Imaging of the Cardiovascular System, Thorax, and Abdomen

Magnetic Resonance Imaging Handbook

Image Principles, Neck, and the Brain
Imaging of the Cardiovascular System, Thorax, and Abdomen
Imaging of the Pelvis, Musculoskeletal System, and Special Applications to CAD

Imaging of the Cardiovascular System, Thorax, and Abdomen

edited by

Luca Saba

CRC Press
Taylor & Francis Group
Boca Raton London New York

CRC Press is an imprint of the
Taylor & Francis Group, an **informa** business

CRC Press
Taylor & Francis Group
6000 Broken Sound Parkway NW, Suite 300
Boca Raton, FL 33487-2742

© 2017 by Taylor & Francis Group, LLC
CRC Press is an imprint of Taylor & Francis Group, an Informa business

No claim to original U.S. Government works

Printed and bound in India by Replika Press Pvt. Ltd.

Printed on acid-free paper
Version Date: 20160328

International Standard Book Number-13: 978-1-4822-1626-4 (Hardback)

Library of Congress Cataloging-in-Publication Data

Names: Saba, Luca, editor.
Title: Magnetic resonance imaging handbook / [edited by] Luca Saba.
Description: Boca Raton : Taylor & Francis, 2016. | Includes bibliographical references and index.
Identifiers: LCCN 2015043923| ISBN 9781482216288 (set : hardcover : alk. paper) | ISBN 9781482216134 (v. 1 : hardcover : alk. paper) | ISBN 9781482216264 (v. 2 : hardcover : alk. paper) | ISBN 9781482216219 (v. 3 : hardcover : alk. paper)
Subjects: | MESH: Magnetic Resonance Imaging.
Classification: LCC RC386.6.M34 | NLM WN 185 | DDC 616.07/548--dc23
LC record available at http://lccn.loc.gov/2015043923

Visit the Taylor & Francis Web site at
http://www.taylorandfrancis.com

and the CRC Press Web site at
http://www.crcpress.com

This book is dedicated to Giovanni Saba.

Thank you.

Start by doing what's necessary; then do what's possible; and suddenly you are doing the impossible.

Francis of Assisi (1181–1226)

The whole is more than the sum of its parts.

Aristotle, Greek philosopher (ca. 384–322 BC)

Contents

Preface

Magnetic resonance imaging (MRI) is a medical imaging technique used in radiology to visualize internal structures of the body in detail. The introduction of MRI resulted in a fundamental and far-reaching improvement of the diagnostic process because this technique provides an excellent contrast between the different soft tissues of the body, which makes it especially useful in imaging the brain, muscles, heart, and cancers compared with other medical imaging techniques such as computed tomography or X-rays.

In the past 20 years, MRI technology has further improved with the introduction of systems up to 7 T and with the development of numerous postprocessing algorithms such as diffusion tensor imaging (DTI), functional MRI (fMRI), and spectroscopic imaging.

From these developments, the diagnostic potentialities of MRI have impressively improved with exceptional spatial resolution and the possibility of analyzing the morphology and function of several kinds of pathology.

The purpose of this book is to cover engineering and clinical benefits in the diagnosis of human pathologies using MRI. It will cover the protocols and potentialities of advanced MRI scanners with very high-quality MR images. Given these exciting developments in the MRI field, I hope that this book will be a timely and complete addition to the growing body of literature on this topic.

Luca Saba
University of Cagliari, Italy

Acknowledgments

It is not possible to overstate my gratitude to the many individuals who helped to produce this book; their enthusiasm and dedication were unbelievable.

I express my appreciation to CRC Press/Taylor & Francis Group for their professionalism in handling this project. Your help was wonderful and made producing this book an enjoyable and worthwhile experience.

Finally, I acknowledge Tiziana.

Editor

Professor **Luca Saba** earned his MD from the University of Cagliari, Italy, in 2002. Currently, he works at the Azienda Ospedaliero Universitaria of Cagliari. His research is focused on multidetector-row computed tomography, magnetic resonance, ultrasound, neuroradiology, and diagnostics in vascular sciences.

Professor Saba has published more than 180 papers in high-impact factor journals such as the *American Journal of Neuroradiology, Atherosclerosis, European Radiology, European Journal of Radiology, Acta Radiologica, Cardiovascular and Interventional Radiology, Journal of Computer Assisted Tomography, American Journal of Roentgenology, Neuroradiology, Clinical Radiology, Journal of Cardiovascular Surgery, Cerebrovascular Diseases, Brain Pathology, Medical Physics,* and *Atherosclerosis.* He is a well-known speaker and has spoken over 45 times at national and international conferences.

Dr. Saba has won 15 scientific and extracurricular awards during his career, and has presented more than 500 papers and posters at national and international congress events (Radiological Society of North America [RSNA], ESGAR, ECR, ISR, AOCR, AINR, JRS, Italian Society of Radiology [SIRM], and AINR). He has written 21 book chapters and is the editor of 10 books in the fields of computed tomography, cardiovascular surgery, plastic surgery, gynecological imaging, and neurodegenerative imaging.

He is a member of the SIRM, European Society of Radiology, RSNA, American Roentgen Ray Society, and European Society of Neuroradiology, and serves as the reviewer of more than 40 scientific journals.

Contributors

Luis Luna Alcalá
Health Time Group
Clínica Las Nieves
Jaén, Spain

Ersan Altun
Department of Radiology
University of North Carolina
Chapel Hill, North Carolina

Ana C. Andrade
Cardiovascular Rehabilitation
    and Exercise Physiology Unit
Heart Institute (InCor)
University of São Paulo
    Medical School
São Paulo, Brazil

Michele Anzidei
Department of Radiological,
    Oncological, and
    Anatomopathological Sciences
Sapienza University of Rome
Rome, Italy

Pablo Bächler
Department of Radiology and
    Biomedical Imaging Center
School of Medicine
Pontificia Universidad Católica
    de Chile
Santiago, Chile

Grzegorz Bauman
Division of Radiological Physics
Department of Radiology
University of Basel Hospital
Basel, Switzerland

Davide Bellini
Department of Radiological,
    Oncological, and Pathological
    Sciences
University of Rome "Sapienza" -
    Polo Pontino
Latina, Italy

Mario Bezzi
Department of Radiological,
    Oncological, and
    Anatomopathological Sciences
Sapienza University of Rome
Rome, Italy

Priya Bhosale
Diagnostic Radiology
University of Texas MD Anderson
    Cancer Center
Houston, Texas

Jürgen Biederer
Translational Lung Research
    Center Heidelberg (TLRC)
University of Heidelberg
Heidelberg, Germany

Michael A. Blake
Department of Radiology
Massachusetts General Hospital
Boston, Massachusetts

Rene M. Botnar
Division of Imaging Sciences
The Rayne Institute
King's College London
London, United Kingdom

Iacopo Carbone
Cardiovascular Imaging Unit
Department of Radiological,
    Oncological, and Pathological
    Sciences
Sapienza University of Rome
Rome, Italy

Carlo Catalano
Department of Radiological,
    Oncological, and
    Anatomopathological Sciences
Sapienza University of Rome
Rome, Italy

Amedeo Chiribiri
Division of Imaging Sciences
The Rayne Institute
King's College London
London, United Kingdom

Maria Ciolina
Department of Radiological,
    Oncological, and Pathological
    Sciences
University of Rome "Sapienza" -
    Polo Pontino
Latina, Italy

Otavio Coelho-Filho
Department of Internal Medicine
State University of Campinas
Campinas, Brazil

Carlo Nicola De Cecco
Department of Radiological,
    Oncological, and Pathological
    Sciences
University of Rome "Sapienza" -
    Polo Pontino
Latina, Italy

and

Department of Radiology and
    Radiological Science
Medical University of South
    Carolina
Charleston, South Carolina

Gianluca De Rubeis
Cardiovascular Imaging Unit
Department of Radiological,
    Oncological, and Pathological
    Sciences
Sapienza University of Rome
Rome, Italy

Domenico De Santis
Department of Radiological,
    Oncological, and Pathological
    Sciences
University of Rome "Sapienza" -
    Polo Pontino
Latina, Italy

Ganeshan Dhakshinamoorthy
Diagnostic Radiology
University of Texas MD Anderson
    Cancer Center
Houston, Texas

Michele Di Martino
Department of Radiological,
    Oncological, and
    Anatomical Pathological Sciences
Sapienza University of Rome
Rome, Italy

Manjiri K. Dighe
Department of Radiology
University of Washington
Seattle, Washington

Christopher J. François
Department of Radiology
School of Medicine and Public
    Health
University of Wisconsin-Madison
Madison, Wisconsin

Marco Francone
Cardiovascular Imaging Unit
Department of Radiological,
    Oncological, and Pathological
    Sciences
Sapienza University of Rome
Rome, Italy

Simon Gabriel
Department of Radiology
University of California
Los Angeles, California

Nicola Galea
Cardiovascular Imaging Unit
Department of Radiological,
    Oncological, and Pathological
    Sciences
Sapienza University of Rome
    Rome, Italy

Verghese George
Department of Radiology
and
Department of Diagnostic and
    Interventional Imaging
University of Texas Health Science
    Center at Houston
Houston, Texas

Abed Ghandour
Department of Radiology
Health Time Group
Clínica Las Nieves
Jaén, Spain

and

Department of Radiology
University Hospitals
Cleveland, Ohio

Giorgia Giustini
Cardiovascular Surgery
    Department
Sapienza University of Rome
Rome, Italy

Markus Henningsson
Division of Imaging Sciences
The Rayne Institute
King's College London
London, United Kingdom

Franco Iafrate
Department of Radiological,
    Oncological, and Pathological
    Sciences
University of Rome "Sapienza" -
    Polo Pontino
Latina, Italy

Michael Jerosch-Herold
Department of Radiology
Brigham and Women's Hospital
Havard Medical School
Boston, Massachusetts

Shelby Kent
Diagnostic Radiology
University of Texas MD Anderson
    Cancer Center
Houston, Texas

Sonja M. Kirchhoff
Institute of Clinical Radiology
University Hospital Munich
    Großhadern
Ludwig Maximilians Universität
    München
Munich, Germany

Sachin Kumbhar
Department of Radiology
University of Washington
Seattle, Washington

Andrea Laghi
Department of Radiological,
    Oncological, and Pathological
    Sciences
University of Rome "Sapienza" -
    Polo Pontino
Latina, Italy

Neeraj Lalwani
Department of Radiology
University of Washington
Seattle, Washington

Diana L. Lam
Seattle Cancer Care Alliance
University of Washington
Seattle, Washington

Pierleone Lucatelli
Department of Radiological,
    Oncological, and
    Anatomopathological Sciences
Sapienza University of Rome
Rome, Italy

Antonio Luna
Department of Radiology
Health Time Group
Clínica Las Nieves
Jaén, Spain

and

Department of Radiology
University Hospitals Cleveland
and
Case Western Reserve University
    School of Medicine
Cleveland, Ohio

Juan Maestre-Antequera
General Surgery Unit
Jesus Uson Minimally Invasive
    Surgery Center
Cáceres, Spain

Beatrice Cavallo Marincola
Department of Radiological,
    Oncological, and
    Anatomopathological Sciences
Sapienza University of Rome
Rome, Italy

Diego Masjoan
Department of Radiology
Health Time Group
Clínica Las Nieves
Jaén, Spain

and

San Miguel Clinic
Cáceres, Spain

Colin J. McCarthy
Department of Radiology
Massachusetts General Hospital
Boston, Massachusetts

Shaunagh McDermott
Department of Radiology
Massachusetts General Hospital
Boston, Massachusetts

Christine Menias
Mayo Clinic LL Radiology
Scottsdale, Arizona

and

Mayo Clinic Hospital
Phoenix, Arizona

and

Department of Radiology
Health Time Group
Clínica Las Nieves
Jaén, Spain

Saeed Mirsadraee
Clinical Research Imaging Centre
University of Edinburgh
Edinburgh, United Kingdom

Francesco Molinari
Department of Radiology
University Hospital of Lille
Lille, France

Justin Morris
Department of Radiology and
    Radiological Science
Medical University of South Carolina
Charleston, South Carolina

Jose Luis Moyano-Cuevas
Bioengineering and Health
    Technology Unit
Jesus Uson Minimally Invasive
    Surgery Center
Cáceres, Spain

Alessandro Napoli
Department of Radiological,
    Oncological, and
    Anatomopathological Sciences
Sapienza University of Rome
Rome, Italy

Yoshiharu Ohno
Advanced Biomedical Imaging
    Research Center
Division of Functional and
    Diagnostic Imaging Research
and
Department of Radiology
Kobe University Graduate School
    of Medicine
Hyogo, Japan

José Blas Pagador
Bioengineering and Health
    Technology Unit
Jesus Uson Minimally Invasive
    Surgery Center
Cáceres, Spain

Claudia Prieto
Division of Imaging Sciences
The Rayne Institute
King's College London
London, United Kingdom

Habib Rahbar
Seattle Cancer Care Alliance
University of Washington
Seattle, Washington

Prabhakar Rajiah
Department of Radiology
University Hospitals Cleveland
and
Case Western Reserve University
    School of Medicine
Cleveland, Ohio

Miguel Ramalho
Department of Radiology
University of North Carolina
Chapel Hill, North Carolina

Marco Rengo
Department of Radiological,
    Oncological, and Pathological
    Sciences
University of Rome "Sapienza" -
    Polo Pontino
Latina, Italy

Stefan G. Ruehm
Department of Radiology
University of California
Los Angeles, California

Tobias Saam
Institute for Clinical Radiology
Ludwig-Maximilians-University
    Hospitals
Munich, Germany

Beatrice Sacconi
Department of Radiological,
    Oncological, and
    Anatomopathological Sciences
Sapienza University of Rome
Rome, Italy

**Francisco Miguel
Sánchez-Margallo**
Department of Radiology
Health Time Group
Clínica Las Nieves
Jaén, Spain

and

Jesus Uson Minimally Invasive
    Surgery Center
Cáceres, Spain

Richard Semelka
Department of Radiology
University of North Carolina
Chapel Hill, North Carolina

Ravi S. Shah
Cardiovascular Division
Department of Medicine
Harvard Medical School
Boston, Massachusetts

Chun Kit Shiu
Department of Radiology
Queen Elizabeth Hospital
Hong Kong, China

Rakesh Sinha
Department of Radiology
South Warwickshire Hospitals
    NHS Trust
Coventry, United Kingdom

Andrew J. Swift
Academic Radiology
and
INSIGNEO
Institute for In Silico Medicine
University of Sheffield
Sheffield, United Kingdom

Miguel Trelles
Russell H. Morgan Department
    of Radiology and Radiological
    Science
Johns Hopkins University School
    of Medicine
Baltimore, Maryland

Sergio Uribe
Department of Radiology and
    Biomedical Imaging Center
School of Medicine
Pontificia Universidad Católica
    de Chile
Santiago, Chile

Israel Valverde
Pediatric Cardiology Unit
Institute of Biomedicine of
    Seville, IBIS
Hospital Virgen de Rocio
Seville, Spain

Edwin J.R. van Beek
Clinical Research Imaging Centre
University of Edinburgh
Edinburgh, United Kingdom

Fernanda Garozzo Velloni
Department of Radiology
University of North Carolina
Chapel Hill, North Carolina

and

Department of Diagnostic Imaging
Federal University of Sao Paulo
Sao Paulo, Brazil

Rafael Andres Vicens
Diagnostic Radiology
University of Texas MD Anderson
    Cancer Center
Houston, Texas

Nicolaus Wagner-Bartak
Diagnostic Radiology
University of Texas MD Anderson
    Cancer Center
Houston, Texas

Mark O. Wielpütz
Translational Lung Research Center
    Heidelberg
University of Heidelberg
Heidelberg, Germany

Jim M. Wild
Academic Radiology
University of Sheffield
Sheffield, United Kingdom

1

Coronary and Perfusion Imaging with Cardiovascular Magnetic Resonance: Current State of the Art

Amedeo Chiribiri, Markus Henningsson, Claudia Prieto, Michael Jerosch-Herold, and Rene M. Botnar

CONTENTS

1.1   Introduction

Despite substantial improvements in prevention and treatment,[1] coronary artery disease (CAD) remains the leading cause of death and disability in the Western world.[2] The current gold standard for the diagnosis of CAD is cardiac catheterization. In the United States alone, 16,300,000 patients are suffering from CAD, and approximately 1,000,000 cardiac catheterizations are performed each year.[2] In up to 40% of examined patients, no significant coronary artery stenoses are diagnosed.[3] Therefore, a noninvasive test that could directly assess the integrity of

the coronary lumen would be desirable.[4] Cardiovascular magnetic resonance (CMR) allows a comprehensive evaluation of myocardial function, perfusion, and morphology in patients with CAD.[5] In addition, magnetic resonance angiography (MRA) can be used for direct visualization of the coronary artery lumen, whereas black blood techniques allow for visualization of the coronary vessel wall,[6] making CMR a tool capable of providing all required diagnostic information in one single examination.

Moreover, the same MRA techniques allow the visualization of the coronary veins anatomy, which is of interest for the optimal placement of pacemaker leads in cardiac resynchronization therapy.[7,8]

This chapter provides an update on current improvements in coronary MRA as well as an overview on its current clinical usage.

1.2 Technical Challenges and General Imaging Strategies for Coronary CMR

Coronary MRA demands dedicated techniques to optimize image contrast and to ensure a high signal-to-noise ratio (SNR), yielding a clear delineation from blood-filled structures or the vessel walls with high spatial resolution. Since high spatial resolution is required for adequate visualization of the coronary vessels, the concomitant intrinsic cardiac and respiratory motions pose a major challenge to coronary MRA and vessel wall imaging.

To overcome this, a standard coronary MRA protocol comprises (1) electrocardiogram (ECG) triggering for synchronization of the heart motion with the data acquisition, (2) respiratory navigation for synchronization/compensation of the respiratory motion, (3) the imaging sequence itself, and (4) prepulses for spin preparation to ensure sufficient image contrast (Figure 1.1).

1.2.1 Compensation of Cardiac Motion: ECG Triggering

To freeze cardiac motion, data acquisition has to be synchronized with the cardiac cycle and be limited to periods of minimal cardiac movement.[9] Resting periods occur in end-systole (approximately 280–350 ms after the R wave) and in mid-diastole (immediately prior to atrial systole). Both acquisition strategies (systolic or diastolic) have advantages and disadvantages (Table 1.1).

The optimal trigger delay and the length of the acquisition window depends on the patients' heart rate, on the type of imaging sequence used, on the structure to visualize (arteries versus veins), and other hemodynamic factors. While the use of a heart-rate dependent formula to identify the mid-diastolic resting period is effective in many subjects, there may be considerable intersubject variation.[10] Therefore, the resting period should be identified for each patient from a free-breathing high temporal resolution cine scan in the four-chamber view performed shortly before the coronary scan.[9] Real-time arrhythmia rejection to exclude irregular heartbeats may further improve coronary MRA image quality.[11–14]

Another important parameter to consider is the duration of the resting period. It is typically longer for the left compared to the right coronary artery system. Thus, the length of the acquisition window of a whole heart acquisition is determined by the duration of RCA diastasis.

1.2.2 Compensation of Respiratory Motion: Navigator

The displacement of the heart due to respiration can exceed 2–3 cm, requiring synchronization of image acquisition with the respiratory cycle. High-resolution three-dimensional (3D) datasets are not compatible

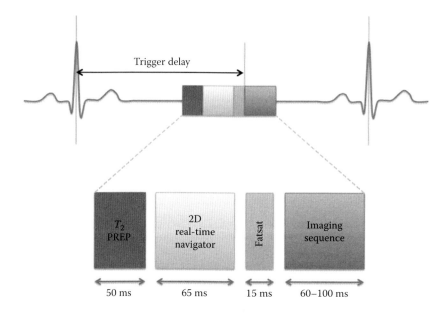

FIGURE 1.1

Schematic representation of pulse sequence elements for MRA. Image acquisition is performed in mid-diastole after a trigger delay from the R wave of the ECG. The imaging block is preceded by a contrast-enhancing spin preparation (T_2 PREP), a 2D selective real-time navigator pulse for respiratory motion compensation, and a frequency-selective fat-suppression prepulse (FATSAT). These sequence blocks are repeated with every heart cycle. (Adapted from Botnar, R.M. et al., *Circulation*, 99, 3139–3148, 1999. With permission.)

TABLE 1.1

Comparison between Systolic and Diastolic Coronary CMR

Diastolic Resting Period	Systolic Resting Period
Advantages	*Advantages*
Longer acquisition window (~100–125 ms/heartbeat)	Less sensitive to heart rate variability
Higher blood flow (higher signal in gradient echo sequences)	Larger diameter of the venous vessels
Disadvantage	*Disadvantage*
Higher sensitivity to heart rate variability	Shorter acquisition window (~50 ms/heartbeat)

with breath-hold acquisitions. Several approaches have been tested to reduce the effect of respiratory motion on image quality. Prospective real-time navigator gating and correction techniques are the current approach to minimize respiratory motion artifacts.[15,16] A pencil beam one-dimensional navigator typically positioned on the dome of the right hemidiaphragm is used to monitor the foot head motion[17] of the diaphragm immediately prior to coronary image acquisition (Figure 1.2). Depending on the position of the diaphragm, the data is either accepted, when the position falls within a certain acceptance window (usually 3–5 mm), or rejected. In the latter case, the data has to be remeasured in the subsequent cardiac cycle. Reduction of the acceptance window reduces the motion artifacts but increases the overall acquisition time because more data is rejected. An acceptance window of 5 mm usually allows an efficiency approaching 50%.[18–20]

To increase the percentage of accepted data, the position of the imaging slice can be prospectively adapted to the measured respiratory position. Normally the movement of the heart due to breathing is less pronounced than the motion of the diaphragm itself, and a scale factor between 0.4 and 0.6 has been used for optimal slice tracking. However, also with the use of these techniques, respiratory induced motion of the heart often cannot be completely modeled by a simple translation along the foot–head direction, and it has been shown that in up to 30% of patients, an affine transformation models the respiratory motion more accurately.[21,22]

Other proposed approaches include the use of image-based navigators that directly track cardiac motion, navigators that monitor the movement of the epicardial fat,[23] scanning in prone position,[24,25] and the use of an abdominal or a thoracic banding.[25,26]

1.2.3 Coronary Artery Imaging

1.2.3.1 Sequences

3D approaches require long acquisition times and were initially not feasible. The first approaches to coronary artery angiography were attempted by Edelman[22] and Manning[27] by two-dimensional (2D) gradient-echo techniques. One slice was acquired in 16 heartbeats (in a single breath-hold) and patients were free to breathe between acquisitions. The introduction of navigator techniques allowed the acquisition of data in free breathing and 3D techniques became feasible.

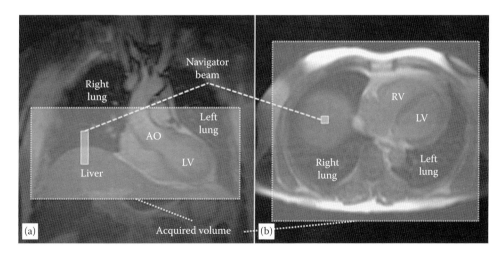

FIGURE 1.2
Planning of the navigator and of the whole heart imaging volume from coronal (a) and axial (b) scout images. A pencil beam one-dimensional navigator is positioned on the dome of the right hemidiaphragm. A 3D volume covering the entire heart is planned in the axial plane for data acquisition and includes 2 cm cranial of the aortic root and caudal to the diaphragm.

TABLE 1.2

Planning of a Coronary MRA Examination

Whole-Heart Imaging	Target-Volume Imaging
Axial-, sagittal-, or coronal-oriented imaging volume	Right and left coronaries are imaged in two individually planned thin slabs
Planning on the set of axial, sagittal, and coronal scout images to include the entire heart	Quicker planning performed on a low-resolution free-breathing coronary scout scan
Preferably isotropic spatial resolution to allow for 3D reformatting along the course of the coronary vessels	The central slice is defined by selection of three points located on the path of the coronary vessel
Full coverage (simultaneous acquisition of the venous system)	Targeted coverage
Long acquisition time	Shorter acquisition time than the whole-heart approach
Limited in-flow effect (less contrast, particularly in gradient echo techniques)	Nonisotropic voxels with in-plane spatial resolution of 1×1 mm^2 and slice thickness of 2–3mm
Typical spatial resolution: isotropic voxels with $1.3 \times 1.3 \times 1.3$-$mm^3$ size	

Two different approaches are currently in use, the whole-heart and the vessel-targeted approach, which require an individual planning procedure. Whole-heart techniques have also been used for coronary vein imaging.

These sequences can be acquired using a whole heart or target volume approach (Table 1.2).

1.2.3.2 *Contrast-Enhancing Spin Preparations*

The use of 3D acquisitions has not only the advantage of improved SNR and higher spatial resolution but also the disadvantage of reduced contrast between blood and the myocardium (less in-flow effects). Therefore, contrast-enhancing spin preparation techniques[28–30] in combination with steady-state-free-precession (SSFP) techniques have been developed to provide sufficient contrast for the delineation of the coronary arteries.[31] SSFP sequences are currently preferred to T_1-weighted gradient-echo sequences on 1.5 T systems due to both higher intrinsic SNR and better contrast between blood and myocardium.[31–33] However, SSFP sequences have limitations at 3 T due to longer repetition times (more stringent specific adsorption rate[SAR] model) and thus increased sensitivity to off-resonances and the need for higher flip angles to obtain enough contrast between blood and myocardium. At 3 T and higher field strengths,[34] gradient-echo techniques appear as promising alternative to SSFP techniques.

For non-contrast agent-enhanced imaging, spin preparation usually includes fat suppression and T_2 preparation. Frequency selective prepulses can be applied to saturate the signal from fat tissue, allowing the visualization of the coronaries.[22,29,30] To enhance the contrast between the coronary lumen and the underlying myocardium, T_2-preparation techniques can be used[28,29] to suppress signal from myocardium, as blood and myocardium have similar T_1 but different T_2 (Figure 1.3).

T_2 preparation also suppresses deoxygenated venous blood and therefore is not suited for coronary vein imaging. A few other techniques to improve contrast between the vessel lumen and the myocardium have been proposed but are not widely adopted. Among

these are spin-locking[35] and magnetization transfer techniques.[30] Magnetization transfer preparation does not affect the signal from deoxygenated venous blood and can therefore be used to image the coronary veins.[36] Further improvement of contrast between blood and the surrounding tissues can be achieved by application of contrast agents.

1.2.3.3 *Contrast Agents*

Extracellular,[37] blood-pool,[38–41] and contrast agents with weak albumin binding[42–44] have been used to improve contrast-to-noise and image quality of coronary MRA. As coronary artery imaging protocols should ideally become part of a routine ischemia/CAD diagnostic imaging protocol, the choice of the contrast agent depends on a balance between increasing contrast between lumen and myocardium while maintaining to ability to provide useful scar information in late gadolinium-enhancement images. Extracellular contrast agents allow limited coronary artery enhancement due to the rapid extravasation in the interstitial space. Blood-pool contrast agents offer the highest contrast between the vessel lumen and the surrounding tissues but may face limitations for late enhancement imaging. Contrast agents with weak albumin binding appear most promising for combined coronary artery and infarct imaging due to their prolonged retention time in blood and their higher relaxivities (useful to improve image quality of coronary MRA),[45] while maintaining good late enhancement properties.[46] To take advantage of contrast agents for coronary MRA, a saturation[47] or inversion prepulse[48] is usually applied instead of T_2 preparation. The rapid T_1 recovery of blood after contrast agent administration (native T_1 of blood = 1200 ms, T_1 of blood with contrast agent = 50–100 ms, native T_1 of myocardium = 850 ms) allows the generation of high contrast between the coronary lumen and myocardium.

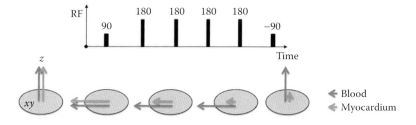

FIGURE 1.3

Endogenous contrast enhancement (T_2 preparation). The T_2 preparation allows suppression of tissues with short T_2 relaxation times (cardiac muscle, cardiac veins, and epicardial fat), while minimally influencing tissues with long T_2 relaxation times, such as arterial blood. A 90° radio-frequency (RF) pulse flips the magnetization vector from the z-direction into the xy plane. An even number of nonselective RF refocusing pulses minimizes flow sensitivity of the T_2 preparation, while allowing T_2-dependent decay of the signal. Finally, a −90° RF tip up pulse restores the magnetization in the z-direction. (Adapted from Brittain, J.H. et al., *Magn. Reson. Med.*, 33, 689–696, 1995. With permission; Botnar, R.M. et al., *Circulation*, 99, 3139–3148, 1999.)

1.2.3.4 Recent Improvements in Acquisition Speed and Resolution

Faster image acquisition can result in better image quality due to reduced sensitivity to motion because of a shorter acquisition window or shorter overall data acquisition time. To accelerate data acquisition and/or reduce motion sensitivity, several approaches have been proposed, such as faster encoding of k-space using echo planar imaging,[49,50] more efficient k-space sampling using spiral,[51] or less motion sensitive k-space sampling using radial trajectories.[52] These techniques have not yet become established standards for coronary MRA due to their off-resonance sensitivity (echo planar imaging and spiral) or signal-to-noise penalty (radial).

More recently developed parallel imaging techniques such as sensitivity encoding (SENSE)[53] or simultaneous acquisition of spatial harmonics (SMASH)[54] have been successfully applied to reduce the overall MRA acquisition time while maintaining image quality. With the advent of 2D coil arrays (e.g., 32 channel coils), acceleration along the slice encoding and phase-encoding direction can be applied, thereby further reducing scan time. However, acceleration factors above 4 can hardly be realized due to SNR limitations and increasing reconstruction artifacts.[54,55]

1.2.4 Coronary Vein Imaging

During the past few decades, coronary imaging has been mainly focused on the coronary arteries. With the advent of resynchronization devices, the assessment of the course of the coronary veins has become increasingly important and has recently gained interest for preinterventional identification of optimal placement site for the left ventricular lead of resynchronization devices. 3D MR coronary vein angiograms can be overlaid onto real-time time acquired X-ray images, to provide improved guidance for catheter implantation.[56,57]

Because of the low oxygen saturation of coronary venous blood (causing a strong reduction of blood T_2 values), T_2 preparation is not suitable for coronary vein imaging. Current approaches include non-contrast-enhanced imaging by magnetization transfer preparation[36,58] or gadolinium contrast, including blood-pool,[7,8,59] extracellular,[60] and contrast agents with weak albumin binding.[61] Of interest is the approach proposed by Duckett of slow infusion of a high relaxivity contrast agent during coronary vein MRA acquisition, which offers the possibility to acquire late gadolinium enhancement images after the redistribution of the contrast agent.[61]

The optimal acquisition window for coronary vein imaging is in end-systole, when the coronary vein diameter is maximal.[36] However, ECG triggering may be difficult in patients with heart failure due to tachycardia, orthopnea, and asynchronous contraction of the left ventricle, making the resting period different in independent segments of the chamber. In these cases, the acquisition parameters should be adapted and data acquisition in end-diastole might be an alternative (Figure 1.4).

1.2.5 Coronary Vessel Wall Imaging

The excellent soft tissue contrast of magnetic resonance imaging (MRI) enables the visualization of the vessel wall. First, *in-vivo* images of coronary vessel wall were obtained by 2D fat-saturated fast-spin echo techniques.[62,63] For improved contrast between the blood and the vessel wall, a double inversion recovery preparation was applied to obtain black-blood images.[64]

Further developments of the technique combined the double inversion recovery prepulse with fast gradient echo readout techniques[65] and recently with spiral[66] and radial acquisition trajectories.[67] Clinical studies demonstrated outward positive remodeling with relative lumen preservation in patients with established CAD

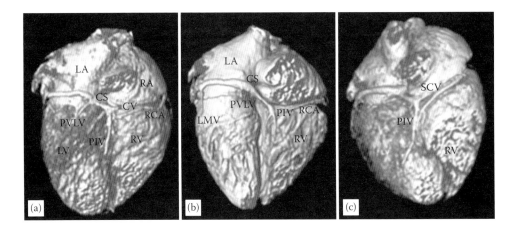

FIGURE 1.4

MRA of the coronary veins obtained using a whole heart approach and an intravascular contrast agent. There is a wide inter-individual variability (a–c) of the anatomy of the cardiac venous system in the terms of presence, relative position, and diameter of the coronary sinus tributaries. The knowledge of the anatomy and variations of the cardiac venous system may facilitate the positioning of the left ventricle lead in patients undergoing cardiac resynchronization therapy. LA, left atrium; LV, left ventricle; LMV, left marginal vein; PIV, posterior interventricular vein; PVLV, posterior vein of the left ventricle; RA, right atrium; RV, right ventricle; RCA, right coronary artery; and SCV, small cardiac vein. (Adapted from Chiribiri, A. et al., *Am. J. Cardiol.*, 101, 407–412, 2008. With permission.)

and increased vessel wall thickness in patients with type I diabetes and renal dysfunction.[68,69]

For selective imaging of fibrous or inflamed plaque, several preclinical and clinical studies have been performed using delayed gadolinium enhancement techniques, showing promising results.[70] *In vivo*, clinically approved contrast agents showed nonspecific uptake in plaques both in patients with chronic angina[71] and in patients with acute coronary syndromes (ACS)[72] (Figure 1.5). Contrast uptake in patients with stable angina was associated with calcified or mixed plaques on multislice computed tomography (MSCT), while contrast uptake in patients with ACS was transient and thus more likely related to inflammation.

In experimental animal models, several target-specific contrast agents have been tested. Accumulation of albumin-binding blood-pool contrast agent indicates increased endothelial permeability and/or increased neovascularization.[73] The accumulation of iron-oxide particles (USPIO) indicates increased endothelial permeability and vessel wall inflammation due to the presence of intraplaque macrophages.[74,75]

Novel molecular contrast agents, which specifically target molecules or cells, have recently been developed allowing selective visualization of inflammatory markers such as intercellular adhesion molecule-1 (ICAM-1), vascular adhesion molecule-1 (VCAM-1), or matrix metalloproteinase (MMP).[76,77]

Recent work also includes the specific labeling of thrombi by a fibrin-specific contrast agent.[78,79]

Molecular contrast agent may provide new opportunities for the characterization of early atherosclerotic lesions as well as for the assessment of plaque vulnerability.

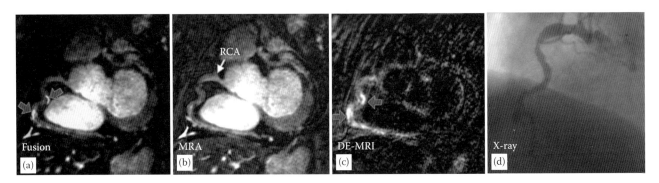

FIGURE 1.5

Localized uptake of extracellular contrast agent (Gd-DTPA) in unstable CAD. Right coronary artery from a subject with unstable angina. (a) Image fusion between coronary MRA (b) and delayed-enhancement (DE)-MRI (c) that shows significantly enhanced vessel wall signal. Comparison with X-ray angiography (d). (Adapted from Ibrahim, T. et al., *JACC Cardiovasc. Imaging*, 2, 580–588, 2009. With permission.)

1.2.6 Special Considerations: Intracoronary Stents

Coronary stents are currently used in a large numbers of patients undergoing percutaneous revascularization. Typically made of high-grade stainless steel, coronary stents cause negligible attractive force and local heating at both 1.5T[80–85] and 3T,[86] with clinical safety demonstrated for imaging early after implantation at 1.5T.[80–82] In the United States, both CYPHER™ (Cordis, Miami Lakes, Florida) and Taxus® Liberté® (Boston Scientific, Natick, Massachusetts) drug-eluting stents are approved for CMR scanning immediately after implantation. However, coronary stents cause local susceptibility artifacts and signal voids that preclude the assessment of peri-stent and intra-stent coronary integrity.

1.3 Clinical Applications

The application of MRA is currently limited to the assessment of anomalies of the coronary arteries (class I indication) and aorto-coronary bypass grafts (class II indication). The use of MRA for the assessment of native coronary arteries has not yet entered clinical routine.[87]

1.3.1 Coronary Artery Angiography for the Detection of CAD

Proximal and medial segments of the main branches of the coronary arteries can routinely be visualized by free-breathing navigator-gated fast gradient-echo or SSFP techniques. Especially the proximal segments are evaluable in nearly 100% of subjects. The left anterior descending (LAD) and the right coronary artery (RCA) are usually imaged with better image quality than the left circumflex (LCX), which runs in the direct vicinity of the myocardium and at a larger distance from the coil elements. The mean length of the vessels that was visible in previous studies was approximately 50 mm for the LAD, 80 mm for the RCA and 40 mm for the LCX,[27,28,73,88–92] with an excellent agreement between the proximal vessel diameters between MRA and conventional angiography.[93] However, despite the most recent technical developments, spatial resolution of coronary MRA is still lower than that of invasive coronary angiography, limiting the size of the branches that can be assessed by coronary MRA. In spite of these limitations, coronary artery stenosis in proximal segments can often be identified by coronary MRA. On *bright-blood* (gradient echo) images, a stenosis appears as a signal void in the otherwise bright vessel (Figure 1.6). This is primarily due to the reduction of the lumen and partly due to the presence of turbulence of the blood flow distal to the stenosis that may result in a dephasing of the signal and thus lead to an overestimation of the severity of the stenosis with MRA when compared to invasive angiography.[94]

An international multicenter study[95] assessing the diagnostic accuracy of coronary MRA revealed a high sensitivity (92%) and a low specificity (59%) for the detection of CAD. The relatively limited spatial resolution of coronary MRA contributed to these results, as demonstrated by the excellent performance observed for the exclusion of left main or three-vessel disease (sensitivity 100%; negative predictive value 100%). These results have been confirmed in a series of smaller single-center studies, with good results for the detection of significant coronary stenosis in the proximal coronary segments.[26,28,96–103]

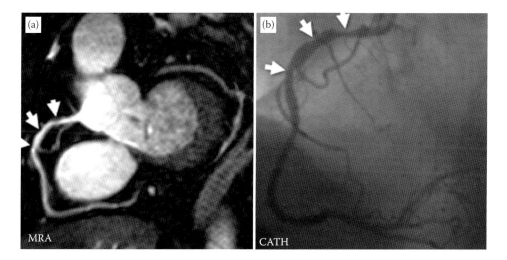

FIGURE 1.6
Coronary artery stenosis visualized on white blood MRA images (a). Comparison with coronary artery angiography (b). Visualization of multiple stenoses and wall irregularities in the proximal RCA (white arrows).

Coronary MRA has recently been proposed in conjunction with late Gadolinium enhancement to rule out ischemic etiology in patients presenting with dilated cardiomyopathy without symptoms of myocardial infarction.[87]

A recent meta-analysis compared coronary MRA and MSCT for ruling out significant CAD in adults.[104] The conclusion was that MSCT is more accurate than MRA and that, therefore, MSCT should be considered as the preferred noninvasive alternative method to coronary catheterization to exclude CAD. However, the added value of a coronary MRA scan integrated into a comprehensive clinical protocol encompassing function, structure, perfusion, and viability scans still needs to be assessed and has the potential to provide a more accurate evaluation of patients with known or suspected CAD. A more recent national multicenter study from Japan demonstrated that non-contrast-enhanced whole heart coronary MRA at 1.5T can noninvasively detect significant CAD with high sensitivity (88%) and moderate specificity (72%). In the study, a negative predictive value (NPV) of 88% indicates that whole heart coronary MRA can rule out CAD.[105] Noteworthy is the fact that the NPV of this MRCA multicenter trial is identical to the NPV of the CORE-64 MSCT multicenter study.[106] Remarkably, the value of MRCA in patients with a low pretest likelihood is similar to CT and can reliable rule out CAD in patients with a pretest likelihood of <20%.[107]

1.3.2 Coronary Anomalies and Aneurysms

Anomalies of the origin and path of the coronary arteries and coronary aneurysms are accurately visualized by coronary MRA, as a result of their higher caliber and preferred location in proximal or ectatic segments. Coronary MRA has the important added benefit of allowing a noninvasive reliable diagnosis without exposing the patient to ionizing radiation (Figure 1.7). This is particularly important in younger patients and children, as well as in younger women.[87,108]

1.3.3 Coronary Bypass Grafts

The visualization of bypass grafts benefits from their rather stationary position, straight and known course, and large diameter as compared to the native coronary arteries. Spin-echo techniques[109–112] as well as gradient-echo techniques have been used. Contrast agents have been applied for enhancement of the blood signal,[113–115] which resulted in sensitivities between 95% and 100%.

A major limitation of bypass imaging results from the presence of metallic clips, which cause signal voids due to susceptibility artifacts. Since clips are often located in the proximity of the bypass grafts and coronary arteries, the resulting signal voids may often not allow the assessment of the respective anatomy.

At specialized centers, coronary MRA may be used to identify coronary arterial stenosis in arterial bypass grafts.[87]

1.3.4 Coronary Vessel Wall Imaging

By means of black-blood techniques, the latest clinical MRI scanners can provide a detailed visualization of the coronary artery wall. The major impediment to clinical application of these techniques is the need for very high spatial resolution and the related long acquisition times.

The vessel wall can be visualized either in a cross-sectional view or along the path of the vessel. Because of reduced partial volume effects, the cross-sectional orientation should provide more accurate quantification of the vessel wall thickness, but it provides limited coverage. A long-axis view of the vessel wall provides a more extensive visualization and typically allows assessment of the proximal 5 cm. The use of contrast agents allows for selective plaque visualization. After injection of, for example, an extracellular contrast agent, delayed enhancement images showed focal or diffuse uptake of contrast agent and enhancement can be due to the presence of a fibrous plaque or due to inflammation. The current application of coronary vessel wall imaging is restricted to research purposes but in future, it may become part of CAD risk assessment and monitoring of treatment response, especially if plaque targeting agents become available.

1.3.5 Coronary Vein Imaging

Imaging of the coronary veins and the integration of coronary vein anatomy and myocardial scar information may help to guide the left ventricular lead implantation to achieve optimal cardiac resynchronization therapy (Figure 1.8). Early studies have shown that contrast agent-enhanced CMR can be used for the assessment of the course of the coronary sinus, the great cardiac vein, and the respective tributaries.[7,8,59] A major limitation in heart failure patients may rise from the asynchronous contraction pattern of the left ventricle, causing constant systolic motion of the heart. This may disqualify the suggested usage of an end-systolic trigger window,[36] and the use of a mid-diastolic triggering window may be more effective.

1.4 Myocardial Perfusion Imaging

The coronary tree can be well visualized with MRA at the level of the coronary arteries and veins, but the coronary resistance vessels and the capillary network

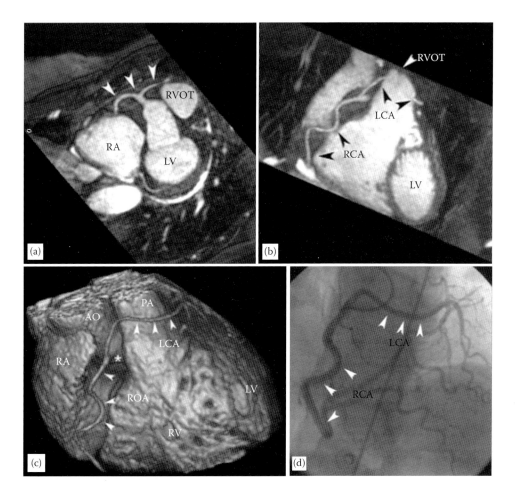

FIGURE 1.7
Coronary MRA for the visualization of the anomalous origin of the coronary arteries. T_2-prepared navigator gated whole heart technique shows the anomalous origin of a single coronary artery from the right sinus of Valsalva in a patient complaining for angina. (a) Maximum intensity projection (MIP) image demonstrating the anomalous origin of the single coronary artery from the right sinus of Valsalva (white arrows); LV, left ventricle; RA, right atrium; and RVOT, right ventricle outflow tract. (b) MIP image showing the path of the right coronary artery (RCA) and of the left coronary artery (LCA), crossing in front of the RVOT. (c) Volume rendering reconstruction of the heart, showing the anomalous origin of the single coronary artery (*) from the aorta (AO), the course of the RCA and the LCA; PA, pulmonary artery. (d) Coronary angiography in right anterior oblique projection, showing the anomalous coronary tree. (From Boffano, C. et al., *Int. J. Cardiol.*, 137, e27–e28, 2009.)

in the myocardium are well beyond the reach of these techniques. Arterioles have diameter of the order 10–100 μm, with smooth muscle layers in the vessel wall that control the vascular resistance, and thereby the flow of blood through the downstream network of capillaries and venules. A reduction of blood flow in the myocardium can cause myocardial ischemia from a lack of oxygen supply. It can occur as a result of an up-stream stenosis in a coronary artery supplying the myocardium, but also from microvascular or smooth muscle cell dysfunction. The focus of myocardial perfusion imaging lies therefore primarily in assessing the level of myocardial blood flow at the level of the microcirculation. Secondary aspects such as capillary density in the myocardium, are potentially also of interest, but assessing myocardial perfusion can be viewed as the physiologically most relevant measure

to assess the risk of myocardial ischemia in a patient under resting and stress conditions.

1.4.1 Techniques for Magnetic Resonance Myocardial Perfusion Imaging

First pass perfusion imaging is the most widely used technique to assess and quantify the function of the coronary microcirculation. This technique relies on the rapid acquisition of T_1-weighted (T_1w) images during the first pass of a contrast bolus through the heart. There are several requirements to enable myocardial perfusion MRI: (1) an image acquisition speed sufficient to avoid artifacts from cardiac motion and fast enough to image the heart within one heart beat, (2) a spatial resolution that allows detection of endo-to-epicardial perfusion gradients, and (3) high sensitivity of the acquired signal

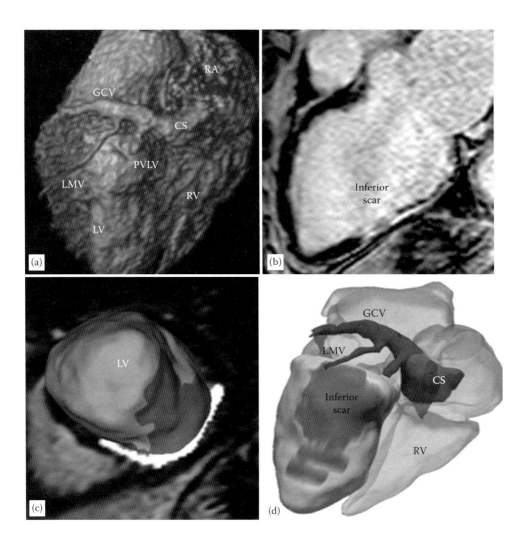

FIGURE 1.8

Visualization of coronary veins and scar in a patient referred for cardiac resynchronization therapy. Figure 1.6. (a) 3D reconstruction of the heart with the coronary venous system. (b) Two chamber late Gadolinium enhancement, with scar in the inferior segments of the left ventricle. (c) Segmented LV registered to the scar imaging in the short axis view. (d) Segmented whole heart with the coronary veins and scar all superimposed. RA, right atrium; LMV, left marginal vein; RV, right ventricle; LV, left ventricle; GCV, great cardiac vein; and PVLV, posterior lateral vein of the LV. (Adapted from Duckett, S.G. et al., *J. Magn. Reson. Imaging*, 33, 87–95, 2011. With permission.)

to contrast enhancement during the transit of a contrast agent through the ventricular cavities and the myocardium. Currently, most cardiac perfusion imaging techniques in MRI are based on ECG-triggered 2D imaging methods, and coverage of the heart is achieved by multislice imaging. The stated requirements entail the need for a rapid, *single-shot*[*] image acquisition of the order 100–150 ms per dynamic view in each slice, and all slices imaged during a heart-beat, an in-plane resolution of 2.5 mm or less, and a pulse sequence that imparts to the acquired signal a strong T_1-weighting (T_1w). The necessary T_1-weighting is generally achieved by preceding the image acquisition by a magnetization preparation in the form of an inversion or saturation pulse, such that the image acquired during the subsequent

magnetization recovery is strongly dependent on the T_1 of the blood or myocardial tissue. The total number of images per slice should be sufficient to cover the first pass and at least early recirculation of contrast in the left ventricle cavity. In addition, imaging should be started before contrast injection to obtain to obtain *baseline* reference images without contrast enhancement—60 images per slice are typical for myocardial perfusion studies. Figure 1.9 illustrates a common scheme for 2D, multi-slice perfusion imaging.

The 2D image readout in a perfusion sequence is generally in the form of a gradient-echo acquisition, with or without SSFP, or a form hybrid of gradient-echo and echo-planar imaging.[116] In both cases, the MR signal data (i.e., a phase- and frequency-encoded

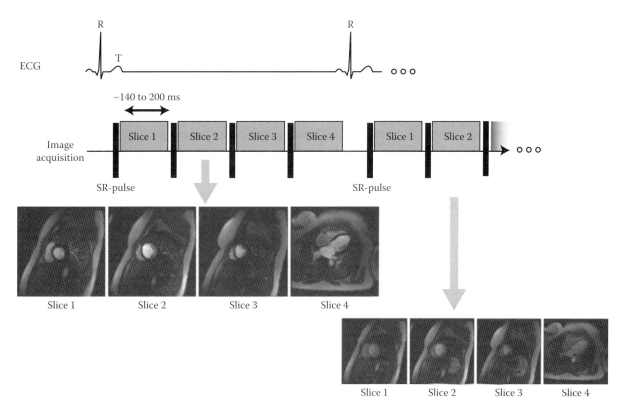

FIGURE 1.9

Multislice, T_1-weighted, and ECG-gated myocardial perfusion imaging. The schematic illustrates a common scheme for acquiring T_1-weighted myocardial perfusion images during a first pass perfusion study over approximately 60 heart beats to capture the first pass and early recirculation of contrast. The narrow black rectangles represent the magnetization preparation (e.g., a saturation pulse) that is applied before acquiring an image to control the T_1-weighting. The grey rectangles represent the 2D image acquisitions, which typically take 150–200 ms each with fast gradient-echo imaging, if possible, parallel-imaging acceleration. Phase-encodings would be applied in a linear order to provide sufficient time for magnetization recovery before the central k-space lines are acquired. The magnetization preparations are normally non-slice selective to avoid inflow-dependent contrast in the blood pool, and they are repeated for each slice to produce the same T_1-weighting. The images show an example from a an acquisition with a combination of three short-axis slices, and one four-chamber long-axis view of the heart, which can be acquired without noticeable interference when a saturation-preparation is used as magnetization preparation.

gradient-echo signal) are readout as lines of a Cartesian grid of k-space data. Image acquisitions, using non-Cartesian data-sampling trajectories, for example, in the form radial lines that represent projection profiles like in computed tomography (CT) imaging,[117,118] or along spiral k-space trajectories[119,120] have in recent years gained increasing attention. These non-Cartesian acquisition methods provide not only a gain in image acquisition speed but have also beneficial characteristics for avoiding certain types of *dark rim* artifacts[121] at the endocardial border[122] that can compromise the diagnostic accuracy of myocardial perfusion imaging by being mistaken for endocardial perfusion defects. A disadvantage of the non-Cartesian acquisition techniques is the more complex nature of the image reconstruction. Data sampled on a non-Cartesian (k-space) trajectory require a relatively time-consuming process of resampling onto a Cartesian grid before a fast Fourier

Transform can be applied. Spiral and radial perfusion imaging therefore entail a considerable latency time for the display of the images, rendering it currently impossible to see the passage of the bolus through the heart in *real time*.

Because of the need for cardiac perfusion imaging to cover the heart, 2D images are readout during almost the entire cardiac cycle. The image acquisition is either triggered by an ECG signal, or the image data are acquired continuously and binned into cardiac phases, using some cardiac-cycle related signal from the image data.[118,123] Most myocardial perfusion occurs during heart relaxation (diastole). This raises the question if perfusion images acquired during systole and diastole carry different information about myocardial blood flow. It is useful to point out here that myocardial perfusion imaging provides an estimate of myocardial perfusion *averaged* over an entire heart cycle, that is, one observes beat-to-beat changes in tissue and blood contrast, and not the instantaneous flow during a specific time point in the cardiac cycle. As images for each slice position are readout during a fixed phase of the cardiac cycle, cardiac motion will appear frozen when the images are played in cine mode, and only breathing motion will normally remain visible, unless there is a cardiac trigger error.

Covering the entire left ventricle in a perfusion scan is currently mostly achieved by multislice coverage, using, for example, 3–4 short axis slices, in combination with one long axis view to assess the heart apex. The limitations of this approach could in principle be overcome by 3D perfusion imaging, as is a standard in nuclear cardiology. For 3D MR myocardial perfusion imaging, it becomes more challenging to freeze cardiac motion than for its 2D multislice analog, because the entire acquisition time for the 3D dataset determines the *shutter speed*, rather than the acquisition time for a single 2D image in a multislice approach. For 3D perfusion imaging, one therefore only uses a short (diastolic) portion of the cardiac cycle to acquire the data, and it becomes necessary to use parallel-imaging acceleration and sparse sampling to achieve sufficiently short acquisition times.[124–126]

1.4.2 Clinical Applications

In the clinical setting myocardial perfusion is assessed both at rest and during stress to detect CAD. For stress, patients undergo most often some form of pharmacologically induced vasodilation. The presence of low limiting epicardial lesions can, with the exception of high-grade lesions (luminal narrowing above 90%), only be detected when the resistance of the coronary microcirculation is minimized by vasodilation. The perfusion study under resting conditions serves as a reference to assess the myocardial blood flow response that can elicited by maximal vasodilation.

Both a short contrast bolus, using a power injector, and imaging with a temporal resolution equivalent to a heart beat duration are particularly important during stress to enable an adequate assessment of myocardial perfusion. Otherwise myocardial contrast enhancement becomes too dependent on the rate at which the contrast agent is injected, rather than myocardial blood flow, and lower temporal sampling renders it more difficult to detect visually, or quantitatively, differences in the *rate* of myocardial contrast enhancement.

For the interpretation of MR perfusion studies, one primarily evaluates the rate of myocardial contrast enhancement during the first pass of the contrast bolus, as illustrated by the example in Figure 1.10. This can be done qualitatively by visual comparison of contrast enhancement in different myocardial segments when the perfusion images are displayed in cine mode, or based on signal-intensity curves that are used to quantify the rate of contrast enhancement, or even to quantify the myocardial perfusion reserve. Determining the perfusion reserve has the advantage in comparison to a visual assessment of providing a measure that can be used to detect both global and regional reductions of the vasodilator response, an aspect that gains particular importance in multivessel disease and microvascular dysfunction. Secondary measures of potential interest are the delay in the arrival of contrast in a myocardial region of interest relative to the blood pool and other segments— significantly longer delays are found in collateral-dependent myocardium.[127–129] Quantitative approaches have been held back by the need to (manually) segment the perfusion images, though recent advances for motion correction[130,131] during image reconstruction have rendered this task much easier.

1.5 Conclusions—Future Perspective

Although the spatial resolution of coronary MRA for the visualization of the coronary lumen is not as high as for MSCT, novel coronary MRI techniques have shown potential in providing additional information including plaque burden and biology by black-blood coronary MRI or use of nonspecific and targeted MR contrast agents. Currently, novel contrast agents targeting specific molecules or cells are undergoing preclinical evaluation, and some have been already tested in humans. In addition, direct thrombus visualization, assessment of endothelial function, and coronary edema has recently been demonstrated.

Finally, ongoing developments in motion compensation techniques will simplify the acquisition of coronary MRA with the advantage of improved image quality and a shorter scan time.

To look in the coronary tree beyond the coronary arteries at the coronary microcirculation has led to the development of myocardial perfusion imaging as a technique to detect myocardium at risk of ischemia. The information obtained by myocardial perfusion imaging is a measure of microcirculatory function, which is assessed both during rest and during stress. While coronary MRA can be applied to assess plaque burden and coronary luminal narrowing, myocardial perfusion imaging can determine their hemodynamical significance, and the presence of collateral circulation, which can reduce the detrimental impact of epicardial lesions. Taken together, these techniques represent a versatile and multifaceted approach to study coronary physiology and disease.

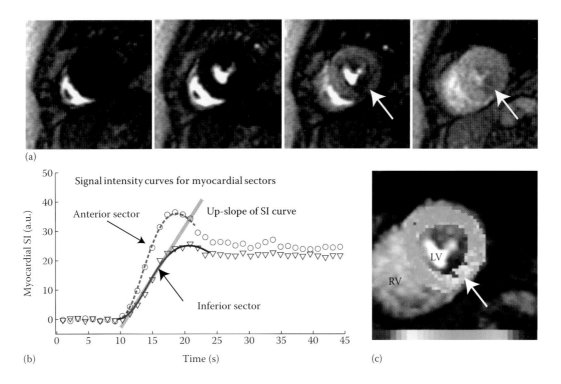

FIGURE 1.10

First-pass myocardial perfusion study in patient with coronary artery disease. (a) Selected perfusion images acquired during maximal vasodilation in a 64-year-old female with coronary artery disease depicting the first pass of a 0.05 mmol/kg bolus of gadolinium contrast through the right and left ventricles. (b) The average signal intensity (SI) in the anterior and inferior myocardial sectors is shown as a function of the time from the start of perfusion imaging. The SI curve for the inferior myocardial sector (arrow in images) shows a reduced rate of myocardial contrast enhancement in comparison to the anterior sector. (c) The linear rate of contrast enhancement (*up-slope*) in the myocardium is color encoded and overlaid an MR perfusion image to show the spatial variation of the up-slope parameter and depict *relative* variations in myocardial blood flow within the heart.

References

1. Ford ES, Ajani UA, Croft JB et al. Explaining the decrease in U.S. deaths from coronary disease, 1980–2000. *New Engl J Med* 2007;356:2388–98.

2. Roger VL, Go AS, Lloyd-Jones DM et al. Heart disease and stroke statistics—2011 update: A report from the American Heart Association. *Circulation* 2011;123:e18–e209.

3. Patel MR, Peterson ED, Dai D et al. Low diagnostic yield of elective coronary angiography. *New Engl J Med* 2010;362:886–95.

4. Kim WY, Danias PG, Stuber M et al. Coronary magnetic resonance angiography for the detection of coronary stenoses. *New Engl J Med* 2001;345:1863–9.

5. Karamitsos TD, Dall'armellina E, Choudhury RP, Neubauer S. Ischemic heart disease: Comprehensive evaluation by cardiovascular magnetic resonance. *Am Heart J* 2011;162:16–30.

6. Spuentrup E, Botnar RM. Coronary magnetic resonance imaging: Visualization of the vessel lumen and the vessel wall and molecular imaging of arteriothrombosis. *Eur Radiol* 2006;16:1–14.

7. Chiribiri A, Kelle S, Gotze S et al. Visualization of the cardiac venous system using cardiac magnetic resonance. *Am J Cardiol* 2008;101:407–12.

8. Chiribiri A, Kelle S, Kohler U et al. Magnetic resonance cardiac vein imaging: Relation to mitral valve annulus and left circumflex coronary artery. *JACC Cardiovasc Imaging* 2008;1:729–38.

9. Kim WY, Stuber M, Kissinger KV, Andersen NT, Manning WJ, Botnar RM. Impact of bulk cardiac motion on right coronary MR angiography and vessel wall imaging. *J Magn Reson Imaging* 2001;14:383–90.

10. Plein S, Jones TR, Ridgway JP, Sivananthan MU. Three-dimensional coronary MR angiography performed with subject-specific cardiac acquisition windows and motion-adapted respiratory gating. *AJR Am J Roentgenol* 2003;180:505–12.

11. Jahnke C, Paetsch I, Nehrke K et al. A new approach for rapid assessment of the cardiac rest period for coronary MRA. *J Cardiovasc Magn Reson* 2005;7:395–9.

12. Leiner T, Katsimaglis G, Yeh EN et al. Correction for heart rate variability improves coronary magnetic resonance angiography. *J Magn Reson Imaging* 2005;22:577–82.

13. Tangcharoen T, Jahnke C, Koehler U et al. Impact of heart rate variability in patients with normal sinus rhythm on image quality in coronary magnetic angiography. *J Magn Reson Imaging* 2008;28:74–9.

14. Ustun A, Desai M, Abd-Elmoniem KZ, Schar M, Stuber M. Automated identification of minimal myocardial motion for improved image quality on MR angiography at 3 T. *AJR Am J Roentgenol* 2007;188:W283–90.

15. Ehman RL, Felmlee JP. Adaptive technique for high-definition MR imaging of moving structures. *Radiology* 1989;173:255–63.

16. Nehrke K, Bornert P, Groen J, Smink J, Bock JC. On the performance and accuracy of 2D navigator pulses. *Magn Reson Imaging* 1999;17:1173–81.

17. Wang Y, Riederer SJ, Ehman RL. Respiratory motion of the heart: Kinematics and the implications for the spatial resolution in coronary imaging. *Magn Reson Med* 1995;33:713–9.

18. Danias PG, McConnell MV, Khasgiwala VC, Chuang ML, Edelman RR, Manning WJ. Prospective navigator correction of image position for coronary MR angiography. *Radiology* 1997;203:733–6.

19. Danias PG, Stuber M, Botnar RM, Kissinger KV, Edelman RR, Manning WJ. Relationship between motion of coronary arteries and diaphragm during free breathing: Lessons from real-time MR imaging. *AJR Am J Roentgenol* 1999;172:1061–5.

20. Nagel E, Bornstedt A, Schnackenburg B, Hug J, Oswald H, Fleck E. Optimization of realtime adaptive navigator correction for 3D magnetic resonance coronary angiography. *Magn Reson Med* 1999;42:408–11.

21. Jahnke C, Nehrke K, Paetsch I et al. Improved bulk myocardial motion suppression for navigator-gated coronary magnetic resonance imaging. *J Magn Reson Imaging* 2007;26:780–6.

22. Edelman RR, Manning WJ, Burstein D, Paulin S. Coronary arteries: Breath-hold MR angiography. *Radiology* 1991;181:641–3.

23. Manke D, Nehrke K, Bornert P. Novel prospective respiratory motion correction approach for free-breathing coronary MR angiography using a patient-adapted affine motion model. *Magn Reson Med* 2003;50:122–31.

24. Huber S, Bornstedt A, Schnackenburg B, Paetsch I, Fleck E, Nagel E. The impact of different positions and thoracial restrains on respiratory induced cardiac motion. *J Cardiovasc Magn Reson* 2006;8:483–8.

25. Stuber M, Danias PG, Botnar RM, Sodickson DK, Kissinger KV, Manning WJ. Superiority of prone position in free-breathing 3D coronary MRA in patients with coronary disease. *J Magn Reson Imaging* 2001;13:185–91.

26. Sakuma H, Ichikawa Y, Chino S, Hirano T, Makino K, Takeda K. Detection of coronary artery stenosis with whole-heart coronary magnetic resonance angiography. *J Am Coll Cardiol* 2006;48:1946–50.

27. Manning WJ, Li W, Boyle NG, Edelman RR. Fat-suppressed breath-hold magnetic resonance coronary angiography. *Circulation* 1993;87:94–104.

28. Botnar RM, Stuber M, Danias PG, Kissinger KV, Manning WJ. Improved coronary artery definition with T2-weighted, free-breathing, three-dimensional coronary MRA. *Circulation* 1999;99:3139–48.

29. Brittain JH, Hu BS, Wright GA, Meyer CH, Macovski A, Nishimura DG. Coronary angiography with magnetization-prepared T2 contrast. *Magn Reson Med* 1995;33:689–96.

30. Li D, Paschal CB, Haacke EM, Adler LP. Coronary arteries: Three-dimensional MR imaging with fat saturation and magnetization transfer contrast. *Radiology* 1993;187:401–6.

31. Deshpande VS, Shea SM, Laub G, Simonetti OP, Finn JP, Li D. 3D magnetization-prepared true-FISP: A new technique for imaging coronary arteries. *Magn Reson Med* 2001;46:494–502.

32. Giorgi B, Dymarkowski S, Maes F, Kouwenhoven M, Bogaert J. Improved visualization of coronary arteries using a new three-dimensional submillimeter MR coronary angiography sequence with balanced gradients. *AJR Am J Roentgenol* 2002;179:901–10.

33. Spuentrup E, Bornert P, Botnar RM, Groen JP, Manning WJ, Stuber M. Navigator-gated free-breathing three-dimensional balanced fast field echo (TrueFISP) coronary magnetic resonance angiography. *Invest Radiol* 2002;37:637–42.

34. Kaul MG, Stork A, Bansmann PM et al. Evaluation of balanced steady-state free precession (TrueFISP) and K-space segmented gradient echo sequences for 3D coronary MR angiography with navigator gating at 3 Tesla. *Rofo* 2004;176:1560–5.

35. Dixon WT, Oshinski JN, Trudeau JD, Arnold BC, Pettigrew RI. Myocardial suppression in vivo by spin locking with composite pulses. *Magn Reson Med* 1996;36:90–4.

36. Nezafat R, Han Y, Peters DC et al. Coronary magnetic resonance vein imaging: Imaging contrast, sequence, and timing. *Magn Reson Med* 2007;58:1196–206.

37. Regenfus M, Ropers D, Achenbach S et al. Noninvasive detection of coronary artery stenosis using contrast-enhanced three-dimensional breath-hold magnetic resonance coronary angiography. *J Am Coll Cardiol* 2000;36:44–50.

38. Huber ME, Paetsch I, Schnackenburg B et al. Performance of a new gadolinium-based intravascular contrast agent in free-breathing inversion-recovery 3D coronary MRA. *Magn Reson Med* 2003;49:115–21.

39. Kelle S, Thouet T, Tangcharoen T et al. Whole-heart coronary magnetic resonance angiography with MS-325 (Gadofosveset). *Med Sci Monit* 2007;13:CR469–74.

40. Li D, Dolan RP, Walovitch RC, Lauffer RB. Three-dimensional MRI of coronary arteries using an intravascular contrast agent. *Magn Reson Med* 1998;39:1014–8.

41. Tang L, Merkle N, Schar M et al. Volume-targeted and whole-heart coronary magnetic resonance angiography using an intravascular contrast agent. *J Magn Reson Imaging* 2009;30:1191–6.

42. Liu X, Bi X, Huang J, Jerecic R, Carr J, Li D. Contrast-enhanced whole-heart coronary magnetic resonance angiography at 3.0 T: Comparison with steady-state free precession technique at 1.5 T. *Invest Radiol* 2008;43:663–8.

43. Nassenstein K, Breuckmann F, Hunold P, Barkhausen J, Schlosser T. Magnetic resonance coronary angiography: Comparison between a Gd-BOPTA- and a

Gd-DTPA-enhanced spoiled gradient-echo sequence and a non-contrast-enhanced steady-state free-precession sequence. *Acta Radiol* 2009;50:406–11.

44. Yang Q, Li K, Liu X et al. Contrast-enhanced whole-heart coronary magnetic resonance angiography at 3.0-T: A comparative study with X-ray angiography in a single center. *J Am Coll Cardiol* 2009;54:69–76.

45. Laurent S, Elst LV, Muller RN. Comparative study of the physicochemical properties of six clinical low molecular weight gadolinium contrast agents. *Contrast Media Mol Imaging* 2006;1:128–37.

46. Krombach GA, Hahnen C, Lodemann KP et al. Gd-BOPTA for assessment of myocardial viability on MRI: Changes of T1 value and their impact on delayed enhancement. *Eur Radiol* 2009;19:2136–46.

47. Goldfarb JW, Edelman RR. Coronary arteries: Breath-hold, gadolinium-enhanced, three-dimensional MR angiography. *Radiology* 1998;206:830–4.

48. Stuber M, Botnar RM, Danias PG et al. Contrast agent-enhanced, free-breathing, three-dimensional coronary magnetic resonance angiography. *J Magn Reson Imaging* 1999;10:790–9.

49. Bhat H, Yang Q, Zuehlsdorff S, Li K, Li D. Contrast-enhanced whole-heart coronary magnetic resonance angiography at 3 T using interleaved echo planar imaging. *Invest Radiol* 2010;45:458–64.

50. Slavin GS, Riederer SJ, Ehman RL. Two-dimensional multishot echo-planar coronary MR angiography. *Magn Reson Med* 1998;40:883–9.

51. Bornert P, Stuber M, Botnar RM et al. Direct comparison of 3D spiral vs. Cartesian gradient-echo coronary magnetic resonance angiography. *Magn Reson Med* 2001;46:789–94.

52. Priest AN, Bansmann PM, Mullerleile K, Adam G. Coronary vessel-wall and lumen imaging using radial k-space acquisition with MRI at 3 Tesla. *Eur Radiol* 2007;17:339–46.

53. Pruessmann KP, Weiger M, Scheidegger MB, Boesiger P. SENSE: Sensitivity encoding for fast MRI. *Magn Reson Med* 1999;42:952–62.

54. Sodickson DK, Manning WJ. Simultaneous acquisition of spatial harmonics (SMASH): Fast imaging with radiofrequency coil arrays. *Magn Reson Med* 1997;38:591–603.

55. Yu J, Schar M, Vonken EJ, Kelle S, Stuber M. Improved SNR efficiency in gradient echo coronary MRA with high temporal resolution using parallel imaging. *Magn Reson Med* 2009;62:1211–20.

56. Duckett SG, Ginks M, Shetty AK et al. Realtime fusion of cardiac magnetic resonance imaging and computed tomography venography with X-ray fluoroscopy to aid cardiac resynchronisation therapy implantation in patients with persistent left superior vena cava. *Europace* 2011;13:285–6.

57. Duckett SG, Ginks MR, Knowles BR et al. Advanced image fusion to overlay coronary sinus anatomy with real-time fluoroscopy to facilitate left ventricular lead implantation in CRT. *Pacing Clin Electrophysiol* 2011;34:226–34.

58. Stoeck CT, Han Y, Peters DC et al. Whole heart magnetization-prepared steady-state free precession coronary vein MRI. *J Magn Reson Imaging* 2009;29:1293–9.

59. Rasche V, Binner L, Cavagna F et al. Whole-heart coronary vein imaging: A comparison between non-contrast-agent- and contrast-agent-enhanced visualization of the coronary venous system. *Magn Reson Med* 2007;57:1019–26.

60. Younger JF, Plein S, Crean A, Ball SG, Greenwood JP. Visualization of coronary venous anatomy by cardiovascular magnetic resonance. *J Cardiovasc Magn Reson* 2009;11:26.

61. Duckett SG, Chiribiri A, Ginks MR et al. Cardiac MRI to investigate myocardial scar and coronary venous anatomy using a slow infusion of dimeglumine gadobenate in patients undergoing assessment for cardiac resynchronization therapy. *J Magn Reson Imaging* 2011;33:87–95.

62. Botnar RM, Stuber M, Kissinger KV, Kim WY, Spuentrup E, Manning WJ. Noninvasive coronary vessel wall and plaque imaging with magnetic resonance imaging. *Circulation* 2000;102:2582–7.

63. Fayad ZA, Fuster V, Fallon JT et al. Noninvasive in vivo human coronary artery lumen and wall imaging using black-blood magnetic resonance imaging. *Circulation* 2000;102:506–10.

64. Edelman RR, Chien D, Kim D. Fast selective black blood MR imaging. *Radiology* 1991;181:655–60.

65. Botnar RM, Stuber M, Lamerichs R et al. Initial experiences with in vivo right coronary artery human MR vessel wall imaging at 3 tesla. *J Cardiovasc Magn Reson* 2003;5:589–94.

66. Botnar RM, Kim WY, Bornert P, Stuber M, Spuentrup E, Manning WJ. 3D coronary vessel wall imaging utilizing a local inversion technique with spiral image acquisition. *Magn Reson Med* 2001;46:848–54.

67. Katoh M, Spuentrup E, Buecker A et al. MRI of coronary vessel walls using radial k-space sampling and steady-state free precession imaging. *AJR Am J Roentgenol* 2006;186:S401–6.

68. Kim WY, Stuber M, Bornert P, Kissinger KV, Manning WJ, Botnar RM. Three-dimensional black-blood cardiac magnetic resonance coronary vessel wall imaging detects positive arterial remodeling in patients with nonsignificant coronary artery disease. *Circulation* 2002;106:296–9.

69. Kim WY, Astrup AS, Stuber M et al. Subclinical coronary and aortic atherosclerosis detected by magnetic resonance imaging in type 1 diabetes with and without diabetic nephropathy. *Circulation* 2007;115:228–35.

70. Kerwin WS, Zhao X, Yuan C, Hatsukami TS, Maravilla KR, Underhill HR. Contrast-enhanced MRI of carotid atherosclerosis: Dependence on contrast agent. *J Magn Reson Imaging* 2009;30:35–40.

71. Yeon SB, Sabir A, Clouse M et al. Delayed-enhancement cardiovascular magnetic resonance coronary artery wall imaging: Comparison with multislice computed tomography and quantitative coronary angiography. *J Am Coll Cardiol* 2007;50:441–7.

72. Ibrahim T, Makowski MR, Jankauskas A et al. Serial contrast-enhanced cardiac magnetic resonance imaging demonstrates regression of hyperenhancement within the coronary artery wall in patients after acute myocardial infarction. *JACC Cardiovasc Imaging* 2009;2:580–8.

73. Lobbes MB, Miserus RJ, Heeneman S et al. Atherosclerosis: Contrast-enhanced MR imaging of vessel wall in rabbit model—Comparison of gadofosveset and gadopentetate dimeglumine. *Radiology* 2009;250:682–91.

74. Kooi ME, Cappendijk VC, Cleutjens KB et al. Accumulation of ultrasmall superparamagnetic particles of iron oxide in human atherosclerotic plaques can be detected by in vivo magnetic resonance imaging. *Circulation* 2003;107:2453–8.

75. Tang TY, Howarth SP, Miller SR et al. The ATHEROMA (Atorvastatin Therapy: Effects on Reduction of Macrophage Activity) Study. Evaluation using ultrasmall superparamagnetic iron oxide-enhanced magnetic resonance imaging in carotid disease. *J Am College Cardiol* 2009;53:2039–50.

76. Nahrendorf M, Jaffer FA, Kelly KA et al. Noninvasive vascular cell adhesion molecule-1 imaging identifies inflammatory activation of cells in atherosclerosis. *Circulation* 2006;114:1504–11.

77. Nahrendorf M, Keliher E, Panizzi P et al. 18F-4V for PET-CT imaging of VCAM-1 expression in atherosclerosis. *JACC Cardiovasc Imaging* 2009;2:1213–22.

78. Botnar RM, Buecker A, Wiethoff AJ et al. In vivo magnetic resonance imaging of coronary thrombosis using a fibrin-binding molecular magnetic resonance contrast agent. *Circulation* 2004;110:1463–6.

79. Botnar RM, Perez AS, Witte S et al. In vivo molecular imaging of acute and subacute thrombosis using a fibrin-binding magnetic resonance imaging contrast agent. *Circulation* 2004;109:2023–9.

80. Gerber TC, Fasseas P, Lennon RJ et al. Clinical safety of magnetic resonance imaging early after coronary artery stent placement. *J Am Coll Cardiol* 2003;42:1295–8.

81. Hug J, Nagel E, Bornstedt A, Schnackenburg B, Oswald H, Fleck E. Coronary arterial stents: Safety and artifacts during MR imaging. *Radiology* 2000;216:781–7.

82. Kramer CM, Rogers WJ, Jr., Pakstis DL. Absence of adverse outcomes after magnetic resonance imaging early after stent placement for acute myocardial infarction: A preliminary study. *J Cardiovasc Magn Reson* 2000;2:257–61.

83. Scott NA, Pettigrew RI. Absence of movement of coronary stents after placement in a magnetic resonance imaging field. *Am J Cardiol* 1994;73:900–1.

84. Shellock FG, Shellock VJ. Metallic stents: Evaluation of MR imaging safety. *AJR Am J Roentgenol* 1999;173:543–7.

85. Strohm O, Kivelitz D, Gross W et al. Safety of implantable coronary stents during 1H-magnetic resonance imaging at 1.0 and 1.5 T. *J Cardiovasc Magn Reson* 1999;1:239–45.

86. Shellock FG, Forder JR. Drug eluting coronary stent: In vitro evaluation of magnet resonance safety at 3 Tesla. *J Cardiovasc Magn Reson* 2005;7:415–9.

87. Hundley WG, Bluemke DA, Finn JP et al. ACCF/ACR/AHA/NASCI/SCMR 2010 expert consensus document on cardiovascular magnetic resonance: A report of the American College of Cardiology Foundation Task Force on Expert Consensus Documents. *J Am Coll Cardiol* 2010;55:2614–62.

88. Hofman MB, Paschal CB, Li D, Haacke EM, van Rossum AC, Sprenger M. MRI of coronary arteries: 2D breath-hold vs 3D respiratory-gated acquisition. *J Comput Assist Tomogr* 1995;19:56–62.

89. Oshinski JN, Hofland L, Mukundan S, Jr., Dixon WT, Parks WJ, Pettigrew RI. Two-dimensional coronary MR angiography without breath holding. *Radiology* 1996;201:737–43.

90. Paschal CB, Haacke EM, Adler LP. Three-dimensional MR imaging of the coronary arteries: Preliminary clinical experience. *J Magn Reson Imaging* 1993;3:491–500.

91. Post JC, van Rossum AC, Hofman MB, Valk J, Visser CA. Three-dimensional respiratory-gated MR angiography of coronary arteries: Comparison with conventional coronary angiography. *AJR Am J Roentgenol* 1996;166:1399–404.

92. Stuber M, Botnar RM, Danias PG et al. Double-oblique free-breathing high resolution three-dimensional coronary magnetic resonance angiography. *J Am Coll Cardiol* 1999;34:524–31.

93. Scheidegger MB, Muller R, Boesiger P. Magnetic resonance angiography: Methods and its applications to the coronary arteries. *Technol Health Care* 1994;2:255–65.

94. Pennell DJ, Bogren HG, Keegan J, Firmin DN, Underwood SR. Assessment of coronary artery stenosis by magnetic resonance imaging. *Heart* 1996;75:127–33.

95. Kim WY, Danias PG, Stuber M et al. Coronary magnetic resonance angiography for the detection of coronary stenoses. *N Engl J Med* 2001;345:1863–9.

96. Bogaert J, Kuzo R, Dymarkowski S, Beckers R, Piessens J, Rademakers FE. Coronary artery imaging with real-time navigator three-dimensional turbo-field-echo MR coronary angiography: Initial experience. *Radiology* 2003;226:707–16.

97. Dewey M, Teige F, Schnapauff D et al. Combination of free-breathing and breathhold steady-state free precession magnetic resonance angiography for detection of coronary artery stenoses. *J Magn Reson Imaging* 2006;23:674–81.

98. Jahnke C, Paetsch I, Nehrke K et al. Rapid and complete coronary arterial tree visualization with magnetic resonance imaging: Feasibility and diagnostic performance. *Eur Heart J* 2005;26:2313–9.

99. Jahnke C, Paetsch I, Schnackenburg B et al. Coronary MR angiography with steady-state free precession: individually adapted breath-hold technique versus free-breathing technique. *Radiology* 2004;232:669–76.

100. Maintz D, Aepfelbacher FC, Kissinger KV et al. Coronary MR angiography: Comparison of quantitative and qualitative data from four techniques. *AJR Am J Roentgenol* 2004;182:515–21.

101. Manning WJ, Li W, Edelman RR. A preliminary report comparing magnetic resonance coronary angiography with conventional angiography. *N Engl J Med* 1993;328:828–32.

102. Ozgun M, Hoffmeier A, Kouwenhoven M et al. Comparison of 3D segmented gradient-echo and steady-state free precession coronary MRI sequences in patients with coronary artery disease. *AJR Am J Roentgenol* 2005;185:103–9.

103. Sakuma H, Ichikawa Y, Suzawa N et al. Assessment of coronary arteries with total study time of less than 30 minutes by using whole-heart coronary MR angiography. *Radiology* 2005;237:316–21.

104. Schuetz GM, Zacharopoulou NM, Schlattmann P, Dewey M. Meta-analysis: Noninvasive coronary angiography using computed tomography versus magnetic resonance imaging. *Ann Intern Med* 2010;152:167–77.

105. Kato S, Kitagawa K, Ishida N et al. Assessment of coronary artery disease using magnetic resonance coronary angiography: A national multicenter trial. *J Am Coll Cardiol* 2010;56:983–91.

106. Miller JM, Rochitte CE, Dewey M et al. Diagnostic performance of coronary angiography by 64-row CT. *New Engl J Med* 2008;359:2324–36.

107. Nagel E. Magnetic resonance coronary angiography: The condemned live longer. *J Am College Cardiol* 2010;56:992–4.

108. Boffano C, Chiribiri A, Cesarani F. Native whole-heart coronary imaging for the identification of anomalous origin of the coronary arteries. *Int J Cardiol* 2009;137:e27–8.

109. Galjee MA, van Rossum AC, Doesburg T, van Eenige MJ, Visser CA. Value of magnetic resonance imaging in assessing patency and function of coronary artery bypass grafts. An angiographically controlled study. *Circulation* 1996;93:660–6.

110. Jenkins JP, Love HG, Foster CJ, Isherwood I, Rowlands DJ. Detection of coronary artery bypass graft patency as assessed by magnetic resonance imaging. *Br J Radiol* 1988;61:2–4.

111. Rubinstein RI, Askenase AD, Thickman D, Feldman MS, Agarwal JB, Helfant RH. Magnetic resonance imaging to evaluate patency of aortocoronary bypass grafts. *Circulation* 1987;76:786–91.

112. White RD, Caputo GR, Mark AS, Modin GW, Higgins CB. Coronary artery bypass graft patency: Noninvasive evaluation with MR imaging. *Radiology* 1987;164:681–6.

113. Vrachliotis TG, Bis KG, Aliabadi D, Shetty AN, Safian R, Simonetti O. Contrast-enhanced breath-hold MR angiography for evaluating patency of coronary artery bypass grafts. *AJR Am J Roentgenol* 1997;168:1073–80.

114. Wintersperger BJ, Engelmann MG, von Smekal A et al. Patency of coronary bypass grafts: Assessment with breath-hold contrast-enhanced MR angiography—Value of a non-electrocardiographically triggered technique. *Radiology* 1998;208:345–51.

115. Wintersperger BJ, von Smekal A, Engelmann MG et al. Contrast media enhanced magnetic resonance angiography for determining patency of a coronary bypass. A comparison with coronary angiography. *Rofo* 1997;167:572–8.

116. Wang Y, Moin K, Akinboboye O, Reichek N. Myocardial first pass perfusion: Steady-state free precession versus spoiled gradient echo and segmented echo planar imaging. *Magn Reson Med* 2005;54:1123–9.

117. Adluru G, McGann C, Speier P, Kholmovski EG, Shaaban A, Dibella EV. Acquisition and reconstruction of undersampled radial data for myocardial perfusion magnetic resonance imaging. *J Magn Reson Imaging* 2009;29:466–73.

118. Sharif B, Arsanjani R, Dharmakumar R, Bairey Merz CN, Berman DS, Li D. All-systolic non-ECG-gated myocardial perfusion MRI: Feasibility of multi-slice continuous first-pass imaging. *Magn Reson Med* 2015;74:1661–74.

119. Salerno M, Sica CT, Kramer CM, Meyer CH. Optimization of spiral-based pulse sequences for first-pass myocardial perfusion imaging. *Magn Reson Med* 2011;65:1602–10.

120. Salerno M, Taylor A, Yang Y et al. Adenosine stress cardiovascular magnetic resonance with variable-density spiral pulse sequences accurately detects coronary artery disease: initial clinical evaluation. *Circ Cardiovasc Imaging* 2014;7:639–46.

121. Di Bella EV, Parker DL, Sinusas AJ. On the dark rim artifact in dynamic contrast-enhanced MRI myocardial perfusion studies. *Magn Reson Med* 2005;54:1295–9.

122. Sharif B, Dharmakumar R, Labounty T et al. Projection imaging of myocardial perfusion: Minimizing the subendocardial dark-rim artifact. *J Cardiovasc Magn Reson* 2012;14 (Suppl 1):P275.

123. DiBella EV, Chen L, Schabel MC, Adluru G, McGann CJ. Myocardial perfusion acquisition without magnetization preparation or gating. *Magn Reson Med* 2012;67:609–13.

124. Manka R, Wissmann L, Gebker R et al. Multicenter evaluation of dynamic three-dimensional magnetic resonance myocardial perfusion imaging for the detection of coronary artery disease defined by fractional flow reserve. *Circ Cardiovasc Imaging* 2015;8:e003061.

125. Motwani M, Kidambi A, Sourbron S et al. Quantitative three-dimensional cardiovascular magnetic resonance myocardial perfusion imaging in systole and diastole. *J Cardiovasc Magn Reson* 2014;16:19.

126. Fair MJ, Gatehouse PD, DiBella EV, Firmin DN. A review of 3D first-pass, whole-heart, myocardial perfusion cardiovascular magnetic resonance. *J Cardiovasc Magn Reson* 2015;17:68.

127. Jerosch-Herold M, Hu XD, Murthy NS, Seethamraju RT. Time delay for arrival of MR contrast agent in collateral-dependent myocardium. *IEEE Trans Med Imaging* 2004;23:881–90.

128. Muehling OM, Cyran C, Jerosch-Herold M et al. Quantitative magnetic resonance first-pass perfusion imaging detects collateral-dependent myocardium. *Circulation* 2005;112:2272.

129. Muehling OM, Huber A, Cyran C et al. The delay of contrast arrival in magnetic resonance first-pass perfusion imaging: A novel non-invasive parameter detecting collateral-dependent myocardium. *Heart* 2007;93:842–7.

130. Xue H, Guehring J, Srinivasan L et al. Evaluation of rigid and non-rigid motion compensation of cardiac perfusion MRI. *Med Image Comput Comput Assist Interv* 2008;11:35–43.

131. Xue H, Zuehlsdorff S, Kellman P et al. Unsupervised inline analysis of cardiac perfusion MRI. *Med Image Comput Comput Assist Interv* 2009;12:741–9.

2

Imaging of the Heart: Myocardial Imaging

Ravi S. Shah, Otavio Coelho-Filho, Ana C. Andrade, and Michael Jerosch-Herold

CONTENTS

2.1 Introduction

Magnetic resonance imaging (MRI) is by now firmly established as a modality to image the heart, not least because of its arguably unrivaled capabilities for control of soft tissue contrast, its versatility in characterizing myocardial tissue properties, and to measure and quantify myocardial motion. The purpose of this chapter is to introduce the reader to the different methods used to achieve comprehensive myocardial imaging.

The vast majority of clinical cardiac MRI studies start with an assessment of ventricular volumes, myocardial mass, and global ventricular function by cine imaging. Hence, this chapter will introduce first this important component of a cardiac MRI exam. Although standard cine MRI of the heart also allows an assessment of regional cardiac function, some rather unique methods can be added to the repertoire, which can allow a highly sophisticated characterization of regional myocardial mechanics.

Assessing myocardial properties generally includes the standard qualitative methods probing the myocardial relaxation properties through T_1-, T_2-, and T_2*-weighted imaging. The resulting forms of soft-tissue contrast are widely used for detecting disease in the heart muscle. In recent years, the field has seen a growing interest in quantification of the relaxographic properties of myocardial tissue, that is, measuring and mapping the relaxation times T_1, T_2, and T_2*, which overcomes some of the limitations of using the relaxation properties to modulate tissue contrast.

2.2 Ventricular Function and Volumes

One of the most well-established applications for cardiac magnetic resonance (CMR) imaging is the precise and accurate quantification of ventricular function and volumes. The prognostic importance of left ventricular volumes is well known, with numerous studies demonstrating the link between myocardial infarct size, LV remodeling [1–3], and progressive ventricular dilatation [4–6]. The prognostic epidemiology of left ventricular structure and function is the subject of multiple reviews [7], and will not be recapitulated here. In this section, we will focus on the technical and clinical aspects of CMR imaging for assessment of structure and function.

Given the widespread use of echocardiography (ECG) for the assessment of cardiac volumes and function throughout cardiovascular medicine, CMR approaches to quantification of ventricular volumes and function have been essentially adapted from echocardiographic approaches. Although an area-length method has been utilized to measure LV volumes [8], the more common approach to assess LV volumes and ejection fraction is the Simpson's method (method of disks; Figure 2.1). In this approach, a stack of short-axis images planned in three dimensions from the base of the LV (at the mitral valve annular plane) to the apex is prescribed and imaged with a *cine* sequence (fast gradient-echo pulse sequence with or without steady-state free precession). The volume is calculated by the summation over all slices from base to apex of the product between the slice thickness and LV endocardial volume. For volume assessment, the papillary muscles are generally included in the LV volume.

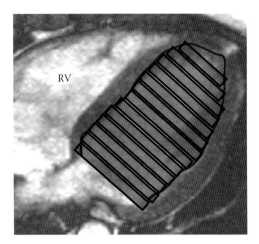

FIGURE 2.1

Illustration of quantification of ventricular volume using Simpson's method to sum the slice volumes falling within the ventricular cavity within each slice of a stack of slices with short-axis orientation that cover the left ventricle from base to apex. The slices in the stack do not have to be contiguous (*no slice gap*), in which case the volumes in the gaps are estimated by interpolation from the two adjacent slices. Cine MR images are acquired for during breath holding, with typically one or two slices acquired within one breath hold. The illustration represents an idealized situation insofar as it is assumed that the patient has suspended breathing such that the position of the heart is the same for each cine acquisition.

Traditionally, an electrocardiographically triggered fast gradient-echo (FGRE) pulse sequence has been used to achieve a temporal resolution of approximately 20 phases or more over the cardiac cycle. This is only possible by splitting the image acquisition into *segments* that define the temporal resolution. The term *segments* refers here to the subset of lines of data that are acquired for each cardiac phase. The image acquisition is therefore spread over a number of cardiac cycles, until enough data have been collected for each cardiac phase to reconstruct an image each phase in the cardiac cycle. During cine MRI, the continuous readout of gradient echoes (lines of data) over the cardiac cycle causes saturation of the myocardial signal, while blood pool in the ventricular cavity appears bright due to the inflow of unsaturated spins. This effect gave rise to the term *bright-blood* cine imaging. Nevertheless, in patients with poor cardiac function, the bright-blood effect is muted, leading to poor definition of the endocardial borders on the cine images. In addition, the continued application of radio-frequency pulses means that their flip angle has to be relatively low (15°–20°), and the resulting signal-to-noise ratio is sometimes suboptimal.

More recently, steady-state free precession (SSFP) cine sequences have been introduced to overcome these limitations and image the heart with a higher signal-to-noise ratio, reduced acquisition times, and better definition of the endocardial border [9]. (It should be noted here that higher signal-to-noise leaves more room for accelerating image acquisitions through parallel imaging, a technique that uses multiple receive-coil with receivers to speed up the image acquisition.) However, some large population studies have utilized FGRE imaging techniques in the past, and the results of SSFP and FGRE have some systematic differences [10]. In general, LV volumes are higher and LV mass lower with an SSFP approach [11]. This probably has to do with the fact that the endocardial border is more sharply defined on SSFP images, that is, a bright-blood pool signal is observed right up to the true myocardial border, shifting the balance between what is considered blood pool and myocardium toward the former.

Unlike the area-length method or visual estimation, a stack of slices yields reproducible and precise quantification of LV volumes, ejection fraction, and end-diastolic mass (if both the epicardial and endocardial borders are traced) without assumptions for cavity geometry [12,13]. Normative ranges have been developed for both right and left ventricular volumes and ejection fraction using this technique, and published evidence suggests that this is highly reproducible, more accurate than ECG [14], and may even lead to smaller sample sizes in studies using LV volumes or function as an endpoint [15]. Left ventricular mass assessment is performed in a similar manner, with the area between the endocardium and epicardium (excluding the papillary muscles in most cases, unless specifically queried) used in a Simpson's calculation from base to apex. The volumes are multiplied by 1.02 g/ml (the approximate density of myocardium) to generate an LV mass from the myocardial volume.

Quantification of right ventricular structure is of paramount concern in several different pathophysiologic states, including pulmonary hypertension, heart failure, and congenital heart disease. CMR is the gold standard for the assessment of RV morphology, structure, and function, given that it is based on tomographic imaging planes. Although approaches that image the RV axially have yielded improved precision [16], current approaches to RV imaging rely on short-axis base-to-apex imaging, with inclusion from the pulmonary valve to the apical region.

An interesting consideration that has helped to establish CMR as the gold standard for quantification of cardiac volumes and function is comparison of *in vivo* with *ex vivo* (postmortem) data. In a study of 15 latex cast postmortem hearts, Rehr and colleagues reported a near perfect correlation (coefficient 0.997) between anatomic cast volume and volumes by CMR [17]. Anatomic validation of left and right ventricular mass has also been performed [18,19], with excellent correlation between *ex vivo* specimens and CMR-determined ventricular mass.

Finally, the improved precision of CMR (e.g., over ECG or nuclear techniques) has established CMR imaging as a mainstay to limit sample size in clinical interventional studies: in one report, CMR assessments of ventricular morphology yielded an over 10-fold decrease in sample size over ECG [20].

Many studies have utilized CMR imaging for mass, function, and volume quantification in different important clinical circumstances, including postmyocardial infarction, valvular heart disease, and heart failure. Importantly, one major advantage of CMR in these varied clinical conditions is the availability of a multiparametric CMR approach, involving infarct size assessment (see Section 2.7), myocardial perfusion, and T_2 characterization, to further interrogate areas with reduced myocardial function. As a result, the scope of this chapter is insufficient to discuss the breadth of data in this field.

Several population-based studies (e.g., the Multi-Ethnic Study of Atherosclerosis) have taken advantage of non-contrast CMR approaches quantifying global and regional function, to shed light in larger patient populations on disease physiology and mechanism [21–24]. In addition, more recent studies (e.g., the United Kingdom BioBank) have embarked on an even larger approach, integrating biomarker and *big data* methodologies to assess of regional myocardial phenotypes, mass, and function, as a means to uncover important physiologic relationships. Ultimately, CMR may be uniquely positioned with these approaches to shed substantial light on the relationships between myocardial biology and clinical outcomes.

2.3 Regional Myocardial Function

In addition to the quantification of *global* LV and RV function, CMR is highly useful to assess *regional* dysfunction and subclinical dysfunction—including in the context of new approaches in CMR to examine regional subclinical myocardial tissue remodeling. Regional myocardial dysfunction may be prognostic beyond global function, especially in preclinical disease (e.g., hypertension, obesity, and diabetes) (MESA REF). As the first responder in this area, ECG has established techniques to examine regional function, including regional wall thickness (in systole and diastole), thickening, and strain and strain rate imaging using speckle tracking. CMR has taken a similar approach by harnessing information about intra-myocardial motion (tagging, phase contrast [25], and regional thickening) to examine regional function. In this section, we will describe the methods used to acquire myocardial tagging images and some of the parameters garnered from

these approaches. The reader is also directed to a recent review that details the history and current state-of-the-art in this field [26].

Several different techniques have emerged for the quantification of regional myocardial function [26]. Initial approaches in CMR utilized a scheme for spatial modulation of magnetization (CSPAMM) to lay down a grid of *taglines* in the myocardium at the beginning of the cardiac cycle, followed by imaging to track the intra-myocardial deformation of these taglines [27,28]. The taglines (e.g., parallel lines or a grid pattern) are created at the beginning of the cardiac cycle, and before starting the cine readout by saturating the magnetization along narrow lines with a combination of radio-frequency pulses and gradient pulses. Figure 2.2 shows an example. Tagline separation is of the order 5–10 mm for human studies, and this effectively defines the spatial resolution of the strain maps that can be calculated from *tagged* MR images. With the parallel imaging, the spatial resolution of the underlying image is between 1 and 2 mm, and the temporal resolution for following the movement of the taglines in the myocardium is near 20–50 ms. Several different image-readout techniques (e.g., echo planar imaging, SSFP, and GRE) have been successfully employed to acquire *tagged* images. Tagged cine MR images are typically acquired for a set of short-axis (or long-axis) myocardial slices allowing quantification of regional strain and strain rate over the course of the cardiac cycle. The fading of tag gridlines is dependent on myocardial T_1 relaxation. For this reason, tagged cine imaging is generally performed before contrast administration (to take advantage of a longer T_1 relaxation time, that is, slower tagline fading over the cardiac cycle).

The analysis of tagged images relies on correct identification of gridlines and tracking those lines over time during the cardiac cycle. In general, four sets of cine tagging images are acquired for analysis, including long-axis (four-chamber), and three short-axis slices (base, mid, and apical segments) for assessment of regional LV function. Circumferential and radial strains are assessed from short-axis images, whereas longitudinal strain is assessed from long-axis views (e.g., radial long axis view rotated about the long axis of the LV). Circumferential, radial, and longitudinal strain components, strain rates, and higher order parameters (e.g., torsion) can be then obtained, using a variety of proprietary software packages. Tissue tagging has been extensively applied in the evaluation of myocardial disease, including aging [29], RV function [30], coronary disease [31], cardiomyopathies, and dyssynchrony [32]. Nevertheless, the clinical application of *quantitative* strain imaging has been hampered by the relatively extensive post-processing requirements, which to date have been reduced to 2–3 min per slice for a trained observer. Future improvements in signal-to-noise ratios and automated console

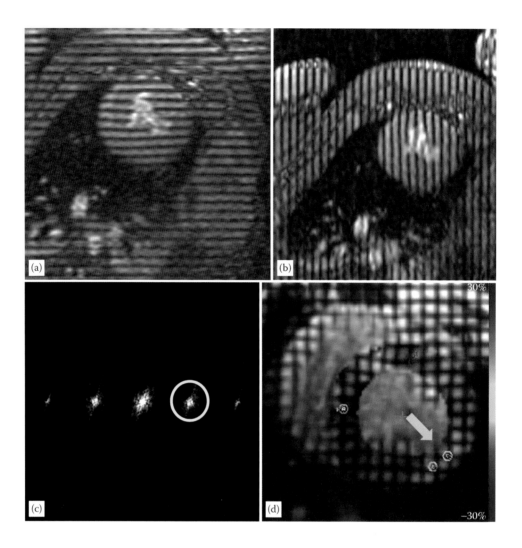

FIGURE 2.2
Selected tagged MR images with (a) horizontal and (b) vertical taglines were acquired with a cine gradient-echo pulse sequence with spatial modulation of the magnetization being applied right after the R-wave on the electrocardiogram. (c) For semi-automatic quantitative detection of tagline deformation, one can take advantage of the characteristic peaks and their harmonics in the Fourier-transformed images, which result from the periodic pattern of parallel taglines in the images. The changes of the first harmonic peak (yellow circle) location and phase from image to image can be used to quantify the tagline deformation. (d) shows a map of circumferential myocardial strain superimposed on combined images from (a) and (b). Alternatively, one can acquire directly tagged images where the spatial modulation of magnetization produces a grid pattern.

analysis will improve clinical translatability of CMR regional myocardial function and its integration as an indispensible aspect of the clinical CMR examination.

2.4 T_1- and T_2-Weighted Myocardial Imaging

The T_2-relaxation time constant refers to the decay of the transverse magnetization created by an initial radio frequency. For T_2 measurements, the transverse magnetization is refocused into spin-echoes, to render the decay rate independent of field inhomogeneities, which

cause a loss of phase coherence. Spin-echoes are created by an initial 90° radio-frequency pulse followed by single (conventional spin-echo) or multiple (*turbo* spin-echo) 180° refocusing pulses. The time delay between the initial 90° radio-frequency pulse and the generation of a spin-echo is called the *echo time* (TE). In addition, there is the repetition time (TR) for repeating the 90° excitation pulse and spin-echo readouts, which is important for the T_1-weighting of the images. Multiple spin-echoes can be produced by a train of 180° pulses, spaced apart by TE. The decay of the amplitudes of the resulting echoes defines the T_2-relaxation time. When the application radio-frequency pulse stops, that is, at the end of the echo train, the 1H spins return to their original orientation along the static magnetic field at a rate determined by the T_1-relaxation time, that is, the longitudinal magnetization component, which was initially converted into transverse magnetization by the first 90° pulse, recovers at a rate determined by T_1. (In reality, the T_1- and T_2-relaxation processes occur in parallel, but T_1 effects during the generation of the echo train can be neglected to a first approximation because T_2 is generally short compared to T_1.) If spin-echoes are created again before full relaxation, the resulting image will acquire T_1-weighting, that is, the signal from different tissues types in such an image will reflect the T_1 of the tissue.

MRI takes advantage of the differences in T_1 and T_2 (or T_2*) relaxation to produce contrast between different soft tissues. In ECG-gated spin-echo imaging, TE determines the T_2-contrast, and the TR for the spin-echo readouts (e.g., between 90° pulses) determines the T_1-weighting. T_1-weighting is increased by reducing this TR, with TE being kept short to avoid a competing T_2-weighting. On T_1-weighted spin-echo images, fat has a bright (high) signal, and myocardium a relatively low signal. For routine cardiac spin-echo imaging, a short TR corresponds to one R-R interval, and a long TR corresponds to two or three R-R intervals. T_2-weighting is obtained with a long TR, and a long TE. T_2-weighted spin-echo images are used for edema imaging because increased fluid content lengthens T_2.

On cardiac (turbo) spin-echo imaging, a bright signal from the blood pool can cause considerable ghosting over the myocardium. As the signal intensity of the blood pool is generally not of particular interest, one resorts to suppressing the signal from the blood pool, giving rise to the so-called T_1- or T_2-weighted *black-blood* (turbo) spin-echo sequences. One example is black-blood spin-echo imaging with a double inversion recovery preparation [33]. The double inversion preparation applies a non-slice-selective inversion pulse to the whole volume to be imaged, followed by a second, slice-selective inversion pulse that restores signal mostly for the myocardium. This means that effectively the myocardial magnetization undergoes a 360° rotation (i.e., returns to its original state), while blood has experienced an inversion, except for a thin slice of spins that has moved out of the image plane by the time the image is acquired. The readout of the spin-echo is timed such that it coincides with the zero-crossing of the blood magnetization during its inversion recovery. This delay is called the *inversion time* (TI). In practice, the double inversion preparation is triggered with a delay after an R-wave on the ECG, and the spin-echoes are produced 400 to 600 ms later, such that they fall into diastole. Turbo spin-echo sequences acquire multiple lines of *k*-space during each heartbeat—one line with each spin-echo—by repeating the 180° refocusing

a fixed number of times, referred to as the *echo train length*. An echo train length of 10 to 20 echoes usually allows the acquisition of one or two slices in a single breath hold, though it will introduce T_2-weighting. For T_1-weighted turbo spin-echo imaging, the T_2-weighting is kept low by acquiring the data lines defining the image contrast of large image features early in the echo train. (For readers familiar with the concept of *k*-space, this can be equivalent states as the central *k*-space lines are acquired early in the echo train.)

A common challenging of double inversion recovery turbo spin-echo imaging is loss of signal of the myocardium as a result of cardiac motion (which cannot be compensated by the 180° refocusing pulses). Structures of interest may move out of the slice, where they do not experience the second inversion pulse. This situation is remedied by increasing the thickness of the slice selective pulse. Signal from fat can be suppressed by adding a third inversion pulse immediately before the spin-echo sequence or by suppressing signal with the resonance frequency of fat molecules. Fat saturation is usually applied in addition to blood suppression to increase the contrast between normal and edematous myocardium.

Double inversion-recovery T_2-weighted black-blood imaging has been used to delineate the area at risk after myocardial infarction and to identify myocardial edema and inflammation. (Figure 2.3 shows an example in a patient with acute chest pain.) T_2-weighted fast spin-echo or gradient-echo techniques have been used to characterize the extent of myocardial edema as the *at-risk region* due to ischemia [34]. Application in the clinical setting where acute coronary syndrome is common has shown encouraging results, though myocardial contrast can be affected by extraneous sources such as coil sensitivity variation over the heart. T_2-weighted fast spin-echo imaging is therefore preferably carried out with the body coil used for radio-frequency transmission and signal reception, rather than using a surface coil for the latter, and even though this results in lower signal-to-noise ratios. Cury and colleagues assessed 62 chest-pain patients in the emergency room and found that the addition of T_2-weighted imaging for myocardial edema improved the differentiation of acute versus chronic coronary artery disease [32]. T_2-weighted technique enhanced the positive predictive value in detecting acute coronary syndrome from 55% to 85%. After an acute MI, infarct remodeling continues to take place over months and is governed by status of coronary reflow, ongoing ischemia, collateral formation, and infarct location. Thus, it is expected that novel CMR techniques that can capture different information of myocardial physiology will continue to shed light on prognostic value of CMR beyond total infarct size, area-at-risk, and left ventricular function.

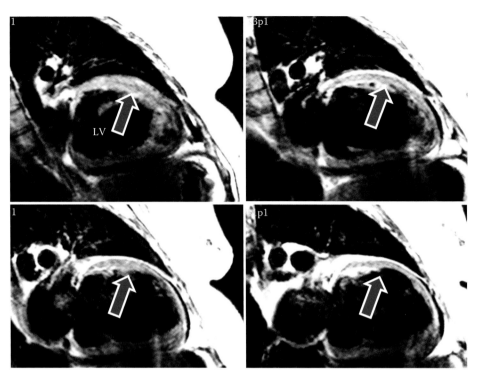

FIGURE 2.3
Black-blood T_2-weighted turbo-spin-echo images of the left ventricle (LV) for four parallel long-axis views in a 26-year-old female patient presenting at the emergency room with shortness of breath and chest pain. With this technique, the signal from the blood pool is suppressed to avoid ghosting artifacts over myocardium, and to highlight intra-myocardial signal-intensity differences. The black blood preparation is achieved with a combination of consecutive slice-selective and non-slice selective inversion pulses, which have no net effect on the myocardium, but inverted the signal from the blood that at ends up in the slice plane when the image is acquired. The time between inversion pulses and image acquisition is adjusted to coincide with the null-crossing of the magnetization in blood.

2.5 T_2^*-Weighted Imaging

T_2^*-weighted images can be acquired with gradient-echo sequences, and they have been used for assessment of myocardial iron content. Iron overload causes signal loss in affected tissues. Iron deposition induces local, microscopic irregularities in the magnetic field that cause water protons around these deposits to lose phase coherence more rapidly. This effect is iron-concentration dependent. T_2^*-weighted images can be obtained for a series of T_2^*-weightings/echo times using a single breath-hold multiecho gradient-echo technique. A single short-axis mid-ventricular slice is acquired at various TEs ranging approximately from 2 to 18 ms. A region of interest (ROI) is then drawn in the interventricular septum to measure the signal intensity at each TE. Plotting the signal intensity against the TE produces an exponential decay curve, from which T_2^* can be estimated.

T_2^* times of the heart (in ms) determined by CMR is inversely related to the amount of iron deposition in the heart. At 1.5 T a myocardial T_2^* less than 20 ms is abnormal and indicates probable iron overload, and a myocardial T_2^* less than 10 ms indicates severe iron overload. Iron overload states can lead to heart failure and mortality. However, there is usually good response to medical management using chelation therapy as indicated by improvement in LV dysfunction. Shorter T_2^* times are associated with worsening LV dysfunction and heart failure [35,36]. Anderson and colleagues reported a threshold value of T_2^*, below which is suggestive of iron-overloading induced cardiac dysfunction in patients with undiagnosed heart failure [37]. Some studies have demonstrated that T_2^* CMR imaging can be used to monitor therapeutic response in patients who are being treated with chelation therapy [38–40]. A recent multicenter study of 650 patients has shown that a cardiac T_2^* of <6 ms is associated with development within 1 year of heart failure in 47% of patients and arrhythmia in 14% [41]. This and other pieces of evidence indicate that cardiac T_2^* measurements are reproducible, transferable, and strong predictors of clinical outcome. This noninvasive technique not only is an instrumental tool to study the efficacy of new chelation therapy, but it should become the standard of care for all patients receiving long-term transfusions. As a result, CMR has become a useful tool in the initial diagnosis as well as follow-up for effectiveness of medical management in this patient population.

T_2 or T_2* imaging is also useful to assess the oxygenation status of the myocardium, as an increase of deoxy-hemoglobin will cause of reduction of T_2 and T_2*. With T_2- or T_2*-weighted imaging, the baseline resting state carries little information by itself. However, if T_2/T_2*-weighted imaging is also carried out during *stress* (generally pharmacological induced stress, e.g., with an inotropic agent such as dobutamine), then it becomes feasible to assess the relative change of the T_2/T_2*-weighted signal. This blood-oxygen-level-dependent (BOLD) contrast change between baseline and stress has proven useful to identify ischemic myocardium.

2.6 Quantitative Myocardial Tissue Characterization by T_1 or T_2 Mapping

Although T_1- or T_2-weighted myocardial imaging have proven exceedingly useful for myocardial characterization, the resulting images in general only allow a qualitative evaluation, and subtle differences in signal intensity, for example, on T_2-weighted images due to myocardial edema, may not be noticeable, or there are confounding effects that also modulate the myocardial signal intensity. For example, the sensitivity of a surface coil varies over the volume of the heart, potentially causing signal variations on T_2-weighted images that are unrelated to edema. However, the argument in favor quantifying T_1 and T_2, and generating parametric T_1 and T_2 maps on the heart can be carried much further, as has been shown over the last couple of years.

T_1 imaging of blood and myocardium before and after administration of an extra-cellular contrast agent can be used to quantify the extracellular distribution volume (ECV) in the myocardium. ECV expansion occurs with the deposition of connective tissue in the interstitial space. ECV has been shown to be a marker sensitive to build-up of diffuse fibrosis in the myocardium. Similarly, ECV is significantly increased by the deposition of amyloid protein, providing a quantitative marker for patients, where late-gadolinium enhancement (LGE) is often very diffuse. In patients with cardiac amyloidosis, the global effect of the disease on the myocardium makes the standard prescription for LGE imaging of nulling the signal in *normal* myocardium problematic, whereas T_1 mapping can extract an objective, quantitative parameters of ECV expansion due to amyloid protein deposition. There has also been increased interest on the assessment of myocardial T_1 without contrast administration, and estimating what is referred to

as the *native* T_1. This parameter has been shown to be quite sensitive to the presence of myocardial edema [42], where it lengthens T_1, and to the presence of iron, where it leads to T_1 shortening [43].

T_1 imaging is carried out by acquiring a series of images after an inversion pulse. The most widely used technique uses single shot gradient-echo-readouts with steady-state free precession for each sampled time after inversion, as this type of image readout has a relatively small disturbing effect on the inversion recovery [44]. Therefore, multiple images can be acquired after an inversion pulse, and the true T_1 can be estimated relatively easily by an exponential fit of signal intensity as a function of the time after inversion (TI). The method of reading out signal intensity after an inversion pulse was originally introduced by Look and Locker before the era of MR imaging [45], when. This technique was later adapted to MR imaging, and more recently modified for cardiac T_1 imaging to acquire images for different times after the inversion pulse (TI) in the *same* phase of the cardiac cycle—it was introduced with the acronym of MOLLI for modified Look–Locker imaging [46]. As all images with different TI times are acquired during the same phase of the cardiac cycle, the heart appears stationary, which makes it relatively straightforward to map T_1 for each pixel location in the MOLLI *source* images [47]. The resulting high-resolution T_1-maps have had a major impact in the research of myocardial pathologies in various diseases.

Analogous methods can be used to generate a series of images with different effective TE weightings, and the generation of T_2 or T_2* maps. Because these parametric maps almost exclusively reflect the variation of T_2/T_2*, they have substantially improved the assessment of tissue pathology that alters T_2 or T_2*, like iron overload. In fact, it can be argued, based on the experiences and results of research studies, that are an essential part of the standard repertoire for tissue characterization in patients with suspected iron-overload [48,49].

The qualitative evaluation of ECV can help in terms of research, disease progression, prognostic, and therapeutic aspects. The amount of fibrosis in myocardial interstitium, its varying degrees in diseases, such as infarction, myocarditis, hypertrophic and dilated cardiomyopathy, cardiac amyloidosis, cardiac involvement in systemic diseases as systemic lupus erythematosus or in patients with human immunodeficiency virus, and aortic stenosis [50].

Noninvasive sampling analysis are as reliable as anatomopathological ones, and there are regions not achievable through *in-vivo* biopsy only with autopsy. Biopsy of certain regions can result in false negative results, while T_1 mapping allows a reproducible exam of the entire myocardium [51].

2.7 Myocardial Viability

A loss of myocardial viability coincides with the breakdown of cardiomyocyte membranes. The detection of this breakdown in membrane integrity in acute myocardial infarction is feasible with MRI through commonly used extra-cellular contrast gadolinium-based contrast agents. Cardiomyocytes account in healthy myocardium for >70% of the myocardial tissue volume. In viable myocardium, a gadolinium-based contrast chelate is excluded from this volume, and with breakdown of cardiomyocyte membrane integrity, the volume of distribution is increased.

Imaging of LGE was introduced as a method to highlight areas with an increased volume of distribution for gadolinium contrast, relative to viable myocardium [52]. Differences in the volume of distribution result after contrast administration in differences of myocardial T_1, which can be dramatically highlighted by adjusting T_1 contrast. During an inversion recovery, the myocardial magnetization crosses during its recovery from the inverted state to the equilibrium state a point where its longitudinal component, that is, the component from which one can generate a signal for an image, is effectively zero. If imaging data are acquired at this particular time point after the magnetization inversion, then any tissue where the longitudinal magnetization crosses the null point appears effectively black. In non-viable myocardial tissue, the relaxation T_1 will be shorter due to

the increased volume of distribution, or in other words, the increased accumulation of contrast agent per voxel volume. On an image that is acquired at the time when the signal intensity of *normal* myocardium is nulled through careful selection of the time after inversion, the non-viable myocardium can have a dramatically higher signal intensity.

For the practical implementation of LGE imaging, a few key challenges had to be overcome: (1) The image contrast should be predominantly T_1-weighted, with as little sensitivity to cardiac motion as possible, and allow for high spatial resolution to allow an accurate assessment of infarct transmurality and (2) image data should only be acquired within a small time window around the time after inversion when the longitudinal magnetization in normal myocardium is nulled. Simonetti and colleagues introduced a highly successful T_1-weighted, segmented gradient-echo imaging technique that met these requirements [53]. The term *segmented* refers here to the fact that the image data are acquired in segments, that is, the acquisition of the image data is distributed over several heartbeats. Each readout of image data is preceded by a non-slice-selective inversion pulse, and the image readout occurs at a time-after inversion that can be adjusted by the operator to null the signal in *normal* myocardium. Figure 2.4 illustrates the technique. As one would like to acquire the image data during a relatively quiet the portion of the cardiac cycle (e.g., early diastole), the implementations of the technique allow one to specify the trigger delay for the image acquisition,

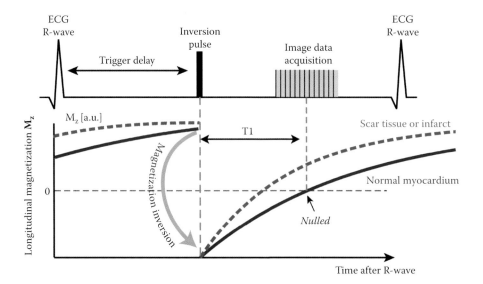

FIGURE 2.4
The technique for imaging of late-gadolinium enhancement is based on an inversion pulse, followed at a user-specified time after the magnetization inversion by the readout of a segment of the image data. This pulse sequence block is typically repeated every second heartbeat to acquire an entire image. The acquisition of the image data segments are timed to coincide with null-crossing of magnetization in normal myocardium after magnetization inversion. Infarcted myocardium or myocardial scar has a shorter T_1, and the magnetization has already passed the null point when the image is acquired, making these areas appear to be bright in the resulting images. The trigger delay is chosen, so that images are acquired during a relatively quiescent phase of the cardiac cycle.

and the inversion pulse is applied at a time TI (time after inversion) before the image read-out—when TI is changed, the position in time of the inversion pulse is shifted relative to the image acquisition time, rather than shifting the time at which the image is acquired in the cardiac cycle.

To make the determination of the optimal TI for nulling the signal in normal myocardium as easy as possible, it has become customary to run a so-called TI scout sequence after contrast administration and before LGE imaging to quickly determine the correct TI value. This is essentially a cine technique, combined with an inversion pulse, so that each image in the cine loop corresponds to a different TI. It is acquired within a single-breath-hold duration, in contrast to the LGE technique that typically acquires only one image for a single TI value within a breath-hold duration. The TI value corresponding to the image in the TI scout cine loop where viable myocardium is nulled is then adopted for the LGE acquisition.

The optimal delay for acquiring LGE images after contrast administration was initially assumed to be of the order 20 min, but with the wide adoption of this technique, it has become clear that at least in acute myocardial infarctions LGE imaging can start as early as approximately 5 min after contrast injection.

The prescription for adjusting TI so that the signal intensity in normal myocardium is nulled may appear to defy the logic of needing to detect non-viable myocardium. However, as LGE imaging is typically performed toward the end of a cardiac study when cine imaging of the heart has already been completed, one can safely assume that myocardial segments that are not hypokinetic can at least partially serve as reference regions to achieve nulling of the myocardial signal. Another reason why cine imaging is useful for LGE imaging relates to the not infrequent difficulty of delineating the endocardial border of infarcted segments, where the enhanced myocardial tissue appears as bright as the signal from the blood pool: On a precontrast cine, the endocardial border can be clearly delineated in the infarct region, thereby providing a reference for determining the transmural extent a myocardial infarct or scar.

LGE imaging provides a reliable quantitation of the presence and the extent of infarcted myocardial tissue [54]. Kim and colleagues showed that LGE technique delineates the transmural extent of infarction and distinguishes between reversible and irreversible myocardial injury regardless of the extent of wall motion at rest, the age of the infarct, or the reperfusion status [52,55]. In a landmark study of patients with coronary artery disease, the probability of improvement in regional contractility after successful coronary revascularization decreased in an inversely proportional fashion to the transmural extent of LGE before revascularization

[54]. In addition, the transmural extent of LGE can provide accurate prediction of the response of LV function from beta-blocker therapy in heart failure patients [56]. In another study, Orn et al. [57] found that scar size assessed by CMR was the strongest independent predictor of ejection fraction and left ventricular volumes in acute myocardial infarction patients and heart failure.

References

1. Pfeffer, M.A. and E. Braunwald, Ventricular remodeling after myocardial infarction. Experimental observations and clinical implications. *Circulation*, 1990. 81(4): 1161–72.
2. Chareonthaitawee, P. et al., Relation of initial infarct size to extent of left ventricular remodeling in the year after acute myocardial infarction. *J Am Coll Cardiol*, 1995. 25(3): 567–73.
3. Van Gilst, W.H. et al., Which patient benefits from early angiotensin-converting enzyme inhibition after myocardial infarction? Results of one-year serial echocardiographic follow-up from the Captopril and Thrombolysis Study (CATS). *J Am Coll Cardiol*, 1996. 28(1): 114–21.
4. Douglas, P.S. et al., Left ventricular shape, afterload and survival in idiopathic dilated cardiomyopathy. *J Am Coll Cardiol*, 1989. 13(2): 311–5.
5. Konstam, M.A. et al., Effects of the angiotensin converting enzyme inhibitor enalapril on the long-term progression of left ventricular dysfunction in patients with heart failure. SOLVD Investigators. *Circulation*, 1992. 86(2): 431–8.
6. Konstam, M.A. et al., Effects of the angiotensin converting enzyme inhibitor enalapril on the long-term progression of left ventricular dilatation in patients with asymptomatic systolic dysfunction. SOLVD (Studies of Left Ventricular Dysfunction) Investigators. *Circulation*, 1993. 88(5 Pt 1): 2277–83.
7. Heckbert, S.R. et al., Traditional cardiovascular risk factors in relation to left ventricular mass, volume, and systolic function by cardiac magnetic resonance imaging: the Multiethnic Study of Atherosclerosis. *J Am Coll Cardiol*, 2006. 48(11): 2285–92.
8. Lawson, M.A. et al., Accuracy of biplane long-axis left ventricular volume determined by cine magnetic resonance imaging in patients with regional and global dysfunction. *Am J Cardiol*, 1996. 77(12): 1098–104.
9. Thiele, H. et al., Functional cardiac MR imaging with steady-state free precession (SSFP) significantly improves endocardial border delineation without contrast agents. *J Magn Reson Imaging*, 2001. 14(4): 362–7.
10. Malayeri, A.A. et al., Cardiac cine MRI: Quantification of the relationship between fast gradient echo and steady-state free precession for determination of myocardial mass and volumes. *J Magn Reson Imaging*, 2008. 28(1): 60–6.

11. Moon, J.C. et al., Breath-hold FLASH and FISP cardiovascular MR imaging: Left ventricular volume differences and reproducibility. *Radiology*, 2002. 223(3): 789–97.

12. Longmore, D.B. et al., Dimensional accuracy of magnetic resonance in studies of the heart. *Lancet*, 1985. 1(8442): 1360–2.

13. Maceira, A.M. et al., Normalized left ventricular systolic and diastolic function by steady state free precession cardiovascular magnetic resonance. *J Cardiovasc Magn Reson*, 2006. 8(3): 417–26.

14. Grothues, F. et al., Comparison of interstudy reproducibility of cardiovascular magnetic resonance with two-dimensional echocardiography in normal subjects and in patients with heart failure or left ventricular hypertrophy. *Am J Cardiol*, 2002. 90(1): 29–34.

15. Bellenger, N.G. et al., Reduction in sample size for studies of remodeling in heart failure by the use of cardiovascular magnetic resonance. *J Cardiovasc Magn Reson*, 2000. 2(4): 271–8.

16. Alfakih, K. et al., Comparison of right ventricular volume measurements between axial and short axis orientation using steady-state free precession magnetic resonance imaging. *J Magn Reson Imaging*, 2003. 18(1): 25–32.

17. Rehr, R.B. et al., Left ventricular volumes measured by MR imaging. *Radiology*, 1985. 156(3): 717–9.

18. Katz, J. et al., Estimation of human myocardial mass with MR imaging. *Radiology*, 1988. 169(2): 495–8.

19. Katz, J. et al., Estimation of right ventricular mass in normal subjects and in patients with primary pulmonary hypertension by nuclear magnetic resonance imaging. *J Am Coll Cardiol*, 1993. 21(6): 1475–81.

20. Bottini, P.B. et al., Magnetic resonance imaging compared to echocardiography to assess left ventricular mass in the hypertensive patient. *Am J Hypertens*, 1995. 8(3): 221–8.

21. Chahal, H. et al., Obesity and right ventricular structure and function: The MESA-Right Ventricle Study. *Chest*, 2012. 141(2): 388–95.

22. Yan, R.T. et al., Regional left ventricular myocardial dysfunction as a predictor of incident cardiovascular events MESA (multi-ethnic study of atherosclerosis). *J Am Coll Cardiol*, 2011. 57(17): 1735–44.

23. Rosen, B.D. et al., Reduction in regional myocardial function is associated with concentric left ventricular remodeling: Multi-ethnic study of atherosclerosis. *J Am College Cardiol*, 2004. 43(5): 221A–221A.

24. Rosen, B.D. et al., Left ventricular concentric remodeling is associated with decreased global and regional systolic function: The Multi-Ethnic Study of Atherosclerosis. *Circulation*, 2005. 112(7): 984–91.

25. Pelc, L.R. et al., Evaluation of myocardial motion tracking with cine-phase contrast magnetic resonance imaging. *Invest Radiol*, 1994. 29(12): 1038–42.

26. Ibrahim el, S.H., Myocardial tagging by cardiovascular magnetic resonance: Evolution of techniques—Pulse sequences, analysis algorithms, and applications. *J Cardiovasc Magn Reson*, 2011. 13: 36.

27. Axel, L. and L. Dougherty, MR imaging of motion with spatial modulation of magnetization. *Radiology*, 1989. 171(3): 841–5.

28. Zerhouni, E.A. et al., Human heart: Tagging with MR imaging—A method for noninvasive assessment of myocardial motion. *Radiology*, 1988. 169(1): 59–63.

29. Fonseca, C.G. et al., Aging alters patterns of regional nonuniformity in LV strain relaxation: A 3-D MR tissue tagging study. *Am J Physiol Heart Circ Physiol*, 2003. 285(2): H621–30.

30. Haber, I., D.N. Metaxas, and L. Axel, Three-dimensional motion reconstruction and analysis of the right ventricle using tagged MRI. *Med Image Anal*, 2000. 4(4): 335–55.

31. Kraitchman, D.L. et al., Quantitative ischemia detection during cardiac magnetic resonance stress testing by use of FastHARP. *Circulation*, 2003. 107(15): 2025–30.

32. Curry, C.W. et al., Mechanical dyssynchrony in dilated cardiomyopathy with intraventricular conduction delay as depicted by 3D tagged magnetic resonance imaging. *Circulation*, 2000. 101(1): E2.

33. Simonetti, O.P. et al., "Black blood" T2-weighted inversion-recovery MR imaging of the heart. *Radiology*, 1996. 199(1): 49–57.

34. Aletras, A.H. et al., Retrospective determination of the area at risk for reperfused acute myocardial infarction with T2-weighted cardiac magnetic resonance imaging: Histopathological and displacement encoding with stimulated echoes (DENSE) functional validations. *Circulation*, 2006. 113(15): 1865–70.

35. Westwood, M. et al., A single breath-hold multiecho T2* cardiovascular magnetic resonance technique for diagnosis of myocardial iron overload. *J Magn Reson Imaging*, 2003. 18(1): 33–9.

36. Tanner, M.A. et al., Myocardial iron loading in patients with thalassemia major on deferoxamine chelation. *J Cardiovasc Magn Reson*, 2006. 8(3): 543–7.

37. Anderson, L.J. et al., Cardiovascular T2-star (T2*) magnetic resonance for the early diagnosis of myocardial iron overload. *Eur Heart J*, 2001. 22(23): 2171–9.

38. Mavrogeni, S.I. et al., A comparison of magnetic resonance imaging and cardiac biopsy in the evaluation of heart iron overload in patients with beta-thalassemia major. *Eur J Haematol*, 2005. 75(3): 241–7.

39. Westwood, M.A. et al., Myocardial biopsy and T2* magnetic resonance in heart failure due to thalassaemia. *Br J Haematol*, 2005. 128(1): 2.

40. Westwood, M.A. et al., Left ventricular diastolic function compared with T2* cardiovascular magnetic resonance for early detection of myocardial iron overload in thalassemia major. *J Magn Reson Imaging*, 2005. 22(2): 229–33.

41. Kirk, P. et al., Cardiac T2* magnetic resonance for prediction of cardiac complications in thalassemia major. *Circulation*, 2009. 120(20): 1961–8.

42. Ferreira, V.M. et al., Non-contrast T1-mapping detects acute myocardial edema with high diagnostic accuracy: A comparison to T2-weighted cardiovascular magnetic resonance. *J Cardiovasc Magn Reson*, 2012. 14: 42.

43. Sado, D.M. et al., Noncontrast myocardial T mapping using cardiovascular magnetic resonance for iron overload. *J Magn Reson Imaging*, 2014. 41: 1505–1511.

44. Scheffler, K. and J. Hennig, T(1) quantification with inversion recovery TrueFISP. *Magn Reson Med*, 2001. 45(4): 720–3.

45. Look, D. and D. Locker, Time saving in measurement of NMR and EPR relaxation times. *Rev Sci Instrum*, 1970. 41: 250–251.

46. Messroghli, D.R. et al., Modified Look-Locker inversion recovery (MOLLI) for high-resolution T1 mapping of the heart. *Magn Reson Med*, 2004. 52(1): 141–6.

47. Messroghli, D.R. et al., Myocardial T1 mapping: Application to patients with acute and chronic myocardial infarction. *Magn Reson Med*, 2007. 58(1): 34–40.

48. Baksi, A.J. and D.J. Pennell, T2* imaging of the heart: Methods, applications, and outcomes. *Top Magn Reson Imaging*, 2014. 23(1): 13–20.

49. Pennell, D.J. et al., Cardiovascular function and treatment in beta-thalassemia major: A consensus statement from the American Heart Association. *Circulation*, 2013. 128(3): 281–308.

50. Ferreira, V.M. et al., Myocardial tissue characterization by magnetic resonance imaging: Novel applications of T1 and T2 mapping. *J Thorac Imaging*, 2014. 29(3): 147–54.

51. Flett, A.S. et al., Equilibrium contrast cardiovascular magnetic resonance for the measurement of diffuse myocardial fibrosis: Preliminary validation in humans. *Circulation*, 2010. 122(2): 138–44.

52. Kim, R.J. et al., Relationship of MRI delayed contrast enhancement to irreversible injury, infarct age, and contractile function. *Circulation*, 1999. 100(19): 1992–2002.

53. Simonetti, O.P. et al., An improved MR imaging technique for the visualization of myocardial infarction. *Radiology*, 2001. 218(1): 215–23.

54. Kim, R.J. et al., The use of contrast-enhanced magnetic resonance imaging to identify reversible myocardial dysfunction. *N Engl J Med*, 2000. 343(20): 1445–53.

55. Kim, R.J. et al., Myocardial Gd-DTPA kinetics determine MRI contrast enhancement and reflect the extent and severity of myocardial injury after acute reperfused infarction. *Circulation*, 1996. 94(12): 3318–26.

56. Bello, D. et al., Gadolinium cardiovascular magnetic resonance predicts reversible myocardial dysfunction and remodeling in patients with heart failure undergoing beta-blocker therapy. *Circulation*, 2003. 108(16): 1945–53.

57. Orn, S. et al., Effect of left ventricular scar size, location, and transmurality on left ventricular remodeling with healed myocardial infarction. *Am J Cardiol*, 2007. 99(8): 1109–14.

3

Computed Topography/Magnetic Resonance Imaging of Pericardial Disease

Marco Francone, Giorgia Giustini, Gianluca De Rubeis, Iacopo Carbone, and Nicola Galea

CONTENTS

3.1 Introduction

Pericardial pathology has attained raising interest due to both increasing comprehension of its complex pathophysiology and the development and widespread diffusion of accurate noninvasive imaging tools, which combine the assessment of both functional and morphological aspects of its abnormalities.

Pericardial disease includes a wide spectrum of congenital and acquired pathological conditions characterized by heterogeneous clinical manifestations, with potential impact on biventricular function and therefore on patient's management and prognosis.

Beyond the isolated primary diseases, it is stated that "everything causes pericardial pathology" to underline the frequency of its secondary involvement in daily clinical practice and the variety of organs and systemic conditions that may potentially affect the pericardium, ranging from infective, autoimmune, neoplastic to iatrogenic processes (e.g., complications of cardiac surgery or post-radiation therapy) [1,2].

In spite of its relatively simple macroscopic and ultrastructural architecture, pericardial sac plays a complex role in cardiac physiology, which is derived by combining mechanical containment and anchoring functions involved in the regulation of normal ventricular compliance. Pericardial sac contributes to minimize ventricular dilatation, to physically protect the heart, through production of fluid and surfactants and limits its displacement in the mediastinum [1,3].

Although in presence of *typical* clinical manifestations (e.g., acute pericarditis with chest pain exacerbated in orthostatic position and by pericardial rub) diagnosis can be established straightforwardly [4,5] with conventional diagnostic tools, there is an increasing number of *atypical* clinical expression of disease or scenarios in which diagnostic contribution of *second-line* imaging modalities such as computed topography (CT) and magnetic resonance (MR) is widely advocated to establish a correct diagnosis and adequate patient's management [6].

Moreover, in a large variety of conditions, conventional transthoracic echocardiography may have limited diagnostic accuracy, such as in patients with suboptimal acoustic window (obese subjects, severe chronic obstructive pulmonary disease, or skeletal malformations), and the recognition of small loculated effusions or characterization of complex high-protein content exudates may result challenging [4,6,7].

The contribution of MR/CT in the presence of suspected neoplastic disease/infiltration of the pericardial layer is now well established, in which the combination of high-resolution imaging and larger field of view enables better depiction and characterization of the underlying process.

The additional value of MR has also been reported in various forms of suspected constrictive pericarditis (CP), such as in the presence of non-thickened pericardium with restrictive pattern of ventricular dysfunction and its differential diagnosis between pericardial constriction and restrictive cardiomyopathy (RC) [8,9].

For these reasons, also supported by the recent technological developments, which range from real-time imaging in cardiac MR to highly increased temporal resolution and functional imaging capability of the latest multidetector computerized topography (MDCT) systems, both techniques have rapidly emerged in a large variety of pathological conditions.

3.1.1 Pericardial Anatomy and Physiology

The pericardium is a thin, flask-shaped, elastic membrane covering the heart, and great vessels (Figure 3.1) and macroscopically composed of two paired layers: the inner serosa (also known as *visceral pericardium* or *epicardium*) and the outer fibrosa (the *parietal pericardium*) [10–12].

The serosa is a complete sac distended by a variable amount of 15–50 mL of plasmatic ultrafiltrate, and it is separated from the heart by loose epicardial connective tissue and a single line of mesothelial cells. The adipose tissue that surrounds the pericardium is divided into two components: subserosal and epicardial (the latter seems to be the most important one due to its paracrine and physical role). Subserosal fat is localized in normal weight people, especially over the interventricular sulcus and over the atrioventricular (AV) sulcus, and is found to increase in obese people [13].

Similar to the visceral abdominal fat, epicardial fat thickness is increased in obesity and could directly influence coronary atherogenesis because there is no fibrous layer to impede diffusion of free fatty acids and adipokines that modulate atherothrombosis to the underlying coronary arteries [13–15].

Anatomically, three- to fourfold more epicardial fat is distributed along the right ventricle than in the left [13], and the highest amount of fat tissue is usually located at the level of the right ventricular free wall [13–15], creating high natural contrast between visceral pericardium and the myocardium, thereby enabling better anatomical delineation of the structure with CT and MR.

The outer layer (fibrous pericardium) is clasped internally to the epicardium and cranially extends above the aortic root, continuing with the deep cervical fascia; it is attached to the sternum and to the diaphragm by loose ligaments, which impede myocardial displacement within the mediastinum and protect the heart from the adjacent structures.

When surgically incised, fibrous retraction of pericardium reflects the mechanical stress that is predominantly applied on thinner wall structures such as the right ventricle and atrium, opposing to the pathological overdistension of those chambers (as in presence of chronic large effusions) and directly influencing endocavitary diastolic pressure.

The relative pericardial *inelasticity* is also necessary to maintain ventricular coupling or interdependence in order to equalize the filling and relaxation between two ventricles, as confirmed in pathological conditions such as CP or cardiac tamponade (see Sections 3.3.2 and 3.3.3) [10,16].

Another property of the pericardium that deserves to be mentioned because of its potential implications in constrictive physiology attains its adaptation to pleural and intrathoracic pressures [2,8,17], in which the membrane acts as a sort of transducer between pleural spaces and cardiac chambers, directly transmitting intrathoracic respiratory changes to the heart. This phenomenon can be physiologically observed during inspiration in which respiratory fall of pleural pressures is transmitted to cardiac chambers as a consequence of the pleura–pericardial interactions causing an increment of venous return and right heart filling and therefore direct increase on right ventricular preload and output.

The pericardium has also two major reflections of the visceral layer arranged around two complex connected

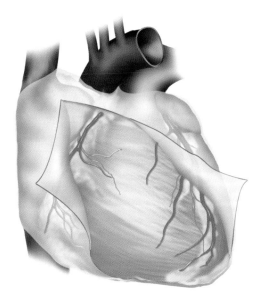

FIGURE 3.1
Anatomic drawing of the normal pericardial sac after cutting of parietal layer; the visceral layer is strictly adherent to the epicardium. (Courtesy of Bettina Conti, MD, Sapienza University, Rome, Italy.)

tubes: the former covers the superior and inferior vena cava and the right pulmonary veins, and the latter covers the left pulmonary veins. These reflections delimit two pocket-like structures described as oblique and transverse sinuses, which include several recesses, and contribute to create the so-called pericardial reserve volume in which fluid can strategically accumulate in presence of increased fluid content [11,18]

> *Transverse sinus* is interposed between the great arteries, atria, and vena cava, and it is characterized by a linear-shaped virtual cavity, which sometimes can be misinterpreted to simulate an aortic focal dissection/thrombus or a pathological lymphadenopathy on CT-MR examination [11].
>
> *Oblique sinus* is located behind the left atrium, and it is *J*-shaped [2,11,12].

As previously mentioned, Pericardium has three main functions:

> *Mechanical:* It ensures the maintenance of normal ventricular compliance to minimize ventricular dilatation; the production of pericardial fluid protects the heart from stretching, underlying some physical stress. Pericardium has a crucial role in the physiological interaction between the right and the left ventricles, defined as *ventricular coupling,* which in normal loading conditions is characterized by a slight prevalence of LV diastolic pressures with respect to RV pressures, determining the typically convex shape of septum with a slight right-sided bulge. Some pericardial diseases may affect the interventricular interaction, such as in a condition where the pericardial compliance or pressure is altered; the RV diastolic pressure is higher than the LV pressure, with subsequent inversion or flattening of interventricular septal shape. The pericardium also regulates the pressure relationships between the heart and the pleura, so that in normal condition, the respiratory changes are transmitted to the heart with an increase of RV filling during inspiration. This important physiological feature could be crucial to differential diagnosis between CP and RC.
>
> *Membranous:* Protecting the heart from infection by contact coming from other neighboring organs such as lungs, physical damage from trauma, and tumoral infiltration coming from adjacent organs.
>
> *Ligamentous:* Fixing the heart in the medium mediastinum to prevent excessive movement caused by no-direct impact.

3.2 CT and MR Imaging Techniques

CT and MR imaging allows excellent anatomical delineation of the pericardial sac in most of the cases, which appears on both techniques as a thin linear band parallel to the myocardium, surrounded by a variable amount of epicardial fat tissue (Figure 3.2).

Pericardial line is better depicted at the level of the right ventricular free wall due to the greater amount of fat surrounding the heart, usually not (or poorly) visible in the lateral and posterior walls of the left ventricle [2,12,19].

Normal pericardial thickness measured with MR and CT ranges, respectively, between 1.2 and 1.7 mm [20–22], and 2.0 ± 0.4 mm [23] and are constantly overestimated when compared to those reported in pathological studies [24], likely as a consequence of the intrinsically limited spatial and temporal resolutions not allowing to fully distinguish between layers and the internal fluid [25].

CT images of the pericardial membrane show a single-layered hyper dense structure, which is easily detectable using or not using contrast media because of its natural contrast against the low attenuation of the surrounding fat; the membrane slightly linearly enhances after iodine administration and is likely due to the presence of small capillary vessels within the fibrous component of the parietal layer [22,26].

CT technique for the depiction of pericardial pathology is usually performed with a simple volumetric high-resolution acquisition technique (i.e., slice thickness <3 mm), which allows excellent delineation of pericardial layers [19,27], whereas the use of ECG-gating or triggering is beneficial to minimize heart-related motion artifacts and to improve depiction of sinuses and recesses at the level of the upper mediastinum, particularly to avoid a misdiagnosis of lymphadenopathy or other mediastinal or hilar disease [18,27].

Retrospectively, ECG-gated MDCT dynamic evaluation of septal motion has also been proposed in the clinical setting of constrictive pericarditis [22,28,29] to evaluate the impact of pericardial pathology on interventricular septal motion and configuration at the price of a non-negligible radiation dose; however, the use of MRI or echocardiogram is preferable for this purpose.

Although limited by the intrinsically low contrast resolution impeding differentiation between layers and the fluid, by its relative invasiveness (ionizing radiations exposure and iodine contrast agent) and by the lack of functional information provided, CT has the advantage of being an extremely fast acquisition technique, offering excellent anatomical outline of the pericardial layers and with high sensitivity for detection of pericardial calcifications (typical morphological characteristic

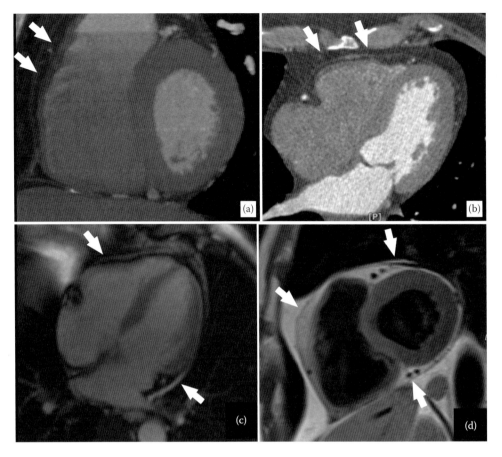

FIGURE 3.2

Normal appearance of the pericardium. Contrast-enhanced CT images of the heart obtained from a dual-source multidetector scanner with 1.0 mm collimation on short-axis (a) and horizontal long-axis (b) views. Pericardium is typically best visible along the free wall of right ventricle as a thin line surrounded by fat (arrows). On horizontal long-axis balanced steady-state free-precession (SSFP) cine MR image (c) pericardium can be easily depicted when a little amount of pericardial effusion is present because of the *chemical shift* artifact, which produce black lines at fat–water interfaces (also known as *black boundary artifact*). On black-blood T_1-weighted fast spin-echo MR images in short axis (d), the pericardium appears as a thin curvilinear hypointense structure (arrows), because of the presence of a small amount of pericardial fluid interposed between two layers, best appreciable when surrounded by fat tissue.

of CP) or hemorrhage, which represents in acute setting a potential life-threatening condition.

A further clinical situation where the use of CT should be preferred is in the presence of suspected secondary pericardial involvement (like metastatic or systemic disease), in which the whole body single-step approach of the technique (including lung parenchyma) allows a faster and better diagnostic workup.

Cardiac MR is the ideal noninvasive imaging modality for the assessment of pericardial pathology, combining excellent morphological depiction and tissue characterization with functional information about the effects of pathology on ventricular filling and specifically on diastolic function and ventricular coupling.

The preferred *morphological* MR technique for imaging of the pericardium is a black-blood segmented T_1-weighted ECG-gated turbo spin-echo sequence, usually acquired with small field of views and saturation blocks to avoid unfolding artifacts and to allow depiction of the pericardium as a thin hypointense curvilinear band surrounded by the high-intensity mediastinal and epicardial fat and the mid-intensity of the myocardium.

The combination of T_2-weighted spin-echo morphological imaging with a triple inversion recovery black-blood acquisition, which allows signal suppression from flowing blood as well as from fat, can be applied to detect fluid and/or edema of the inflamed pericardial layers in the presence of pericarditis [25,30].

Functional imaging still relies on balanced steady-state free precession (SSFP) sequences, which are not only the reference standard for quantification of regional and global ventricular function, but can also be used in pericardial imaging for the assessment of layers mobility, providing indirect relevant information about stiffness of the pericardial sac and its potential impact on diastolic filling.

Data from cine-SSFP images could also be integrated with tagged cine-MR sequences, which can be helpful to recognize fibrotic adhesion of the thickened layers to the myocardium indicated by the persistent appearance of tagged signals between the pericardium/myocardium interface throughout the cardiac cycle [31].

A further MR functional technique that can be adopted in presence of suspected CP is the real-time non-triggered imaging approach, which is based on a high temporal resolution sequence using the SSFP technique in combination with parallel imaging. This technique enables to assess the effects of respiratory activity on interventricular septal motion as explained above [8,17].

Contrast-enhanced T_1-weighted images, using both turbo spin-echo (TSE) or inversion recovery sequences, may highlight inflamed-enhancing pericardial layers in the clinical setting of inflammatory pericardial disease, offering indirect information about disease activity and allowing discrimination of the different components of pericardial and myocardial inflammation (e.g., fluid vs. layers) [32] (see Figure 3.3).

Paramagnetic contrast agent is also recommended when pericardial masses are suspected and in case of underlying myocardial involvement (e.g., suspected pericardio-myocarditis).

Finally, use of velocity encoded cine MRI (VENC) or phase contrast imaging is mandatory to assess impact of pericardial pathology on diastolic function by measuring velocities and inflow patterns at the level of AV valves and pulmonary and/or systemic veins [33–35].

3.3 Congenital Disease

3.3.1 Pericardial Cysts and Diverticula

Pericardial cysts and diverticula are rare congenital diseases with a prevalence of 1/100,000 individuals and represent 13%–17% of all mediastinal cysts [36].

Congenital cysts are benign unilocular masses of coelomic origin during the embryogenesis, [37] usually detected on a routine chest X-ray or transthoracic echocardiography examination in an asymptomatic individual, whereas symptoms may occur in the presence of compression due to adjacent structures [1].

They usually present on both CT and MR imaging techniques as encapsulated structures containing clear fluid, without internal septa or nodules, and their most common locations are in the right cardiophrenic angle (30% of cases) and in the left cardiophrenic angle (20% of cases) [38], directly attached to the pericardium or rarely by a peduncle (Figures 3.3 and 3.4).

Pericardial diverticula are focal extroflexions of the pericardial sac characterized by direct communication with the pericardial cavity, unlike congenital cysts, as demonstrated by their changing size according to the body position [18,36].

3.3.2 Pericardial Defects

Congenital defects of the pericardium are extremely rare anomalies due to the premature atrophy of the cardinal veins supplying the pleuro-pericardial folds during embryogenesis, resulting in developmental arrest of the membrane. The extent of the defect, likely determined by the timing of vascular degeneration, ranges from small foramen in the pericardium to complete absence of the pericardial sac [39,40]. Most pericardial defects are asymptomatic and are an unexpected finding at surgery or postmortem [39], although a partial defect may have an acute presentation with chest pain, dyspnea, dysrhythmias, syncope, and even sudden death when complicated by herniation and entrapment of cardiac structures [41].

Partial defects are predominantly on the left side of the heart, but they can be located anywhere in the pericardium. Pericardial abnormalities may be associated with other congenital cardiac abnormalities such as atrial or ventricular septal defect, patent ductus arteriosus, bicuspid aortic valve, diaphragmatic hernia, or pulmonary malformations [39].

Mobilization of cardiac structures as well as lung parenchyma can cause the typical *levo-displacement* of the heart and aortic knob on chest X-ray (in the case of left absence), with the trachea remaining midline.

Cross-sectional (CT/MR) imaging techniques are usually performed as second-line modalities after chest X-ray recognition, even though it has the relevant diagnostic limitation that pericardial line is usually not visualized along left chambers, which largely represent the most frequent location of the defect, as mentioned above.

Therefore, indirect signs of an absent pericardium are usually necessary for establishing a diagnosis, consisting in a variable leftward displacement of the heart within the chest cavity or cardiac indentation at the location of the defect (in partial absence), usually emphasized with positional changes and associated with increased apical mobility observed during cine images [42–44].

3.3.2.1 Pericardial Effusion

Pericardial effusion occurs when there is an increased production of fluid within the visceral layers or obstruction of lymphatic and venous drainage from the heart, which may be caused by a large number of local and systemic conditions, including heart failure, neoplastic

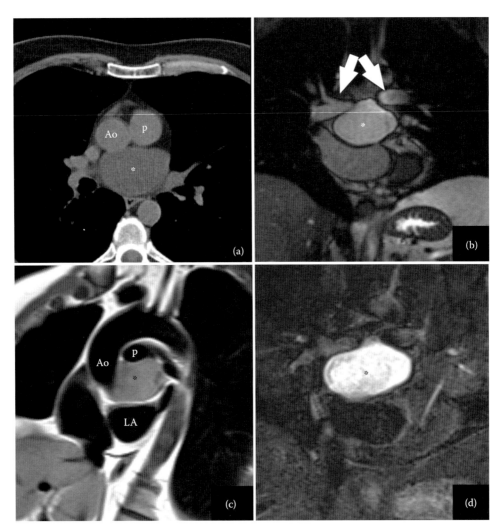

FIGURE 3.3
Typical pericardial cyst of transverse sinus. Axial unenhanced computed tomography (a) and coronal bSSFP cine MR (b) images show a mediastinal sharply edged oval-shaped cystic structure (*) interposed between pulmonary trunk (p) and its main branches (arrows), ascending aorta (Ao), and left atrium (LA). The cyst has low density (attenuation: 7 HU) and no wall thickening on CT scan (a). SSFP cine MR image (b) is helpful to confirm thin and regular margins and to exclude the presence of clot or any corpuscolar content. Low signal intensity in spin echo T_1-weighted sequence (c) and high signal intensity in coronal STIR T_2-weighted sequence (d) suggest liquid content.

diseases, inflammation, renal and liver insufficiency, traumas, and myocardial infarction [1–3].

Transthoracic echocardiography is the preferred imaging modality to detect pericardial effusion and to address a correct diagnosis in most of the cases, by being a rapid, widely available, and a relatively inexpensive diagnostic tool. However, echocardiography is limited in patients with unfavorable acoustic windows or with small focal effusions [6,16]; furthermore, pericardial thickness is usually nonassessable with ultrasound, with obvious limited accuracy in the presence of suspicious overlapping inflammation [16].

Both CT and MR are highly accurate to confirm presence of the pericardial fluid, to quantify its amount, to provide a certain content characterization (transudates vs. exudates), to detect pericardial inflammation, to assess its

hemodynamic impact on ventricular filling, and eventually to guide pericardiocentesis (Figures 3.5 and 3.6).

CT and MR imaging techniques are usually required also to rule out underlying secondary causes of effusion from pericarditis to neoplastic lesions, or when a complex inhomogeneously echoic exudate is observed at echocardiography [5].

Both modalities allow accurate evaluation of pericardial width and detection of even minimal amounts of fluid (up to 10–40 mL representing the physiological plasmatic ultrafiltrate [45,46]), which can strategically accumulate in those anatomical recesses usually nonassessable with ultrasound, such as the interior-basal wall of the right atrium or the oblique sinus.

The amount of effusion can be measured semi-quantitatively, assuming that a pericardial space greater

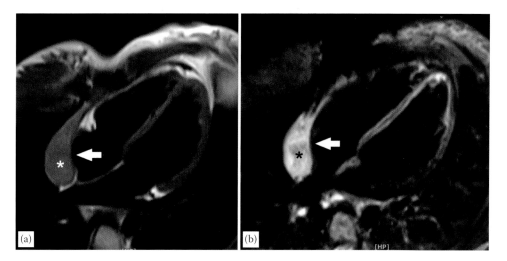

FIGURE 3.4
Congenital cyst. (a) TSE T_1-weighted and (b) STIR T_2-weighted images show a cystic elongated formation (*) located within the pericardial sac at the level of the lateral atrial wall (arrows). Signal intensity is compatible with inhomogeneous corpuscolated content.

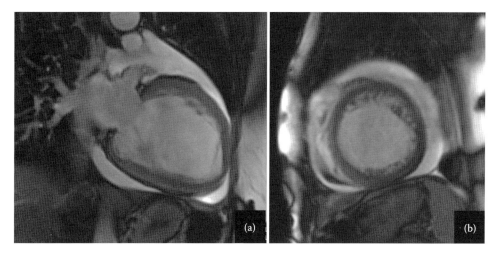

FIGURE 3.5
Balanced steady-state free precession cine MR images acquired on longitudinal (a) and horizontal (b) long-axis views show a large pericardial transudate effusion in patients with dilated cardiomyopathy. The slow and gradual accumulation of a large amount of pericardial effusion stretches the pericardial sac and increases compliance without increasing the pericardial pressure.

than 5 mm anterior to the right ventricle is equated to at least a moderate effusion (i.e., 100–500 mL) or using a volumetric approach based on a multi-slice manual contouring of fluid, like for ventricular volumes [47].

Generally, the pericardial space is filled with about 25–30 mL and the negative pressure regime present therein, causes the phenomenon *ventricular coupling*; the presence of effusion increases the intrapericardial pressure and hampers venous return, resulting in a disturbance of cardiac hemodynamics.

Analysis of fluid attenuation/signal intensity also enables differentiation between transudates (hydropericardium) and exudates; however, reliable characterization of fluid content is not always feasible because of the extreme variability in the exudates characteristics, the

artifacts related to cardiac motion, and the presence of surrounding fat tissue (Figure 3.6).

Hydropericardium typically present a water-like attenuation, in contrast to hemopericardium or purulent exudates, that may be easily suspected on MDCT when a high-attenuation effusion is depicted, or to chylopericardium, characterized by inhomogeneous low-attenuation pericardial fluid [19]; presence of fluid–gas levels within visceral layers can also be identified in patients with pneumopericardium or purulent anaerobic infections [22].

MR imaging shows a peculiar homogeneous T_1-weighted low signal and T_2-weighted high signal intensity of transudates as compared to the extremely variable shortening of T_1- and T_2-relaxation of the

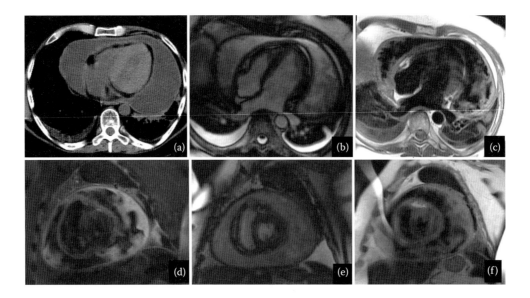

FIGURE 3.6
Pericardial effusion. Axial unenhanced CT image (a) shows a huge pericardial effusion (approximately 1100–1200 mL) in a patient with bone marrow aplasia, with low-intermediate attenuation content (15–25 Hounsfield Units). bSSFP cine MR (b and e) and black-blood TSE T_1-weighted (c and f) images, respectively, acquired on long-axis and short-axis views demonstrate an inhomogeneous fluid-corpuscolated mixed content suggesting exudative origin. Please note how ECG-gated short-tau inversion recovery (STIR) T_2-weighted sequence (d) is not very useful to characterize pericardial fluid content, because it is prone to flow artifact due to cardiac beat, especially for great amount of effusion. Bilateral small pleural effusion can be also detected.

highly proteinaceous effusions. Hemopericardium can be suspected in patients who have previously undergone aortic/cardiac surgery, who have a history of trauma, neoplastic disease, tuberculosis or in the case of Dressler's syndrome, and is characterized by high signal intensity on T_1-weighted images and inhomogeneous low signal intensity on cine SSFP images [19,25,26].

However, it is also remarkable to note that signal intensity of the fluid may vary on MR (and mostly on T_1-weighted sequences) due to both nonlaminar motion of the water and the presence of chemical shift artifacts at the level of the fluid–fat interface surrounding the pericardium itself [2,12,26]. Bright-blood dynamic cine MR imaging often enables the recognition of intrapericardial contents, such as the visualization of fibrinous strands or of the presence of coagulated blood [25].

Finally, a central feature of CT/MR imaging is the capability to accurately assess the pericardial layer thickness and composition, allowing the differentiation of simple pericardial effusions from inflammatory effusive pericarditis or malignant pericardial diseases [12].

3.3.3 Cardiac Tamponade

The effects of effusion on diastolic function are largely dependent on the mechanical proprieties of the pericardial sac, which normally allows equalization of transmural end-diastolic pressures throughout the ventricles without remarkable effects on ventricular filling.

Generally, the elastic properties of the pericardium allow to tolerate a large amount of fluid (up to 1000 mL) [48], particularly when slowly accumulated (Figure 3.6). However, stretching capacities may be overcome when a quick and considerable increase in pericardial fluid volume occurs and intrapericardial pressure increases, resulting in compression of cardiac chambers, compromission of diastolic filling, inversion of the interventricular septum, and sudden reduction of cardiac output with subsequent hypotension, tachycardia, and progression to cardiogenic shock [49] (Figures 3.7 and 3.8).

This condition, called *cardiac tamponade*, configures a hemodynamic potentially life-threatening condition in which a rapid diagnosis must be established followed by urgent therapy [71].

A sign of mechanical impairment due to cardiac tamponade is the diastolic collapse of right ventricle, which occurs once intrapericardial pressure exceeds intracavitary pressure, subsequent distension of the vena cava, prominence of diastolic reversals in inferior caval system and hepatic veins, and enhanced respiratory variation in mitral and tricuspid inflow as a consequence of increased interventricular dependence [5,50].

Cardiac tamponade can be triggered by multiple causes, including trauma, inflammation, aortic dissection, neoplastic invasion of the pericardial sac, acute myocardial infarction, and cardiac surgery [49,50].

The clinical presentation (acute vs. chronic) depends from absolute volume of fluid, the speed of fluid accumulation, and the elasticity of the pericardium, so that

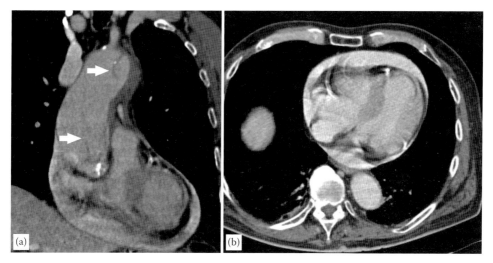

FIGURE 3.7

Acute cardiac tamponade due to ruptured aortic dissection. Coronal (a) and axial (b) contrast-enhanced CT images show intimal flap in the ascending aorta extending into the aortic arch (arrows), corresponding to Stanford type A aortic dissection, and large high-attenuation pericardial effusion (asterisks) corresponding to abundant hemopericardium. The patient died immediately after CT angiography. (Courtesy of G. F. Gualdi, MD, and C. Valentini, MD, DEA Policlinico Umberto I, Rome, Italy.)

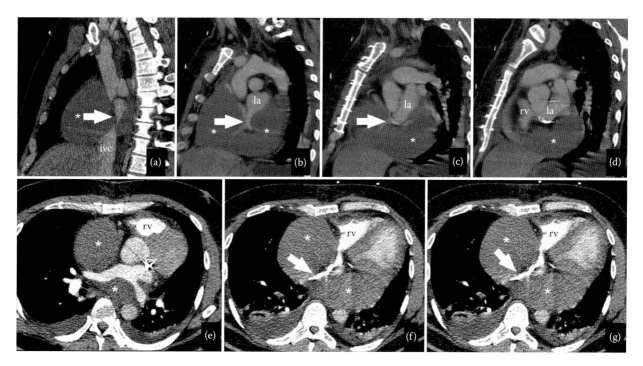

FIGURE 3.8

Pericardial hematoma causing cardiac tamponade. Contrast-enhanced CT images reconstructed on oblique sagittal (a–d) and axial (e–g) views show a huge pericardial low-attenuated fluid collection (*) in a 53-year-old male patient who had recently (4 days prior) undergone a cardiac surgery for mitral valve replacement, presenting worsening dyspnea, hypoxia, hypercapnia, and signs of splanchnic and peripheral engorgement (e.g., pedal edema and jugular distension). Pericardial hematoma is located on cardiac basis and compress superior (svc) and inferior (ivc) vena cava, right atrium (arrow), and left atrium (la); this condition determines a significant obstacle to systemic venous return and distolic filling of the right ventricle (rv), which is of reduced dimensions.

a rapid onset of a 200–250 mL amount of pericardial effusion may be lethal, whereas slow accumulation of 2 L can be asymptomatic and devoid of hemodynamic effects.

Transthoracic echocardiography is the first-line modality to assess the impairment of cardiac hemodynamic, given the need to rapidly obtain a morphological and functional assessment to establish a diagnosis; however CT and MR may be particularly beneficial in hemodynamically stable forms of tamponade, in which the superior resolution and tissue characterization of these

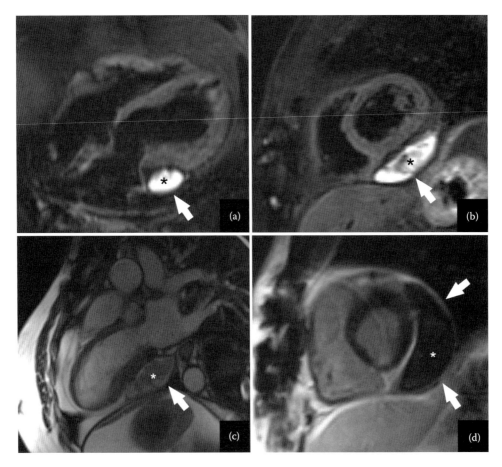

FIGURE 3.9
Pericardial loculated organized hematoma following cardiac surgery in a 74-year-old male. CMR images show an ovoid pericardial fluid collection (*) slightly compressing left ventricular inferior wall characterized by inhomogeneous high-signal intensity on T_2-weighted STIR (a and b) and intermediate signal intensity in SSFP image (c). CE-IR T_1-weighted short-axis image (d) shows hypointense fluid content bordered by thick and enhancing pericardial layers (arrows).

techniques may be helpful to characterized fluid content (or other content such as air) (Figure 3.9).

Furthermore, cardiac tamponade should be differentiated from CP with effusion, where the disorder is caused by a pathologic noncompliant pericardium rather than by the effusion itself.

3.3.3.1 Acute and Chronic Pericarditis

The terms *acute* and *chronic* pericarditis (inflammatory pericarditis) refer to different clinical manifestations of a large spectrum of pathological and etiological entities, ranging from idiopathic forms to infective and autoimmune diseases, postinfarction, postradiation, and paraneoplastic conditions [16].

Excluding idiopathic forms, whose prevalence ranges between 30% and 85% of the cases, the most common etiology of pericardial inflammation is viral infection (coxackie A and B, echovirus, mumps virus, cytomegalovirus, herpes simplex virus, herpes zoster virus, adenovirus, and Epstein–Barr virus) followed by bacterial

and tubercular infective disease; among noninfectious (generally autoimmune disease such as rheumatoid arthritis, systemic lupus erythematosus, and progressive systemic sclerosis) forms, myocardial infarction and radiation therapy represent two additional not infrequent causes of pericarditis [16,51].

Great emphasis has been recently given to the growing number of cases of tuberculous-related pericarditis reported in developed countries, as a consequence of immigration flows, or in association with human immunodeficiency virus infection [52–55] (Figure 3.10).

Other leading causes of pericarditis include iatrogenic post-traumatic pericarditis after radiation therapy for breast cancer and mediastinal tumors and cardiac interventions such as cardiac surgery, percutaneous coronary interventions, pacemaker insertion, and catheter ablation [56].

Postinfarction pericarditis occurs in approximately 10% of patients with transmural acute myocardial infarction [57] as a consequence of direct epicardial propagation (*epistenocardiac* pericarditis) of the infarct-related inflammation (Figure 3.11).

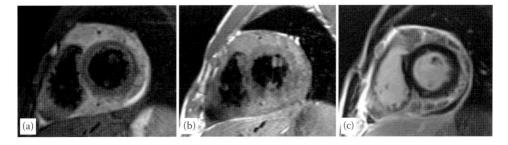

FIGURE 3.10
Tuberculous pericarditis. STIR T_2-weighted (a) and black-blood TSE T_1-weighted (b) images acquired on short axis view show an abundant amount of pericardial effusion with inhomogeneous signal with mixed fluid-solid composition, hyperintense both on T_1- and T_2-weighted sequence as caseous content. Corresponding short-axis LGE image (c) reveals thickening of pericardial layers, particularly of visceral layer, and irregular streaks of enhancement for bridges, adhesions, and fibrinous strands internal to organized effusion.

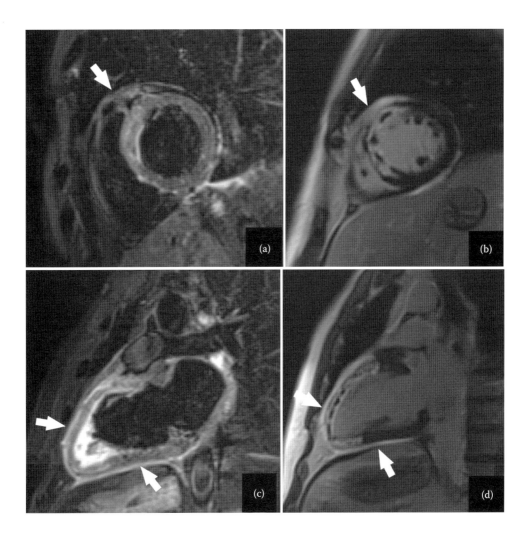

FIGURE 3.11
Epistenocardiac pericarditis. Early postinfarction pericarditis in a 65-year-old patient with a recent large anteroapical myocardial infarction. (a) Short-axis and (c) vertical long-axis STIR T_2-weighted images and corresponding (b) short-axis and (d) vertical long-axis LGE images show a large transmural acute myocardial infarction involving ventricular septum, LV anterior wall, and apex, centrally containing zone of intra-myocardial hemorrhage (on STIR T_2-weighted images) and microvascular obstruction (on LGE images). Thickening of pericardial layers of the entire sac with edematous signal is shown in STIR T_2-weighted images (a, c), which corresponds intense late enhancement (b, d), reflecting active pericardial inflammation (arrows).

The process should be clinically differentiated from late postinfarction pericarditis (or Dressler's syndrome) representing a rare autoimmune process without a close temporal relation with the infarction itself (2–3 weeks following the infarction), unlike the postinfarction form (immediately after the event).

The classic acute presentation of pericardial inflammation, characterized by acute chest pain, pericardial rubs, and widespread ST-elevation with pathognomonic absence of Q wave of necrosis on ECG, is a main differential diagnosis of acute chest pain observed in about 5% of patients admitted to emergency department and reported as a main differential in the *triple rule-out* protocols for CT angiography [58–60] (Figure 3.7).

Depending on the pathological substrate, imaging findings may vary from the presence of a simple pericardial effusion to more exudative or purulent forms with abundant highly proteinaceous effusion potentially causing tamponade or evolving to pericardial constriction.

There is also a clinical entity called *dry pericarditis* (or *pericarditis sicca*) that describes pure pericardial inflammation without associate effusion, pathologically characterized by fibrinous inflammatory exudate producing the typical pericardial friction rub [61].

In the acute setting, the inflammatory process is characterized by formation of a highly vascularized granulation tissue with a relative presence of fluid in pericardial sac, which appears on CT as a diffuse and irregular thickening of pericardial line including both the fluid and fibrinous-inflammatory components [32].

As previously mentioned, a limitation of CT imaging in this condition is the limited capability to discriminate small effusions from pericardial thickening [5].

Echocardiography remains the modality of choice for diagnosis, and generally it is sufficient to support the clinical management of the patient; however, in some cases, the use of cross-sectional methods may provide additional information for pathology characterization or guide diagnostic or therapeutic procedures (pericardiocentesis or pericardial biopsy).

MR features of acute/active pericarditis include presence of a diffuse edematous imbibition of visceral layers, hyperintense on T_2-weighted short-tau inversion recovery (STIR) sequences, usually associated with a variable amount of effusion and irregular thickening of the membrane that depicted with T_1-weighted TSE and cine-SSFP sequences (Figure 3.12).

Active inflammation is also characterized by a certain degree of pericardial contrast enhancement, such as that described in various reports in patients with pure acute inflammatory forms and with clinically recurrent chronic pericarditis using both T_1-weighted TSE or inversion-recovery sequences with late enhancement acquisitions [23,32,62] (Figure 3.13).

Contrast-enhanced MR imaging, moreover, allows evaluation of the extent of inflammation into the surrounding fat and adjacent myocardial tissue [12].

Acute pericarditis heals without detectable sequelae in most of the cases or may eventually leave late fibrous or fibrogranulomatous residual adhesions [61]. However, despite the natural history is usually benign with a good response to nonsteroidal anti-inflammatory drugs, in some cases, it can progress to chronic sclerosing pericarditis, with extensive progressive fibroblasts proliferation and collagen deposition, which ultimately conduct to a thickened fibrotic stiff pericardium that constricts the heart (CP).

Chronic fibrosing pericarditis is characterized on both CT and MR imaging by the presence of a nonenhancing focally or diffusely thickened pericardium very often showing loculated effusions caused by the presence of focal adhesions [32,61].

As mentioned above, the presence of pericardial adhesions can be depicted using MR tagging techniques, visualized as persistence of tagged signals between the pericardium/myocardium interface throughout the cardiac cycle [31].

Finally, positron emission tomography (PET)/CT can offer some advantages, particularly with the presence of *complex pericarditis*, because its capability to depict metabolic activity in the pericardium or in pericardial fluid, distinguishing an infectious or malignant process from a pure inflammatory process [23,63].

3.3.3.2 Pericardio-Myocarditis (or Myocardio-Pericarditis)

Pericardial inflammation is often accompanied by a variable degree of myocarditis (and vice versa), which occurs as a consequence of direct epicardial propagation (or vice versa) of the inflammatory process and because of their common etiologic agents, mainly cardiotropic viruses.

Usually the terms *pericardio-myocarditis* or *myocardio-pericarditis* are interchangeably used to emphasize the dominant clinical and pathological expression of one of the two entities, which are rarely equally represented [64]. In particular, *myopericarditis* refers primarily to a pericarditis with lesser myocarditis, as opposed to a *perimyocarditis*, which indicates a predominantly myocarditic syndrome [51].

Recognition of overlapping myocarditis may be clinically relevant representing a negative prognostic predictor in patients with pericarditis, often requiring hospitalization and a full etiologic search [12,53,64,65].

Myocardial involvement can be suspected in acute pericarditis in the presence of atypical ECG changes (e.g., new onset of arrhythmias, sinus tachycardia, and

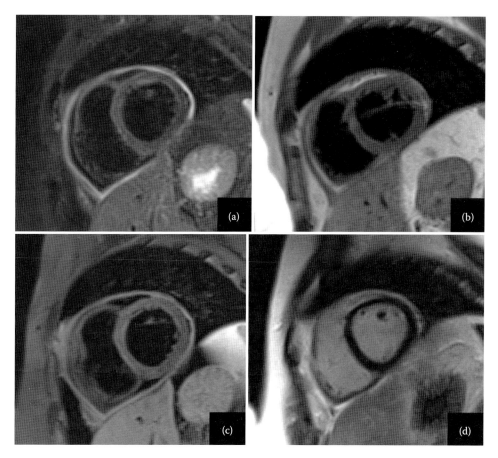

FIGURE 3.12

Diffuse active pericarditis. STIR spin-echo T_2-weighted (a) and unenhanced black-blood TSE T_1 weighted (b) images on short axis view well demonstrate a diffuse pericardial thickening in viral pericarditis. The contrast-enhancement ECG-gated fat-suppressed TSE T_1-weighted image (c) on the corresponding short-axis view excellently highlights the strong homogeneous enhancement of the entire pericardium following gadolinium administration reflecting the inflammation activity, even if compared to similar image obtained with the traditional late gadolinium-enhancement technique (d), which poorly distinguish between pericardial enhancement and epicardial fat tissue.

QT prolongation) associated with transient regional and global wall motion abnormalities, and raise of cardiac enzymes.

Diagnostic contribution of echocardiography in this clinical setting is limited by the fact that many patients with nonsevere forms of myocardial involvement have a normal echocardiogram or may exhibit nonspecific signs and extremely variable evidence of disease even in the presence of severe myocardial inflammation [66].

The role of CT is also limited in this condition, although epicardial late enhancement has been described in patients with acute myocarditis and correlated with MR features [67].

Cardiac MR has become the reference for noninvasive assessment of myocarditis due to its unique capabilities of combining tissue characterization and functional assessment [7,68–70]. It usually shows an association of pericardial inflammation with typical signs of acute myocarditis characterized by predominantly meso-epicardial late enhancement with associated focal edema on T_2-weighted STIR and a variable degree of regional dysfunction [7,68–70] (Figure 3.14).

3.3.3.3 Constrictive Pericarditis

The main hallmark of CP is characterized by the formation of a thick, anelastic fibrotic or fibro-calcified sac surrounding the heart and impeding physiological biventricular relaxation with progressively severe diastolic dysfunction [1].

Fibrous pericardial replacement is usually the consequence of an exaggerated healing process, potentially following any form of acute inflammation, and resulting in a thick scar tissue formation causing total or subtotal obliteration of the pericardial space, which can be partially filled with lacunae of exudative/purulent material [1,54,61].

Calcium deposition is also commonly observed in this condition and may contribute to stiffening of the

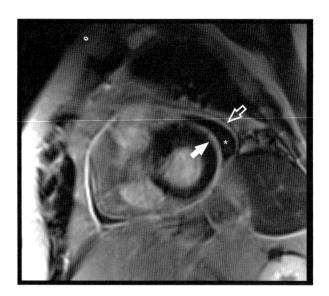

FIGURE 3.13
Diffuse active pericarditis. CE-IR T_1-weighted image on short-axis view shows a diffuse thickening and gadolinium enhancement of both parietal and visceral (white arrow) pericardial layers spaced by a small amount of pericardial effusion (*). Presence of contrast enhancement is a feature of active inflammation.

The risk of CP after acute viral or idiopathic pericarditis is quite low (< 0.5%) but is relatively frequent in purulent pericarditis with an incessant course and large pericardial effusions [72,73].

Clinical symptoms usually are due to the consequent impaired ventricular filling, severe diastolic dysfunction, and the right heart failure, and start with signs of splanchnic and peripheral engorgement (e.g., pedal edema, ascites, and jugular distension), which later progress to more central symptoms such as dyspnea, cough, and fatigability [16,61].

CP is usually a chronic process that slowly progresses over time and becomes clinically manifest several years after the initial pathologic trigger; however, subacute (i.e., less than 1 year) and even acute (i.e., days) forms of disease have also been observed in some specific conditions like after extensive cardiac tamponade or following surgical maneuvers, causing focal scarring with band-like fibrous adherences (e.g., after pericardial drainage removal) [16,61].

Chronic CP is usually associated with predominant fibrous pericardial tissue with few inflammatory cells, whereas acute and subacute processes are generally histologically characterized by a high prevalence of active inflammatory elements with minimal amount of connective tissue [61,72].

Despite the progress in understanding the complex hemodynamics of pericardial constriction and the advancement of noninvasive diagnostic tools, diagnosis of CP remains challenging, particularly in the differentiation from RC and other causes of systemic venous

membrane as a consequence of the chronic inflammatory stimulus (Figure 3.15).

The spectrum of possible etiologies causing pericardial constriction has progressively changed over time moving from pure infectious (particularly tubercular) to postradiation and postsurgical forms, which have become the most frequent causes of disease [71,72].

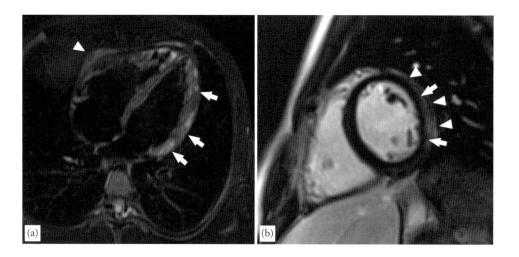

FIGURE 3.14
Acute perimyocarditis in a 54-year-old woman with recent onset of chest pain and mild left ventricular dysfunction. Short-axis STIR T_2-weighted image (a) shows subepicardial patchy focal areas of myocardial edema on the LV lateral wall matching a thick subepicardial band-like area of myocardial enhancement on LGE T_1-weighted image (b). A minimal pericardial thickening is visible both on STIR T_2-weighted image near the RV free wall (a, arrowhead) and on LGE T_1-weighted image on LV lateral wall (b, arrowheads). A LV myocardial biopsy confirmed the myocardial inflammation.

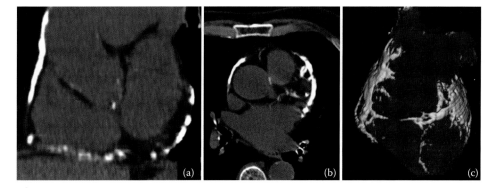

FIGURE 3.15
Extensive pericardial calcifications detected on calcium-scoring scan as *incidental finding* in a patient with no history of previous reported nonspecific cardiac diseases. Short-axis (a) and axial (b) unenhanced CT image shows thickened and heavily calcified pericardium. Three-dimensional volume rendered image (c) of unenhanced CT scan is useful to best depict global extent of pericardial calcifications.

congestion, including right heart failure and restrictive disease, and in the presence of atypical onset of disease.

Differential diagnosis between RC and CP in particular has been defined as a *diagnostic dilemma* as restrictive and constrictive pathophysiology are characterized by similar diastolic abnormalities and share common clinical features [71,74].

The most relevant difference between these two entities is that constriction is potentially surgically corrected, whereas in RC the primary abnormalities consist in the hampered myocardial relaxation and compliance and therefore therapy is largely palliative with poor prognosis [16,75].

Clinical recognition of the CP should also be done as early as possible because of the beneficial clinical and prognostic effects of pericardiectomy, which becomes less effective in the presence of extensive calcifications and significant myocardial involvement [71,74,76].

Since no single imaging modality could be considered *comprehensive* or completely exhaustive to recognize all cases of CP, diagnostic strategy should be necessarily tailored according to each patient [16,77,78].

3.3.3.4 Morphological Abnormalities

Focal or diffuse pericardial thickening (i.e., >4 mm) is the typical morphological sign of CP, which is usually more pronounced at the right ventricular free wall and anterior AV groove where pericardial line appears irregular and often shows some degree of calcification (Figure 3.16).

In a retrospective analysis with surgical correlation, Young PM and colleagues [23] found an average pericardial thickness of 4.6 ± 2.1 cm in patients with noncalcified CP and 9.2 ± 7.0 cm in the presence of constriction with calcifications.

It has been reported that a pericardial thickness of >6 mm has a high specificity for detection of pericardial

constriction, whereas at 4–6 mm, diagnosis can be suspected in the presence of a typical clinical presentation and becomes very unlikely when the pericardium is less than 2 mm thick [4,54,79].

However, it was recently demonstrated that, even though fibrosis with or without calcification is almost always present at histology (96%), in cases of pericardial constriction, the maximal pericardial thickness may vary considerably (1–17 mm; mean, 4 mm), and it is only weakly related to the degree of cardiac constriction, with up to 20% of patients showing a normal (<2 mm) thickness [16,54,80].

The absence of membrane thickening (even focal), however, does not rule out the presence of constriction; previous studies have shown that a normally thickened but stiff pericardium may also be responsible of a severe diastolic impairment impeding physiological ventricular filling [8,80].

Therefore, function assessment with cardiac MR is highly recommended in cases where symptoms of constriction are not coupled with evidence of pericardial thickening [8,80]. Moreover, some authors showed that evolution toward an end-stage irreversible fibrosing pericarditis is characterized by a thinning of the chronically inflamed pericardium [81–83].

Cardiac CT is the technique of choice to depict the presence of calcium allowing to identify even minimal focal deposits, whereas extensive calcification may be potentially missed with MR [12,22,29,45].

Presence of pericardial calcifications has been observed in up to 50% of patients with evidence of constriction and usually allows to exclude underlying RC with high degree of confidence [4,22,84].

Although pericardial calcifications typically characterize tuberculous pericarditis, they have been reported in approximately 28% of patients with histologically confirmed CP in which tuberculosis

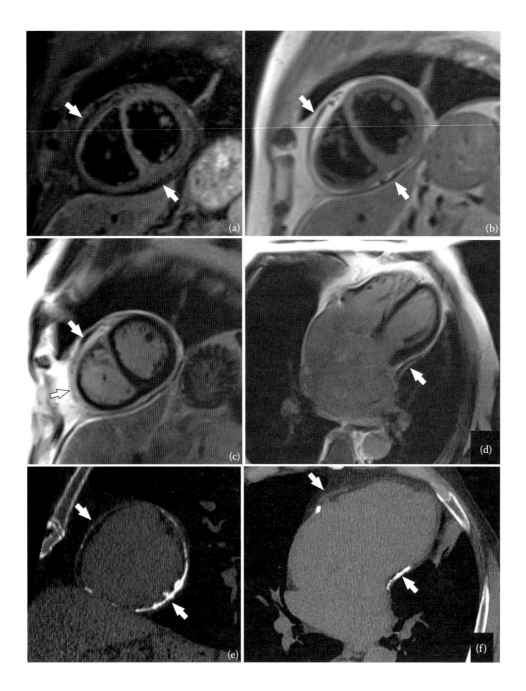

FIGURE 3.16
Calcified constrictive pericarditis. Short-axis ECG-gated (a) STIR T_2-weighted and (b) TSE T_1-weighted images show irregularly diffuse thickened hypointense appearance of the pericardium, predominantly involving the basal part of ventricles (arrows), with subsequent flattening of septum. On LGE short-axis (c) and horizontal long-axis (d) images, the enhancement of the thickened non-calcified pericardial layers is hardly distinguishable from the epicardial and adjacent paracardiac adipose tissue. (b) Short-axis T_1-weighted spin-echo MR image shows the extent of pericardial abnormalities along inferior part of the heart (arrows). Sagittal (e) and axial (f) CT images are given and the calcifications are visible.

was excluded [76]; therefore, calcifications should be considered a nonspecific response to chronic inflammation.

In prediction of pericardial surgery, preoperative CT examination may be used for detailed description of both severity of thickening and the presence and location of calcifications, allowing better surgical planning and stratification of procedural risk [26].

On MR imaging, the thickened fibrotic and/or calcified pericardial layers have low signal intensity on T_1- and T_2-weighted MR images and at cine imaging and are easily distinguishable from effusion.

Inter-ventricular septum may exhibit a sinuous or flattened morphology or may appear leftward bowed as a consequence of increased trans-septal pressures.

Additional indirect signs of constriction include dilatation of the caval and sovra-hepatic venous systems, unilateral or bilateral atrial enlargement (particularly the left), and ventricular tubular morphology caused by direct compression from the thickened pericardial sac.

Bilateral pleural effusion and ascites may also be observed. Typical morphological features of CP may also lack in various conditions, making diagnosis more challenging with need to combined clinical data with functional evaluation [74,80].

Pericardial constriction may be determined by a focal, strategically located, thickening of the serosa, like in proximity to the right AV groove, causing severe hemodynamic impairment. Panoramic assessment of pericardial line with cross-sectional imaging modalities is extremely useful in these cases [2,9,25] and tagging MR may be particularly helpful to directly identify focal adhesions.

After gadolinium administration, *end-stage* chronic fibrosing CP do not present significant contrast enhancement, whereas pericardial enhancement is suggestive of residual inflammation [32,83].

The evidence of persistent chronic inflammation is associated with *reversible* or *transient* forms of CP that respond to anti-inflammatory treatment. In this clinical setting, contrast-enhanced MR imaging can be helpful to identify patient with residual inflammatory activity that can be benefited by medical therapy, without undergoing to pericardiectomy [81–83].

Finally, there is a further entity called *occult pericarditis* (or volume-depleted CP) [25] in which clinical symptoms and hemodynamic manifestations of disease are absent in basal conditions and become clinically manifested in the presence of increased venous return, like during deep inspiration or intense physical activity. In such cases, use of dynamic maneuvers during diagnostic imaging like the Valsalva's one may be helpful to recognize underlying diseases, showing the functional diastolic changes associated with the increased right ventricular inflow [17,25].

3.3.3.5 Function Abnormalities

Functional abnormalities of CP obviously reflect the anatomical changes induced by constriction and are mainly caused by the hampered diastolic biventricular expansion with accelerated filling rate.

A thickened pericardium, in fact, increases ventricular coupling, causing accentuated ventricular volume/ pressure relation, and induces an increase of diastolic pressures, which are equalized or nearly equalized in all cardiac chambers.

The effects of increased filling pressures are particularly evident at the level of interventricular septum, which shows abnormal motion with leftward displacement during early diastolic phase as a consequence of increased trans-septal filling pressures. This phenomenon, also called *paradoxical septal motion*, represents a typical feature of CP, which enables its differentiation from RC [8,9].

A further relevant sign of CP is the dissociation between intracardiac and intrathoracic pressures during respiration, which was initially described by Hatle and colleagues [85], and it causes pressure isolation of cardiac cavities during normal respiratory changes. During deep inspiration, if the heart is totally encased, intrathoracic respiratory fall is not transmitted to the high-pressure cardiac cavities, causing a decreased pulmonary venous return and a consequent decreased left ventricular filling; conversely, systolic function is generally well preserved.

It should be noted that all these key functional and hemodynamic features can be detected with transthoracic echocardiography and cardiac catheterization, which represent robust and widely utilized tools in this clinical setting [75,86].

Cardiac CT has very limited role in the assessment of CP-related functional changes, although dynamic evaluation of interventricular septal motion is theoretically possible using a retrospective-ECG triggered technique with a high temporal resolution scanner [28].

The role of cardiac-MR is more comprehensive and allows to identify most of the aforementioned functional features of constriction using a combination of cine-SSFP, phase-contrast, and eventually real-time cine-MR imaging and tagging sequences.

Use of cine-SSFP imaging for depiction of abnormal septal motion has already been widely described in literature and offers the advantage toward echocardiography of a superior contrast resolution [9].

Giorgi and colleagues [9], using the conventional cine-SSFP breath-hold technique, found septal flattening or inversion in most patients with CP during the early filling phase. This phenomenon was more pronounced at basal level, and it was characterized by a sigmoid distortion of septal shape during cardiac cycle and was absent in patients with RC.

The typical diastolic transtricuspid early-wave ratio is decreased compared with healthy subjects on velocity-encoded cine-MR sequences with a decreased or absent A-wave (atrial filling) [87].

Reduced or even reverse systolic forward flow can also be depicted at the inferior vena cava level and can be associated with increased early diastolic forward flow of restrictive pathophysiology.

The effects of respiratory activity on ventricular septal motion can be assessed with real-time MR techniques, which allow the identification of early mid-diastolic respiratory bounce of interventricular septum in the presence of CP using a SSFP nontriggered cine MRI technique in combination with parallel imaging [8,17].

In normal loading conditions, interventricular septum has a convex profile toward the right ventricle as a consequence of the left to right cavities positive pressure gradient, and this configuration in normal subjects is minimally affected by respiratory activity as a consequence of pleuro–pericardial interactions with pressures equalization.

Conversely, in CP patients, the outward right ventricular expansion is restricted by the pathological pericardium, resulting in a sudden increase of the RV pressure, with a subsequent sudden leftward displacement with flattening or inversion of the ventricular septum.

Impact of respiratory activity on ventricular septal motion and configuration was previously described by Francone using a real-time breath-hold technique in a study population of patients with CP, RC, and inflammatory pericarditis [8]. Results of the study showed typical septal flattening or inversion in CP patients occurring in early diastole, which was more pronounced at the onset of deep inspiration and rapidly disappearing in late filling phases (Figure 3.17).

This phenomenon was absent in patients with RC and was proposed to be helpful for differentiating between CP and RC patients, especially in the presence of a normal or minimally thickened pericardium.

Pericardial mobility and stiffness can also be assessed by conventional ECG-gated tagged cine-MRI, which enable to identify focal fibrotic adhesions between inner and outer pericardial layers as persistence of taglines during the cardiac cycle [31].

3.3.4 Pericardial Masses

The heterogeneous group of pericardial masses includes cysts, hematomas, complex *organized* effusions, and primary and secondary malignancies affecting the pericardium.

CT and MR imaging techniques are usually required to characterize a mass detected in a preliminary echocardiographic exam or as incidental finding on a chest-ray or CT examination. Cross-sectional imaging allows (1) to distinguish between a *true* pericardial mass from a so-called pseudo-mass, (2) to accurately define the location and extent of the lesion and its relationships with surrounding structures, and (3) to provide a certain degree of tissue characterization.

Organized hematoma, cystic collections or pseudo masses are generally well discriminated from neoplasms using CT and MRI (Figures 3.3 and 3.9). Abundant epicardial fat may be misinterpreted for a pericardial mass at transthoracic echocardiography, even by experienced operators and easily recognized on MR, using fat-saturated sequences, or CT imaging.

Pseudoaneurysm occurs when a haemopericardium caused by myocardial wall rupture is contained by thickened pericardium [88].

Primary neoplastic disease of the pericardium is extremely rare, whereas its secondary involvement has been described in up to 10%–12% of patients with known neoplasia [89,90].

It may be helpful to remember that pericardial masses are characterized by a variable amount of effusion, which is disproportionally greater to the size of the solid lesion, very often hemorrhagic, and therefore characterized by inhomogeneous highly proteinaceous signal intensity on both T_1- and T_2-weighted CMR sequences and high density on unenhanced CT exam (Figure 3.18).

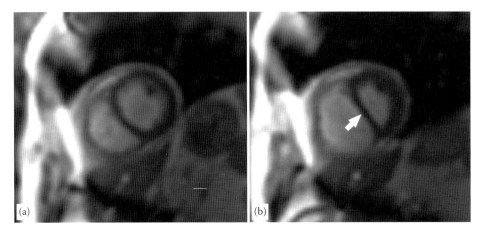

FIGURE 3.17

Constrictive pericarditis (same cases of Figure 3.13). Short-axis real time SSFP cine MR images obtained (a) at the end-expiration and (b) during inspiration show inversion of septal shape (arrow) due to a rapid increase of right ventricular filling not compensated by pericardial distensibility.

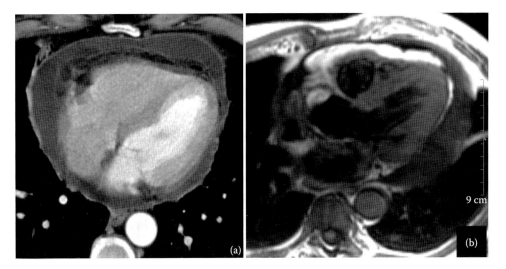

FIGURE 3.18
Pericardial mesothelioma. Axial contrast-enhanced computed tomography (a) and black-blood TSE T_1-weighted (b) images show a diffuse contrast enhancement of pericardial layers with some soft tissue attenuation nodular thickening and coexistence of moderate pericardial effusion with mixed corpuscular-fluid content. Pericardial fluid cytological analysis performed after pericardiocentesis diagnosed a primary malignant pericardial mesothelioma.

Benign tumors of the pericardium such as fibroma, teratoma, haemangioma, and lipoma are overall extremely rare.

Primary lesion is more frequently malignant and includes mesothelioma, sarcoma, liposarcoma, and lymphoma (Figure 3.19). Malignant mesotheliomas are the most common primary lesions that accounts for 50% of all pericardial tumors and are usually characterized by the presence of effusion with pericardial nodular lesions or plaques [91,92].

Metastatic pericardial disease can occur by direct invasion of the membrane from neoplasms originated from nearby anatomical structures such as the lung, mediastinum (lymphoma), or the myocardium (angiosarcoma); through hematic spread (more frequently malignant melanoma, lymphoma, and breast neoplasms) or venous diffusion (by infiltrating the right cavities, usually renal cell or hepatocellular carcinomas) (Figure 3.20).

The typical sign of malignancy is the focal interruption of pericardial line from the epicardial or mediastinal side, which can be easily appreciated at the level the right ventricular free wall, because of the pericardial invasion from primary lesions.

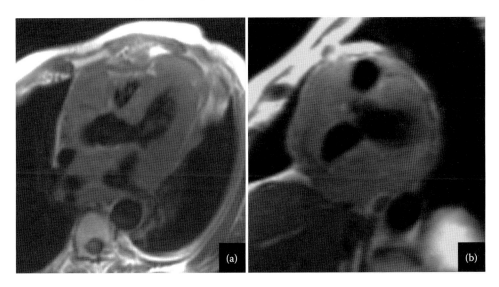

FIGURE 3.19
Primary non-Hodgkin's lyphoma. Unenhanced black-blood TSE T_1-weighted images acquired on long axis (a) and short-axis (b) views demonstrate a severe diffuse thickening of pericardium with poorly demarcated soft tissue, invading the epicardial adipose space and completely surrounding cardiac cavities.

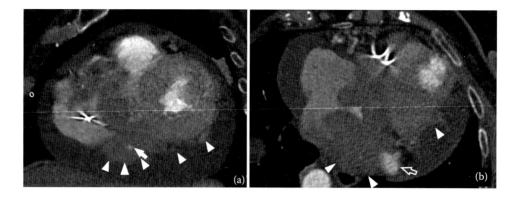

FIGURE 3.20
Pericardial metastasis with myocardial invasion in a patient with metastatic renal cell cancer. ECG-gated contrast-enhanced CT images reconstructed on short-axis (a) and axial (b) views show a large infiltrating solid tissue (arrowheads) surrounding the posterior descending artery (arrow) with diffuse invasion of the pericardial sac and right ventricular cavity. The pathological tissue spreads along the pericardium covering the left ventricular inferolateral wall, infiltrating and causing stenosis of coronary sinus (curved arrow). A moderate size pericardial effusion and presence of electrode in right ventricle are also seen.

Direct invasion of the visceral layer can be observed, for example, in the presence of primary angiosarcomas, which generally arise from the right atrium, and rapidly grow toward the causing infiltration and extensive exudative effusion. Parietal pericardial layer is often infiltrated by an aggressive primary lung perihilar neoplasia.

Although the information obtainable with CT and MRI are not little, biopsy and histopathologic analysis are still necessary to achieve a definitive diagnosis of most pericardial tumors [92].

References

1. Spodick DH. *The Pericardium: A Comprehensive Textbook.* New York: Marcel Dekker, Inc., 1997.
2. Francone M, Dymarkowski S, Kalantzi M, Bogaert J. Magnetic resonance imaging in the evaluation of the pericardium: A pictorial essay. *La Radiologia medica* 2005;109:64–74; quiz 75–76.
3. Ivens EL, Munt BI, Moss RR. Pericardial disease: What the general cardiologist needs to know. *Heart* 2007;93:993–1000.
4. Maisch B, Seferovic PM, Ristic AD et al. Guidelines on the diagnosis and management of pericardial diseases executive summary; The Task force on the diagnosis and management of pericardial diseases of the European Society of Cardiology. *European Heart Journal* 2004;25:587–610.
5. Verhaert D, Gabriel RS, Johnston D, Lytle BW, Desai MY, Klein AL. The role of multimodality imaging in the management of pericardial disease. *Circulation: Cardiovascular Imaging* 2010;3:333–343.
6. Imazio M, Spodick DH, Brucato A, Trinchero R, Adler Y. Controversial issues in the management of pericardial diseases. *Circulation* 2010;121:916–928.
7. Pennell DJ, Sechtem UP, Higgins CB et al. Clinical indications for cardiovascular magnetic resonance (CMR): Consensus Panel report. *European Heart Journal* 2004;25:1940–1965.
8. Francone M, Dymarkowski S, Kalantzi M, Rademakers FE, Bogaert J. Assessment of ventricular coupling with real-time cine MRI and its value to differentiate constrictive pericarditis from restrictive cardiomyopathy. *European Radiology* 2006;16:944–951.
9. Giorgi B, Mollet NR, Dymarkowski S, Rademakers FE, Bogaert J. Clinically suspected constrictive pericarditis: MR imaging assessment of ventricular septal motion and configuration in patients and healthy subjects. *Radiology* 2003;228:417–424.
10. Spodick DH. Macrophysiology, microphysiology, and anatomy of the pericardium: A synopsis. *American Heart Journal* 1992;124:1046–1051.
11. Groell R, Schaffler GJ, Rienmueller R. Pericardial sinuses and recesses: Findings at electrocardiographically triggered electron-beam CT. *Radiology* 1999;212:69–73.
12. Bogaert J, Francone M. Pericardial disease: Value of CT and MR imaging. *Radiology* 2013;267:340–356.
13. Rabkin SW. Epicardial fat: Properties, function and relationship to obesity. Obesity reviews : An official journal of the International Association for the *Study of Obesity* 2007;8:253–261.
14. Iacobellis G, Corradi D, Sharma AM. Epicardial adipose tissue: Anatomic, biomolecular and clinical relationships with the heart. *Nature Clinical Practice Cardiovascular Medicine* 2005;2:536–543.
15. Sacks HS, Fain JN. Human epicardial adipose tissue: A review. *American Heart Journal* 2007;153:907–917.
16. Troughton RW, Asher CR, Klein AL. Pericarditis. *Lancet* 2004;363:717–727.
17. Francone M, Dymarkowski S, Kalantzi M, Bogaert J. Real-time cine MRI of ventricular septal motion: A novel approach to assess ventricular coupling. *Journal of Magnetic Resonance Imaging: JMRI* 2005;21:305–309.

18. Truong MT, Erasmus JJ, Gladish GW et al. Anatomy of pericardial recesses on multidetector CT: Implications for oncologic imaging. *AJR American Journal of Roentgenology* 2003;181:1109–1113.

19. Wang ZJ, Reddy GP, Gotway MB, Yeh BM, Hetts SW, Higgins CB. CT and MR imaging of pericardial disease. *RadioGraphics: A Review Publication of the Radiological Society of North America, Inc* 2003;23(Spec. No.): S167–S180.

20. Sechtem U, Tscholakoff D, Higgins CB. MRI of the normal pericardium. *AJR American Journal of Roentgenology* 1986;147:239–244.

21. Bogaert J, Duerinckx AJ. Appearance of the normal pericardium on coronary MR angiograms. *Journal of Magnetic Resonance Imaging: JMRI* 1995;5:579–587.

22. O'Leary SM, Williams PL, Williams MP et al. Imaging the pericardium: Appearances on ECG-gated 64-detector row cardiac computed tomography. *The British Journal of Radiology* 2010;83:194–205.

23. Young PM, Glockner JF, Williamson EE et al. MR imaging findings in 76 consecutive surgically proven cases of pericardial disease with CT and pathologic correlation. *The International Journal of Cardiovascular Imaging* 2012;28:1099–1109.

24. Ferrans VJ IT, Roberts WC. Anatomy of the pericardium. In: Reddy PS LD, Shaver JA, eds., *Pericardial Disease*. New York: Raven, 1982: pp. 77–92.

25. Bogaert J, Francone M. Cardiovascular magnetic resonance in pericardial diseases. *Journal of Cardiovascular Magnetic Resonance: Official Journal of the Society for Cardiovascular Magnetic Resonance* 2009;11:14.

26. Yared K, Baggish AL, Picard MH, Hoffmann U, Hung J. Multimodality imaging of pericardial diseases. *JACC Cardiovascular Imaging* 2010;3:650–660.

27. Kodama F, Fultz PJ, Wandtke JC. Comparing thin-section and thick-section CT of pericardial sinuses and recesses. *AJR American Journal of Roentgenology* 2003;181:1101–1108.

28. Ghersin E, Lessick J, Litmanovich D et al. Septal bounce in constrictive pericarditis. Diagnosis and dynamic evaluation with multidetector CT. *Journal of Computer Assisted Tomography* 2004;28:676–678.

29. Belgour A, Christiaens LP, Varroud-Vial N, Vialle R, Tasu JP. Chronic pericarditis: CT and MR imaging features. *Journal de radiologie* 2010;91:615–622.

30. Francone M, Carbone I, Agati L et al. Utility of T2-weighted short-tau inversion recovery (STIR) sequences in cardiac MRI: An overview of clinical applications in ischaemic and non-ischaemic heart disease. *La radiologia medica* 2011;116:32–46.

31. Kojima S, Yamada N, Goto Y. Diagnosis of constrictive pericarditis by tagged cine magnetic resonance imaging. *The New England Journal of Medicine* 1999;341:373–374.

32. Taylor AM, Dymarkowski S, Verbeken EK, Bogaert J. Detection of pericardial inflammation with late-enhancement cardiac magnetic resonance imaging: Initial results. *European Radiology* 2006;16:569–574.

33. Rademakers FE, Bogaert J. Cardiac dysfunction in heart failure with normal ejection fraction: MRI measurements. *Progress in Cardiovascular Diseases* 2006;49:215–227.

34. Paelinck BP, Lamb HJ, Bax JJ, Van der Wall EE, de Roos A. Assessment of diastolic function by cardiovascular magnetic resonance. *American Heart Journal* 2002;144:198–205.

35. Mohiaddin RH, Wann SL, Underwood R, Firmin DN, Rees S, Longmore DB. Vena caval flow: Assessment with cine MR velocity mapping. *Radiology* 1990;177:537–541.

36. Akiba T, Marushima H, Masubuchi M, Kobayashi S, Morikawa T. Small symptomatic pericardial diverticula treated by video-assisted thoracic surgical resection. *Annals of Thoracic and Cardiovascular Surgery: Official Journal of the Association of Thoracic and Cardiovascular Surgeons of Asia* 2009;15:123–125.

37. Feigin DS, Fenoglio JJ, McAllister HA, Madewell JE. Pericardial cysts. A radiologic-pathologic correlation and review. *Radiology* 1977;125:15–20.

38. Jeung MY, Gasser B, Gangi A et al. Imaging of cystic masses of the mediastinum. *RadioGraphics: A Review Publication of the Radiological Society of North America, Inc* 2002;22(Spec. No.):S79–S93.

39. Drury NE, De Silva RJ, Hall RM, Large SR. Congenital defects of the pericardium. *The Annals of Thoracic Surgery* 2007;83:1552–1553.

40. Yamano T, Sawada T, Sakamoto K, Nakamura T, Azuma A, Nakagawa M. Magnetic resonance imaging differentiated partial from complete absence of the left pericardium in a case of leftward displacement of the heart. *Circulation Journal: Official Journal of the Japanese Circulation Society* 2004;68:385–388.

41. Peebles CR, Shambrook JS, Harden SP. Pericardial disease—Anatomy and function. *The British Journal of Radiology* 2011;84(Spec. No. 3):S324–S337.

42. Scheuermann-Freestone M, Orchard E, Francis J et al. Images in cardiovascular medicine. Partial congenital absence of the pericardium. *Circulation* 2007;116:e126–e129.

43. Psychidis-Papakyritsis P, de Roos A, Kroft LJ. Functional MRI of congenital absence of the pericardium. *AJR American Journal of Roentgenology* 2007;189:W312–W314.

44. Abbas AE, Appleton CP, Liu PT, Sweeney JP. Congenital absence of the pericardium: Case presentation and review of literature. *International Journal of Cardiology* 2005;98:21–25.

45. Ovchinnikov VI. Computerized tomography of pericardial diseases. *Vestnik rentgenologii i radiologii* 1996:10–15.

46. Mulvagh SL, Rokey R, Vick GW, 3rd, Johnston DL. Usefulness of nuclear magnetic resonance imaging for evaluation of pericardial effusions, and comparison with two-dimensional echocardiography. *The American Journal of Cardiology* 1989;64:1002–1009.

47. Frank H, Globits S. Magnetic resonance imaging evaluation of myocardial and pericardial disease. *Journal of Magnetic Resonance Imaging: JMRI* 1999;10:617–626.

48. Breen JF. Imaging of the pericardium. *Journal of Thoracic Imaging* 2001;16:47–54.

49. Spodick DH. Acute cardiac tamponade. *The New England Journal of Medicine* 2003;349:684–690.

50. Restrepo CS, Lemos DF, Lemos JA et al. Imaging findings in cardiac tamponade with emphasis on CT. *RadioGraphics: A Review Publication of the Radiological Society of North America, Inc* 2007;27:1595–1610.

51. Imazio M, Trinchero R. Myopericarditis: Etiology, management, and prognosis. *International Journal of Cardiology* 2008;127:17–26.

52. Restrepo CS, Diethelm L, Lemos JA et al. Cardiovascular complications of human immunodeficiency virus infection. *RadioGraphics: A Review Publication of the Radiological Society of North America, Inc* 2006;26:213–231.

53. Little WC, Freeman GL. Pericardial disease. *Circulation* 2006;113:1622–1632.

54. Oh KY, Shimizu M, Edwards WD, Tazelaar HD, Danielson GK. Surgical pathology of the parietal pericardium: A study of 344 cases (1993-1999). *Cardiovascular Pathology: The Official Journal of the Society for Cardiovascular Pathology* 2001;10:157–168.

55. Syed FF, Ntsekhe M, Gumedze F, Badri M, Mayosi BM. Myopericarditis in tuberculous pericardial effusion: Prevalence, predictors and outcome. *Heart* 2014;100:135–139.

56. Imazio M, Brucato A, Derosa FG et al. Aetiological diagnosis in acute and recurrent pericarditis: When and how. *Journal of Cardiovascular Medicine* 2009;10:217–230.

57. Doulaptsis C, Cazacu A, Dymarkowski S, Goetschalckx K, Bogaert J. Epistenocardiac pericarditis. *Hellenic Journal of Cardiology: HJC = Hellenike kardiologike epitheorese* 2013;54:466–468.

58. Halpern EJ. Triple-rule-out CT angiography for evaluation of acute chest pain and possible acute coronary syndrome. *Radiology* 2009;252:332–345.

59. Takakuwa KM, Halpern EJ, Shofer FS. A time and imaging cost analysis of low-risk ED observation patients: A conservative 64-section computed tomography coronary angiography "triple rule-out" compared to nuclear stress test strategy. *The American Journal of Emergency Medicine* 2011;29:187–195.

60. Frauenfelder T, Appenzeller P, Karlo C et al. Triple rule-out CT in the emergency department: Protocols and spectrum of imaging findings. *European Radiology* 2009;19:789–799.

61. Spodick DH. Pericarditis, pericardial effusion, cardiac tamponade, and constriction. *Critical Care Clinics* 1989;5:455–476.

62. Yelgec NS, Dymarkowski S, Ganame J, Bogaert J. Value of MRI in patients with a clinical suspicion of acute myocarditis. *European Radiology* 2007;17:2211–2217.

63. Strobel K, Schuler R, Genoni M. Visualization of pericarditis with fluoro-deoxy-glucose-positron emission tomography/computed tomography. *European Heart Journal* 2008;29:1212.

64. Spodick DH. Risk prediction in pericarditis: Who to keep in hospital? *Heart* 2008;94:398–399.

65. Lange RA, Hillis LD. Clinical practice. Acute pericarditis. *The New England Journal of Medicine* 2004;351:2195–2202.

66. Pinamonti B, Alberti E, Cigalotto A et al. Echocardiographic findings in myocarditis. *The American Journal of Cardiology* 1988;62:285–291.

67. Brett NJ, Strugnell WE, Slaughter RE. Acute myocarditis demonstrated on CT coronary angiography with MRI correlation. *Circulation Cardiovascular Imaging* 2011;4: e5–e6.

68. Friedrich MG, Sechtem U, Schulz-Menger J et al. Cardiovascular magnetic resonance in myocarditis: A JACC White Paper. *Journal of the American College of Cardiology* 2009;53:1475–1487.

69. Hundley WG, Bluemke DA, Finn JP et al. ACCF/ACR/AHA/NASCI/SCMR 2010 expert consensus document on cardiovascular magnetic resonance: A report of the American College of Cardiology Foundation Task Force on Expert Consensus Documents. *Journal of the American College of Cardiology* 2010;55:2614–2662.

70. Francone M, Chimenti C, Galea N et al. CMR sensitivity varies with clinical presentation and extent of cell necrosis in biopsy-proven acute myocarditis. *JACC Cardiovascular Imaging* 2014;7:254–263.

71. Bertog SC, Thambidorai SK, Parakh K et al. Constrictive pericarditis: Etiology and cause-specific survival after pericardiectomy. *Journal of the American College of Cardiology* 2004;43:1445–1452.

72. Cameron J, Oesterle SN, Baldwin JC, Hancock EW. The etiologic spectrum of constrictive pericarditis. *American Heart journal* 1987;113:354–360.

73. Imazio M, Brucato A, Maestroni S et al. Risk of constrictive pericarditis after acute pericarditis. *Circulation* 2011;124:1270–1275.

74. Schwefer M, Aschenbach R, Heidemann J, Mey C, Lapp H. Constrictive pericarditis, still a diagnostic challenge: Comprehensive review of clinical management. *European Journal of Cardio-Thoracic Surgery: Official Journal of the European Association for Cardio-Thoracic Surgery* 2009;36:502–510.

75. Nishimura RA. Constrictive pericarditis in the modern era: A diagnostic dilemma. *Heart* 2001;86:619–623.

76. Ling LH, Oh JK, Breen JF et al. Calcific constrictive pericarditis: Is it still with us? *Annals of Internal Medicine* 2000;132:444–450.

77. DeValeria PA, Baumgartner WA, Casale AS et al. Current indications, risks, and outcome after pericardiectomy. *The Annals of Thoracic Surgery* 1991;52:219–224.

78. Uchida T, Bando K, Minatoya K, Sasako Y, Kobayashi J, Kitamura S. Pericardiectomy for constrictive pericarditis using the harmonic scalpel. *The Annals of Thoracic Surgery* 2001;72:924–925.

79. Soulen RL, Stark DD, Higgins CB. Magnetic resonance imaging of constrictive pericardial disease. *The American Journal of Cardiology* 1985;55:480–484.

80. Talreja DR, Edwards WD, Danielson GK et al. Constrictive pericarditis in 26 patients with histologically normal pericardial thickness. *Circulation* 2003;108:1852–1857.

81. Zurick AO, Bolen MA, Kwon DH et al. Pericardial delayed hyperenhancement with CMR imaging in patients with constrictive pericarditis undergoing surgical pericardiectomy: A case series with histopathological correlation. *JACC Cardiovascular Imaging* 2011;4:1180–1191.

82. Haley JH, Tajik AJ, Danielson GK, Schaff HV, Mulvagh SL, Oh JK. Transient constrictive pericarditis: Causes and natural history. *Journal of the American College of Cardiology* 2004;43:271–275.

83. Feng D, Glockner J, Kim K et al. Cardiac magnetic resonance imaging pericardial late gadolinium enhancement and elevated inflammatory markers can predict

the reversibility of constrictive pericarditis after antiin-flammatory medical therapy: A pilot study. *Circulation* 2011;124:1830–1837.

84. Masui T, Finck S, Higgins CB. Constrictive pericarditis and restrictive cardiomyopathy: Evaluation with MR imaging. *Radiology* 1992;182:369–373.

85. Hatle LK, Appleton CP, Popp RL. Differentiation of constrictive pericarditis and restrictive cardiomyopathy by Doppler echocardiography. *Circulation* 1989;79:357–370.

86. Rajagopalan N, Garcia MJ, Rodriguez L et al. Comparison of new Doppler echocardiographic methods to differentiate constrictive pericardial heart disease and restrictive cardiomyopathy. *The American Journal of Cardiology* 2001;87:86–94.

87. Bauner K, Horng A, Schmitz C, Reiser M, Huber A. New observations from MR velocity-encoded flow measurements concerning diastolic function in constrictive pericarditis. *European Radiology* 2010;20:1831–1840.

88. Mangia M, Madeo A, Conti B, Galea N. Giant left ventricular pseudoaneurysm following coronary artery bypass graft surgery. *European Journal of Cardio-Thoracic Surgery: Official Journal of the European Association for Cardio-Thoracic Surgery* 2012;41:e21.

89. Meleca MJ, Hoit BD. Previously unrecognized intrapericardial hematoma leading to refractory abdominal ascites. *Chest* 1995;108:1747–1748.

90. Brown DL, Ivey TD. Giant organized pericardial hematoma producing constrictive pericarditis: A case report and review of the literature. *The Journal of Trauma* 1996; 41:558–560.

91. Hoffmann U, Globits S, Frank H. Cardiac and paracardiac masses. Current opinion on diagnostic evaluation by magnetic resonance imaging. *European Heart journal* 1998;19:553–563.

92. Grebenc ML, Rosado de Christenson ML, Burke AP, Green CE, Galvin JR. Primary cardiac and pericardial neoplasms: Radiologic-pathologic correlation. *RadioGraphics: A Review Publication of the Radiological Society of North America, Inc* 2000;20:1073–1103; quiz 1110–1111, 1112.

4

Vascular Imaging of the Head and Neck

Miguel Trelles and Tobias Saam

CONTENTS

4.1 Vascular Anatomy

Magnetic resonance imaging (MRI) allows for excellent depiction of the vascular anatomy through routine planar imaging, contrast-enhanced MR angiography (MRA) and non-contrast-enhanced MRA. Conventional angiography should be relegated to cases where depiction of small vessels is necessary such as in the evaluation of vasculitis and to therapeutic interventions. Figure 4.1a shows a volume rendered image of the aorta showing the most common vascular distribution: left-sided aortic arch, right brachiocephalic artery, left common carotid, and left subclavian artery, arising from the aorta and both vertebral arteries arising from the subclavian arteries. Figure 4.1b and c shows the most common variations including *bovine* arch (Figure 4.1b),

where both common carotid arteries arise from a common trunk (which is actually a misnomer as the bovine anatomy is different) and an aberrant right subclavian artery in the setting of a left aortic arch (Figure 4.1c), potentially creating a vascular compression on the esophagus, which can be seen on fluoroscopy. This artery is also called *lusoria artery*. In these patients, the right subclavian artery can have a dilated origin from the aortic arch named the diverticulum of Kommerell. Important anomalies to keep in mind in the evaluation of the aortic arch include right aortic arch with mirror image branching, which is associated with cyanotic congenital heart disease, and right aortic arch with aberrant left subclavian artery, which, although not as strongly related with congenital heart disease, causes a vascular ring and its associated symptoms from compression of the trachea and esophagus, such as inspiratory stridor,

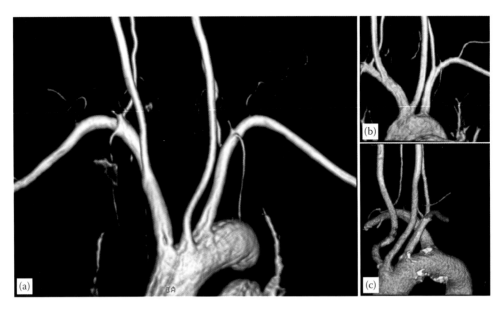

FIGURE 4.1
Volume-rendered CTA images of (a) a normal aortic arch, (b) a bovine type aortic arch, and (c) an aberrant right subclavian artery.

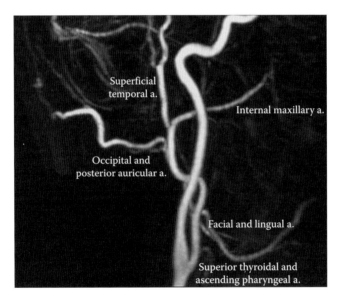

FIGURE 4.2
MRA showing the major branches of the external carotid artery.

wheezing, dyspnea, cough, dysphagia, and recurrent respiratory tract infections [1].

The common carotid artery bifurcates into external and internal carotid arteries at the level of C3–C4 at the carotid bulb. As we will see later, this is a favorite site for atheromatous plaque formation. The external carotid artery subsequently divides into the superior thyroidal, ascending pharyngeal, lingual, facial, occipital, posterior auricular, superficial temporal, and maxillary arteries, and remembering these branches is the nightmare of all starting residents; we recommend using the mnemonic "She always likes friends over

papa, sister and mamma" [2]. The main branches can be seen in Figure 4.2.

The internal carotid artery has seven segments (C1–C7) [3]: cervical (which has no branches easily distinguishing it from the external carotid artery), petrous, lacerum, cavernous, clinoid (piercing the dura and becoming intracranial at this level), ophthalmic, and communicating segments. The first commonly visualized branch by MRA is the ophthalmic artery (at the ophthalmic segment). Small branches such as the carotid cotympanic, vidian, meningohypopheseal trunks, inferolateral trunk, and anterior choroidal arteries are not

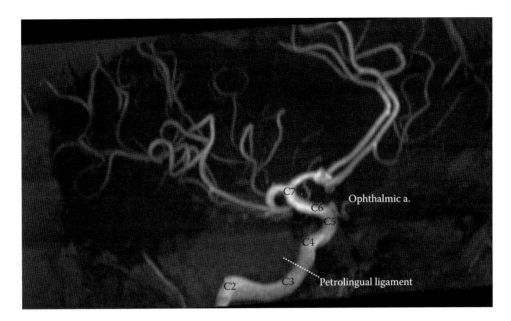

FIGURE 4.3
Lateral view of a maximum intensity projection (MIP) of a time-of-flight MRA of the brain showing the intracranial segments or the ICA: C1, cervical (not shown); C2, petrous; C3, lacerum; C4, cavernous; C5, clinoid; C6, opthalmic; C7, communicating segments.

routinely visualized on MRA if vessels are of normal size (Figure 4.3).

It is important to realize that multiple anastomotic channels exist between the extracranial and intracranial circulation [4], which are beyond the scope of this chapter. These branches become clinically significant if embolization to the extracranial circulation is attempted, as they may provide a pathway for intracranial spill. Likewise, they may provide collateral flow in cases of obstruction of the intracranial circulation.

The Circle of Willis is a system of anastomosing vessels in the base of the skull that can provide collateral flow maintaining cerebral blood flow and life in the presence of single or even dual vessel stenosis. It forms a vascular ring or pentagon made up of the terminal branches of both internal carotid arteries and basilar artery (Figure 4.4); specifically both A1 segments of the anterior cerebral arteries (ACA), anterior communicating artery (AComm), both posterior communicating arteries (PComm), and the basilar tip. Anatomical variants with hypoplastic segments and incomplete rings are the norm, less than 50% of the population has a complete Circle of Willis ring [5,6].

The vertebral arteries normally arise from the subclavian arteries. They are divided into four segments. Proximally, they course between the longus coli and anterior scalene muscles in the neck (V-1); they enter and course through the transverse foramina from C6 to C2 (V-2); subsequently, they exit the transverse foramen of C2 and pass through the transverse foramen of C1, and make a curve behind the posterior articular process of the atlas to enter the spinal canal (V-3); finally, they pierce

through the dura (V-4) and join with its contralateral partner to form the basilar artery. Anatomical variations include origin of the vertebral arteries from the aortic arch or rarely from the common carotid arteries [7]. Important branches include the anterior spinal artery and postero-inferior cerebellar arteries (PICA). The basilar artery gives rise to the anterior inferior cerebellar artery (AICA), superior cerebellar artery (SCA) and pontine perforators and terminates in the posterior cerebral artery (PCA). An anatomical pearl is the artery of Percheron, which originates from the proximal PCA and may supply both medial thalami [8].

4.1.1 Vessel Wall Histology

The vessel wall is composed of three layers: the *intima, media,* and *adventitia*. The intima constitutes the innermost layer and is made up of endothelium and a thin supportive connective tissue. The media consists of smooth muscle cells and elastic tissue in varying proportions. Elastic arteries have the highest proportion of elastic tissue, whereas muscular arteries have the highest proportion of smooth muscle. A continuous layer of elastic tissue, called the *internal elastic lamina*, forms the boundary between the media and the intima. The adventitia consists of fibrous connective tissue, which may be continuous with the stromal connective tissue of the organ, in which the vessel is found. The adventitia may also contain numerous elastin fibers. The distinct arteries, arterioles, capillaries, venules, and veins have distinct compositions that also depend on their location, which is beyond the scope of this book.

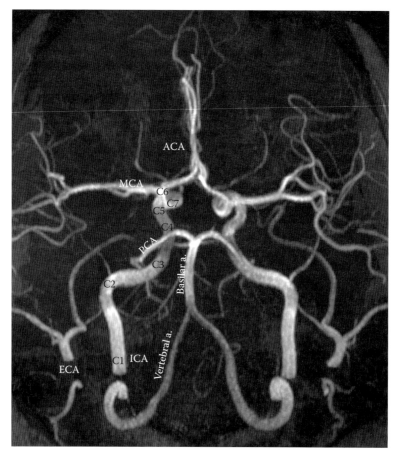

FIGURE 4.4
Superior MIP of a time-of-flight MRA of the brain showing the Circle of Willis. The main arteries and segments of the ICA are labeled on the right.

4.1.2 Vascular Imaging

New advances in endovascular surgery using stent-graft technology require exact vascular measurements [9] such as location and length of stenosis, percentage of stenosis based on diameter, area or flow, distance from landmarks, length, and size from the normal vessel, among others. MRI offers a lot of different possibilities to perform vascular imaging. The vessel lumen can be visualized using contrast-enhanced and non-contrast-enhanced techniques such as time-of-flight (TOF) and phase contrast MRA. The vessel wall may be studied using black-blood imaging. Flow velocity can be analyzed using phase contrast imaging. Before deciding which specific technique to use, it is important to know what it is that one wants to image. The following methods are briefly described.

4.1.2.1 Time-of-Flight MRA

TOF MRA is a non-contrast-enhanced method to visualize moving blood that may be used for angiographic purposes. Briefly, blood flowing into the imaging plane shows increased T_1 signal on gradient echo imaging,

because it has not been subjected to an RF pulse and retains its complete longitudinal magnetization. This flow-related enhancement is optimized in TOF imaging by using a short repetition time (TR) and a high flip angle, which reduces signal from background (stationary) tissue enhancing flow-related signal. It is important to realize that TOF imaging is T_1 weighted and as such methemoglobin and, in case fat suppression is not applied, fat will also have high signal.

TOF imaging may be performed by acquiring one image at a time (2D) or by acquiring one slab of multiple images at a time (3D). The latter is the preferred method because it reduces spatial misregistration from motion between images and allows for an even shorter TE, reducing background signals and artifacts. Further improvements in signal-to-noise ratio are experienced at high field strengths. A major drawback is that complex flow as seen next to areas of stenosis or bifurcations will produce artifactual signal loss, which may lead to underestimation or overestimation and misdiagnosis. 3D time of flight is the ideal first line method for the noninvasive evaluation of the intracranial arteries due to its ease of use and widespread availability.

4.1.2.2 Contrast-Enhanced MRA

Contrast-enhanced MRA is acquired by imaging a tissue volume before administration of contrast and during the maximal arterial concentration of contrast. This is obtained in similar fashion to CTA with careful tracking of the contrast agent bolus and a power injector. Correct timing of the imaging sequence is of paramount importance since final imaging contrast will depend on the concentration of gadolinium within the arteries at the time the center of k-space is being sampled; remember that this does not always occur at the middle of the imaging sequence acquisition.

Contrast-enhanced MRA permits a higher spatial resolution than TOF MRA and suffers from less flow-related artifacts. However, it requires contrast administration, an IV line, and a power injection with its associated nuances and risks. This is the method of choice for evaluating stenosis at the carotid bulb, the aorta, and its other branches [10].

Dynamic, time-resolved contrast-enhanced MRA is a novel development that allows for ultrafast 3D imaging during a contrast bolus. It works by sampling the center of k-space and only a fraction of the periphery. The complete information for the periphery is reconstructed from multiple time points resulting in great reduction of imaging time per volumetric acquisition. It may be used for initial noninvasive evaluation or follow-up of intracranial or extracranial vascular malformations and fistulas [11,12], although DSA remains the gold standard.

4.1.2.3 Phase Contrast MRA

Phase contrast MRA permits the evaluation of flow velocity in the plane of imaging. In order to perform phase contrast MRA, a magnetic field gradient is applied on the direction in which blood flow is to be measured. Blood flowing through this magnetic gradient will experience different precessional speeds depending on the local magnetic field at each particular location. The change in local magnetic field will cause a phase shift that can be measured. The final phase shift depends on the distance traveled, which in turn depends on its velocity. If the area of the vessel is also known, one can use this information to calculate the flow and obtain a pressure gradient from this. Phase contrast MRA is not limited to evaluation of vascular flow, in the central nervous system (CNS), it may be used to evaluate the flow of CSF fluid in the setting of aqueductal or foramen magnum stenosis [13].

4.1.3 Black-Blood Imaging

Vascular pathology has its origin in the vessel wall. However, the previously described imaging methods are primarily luminographic. Such methods are often insufficient to identify or differentiate vascular diseases. Furthermore, contrast enhancement of the vessel wall may help discriminate inflammatory from noninflammatory arteriopathies, which is vital for proper diagnosis and therapy.

Most MRA techniques aim to have high signal intensity within the vessel lumen while nulling the background. Black-blood imaging aims at the opposite; the objective is to null signal from the vessel lumen in order to be able to visualize the vessel wall without inflow or pulsation artifacts. Additionally, fat suppression should be used in order to better delineate hemorrhage, edema, or inflammation. This is achieved through a series of inversion pulses (Figure 4.5) taking advantage of the fact that blood keeps moving throughout the imaging time. Figure 4.5a shows a coronal plane through the neck, arterial vessels are shown in red for illustrative purposes. An initial non-selective RF pulse excites all spins within the imaging field including both tissue and blood. In Figure 4.5b, the excited volume is highlighted. A second slice-selective inversion RF pulse is then applied to re-invert the spins within the desired imaging slice into the transverse plane preparing it for the normal imaging acquisition portion of the sequence (Figure 4.5c), the tissue that has received both pulses is shown with increased brightness. Normal imaging is performed after a small pause to allow the excited blood to flow out of the slice plane, which is replaced by inverted blood (from the initial non-selective RF pulse) (Figure 4.5d). Only stationary tissues receive all pulses and are imaged. This results in the appearance that moving blood has been removed from the image when in actuality it simply has not been encoded and does not produce a signal.

4.2 Supra-Aortic Arteriopathies

A myriad of vascular diseases affect the head and neck. Although the clinical presentation is highly variable, including headache, encephalopathy, seizures, dizziness, and visual disturbances, all of them may cause severe morbidity and mortality in the form of blindness, transient ischemic attack (TIA) or stroke [14]. They can be classified into inflammatory causes such as CNS arteritis, Takayasu arteritis, Giant cell arteritis, and carotidynia and noninflammatory causes such as fibromuscular dysplasia, reversible vasoconstriction syndrome, and CADASIL. Atherosclerosis is a chronic inflammatory disease that can show features from inflammatory and noninflammatory diseases. Additionally, any diseased vascular wall may be subject to dissection and aneurysm formation. Intracranial berry aneurysms will be discussed in Chapter 24 in *Image Principles, Neck, and the Brain*.

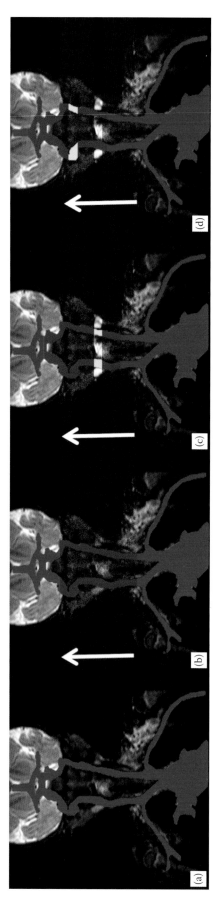

FIGURE 4.5

Dark-blood imaging: (a) coronal plane through the neck; arterial vessels are shown in red. (b) An initial non-selective RF pulse excites all spins within the imaging field including both tissue and blood. The excited volume is shown in grey. (c) A second slice-selective inversion RF pulse is then applied designed to re-invert the spins within the desired imaging slice into the transverse plane preparing it for the normal imaging acquisition portion of the sequence. The tissue that has received both pulses is shown with increased brightness. (d) Normal imaging is performed after a small pause to allow the excited blood to flows out of the slice plane, which is replaced by inverted blood (from the initial non-selective RF pulse). Only stationary tissues receive all pulses and are imaged. This results in the appearance that moving blood has been removed from the image when in actuality it simply has not been encoded and does not produce a signal.

4.2.1 Atherosclerosis

4.2.1.1 Carotid Bulb Atherosclerosis

Atherosclerosis is characterized by slow chronic buildup of lipids such as cholesterol within the intimal layer of the vessel wall; once this lipid collection coalesces it is called an *atheroma*, which causes a surrounding inflammatory response within the intima layer of the arterial wall. Figure 4.6 shows histological slides with the different types of atheromatous plaque formation. The atheroma may progress in different pathways. The best clinical outcome is achieved when the plaque calcifies and is assimilated into the vessel wall (Figure 4.6), which remodels resulting in a normal lumen [15]. The atheroma may grow into the lumen and compress the vessel causing progressive stenosis. If this process is slow enough, it will permit the development of collateral circulation preserving downstream perfusion. However, in some cases the wall that covers the atheroma ruptures exposing the pro-thrombotic components to the lumen (Figure 4.6),

which may in turn embolize downstream, causing embolic strokes and/or generate a clot at the level of the plaque causing an acute complete occluding stenosis reducing downstream perfusion pressure. Depending on the level of collateral flow (and patency of the Circle of Willis), this may result in watershed strokes or territorial infarction. Of all ischemic strokes, approximately 17%–20% of strokes are due to large vessel atherosclerosis, 25% are due to small vessel disease and 15%–27% are cardioembolic, and 35% are undetermined according to the TOAST method [16–18]. Of those undetermined strokes a substantial proportion might be caused by non-stenosing ruptured plaques [19]. Approximately 80%–90% of strokes due to large artery atherosclerosis are caused by plaque rupture, and 10%–20% are caused by watershed infarcts due to high-grade stenosis or vessel occlusion.

Normal luminographic methods are only capable of diagnosing stenosis, black-blood imaging permits detailed evaluation of plaque composition and morphology. NASCET-type vessel wall stenosis is the gold standard for evaluation of risk for thromboembolic event and dictates the need for treatment (endarterectomy) (Figure 4.7). It relies on the percent stenosis when comparing the most narrow segment with a normal distal section of the internal carotid artery, which has shown to correlate with future risk of stroke [20]. Stenosis measurements can also be done as area measurements in analogy to the NASCET method. However, keep in mind that the degree of stenosis calculated by an area

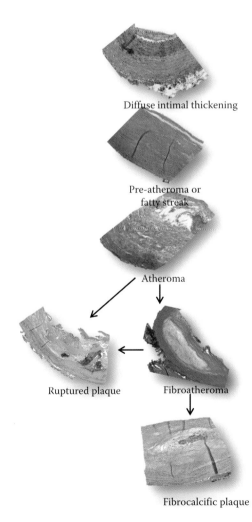

Diffuse intimal thickening

Pre-atheroma or fatty streak

Atheroma

Ruptured plaque   Fibroatheroma

Fibrocalcific plaque

FIGURE 4.6
Atherosclerotic disease progression.

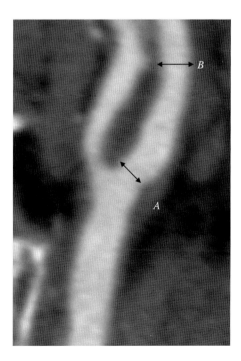

FIGURE 4.7
NASCET-type vessel wall stenosis measurement = $(B-A)/B$. A, diameter at zone of greatest stenosis; B, normal vessel diameter distal to the area of stenosis.

measurement yields higher values than for a 2D diameter measurement when the same formula is used. Contrast-enhanced MRA is the method of choice for the evaluation of carotid bulb stenosis by MRI [10].

High-resolution MRI is capable of identifying the different stages of atherosclerotic buildup including plaque complication and will contribute to the evaluation of atherosclerotic disease. Cai et al. [21] defined the MRI appearance of the distinct stages of atherosclerotic plaque buildup in the vessel wall as defined by the AHA (AHA) (Table 4.1). Any given plaque or vessel may have multiple areas with different classifications, thus the most severe finding must be described. Intraplaque hemorrhage, fibrous cap rupture and mural thrombus define the complicated plaque AHA lesion type VI (AHA LT: VI), and are related to increased risk of stroke independently of the severity of luminal stenosis [22,23].

The different imaging characteristics of plaque components are shown in Table 4.2. The normal vessel wall

TABLE 4.1

Modified AHA Classification for MRI for the Stages of Atherosclerotic Plaque Buildup in the Vessel Wall

Conventional AHA Classification	Modified AHA Classification for MRI
Type 1: Initial lesion with foam cells	Type I–II: Near-normal wall thickness, no calcification
Type II: Fatty streak with multiple foam cell layers	
Type III: Preatheroma with extracellular lipid pools	Type III: Diffuse intimal thickening or small eccentric plaque with no calcification
Type IV: Atheroma with a confluent extracellular lipid core	Type IV–V: Plaque with a lipid or necrotic core surrounded by fibrous tissue with possible calcification
Type V: Fibro-atheroma	
Type VI: Complex plaque with thrombus, hemorrhage or a ruptured fibrous cap	Type VI: Complex plaque with thrombus, hemorrhage or a ruptured fibrous cap
Type VII: Calcified plaque	Type VII: Calcified plaque
Type VIII: Fibrotic plaque without lipid core	Type VIII: Fibrotic plaque without lipid core and with possible small calcifications

Cai J M et al., *Circulation,* 106(11), 1368–1373, 2002.

TABLE 4.2

MRI Characteristics of Plaque Components

	TOF	T_1	PD	T_2	Contrast Enhancement
Normal	o	o	o	o	−/faint
Preatheroma	o	o	o	o	−/faint
Atheroma	o/+	o/+	−	−	None
Hemorrhage type 1	+	+	−/o	−/o	None
Hemorrhage type 2	+	+	+	+	None
Calcification	−	−			None

as well as the fibrotic portions of the diseased vessel wall may show enhancement. In high-resolution black-blood MRI with fat suppression, the lipid-rich necrotic core (LR-NC) without fat suppression appears as an isointense area compared to the normal vessel wall in T_1- and T_2-weighted sequences and can be most easily depicted and quantified on postcontrast T_1-weighted images as hypointense areas compared to the surrounding enhancing fibrous tissue. LR-NC with hemorrhage has increased signal intensity on TOF and T_1-weighted images. Signal intensity on T_2-weighted images depends on stage of hemorrhage with intracellular methemoglobin being low signal (plaque hemorrhage type I or early subacute) and extracellylarmethemoglobin high signal (plaque hemorrhage type II or late subacute) [24]. Calcifications are depicted as an area with decreased signal on all sequences and no peripheral enhancement.

Normal vessels (AHA LT: 1-II) show a normal thin wall (Figure 4.8). Pre-atheromatous plaques (AHA LT: III) have minimal (3–4 mm) diffuse or focal vessel wall thickening without a lipid necrotic core as shown in Figure 4.9. Atheromatous plaques (AHA LT: IV–V) have a lipid-rich necrotic core signal intensities depend on its content (white arrow in Figure 4.10) but which is isointense to the normal vessel wall if hemorrhage is not present. In addition, it will not show any enhancement on postcontrast images (arrowhead Figure 4.10f), which is the single best differentiating factor from fibrous tissue.

CT will show the soft plaque component (black arrowhead Figure 4.10a); however, it is unable to differentiate between AHA LT VI–V and VI, although a large soft plaque makes it more likely that this lesion is a complicated lesion [25].

The complicated plaque (AHA LT: VI) is characterized by hemorrhage, a prothrombotic surface defect (ruptured fibrous cap) or thrombus. Hemorrhage appears as an area of high T_1 signal that is more evident on the precontrast scan. It may demonstrate hypo-, iso-, or slightly hyperintense signal in T_2 and PD, depending on its age and composition. Figure 4.11 shows increased TOF and T_1 (white arrows) signal of the soft plaque component without enhancement representing hemorrhage. Thrombus is seen as a thin area of increased signal next to the vessel lumen, which can show contrast enhancement. Figure 4.12 demonstrates almost complete occlusion (* denotes the lumen) of the lumen by a structure showing increased T_2 signal and enhancement compatible with a thrombus (white arrow). Disruption of the fibrous cap can be diagnosed by (1) the absence of a dark band on TOF images and/or an absence of a fibrous cap on postcontrast T_1-weighted sequences, (2) lumen surface irregularity, and (3) a continuous high signal between the lumen and the vessel wall in TOF MRA. Ideally, all three of these features should be depicted in order to diagnose a fibrous cap rupture. Figures 4.13 and 4.14

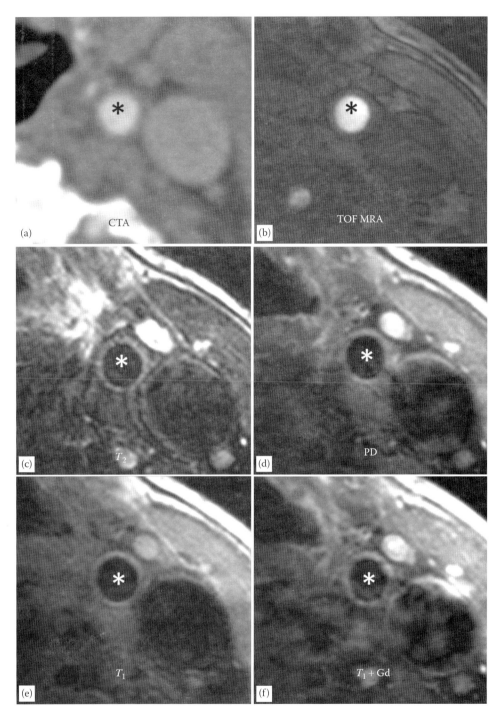

FIGURE 4.8
Axial CTA (a), TOF MRA (b) images as well as high-resolution fat-suppressed dark-blood images with T_2 (c), PD (d), T_1 (e), and T_1 + Gd (f) weighting of the distal left CCA in an asymptomatic 55-year-old patient. Normal vessel wall thickness and expected minimal normal vessel wall enhancement is present. * denotes the vessel lumen.

show disruption of the hypointense fibrous band seen between the hemorrhagic plaque and the vessel lumen (white arrowheads in Figure 4.13).

The calcified plaques (AHA LT: VII) are difficult to see on MRI. Most plaques will have a calcified portion, and it is important in evaluating any single lesion to consider the most pathologic segment. Figure 4.15 shows a calcified plaque easily seen on CTA; however, the signal void is harder to appreciate in MRI. It is easy to confuse the calcification with part of the lumen or a lipid-rich necrotic plaque; therefore, a combination of white- and black-blood images is mandatory. On the T_1 image

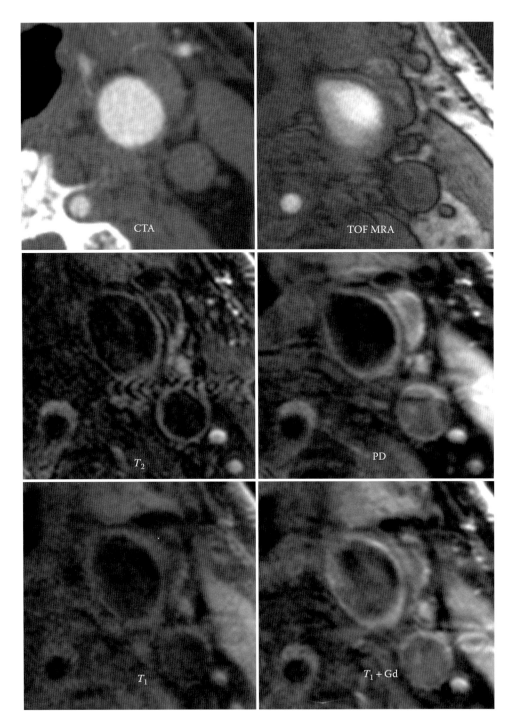

FIGURE 4.9
Axial images just below the level of the left bifurcation show minimal vessel wall thickening at 5 o'clock representing intimal thickening.

(Figure 4.15e) one could mistakenly think that the juxta-luminal area of signal void is part of the lumen, likewise on the PD and T_2 images (Figure 4.15c and d), the plaque resembles a lipid-rich necrotic core or atheroma. If available, always correlate with CTA or TOF images.

Fibrous plaques (AHA LT: VIII) are quite uncommon in the carotid arteries. They are characterized by circumferential vessel wall thickening and fibrosis of the plaque, which causes high-grade stenosis or occlusion (Figure 4.16). This is a late phase in the atherosclerotic spectrum, and it is believed to be due to chronic inflammation with scaring. Calcifications (not shown) and slight circumferential luminal enhancement in the usually highly stenosed vessel are common.

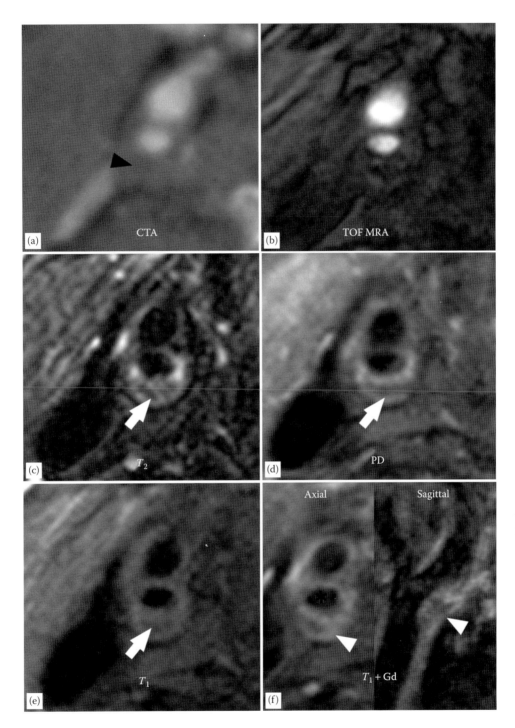

FIGURE 4.10
Images at the level of the right carotid bifurcation in a 54-year-old male with hypercholesterolemia and hypertension show vessel wall thickening with a large plaque showing decreased CT density (black arrowhead in a), decreased signal intensity (white arrow in c–e), increased peripheral T_2 signal and subtle peripheral enhancement (arrowheads in f) representing an atheroma.

The treatment of carotid artery stenosis is based on the degree of stenosis, symptom status, gender, and other risk factors and guidelines may differ across different countries. In general, surgical carotid endarterectomy or carotidartery stenting are the treatments of choice for lesions with a degree of stenosis higher than 70% by NASCET criteria. Nonstenotic complicated atherosclerotic plaques should be treated with best medical treatment and interventions can currently not be recommended for these patients.

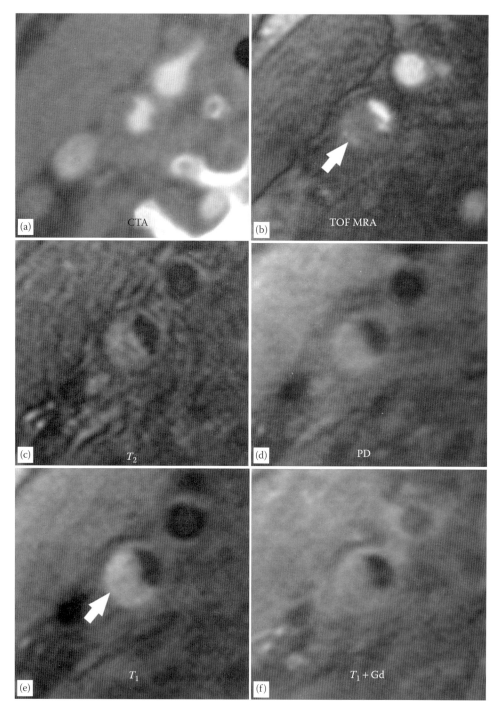

FIGURE 4.11

Axial images just above the right carotid bifurcation from a 64-year-old male with right anterior circulation stroke. CTA shows large soft plaque with ulceration (better seen on sagittal reformat). MRI shows increased TOF and T_1 (white arrows in b and e) signal of the soft plaque component without enhancement representing hemorrhage within a lipid-rich/necrotic core, consistent with a complicated AHA lesion type VI plaque.

4.2.1.2 Intracranial Atherosclerosis

Atherosclerotic disease is a systemic disease affecting all vascular beds. Atherosclerosis of the intracranial vessels may be asymptomatic orcause stenosis, obstruction, and emboli formation. As with other vasculopathies, the most important end point is TIA and stroke. Vascular dementia can also occur and is an increasing epidemiological problem in the aging western population.

Intracranial atherosclerosis accounts for approximately 5%–10% of strokes [18], is much more common in the Asian compared to non-Asian populations, and is also

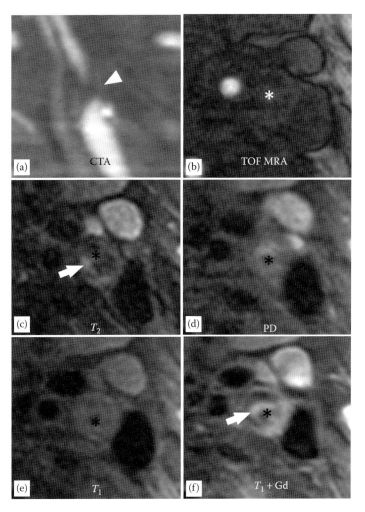

FIGURE 4.12
Sagittal CTA and axial MRI images of the proximal left ICA of a 78-year-old female with left anterior circulation stroke. CTA shows almost complete occlusion at the proximal ICA by a hypodense band (arrowhead) MRI again shows almost complete occlusion (* denotes the lumen) by a structure showing increased T_2 signal and enhancement compatible with a juxtaluminal thrombus (white arrows in c and f).

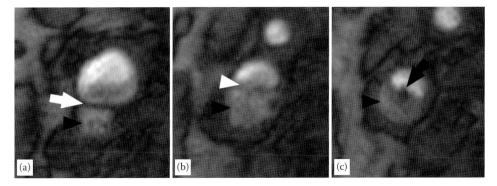

FIGURE 4.13
Sequential axial TOF MRI images from a 67-year-old male with right anterior circulation TIA. Increased signal of vessel wall is due to intra-plaque hemorrhage (black arrowhead). The proximal ICA shows an intact fibrous band (white arrow in a). Subsequent image shows rupture of the fibrous band (white arrowhead in b). Distal to the site of plaque rupture (c) turbulent flow can be appreciated (black arrow).

related to a high risk of recurrent stroke. The mechanisms underlying MCA stenosis-related stroke may include thrombosis formation in situ, artery-to-artery embolism, occlusion of penetrating arteries, and hypoperfusion [26].

TOF MRA is the primary screening method to study the intracranial vascular supply, because it is a fast and noninvasive. Both signal-to-noise ratio and spatial resolution are improved at higher imaging fields. The major

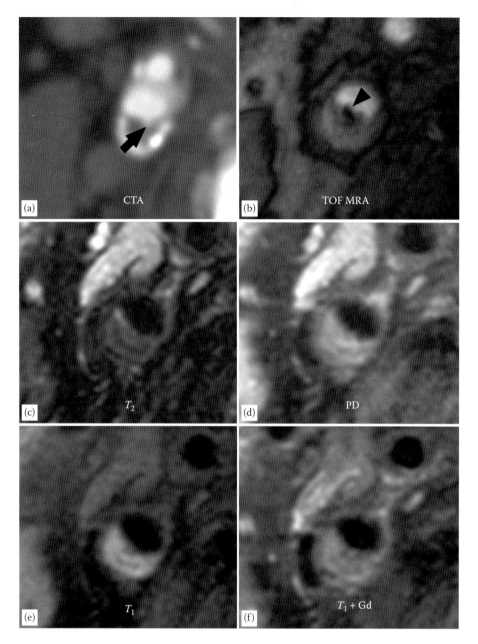

FIGURE 4.14
Same patient as before, axial images just above the right carotid bifurcation from a 67-year-old male with right anterior circulation TIA. CTA shows small irregularity at 7 o'clock (may represent irregular lumen or calcification) (black arrow in a). Turbulence artifact on TOF-MRA image (black arrowhead in b) and lack of blooming on PD image (d) rule out a calcification. Note that this plaque also shows intraplaque hemorrhage (characterized by increased precontrast T_1 and TOF signal).

pitfalls are that small vessels and vessels with slow flow distal to areas of stenosis are not well depicted, which may cause an over estimation of stenosis percentage. Thus, any abnormality should be confirmed by contrast-enhanced MRA, CTA, or DSA, especially if an intervention is contemplated or the diagnosis will alter clinical management.

High-resolution black-blood MRI is additionally able to depict plaque buildup in the intracranial vessels as a focal area of usually eccentric irregular vessel wall thickening, which may cause stenosis. Enhancement of the plaque has

been related to plaque instability and ischemic events [27]. However, the depiction of the atheroma or plaque complication such as hemorrhage is currently close to the resolution boundaries of the technique [28]. Currently, the study of the progression of intracranial atherosclerotic disease is limited; it may be distinct from progression in other vascular beds due to lack of vasa vasorum at birth [29].

Treatment for intracranial atherosclerosis remains controversial, recent studies suggest that aggressive medical therapy appears to be superior to angioplasty and stenting [30].

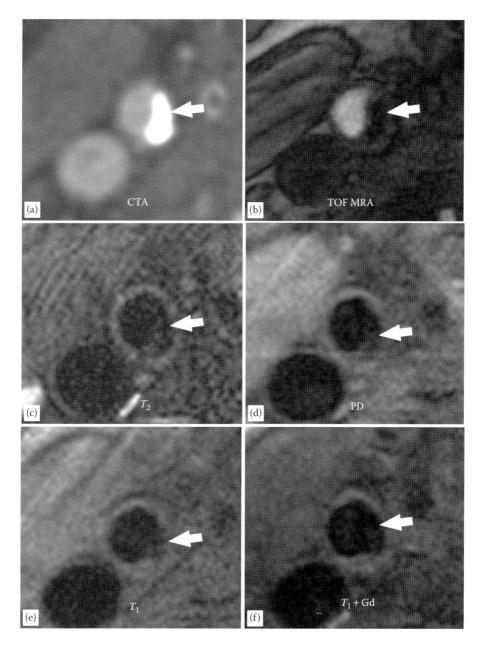

FIGURE 4.15

Axial images through the proximal right ICA of a 67-year-old male with diabetes shows a calcified plaque. Note how the calcification is easily see on CTA (white arrow in a); however, the signal void (white arrows in b–f) is harder to appreciate in MRI, although it can be identified by the combined use of white- and black-blood images.

Figure 4.17 shows a left MCA intracranial atherosclerotic plaque showing contrast enhancement on black-blood images suggestive of plaque instability [27].

A new development is the concept of the brain stress test, used to evaluate the ability of the brain to maintain normal cerebral blood flow by autoregulatory mechanisms [31]. In analogy to a cardiac stress test, a stimulus is given to cause vasodilation, increasing perfusion in areas irrigated by normal vessels, which itself causes a steal phenomenon from areas with arterial stenosis or decrease capacity for further vasodilation and results in decreased perfusion of these vascular beds. The brain stress test may be performed in multiple ways; the stimulus may be carbon dioxide or acetazolamide, which act as intracranial vasodilators [32]. Response to regional perfusion levels may be measured using perfusion MRI or BOLD imaging (Chapters 1 and 7). Comparison of the results with and without vasodilators provides an idea of the cerebral flow reserve and may aid to determine the risk of future ischemic events and guide therapy. It is still unclear if the information obtained will be used in the clinical practice or whether it will be part of our standard future clinical armamentarium.

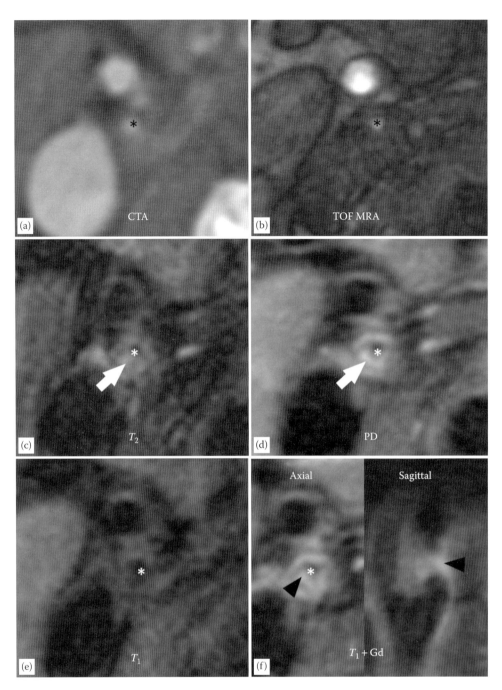

FIGURE 4.16

Shows circumferential narrowing of the vessel lumen (*) with a T_2 and PD hyperintense (white arrow), enhancing vessel wall. Note how a small area of increased T_2 signal and enhancement (black arrowheads in f) is seen at 7 o'clock suggesting loose fibrous matrix or thrombus.

4.2.2 Noninflammatory

4.2.2.1 Fibromuscular Dysplasia

Fibromuscular dysplasia is an uncommon, frequently asymptomatic noninflammatory, non-atherosclerotic arteriopathy, affecting small- to medium-sized vessels. It is more frequent in women, often diagnosed between ages 20 and 40. The renal arteries are most commonly affected (60%–75% of cases), followed by the carotid arteries (25%–30% of cases). However, it has been described in almost all vascular territories. Although its pathogenesis remains unknown multiple factors, including genetic, mechanical, hormonal, and toxics such as cigarette smoking have been implied. It is classified according to the site of abnormal fibrous/collagen proliferation [33]. Type 1 or media fibroplasia is the most common type (85% of cases), which is affecting the media. Figure 4.18 shows typical DSA findings

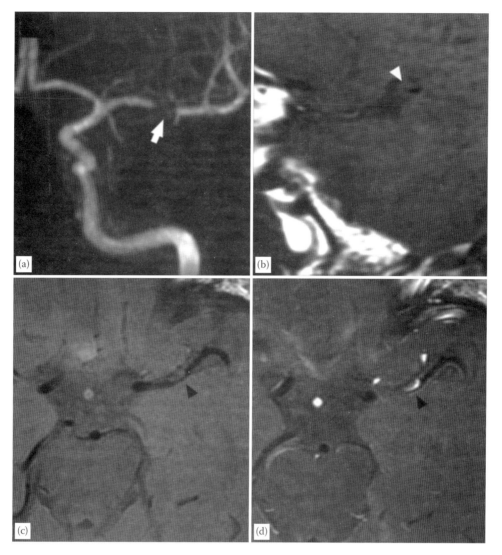

FIGURE 4.17
(a) MIP TOF MRA shows a focal stenosis in the left MCA (white arrow) (b) sagittal T_1 FLAIR image shows eccentric thickening of the antero-superior wall of the left MCA (white arrowhead). Axial T_1 black-blood FLAIR before (c) and after (d) contrast administration show an eccentric enhancing plaque on the anterior wall of the left MCA (black arrowhead), likely representing a plaque instability. (Courtesy of Dr. D. Mikulis; from Swartz RH, Bhuta SS, Farb RI et al. Intracranial arterial wall imaging using high-resolution 3-tesla contrast-enhanced MRI. *Neurology.* 2009;72:627–634. With permission.)

with a long segment area of multifocal stenosis and dilatations, the so-called string of beads sign. Type 2 or intimal fibroplasia accounts for less than 10% of cases. It typically appears as focal concentric stenosis. Long segments with smooth narrowing may also be present, which may be mistaken for large artery vasculitis, in particular Takayasu arteritis. Type 3 or adventitia hyperplasia is the rarest type accounting for less that 5% of cases. Abnormal vessel wall morphology increases risk of aneurysm, stenosis, dissection and arterio-venous fistula, or embolus formation [34].

All three types of cervical fibromuscular dysplasia tend to spare the great vessel origin and carotid bifurcation. The mid-vessel segments are more commonly involved, especially at the C1–C2 level in the internal carotid artery. Vasculitis may be differentiated from fibromuscular dysplasia by concentric wall thickening and the presence of enhancement and perivascular edema on dark blood imaging.

Treatment depends on the presence and severity of stenosis, or aneurysms as well as the extent of disease, symptoms, age, and comorbidities. It varies from observation through antiplatelet medication, anticoagulation, percutaneous angioplasty, and surgery [35]. No clear randomized trials are available comparing treatment methods and the treating physician, as well as the institution's experience has an important role in treatment decision.

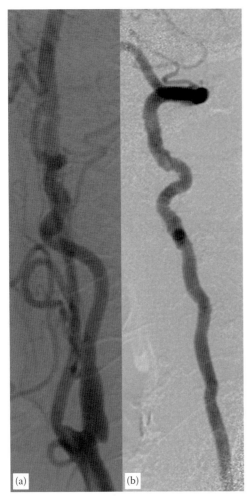

FIGURE 4.18
Two different patients with type 1 ICA (a) and vertebral artery (b) fibromuscular dysplasia showing long segment of multifocal stenosis and dilations *string of beads* sign.

4.2.2.2 Arterial Dissection

An arterial dissection is characterized by a break in the intimal layer of the arterial wall with flow of blood into a false lumen created within the media and constrained between the intimal and adventitia layers. This false lumen expands and may balloon toward the true lumen occluding it and decreasing distal perfusion pressure; alternatively, it may balloon toward the adventitia creating an aneurysmal outpouching, termed as *dissecting aneurysm*, which may cause mass effect on adjacent structures including nerve impingement and might be a source of emboli. Additionally, thrombus formation at the site of dissection may embolize causing downstream ischemia [36].

The pathophysiology is poorly understood; however, constitutional factors such as connective tissue disorders, genetic predisposition as well as environmental factors such as major or minor trauma and infection and vascular inflammation in a subset of patients [37]

have been implicated in predisposing factors or factors actually causing arterial dissection [37].

Arterial dissection is increasingly recognized as a cause of TIA and stroke. It may occur spontaneously or after trauma. Although spontaneous dissection is uncommon, it is an important cause of stroke in persons <50 years of age [38]. Vertebral artery dissection should always be suspected in any patients with cervical spine fracture involving the transverse foramen.

Imaging findings include long segment stenosis, the so-called string sign and high signal in T_1-weighted fat-suppressed images from methemoglobin in the false lumen or in the vessel wall. Recent studies have shown that focal contrast enhancement is a common finding at the site of the dissection [37]. In TOF MRA, special care must be employed not to confuse the increased signal from methemoglobin in the false lumen, which can be falsely attributed to normal flow (Figure 4.19). Figure 4.20 shows a long segment complete occlusion of the right ICA with increased T_1 signal of the wall (red arrow) representing clot within the false lumen. Dissecting aneurysmal outpunching will have similar appearance to aneurysms and may cause mass effect on adjacent structures.

Treatment remains controversial and depends on the clinical scenario and patient characteristics and may range from conservative measures and observation to anticoagulation and stent placement [39,40].

4.2.2.3 Reversible Vasoconstriction Syndrome

Reversible cerebral vasoconstriction syndrome (RCVS), also called Call–Fleming syndrome, is a rare syndrome characterized by severe *thunderclap* headache, transient multifocal segmental vasoconstriction of cerebral arteries, and focal neurologic symptoms with or without strokes. The diagnosis is made by excluding other more common causes such as ruptured aneurysms or vasculitis and spontaneous resolution of clinical and imaging findings in weeks to months. Convexal SAH is a common imaging finding at presentation and should alert

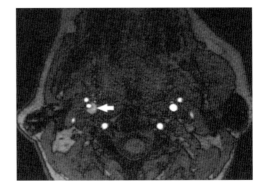

FIGURE 4.19
TOF angiogram of the upper neck showing spontaneous dissection of the right ICA, note increased T_1 signal within the vessel wall in the false lumen (white arrow), which can be mistaken for flow on MIP images.

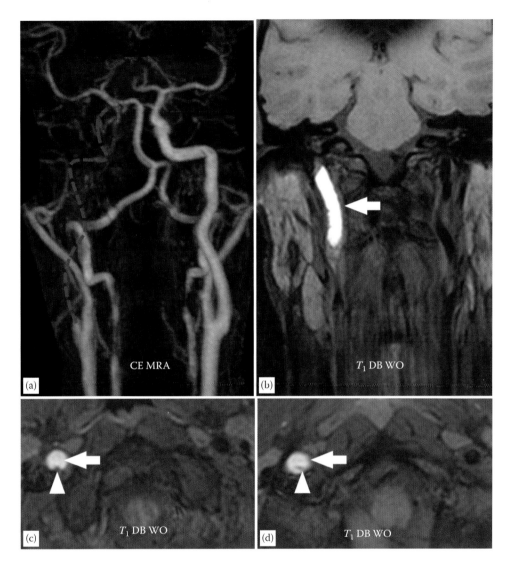

FIGURE 4.20

Images of a 26-year-old female show a long segment stenosis of the right ICA seen on subtracted contrast-enhanced MRA MIP (a); the dashed line represents the site of the missing vessel. Increased T_1 signal within the vessel wall (white arrow in b–d) represents clot within the false lumen. On axial images (c and d) the high signal takes a semilunar shape. The true lumen is seen as a dark dot on the dark-blood images (arrowhead in c and d).

the radiologist to this possibility. The pathogenesis of intracranial vasoconstriction is unknown but certain drugs have been postulated as triggers [41,42]. Females are affected more often than males.

Initial imaging will often only show convexal SAH [43]. Areas of ischemia may be noted on MRI within a watershed distribution. TOF MRA of the circle of Willis is a good noninvasive first-line test; however, any abnormality should be confirmed with DSA. The diagnostic hallmark is areas of vasoconstriction in a *sausage-string* or *beading* pattern on angiography. The main differential consideration is CNS vasculitis, differentiation by luminographic imaging may not be possible. Recent studies have shown that vessel wall enhancement as depicted by black-blood imaging may allow differentiation between reversible vasocontriction syndrome and central nervous system vasculitis [44]. Clinical history

and laboratory analysis may help differentiate both. A novel method includes black-blood imaging of the intracranial vessels to detect vessel wall enhancement, which in conjunction with clinical and laboratory data help to establish the diagnosis.

Figure 4.21 shows a typical case of RCVS with convexal hemorrhage, abnormal FLAIR signal in a watershed distribution corresponding to ischemia. Figure 4.22 shows mild generalized vasoconstriction of the distal internal carotid arteries, MCAs, and ACAs that is confirmed with DSA. Follow-up imaging a month later shows resolution of vasoconstriction.

Treatment is controversial, by definition, this is a self-limiting disease; however, intra-arterial administration of calcium channel blockers as vasodilators such as nimodipine, nifedipine, or verapamil and angioplasty have been performed with different degrees of success [45,46].

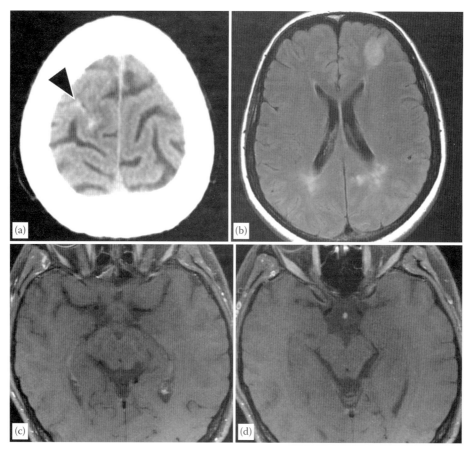

FIGURE 4.21
Images of a 51-year-old female presenting to the emergency room with the worst headache of her life. Image in (a) shows a nonenhanced head CT with convexal hemorrhage (arrowhead), while (b) shows an axial FLAIR image demonstrating multiple areas of abnormal signal representing ischemia in a watershed distribution. Images (c) and (d) show contrast-enhanced dark-blood images through the base of the skull demonstrating no abnormal contrast enhancement.

4.2.2.4 CADASIL Syndrome

Cerebral *A*utosomal *D*ominant *A*rteriopathy with *S*ubcortical *I*nfarcts and *L*eukoencephalopathy (CADASIL) is the most common hereditary cause of stroke and vascular dementia in adults. It is characterized by migraine headaches, TIAs and ischemic strokes, depression, progressive dementia evolving into inability to walk and death. The disease is usually diagnosed by 40–60 years. The mean duration from diagnosis to death is 23 years.

Imaging abnormalities on MRI usually precede the development of symptoms and may thus help the clinician make an early diagnosis. The most important imaging characteristic is diffuse white matter involvement more than expected for age not typical of demyelinating diseases. Diagnosis is given by ruling out more frequent causes of white matter disease and the relentless pattern of disease progression. Involvement of the anterior temporal and superior frontal subcortical white matter (Figure 4.23a and b) and external capsule-insula are characteristic [47]. In the initial stage of the disease, the subcortical µ-fibers are spared. Microbleeds may also be seen

with gradient-echo or susceptibility-weighted imaging (SWI) sequences [48]. Late in the disease decreased T_2 and FLAIR signal may be seen in the subcortical nuclei, which is thought to be due to iron deposition (Figure 4.23c) [49]. Although these findings are helpful in making the diagnosis none are specific of CADASIL. The role of the radiologist is to bring this possibility to the clinician's mind, especially in young patients with typical imaging findings in which no other definite diagnosis can be made.

Figure 4.24 shows disease progression through 15 years. So far no treatment has proven efficacy in delaying disease progression.

4.2.3 Inflammatory

4.2.3.1 CNS Vasculitis

CNS vasculitis presents a heterogeneous group of inflammatory diseases that primarily affect the small leptomeningeal and parenchymal blood vessels of the brain, spinal cord, and leptomeninges. Classification remains difficult due to multiple overlapping clinical, laboratory, and

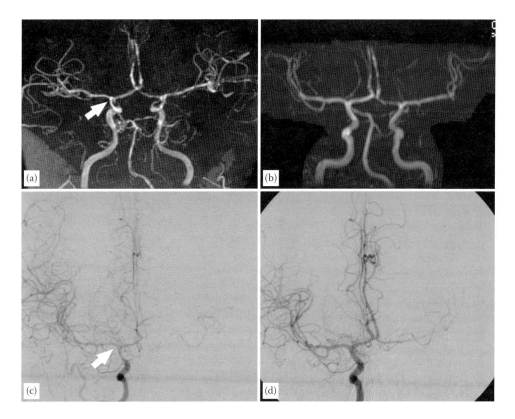

FIGURE 4.22
Same patient as before. TOF MRA of the COW (a, b) and right ICA DSA (c, d) images during the acute onset of symptomatology (a, c) and after resolution of symptoms (b, d). Vasoconstriction of the distal ICAs, MCAs, and ACAs is present on initial imaging (white arrows). Follow-up imaging a month later after the resolution of symptoms shows normal vessel caliber.

pathological findings and thus is controversial. The cardinal histopathological feature is inflammation affecting the blood vessel wall. Vasculitis affecting exclusively the CNS is referred to as primary CNS vasculitis (PCNSV). Secondary vasculitis may occur in the setting of other autoimmune diseases or after a variety of insults such as infection, malignancy, ionizing radiation, drugs, and autoimmune disease. PCNSV, systemic lupus erythematosus, polyarteritisnodosa, giant cell arteritis, and Sjögren syndrome comprise the majority of autoimmune conditions associated with CNS vasculitis [50].

The most common clinical presentation is stroke, encephalopathy, and seizures. Diagnosis depends on clinical, laboratory, and imaging findings. In many cases, biopsy must be performed to confirm the disease as imaging and symptomatology are nonspecific [51]. Immunosuppressive agents are the cornerstones of treatment.

Imaging findings between the different etiologies is indistinct and is characterized by parenchymal ischemic lesions in the affected vascular beds, which commonly correspond to multiple vascular territories. In a small number of cases, convexal SAH may be the first imaging finding on initial CT.

Conventional MRI is nonspecific; however, involvement of the cortical and subcortical white matter as well as the basal ganglia in a nonvascular distribution is highly suggestive. T_2/FLAIR imaging is more sensitive

for detecting abnormalities. GRE and SWI imaging may additionally show small parenchymal microhemorrhages and/or SAH. Patchy parenchymal enhancement with some areas of more focal or linear enhancement may be present; additionally, areas with subacute infarction may also enhance making the overall picture confusing.

Multifocal irregularities with stenosis and dilation in a *string of beads* pattern are the most important imaging finding. Inflammatory changes to the vessel wall increase the prevalence of complications such as dissection and fusiform aneurysm formation. On axial imaging, vessel wall thickening is a subtle finding, which may sometimes be identified.

As the presentation is characterized by broad symptomatology, the main role of the radiologist is to identify patients in which vasculitis could be the causative agent and bring this possibility into the clinician's mind. MRA is a good screening test, but it should not be used to diagnose or rule out vasculitis, as it does not have the spatial resolution to diagnose small vessel stenosis and dilation. Although CTA has a slightly higher spatial resolution, in all cases with high clinical suspicion of vasculitis, DSA should be ultimately performed.

Black-blood MRI is evolving into a new powerful tool for the evaluation of vasculitis. It may be performed to visualize the vessel wall and identify areas of concentric

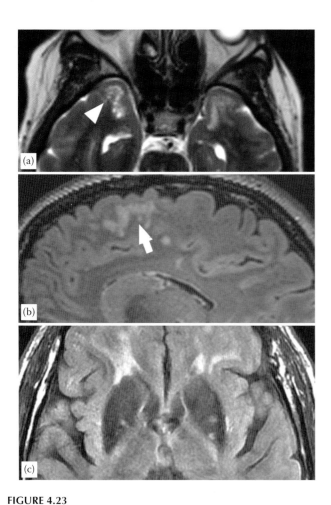

FIGURE 4.23

A 52-year-old female with progressive dementia. T_2 (a) and FLAIR (b, c) hyper-intensities are present in the anterior temporal (arrowhead in a) and superior frontal (white arrow in b) white matter. Low signal intensity of subcortical nuclei in T_2 and FLAIR imaging (c) is typical of late disease, thought to be due to iron deposition.

wall thickening and more importantly enhancement that has been related to active inflammation [52]. Figure 4.25 shows an example of contrast enhancement in the vessel wall of the distal ICA and proximal A1 and M1 and PCOM arteries in the setting of acute left-sided infarctions, which was treated with immunosuppressive therapy without progression of symptoms. This patient did not have any other symptomatology outside of the CNS and was empirically classified as PCNSV. However, black-blood imaging is currently limited to the evaluation of the larger and medium-sized intracranial vessels.

4.2.3.2 Giant Cell Arteritis

Giant cell arteritis is a granulomatous vasculitis of medium and large arteries. It predominantly affects the superior temporal arteries, less common presentations include involvement of the main branches of the aorta, its primary and secondary branches including subclavian and axillary arteries, superficial temporal, ophthalmic,

posterior ciliary, and vertebral arteries [53]. However, any vessel may be affected. The most common presentation includes headache, scalp tenderness, jaw claudication, and systemic inflammatory manifestations; nevertheless, involvement of the cervical or cranial vessels may present as stroke. The most dreaded complication is painless blindness from arterial involvement.

Diagnosis is centered on clinical and laboratory findings with temporal artery biopsy being the gold standard [54]. Imaging with ultrasound or MRI plays a role in cases with atypical presentation, where confirming the presence of vessel wall inflammation is desired before undertaking an invasive procedure such as a biopsy. Conventional T_2 imaging demonstrates vessel wall thickening. High-resolution black-blood MRI may show vessel wall thickening and enhancement, also called *target sign* [55,56]. Figure 4.26 shows dark-blood imaging through the left temple with superficial temporal artery vessel wall thickness, increased T_2 signal and contrast enhancement in a patient with biopsy proven Giant cell arteritis. Systemic corticosteroid therapy is the mainstay of treatment.

4.2.3.3 Takayasu Arteritis

Takayasu arteritis is a chronic granulomatous vasculitis mainly affecting large vessels, mainly the aorta and its main branches. The disease is trimodal with an initial systemic phase characterized by constitutional symptoms; an intermediate phase with addition of vascular involvement such as angiodynia; and a late phase in which stenosis and occlusion occur [57]. This late phase shows the characteristic features of Takayasu arteritis *pulseless disease."* Symptomatology depends on the vascular territory affected.

Imaging with US, CT, or MRI shows vessel wall thickening and inflammation. Angiography in early stages may be normal, and in later stages, all imaging methods will show stenosis. MRI may additionally be used to monitor the disease activity by evaluating the degree of wall thickening and enhancement and increased signal in fluid sensitive sequences, also called *target sign*. In the late or the occlusive phase, contrast-enhanced CTA or MRA allows noninvasive evaluation of the degree and extent of stenosis and collateral vessel formation as well as associated complications such as intraluminal thrombus and rarely aneurysm formation or dissection [58]. Although the gold standard remains digital subtraction angiography, it is now mainly used in the setting of an invasive procedure and in selected cases for treatment planning.

Figure 4.27 shows images of Takayasu arteritis affecting the aortic arch with complete occlusion of the left common carotid and left subclavian arteries and abundant wall thickening and enhancement surrounding the aortic arch and proximal great vessels. Figure 4.28 shows

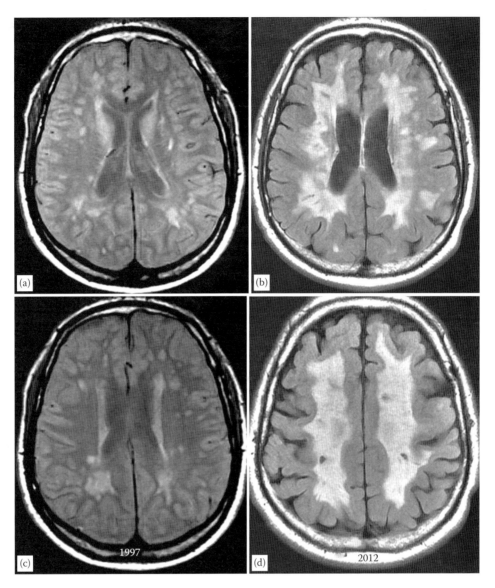

FIGURE 4.24
Axial FLAIR images show initial scan of a 45-year-old male from 1997 (a, c), showing multiple subcortical strokes, more than expected for age. Note early sparing of subcortical arcuate μ-fibers. Follow-up imaging from 2012 (b, d), when the patient was 60, shows confluent lesions with cystic changes and late involvement of arcuate μ-fibers.

biopsy proven Takayasu arteritis affecting the left vertebral artery with vessel wall thickening, enhancement, and luminal stenosis.

Treatment is centered on immunosuppression, and in some instances, oral anticoagulation can also be added. Surgical or endovascular procedures may also be needed in order to treat complications and preserve distal perfusion [59].

4.2.3.4 Carotidynia

Carotidynia is an idiopathic neck pain syndrome associated with tenderness to palpation over the carotid artery bifurcation and ipsilateral neck pain [60]. Although

diagnosis is clinical, a variety of diseases may present with similar symptoms such as vasculitis, fibromuscular dysplasia, complicated atherosclerosis with aneurysm or dissection, or other nonvascular diseases such as lymphadenitis, spinal degeneration, or syaloadenitis [61]. Imaging aids diagnosis and rules out confounding entities by demonstrating carotid wall thickening, perivascular contrast enhancement, and edema, consistent with inflammation. High-resolution MRI of the carotid bulb shows amorphous vessel wall thickening and enhancement, which is usually focal and eccentric but may also be circumferential [62]. Stenosis is not commonly present [63,64]. Increased T_2 signal is also present, and it has been postulated as a finding of edema

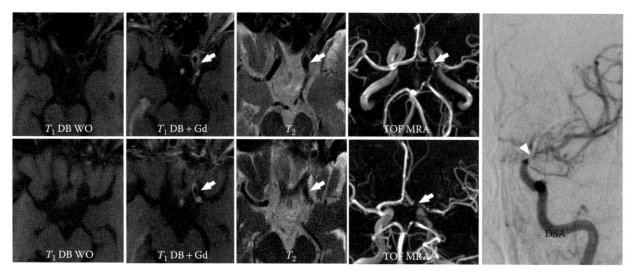

FIGURE 4.25

A 60-year-old female presents with stroke in the left anterior circulation. Black-blood MRI shows vessel wall thickening with stenosis and enhancement (white arrows) of the distal carotid and proximal ACA and MCA suggesting CNS vasculitis. Findings where confirmed on DSA (arrowhead). The patient was treated with cortisone without progression of symptoms.

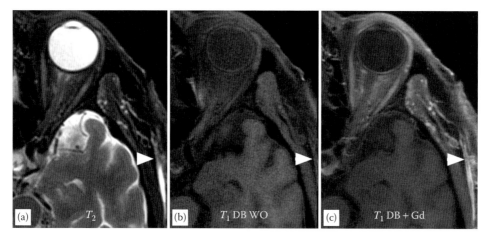

FIGURE 4.26

A 77-year-old male presenting with headaches and tenderness to the left temple. Axial DB images of the left temple show increased T_2 signal representing vessel wall edema (a) and contrast enhancement (b, c) of left temporal artery (arrowheads). Biopsy demonstrated Giant cell arteritis successfully treated with steroids.

related to inflammation. Figure 4.29 shows typical findings of carotidynia with focal enhancing carotid wall that partially resolved after corticosteroid treatment.

4.2.4 Unknown

4.2.4.1 *Moyamoya Disease and Moyamoya Syndrome*

Moyamoya disease is an idiopathic cerebrovascular occlusive disorder characterized by progressive stenosis of the intracranial internal carotid arteries and their proximal branches with associated compensatory collateral vasculature formation by small vessels near the apex of the carotid, cortical brain surface, leptomeninges and branches of the external carotid artery supplying the dura and skull base. By definition, moyamoya disease is bilateral, although it may be asymmetric [65]. In rare cases, this process may also involve the posterior circulation [66]. The incidence peaks in children aged around 5 and in adults aged around 40.

Patients with this characteristic vasculopathy and no known associated entity are said to have moyamoya disease. Patients with intracranial vessel stenosis and occlusion and associated collateral formation in the setting of another primary disease, such as atherosclerosis, Down's syndrome, neurofibromatosis, sickle cell anemia, among others, are classified as moyamoya syndrome.

Symptomatology relates to two underlying physiopathologic processes, progressive stenosis, and fragile

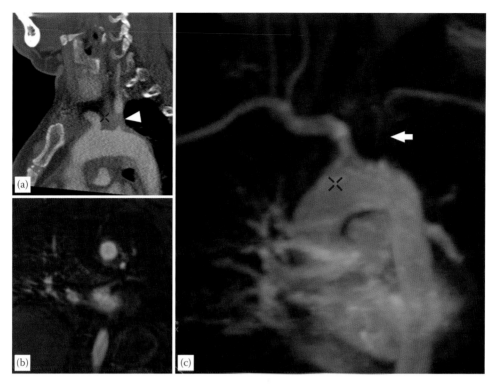

FIGURE 4.27

A 21-year-old female with Takayasu arteritis: (a) oblique reformat of a CTA showing a candycane view of the thoracic aorta with vessel wall thickening and complete occlusion of the proximal left common carotid artery (white arrowhead); (b) shows a coronal postcontrast MRI with abundant enhancement and wall thickening surrounding the aortic arch; and (c) corresponding MRA shows complete occlusion of the left common carotid and subclavian arteries (white arrow).

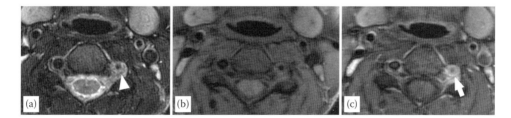

FIGURE 4.28

A 19-year-old female with known diagnosis of Takayasu arteritis presenting with vertebrobasilar symptoms. MRI shows increased vessel wall thickening and T_2 signal representing edema (white arrowhead in a) and enhancement (white arrow in c) of the left vertebral artery causing luminal narrowing.

collateral vessel formation. Progressive vessel stenosis of the distal ICA and proximal MCA and ACA, together with insufficient collateral flow produces reduced perfusion pressure, which results into parenchymal ischemia which then may lead to stroke and TIA or other nonspecific symptoms such as focal neurological deficits, nausea/vomits, dizziness, or even seizures [66]. Formation of a fragile network of collateral vessels predisposes to micro- and macrohemorrhage and headache from dilated transdural collaterals [66,67]. There is great individual variation in clinical presentation due to variations in involvement of different vascular beds, the speed of stenosis progression, and regions of ischemic cortex involved. However, the posterior circulation including the posterior cerebral arteries is usually not involved.

There are three basic collateral pathways shown in Figure 4.30: basal collateral vessels from perforators (orange), leptomeningeal collateral vessels from the PCA (blue) and transdural collateral vessels from the external carotid artery (green). On imaging, these vessels may be seen as a puff of smoke on angiography, termed the *puff of smoke* sign (*moyamoya* in Japanese) [68], which represents a tangle of tiny vessels (Figures 4.31c and 4.32e and f). On MRI, decrease in the flow voids of the distal internal carotid arteries together with prominent linear signal or even flow voids in the basal ganglia and thalamus can be seen secondary to collateral vessel formation.

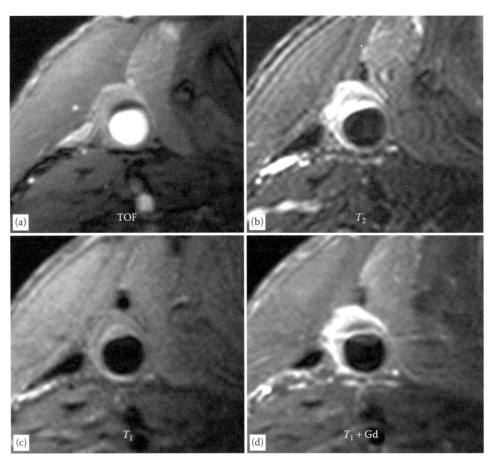

FIGURE 4.29
A 44-year-old male patient with right-sided neck pain for 8 days. TOF (a) and high-resolution black-blood MRI images (b–d) show localized vessel wall thickening at 10–12 o'clock with increased T_2 signal (b) and peripheral enhancement (c, d) in the distal right CCA just below the bifurcation. No stenosis was identified.

On contrast-enhanced MRI, the collaterals may be seen as abnormal linear signal within the leptomeninges with associated contrast enhancement, transdural collaterals appear as leptomeningeal enhancement close to the convexity and are termed *ivy* sign (Figure 4.31c) [69]. SWI and gradient-echo images may be used to evaluate for microhemorrhages, which are commonly seen. Additionally, SWI may better show intraparenchymal collateral vessels (termed *brush sign*) and has been postulated as a way to stage for moyamoya progression [70].

Disease staging was initially described by Suzuki and Takaku in 1969 [65] and evolves from initial minimal distal carotid artery stenosis (stage I) into progressive narrowing with initial collateral vessel formation and appearance of the puff of smoke sign (stage II), further narrowing and intensification of the puff of smoke associated collaterals (stage III), initial ischemic symptomatology as collateral vessel formation is unable to compensate and development of extracranial collaterals (stage IV), and continued narrowing of internal carotid branches, reducing the size of the puff of smoke sign

together with incremental flow from collateral vessels from the extracerebral circulation (stage V). The final stage VI is characterized by complete occlusion of the major intracranial arteries and thorough reliance in collateral flow; the puff of smoke sign disappears. Disease progression goes together with increased likelihood of ischemic symptoms from insufficient perfusion pressure and hemorrhagic complications due to incremental collateral vessel formation.

Besides diagnosis and disease progression, imaging also has a crucial role in the evaluation of complications, such as stroke and hemorrhage. Additionally, the weak collateral vessels are prone to aneurysm and dissection [71] formation. In extremely rare situations, AVMs have also been seen in the setting of moyamoya; it is unclear if there is a physiopathologic relationship such as acquired arteriovenous shunts that developed as a consequence of angiogenic failure or mere coincidence [72].

In moyamoya syndrome, the therapy is aimed at the underlying disease process. Treatment in moyamoya disease is still controversial, and no known treatment

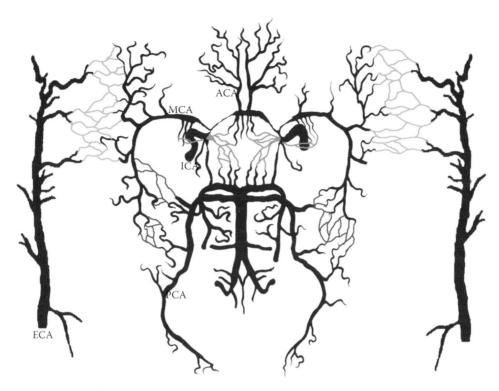

FIGURE 4.30
Diagrammatic depiction showing collateral vessel formation in moyamoya disease. Basal collateral vessels from perforators are shown in orange, leptomeningeal collateral vessels from the posterior cerebral artery in blue and transdural collateral vessels from the external carotid artery in green.

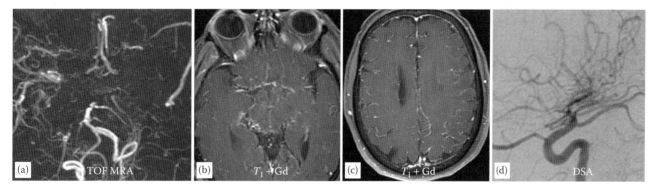

FIGURE 4.31
(a) Axial TOF-MRA MIP shows occlusion of both distal ICAs with abnormal linear signal representing an abnormal collateral network of small friable vessels. (b) Axial CE-T_1 shows abnormal linear enhancement corresponding to these basal and leptomeningeal collaterals. (c) Ivy sign, CE-T_1 axial image through the cerebral vertex showing serpinginous meningeal enhancement, representing abnormal meningeal collaterals from internal meningeal artery. (d) Right internal carotid injection DSA confirms almost complete occlusion of the distal ICA with an adjacent *puff of smoke* sign.

method will reverse the primary disease process. Current treatment aims to improve perfusion to the affected cerebral parenchyma, which will reduce collateral vessel formation, and thus bring reduced risk of both ischemic and hemorrhagic strokes [73]. Medical therapy with antiplatelete and anticoagulant medication has been used, but surgical revascularization remains the treatment of choice. Two general methods are used. In direct revascularization, a branch of the external carotid artery such as the superficial temporal artery is anastomosed into a cortical artery through a small cranial burr hole. In indirect techniques, tissue supplied by the external carotid artery is placed together with its vascular stump in direct contact with the brain, hoping for ingrowth of new blood vessels, which will supply the cortex. Both techniques have their merits and shortcomings; a combined approach is used in multiple institutions [74].

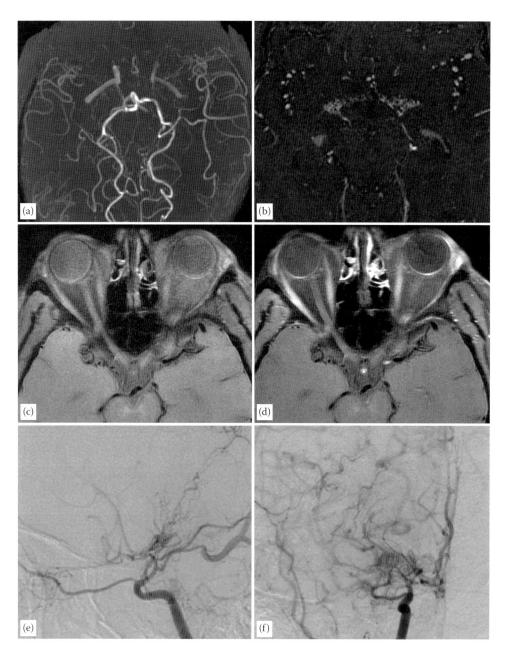

FIGURE 4.32
Companion case, a 15-year-old male presenting with tinnitus for 2–3 years. TOF MRA MIP (a) and axial image (b) show severe stenosis of both distal ICAs and a complex network of reticular collateral vessels at the skull base. High-resolution black-blood MRI (c, d) demonstrates no abnormal enhancement of these vessels to suggestive of active inflammation. DSA (e, f) confirms severe stenosis of the distal ICAs and shows the typical *puff of smoke* sign corresponding to these collateral vessels.

4.3 Conclusion

Traditional luminographic methods such as CTA and conventional MRA are optimized to demonstrate the vessel lumen and to try to copy digital subtraction angiography results. Newer methods, however, permit imaging and visualization of the vessel wall, allowing further characterization of disease processes affecting vessels. Two major breakthroughs in this area include the ability to characterize atherosclerotic plaque composition and morphology in the carotid bulb and the ability to differentiate inflammatory from noninflammatory causes of vessel stenosis in the remaining circulation by evaluation of contrast enhancement and concentric wall thickening. Table 4.3 summarizes the clinical history and the typical imaging features of the most common arteriopathies. Future developments

TABLE 4.3

Summary of the Clinical History and the Typical Imaging Features of the Most Common Arteriopathies

	Atherosclerosis	Fibromuscular Dysplasia	Dissection	Reversible Vasoconstriction Syndrome	CADASIL	CNS Vasculitis	Giant Cell Arteritis	Takayasu	Carotidynia	Moyamoya Disease
History	>50 years old, prevalence increases with age. TIA, stroke or recurrent stroke in one hemisphere	Females 20–40 years old	Rare after 65 years, peak age is 40–45. May be post traumatic or spontaneous	Severe *thunderclap* headache, focal neurologic symptoms and spontaneous resolution. Diagnosis of exclusion	40–60 years, hereditary cause of early vascular dementia	Heterogeneous group of diseases presenting with stroke, encephalopathy, and seizures	>50 years, headache, scalp tenderness, jaw claudication and systemic inflammatory manifestations	<40 years old, female, pulseless disease	Tenderness to palpation over the carotid artery bifurcation and ipsilateral neck pain	Bimodal (5 and 40 years), headache, stroke, TIA, focal neurological deficit and micro and macro-hemorrhage
Localization	Carotid bulb, intracranial, branching of vertebral arteries	60%–75% renal arteries, 25%–30% carotid arteries	Internal carotid and vertebral arteries long segment	Any intracranial vessel may be affected, may involve single segment, whole artery or multiple vessels	Diffuse white matter involvement, predilection of anterior temporal and superior frontal subcortical white matter and external capsule-insula	Any intracranial vessel may be affected, may involve single segment, whole artery or multiple vessels	Superior temporal arteries, less common: main branches of the aorta, ophthalmic, posterior ciliary, and vertebral arteries	Aorta and its main branches	Carotid bulb	Distal ICA and proximal MCA and ACA stenosis
Morphology	Eccentric short segment wall thickening	Long segment area of multifocal stenosis and dilations	Intramural hematoma in semilunar shape	Variable size area of decreased luminal diameter	Small vessels and capillaries affected	Variable size area of decreased luminal diameter with areas of dilation	Concentric and long segment wall thickening with decrease in vessel caliber	Concentric and long segment wall thickening with decrease in vessel caliber	Eccentric or circumferential wall edema	Stenosis without wall thickening

(Continued)

TABLE 4.3 (*Continued*)
Summary of the Clinical History and the Typical Imaging Features of the Most Common Arteriopathies

	Atherosclerosis	Fibromuscular Dysplasia	Dissection	Reversible Vasoconstriction Syndrome	CADASIL	CNS Vasculitis	Giant Cell Arteritis	Takayasu	Carotidynia	Moyamoya Disease
Typical imaging characteristic	Nonenhancing atheroma, check for findings to suggest a complicated plaque	String of beads sign	Semilunar area of increased signal in the vessel wall on fat sat T_1 sequences.	Sausage-string or beading of intracranial vessels	Diffuse white matter disease, more than expected for age	String of beads	Target sign	Target sign	Enhancing eccentric or circumferential thickening	Stenosis with collateral vessel formation (Puff of smoke sign on angiography, Ivy and brush signs on MRI)
Stenosis	Variable	+, with areas of dilatation	>70% initially, ++ commonly regresses with time	+++	–	+++, with areas of dilation	Variable	Variable	Uncommon	+++
Black-Blood Imaging:										
Vessel wall-contrast enhancement	–	Never	+	–	–	+++	+++	+++	+++	–
Vessel wall vessel edema	Never	Never	+	–	–	++	+++	+++	+++	–

might include (1) the establishment of new biomarkers for atherosclerotic disease, such as intraplaque hemorrhage and thin/ruptured fibrous cap for identification of patients at increased risk of stroke; (2) further evaluation of nonstenotic vascular disease that may permit earlier diagnosis and might improve our understanding of vascular disease in general; and (3) higher spatial resolution that will permit the study of smaller, more distal vascular beds.

References

1. Humphrey C, Duncan K, Fletcher S. Decade of experience with vascular rings at a single institution. *Pediatrics.* 2006;117(5):e903–e908.

2. Yousem DM, Grossman RI. *Neuroradiology: The Requisites.* 3rd ed. Philadelphia, PA: Mosby/Elsevier, 2010.

3. Bouthillier A, van Loveren HR, Keller JT. Segments of the internal carotid artery: A new classification. *Neurosurgery.* 1996;38(3):425–432; discussion 32–33.

4. Liebeskind DS. Collateral circulation. *Stroke.* 2003;34(9):2279–2284.

5. Riggs HE, Rupp C. Variation in form of circle of Willis. The relation of the variations to collateral circulation: Anatomic analysis. *Arch Neurol.* 1963;8:8–14.

6. Eftekhar B, Dadmehr M, Ansari S, Ghodsi M, Nazparvar B, Ketabchi E. Are the distributions of variations of circle of Willis different in different populations?—Results of an anatomical study and review of literature. *BMC Neurol.* 2006;6:22.

7. Chen CJ, Wang LJ, Wong YC. Abnormal origin of the vertebral artery from the common carotid artery. *AJNR Am J Neuroradiol.* 1998;19(8):1414–1416.

8. Lazzaro NA, Wright B, Castillo M et al. Artery of percheron infarction: Imaging patterns and clinical spectrum. *AJNR Am J Neuroradiol.* 2010;31(7):1283–1289.

9. Finlay A, Johnson M, Forbes TL. Surgically relevant aortic arch mapping using computed tomography. *Ann Vasc Surg.* 2012;26(4):483–490.

10. Lim RP, Shapiro M, Wang EY et al. 3D time-resolved MR angiography (MRA) of the carotid arteries with time-resolved imaging with stochastic trajectories: Comparison with 3D contrast-enhanced Bolus-Chase MRA and 3D time-of-flight MRA. *AJNR Am J Neuroradiol.* 2008;29(10):1847–1854.

11. Meckel S, Maier M, Ruiz DS et al. MR angiography of dural arteriovenous fistulas: Diagnosis and follow-up after treatment using a time-resolved 3D contrast-enhanced technique. *AJNR Am J Neuroradiol.* 2007;28(5):877–884.

12. Kramer U, Ernemann U, Fenchel M et al. Pretreatment evaluation of peripheral vascular malformations using low-dose contrast-enhanced time-resolved 3D MR angiography: Initial results in 22 patients. *AJR Am J Roentgenol.* 2011;196(3):702–711.

13. Alperin N, Ranganathan S, Bagci AM et al. MRI evidence of impaired CSF homeostasis in obesity-associated idiopathic intracranial hypertension. *AJNR Am J Neuroradiol.* 2013;34(1):29–34.

14. Birnbaum J, Hellmann DB. Primary angiitis of the central nervous system. *Arch Neurol.* 2009;66(6):704–709.

15. Varnava AM, Mills PG, Davies MJ. Relationship between coronary artery remodeling and plaque vulnerability. *Circulation.* 2002;105(8):939–943.

16. Kolominsky-Rabas PL, Weber M, Gefeller O, Neundoerfer B, Heuschmann PU. Epidemiology of ischemic stroke subtypes according to TOAST criteria: Incidence, recurrence, and long-term survival in ischemic stroke subtypes: A population-based study. *Stroke.* 2001;32(12):2735–2740.

17. Kizer JR. Evaluation of the patient with unexplained stroke. *Coron Artery Dis.* 2008;19(7):535–540.

18. Sacco RL, Kargman DE, Gu Q, Zamanillo MC. Race-ethnicity and determinants of intracranial atherosclerotic cerebral infarction. The Northern Manhattan Stroke Study. *Stroke.* 1995;26(1):14–20.

19. Freilinger TM, Schindler A, Schmidt C et al. Prevalence of nonstenosing, complicated atherosclerotic plaques in cryptogenic stroke. *JACC Cardiovasc Imaging.* 2012;5(4):397–405.

20. Clinical alert: Benefit of carotid endarterectomy for patients with high-grade stenosis of the internal carotid artery. National Institute of Neurological Disorders and Stroke Stroke and Trauma Division. North American Symptomatic Carotid Endarterectomy Trial (NASCET) investigators. *Stroke.* 1991;22(6):816–817.

21. Cai JM, Hatsukami TS, Ferguson MS, Small R, Polissar NL, Yuan C. Classification of human carotid atherosclerotic lesions with in vivo multicontrast magnetic resonance imaging. *Circulation.* 2002;106(11):1368–1373.

22. Saam T, Hetterich H, Hoffmann V et al. Meta-analysis and systematic review of the predictive value of carotid plaque hemorrhage on cerebrovascular events by magnetic resonance imaging. *J Am Coll Cardiol.* 2013; 62(12):1081–1091.

23. Gupta A, Baradaran H, Schweitzer AD et al. Carotid plaque MRI and stroke risk: A systematic review and meta-analysis. *Stroke.* 2013;44(11):3071–3077.

24. Underhill HR, Yuan C, Terry JG et al. Differences in carotid arterial morphology and composition between individuals with and without obstructive coronary artery disease: A cardiovascular magnetic resonance study. *J Cardiovasc Magn Reson.* 2008;10:31.

25. Trelles M, Eberhardt KM, Buchholz M et al. CTA for screening of complicated atherosclerotic carotid plaque— American Heart Association Type VI lesions as defined by MRI. *AJNR Am J Neuroradiol.* 2013;34(12):2331–7.

26. Wong KS, Gao S, Chan YL et al. Mechanisms of acute cerebral infarctions in patients with middle cerebral artery stenosis: A diffusion-weighted imaging and microemboli monitoring study. *Ann Neurol.* 2002;52(1):74–81.

27. Swartz RH, Bhuta SS, Farb RI et al. Intracranial arterial wall imaging using high-resolution 3-tesla contrast-enhanced MRI. *Neurology.* 2009;72(7):627–634.

28. Turan TN, Bonilha L, Morgan PS, Adams RJ, Chimowitz MI. Intraplaque hemorrhage in symptomatic intracranial atherosclerotic disease. *J Neuroimaging.* 2011;21(2):e159–e161.

29. Portanova A, Hakakian N, Mikulis DJ, Virmani R, Abdalla WM, Wasserman BA. Intracranial vasa vasorum: Insights and implications for imaging. *Radiology.* 2013;267(3):667–679.

30. Chimowitz MI, Lynn MJ, Derdeyn CP et al. Stenting versus aggressive medical therapy for intracranial arterial stenosis. *N Engl J Med.* 2011;365(11):993–1003.

31. Vagal AS, Leach JL, Fernandez-Ulloa M, Zuccarello M. The acetazolamide challenge: Techniques and applications in the evaluation of chronic cerebral ischemia. *AJNR Am J Neuroradiol.* 2009;30(5):876–884.

32. Gambhir S, Inao S, Tadokoro M et al. Comparison of vasodilatory effect of carbon dioxide inhalation and intravenous acetazolamide on brain vasculature using positron emission tomography. *Neurol Res.* 1997;19(2):139–144.

33. Harrison EG, Jr., McCormack LJ. Pathologic classification of renal arterial disease in renovascular hypertension. *Mayo Clin Proc.* 1971;46(3):161–167.

34. Slovut DP, Olin JW. Fibromuscular dysplasia. *N Engl J Med.* 2004;350(18):1862–1871.

35. Begelman SM, Olin JW. Fibromuscular dysplasia. *Curr Opin Rheumatol.* 2000;12(1):41–47.

36. Lucas C, Moulin T, Deplanque D, Tatu L, Chavot D. Stroke patterns of internal carotid artery dissection in 40 patients. *Stroke.* 1998;29(12):2646–2648.

37. Pfefferkorn T, Saam T, Rominger A et al. Vessel wall inflammation in spontaneous cervical artery dissection: A prospective, observational positron emission tomography, computed tomography, and magnetic resonance imaging study. *Stroke.* 2011;42(6):1563–1568.

38. Ducrocq X, Lacour JC, Debouverie M, Bracard S, Girard F, Weber M. Cerebral ischemic accidents in young subjects. A prospective study of 296 patients aged 16 to 45 years. *Rev Neurol (Paris).* 1999;155(8):575–582.

39. Norris JW. Extracranial arterial dissection: Anticoagulation is the treatment of choice: For. *Stroke.* 2005;36(9):2041–2042.

40. Lyrer PA. Extracranial arterial dissection: Anticoagulation is the treatment of choice: Against. *Stroke.* 2005;36(9):2042–2043.

41. Meschia JF, Malkoff MD, Biller J. Reversible segmental cerebral arterial vasospasm and cerebral infarction: Possible association with excessive use of sumatriptan and Midrin. *Arch Neurol.* 1998;55(5):712–714.

42. Singhal AB, Caviness VS, Begleiter AF, Mark EJ, Rordorf G, Koroshetz WJ. Cerebral vasoconstriction and stroke after use of serotonergic drugs. *Neurology.* 2002;58(1):130–133.

43. Ducros A, Fiedler U, Porcher R, Boukobza M, Stapf C, Bousser MG. Hemorrhagic manifestations of reversible cerebral vasoconstriction syndrome: Frequency, features, and risk factors. *Stroke.* 2010;41(11):2505–2511.

44. Mandell DM, Matouk CC, Farb RI et al. Vessel wall MRI to differentiate between reversible cerebral vasoconstriction syndrome and central nervous system vasculitis: Preliminary results. *Stroke.* 2012;43(3):860–862.

45. Sattar A, Manousakis G, Jensen MB. Systematic review of reversible cerebral vasoconstriction syndrome. *Expert Rev Cardiovasc Ther.* 2010;8(10):1417–1421.

46. Farid H, Tatum JK, Wong C, Halbach VV, Hetts SW. Reversible cerebral vasoconstriction syndrome: Treatment with combined intra-arterial verapamil infusion and intracranial angioplasty. *AJNR Am J Neuroradiol.* 2011;32(10):E184–E187.

47. Auer DP, Putz B, Gossl C, Elbel G, Gasser T, Dichgans M. Differential lesion patterns in CADASIL and sporadic subcortical arteriosclerotic encephalopathy: MR imaging study with statistical parametric group comparison. *Radiology.* 2001;218(2):443–451.

48. Dichgans M, Holtmannspotter M, Herzog J, Peters N, Bergmann M, Yousry TA. Cerebral microbleeds in CADASIL: A gradient-echo magnetic resonance imaging and autopsy study. *Stroke.* 2002;33(1):67–71.

49. Liem MK, Lesnik Oberstein SA, Versluis MJ et al. 7 T MRI reveals diffuse iron deposition in putamen and caudate nucleus in CADASIL. *J Neurol Neurosurg Psychiatry.* 2012;83(12):1180–1185.

50. Fieschi C, Rasura M, Anzini A, Beccia M. Central nervous system vasculitis. *J Neurol Sci.* 1998;153(2):159–171.

51. Marsh EB, Zeiler SR, Levy M, Llinas RH, Urrutia VC. Diagnosing CNS vasculitis: The case against empiric treatment. *Neurologist.* 2012;18(4):233–238.

52. Saam T, Habs M, Pollatos O et al. High-resolution black-blood contrast-enhanced T1 weighted images for the diagnosis and follow-up of intracranial arteritis. *Br J Radiol.* 2010;83(993):e182–e184.

53. Kale N, Eggenberger E. Diagnosis and management of giant cell arteritis: A review. *Curr Opin Ophthalmol.* 2010;21(6):417–422.

54. Borchers AT, Gershwin ME. Giant cell arteritis: A review of classification, pathophysiology, geoepidemiology and treatment. *Autoimmun Rev.* 2012;11(6–7):A544–A554.

55. Saam T, Habs M, Cyran CC et al. New aspects of MRI for diagnostics of large vessel vasculitis and primary angiitis of the central nervous system. *Radiologe.* 2010;50(10):861–871.

56. Bley TA, Uhl M, Carew J et al. Diagnostic value of high-resolution MR imaging in giant cell arteritis. *AJNR Am J Neuroradiol.* 2007;28(9):1722–1727.

57. Miller DV, Maleszewski JJ. The pathology of large-vessel vasculitides. *Clin Exp Rheumatol.* 2011;29(1 Suppl. 64):S92–S98.

58. Khalife T, Alsac JM, Lambert M et al. Diagnosis and surgical treatment of a Takayasu disease on an abdominal aortic dissection. *Ann Vasc Surg.* 2011;25(4):556.e1–556.e5.

59. Sparks SR, Chock A, Seslar S, Bergan JJ, Owens EL. Surgical treatment of Takayasu's arteritis: Case report and literature review. *Ann Vasc Surg.* 2000;14(2):125–129.

60. Roseman DM. Carotidynia. A distinct syndrome. *Arch Otolaryngol.* 1967;85(1):81–84.

61. Schaumberg J, Eckert B, Michels P. Carotidynia: Magnetic resonance imaging and ultrasonographic imaging of a self-limiting disease. *Clin Neuroradiol.* 2011;21(2):91–94.

62. Comacchio F, Bottin R, Brescia G et al. Carotidynia: New aspects of a controversial entity. *Acta Otorhinolaryngol Ital.* 2012;32(4):266–269.

63. da Rocha AJ, Tokura EH, Romualdo AP, Fatio M, Gama HP. Imaging contribution for the diagnosis of carotidynia. *J Headache Pain*. 2009;10(2):125–127.

64. Burton BS, Syms MJ, Petermann GW, Burgess LP. MR imaging of patients with carotidynia. *AJNR Am J Neuroradiol*. 2000;21(4):766–769.

65. Suzuki J, Takaku A. Cerebrovascular "moyamoya" disease. Disease showing abnormal net-like vessels in base of brain. *Arch Neurol*. 1969;20(3):288–299.

66. Scott RM, Smith ER. Moyamoya disease and moyamoya syndrome. *N Engl J Med*. 2009;360(12):1226–1237.

67. Sun W, Yuan C, Liu W et al. Asymptomatic cerebral microbleeds in adult patients with moyamoya disease: A prospective cohort study with 2 years of follow-up. *Cerebrovasc Dis*. 2013;35(5):469–475.

68. Ortiz-Neira CL. The puff of smoke sign. *Radiology*. 2008;247(3):910–911.

69. Yoon HK, Shin HJ, Chang YW. "Ivy sign" in childhood moyamoya disease: Depiction on FLAIR and contrast-enhanced T1-weighted MR images. *Radiology*. 2002;223(2):384–389.

70. Horie N, Morikawa M, Nozaki A, Hayashi K, Suyama K, Nagata I. "Brush Sign" on susceptibility-weighted MR imaging indicates the severity of moyamoya disease. *AJNR Am J Neuroradiol*. 2011;32(9):1697–1702.

71. Nagamine Y, Takahashi S, Sonobe M. Multiple intracranial aneurysms associated with moyamoya disease. Case report. *J Neurosurg*. 1981;54(5):673–676.

72. Nakashima T, Nakayama N, Furuichi M, Kokuzawa J, Murakawa T, Sakai N. Arteriovenous malformation in association with moyamoya disease. Report of two cases. *Neurosurg Focus*. 1998;5(5):e6.

73. Ikezaki K. Rational approach to treatment of moyamoya disease in childhood. *J Child Neurol*. 2000;15(5):350–356.

74. Fung LW, Thompson D, Ganesan V. Revascularisation surgery for paediatric moyamoya: A review of the literature. *Childs Nerv Syst*. 2005;21(5):358–364.

5

Magnetic Resonance Imaging: Aorta and Splanchnic Vessels

Christopher J. François

CONTENTS

5.1 Introduction

Magnetic resonance angiography (MRA) is increasingly used clinically in the workup of patients with known or suspected cardiovascular diseases as a result of technological advances in both hardware and imaging techniques. This is particularly true as the public becomes more aware of the potential risks associated with ionizing radiation and nephrotoxic contrast agents required for computed tomography angiography (CTA). This chapter will briefly review recent developments in contrast-enhanced (CE)-MRA and non-contrast-enhanced (NCE) MRA techniques. This will be followed by more detailed presentation of the use of CE-MRA and NCE-MRA in the evaluation of patients with aortic and splanchnic vascular disease.

5.2 MRA Techniques

5.2.1 CE-MRA

CE-MRA techniques are used more frequently because acquisition times are shorter with much greater field-of-view than NCE-MRA techniques (Figure 5.1).

In addition, inflow and pulsatility artifacts are largely overcome with CE-MRA techniques. Gadolinium-based contrast agents (GBCA) are used to increase the signal of the vasculature relative to surrounding soft tissues. Broadly speaking, CE-MRA can be performed with *static* three-dimensional (3D) CE-MRA sequences or time-resolved CE-MRA sequences. With static CE-MRA, the image acquisition is started when there is maximum enhancement of the vessels of interest—in real time with bolus tracking methods or using a prior knowledge of the contrast arrival time from a test bolus injection. With time-resolved CE-MRA, no specific timing of image acquisition is necessary. As with digital subtraction angiography, multiple 3D CE-MRA datasets are acquired throughout the passage of contrast through the vasculature of interest, ensuring the acquisition of at least one dataset during maximum enhancement of the vessels of interest.

Due to the compromises that need to be made between acquisition time, spatial resolution and temporal resolution, static 3D CE-MRA techniques have greater volumetric coverage and higher spatial resolution than time-resolved CE-MRA sequences. This is particularly true with parallel imaging techniques. Time-resolved CE-MRA, on the other hand, provides dynamic information on blood flow that is less apparent on static 3D CE-MRA.

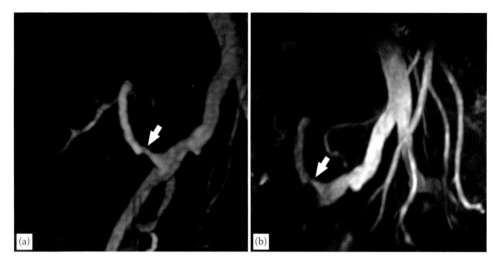

FIGURE 5.1
(a) CE-MRA and (b) NCE-MRA of a right iliac fossa renal transplant. Both techniques clearly show severe stenosis of the renal transplant artery (closed arrow). However, the acquisition time of the CE-MRA was less than 20 s, while the acquisition time of the NCE-MRA was greater than 3 min.

5.2.1.1 Static 3D CE-MRA

High spatial resolution CE-MRA is based on a 3D spoiled gradient-recalled echo sequence. The image acquisition is timed such that the central k-space lines are obtained when the GBCA maximally enhance the vessels of interest. Using currently available sequences, spatial resolution for thoracic and abdominal CE-MRA is typically ≤1.5mm isotropic. This enables CE-MRA to be used to delineate small and complex anatomical structures in patients with a variety of cardiovascular diseases (Figure 5.2). Furthermore, accelerated image acquisition methods allow for greater volumetric coverage in the same time as non-accelerated methods. As a result, it is now possible to obtain high-resolution 3D CE-MRA images of the entire thorax or thorax and abdomen (Figure 5.3) within a single breath hold.

A single dose of extracellular GBCA (0.1 mmol/kg) is usually sufficient for most CE-MRA applications. GBCA are injected intravenously (IV) through an 18–22 gauge catheter (preferably in an antecubital vein). The size of the catheter, quality of the patient's veins, and length of the acquisition will affect the rate of GBCA administration (i.e., 0.5–4.0 mL/s). The timing of image acquisition is such that the lower order k-space frequencies are acquired during greater enhancement of the vessels of interest. This is due to the fact that the center of k-space primarily contributes to the signal and enhancement of the image, while the higher order k-space frequencies contribute to the sharpness of the images (Figure 5.4). A test-bolus injection with a small amount of GBCA can be used to determine the time-to-peak enhancement of the

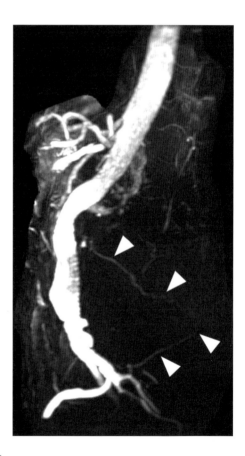

FIGURE 5.2
High-resolution 3D CE-MRA of a type II endoleak in a patient with endovascular abdominal aortic aneurysm repair. The high spatial resolution of the acquisition allows for the clear delineation of the feeding vessel (arrowheads).

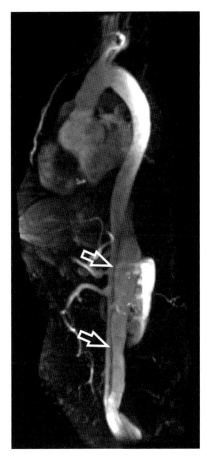

FIGURE 5.3

3D CE-MRA of the entire aorta in a patient with aortic dissection (open arrows). Using parallel imaging, the images were acquired in 18 s following the IV administration of 0.1 mmol/kg of gadobenate dimeglumine (Multihance®, Bracco).

vessels of interest or, with centric *k*-space ordering, real-time bolus-tracking can be used to start CE-MRA acquisition when the contrast bolus arrives in the vessels of interest. Artifacts occur when CE-MRA acquisition is not timed appropriately to the passage of the GBCA bolus. These include lower signal-to-noise ratio (SNR), edge blurring, and truncation artifact [1].

Most GBCA used for CE-MRA distribute throughout the extracellular space. Although not usually a problem, this can be a problem for some vascular applications when trying to achieve the maximum contrast between the vessels and surrounding soft tissues. With intravascular GBCA, scan times can be extended to increase spatial resolution and improve contrast resolution. Gadofosveset trisodium is a protein-binding intravascular contrast agent that has been approved for the evaluation of occlusive aortoiliac disease [2]. The protein binding extends the intravascular half-life of gadofosveset trisodium. In addition, gadofosveset trisodium has a higher relaxivity than other GBCA [3]. Therefore, CE-MRA images of similar quality can be obtained with smaller volumes of gadofosveset trisodium compared to other GBCA [4,5]. Very high spatial resolution CE-MRA images are feasible with intravascular GBCA by acquiring data during the steady state. Because there is less concern with acquiring images quickly during the first pass of contrast through the circulation to avoid contamination from enhancement of the surrounding tissues, it is possible to perform the steady-state imaging during free breathing [6].

For thoracic 3D CE-MRA applications, compensation for cardiac motion is needed to optimize image quality

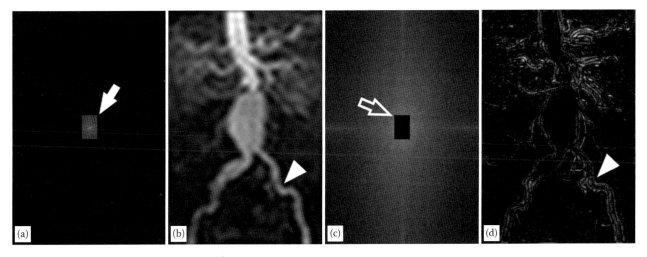

FIGURE 5.4

CE-MRA of the abdominal aorta. When only the inner 10% of *k*-space data (a) are used to reconstruct the image (b), the vessels are clearly depicted but the vessel edges are blurry (arrowhead). When the inner 10% of *k*-space data are removed (c), the resulting image (d) has very little signal within the vessels, but the edges are well delineated (arrowhead).

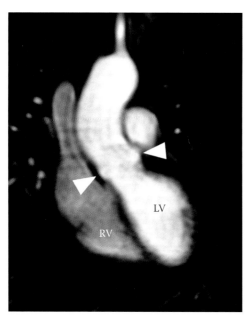

FIGURE 5.5
ECG-triggered, 3D CE-MRA is used in patients with aortic root or ascending aortic aneurysms to minimize blurring from cardiac motion. In this case, the aortic root (arrowheads) is sharp and well delineated.

of the ascending aorta. This is done using electrocardiographic (ECG) triggering and adjusting the acquisition to occur only during diastole (Figure 5.5). Because data are acquired during diastole, ECG-triggered CE-MRA techniques are longer than non-ECG-triggered CE-MRA techniques. Therefore, the field of view is typically restricted to just the aorta to ensure acquisition in a single breath hold.

Scanning at higher field strengths (i.e., 3 T and higher) is beneficial to CE-MRA because the SNR is proportional to the field strength. At 3 T, the SNR is approximately double the SNR at 1.5 T. The increase in SNR at higher field strengths is used to decrease scan time and/or increase spatial resolution while obtaining images of relatively similar or better quality compared to those obtained at lower field strength [7]. Another benefit of performing CE-MRA at higher field strengths is that the T_1-shortening effects of GBCA are different than at lower field strengths. This results in greater contrast between the enhancing vessels and the surrounding soft tissues. As a result, CE-MRA at 3 T can be performed with smaller doses of GBCA [8–11].

However, CE-MRA at 3 T is not without its challenges and limitations due to the stronger magnetic field. In particular, CE-MRA at 3 T is more susceptible to radiofrequency field inhomogeneity. The radiofrequency field inhomogeneities, which result in areas of decreased or increased signal, can be mitigated by using multi-coil transmit coils to suppress eddy currents [12] or using newer 3D radiofrequency pulses [13].

Another consequence of performing CE-MRA at 3 T is the increased specific absorption rate (SAR) due to constructive and destructive interference due to radiofrequency field inhomogeneity. The SAR is an approximation of the energy transferred to tissues by the radiofrequency pulse and is proportional to the square of the resonance frequency. Compared to 1.5 T, the SAR at 3 T would increase fourfold for the same sequence at 1.5, assuming all parameters are equal [14]. Therefore, modifications are made to the pulse sequences, acquisition techniques, and hardware designs to ensure that scans remain within constraints of the allowable SAR.

5.2.1.2 Time-Resolved CE-MRA

As indicated previously, an advantage of time-resolved CE-MRA relative to static 3D CE-MRA is that precise timing of the acquisition relative to the contrast bolus is not necessary due to the continuous acquisition of data as contrast passes through the cardiovascular system. Multiple methods of accelerating image acquisition to achieve high temporal resolution CE-MRA have been developed. This can be done with 3D, asymmetric k-space undersampling combined with thicker slice thickness in the slice encoding direction [15]. However, multiplanar reformations are limited because of the relatively thick slice thickness. Other approaches are necessary to acquire time-resolved data with relatively high spatial resolution. These are based on various methods of undersampling k-space. In short, the higher spatial frequencies are acquired less frequently [16–19] while the lower k-space frequencies are updated more frequently to observe the passage of contrast through the imaged vasculature (Figure 5.6). By updating the center of k-space frequently and *sharing* the higher frequencies of k-space between multiple images, it is possible to generate images with high temporal and spatial resolution. By subtracting an initial mask image obtained prior to the administration of GBCA from subsequent images acquired after the administration of GBCA, it is possible to generate images with purely vascular signal, analogous to digital subtraction angiography.

5.2.2 NCE-MRA

NCE-MRA for aorta and splanchnic vessel imaging has been limited by long acquisition times and cardiac and respiratory motion artifacts. However, software and hardware improvement have helped renew the interest in NCE-MRA, particularly in patients in whom GBCA would be contraindicated. Data demonstrating a link between the development of nephrogenic systemic fibrosis and exposure to GBCA in patients with reduced renal function [20–22] have been a primary motivation for expanding the use of NCE-MRA for a variety of

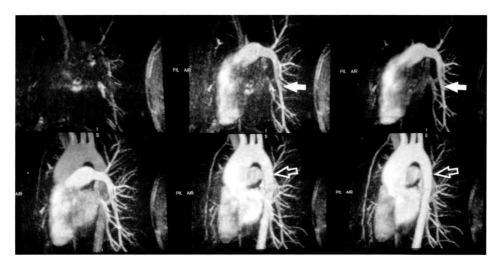

FIGURE 5.6
Time-resolved CE-MRA demonstrated the dynamic passage of contrast through the thoracic circulation. As expected, there is earlier enhancement of the pulmonary arteries (solid arrows) than the aorta (open arrows).

applications. Other contraindications to GBCA include allergy to GBCA, pregnancy, and poor IV access. Time-of-flight sequences are still routinely used clinically for cervical and intracranial studies, with diagnostic capabilities comparable to CTA and digital subtraction angiography [23,24]. However, the use of time-of-flight sequences is markedly limited in the thorax. In the subsequent paragraphs, the two main NCE-MRA sequences used for aortic and splanchnic vasculature imaging will be reviewed.

5.2.2.1 3D Balanced Steady-State Free Precession

3D balanced steady-state free-precession (bSSFP) sequences are primarily used for thoracic and abdominal NCE-MRA. This is because 3D bSSFP has inherently high blood pool signal intensity and its signal is relatively flow independent [25]. The image contrast in 3D bSSFP sequences is determined by both the T_2 and T_1 characteristics of the tissues, resulting in high blood pool signal. 3D bSSFP NCE-MRA techniques result in images with bright arterial and venous signal (Figure 5.7). For aortic imaging [26–29], this is not a problem because of its large size. In fact, 3D bSSFP NCE-MRA can be used to reliably measure aortic size in patients suspected or known to have aortic aneurysms and may even better depict the aortic root than CE-MRA techniques [26]. 3D bSSFP NCE-MRA is also extensively used to assess the renal and mesenteric vasculature (Figure 5.1b).

3D bSSFP sequences are acquired during free breathing, using respiratory triggering with bellows or with navigation, to increase spatial resolution and coverage. 3D bSSFP NCE-MRA is also enhanced by using ECG-triggering to acquire data during diastole, which minimizes cardiac motion artifacts. As with bSSFP

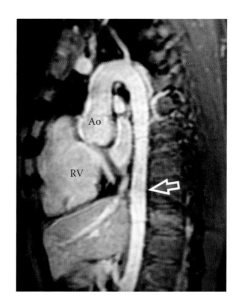

FIGURE 5.7
Sagittal oblique reformatted image from a free-breathing, ECG-triggered, 3D bSSFP NCE-MRA in a 15-year old with repaired tetralogy of Fallot clearly demonstrates the cardiovascular anatomy, including the right ventricle (RV), ascending aorta (Ao), and descending aorta (open arrow).

sequences in other applications, 3D bSSFP NCE-MRA is susceptible to degraded image quality due to field inhomogeneities, such as at lung–soft tissue interfaces and in the vicinity of metallic devices and implants.

5.2.2.2 4D-Flow MRI

The signal intensity in flow-sensitive, phase-contrast (PC) MRI is proportional to the velocities of flowing protons in the imaging field of view and to velocity encoding chosen during the scan prescription.

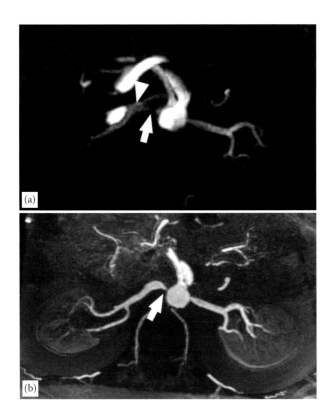

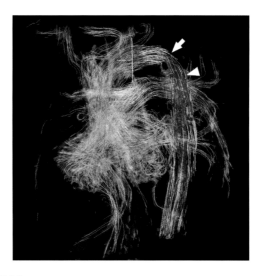

FIGURE 5.9
4D-flow MRI of the thoracic circulation in a patient with a history of aortic coarctation (solid arrow). Streamlines are color coded to the average velocity throughout the cardiac cycle with oranges and yellows indicating lower velocities and blues indicating higher velocities. The higher velocities are present distal to the coarctation (arrowhead).

FIGURE 5.8
(a) 3D PC NCE-MRA in a patient with severe right renal artery stenosis (solid arrow). Distal to the stenosis, the signal in the right renal artery is decreased (arrowhead) due to slow flow. Severe right renal artery stenosis is confirmed on the 3D CE-MRA (b).

Time-resolved, ECG-triggered, two-dimensional (2D) flow-sensitive MRI is routinely used in cardiac MRI studies to quantify flow through the great vessels and cardiac valves [30,31]. Flow-sensitive PC MRI can also be used for NCE-MRA (Figure 5.8). This is done by using a 3D PC sequence that is sensitive to flow in all three directions [32]. Historically, the temporal information of flow was ignored when using this sequence to perform NCE-MRA. As a result, the images obtained represented an average flow within the vasculature during the time of acquisition. Adding the temporal and directional flow data to the anatomical information results in what has become commonly referred to as four-dimensional (4D) flow MRI (three directional velocity encoding + time, in additional to the three spatial dimensions) [33]. Substantial data acquisition acceleration using parallel imaging [34] or 3D radial undersampling [35] is necessary to shorten scan time for 4D-flow MRI to a point where these become clinically feasible. In addition to their use for NCE-MRA, 4D-flow MRI can also be used to quantify a variety of hemodynamic parameters that characterize cardiovascular disease.

4D-flow MRI simplifies flow quantification because the analysis is performed *post priori* through any number of vessels of interest included in the imaging volume [36]. In addition to quantifying blood flow velocities and volumes, 4D-flow MRI data can be used to derive more complex hemodynamic parameters that can indicate the presence or severity of disease. For example, initial studies have demonstrated the ability to use 4D-flow MRI to assess changes in flow patterns in patients with cardiovascular disease (Figure 5.9) and how these alterations in flow patterns are related to changes in wall shear stress [33,34], pulse wave velocity [33], and pressure gradients [37–41].

5.3 Aorta

5.3.1 Acute Aorta Syndrome

Acute aorta syndrome refers to a spectrum of acute pathologies (penetrating atherosclerotic ulcer, intramural hematoma, and aortic dissection) affecting the aorta that frequently have a similar presentation. Patients often experience acute onset of severe chest and/or back pain. Additional symptoms may include dyspnea, diaphoresis, or abdominal pain when the abdominal vasculature is involved. Predisposing conditions that raise the risk of acute aortic syndrome include hypertension, Marfan syndrome, bicuspid aortic valve, or other conditioning resulting in weakening of the aortic wall.

In all three causes of acute aorta syndrome, there is disruption to one or more layers of the wall of the aorta. In intramural hematoma, there is hemorrhage within the media of the aortic wall without disruption of the intima or adventitia [42]. In aortic dissection, a tear is present in the intima and the injury to the aorta wall extends along a variable length of the aortic wall, within the media. The adventitia is intact with aortic dissection, unless there is rupture. Penetrating atherosclerotic ulcers are localized lesion affecting the intima and media of the aortic wall, adjacent to an atherosclerotic plaque [43].

Patients suspected of having an acute aortic dissection require accurate and prompt diagnosis. As a result, CTA is the primary diagnostic test of choice because its accuracy is very close to 100%. In patients with severe allergy to iodinated contrast, in whom there is not time for pre-medication to be given, MRA is also reliable to determining whether or not a patient has aortic dissection [44,45]. In general, MRA has a greater role in following patients with known aortic dissection, particularly younger patients, or renal insufficiency.

The CE-MRA and NCE-MRA methods described previously can all be used. Given the more rapid acquisition times for CE-MRA techniques, these would be preferred in the acute setting, while in follow-up both CE-MRA and NCE-MRA methods can be beneficial. Previously published reports on the use of MRA for the diagnosis of aortic dissection have successfully used a wide variety of GBCA, differing doses, and contrast injection rates. The sensitivities and specificities of MRA for the diagnosis of aortic dissection are between 90% and 100%.

The criteria for establishing aortic dissection with MRA are very similar to those used with CTA. Aortic dissections are characterized by a curvilinear intimal flap that extends along the aorta, separating the true and false lumina (Figure 5.10). Intimal tears may also be present, helping to confirm the diagnosis of aortic dissection. The presence of the intimal flap and length of involvement help distinguish aortic dissection from other causes of acute aortic syndrome, intramural hematoma, and penetrating atherosclerotic ulcer. The role of MRA in management of patients with acute aortic syndrome is to assess the location and extent of dissection, the location of intimal tears (Figure 5.11), relative filling of the lumina, and size of the aorta. Time-resolved CE-MRA can be helpful in distinguishing slow filling of the false lumen from a thrombosed false lumen (Figure 5.12).

5.3.2 Aortic Aneurysm

Aortic aneurysms are usually asymptomatic and detected incidentally on diagnostic imaging studies, including X-rays, echocardiography, ultrasonography, and computed tomography. In patients with mildly

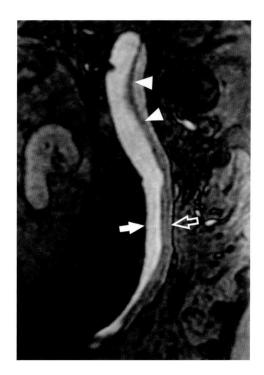

FIGURE 5.10
CE-MRA in a patient with aortic dissection confirmed by a thin, curvilinear intimal flap (arrowheads) separating the true lumen (solid arrow) from the false lumen (open arrow).

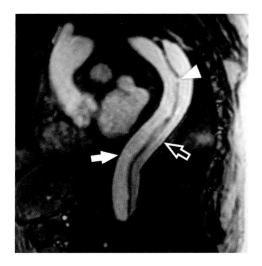

FIGURE 5.11
CE-MRA in a patient with aortic dissection involving the descending thoracic aorta. The true lumen is anterior (closed arrow) and the false lumen is posterior (open arrow). A small intimal tear (arrowhead) is present in the mid-descending aorta.

enlarged thoracic or abdominal aortas, echocardiography and ultrasound are both sufficient for routine follow-up, respectively. In addition, current evidence supports the use of ultrasonography for the screening for abdominal aortic aneurysms (AAA) in males older than 65 years of age with a history of smoking. However,

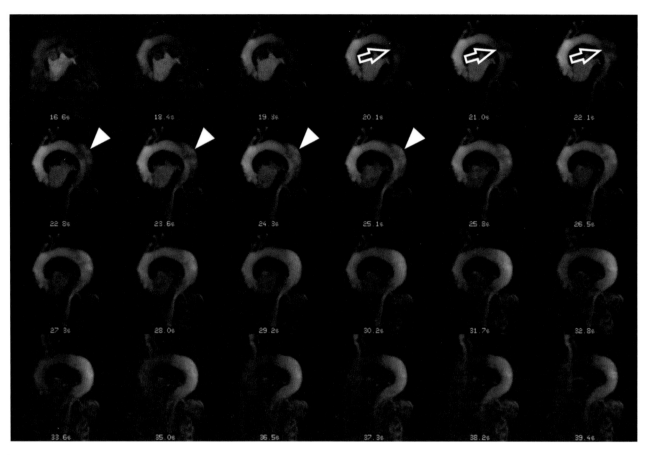

FIGURE 5.12
Time-resolved CE-MRA in patient with descending thoracic aortic dissection. There is earlier enhancement of the true lumen (open arrows) than the false lumen (arrowheads).

echocardiography and ultrasonography tend to underestimate the size of the aorta, compared to CTA and MRA. Patients that are obese or other causes resulting in poor acoustic windows are incompletely evaluated with echocardiography and ultrasonography. In these patients, CTA and MRA are preferred for more accurate characterization of the aortic aneurysm size and extent [46]. This is especially true in patients with tortuous aortas or complex aneurysms in whom endovascular repair is being considered.

The risk of aortic aneurysms is continued growth of the aorta leading to dissection or rupture. The risk of dissection and rupture is directly related to the size of the aorta. Thoracic aortic aneurysms are repaired when they are larger than 5–6 cm in maximum diameter. In patients with an underlying connective tissue disorder, repair may be considered earlier. AAA are usually repaired when larger than 5.5 cm in maximum diameter. The repair of ascending thoracic aortic aneurysms is almost always with an open surgical approach. Descending thoracic, thoraco-abdominal, and AAA are increasingly being treated using endovascular approaches for placement of stent grafts. The decision

on the approach to repair is based on multiple factors, particularly location and extent of the aneurysm.

CTA and MRA have similar results for determining aortic size and extent of aneurysm [47]. CE- and NCE-MRA are viable alternatives for the follow-up of patients with known aortic aneurysm. This is particularly true with the newer 3D CE-MRA methods that enable imaging of the entire chest and abdomen within a short breath hold. Compared to CTA, MRA avoids the use of nephrotoxic contrast agents and ionizing radiation.

Localized or diffuse dilation of the aorta, greater than 50% larger than the normal size of the aorta, is considered aneurysmal. Aneurysms that extend peripherally from the wall of the aorta are called saccular aneurysms (Figure 5.13), while aneurysms that result in circumferential dilation of the vessel are called fusiform (Figure 5.14). Aneurysms can also be characterized based on the integrity of the vessel wall. In true aneurysms, all three layers of the vessel wall are intact, while in false aneurysms (or pseudoaneurysms), there is disruption of the intima and part of the media. Due to the disruption of the vessel wall, pseudoaneurysms are less stable than true aneurysms and require more urgent repair.

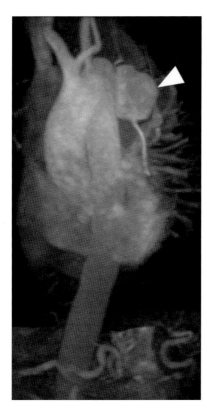

FIGURE 5.13
CE-MRA in a patient with a saccular aneurysm arising from the aortic arch (arrowhead).

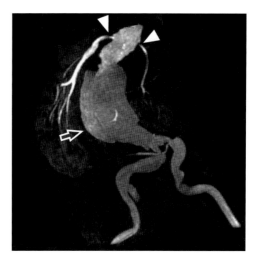

FIGURE 5.14
CE-MRA in a patient with a large fusiform abdominal aortic aneurysm (open arrow). CE-MRA was performed because of the patient's renal insufficiency secondary to bilateral renal artery stenosis (arrowheads).

Due to an association between thoracic aortic aneurysms and AAA, it is important to evaluate the thoracic aorta as well in patients with known AAA. Women and elderly patients have a higher association between thoracic and AAA than men and younger patients.

Large aortic aneurysms can develop intraluminal thrombus due to slow, helical flow. Therefore, it is critical that aortic measurements be made from source data from outer wall to outer wall of the aorta. Reformatted or maximum intensity projection images can underestimate the true size of the aorta when there is thrombus within the aneurysm sac.

5.3.3 Aortitis

Systemic vasculitides are currently categorized based on the size of the affected vessels into small vessel vasculitis, medium vessel vasculitis, and large vessel vasculitis [48]. MRA primarily has an important role in the diagnosis and follow-up of patients with large vessel vasculitides and, to a lesser extent, medium vessel vasculitides. The large vessel vasculitides include giant cell arteritis and Takayasu's arteritis. The medium vessel vasculitides include polyarteritis nodosa and Kawasaki disease. MRA has a limited role in the evaluation of the small vessel vasculitis—granulomatous

polyangiitis, Churg–Strauss vasculitis, microscopic polyangiitis, and small vessel vasculitides associated with immune complexes. Non-infectious aortitis may also be a component of other systemic inflammatory diseases such as rheumatoid arthritis, sarcoidosis, polychondritis, spondyloarthritis, and Behçet's disease.

Due to the more limited role of MRA in the evaluation of patients with small and medium vessel vasculitides, this section will focus on the large vessel vasculitides. High resolution MRA depicts the intramural inflammatory changes within the large vessels affected [49,50]. In medium and small vessel vasculitides, the vessels are too small to be able to detect changes in vessel thickness. Therefore, MRA and MRI are used to detect the secondary changes in end organs supplied by the affected vessels.

MRI is frequently preferred to other imaging modalities for the detection of vessel involvement in aortitis due to the multiple tissue contrasts available with the different MRA sequences. Both NCE- and CE-MRA sequences are used to assess the vessel wall thickness and the presence of active inflammation. Features of active inflammation include mural enhancement and edema (Figure 5.15). With treatment, these changes can rapidly resolve [51]. Therefore, it is important to scan patients with aortitis prior to or as soon as possible after initiation of therapy. Alternatively, MRA can be used to assess disease activity or evaluate response to therapy. Patients with both giant cell arteritis and Takayasu's arteritis are at risk of developing stenosis and aneurysm (Figure 5.16).

In patients with giant cell arteritis, the addition of MRA and MRI of the circulation in the head and neck is critical due to the high incidence of involvement of these vessels. With MRA, this can be easily accomplished within the same clinical visit. This combined approach

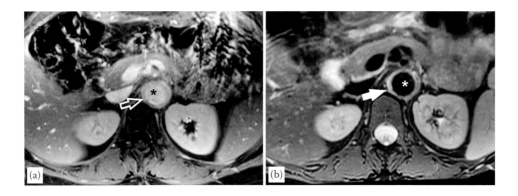

FIGURE 5.15
A 39-year-old female with Takayasu's arteritis involving the abdominal aorta (*). (a) T_1-weighted fast spoiled gradient-recalled echo image obtained after the administration of IV contrast showing circumferential thickening (open arrow) of the wall of the aorta. (b) T_2-weighted image with fat saturation reveals circumferential high signal within the wall of the aorta (closed arrow) indicative of inflammatory mural edema.

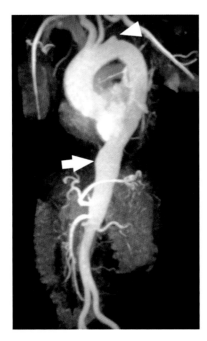

FIGURE 5.16
CE-MRA in a patient with Takayasu's arteritis resulting in a fusiform thoraco-abdominal aortic aneurysm (closed arrow). The left subclavian artery is occluded as well (arrowhead).

offers the incremental value for the patient and the practitioner to understand the extent and activity of this systemic vasculitis.

5.4 Splanchnic Arteries

5.4.1 Mesenteric

Although mesenteric ischemia is uncommon, patients who do present with acute mesenteric ischemia suffer from a high rate of morbidity and mortality. The signs and symptoms of mesenteric ischemia are nonspecific and the diagnosis is frequently delayed, resulting in greater risk of complications. The differential diagnosis of acute abdominal pain is broad and includes both vascular and nonvascular etiologies. Chronic mesenteric ischemia, on the other hand, typically presents with weight loss and postprandial abdominal pain.

Abdominal radiographs are usually the first imaging studies obtained in patients presenting with acute abdominal pain. However, these are both nonspecific and insensitive until signs of bowel infarction are present [52]. As a result, there is a low threshold for performing CTA in patients suspected of having acute mesenteric ischemia due to its availability and rapid acquisition. CTA is also the preferred means of evaluating suspected acute mesenteric ischemia due to its superior ability to detect changes in the bowel wall indicative of intestinal ischemia [53].

Studies have demonstrated the high sensitivity and specificity of MRA for the diagnosis of mesenteric vessel stenosis and occlusion, including the vessels supplying the gastrointestinal tract [54]. However, there is a very limited role for CE-MRA and NCE-MRA in the acute diagnosis and management of acute mesenteric ischemia [55,56]. This is primarily due to the longer scan times, decreased availability, and lower sensitivity for detecting the secondary signs of bowel ischemia. In patients with contraindications for CTA, CE-MRA would be appropriate to assess the vasculature in patients found to have bowel ischemia on non-contrast CT (Figure 5.17).

As indicated previously, postprandial abdominal pain and weight loss are the most frequent presenting symptom in patients with chronic mesenteric ischemia. In older patients, this is usually secondary to atherosclerotic disease. Fibromuscular dysplasia, vasculitis, and median arcuate ligament syndrome are less frequent causes of chronic mesenteric ischemia seen in younger patients.

The accuracy of MRA for the diagnosis of celiac and superior mesenteric artery stenosis is greater than 90%, similar to CTA. Findings supporting the diagnosis include stenosis or occlusion in the proximal celiac and superior mesenteric arteries (Figure 5.18) [54]. In cases where the stenosis is non-occlusive, the addition of a meal challenge to detect appropriate or inappropriate increases in mesenteric blood flow can be extremely helpful [52,57]. In healthy individuals, blood flow through the gastrointestinal tract will usually double following a meal. In patients with mesenteric ischemia, however, the flow through the mesenteric vessels increases much less.

For patients with suspected median arcuate ligament syndrome, it is important to perform dynamic MRA imaging to evaluate the effect of the median arcuate ligament on the compression of the celiac artery. The presence of compression of the celiac artery during expiration that diminishes with inspiration in patients with signs of chronic mesenteric ischemia is diagnostic of median arcuate ligament syndrome (Figure 5.19).

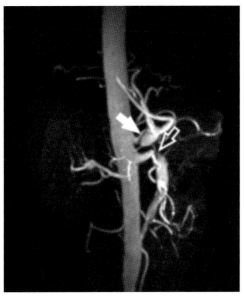

FIGURE 5.17
Celiac (solid arrow) and superior mesenteric (open arrow) artery dissections in a patient with segmental arterial mediolysis. CE-MRA was performed due to contraindications for iodinated contrast administration.

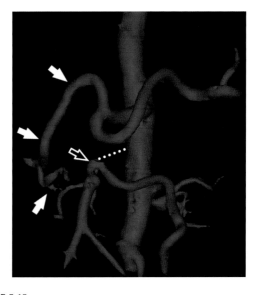

FIGURE 5.18
MRA in patient with chronic mesenteric ischemia secondary to occluded superior mesenteric artery (dashed line). The superior mesenteric artery is reconstituted (open arrow) by gastroduodenal and pancreaticoduodenal artery collaterals (solid arrows).

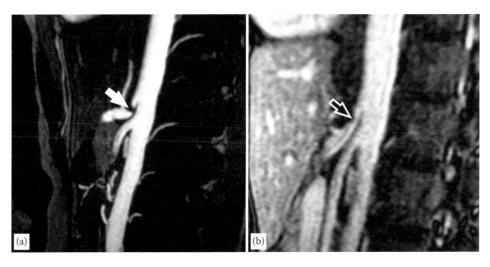

FIGURE 5.19
A 29-year-old patient with chronic mesenteric ischemia secondary to median arcuate ligament syndrome. The celiac artery is nearly completely occluded (solid arrow) during expiration (a) and has much less severe narrowing (open arrow) during inspiration (b).

5.4.2 Renal

Hypertension is a common condition that increases in prevalence with age. Greater than 90% of patients with hypertension have essential hypertension that has not clear etiology and may be due to genetics, obesity, poor diet, or insufficient exercise. Secondary hypertension is that which is related to another medical condition. Causes of secondary hypertension are numerous, but include renal artery stenosis (RAS). Patients are suspected of having renovascular hypertension based on the presence of an abdominal bruit, rapidly rising hypertension, greater than 50 years of age at onset of hypertension, or hypertension that fails to respond to medical management [58]. RAS is most commonly due to chronic atherosclerotic disease in patients over 50 years of age. Acute RAS or RAS secondary to vasculitis or fibromuscular dysplasia is less common. Other causes of RAS, including vasculitis, fibromuscular dysplasia, and neurofibromatosis (Figure 5.20), would be more common in patients under 30 years of age [58].

Doppler ultrasound remains the primary initial diagnostic study in patients suspected to have RAS due to its low costs and accessibility [59]. However, Doppler ultrasound is not sensitive enough to completely exclude RAS in patients with a high pretest probability of disease. This is especially true in patients with poor acoustic windows. In patients with a positive Doppler ultrasound or in patients with a high-pretest probability of RAS, both CTA and MRA are viable alternatives. However, because renal function is frequently impaired or at risk of impairment in patients with secondary hypertension, MRA is preferred to avoid the use of nephrotoxic contrast agents.

Renal artery MRA is accurate and safe in all patients without contraindication for MRI. In fact, the sensitivity and specificity of MRA are greater than 95% and 85%, respectively. Both CE-MRA (Figures 5.1a, 5.20, and 5.21) [60] and NCE-MRA (Figures 5.1b and 5.22) [25,61–63] of the renal arteries are used clinically with essentially equal accuracy. The diagnosis of RAS is primarily based on the presence of narrowing within the renal artery on the anatomical MRA images (Figures 5.1 and 5.14). Secondary signs of significant RAS include delayed enhancement (Figure 5.21) or atrophy of the affected kidney relative to the contralateral normal kidney. For stenoses that are borderline between moderate and severe, it can be difficult to confirm the hemodynamic significance of the stenosis based on the anatomy alone. In these cases, the PC MRA techniques described earlier in this chapter can be beneficial in identifying increased turbulence or flow acceleration distal to the stenosis (Figure 5.8), findings that would confirm a hemodynamically significant lesion. More recent data even suggest that newer 4D-flow MRI techniques may be able to accurately quantify the pressure gradient across the stenosis noninvasively.

Although MRA is usually sufficient to confirm or exclude the presence of RAS in the proximal arteries, assessment of the smaller renal artery branches can be limited due to their small size. For example, in patients with fibromuscular dysplasia [64], the multifocal areas of stenosis and dilation can be difficult to ascertain with MRA [65]. In these

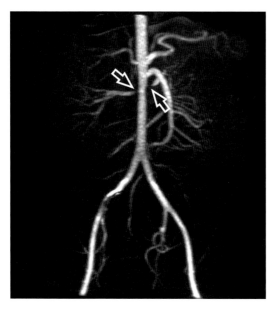

FIGURE 5.20
Bilateral severe RAS (open arrows) secondary to neurofibromatosis.

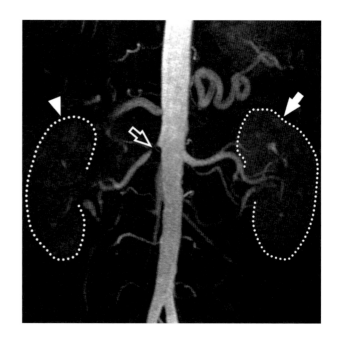

FIGURE 5.21
CE-MRA in a patient with hypertension secondary to severe right renal artery stenosis (open arrow). In addition, there is delayed enhancement of the right renal parenchyma (arrowhead) compared to the left renal parenchyma (closed arrow).

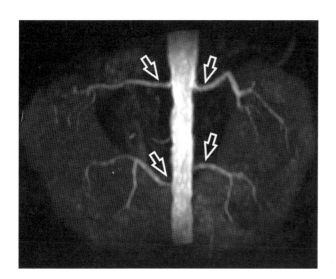

FIGURE 5.22
NCE-MRA in a patient with four renal arteries (open arrows) using an inflow inversion recovery prepared 3D bSSFP technique to enable high-quality imaging of the aorta and renal arteries. The inversion recovery preparation pulse nulls the signal from the surrounding soft tissues and veins.

situations, CTA should be considered if clinically appropriate. CTA would also be appropriate in evaluating the patency of renal artery stents in patients that have already undergone endovascular treatment of RAS.

5.5 Conclusion

The improvements in hardware and software have resulted in dramatic improvements in the quality of CE- and NCE-MRA for the assessment of thoracic and abdominal cardiovascular diseases. Major advances include (a) accelerated image acquisition methods to expand the field of view or improve temporal resolution for time-resolved CE-MRA methods, (b) increased use of higher magnetic field strengths to improve vascular signal and contrast between vasculature and surrounding soft tissues, and (c) renewed interest in higher spatial resolution and faster NCE-MRA methods based on 3D bSSFP and PC acquisitions. Although beyond the scope of this chapter, the 4D-flow MRI techniques have great potential for impacting how patients with cardiovascular disease are evaluated and managed in the future due to their ability to provide exquisite anatomical detail and a broad spectrum of hemodynamic parameters.

Although MRA has a limited role to play in the evaluation of patients with acute cardiovascular pathology, such as acute aortic syndrome or acute mesenteric ischemia, MRA is integral to the clinical management of patients with chronic cardiovascular disease due to its improving

spatial resolution, quicker acquisition times, and safety. Younger patients, in particular, benefit from the use of MRA to detect and follow aortic aneurysms, chronic aortic dissections, chronic mesenteric ischemia, and RAS.

References

1. Maki JH, Prince MR, Londy FJ, Chenevert TL. The effects of time varying intravascular signal intensity and k-space acquisition order on three-dimensional MR angiography image quality. *J Magn Reson Imaging* 1996;6(4):642–651.
2. Rapp JH, Wolff SD, Quinn SF et al. Aortoiliac occlusive disease in patients with known or suspected peripheral vascular disease: Safety and efficacy of gadofosveset-enhanced MR angiography—Multicenter comparative phase III study. *Radiology* 2005;236(1):71–78.
3. Rohrer M, Bauer H, Mintorovitch J, Requardt M, Weinmann HJ. Comparison of magnetic properties of MRI contrast media solutions at different magnetic field strengths. *Invest Radiol* 2005;40(11):715–724.
4. Klessen C, Hein PA, Huppertz A et al. First-pass whole-body magnetic resonance angiography (MRA) using the blood-pool contrast medium gadofosveset trisodium: Comparison to gadopentetate dimeglumine. *Invest Radiol* 2007;42(9):659–664.
5. Maki JH, Wang M, Wilson GJ, Shutske MG, Leiner T. Highly accelerated first-pass contrast-enhanced magnetic resonance angiography of the peripheral vasculature: Comparison of gadofosveset trisodium with gadopentetate dimeglumine contrast agents. *J Magn Reson Imaging* 2009;30(5):1085–1092.
6. Naehle CP, Kaestner M, Muller A et al. First-pass and steady-state MR angiography of thoracic vasculature in children and adolescents. *JACC Cardiovasc Imaging* 2010;3(5):504–513.
7. Nael K, Fenchel M, Krishnam M, Finn JP, Laub G, Ruehm SG. 3.0 Tesla high spatial resolution contrast-enhanced magnetic resonance angiography (CE-MRA) of the pulmonary circulation: Initial experience with a 32-channel phased array coil using a high relaxivity contrast agent. *Invest Radiol* 2007;42(6):392–398.
8. Huang BY, Castillo M. Neurovascular imaging at 1.5 tesla versus 3.0 tesla. *Magn Reson Imaging Clin N Am* 2009;17(1):29–46.
9. Tomasian A, Salamon N, Lohan DG, Jalili M, Villablanca JP, Finn JP. Supraaortic arteries: Contrast material dose reduction at 3.0-T high-spatial-resolution MR angiography—Feasibility study. *Radiology* 2008;249(3):980–990.
10. Habibi R, Krishnam MS, Lohan DG et al. High-spatial-resolution lower extremity MR angiography at 3.0 T: Contrast agent dose comparison study. *Radiology* 2008;248(2):680–692.
11. Attenberger UI, Michaely HJ, Wintersperger BJ et al. Three-dimensional contrast-enhanced magnetic-resonance angiography of the renal arteries: Interindividual

comparison of 0.2 mmol/kg gadobutrol at 1.5 T and 0.1 mmol/kg gadobenate dimeglumine at 3.0 T. *Eur Radiol* 2008;18(6):1260–1268.

12. Vaughan JT, Adriany G, Snyder CJ et al. Efficient high-frequency body coil for high-field MRI. *Magn Reson Med* 2004;52(4):851–859.

13. Saekho S, Yip CY, Noll DC, Boada FE, Stenger VA. Fast-kz three-dimensional tailored radiofrequency pulse for reduced B1 inhomogeneity. *Magn Reson Med* 2006;55(4):719–724.

14. Barth MM, Smith MP, Pedrosa I, Lenkinski RE, Rofsky NM. Body MR imaging at 3.0 T: Understanding the opportunities and challenges. *RadioGraphics* 2007;27(5):1445–1462; discussion 1462–1464.

15. Finn JP, Baskaran V, Carr JC et al. Thorax: Low-dose contrast-enhanced three-dimensional MR angiography with subsecond temporal resolution—Initial results. *Radiology* 2002;224(3):896–904.

16. van Vaals JJ, Brummer ME, Dixon WT et al. "Keyhole" method for accelerating imaging of contrast agent uptake. *J Magn Reson Imaging* 1993;3(4):671–675.

17. Korosec FR, Frayne R, Grist TM, Mistretta CA. Time-resolved contrast-enhanced 3D MR angiography. *Magn Reson Med* 1996;36(3):345–351.

18. Lim RP, Shapiro M, Wang EY et al. 3D time-resolved MR angiography (MRA) of the carotid arteries with time-resolved imaging with stochastic trajectories: Comparison with 3D contrast-enhanced Bolus-Chase MRA and 3D time-of-flight MRA. *AJNR Am J Neuroradiol* 2008;29(10):1847–1854.

19. Haider CR, Hu HH, Campeau NG, Huston J, 3rd, Riederer SJ. 3D high temporal and spatial resolution contrast-enhanced MR angiography of the whole brain. *Magn Reson Med* 2008;60(3):749–760.

20. Perazella MA. Advanced kidney disease, gadolinium and nephrogenic systemic fibrosis: The perfect storm. *Curr Opin Nephrol Hypertens* 2009;18(6):519–525.

21. Perazella MA, Rodby RA. Gadolinium-induced nephrogenic systemic fibrosis in patients with kidney disease. *Am J Med* 2007;120(7):561–562.

22. Sadowski EA, Bennett LK, Chan MR et al. Nephrogenic systemic fibrosis: Risk factors and incidence estimation. *Radiology* 2007;243(1):148–157.

23. Provenzale JM, Sarikaya B. Comparison of test performance characteristics of MRI, MR angiography, and CT angiography in the diagnosis of carotid and vertebral artery dissection: A review of the medical literature. *AJR Am J Roentgenol* 2009;193(4):1167–1174.

24. Buhk JH, Kallenberg K, Mohr A, Dechent P, Knauth M. Evaluation of angiographic computed tomography in the follow-up after endovascular treatment of cerebral aneurysms—A comparative study with DSA and TOF-MRA. *Eur Radiol* 2009;19(2):430–436.

25. Miyazaki M, Lee VS. Nonenhanced MR angiography. *Radiology* 2008;248(1):20–43.

26. Francois CJ, Tuite D, Deshpande V, Jerecic R, Weale P, Carr JC. Unenhanced MR angiography of the thoracic aorta: Initial clinical evaluation. *AJR Am J Roentgenol* 2008;190(4):902–906.

27. Francois CJ, Tuite D, Deshpande V, Jerecic R, Weale P, Carr JC. Pulmonary vein imaging with unenhanced three-dimensional balanced steady-state free precession MR angiography: Initial clinical evaluation. *Radiology* 2009;250(3):932–939.

28. Krishnam MS, Tomasian A, Malik S, Desphande V, Laub G, Ruehm SG. Image quality and diagnostic accuracy of unenhanced SSFP MR angiography compared with conventional contrast-enhanced MR angiography for the assessment of thoracic aortic diseases. *Eur Radiol* 2010;20(6):1311–1320.

29. Pasqua AD, Barcudi S, Leonardi B, Clemente D, Colajacomo M, Sanders SP. Comparison of contrast and noncontrast magnetic resonance angiography for quantitative analysis of thoracic arteries in young patients with congenital heart defects. *Ann Pediatr Cardiol* 2011;4(1):36–40.

30. Pelc NJ, Herfkens RJ, Shimakawa A, Enzmann DR. Phase contrast cine magnetic resonance imaging. *Magn Reson Q* 1991;7(4):229–254.

31. Chai P, Mohiaddin R. How we perform cardiovascular magnetic resonance flow assessment using phase-contrast velocity mapping. *J Cardiovasc Magn Reson* 2005;7(4):705–716.

32. Pelc NJ, Bernstein MA, Shimakawa A, Glover GH. Encoding strategies for three-direction phase-contrast MR imaging of flow. *J Magn Reson Imaging* 1991;1(4):405–413.

33. Markl M, Frydrychowicz A, Kozerke S, Hope M, Wieben O. 4D flow MRI. *J Magn Reson Imaging* 2012;36(5):1015–1036.

34. Markl M, Harloff A, Bley TA et al. Time-resolved 3D MR velocity mapping at 3T: Improved navigator-gated assessment of vascular anatomy and blood flow. *J Magn Reson Imaging* 2007;25(4):824–831.

35. Gu T, Korosec FR, Block WF et al. PC VIPR: A high-speed 3D phase-contrast method for flow quantification and high-resolution angiography. *AJNR Am J Neuroradiol* 2005;26(4):743–749.

36. Stalder AF, Russe MF, Frydrychowicz A, Bock J, Hennig J, Markl M. Quantitative 2D and 3D phase contrast MRI: Optimized analysis of blood flow and vessel wall parameters. *Magn Reson Med* 2008;60(5):1218–1231.

37. Lum DP, Johnson KM, Paul RK et al. Transstenotic pressure gradients: Measurement in swine—Retrospectively ECG-gated 3D phase-contrast MR angiography versus endovascular pressure-sensing guidewires. *Radiology* 2007;245(3):751–760.

38. Turk AS, Johnson KM, Lum D et al. Physiologic and anatomic assessment of a canine carotid artery stenosis model utilizing phase contrast with vastly undersampled isotropic projection imaging. *AJNR Am J Neuroradiol* 2007;28(1):111–115.

39. Bley TA, Johnson KM, Francois CJ et al. Noninvasive assessment of transstenotic pressure gradients in porcine renal artery stenoses by using vastly undersampled phase-contrast MR angiography. *Radiology* 2011;261(1):266–273.

40. Tyszka JM, Laidlaw DH, Asa JW, Silverman JM. Three-dimensional, time-resolved (4D) relative pressure mapping using magnetic resonance imaging. *J Magn Reson Imaging* 2000;12(2):321–329.

41. Bock J, Frydrychowicz A, Lorenz R et al. In vivo noninvasive 4D pressure difference mapping in the human aorta: Phantom comparison and application in healthy volunteers and patients. *Magn Reson Med* 2011;66(4):1079–1088.

42. Evangelista A, Mukherjee D, Mehta RH et al. Acute intramural hematoma of the aorta: A mystery in evolution. *Circulation* 2005;111(8):1063–1070.

43. Hayashi H, Matsuoka Y, Sakamoto I et al. Penetrating atherosclerotic ulcer of the aorta: Imaging features and disease concept. *RadioGraphics* 2000;20(4):995–1005.

44. Klem I, Heitner JF, Shah DJ et al. Improved detection of coronary artery disease by stress perfusion cardio-vascular magnetic resonance with the use of delayed enhancement infarction imaging. *J Am Coll Cardiol* 2006;47(8):1630–1638.

45. Panting JR, Norell MS, Baker C, Nicholson AA. Feasibility, accuracy and safety of magnetic resonance imaging in acute aortic dissection. *Clin Radiol* 1995;50(7):455–458.

46. Desjardins B, Dill KE, Flamm SD et al. ACR Appropriateness Criteria((R)) pulsatile abdominal mass, suspected abdominal aortic aneurysm. *Int J Cardiovasc Imaging* 2013;29(1):177–183.

47. Atar E, Belenky A, Hadad M, Ranany E, Baytner S, Bachar GN. MR angiography for abdominal and thoracic aortic aneurysms: Assessment before endovascular repair in patients with impaired renal function. *AJR Am J Roentgenol* 2006;186(2):386–393.

48. Gornik HL, Creager MA. Aortitis. *Circulation* 2008;117(23):3039–3051.

49. Bley TA, Uhl M, Venhoff N, Thoden J, Langer M, Markl M. 3-T MRI reveals cranial and thoracic inflammatory changes in giant cell arteritis. *Clin Rheumatol* 2007;26(3):448–450.

50. Bley TA, Uhl M, Carew J et al. Diagnostic value of high-resolution MR imaging in giant cell arteritis. *AJNR Am J Neuroradiol* 2007;28(9):1722–1727.

51. Bley TA, Ness T, Warnatz K et al. Influence of corticosteroid treatment on MRI findings in giant cell arteritis. *Clin Rheumatol* 2007;26(9):1541–1543.

52. Burkart DJ, Johnson CD, Reading CC, Ehman RL. MR measurements of mesenteric venous flow: Prospective evaluation in healthy volunteers and patients with suspected chronic mesenteric ischemia. *Radiology* 1995;194(3):801–806.

53. Bowersox JC, Zwolak RM, Walsh DB et al. Duplex ultrasonography in the diagnosis of celiac and mesenteric artery occlusive disease. *J Vasc Surg* 1991;14(6):780–786; discussion 786–788.

54. Meaney JF, Prince MR, Nostrant TT, Stanley JC. Gadolinium-enhanced MR angiography of visceral arteries in patients with suspected chronic mesenteric ischemia. *J Magn Reson Imaging* 1997;7(1):171–176.

55. Shih MC, Angle JF, Leung DA et al. CTA and MRA in mesenteric ischemia: Part 2, Normal findings and complications after surgical and endovascular treatment. *AJR Am J Roentgenol* 2007;188(2):462–471.

56. Shih MC, Hagspiel KD. CTA and MRA in mesenteric ischemia: Part 1, Role in diagnosis and differential diagnosis. *AJR Am J Roentgenol* 2007;188(2):452–461.

57. Li KC, Hopkins KL, Dalman RL, Song CK. Simultaneous measurement of flow in the superior mesenteric vein and artery with cine phase-contrast MR imaging: Value in diagnosis of chronic mesenteric ischemia. Work in progress. *Radiology* 1995;194(2):327–330.

58. Safian RD, Textor SC. Renal-Artery Stenosis. *N Engl J Med* 2001;344(6):431–442.

59. De Cobelli F, Venturini M, Vanzulli A et al. Renal arterial stenosis: Prospective comparison of color Doppler US and breath-hold, three-dimensional, dynamic, gadolinium-enhanced MR angiography. *Radiology* 2000;214(2):373–380.

60. Vasbinder GBC, Nelemans PJ, Kessels AGH et al. Accuracy of computed tomographic angiography and magnetic resonance angiography for diagnosing renal artery stenosis. *Ann Intern Med* 2004;141(9):674–682.

61. Maki JH, Wilson GJ, Eubank WB, Glickerman DJ, Pipavath S, Hoogeveen RM. Steady-state free precession MRA of the renal arteries: Breath-hold and navigator-gated techniques vs. CE-MRA. *J Magn Reson Imaging* 2007;26(4):966–973.

62. Wyttenbach R, Braghetti A, Wyss M et al. Renal artery assessment with nonenhanced steady-state free precession versus contrast-enhanced MR angiography. *Radiology* 2007;245(1):186–195.

63. Spuentrup E, Manning WJ, Bornert P, Kissinger KV, Botnar RM, Stuber M. Renal arteries: Navigator-gated balanced fast field-echo projection MR angiography with aortic spin labeling: Initial experience. *Radiology* 2002;225(2):589–596.

64. Slovut DP, Olin JW. Fibromuscular dysplasia. *N Engl J Med* 2004;350(18):1862–1871.

65. Willoteaux S, Faivre-Pierret M, Moranne O et al. Fibromuscular dysplasia of the main renal arteries: Comparison of contrast-enhanced MR angiography with digital subtraction angiography. *Radiology* 2006;241(3):922–929.

6

Peripheral Magnetic Resonance Angiography

Michele Anzidei, Beatrice Sacconi, Beatrice Cavallo Marincola, Pierleone Lucatelli, Alessandro Napoli, Mario Bezzi, and Carlo Catalano

CONTENTS

The angiographic effect in magnetic resonance (MR) imaging is created in a variety of ways, including techniques without contrast agent enhancement such as time-of-flight (TOF) and phase contrast angiography, or with contrast-enhanced MR angiography (MRA) using a gadolinium chelating agent. The non-contrast agent techniques exploit the dynamic nature of flowing blood and produce bright-blood images. However, these techniques are time consuming and often overestimate stenosis, especially in the setting of complex anatomy, tortuous vessels, and abnormal blood flow.

Contrast-enhanced MRA has emerged as a robust and reliable alternative to flow-dependent imaging techniques, and it can be performed in seconds rather than minutes. Contrast between blood vessels and background is achieved by shortening the T_1 relaxation time of blood with a paramagnetic contrast material. However, one of the greatest challenges of contrast-enhanced MRA is achieving optimal timing of the contrast agent bolus and limiting venous contamination. In the peripheral extremities, timing of the bolus is particularly challenging. Test bolus timing examinations and bolus tracking techniques may be limited because small vessel opacification may be obscured by inflow effects.

6.1 Magnetic Resonance Technique

The following points discuss the patient positioning technique for the study of the upper limbs:

- Place the patient in supine position in the magnet, with the arm at side; if only the forearm or hand is to be imaged, Superman position should be considered.

- Use multichannel phased-array coil or extremity coil if available; the high-density coil has to be placed in the region of interest.

- Contrast media has to be injected in the contralateral arm.

- Remove any clothing with any metal.

The following points discuss the patient positioning technique for the study of the lower limbs:

- Place the patient in supine position in the magnet, with the lower limbs close together facing the gantry and the feet slightly internally rotated.
- Use multichannel phased array coil for the abdomen and extremity coil.
- Intravenous access.
- Remove any clothing with any metal.

6.2 Unenhanced MRA: Time-of-Flight and Steady-State Free-Precession Sequences

TOF sequences are one of the most commonly used sequences among the noncontrast MRA sequences but they are usually used in patients with adverse reactions to contrast material and for evaluation of fistula in hemodialysis patients.

Two-dimensional (2D) and three-dimensional (3D) TOF techniques are used to differentiate the stationary tissue and moving protons inside the blood vessels. This can be accomplished by saturating the background signal in the slice that is to be imaged with rapid radio frequency (RF) pulses. Since the fresh incoming blood supplies a significant amount of spins, which has not seen any of the rapid RF pulses, the blood vessels would have a quite bright signal, whereas the stationary tissue signal is almost completely suppressed. TOF sequences use gradient-echo (GRE) pulse sequence family with relatively small flip angle to be able to create a series of rapid RF pulses. We can maximize the signal from blood if we choose the slice orientation perpendicular to the blood vessels. One of the biggest advantages is that the TOF technique has been used to visualize arteries, veins, or both; to selectively visualize arteries or veins, the user is requested to place a saturation band.

The TOF technique also suffers from some major drawbacks; it usually requires longer acquisition time and is sensitive to blood vessel orientation, susceptibility, and flow turbulences in stenosis or bifurcations [1,2] (Tables 6.1 and 6.2).

Steady-state free-precession (SSFP) sequences may generate rapid, bright-blood images without the requirement for gadolinium (Gd) administration. In fact, the SSFP technique operates at very short repetition time (TR) (3 ms) and large flip angles, generating 2D images in less than 1 s. For this reason, breath holding is not required for single shot SSFP, and this sequence may be used effectively in a number of clinical scenarios, including aortic, renal, carotid, and peripheral arterial imaging.

TABLE 6.1

Technical Parameters of TOF Sequences for the Upper Limbs

MRI Parameters	Values
TR (ms)	11
TE (ms)	6.9
FoV (mm)	450×450
Slice thickness (mm)	5
Matrix	384×384
Acquisition time (s)	135

TABLE 6.2

Technical Parameters of TOF Sequences for the Lower Limbs

MRI Parameters	Values
TR (ms)	30
TE (ms)	3.9
FoV (mm)	180×140
Slice thickness (mm)	1
Matrix	256×224
Acquisition time (s)	5

6.3 Contrast-Enhanced MRA

Unlike TOF MRA, contrast-enhanced MRA does not rely on blood flow but on the T_1 shortening of blood by an intravenous gadolinium chelating agent. This allows considerably faster acquisition of data. Accurate timing of the contrast agent bolus is required to capture the arterial phase and reduce venous contamination. Fast-interpolated 3D acquisitions permit near-isotropic acquisitions that can improve depiction of complex vascular anatomy. These sequences are not dependent on blood flow. High temporal and spatial resolution may be achieved with minimal flow-related artifacts. Arteriography and venography may be performed, depending on the timing of the contrast agent bolus. Image quality may be limited by artifacts related to calcifications, metallic devices, and stents and by signal loss due to high gadolinium chelate concentrations at which T_2^* effects may dominate [1–5] (Tables 6.3 and 6.4).

TABLE 6.3

Technical Parameters of 3D T_1-Weighted Gradient-Echo for the Upper Limbs

MRI Parameters	Values
TR (ms)	3.5
TE (ms)	1.2
FoV (mm)	140×180
Slice thickness (mm)	1
Matrix	384×384
Acquisition time (s)	14

TABLE 6.4

Technical Parameters of 3D T_1-Weighted Gradient-Echo for the Lower Limbs

	Abdomen	Coscia	Gamba
TE (ms)	3	3	3
FoV (mm)	0.9	0.9	1.4
Slice thickness (mm)	1.4	1.4	1.2
Matrix	384 × 384	384 × 384	512 × 512
Acquisition time (s)	15	16	21

The protocol includes the following:

- Coronal localizer
- GRE 3D mask on coronal plane
- Visualization of contrast media flow through MR fluoroscopic technique (acquisition of multiple low-resolution images at thoracic aorta level)
- Acquisition of GRE 3D sequence on coronal plane (identical to the mask)
- Processing of subtracted images

6.4 Time-Resolved or Four-Dimensional MRA

Techniques such as time-resolved imaging of contrast kinetics (TRICKS) help achieve high temporal resolution without compromising spatial information. An initial mask, a dataset obtained before contrast agent administration, is obtained for subtraction from contrast-enhanced datasets. The sequence is then repeated multiple times after the administration of contrast agent to get snapshots of the vascular system at multiple time points. This allows visualization of, for example, the gradual enhancement of small, diseased vessels or vessels that are reconstituted by collateral vessels. Both morphologic and kinetic information are obtained, with a low dose (<0.1 mmol/kg) of contrast agent [6–12] (Table 6.5).

These sequences can be used together with GRE 3D in a *hybrid protocol*; this protocol is useful in patients with flow asymmetry in extremities, such as patients with bypass or diabetic patients. It can also permit the

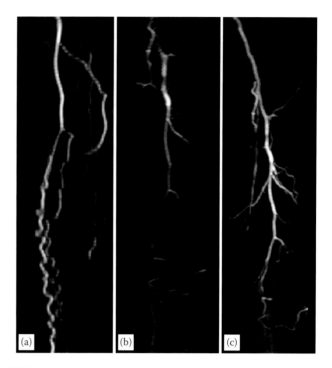

FIGURE 6.1
A 74-year-old male with occlusion of the left superficial femoral artery. (a) Time-of-flight sequence (TOF) demonstrates vessel obstruction as well as patency of collateral branches from the deep femoral artery, with some stairstep artifacts due to non-isotropic voxel size. (b) Steady-state free-precession (SSFP) sequence shows similar findings with some residual signal from venous vessels. (c) Contrast-enhanced MRA (CE-MRA) confirms the findings of both unenhanced techniques.

visualization of smaller vessels, which cannot be visualized only with GRE 3D (Figure 6.1).

The protocol includes the following:

- Time-resolved MRA for the study of distal vessels (leg and foot) using a first bolus of contrast media (8–10 mL)
- Conventional GRE 3D for the study of vessels of the abdomen and thigh using a second bolus of contrast media (10–15 mL)

6.5 Pathology of the Upper Limbs

Vascular disorders of the hand and upper extremity encompass a broad spectrum of diseases, ranging from acute limb-threatening ischemia to chronic disabling disease. Although less common than lower extremity vascular disease, upper extremity disease affects as much as 10% of the population.

For optimal imaging, protocols should be tailored for specific clinical concerns and symptoms. If subclavian steal syndrome is suspected, a 2D TOF sequence

TABLE 6.5

Technical Parameters of Time-Resolved MRA

	Twist (Siemens)	Tricks (GE)	4D-Trak (Philips)
TR (ms)	2.7	5.5	2.2
TE (ms)	1.1	1.2	0.9
Flip angle	20°	30°	15°
Slice thickness	1.8	1.8	1.1
Matrix	384 × 384	384 × 384	224 × 178

in the neck (five to six sections), with a saturation band placed above and then below the imaging slabs to assess direction of flow in the vertebral arteries, may be used. Alternatively, 2D phase contrast or time-resolved imaging may be used for this purpose.

If vasculitis is suspected, a fat-suppressed T_2-weighted sequence such as an inversion-recovery fast-spin-echo sequence in addition to the postcontrast T_1-weighted GRE sequence is recommended to examine for mural disease.

If thoracic outlet syndrome is suspected, the protocol should initially be performed with the patient's arms up. If there is abnormal compression of the vasculature, repeat MRA with the arms down.

6.6 Atherosclerosis

In the upper limbs, atherosclerosis is far less common than in the lower limbs; when present, atherosclerotic lesions may involve the innominate artery and the origin of the subclavian artery. Stenosis or the occlusion of these arteries may be asymptomatic due to the development of collateral vessels, or lead to upper extremity claudication or ischemia. Symptomatic patients can undergo various procedures, such as carotid-subclavian bypass or subclavian transposition; in these cases, MR can be used to assess the graft location, anastomoses, lumen patency, and potential restenosis.

It is strictly necessary to follow a precise pattern in reporting MR images:

- Localize and characterize lesions, evaluating the degree of stenosis.
- Describe collateral vessels in case of obstruction.
- Describe anatomical abnormalities and vessel condition above and below the lesions (including caliber and wall status).
- Evaluate arm circulation (including flow asymmetry) and the presence of additional parietal lesions, although of lesser magnitude.
- Pay attention to nonvascular findings.

6.7 Acute Arterial Occlusion

Arterial occlusive disease of the upper extremity may represent either local or systemic disease. The pattern of arterial disease varies according to etiology. Diseases that affect the brachiocephalic vessels include atherosclerosis, arteritis, congenital anomalies, trauma, and fibromuscular dysplasia. In the United States, atherosclerosis

is the most common cause of subclavian artery stenosis; outside of the United States, Takayasu's arteritis is more common. The axillary and brachial arteries are common sites of injury. One-third of peripheral emboli lodge in the upper extremity, producing acute arterial occlusion. Radiation therapy of the chest or breast may induce subclavian artery disease. Common causes of acute arterial occlusion are as below:

Embolism

Acute thrombosis

Arterial spasm

Compressive mechanism

Dissection

Reports should contain the same information described in case of atherosclerotic disease; it is to be noted that in acute obstruction, collateral blood vessels have not formed.

6.8 Raynaud's Phenomenon

Raynaud's phenomenon is a vasospastic disorder causing discoloration of the fingers, toes, and occasionally other areas. Raynaud's phenomenon includes Raynaud's disease (also known as *primary Raynaud's phenomenon*) where the phenomenon is idiopathic, and Raynaud's syndrome. Raynaud's syndrome can be classified as follows:

- Primary, which is a functional microangiopathy caused by some investigating factor, most commonly connective tissue disorders such as systemic lupus erythematosus.
- Secondary, which is due to compressive mechanism and is included in the thoracic outlet syndrome.

Considering the small caliber of vessels involved, MRI is employed to exclude other diagnosis, such as compression of larger vessels, rather than to specifically evaluate the involved small vessels (Figure 6.2).

6.9 Thoracic Outlet Syndrome

Dynamically induced compression of the neural, arterial, and, more rarely, venous structures crossing the tunnel of the thoracic outlet leads to thoracic outlet syndrome. The diagnosis is based on the results of clinical

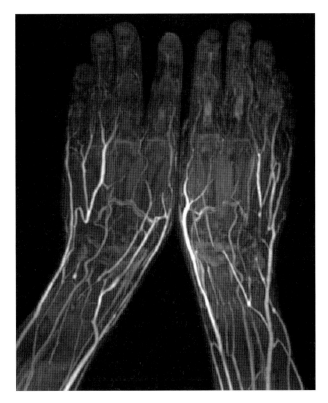

FIGURE 6.2
A 52-year-old female with Raynaud's phenomenon. CE-MRA images obtained in the equilibrium phase demonstrate bilateral and progressive caliber reduction of finger arteries.

TABLE 6.6

Principal Causes of Thoracic Outlet Syndrome

General Condition	Specific Pathology
Skeletal and bone abnormalities	Cervical rib elongated C7 transverse process
	Exostosis or tumor of the I rib or clavicle
	Excess callus of the I rib or clavicle
Soft tissue abnormalities	Fibrous band
	Congenital muscle abnormalities
Acquired soft tissue abnormalities	Posttraumatic fibrous scarring
	Postoperative scarring
Posture-predisposing morphotype	Poor posture-weak muscular support

evaluation, particularly if the patient's symptoms can be reproduced when various dynamic maneuvers, including elevation of the arm, are undertaken. The thoracic outlet, or cervicothoracobrachial junction, includes three confined spaces, extending from the cervical spine and the mediastinum to the lower border of the pectoralis minor muscle, which are potential sites of neurovascular compression. These three compartments are the interscalene triangle, the costoclavicular space, and the retropectoralis minor space (Table 6.6).

Although diagnosis is usually clinical, MR and computed tomography (CT) permit to

- Localize the correct site of compression.
- Individuate the cause and the structures that are involved.
- Evaluate the extension of the compression.

CT can adequately depict vascular compression, but a detailed analysis of the brachial plexus branches is more difficult and should be performed using MRI [13].

6.10 Hemodialysis Fistulae

Hemodialysis requires well-functioning vascular access to allow sufficient blood flow for adequate clearance and dialysis dosing. This is achieved temporarily by central vein catheterization and subsequently by surgical creation of an arteriovenous fistula or graft for more permanent and durable access. An upper extremity arteriovenous fistula is commonly created by surgical end-to-side anastomosis of the radial or brachial artery to an adjacent basilic, cephalic, or medial antecubital vein. Complications of hemodialysis fistulae include venous stenosis, thrombosis, infection, aneurysms, and, rarely, arterial steal syndrome. MR has proven to be an accurate imaging modality for the comprehensive evaluation of the anatomy and function of hemodialysis fistulae, as well for the diagnosis of related complications. It allows noninvasive monitoring and early detection of hemodynamically significant stenoses treatable with elective angioplasty, thereby substantially reducing the frequency of subsequent thrombosis and access failure [14–17].

6.11 Aneurysm of the Upper Limbs

An aneurysm is a localized, blood-filled balloon-like bulge in the wall of a blood vessel. A true aneurysm is one that involves all three layers of the wall of an artery (intima, media, and adventitia); they include atherosclerotic, syphilitic, and congenital aneurysms. A false aneurysm, or pseudoaneurysm, is a collection of blood leaking completely out of an artery or vein, but confined next to the vessel by the surrounding tissue. This blood-filled cavity will eventually either thrombose (clot) enough to seal the leak, or rupture out of the surrounding tissue. Pseudoaneurysms can be caused by trauma that punctures the artery, such as knife and bullet wounds, because of percutaneous surgical procedures such as coronary angiography or arterial grafting, or

TABLE 6.7

Main Causes of Aneurysms

Vessel Involved	Subsegment	Cause
Subclavian artery	Prescalenic portion	Atherosclerotic micotic
Subclavian artery	Distal portion	Poststenotic traumatic
Axillar artery		Traumatic
Radial and ulnar artery		Chronic traumas

use of an artery for injection [18]. Aneurysms can also be classified by their macroscopic shape and size, and are described as either saccular or fusiform. The shape of an aneurysm is not specific for a specific disease. Saccular aneurysms are spherical in shape and involve only a portion of the vessel wall; they vary in size from 5 to 20 cm in diameter, and they are often filled, either partially or fully, by a thrombus. Fusiform aneurysms are variable in both their diameter and length (Table 6.7).

It is strictly necessary to follow a precise pattern in reporting MR images:

- Describe the location, size, and longitudinal extension of the aneurysm.
- Describe the thrombotic apposition in the aneurysm (concentric/eccentric).
- Provide the distance from the normal vessel above and below the aneurysm.
- Identify the incipient rupture or occlusion signs.
- Describe the relationship with the surrounding vascular and nonvascular structures.

6.12 Subclavian Steal Syndrome

Subclavian steal syndrome results from the occlusion or severe stenosis of the left subclavian artery or the brachiocephalic trunk proximal to the origin of the ipsilateral vertebral artery. As the demand for blood supply to the upper extremity increases (overhead activity and muscular stress), blood flow is diverted or *stolen* from the vertebrobasilar system and bypassed to the arm via retrograde flow in the ipsilateral vertebral artery. Stenosis or occlusion of the subclavian artery is most commonly caused by atherosclerotic disease, but the differential diagnosis includes trauma, vasculitis, dissection, and congenital anomaly. Subclavian steal affects the left subclavian artery three times more commonly than the right. In patients affected by this condition, MR and CT can confirm the stenosis or obstruction of the proximal subclavian artery, and give information about the length of the stenosis and allows an evaluation of the ipsilateral reperfusion through the vertebral artery (Figure 6.3).

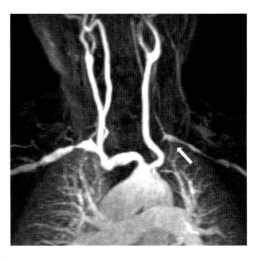

FIGURE 6.3
A 67-year-old male with left upper limb cramps. CE-MRA image shows a severe stenosis of the left subclavian artery (arrow); the blood flow is diverted from the ipsilateral vertebrobasilar system.

6.13 Pathology of the Lower Limbs

6.13.1 Atherosclerosis

Atherosclerotic disease is the main cause of peripheral arterial disease, producing a reduction in arterial blood supply to the lower vascular district by major vessels stenosis or occlusion. The early stage is clinically characterized by chronic cramping and soreness in the posterior compartment of the leg, mainly during exercise, while acute worsening of the symptoms occurs more commonly in the advanced stage of the disease. Critical limb ischemia is the most severe evolution of peripheral arterial occlusive disease (PAOD). The main risk factors are age, male sex, cigarette smoke, diabetes, and dyslipidemia [7,19].

From a radiological point of view, the current guidelines for the management of peripheral arterial disease classified the lesions according to their extension and possible surgical or endovascular approach. In particular, atherosclerotic lesions are divided into *inflow lesions* (those involving the aorta and the iliac arteries) and *runoff lesions* (involving the femoral and popliteal arteries). In these regions, atherosclerotic lesions are classified using a four-point scale [20]:

1. Lesions that should be treated preferably with an endovascular approach.
2. Lesions that are treated with sufficiently good results using endovascular methods; surgical treatment should be limited to selected case.
3. Lesions that are treated surgically with superior long-terms results; endovascular treatment

should be preferred only in patients with high surgical risks.

4. Lesions that should be treated preferably with a surgical approach.

The role of imaging is fundamental in the diagnostic-therapeutic process for the quantification and localization of the stenosis and in treatment planning, intended as analysis of the vascular anatomy and the choice of a more effective therapeutic approach.

It is strictly necessary to follow a precise pattern in reporting MR images:

- Localize and characterize lesions, evaluating the degree of stenosis
- Describe collateral vessels in case of obstruction
- Describe anatomical abnormalities and vessel condition above and below the lesions (including caliber and wall status)
- Evaluate arm circulation (including flow asymmetry) and the presence of additional parietal lesions, although of lesser magnitude
- Pay attention to nonvascular findings

Interventional and surgical therapy for PAOD includes percutaneous transluminal angioplasty, thrombolytic infusion, bypass, and stent placement; with the improvement of new angioplastic techniques, endovascular treatment is becoming an effective alternative to surgery.

MRA is very useful in posttreatment evaluation of peripheral vasculature.

Percutaneous transluminal angioplasty is often used as first-line technique; posttreatment imaging should identify possible restenosis of the stent. However, several types of stent can produce artifacts during MR scan, so CT is usually preferred. Early complications include hemorrhage and distal embolization; in the first case, MR can demonstrate both the hematoma and the bleeding site. In the second case, MR may be helpful in the identification of distal embolization (mostly when a pretreatment examination is available for comparisons) and can demonstrate the disappearance of previously normal distal arterial branches.

Late complications include pseudoaneurysm and stent occlusion/restenosis, commonly caused by progression of the underlying atherosclerotic disease and by intimal hyperplasia.

Endarterectomy is a surgical procedure to remove the atheromatous plaque along the intima; restenoses after the treatment are very frequent.

Surgical bypass consists in deploying a synthetic graft from above the obstruction segment to the distal patent branches. This technique is most durable even

if various complications might occur. Early complications include the formation of hematoma/seroma/limphocele; all these entities usually appear as well-capsulated fluid collections, appearing with fluid signal on MRA; thrombus, blood, or clots in the collection may determine variable signal alterations acute occlusion of the graft can also occur; on MR, bypass obstruction is easily diagnosed when failed opacification occurs. Graft infection can be diagnosed when fluid collection at the anastomotic sites or along the graft body is visible, with surrounding gas bubbles and soft tissue swelling [21,22].

Late complications include graft stenosis occlusion secondary to recurrent obliterative disease; stenosis can involve the anastomosis or the mid-graft region. The appearance of graft stenosis on MR is similar to that of stenosis of native vessels. Thrombosis or stenosis in extranatomical bypass is also possible, due to external compression, kinking, or progression of distal and proximal disease. Other complications are pseudoaneurysms and subsequent anastomotic dehiscence, whose main risk factors are represented by excessive tension of the graft; on MR, pseudoaneurysm is usually identified as a focal outpouching of the vessel walls in the proximity of an anastomotic sites. Aortoenteric fistula is a life-threatening complication with high mortality rates, even if promptly diagnosed (Figures 6.4 through 6.6).

6.13.2 Aneurysm of the Lower Limbs

The cut off caliber ratio with contralateral vessels for the diagnoses of aneurysm is >0.7 for popliteal arteries and >1.5 for femoral arteries.

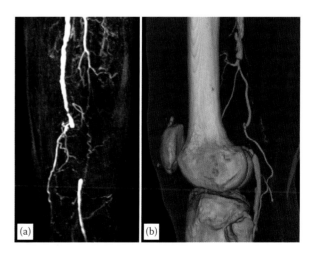

FIGURE 6.4

A 79-year-old male with acute claudication of the left lower limb. (a) CE-MRA image demonstrates occlusion of the distal superficial femoral artery and proximal popliteal artery. (b) Volume rendering of peripheral computed tomography angiography (CTA) confirms MRI findings, also showing distal revascularization from a collateral branch of the superficial femoral artery.

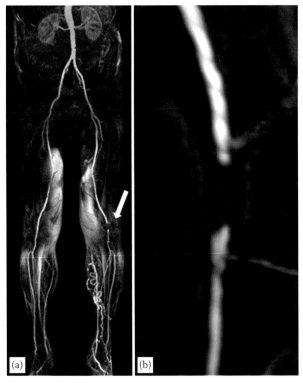

FIGURE 6.5
An 81-year-old male with acute claudication of the left lower limb. (a) Panoramic minimum intensity projection (MIP) reconstruction of CE-MRA demonstrates segmental obstruction of the left supra-articular popliteal artery (arrow). (b) Regional close-up confirms the previous findings.

True aneurysm, mainly associated with atherosclerotic disease, occurs mostly in the popliteal region. False aneurysm is due to a defect in the arterial wall related to trauma (most commonly iatrogenic) or infection.

Reports must

- Describe the location, size, and longitudinal extension of the aneurysm.
- Describe the thrombotic apposition in the aneurysm (concentric/eccentric).
- Provide the distance from the normal vessel above and below the aneurysm.
- Identify the incipient rupture or occlusion signs.
- Describe the relationship with the surrounding vascular and nonvascular structures.

MR technique should be adapted by increasing the delay between the contrast agent administration and the scan start, especially in case of large aneurysms; in fact, the blood flow at the aneurysm level can be very turbulent, causing the dilution of contrast media [5,9,23] (Figure 6.7).

6.13.3 Entrapment

Popliteal artery entrapment syndrome (PAES) is a developmental abnormality that results from an abnormal relationship of the popliteal artery with the gastrocnemius muscle or, rarely, an anomalous fibrous band or

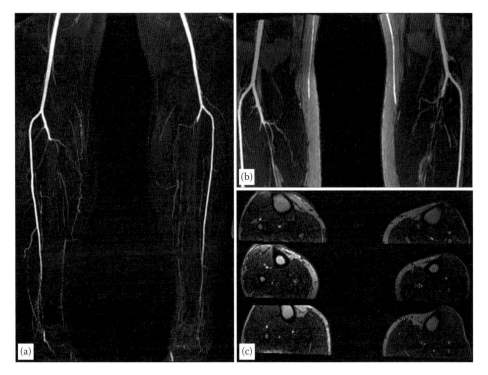

FIGURE 6.6
A 78-year-old male with bilateral acute claudication and foot ulcers. (a) CE-MRA shows bilateral occlusion of the tibial-peroneal trunk. (b and c) High-resolution sequences obtained at the equilibrium phase with axial reformation confirm the findings.

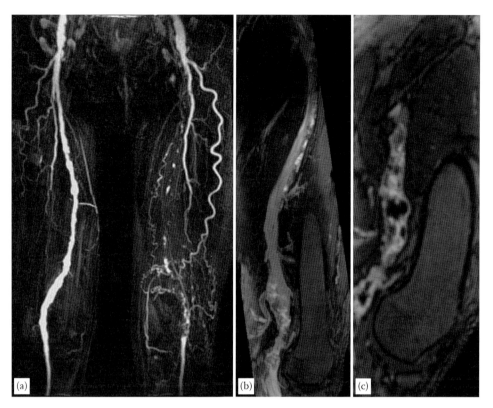

FIGURE 6.7
A 74-year-old male with bilateral claudication. (a) CE-MRA demonstrates bilateral arteriomegaly involving the superficial femoral artery and popliteal artery, with thrombotic occlusion of the distal superficial femoral artery and proximal popliteal artery in the left limb. (b and c) High-resolution sequences obtained at the equilibrium phase confirm the findings.

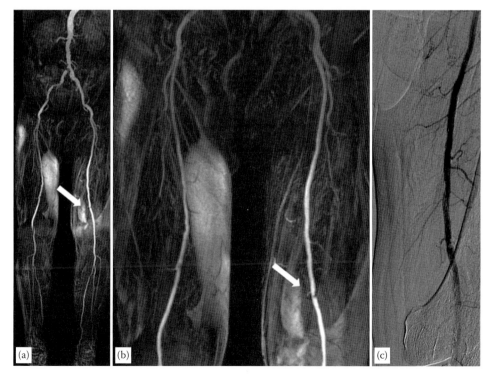

FIGURE 6.8
A 68-year-old male. (a) CE-MRA demonstrates focal dissection of the left distal superficial femoral artery (arrow). (b) Regional close-up confirms the previous finding. (c) Femoral angiography of the same patient.

the popliteus muscle, causing a deviation and compression of the artery. There are essentially four anatomic variants of PAES.

Stress MR, with scan performed at rest and in evocative position, with the foot in either dorsiflexed position to elicit vascular compression, is usually preferred, also because of the young age of the patients [5,9,23].

6.13.4 Dissection

Dissection of peripheral vessels is usually the extension of a dissection of abdominal aorta. We can observe a focal dissection of the femoral artery due to a percutaneous procedure less frequently.

MR imaging requires, in addition to conventional protocol, T_1- and T_2-weighted sequences before contrast media administration, to identify a possible intimal flap (Figure 6.8).

References

1. Lee VS. *Cardiovascular MRI: Physical Principles to Practical Protocols*. Philadelphia, PA: Lippincott Williams & Wilkins, 2006; p. 14.
2. Krinsky G, Rofsky NM. MR angiography of the aortic arch vessels and upper extremities. *Magn Reson Imaging Clin N Am* 1998; 6(2): 269–292.
3. Prince MR, Yucel EK, Kaufman JA, Harrison DC, Geller SC. Dynamic gadolinium-enhanced three-dimensional abdominal MR arteriography. *J Magn Reson Imaging* 1993; 3(6): 877–881.
4. Lee VS, Lee HM, Rofsky NM. Magnetic resonance angiography of the hand: A review. *Invest Radiol* 1998; 33(9): 687–698.
5. Demondion X, Herbinet P, Van Sint Jan S, Boutry N, Chanteot C, Cotten A. Imaging assessment of thoracic outlet syndrome. *RadioGraphics* 2006; 26: 1735–1750.
6. Korosec FR, Frayne R, Grist TM, Mistretta CA. Time-resolved contrast-enhanced 3D MR angiography. *Magn Reson Med* 1996; 36(3): 345–351.
7. Van Rijswijk CS, van der Linden E, van der Woude HJ, van Baalen JM, Bloem JL. Value of dynamic contrast-enhanced MR imaging in diagnosing and classifying peripheral vascular malformations. *AJR Am J Roentgenol* 2002; 178(5): 1181–1187.
8. Konez O, Burrows PE. Magnetic resonance of vascular anomalies. *Magn Reson Imaging Clin N Am* 2002; 10(2): 363–388, vii.
9. Connell DA, Koulouris G, Thorn DA, Potter HG. Contrast-enhanced MR angiography of the hand. *RadioGraphics* 2002; 22(3): 583–599.
10. Choe YH, Han BK, Koh EM, Kim DK, Do YS, Lee WR. Takayasu's arteritis: Assessment of disease activity with contrast-enhanced MR imaging. *AJR Am J Roentgenol* 2000; 175(2): 505–511.
11. Herborn CU, Goyen M, Lauenstein TC, Debatin JF, Ruehm SG, Kroger K. Comprehensive time-resolved MRI of peripheral vascular malformations. *AJR Am J Roentgenol* 2003; 181(3): 729–735.
12. Shah DJ, Brown B, Kim RJ, Grizzard JD. Magnetic resonance evaluation of peripheral arterial disease. *Cardiol Clin* 2007; 25(1): 185–212.
13. Demondion X, Herbinet P, Van Sint Jan S. et al. Stenosis detection in forearm hemodialysis arteriovenous fistulae by multiphase contrast-enhanced magnetic resonance angiography: Preliminary experience. *J Magn Reson Imaging* 2003; 17(1): 54–64.
14. Zhang J, Hecht EM, Maldonado T, Lee VS. Time-resolved 3D MR angiography with parallel imaging for evaluation of hemodialysis fistulas and grafts: Initial experience. *AJR Am J Roentgenol* 2006; 186(5): 1436–1442.
15. Fayad L, Hazirolan T, Bluemke D, Mitchell S. Vascular malformations in the extremities: Emphasis on MR imaging features that guide treatment options. *Skeletal Radiol* 2006; 35(3): 127–137.
16. Norgren L, Hiatt WR, Dormandy JA. et al. Intersociety consensus for the management of peripheral arterial disease. *Int Angiol* 2007; 26(2): 81–157.
17. Cavagna E, D'Andrea P, Schiavon F, Tarroni G. Failing hemodialysis arteriovenous fistula and percutaneous treatment: Imaging with CT, MRI and digital subtraction angiography. *Cardovasc Intervent Radiol* 2000; 23: 262–265.
18. Chaudhry N, Salhab KF. Hand, upper extremity vascular injury. *Emedicine* 2003; http://www.emedicine.com/plastic/topic461.htm. Accessed August 3, 2006.
19. Prince MR, Narasimham DL, Stanley JC. et al. Breath-hold gadolinium-enhanced MR angiography of the abdominal aorta and its major branches. *Radiology* 1995; 197(3): 785–792.
20. Ichihashi S, Higashiura W, Itoh H, Sakaguchi S, Nishimine K, Kichikwa K. Long-term outcomes for systematic primary stent placement in complex iliac artery occlusive disease classified according to Trans-Atlantic Inter-Society Consensus (TASC)-II. *J Vasc Surg* 2011; 53(4): 992–999.
21. Ye W, Liu CW, Ricco JB, Mani K, Zeng R, Jiang J. Early and late outcomes of percutaneous treatment of Trans-Atlantic Inter-Society Consensus class C and D aorto-iliac lesions. *J Vasc Surg* 2011; 53(6): 1728–1737.
22. Sultan S, Hynes N. Five-year Irish trial of CLI patients with TASC II type C/D lesions undergoing subintimal angioplasty or bypass surgery based on plaque echolucency. *J Endovasc Ther* 2009; 16(3): 270–283.
23. Lee VS, Martin DJ, Krinsky GA, Rofsky NM. Gadolinium-enhanced MR angiography: Artifacts and pitfalls. *AJR Am J Roentgenol* 2000; 175(1): 197–205.

7

Magnetic Resonance Venography

Simon Gabriel, Chun Kit Shiu, and Stefan G. Ruehm

CONTENTS

7.1 Introduction

Magnetic resonance venography (MRV) has evolved into a clinically proven, reliable, and effective imaging modality for the evaluation of the venous system. Many of the techniques described in this chapter are generally used for imaging of both the arterial and venous systems.[1] Compared to arterial imaging, MRV tends to be less technically demanding due to slower blood flow and more homogenous flow profiles. The purpose of this chapter is to describe theory, techniques, and current applications of MRV in clinical practice.

7.2 Overview of MRV Techniques

The ability of MRI to depict blood flow coupled with its inherent exceptional soft tissue contrast has led to its implementation as a clinical problem-solving tool or a first-line diagnostic study at certain institutions.

MRV techniques can be broadly categorized into non-contrast-enhanced and contrast-enhanced techniques (Table 7.1). Sequences used for non-contrast imaging include spin-echo (SE)/fast spin-echo (FSE), conventional time-of-flight (TOF), phase contrast (PC), balanced steady-state free precession (bSSFP),

TABLE 7.1

Commonly used Techniques for Magnetic Resonance Venography

	Key Features	Benefits	Limitations
Non-Contrast Techniques			
TOF	Imaging of through-plane blood flow, optional saturation bands, GRE based	First line non-enhanced sequence, commonly used in brain	Long acquisition time, in-plane flow saturation artifacts
Phase contrast	Signal dependent on phase-shift of moving spins (blood)	Quantification of flow velocity including gradient across stenosis	Estimation of flow velocity for correct parameter setting (VENC), long acquisition time
T_1-3D GRE	Three-dimensional direct clot imaging	Non-contrast, direct clot visualization	Dependent on T_1-shortening of clot, limited accuracy
SWI	Frequency shift of paramagnetic deoxyhemoglobin	Signal not flow dependent	High field strength beneficial
SSFP	Signal intrinsic to blood	High signal-to-noise ratio, fast, robust, and lack of motion artifacts	Insensitive to flow and direction, susceptible to field inhomogeneity, prone to off-resonance artifacts, particularly at high field strengths (3 T)
Contrast-Enhanced Techniques			
CE 3D MRV	Three-dimensional, contrast dependent	Spatial resolution, versatility, robustness, and speed	Contraindication to contrast (requires IV injection)
Direct MRV	Direct downstream venous imaging	Selective venous display, low contrast volume, high vascular contrast	Limited display of venous territory, injection of affected venous system

TABLE 7.2

Generic *Non-Contrast* Sequences along with the Vendor-Specific Names

	GE Healthcare	Siemens Medical Solutions	Toshiba	Philips Medical Systems	Hitachi Medical Corporation
SE/FSE	SE/FSE	SE/TSE	SE/FSE	SE/TSE	SE/FSE
TOF	TOF	TOF	TOF	Inflow MRA	TOF
PC	PC	PC	PC	PC	VNEC
bSSFP	FIESTA	True FISP	True SSFP	Balanced FFE	BASG (balanced SARGE)
SWI	SWAN	SWI	Flow-sensitive black blood (FSBB)	Venous bold	BSI

SE, spin echo; FSE, fast spin echo; TOF, time of flight; PC, phase contrast; bSSFP, balanced steady-state free precession; SWI, susceptibility-weighted imaging.

TABLE 7.3

Generic *Contrast* Sequences along with the Vendor-Specific Names

	GE Healthcare	Siemens Medical Solutions	Toshiba American Medical Systems	Phillips Medical Systems	Hitachi Medical Corporation
Time-resolved MR angiography	TRICKS (time-resolved imaging of contrast kinetics)	TWIST (time-resolved angiography with interleaved stochastic trajectories)	Freeze frame/DRKS	4D TRAK (time-resolved angiography using keyhole)	TRAQ
Conventional timed angiography	Smart prep; fluoro triggered MRA	Care bolus	Visual prep	Bolus trak	FLUTE

susceptibility-weighted imaging (SWI), signal targeting using alternative radio frequency (RF) and flow-independent relaxation enhancement, and direct thrombus imaging (Table 7.2). The contrast-enhanced sequences include time-resolved MRV and conventional timed MRV (Table 7.3). Although there are several technique variations, MRI scanner manufacturer-specific options, MRI platforms, and contrast agents the fundamentals of MRV remain universal. Table 7.1

summarizes the various MRV techniques discussed in this chapter. Tables 7.2 and 7.3 summarize the vendor-specific names for the sequences typically used.

Compared to traditional non-contrast techniques, contrast-enhanced techniques are generally preferred due to superior image resolution, superior contrast-to-noise ratio, faster acquisition speed, and blood flow independence. At most institutions, non-contrast techniques or alternative modalities are typically used when there are contraindications to the use of intravenous gadolinium, for example, in patients with severely impaired renal function. At our institution, we favor Gd-based contrast agents with a various degree of albumin binding, which prolongs intravascular half-life and increases T_1-shortening, resulting in higher intravascular signal. Occasionally, we consider a newly resurgent class of non-gadolinium contrast agent known as *ultrasmall iron-oxide particles* (USPIO) in patients with contraindications to gadolinium administration.

7.3 Non-Contrast-Enhanced Sequences

7.3.1 Spin-Echo and Fast Spin-Echo

SE and FSE sequences are frequently used in cardiac imaging and are often referred to as *black-blood* imaging techniques, because the rapidly flowing blood produces a flow void, resulting in a black/hypointense appearance of the lumen. Pre-saturation pulses can be applied proximally to the vessel segment of interest to more effectively null the signal from moving spins while retaining signal from stationary tissue spins. Intravascular enhancement, during black-blood imaging, could be related to factors such as turbulent blood flow, prolonged duration of in-slice plane flow, or slow flow.[2] More robust sequence designs have been proposed to overcome some of the inherent technical limitations of traditional black-blood techniques.[3]

Conventional black-blood imaging is well suited for the evaluation of the tissues surrounding vessels and vessel wall morphology, especially when both non-fat saturation and fat-saturation images are acquired. However, in the modern imaging era, the traditional SE and FSE techniques have limited utility in the evaluation of intraluminal venous pathology.

7.3.2 Time-of-Flight

TOF MRV utilizes a short repetition time (TR) gradient-recalled-echo pulse sequence to accentuate flow-related enhancement. The signal of stationary tissue is suppressed, whereas inflowing blood outside the imaging slice is not exposed to the RF pulse and thus retains full longitudinal relaxation. Upon entering the slice, the longitudinal relaxation is tipped into the transverse plane by an excitation pulse, thereby producing high single intensity. Two- (2D) or three-dimensional (3D) TOF imaging is technically feasible. In 2D imaging, each thin section image is acquired as a result of sequential slice-by-slice excitations. In 3D imaging, a volume of tissue is excited and then later partitioned into thin sections. 2D TOF is ideal for venous imaging because it can detect slow flow, covers a large body area, and has sensitivity

to T_1 effects.[4] Disadvantages of 2D TOF imaging include limited resolution, saturation of in-plane flow, and potential respiratory artifacts.[4]

When blood flows within the image plane long enough to receive multiple RF excitations, the end result will be signal loss from spin saturation. This effect occurs in vessels that are oriented parallel to the image acquisition plane.

TOF imaging generally results in both arteries and veins appearing bright. However, signal from blood flow in a particular direction can be saturated by using flow pre-saturation bands (Figure 7.1), which nulls the specific inflow of blood into the region of interest. For venous MR imaging (MRI), the saturation band is placed proximally to the anatomical region being examined. As a consequence selective TOF venograms can be obtained.

With TOF imaging, blood flow signal is intensified by using longer TR, allowing an increased number of relaxed spins entering the imaging plane, which, however, results in longer acquisition times. The echo time (TE) should be short, a TE of 8 ms (on a 1.5 T magnet) has been proposed, so that the signal from water and fat are out of phase to enable the signal from fat to be reduced.

Selection of the appropriate flip angle is dependent on both slice orientation and axis of vessel under investigation. For longitudinal/in-plane flow a flip angle of 20°–25° tends to be optimal, whereas an angle of 45° should be chosen for imaging of transverse/through plane flow. It is important to note that decreasing the flip angle results in reduced in-plane flow saturation, and more RF pulses are needed to drive the longitudinal magnetization to equilibrium. Too large a flip angle may lead to saturation of the venous signal, whereas too small a flip angle may result in increased image noise.

To enhance the quality of maximum intensity projections (MIP) image reconstructions, partitions should be contiguous or even overlapping in order to reduce, for example, stair-step reconstruction artifacts (Figure 1d).

7.3.3 Phase Contrast MRV

PC imaging is based on phase changes that occur in moving protons. The application of a bipolar magnetic gradient results in moving spins acquiring a different phase (known as a *phase shift*) in relation to stationary spins. The bipolar gradient is generated by one gradient with positive polarity followed by a second gradient with negative polarity. The gradient with reversed polarity results in a phase shift of flowing blood, whereas stationary tissue experiences no net phase change. Image subtractions are typically used to minimize the signal from stationary spins, whereas the magnetization of moving spins is enhanced.[5] PC imaging results in a velocity map with the voxel intensity values proportional to the actual flow velocity in a particular

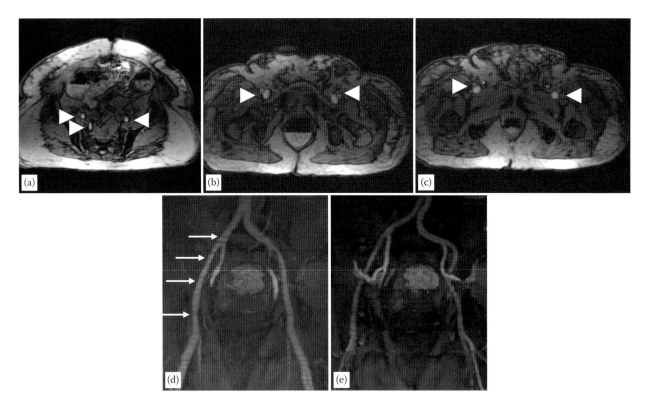

FIGURE 7.1
TOF: 2D TOF MRV of pelvic veins. Sequential slice-by-slice excitations of the pelvis with pre-saturation bands set proximally to nullify arterial flow signal resulted in only venous blood (arrowheads) appearing bright (a–c). Coronal MIP image reconstructed from the axial venous (d) and axial arterial (not shown) datasets (e). Please note that the horizontal bands (white arrows) represent spin saturation artifacts.

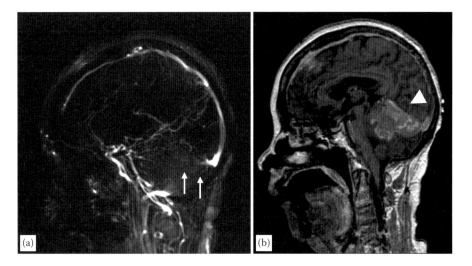

FIGURE 7.2
2D PC MRV. (VENC = 10 cm/s.) Summed magnitude velocity-encoded images generating the venogram at midline sagittal view (a). Note non-filling of the straight sinus (arrows) secondary to invasion by a meningioma (arrowhead) seen on the (b) contrast-enhanced MRI.

flow direction. The measured phase difference in each individual pixel with flow is directly related to the flow velocity along the direction of the first moment change and the velocity encoding time (ΔT). PC-MRV is traditionally performed as a 2D technique (Figure 7.2). Historically, 3D sequences were used for intracranial venous imaging; it has limited utility in modern imaging because of its suboptimal image quality compared to other acquisition techniques.

Intensity variations are dependent on the amount of phase shift: the brighter or the darker the signal, the larger the phase shift. The flow sensitivity can be adjusted by a variable known as the *velocity encoding value* (VENC), which helps determine the largest measurable velocity. The appropriate VENC value should be chosen to exceed the maximum expected velocity by about 25%. If the VENC is too low, aliasing or wrap-around artifacts occur. If the VENC is too high, sensitivity to slow blood flow and accuracy of quantitative analysis may be decreased. A VENC of less than 20 cm/s is typically recommended for venous imaging.[4] The quantification of blood flow is computed by software that often requires the operator to manually outline the vessel of interest for each individual phase. Outlined images should always be checked by the interpreting radiologist for accuracy. Free-breathing PC-MRV has been shown to be a noninvasive technique that can be used to measure hepatic blood flow with lower variability and higher reproducibility than Doppler ultrasound.[6]

7.3.4 Direct Thrombus Imaging

In contrast to most imaging techniques, which delineate thrombus as flow void or contrast filling defect, MR direct thrombus imaging uses a T_1-3D gradient-echo (GRE) sequence, which enables visualization of the thrombus against a suppressed background[7] (Figures 7.3b and 7.6f later in the chapter). During the process of thrombus formation, a predictable reduction in the T_1 value of the clot occurs reflecting the presence of methemoglobin, which possesses strong paramagnetic properties consequent upon the 3+ ferric state of iron. High signal intensity occurs initially at the periphery of the clot with extension toward the center overtime.

In addition to the signal generated by the thrombus itself, nulling the unclotted blood signal using an inversion recovery pulse can create further contrast of a clot against the surrounding blood. Background signal on the T_1-weighted image can be further suppressed through selective RF excitation of water molecules to reduce the fat signal. Magnetic resonance direct thrombus imaging has been shown to be accurate in the detection of acute deep venous thrombosis (DVT) in patient with clinically suspected DVT.[8] It has been proposed that since the T_1 signal normalizes over a period of about 6 months, this modality may be superior to compression ultrasound for the detection of recurrent DVT.[9]

7.3.5 Susceptibility-Weighted Imaging

SWI is a high-resolution 3D GRE sequence that combines magnitude and phase signal resulting in high sensitivity to tissue magnetic susceptibility differences.[10,11]

In SWI, hypointense signal can be produced with deoxyhemoglobin, hemosiderin, ferritin, and calcium. In veins, deoxyhemoglobin acts as a contrast agent causing T_2*-related signal intensity losses in the magnitude image and a shift in the phase image caused by susceptibility differences. In arteries, TOF effects and lack of T_2* effects result in hyperintense endoluminal signal.

Retrograde leptomeningeal venous drainage in dural arteriovenous fistulas is associated with increased morbidity and mortality. It has been shown that SWI hyperintensity within intracerebral venous structures accurately identified patients with retrograde leptomeningeal venous drainage in dural arteriovenous

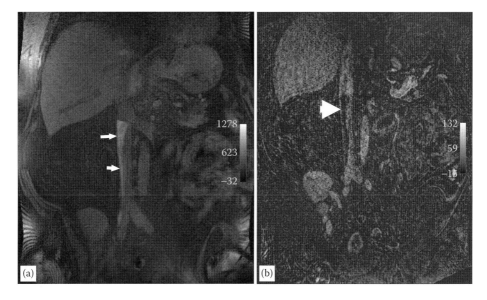

FIGURE 7.3
Coronal T_1 gradient-echo sequence (FLASH) with fat suppression (a) displayed a thrombus, which was bright (arrows), extending from IVC to right internal iliac vein. (b) Corresponding indirect CEMRV with subtraction demonstrated the thrombus as black filling defect/hypointensity (arrowhead).

fistulas in both pre- and posttreatment follow-up evaluations when compared to digital subtraction angiography.[12,13] Furthermore, SWI was also helpful in assessing posttreatment improvement in cerebral circulation as evidenced by decreased venous congestion.[13] Thus, SWI may provide an alternative to conventional angiography, which is widely regarded as the gold standard for diagnosis and assessment of RLVD in DAVFs (Figure 7.4).

Because of the sensitivity of 3D GRE sequence to breathing artifacts SWI is primarily reserved for intracranial imaging; however, recent research suggests that 2D SWI may as well be applicable for abdominal imaging.[14]

7.3.6 Balanced Steady-State Free Precession

bSSFP (FIESTA [General Electric]; True FISP [Siemens]) is a type of bright-blood imaging technique, which generates high signal within both arteries and veins. It is used extensively in cinematic cardiac imaging.

bSSFP is a rapid GRE pulse sequence in which the longitudinal and transverse components of the magnetization are maintained in a steady-state condition. In bSSFP, the net phase imparted is zero. The signal intensity produced is proportional to T_2/T_1 with long T_2 tissues and short T_1 tissues both resulting in hyper-intense signal. Datasets can be acquired as a 2D stack or in 3D with respiratory gating where applicable. The acquisition time for SSFP is considerably fast, particularly in comparison to other non-contrast MRV techniques such as TOF. bSSFP has been proven efficacious in imaging of the pulmonary veins[15] (Figure 7.5), thoracic central veins,[16] and portal veins.[17]

7.3.7 Signal Targeting Using Alternative RF and Flow-Independent Relaxation Enhancement

Signal targeting using alternative RF and flow-independent relaxation enhancement (STARFIRE) is a novel non-contrast technique for flow-independent magnetic resonance angiography (MRA) and MRV with homogenous fat and muscle background suppression.[18] STARFIRE technique produces hyper-intense images of both arteries and veins similar to bSSFP; however, it allows imaging of small peripheral veins and arteries by homogenous suppression of fat and muscle signal.

The technique employs fat suppression by a T_1-dependent mechanism; muscle suppression is achieved by a bSSFP pulse sequence with linear phase-encode

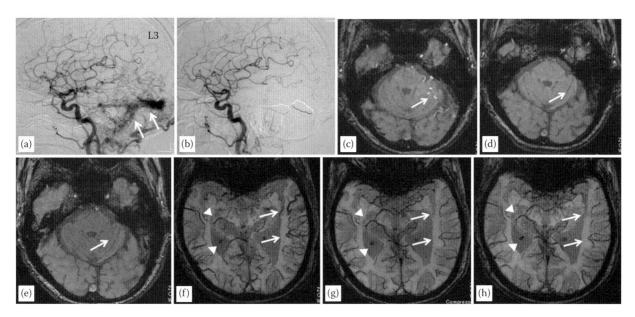

FIGURE 7.4
76-year-old woman with tinnitus. (a) preoperative lateral projection of left common carotid angiography showing Cognard type IIa + b fistula (arrows) affecting left transverse-sigmoid sinus with severe venous congestion in left temporal region. (b) Lateral projection of left common carotid angiography immediately after treatment showing complete obliteration of fistula. (c through e) Susceptibility-weighted MR images from before treatment (c) and 12 (e) months after treatment with transvenous embolization. Note venous hyperintensity present pretreatment (arrow in c), is absent at 3 and 12 months post treatment (arrows in d and e, respectively). (f through h) Susceptibility-weighted MR image venograms from before treatment (c) and 12 (e) months after treatment with transvenous embolization. Note venous congestion in left temporal lobe before treatment (arrow in f), normalized 12 months after treatment (arrow in h). In contrast, there was no change in venous caliber before and after treatment in right temporal lobe (arrowhead in f and h).

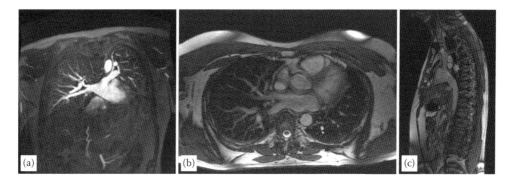

FIGURE 7.5
Conventional pulmonary venous return: MR pulmonary venography using contrast enhanced (a) and bSSFP (b and c) techniques. (a) The high-resolution 3D MRV clearly depicted pulmonary venous anatomy and allowed accurate measurement using multi-planar reconstruction. (b) Axial and (c) sagittal 2D bSSFP for the same patient. The right superior and inferior pulmonary veins are clearly noted.

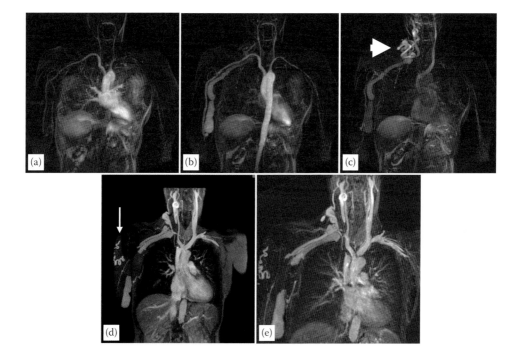

FIGURE 7.6
(a–c) Early, mid, and late time-resolved images demonstrate a right brachial artery to brachial vein fistula with venous outflow aneurysms and an occluded right innominate vein, inferior right internal jugular vein, and proximal right subclavian vein. The right external jugular vein gives rise to numerous regional collaterals (arrowhead). (d) High-resolution contrast-enhanced MRA 17.8 mm MIP confirms findings of occluded inferior right internal jugular vein, right innominate, and proximal right subclavian with right arm venous collaterals (arrow). (e) Subtraction image created by fusing a non-contrast mask and a contrast-enhanced steady-state image. Note the relatively decreased intensity of muscles and soft tissues (arrow) compared to (e).

ordering and large flip angle excitation.[18] To select enhance venous signal, a series of spatially selective pre-saturation RF pulses are applied every 50 ms to inflowing blood. This technique allows imaging of a large field of view and is less susceptible to regional magnetic field inhomogeneities. In a qualitative analysis with a small group of patients, STARFIRE and contrast-enhanced 3D MRA were found to have similar consistency in image quality and similar visualization of superficial venous structures; both were superior to 2D TOF.[18]

7.3.8 3D Contrast-Enhanced MRV Techniques

To overcome the relative limitations of lower resolution and lengthy acquisition times seen in non-contrast-enhanced pulse sequences, contrast-enhanced 3D contrast-enhanced MRV (CEMRV) has been advocated as the mainstay technique for the evaluation of the venous system (Figures 7.6 and 7.7).

CEMRV uses essentially the same techniques as contrast-enhanced MR arteriography, except that images

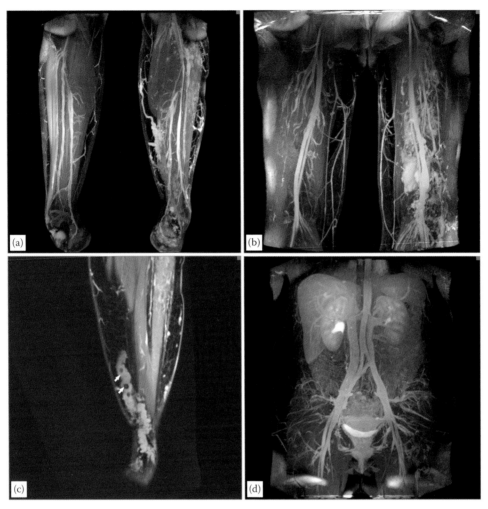

FIGURE 7.7
High spatial resolution 3D MRV following intravenous injection of 0.05 mmol/kg of (Ablavar) intravascular gadolinium in a patient with known Klippel Trenaunay syndrome. Extensive venous malformation is present extending from the foot to upper thigh of left lower extremity with muscle infiltration (a, b). Multiple varicosities are also present in the superficial subcutaneous fat of left leg, with several discrete thrombi identified (c) in the 3D fat-saturated GRE post (arrows). (d) Maximum intensity projection demonstrates a duplicated IVC

are acquired during an equilibrium or venous contrast phase. Alternatively, CEMRV can be performed using a time-resolved technique. The contrast agent can be administered using an indirect or a direct venous approach: For the indirect approach, contrast is administered via a nontarget peripheral vein with images acquired at early equilibrium phase. Usually a large concentrated dose of contrast is needed since significant dilution can occur before it reaches the venous territory under investigation. A major advantage of the indirect approach is that direct cannulation of a vein upstream of the affected extremity is not required.

Image subtraction is typically performed to improve the contrast-to-noise ratio of the opacified vasculature (Figure 7.6f). Image subtraction requires image acquisition before (mask) and after contrast administration. In the chest and abdomen, respiratory motion can potentially result in spatial mis-registration artifacts on the subtracted images.

Direct MRV requires diluted contrast agent to be continuously injected upstream to the vessel of interest. The direct technique ideally results in a full display of the deep and superficial venous system in a manner similar to conventional venography. Compared to the indirect approach, the direct injection technique results in superior regional contrast to noise values while requiring less contrast agent. A dilution factor of 1:10–20 (contrast: saline) is typically used to avoid marked hypointense intravascular signal from T_2-shortening effects. A pre-contrast mask is usually obtained followed by a 3D dataset during slow infusion of diluted contrast at a flow rate of approximately 1 mL/s. Repeated acquisitions can be performed, with dynamic maneuvers to evaluate for functional obstruction of veins (Figure 7.8).[19]

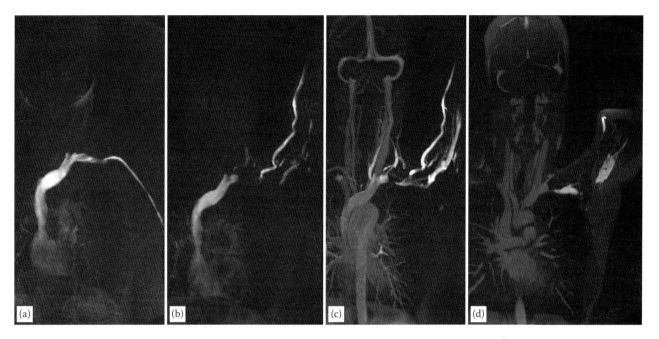

FIGURE 7.8

Direct time-resolved MRV: (a) Direct left upper extremity venous injection of 2 cc Ablavar with the arms in neutral position demonstrates no significant venous stenosis. (b) Reinjection of 2 cc Ablavar with the arms in abduction during early venous phase acquisition demonstrates high grade stenosis at the mid-subclavian vein with regional collaterals. (c) Equilibrium time-resolved images from the same injection demonstrate innumerable regional collaterals without filling defect or thrombus. (d) High-resolution equilibrium imaging confirms findings from direct time-resolved MRV.

In order to increase signal-to-noise ratio and resolution a surface coil should be used.

Acquisition parameters with very short TR and TE values and a flip angle of 30°–40° are typically recommended. Contrast should be administered continuously during image acquisition to avoid artifacts from fluctuating contrast concentrations during the acquisition of the center of k-space. Continuous injection also allows for filling of venous collaterals in the presence of a venous occlusion.

7.3.9 Time-Resolved MRV

Time-resolved MRV is a relatively new concept where rapid and dynamic image acquisitions of a prescribed volume provide information about both blood flow and anatomy. The accelerated image acquisition during contrast injection allows for high temporal resolution at the expense of spatial resolution. Time-resolved MRV benefits from parallel imaging, reordering of k-space acquisition, accelerated techniques of k-space filling, and novel image reconstruction algorithms. Images are acquired at roughly 1–6 seconds per frame in real time. This technique can be helpful for the evaluation of vascular malformations[20] (Figures 7.6, 7.9, and 7.10), venous reflux,[21] or complex congenital cardiovascular disease involving both venous and arterial vessels. Time-resolved imaging requires roughly 10-fold less contrast when compared to conventional CEMRV.

7.3.10 Contrast Agents

7.3.10.1 Extracellular Gadolinium Contrast Agents

Extracellular gadolinium contrast agents are most commonly employed for CEMRV and contrast-enhanced magnetic resonance angiography (CEMRA) studies. These extracellular agents have significantly more favorable safety profiles when compared to iodinated contrast agents. There are currently nine gadolinium-based contrast agents FDA approved for MRI, two of these agents have specific approval for MR angiography (Table 7.4). It should be noted that none of these agents are specifically approved for venous imaging, as such it is an off-label use. These contrast agents have a relatively rapid extravasation from the capillaries into the interstitium, which results in decreasing vascular enhancement over time and increasing background signal. Time-resolved MRV may help find the optimal phase for timing by reviewing the different phases of the dynamic dataset. The advantage of extracellular gadolinium contrast agents are the favorable safety profiles and decreased cost compared to blood pool agents.[22]

7.3.10.2 Blood-Pool Contrast Agents

Blood-pool contrast agents represent a new generation of contrast agents with longer intravascular distribution compared to extracellular contrast agents. Blood-pool contrast agents include a variety of compounds,

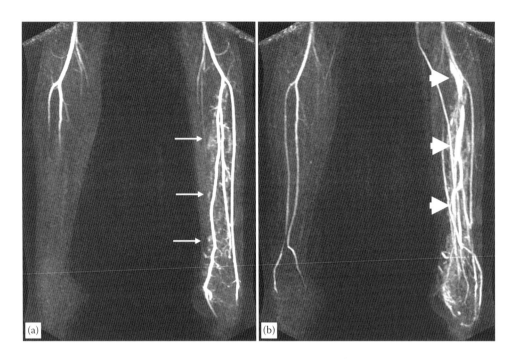

FIGURE 7.9
Dynamic contrast-enhanced MR angiography of legs in a patient with microfistulous vascular malformation using time-resolved angiography with interleaved stochastic trajectories (TWIST) after intravenous injection of 1mL of gadolinium contrast. In the time-resolved images, there is hyperdynamic circulation of left leg with early arterial enhancement (a). There were multiple microfistulous communications (arrows in a). In the calf, early venous draining with enlarged deep veins was present (arrowheads in b).

including ultrasmall supraparamagentic iron-oxide particles, gadolinium-based macromolecules, and gadolinium-based small molecules with strong reversible binding to plasma proteins.[22,23]

Gadofosveset trisodium (Ablavar®) is currently the only FDA-approved blood-pool contrast agent. Gadofosveset tridosidum is approved by the FDA for intravenous use in MR angiography to evaluate aortoiliac occlusive disease in adults with known or suspected peripheral vascular disease (FDA website, http://www.fda.gov/Drugs, accessed 2/15/2014). Thus, using gadofosveset trisodium for MRV is an off-label use. Nevertheless, our experience shows that this agent has the advantage of visualizing smaller vessels and vessels with slow or complex flow.[23] Gadofosveset trisodium is non-covalently bound to albumin and is excreted by the kidneys. The reversible albumin binding of gadofosveset trisodium allows imaging with significant lower gadolinium doses than conventional extracellular contrast agents. It also enhances the paramagenetic effectiveness of gadolinium.[23] Gadofosveset trisodium at 1.0 T exhibits a relaxivity that is 6–10 times that of gadopenteteate-diethylene triamine pentaacetic acid (DTPA).[24] This contrast agent can be used for first pass and steady-state imaging.

7.3.10.3 Pitfalls and Technical Limitations

Imaging of the venous anatomy may be challenging when collateral veins need to be included in the field of view, for example, in the presence of post-thrombotic changes. Data acquisition with thicker prescribed 3D volumes may therefore be necessary.

Conventional clot imaging utilizing CEMRV relies on the depiction of dark intraluminal defects reflecting thrombus. On MIP images the bright signal of blood surrounding the thrombus can potentially mask a central filling defect or thrombus. On TOF MRV, reduced intravascular intensity due to saturation effects may be interpreted as thrombus on MIP reconstructions if the vein enters the acquisition slice obliquely. Therefore, it is always mandatory to scrutinize the source images rather than to simply rely on MIP reconstructions when assessing the venous system for the presence of clot.

During image acquisition, some patients may show accentuated respiratory variations in venous flow. If flow sensitive TOF MRV is used, the vein may appear dark if the central part of the acquisition is obtained while the venous flow is substantially reduced during exhalation. In this circumstance, data acquisition should be performed during breath-hold.

In conventional CEMRV, in scenarios with complete venous occlusion, the time interval might not be optimal to demonstrate collateral veins and reconstitution of the vein distal to the occlusion. To overcome this pitfall, multiple datasets should be obtained during and shortly after infusion of the contrast. If the acquired images are inadequate, longer infusion duration in conjunction with a larger total contrast volume should be

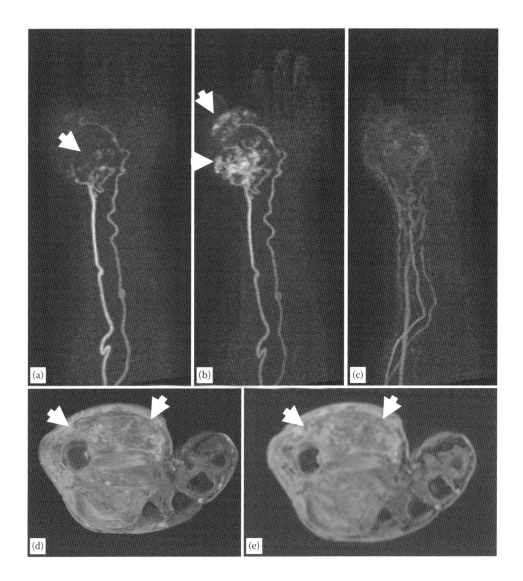

FIGURE 7.10

(a, b) Early time-resolved images demonstrate supply of large vascular malformation (arrowheads) by radial and ulnar artery. (c) Slightly delayed time-resolved image demonstrates drainage mainly via radial veins (d) T_1 FLASH 2D axial image and (e) postcontrast T_1 VIBE with fat saturation clearly demonstrate large arteriovenous malformation intimately associated with thenar and left lateral hand muscles and soft tissues (arrowheads).

used. During indirect conventional timed MRV, the 3D acquisition needs to be acquired so that the acquisition of the center of k-space coincides with the venous phase of the contrast agent bolus. If timing is suboptimal, edge or ringing artifacts may occur.

Depending on the composition, intravascular metallic devices or leads (i.e., stents, vena cava filters, or pacemaker leads) may produce regional susceptibility artifact and limit visualization of the vessel lumen on most MRI sequences (Figures 7.11 and 7.12). In scenarios where

TABLE 7.4

FDA-Approved Contrast Agents for MR Angiography

Agent Name	Trade Name	Company	C (molar) (mmol/mL)	MW (g/mol)	Approved Indication	Primary Excretion	Distribution
Gadobenate dimeglumine	MultiHance®	Bracco	0.5	1058	Renal/aorto-ilio-femoral occlusive disease	Renal	Extracellular
Gadofosveset trisodium	Ablavar	Bayer health care	0.25	957	Aorto-iliac occlusive disease	Renal	Blood-pool agent/ intravascular

FDA approved nine gadolinium-based contrast agents (GBCAs): (1) gadoterate meglumine (Dotarem®), (2) gadodiamide (Omniscan), (3) gadobenate (MultiHance), (4) gadopentetate (Magnevist®), (5) gadoteridol (ProHance®), (6) gadofosveset (Ablavar, formerly Vasovist), (7) gadoversetamide (OptiMARK™), (8) gadoxetate disodium (Eovist®), (9) gadobutrol (Gadavist®).

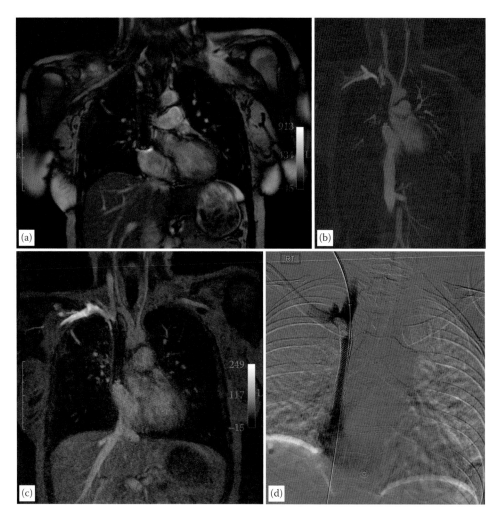

FIGURE 7.11
SVC stent: A 37-year-old female with history of mediastinal fibrosis and previous SVC narrowing and subsequent stent placement (a) true FISP coronal acquisition demonstrates significant regional susceptibility artifact from an SVC stent (b) coronal 10 mm MIP demonstrating regional susceptibility from the stent within the SVC and right innominate vein (c) source CEMRA demonstrating superiorly patent stent (d) DSA showing a widely patent right innominate vein and SVC stent.

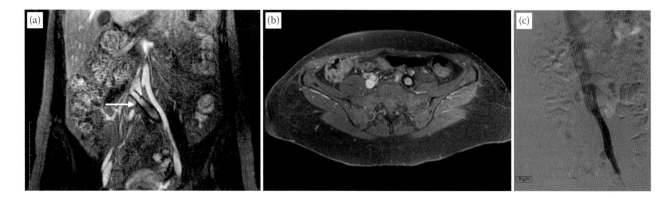

FIGURE 7.12
External iliac vein nitinol stent: (a) endovascular stent within the left common iliac vein, extending proximally into the IVC (arrow) and distally as far as the left external iliac vein, only minimal susceptibility artifact is seen from the nitinol stent and the lumen is widely patent. (b) T_1 fat saturation vibe postcontrast with a widely patent left external iliac vein (c) that was confirmed with digital subtraction venography. Nitinol is a metal alloy consisting of nickel and titanium and is known to have more favorable MRI characteristics.

the high-resolution contrast-enhanced images are non-diagnostic, it is often helpful to refer to the time-resolved images to evaluate for stent patency. With the evaluation for the presence or absence of regional venous collaterals and the temporal sequence of contrast flowing proximal and distal to the stent, one can infer stent patency. However, in most circumstances, endothelialization and stent luminal diameter quantification are significantly limited. As an alternative, single or double extremity direct MRV may provide more diagnostic images. In these predicaments, computed tomography venography or duplex and color ultrasound may be more helpful.

7.4 Clinical Applications

MRV has numerous clinical applications and may be considered a first-line study or complimentary to ultrasound or computed tomography. In Section 7.4.1, the relevant anatomy, imaging techniques, and clinical applications are discussed.

7.4.1 Upper Extremity and Thoracic Systemic Veins

7.4.1.1 Anatomy

The superior vena cava (SVC) is a major systemic vein that is short in length (roughly 7 cm) and wide in diameter; it returns deoxygenated blood from all structures superior to the diaphragm (except the lungs and heart) to the right atrium. The SVC is formed by the confluence of the right and left brachiocephalic veins (also known as innominate veins). The proximal portion of the SVC is found in the right side of the superior mediastinum at the level of the right first costal cartilage. The inferior portion of the SVC is ensheathed by pericardium. Posteriorly, at the level of the second costal cartilage just before being ensheathed by pericardium, the azygos vein runs over the root of the right lung to merge with the posterior aspect of the SVC. The SVC ends at the superior vena caval orifice, where it is in continuity with the right atrium at the level of the third right costal cartilage in the middle mediastinum.

The SVC and two brachiocephalic veins are the major systemic veins found in the superior mediastinum. They receive venous blood from both the subclavian and internal jugular veins, thus providing venous return from both upper extremities and the head and neck.

The left brachiocephalic vein is over twice as long as the right brachiocephalic vein because it passes from left to right side, passing anterior to the roots of the proximal aortic branches to form the SVC. At the origin of the right brachiocephalic vein, where the right internal jugular vein and right subclavian converge, the right lymphatic duct provides afferent lymph drainage. Similarly, at the origin of the left brachiocephalic vein, where the left internal jugular vein and left subclavian veins converge, the thoracic duct provides afferent lymph drainage.

The subclavian vein extends to the first rib where distally it is named the axillary vein. The axillary vein continues distally to form the basilic vein of the arm. The cephalic vein belongs to the superficial venous system and merges with the axillary vein proximally before it continues as the subclavian vein.

7.4.1.2 Imaging Techniques

CEMRV is well suited for this anatomic region. The indirect approach is the most commonly employed technique. It can be performed following a single injection of contrast in an antecubital vein or in combination with the acquisition of the 3D dataset in the equilibrium contrast phase (Figures 7.6 and 7.13). The indirect approach is especially useful in patients where information on both arterial and venous vessels are needed (e.g., anastomosis of a dialysis fistula needs to be displayed).

The less commonly employed direct approach can have advantages in patients with impaired renal function (since dose of contrast is reduced), dialysis shunts, fistulas or long-term central catheter placements. In order to simultaneously visualize the bilateral axillary, subclavian and brachiocephalic veins, and the SVC, simultaneous injection into the right and left upper extremities must be performed. Utilizing two operators, each with two syringes filled with diluted contrast agent, the injection is coordinated. The objective is that both operators complete the injection of the first syringe at roughly the same time. Acquisition of the imaging data should be initiated as soon as half of the contrast volume of the more delayed syringe has been injected.

Because of the variable course of the chest and upper extremity veins 2D TOF, MRV image data must be acquired in variable planes in order to ensure that the scan plane is oriented perpendicular to the veins. This means that a majority of the acquisition will be in the axial and for this specific reason, 2D TOF acquisition can be time consuming.

7.4.1.3 Clinical Applications

MRV may be performed to investigate SVC syndrome, Paget–Schroetter syndrome, deep thrombus of central veins, mediastinal abnormalities with potential vascular involvement, and evaluation of anatomical variants (Figures 7.8 and 7.14). MRV can also be employed to monitor short- or long-term therapeutic outcome in patients with DVT. Both non-contrast and contrast-enhanced MRV techniques are of value for the work up of the above-mentioned clinical scenarios.

SVC syndrome is characterized by any condition that obstructs blood flow through the SVC. The SVC obstruction can result from narrowing or blockage as a result of

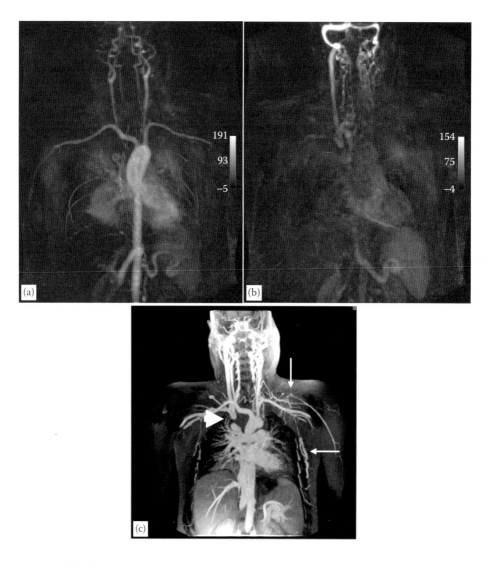

FIGURE 7.13

Indirect contrast-enhanced MRV for the central thoracic veins. Dynamic CEMRA using TWIST for timing was performed following admin-istration of 1 mL of intravascular gadolinium contrast agent. Coronal datasets were acquired in (a) arterial and (b) venous phases. There was non-opacification of left internal jugular vein as shown in (b). (c) High spatial resolution 3D MRV acquired following intravenous injection of an additional 10 mL of intravascular gadolinium. Both the arterial and venous systems were opacified. Left internal jugular vein, left innomi-nate vein, left axillary vein, and left subclavian vein were occluded. There were extensive venous collaterals (arrows) around left shoulder and left upper chest. The right innominate vein was also occluded (arrowhead).

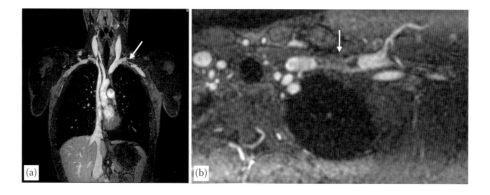

FIGURE 7.14

Reconstructed coronal (a) and oblique axial (b) images from 3D MRV at thoracic outlet shows a near occlusive thrombus (arrow) within the left subclavian vein without venous collaterals consistent with Paget–Schroetter syndrome.

thrombosis or external compression. SVC syndrome is characterized by edema, venous congestion, and swelling of the head, neck, and arm. In the 1980s, malignancy was by far the most common cause of SVC syndrome, resulting in extrinsic compression or invasion of the SVC and accounting for approximately 85% of SVC syndrome cases.[25] Currently, with the rise of intravascular treatment devices, malignancy still accounts for a majority of 60% of patients, while 40% are due to benign processes.[26] Benign etiologies include mediastinal fibrosis and thrombosis secondary to central venous catheters or trans-venous pacing wires.

Obstruction of the subclavian vein may occur as part of a thoracic outlet syndrome (TOS). TOS results from a compression of the subclavian artery, subclavian vein, brachial plexus, or a combination of the above in the cervicothoracic region. The compression may be due to a cervical or anomalous first rib, long C7 transverse process, muscle anomalies, short scalene muscles, exostosis, and vessels passing through the substance of muscles.[27] To exclude vascular TOS, we perform time-resolved MRA and high-resolution contrast-enhanced MRV in both abduction and neutral positions with Ablavar (Figure 7.8a, b).

Persistent left SVC (PLSVC) is the most common congenital thoracic venous anomaly and has a prevalence of 0.3%–0.5%.[28] In most cases, both the left SVC and right SVC are both present. About 40% patients with PLSVC have associated cardiac anomalies.[28] The left SVC commonly drains into the right atrium via the coronary sinus (Figure 7.15); less frequently, the left SVC drains into the left atrium resulting in a right to left shunt.

7.4.2 Inferior Vena Cava and Lower Extremity

7.4.2.1 Anatomy

The inferior vena cava (IVC) begins anterior to the L5 vertebral body level and is formed by the union of both iliac veins. The IVC is located 2.5 cm to the right of the median plane and inferior to the bifurcation of the aorta and posterior to the proximal part of the right common iliac artery. The IVC ascends on the right of L3–L5 vertebral bodies of vertebrae and anterior to the right psoas major. The IVC exits the abdomen by passing through the caval foramen in the diaphragm at the T8 vertebral body level. After entering the thorax, it connects to the right atrium. The abdominal part of the IVC collects poorly oxygenated blood from the lower limbs and nonportal blood from the abdomen and pelvis.

Anomalies of the IVC occur in less than 1% of the population, although the incidence is slightly higher in patients with other cardiovascular anomalies.[29] IVC anomalies include left sided IVC, double IVC (Figure 7.7d), azygous continuation of the IVC (Figure 7.16), circumaortic left

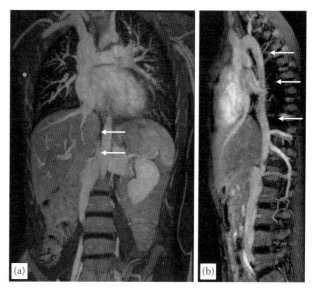

FIGURE 7.16
3D MRV of thorax and abdomen. Coronal image (a) shows interrupted hepatic IVC (arrows) with supra-hepatic reconstitution draining into right atrium (arrowhead). Sagittal image (b) demonstrates azygous continuation of infrahepatic IVC (arrows) draining into SVC. Note also the enlarged intercostal veins (asterisks).

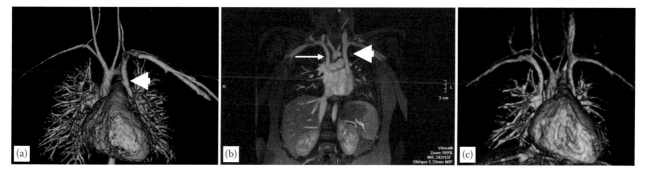

FIGURE 7.15
Left side SVC: (a) 3D volume rendered image during early arterial phase demonstrating a left-sided SVC from a left upper extremity venous injection (arrowhead). (b) oblique coronal MIP demonstrating the continuation of the LSVC to the coronary sinus and right atrium (arrowhead), right SVC is also seen in this later phase steady-state image (arrow) (c) 3D volume rendered image during steady-state demonstrating right SVC and persistent, left-sided SVC.

renal vein, absence of the infrarenal IVC, and absence of the entire IVC. A left-sided IVC is the commonest of these anomalies occurring with a prevalence of 0.2%–0.5%.[30] In these patients the left-sided IVC terminates in the left renal vein, which then crosses anterior to the aorta uniting with the right renal vein to form a normal right sided pre-renal IVC.

The lower extremity veins can be divided into a superficial and deep system. The deep veins usually follow the same course as the named main arteries.

In the lower extremity, the deep venous system includes the femoral and deep femoral veins, popliteal vein, and anterior tibial, posterior tibial, and peroneal veins. The veins are commonly paired at the calf level.

The greater saphenous vein is the longest vein in the human body. It courses from the dorsal arch of the foot medial to the tibia, up the medial thigh to connect to the femoral vein. The lesser saphenous vein courses from the lateral arch of the foot posterolaterally in the calf to connect to the popliteal vein. The superficial venous system is an important collateral venous pathway in the event of DVT.

7.4.2.2 Imaging Techniques

Early after the introduction of GRE techniques, TOF MRV evolved as the primary MR technique for the evaluation of DVT in the pelvic and lower extremity veins.[31–33] TOF venography can adequately evaluate DVT in femoral and calf veins; however, its utility in the pelvis is limited due to inherent disadvantages described previously, for example, orientation of veins relative to the acquisition plane and potential respiratory motion artifacts. Similarly, TOF MRV is not ideal for assessing varicose veins or complex post-thrombotic changes.

Contrast-enhanced 3D MRV overcomes inherent limitations of TOF venography. Conventional CEMRA has the advantage of ultra-fast high-resolution acquisitions, which can be obtained without breath hold. In addition,

direct MRV with injection of diluted paramagnetic contrast agent allows visualization of all vessels, regardless of the underlying flow characteristics and the orientation of the vessel. Small perforating and superficial veins containing slow or even retrograde flowing blood are fully depicted.

Overall, indirect CEMRV is widely regarded the standard imaging technique due to significant advantages in terms of spatial resolution, image reconstruction, and data acquisition times.

7.4.2.3 Clinical Applications

MRV of the IVC and lower extremities may be performed for evaluation of obstruction secondary to intrinsic or extrinsic processes (Figure 7.17), assessment of congenital anatomic variations, or part of a preoperative or preinterventional review of venous anatomy and dimensions.

Hemodynamically significant thrombosis of the IVC or pelvic veins usually results in the development of multiple collateral vessels. In unilateral iliac vein thrombosis (Figure 7.18) or occlusion, blood may drain to the contralateral side by collaterals via the sacral, rectal, uterine, vesical, or prostatic plexus. Complete thrombosis or occlusion of the IVC leads to drainage of blood via the hemiazygous and azygous systems along with collateral veins, which include abdominal epigastric veins, thoracic epigastric, or vertebral venous plexus that finally draining into the SVC.

Non-contrast MRV techniques, described earlier in Section 7.2, are able to demonstrate basic anatomic variations and pathologies of the IVC and pelvic and lower extremity veins. Although, conventional venography is widely regarded as the gold standard for the evaluation of DVT, duplex ultrasonography is now widely used at many institutions as a first-line imaging tool, particularly due to availability and low cost. It is noteworthy that the frequency of isolated DVT detected with MRV (Figure 7.19) was higher than with sonography or conventional ascending venography.[34]

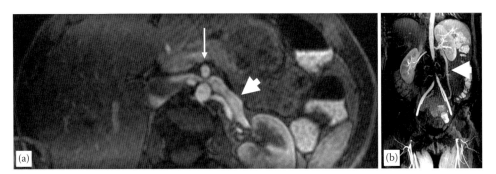

FIGURE 7.17

Nutcracker syndrome: (a) High-resolution 3D MRV of the abdomen in a patient with hematuria reveals extrinsic compression of the left renal vein by the superior mesenteric artery (arrow) with proximal left renal vein dilatation (arrow head). (b) Thick MIP image demonstrates enlarged left gonadal vein (arrowhead).

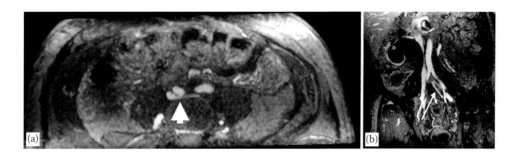

FIGURE 7.18
May–Thurner syndrome. MRV of IVC and pelvis. (a) Extrinsic high-grade compression of the proximal left common iliac vein (arrowhead) by the right common iliac artery, with (b) distal expansion and endoluminal thrombus (arrow) consistent with May–Thurner syndrome. The thrombus was undetected by duplex ultrasonography (not shown). Incidental right hydronephrosis (*).

MRV appears to offer advantages for the differentiation between bland versus tumor thrombi, particularly in the renal veins or IVC, due to its superior soft tissue contrast resolution and utilization of contrast (Figures 7.20 and 7.21). This enables better characterization of thrombus composition,[35] especially based on contrast enhancement characteristics: In general, tumor thrombus demonstrates various degrees of enhancement, whereas bland thrombus typically does not. Thus, MRV has an important role in the staging of renal cell carcinoma and is well suited for evaluating patients with suspected renal vein thrombosis.

To perform peripheral high-resolution MRV, direct MRV technique should be employed by injecting diluted contrast agent into a foot vein. Direct MRV is a robust and reliable technique for the accurate display of deep and superficial venous morphology, postthrombotic changes, and varicosities affecting the lower extremity.[19]

7.4.2.4 Portal, Hepatic, and Mesenteric Veins

MRI/MRV may be used to evaluate portal and hepatic vein anatomy, evaluate for thrombosis or occlusion

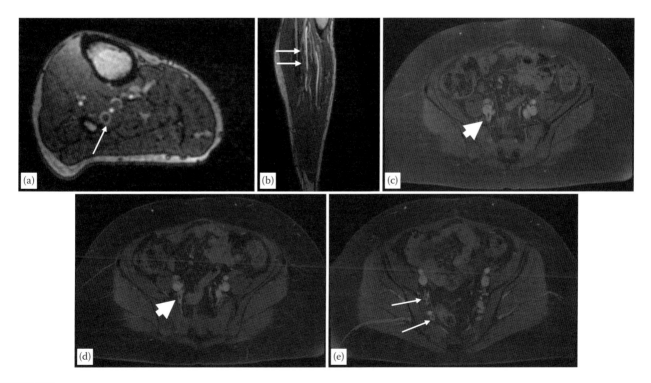

FIGURE 7.19
DVT. High spatial resolution 3D MRV with axial (a) and coronal (b) reconstruction. Intraluminal filling defects (arrows), compatible with thrombus, visualized in peroneal vein. (c–e) Axial T_1 FLASH 2D demonstrates thrombus in right internal iliac vein (arrowhead) and its distal branches (arrows). Compare to patent contralateral left internal iliac vein branches.

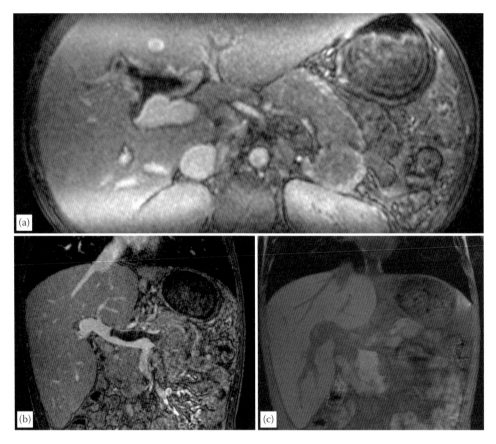

FIGURE 7.20
Bland portal vein thrombus: (a) axial CEMRA demonstrating thrombus within the right anterior, right posterior portal veins, and confluence. (b) Segment of non-occlusive thrombus within the extrahepatic main portal vein and splenic vein stump (spleen has been surgically removed). (c) T_1 FLASH 2D precontrast image demonstrates corresponding mild heterogeneous T_1 hyperintensity coinciding to the area of thrombosis on the CEMRA.

(Figures 7.20 and 7.22), and as part of a preoperative and/or postoperative evaluation for hepatic transplant patients. Applications also exist for the evaluation of regional tumor spread, of importance in pancreatic cancer staging, hepatic cancer staging, pre-liver transplant planning and post-transplant anastomotic evaluation (Figure 7.20). 3D CEMR portography has been shown to be as effective as digital subtraction angiography for the assessment of the portal vein patency and detection of thrombosis in patients with portal hypertension.[36]

7.4.2.5 Pulmonary Venous System

Conventional CEMRV is a proven modality for the evaluation of the pulmonary veins (Figure 7.5) prior to and following RF ablation in patient with atrial fibrillation. In most instances, additional cardiac MRI with delayed enhancement is performed to assess for myocardial scar. If absolute or relative contraindications to MRI or

Gd-based contrast agents exist, non-contrast MRV techniques should be considered.

As a non-contrast technique, bSSFP imaging provides excellent signal to contrast and signal-to-noise ratios. For the assessment of venous structures adjacent to the heart 3D SSFP techniques with cardiac gating are preferable. A study performed by Francois et al., correctly identified all pulmonary vein variants detected by CEMRA using bSSFP and there was no significant difference in the pulmonary vein ostial measurements between the two techniques.[37] In addition, bSSFP has the advantage of defining nonvascular structures. This includes defining the relationship and distance between the left atrium and esophagus in order to prevent a dreaded complication of RF ablation-atrioesophageal fistulas. Using 3D SSFP Francois et al. detected four hiatal hernias, two lung nodules/masses, one bronchogenic cyst, and two cases of lymphadenopathy, although none of these findings were elucidated with the CEMRA sequence.[37] The main disadvantage of this technique is longer acquisition time compared to CEMRA.

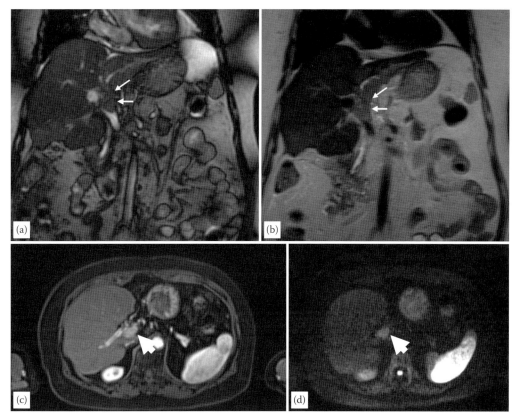

FIGURE 7.21

Coronal trufi (a) and T_2 haste (b) of the portal vein in a patient with resected hepatocellular carcinoma. There is a thrombus extending from the left portal vein to the main portal vein (arrows). The thrombus (arrowhead) demonstrates enhancement in CEMR (c) and restricted diffusion in diffusion-weighted image (d), consistent with tumor thrombus.

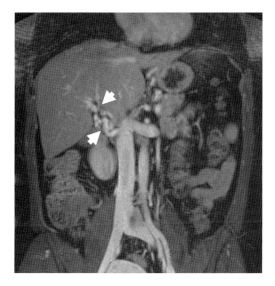

FIGURE 7.22

Contrast enhanced 3D MRV of abdomen in a patient with idiopathic portal vein thrombosis. There was complete occlusion of the extra-hepatic portal vein with extensive portal venous collaterals at porta hepatis (cavernous transformation, arrowheads).

7.4.2.6 Future Advancements

The use of traditional gadolinium-based contrast agents described previously are contraindicated in patients with stage 4 or 5 chronic kidney disease (glomerular filtration rate <30 cc/min/1.73 m²), acute kidney injury, and kidney liver transplant recipients with liver dysfunction, because of the risk nephrogenic systemic fibrosis (NSF).[38,39] Although NSF is not common, it can be severely debilitating and potentially fatal. Non-contrast MRV techniques described previously can provide an alternative in patients with contraindications to gadolinium; however, in the typical clinical scenario high-resolution 3D imaging with fast acquisition times and advanced tissue characterization based on enhancement characteristics are preferred.

Ultrasmall superparamagnetic iron-oxide particles (USPIO) may serve as potential alternative to traditional gadolinium-based contrast compounds in patients with compromised renal function.[40] Of the numerous iron oxide nanoparticles available, ferumoxytol appears to be an appropriate and safe alternative to Gd, and has been proposed as a vascular imaging contrast agent in patients with compromised renal function.[40] Ferumoxytol (Feraheme™; AMAG Pharmaceuticals, Inc., Cambridge, MA) is an intravenous iron preparation consisting of carbohydrate-coated superparamagnetic iron oxide nanoparticles, which was approved by the U.S. Food and Drug Administration in 2009 for IV iron replacement therapy for anemia of chronic kidney disease. It has a high T_1 relaxivity and is not excreted by the kidneys. For the purposes of vascular imaging, ferumoxytol can be injected as a blood pool agent, where it can be administered as a bolus for MRA and dynamic MR. In addition, this agent has a long plasma half-life of about 15 hours, allowing a large temporal window for high-resolution steady-state venous imaging (Figures 7.23 and 7.24).

Another possible future tool in the vast arsenal of MRV may involve clot imaging with targeted contrast agents. These typically gadolinium-based agents can potentially supplement non-contrast MR techniques or serve as alternatives to CEMRV techniques. Flacke et al. described a ligand-directed, lipid-encapsulated liquid perfluorocarbon nanoparticle that had high avidity, prolonged half-life, and carried high Gd-DTPA payloads for high detection sensitivity of fibrin.[41] At field strength of 1.5 T, the agent allowed targeted detection of clots *in vitro* and *in vivo* with detection of intravascular clots as small as 500 nm. Despite early promising reports of targeted clot agents more than a decade ago, none of the agents has passed an experimental preclinical stage to date.

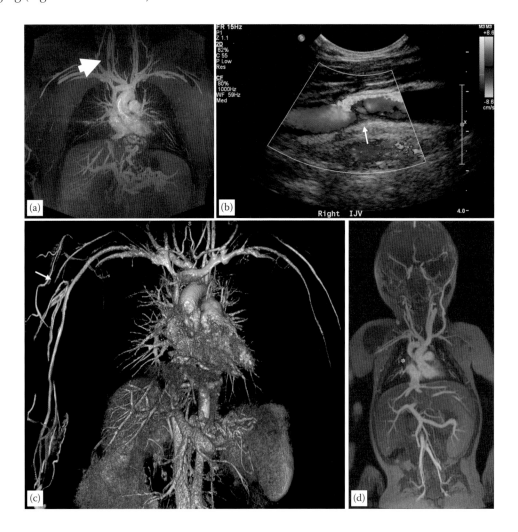

FIGURE 7.23
a) 46 mm subvolume MIP obtained post 15 mL ferumoxytol administered in a 39-year-old patient with end stage renal failure. Patient has moderate right inferior IJ stenosis (arrowhead), caused by a (b) web as demonstrated on color Doppler ultrasound (arrow). Right cephalic vein is also occluded in its mid-portion (arrow), as visualized on volume rendered 3D reconstruction (c). (d) Whole body 3D MRV at steady state in a 3-year-old patient with end stage renal failure. A total dose of 1.6 cc ferumoxytol was administered. Note the occlusion of right internal jugular vein and SVC at cavoatrial junction (asterisk).

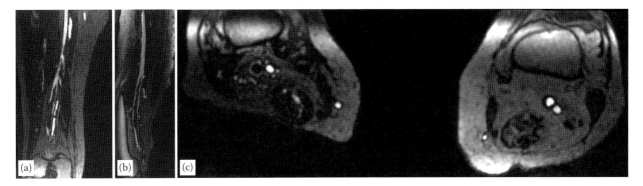

FIGURE 7.24

FH2: A 93-year-old female with impaired renal function and gangrene of the fourth right toe. Vascular imaging was performed with 6.5 mL ferumoxytol (Feraheme, AMAG Pharmaceutricals, Waltham, MA). (a–c) Coronal, sagittal, and axial steady-state reconstructions of the right thigh demonstrate occlusive thrombus within the right distal femoral vein and popliteal vein with regional venous collaterals.

References

1. Spitzer, CE (2009) Progress in MR imaging of the venous system. *Perspect Vasc Surg Endovasc Ther* 21(2):105–116
2. Bradley WG Jr, Waluch V, Lai KS et al. (1984) The appearance of rapidly flowing blood on magnetic resonance images. *AJR Am J Roentgenol* 143(6):1167–1174
3. Liu CY, Bley TA, Wieben O et al. (2010) Flow-independent T2-prepared inversion recovery black-blood MR imaging. *J Magn Reson Imaging* 31(1):248–254
4. Butty S, Hagspiel KD, Leung DA et al. (2002) Body MR venography. *Radiol Clin North Am* 40(4):899–919
5. Constantinesco A, Mallet JJ, Bonmartin A et al. (1984) Spatial or flow velocity phase encoding gradients in NMR imaging. *Magn Reson Imaging* 2(4):335–340
6. Yzet T, Bouzerar R, Allart JD et al. (2010) Hepatic vascular flow measurements by phase contrast MRI and Doppler echography: A comparative and reproducibility study. *J Magn Reson Imaging* 31(3):579–588
7. Kelly J, Hunt BJ, Moody A et al. (2003) Magnetic resonance direct thrombus imaging: A novel technique for imaging venous thromboemboli. *Thromb Haemost* 89(5):773–782
8. Fraser DG, Moody AR, Morgan PS et al. (2002) Diagnosis of lower-limb deep venous thrombosis: A prospective blinded study of magnetic resonance direct thrombus imaging. *Ann Intern Med* 136(2):89–98
9. Westerbeek RE, Van Rooden CJ, Tan M et al. (2008) Magnetic resonance direct thrombus imaging of the evolution of acute deep vein thrombosis of the leg. *J Thromb Haemost* 6(7):1087–1092
10. Haacke EM, Mittal S, Wu Z et al. (2009) Susceptibility-weighted imaging: Technical aspects and clinical applications, part 1. *AJNR Am J Neuroradiol* 30(1):19–30
11. Mittal S, Wu Z, Neelavalli J, Haacke EM (2009) Susceptibility-weighted imaging: Technical aspects and clinical applications, part 2. *AJNR Am J Neuroradiol* 30(2):232–252
12. Letourneau-Guillon L, Krings T (2012) Simultaneous arteriovenous shunting and venous congestion identification in dural arteriovenous fistulas using susceptibility-weighted imaging: Initial experience. *AJNR Am J Neuroradiol* 33:301–307
13. Nakagawa I, Taoka T, Wada T (2013) The use of susceptibility-weighted imaging as an indicator of retrograde leptomeningeal venous drainage and venous congestion with dural arteriovenous fistula: Diagnosis and follow-up after treatment. *Neurosurgery* 72:47–55
14. Dai Y, Zeng M, Li R et al. (2011) Improving detection of siderotic nodules in cirrhotic liver with a multi-breath-hold susceptibility-weighted imaging technique. *J Magn Reson Imaging* 34(2):318–325
15. Krishnam MS, Tomasian A, Malik S et al. (2009) Three-dimensional imaging of pulmonary veins by a novel steady-state free-precession magnetic resonance angiography technique without the use of intravenous contrast agent: Initial experience. *Invest Radiol* 44(8):447–453
16. Tomasian A, Lohan DG, Laub G et al. (2008) Noncontrast 3D steady state free precession magnetic resonance angiography of the thoracic central veins using nonselective radiofrequency excitation over a large field of view: Initial experience. *Invest Radiol* 43(5):306–313
17. Wilson MW, LaBerge JM, Kerlan RK et al. (2002) MR portal venography: Preliminary results of fast acquisition without contrast material or breath holding. *Acad Radiol* 9(10):1179–1184
18. Edelman RR, Koktzoglou I (2009) Unenhanced flow-independent MR venography by using signal targeting alternative radiofrequency and flow-independent relaxation enhancement. *Radiology* 250(1):236–245
19. Ruehm SG, Zimny K, Debatin JF (2001) Direct contrast-enhanced 3D MR venography. *Eur Radiol* 11(1):102–112
20. Herborn CU, Goyen M, Lauenstein TC (2003) Comprehensive time-resolved MRI of peripheral vascular malformations. *AJR Am J Roentgenol* 181(3):729–735

21. Dick EA, Burnett C, Anstee A (2010) Time-resolved imaging of contrast kinetics three-dimensional (3D) magnetic resonance venography in patients with pelvic congestion syndrome. *Br J Radiol* 83(994):882–887

22. Bellin MF, Van Der Molen AJ (2008) Extracellular gadolinium-based contrast media: An overview. *Eur J Radiol* 66(2):160–167

23. Bremerich J, Bilecen D, Reimer P (2007) MR angiography with blood pool contrast agents. *Eur Radiol* 17(12):3017–3024

24. Lauffer RB, Parmelee DJ, Dunham SU (1998) MS-325: Albumin-targeted contrast agent for MR angiography. *Radiology* 207(2):529–38

25. Schraufnagel DE, Hill R, Leech JA et al. (1981) Superior vena caval obstruction: Is it a medical emergency? *Am J Med* 70:1169–1174

26. Rice TW, Rodriguez RM, Light RW (2006) The superior vena cava syndrome: Clinical characteristics and evolving etiology. *Medicine (Baltimore)* 85(1):37–42

27. Aralasmak A, Cevikol C, Karaali K et al. (2012) MRI findings in thoracic outlet syndrome. *Skeletal Radiol* 41(11):1365–1374

28. Goyal SK, Punnam SR, Verma G et al. (2008) Persistent left superior vena cava: A case report and review of literature. *Cardiovasc Ultrasound* 10;6:50

29. Gayer G, Luboshitz J, Hertz M et al. (2003) Congenital anomalies of the inferior vena cava revealed on CT in patients with deep vein thrombosis. *AJR Am J Roentgenol* 180(3):729–732

30. Bass JE, Redwine MD, Kramer LA (2010) Spectrum of congenital anomalies of the inferior vena cava: cross-sectional imaging findings. *RadioGraphics* 20(3):639–652

31. Lanzer P, Gross GM, Keller FS et al. (1991) Sequential 2D inflow venography: Initial clinical observations. *Magn Reson Med* 19:470–476

32. Spritzer CE, Sostman HD, Wilkes DC et al. (1990) Deep venous thrombosis: Experience with gradient echo MR imaging in 66 patients. *Radiology* 177:235–241

33. Erdman WA, Weinreb JC, Cohen JM et al. (1986) Venous thrombosis: Clinical and experimental MR imaging. *Radiology* 161:233–238

34. Spritzer CE, Arata MA, Freed KS (2001) Isolated pelvic deep venous thrombosis: Relative frequency as detected with MR imaging. *Radiology* 219(2):521–525

35. Engelbrecht M, Akin O, Dixit D et al. (2011) Bland and tumor thrombi in abdominal malignancies: Magnetic resonance imaging assessment in a large oncologic patient population. *Abdom Imaging* 36(1):62–68

36. Kreft B, Strunk H, Flacke S et al. (2000) Detection of thrombosis in the portal venous system: Comparison of contrast-enhanced MR angiography with intra-arterial digital subtraction angiography. *Radiology* 216(1):86–92

37. François CJ, Tuite D, Deshpande V et al. (2009) Pulmonary vein imaging with unenhanced three-dimensional balanced steady-state free precession MR angiography: Initial clinical evaluation. *Radiology* 250(3):932–939

38. Cowper SE, Robin HS, Steinberg SM et al. (2000) Scleromyxoedema-like cutaneous diseases in renal-dialysis patients. *Lancet.* 356(9234):1000–1001

39. Rydahl C, Thomsen HS, Marckmann P et al. (2008) High prevalence of nephrogenic systemic fibrosis in chronic renal failure patients exposed to gadodiamide, a gadolinium-containing magnetic resonance contrast agent. *Invest Radiol.* 43(2):141–144

40. Neuwelt EA, Hamilton BE, Varallyay CG et al. (2009) Ultrasmall superparamagnetic iron oxides (USPIOs): A future alternative magnetic resonance (MR) contrast agent for patients at risk for nephrogenic systemic fibrosis (NSF)? *Kidney In* 75:465–474.

41. Flacke S, Fischer S, Scott MJ et al. (2001) Novel MRI contrast agent for molecular imaging of fibrin: Implication for detecting vulnerable plaques. *Circulation* 104:1280–1285

8

Vascular Malformations

Michele Anzidei, Beatrice Sacconi, Beatrice Cavallo Marincola, Pierleone Lucatelli,
Alessandro Napoli, Mario Bezzi, and Carlo Catalano

CONTENTS

8.1 Introduction

Vascular malformations comprise a heterogeneous spectrum of lesions that involve all parts of the body and can cause significant morbidity and even mortality in both adults and children. Confusion about terminology and imaging guidelines can be responsible for improper diagnosis and treatment [1], because therapeutic strategy depends on the type of malformation. Several classification systems have been proposed for vascular anomalies.

In 1982, Mulliken and Glowacki described the most widely accepted classification. This biological classification is based on cellular turnover, histologic features, natural history, and physical findings [2]. They classified vascular anomalies as either hemangiomas or vascular malformations. Hemangiomas are benign vascular tumors of infancy and childhood consisting of cellular proliferation and hyperplasia, characterized by a rapid early proliferative stage and a later involuting stage. Vascular malformations arise from dysplastic vascular channels and exhibit normal endothelial turnover, growing with the child without regression [2–4].

In 1993, Jackson et al. [5] proposed a radiologic classification of vascular malformations, based on flow dynamics; according to this classification, we can distinguish low- and high-flow malformations.

In 1996, these systems were adopted and expanded by the International Society for the Study of Vascular Anomalies [6]. Two categories of vascular anomalies are considered: vascular tumors (with infantile hemangioma being the most common) and vascular malformations.

Vascular malformations are subcategorized according to their flow dynamics as low-flow malformations (venous, lymphatic, capillary, capillary-venous, and capillary-lymphatic-venous) and high-flow malformations (arteriovenous malformations [AVMs] and arteriovenous fistulas [AVFs]). Any malformation with an arterial component is considered high flow, whereas those without an arterial component are considered low flow. The classification of vascular anomalies and their main clinical and magnetic resonance imaging (MRI) features are summarized in Table 8.1.

8.1.1 Classification

Vascular malformations are congenital anomalies, that is, present at birth, although not always evident. They usually grow proportionally with the child and show no regression. Their growth can be exacerbated due to

TABLE 8.1

Features of Vascular Malformations

Vascular Anomalies	Clinical Features	Morphologic Features	MR Features
Vascular Tumors			
Hemangiomas	Occurs in first weeks of life; rapid growth	Strawberry-like, pulsatile, warm mass	Proliferating phase: Well defined, lobulated mass, low SI in T_1-weighted image, high SI on T_2-weighted image, flow-voids on spin-echo images, early homogeneous c.e. Involuting phase: Fat replacement, decreased c.e.
Low-Flow Vascular Malformations			
Venous	Occurs in childhood or early adulthood	Blue, soft, noncompressible, nonpulsatile mass	Septated, lobulated mass without mass effect, phleboliths, fluid-fluid levels, no flow-void, low SI on T_1-weighted image, high SI on T_2-weighted image, slow gradual c.e.
Lymphatic	Occurs in childhood, grows with the child without regression	Smooth, noncompressible, rubbery mass	Septated, lobulated mass without mass effect, fluid-fluid levels, no flow-void, low SI on T_2-weighted image, high SI on T_2-weighted image; if macrocystic has rim c.e., if microcystic has no significative c.e.
Capillary	Occurs in childhood, grows with the child without regression	Cutaneous red discoloration	Skin-thickened lesion
High-Flow Vascular Malformations			
AVM	Occurs in childhood, grows with the child without regression	Red warm, pulsatile mass with a thrill	No well-defined mass, enlarged feeding arteries and draining veins; flow voids; early enhancement of feeding arteries and nidus with shunting to draining veins

hormonal changes during puberty or pregnancy or as a result of thrombosis, infection, trauma, or incomplete treatment [4]. Unlike hemangiomas, they may be infiltrative and usually involve multiple tissue planes.

Vascular malformations are classified into high flow and low flow. This differentiation based on flow dynamics is vital to planning surgical or image-guided treatment procedures.

Complex-combined malformations are found in syndromes such as Klippel–Trénaunay, Sturge–Weber, Parkes Weber, blue rubber bleb, Proteus, and Maffucci [3].

8.1.1.1 Low-Flow Vascular Malformations

8.1.1.1.1 Venous Malformation

Venous malformations are the most common peripheral vascular malformation [7–9], usually located in the head and neck (40% of cases), trunk (20%), and extremities (40%), where they account for almost two-thirds of vascular malformations [7,10].

A venous malformation is defined as a simple malformation with slow flow and an abnormal venous network [9]. They are composed of small and large dysplastic, postcapillary, thin-walled vascular channels with sparse smooth muscle and variable amounts of hamartomatous stroma, thrombi, and phleboliths. The dysplastic venous channels usually connect with adjacent physiologic veins via narrow tributaries [7]. A mural muscular anomaly is probably responsible for the gradual expansion of these lesions [9,11].

Venous malformations are present at birth, but patients usually have symptoms in late childhood or early adulthood. The clinical presentation depends on the depth and extent of the lesion. Skin and subcutaneous venous malformations appear as faint blue, soft, easily compressible, and non-pulsatile masses [7,12] that characteristically enlarge with the Valsalva maneuver and in dependent positions and decompress with extremity elevation and local compression. Venous malformations may permeate across tissue planes and invade multiple adjacent tissues (fat, muscle, tendon, and bone) causing pain, impaired mobility, and skeletal deformity [7,13].

Venous malformations or combined lymphatic-venous malformations are found in different syndromes. Blue rubber bleb nevus syndrome is a familial condition characterized by development of multiple cutaneous, musculoskeletal, and gastrointestinal venous malformations. Patients can present with chronic blood loss and intermittent small bowel obstruction due to chronic bleeding, intussusception, or volvulus of gastrointestinal venous malformations [3,13]. Proteus syndrome includes cutaneous and visceral combined lymphatic-venous malformations with multiple subcutaneous hamartomas, pigmented nevi, hemihypertrophy, hand or foot overgrowth, bone exostoses, and lipomatosis [4,13]. Maffucci syndrome consists of diffuse enchondromatosis involving the phalanges of the hands and feet in association with multiple venous or lymphatic malformations [3].

8.1.1.1.2 Lymphatic Malformation

Lymphatic malformations are the second most common type of vascular malformation after venous malformations [14], usually located in the neck (70%–80%), especially in the posterior cervical triangle, and in the axillary region (20%) and rarely found in the extremities [4,15].

Lymphatic malformations consist of chyle-filled cysts lined with endothelium [4,6]. They result from sequestered lymphatic sacs that fail to communicate with peripheral draining channels [15]. These malformations can be divided into microcystic (multiple cysts smaller than 2 mm in a background of solid matrix) and macrocystic types (larger cysts of variable sizes) [4]. Lymphatic malformations are commonly associated with other vascular malformations [16].

Most lymphatic malformations are usually observed in the first 2 years of life as smooth soft-tissue masses with a rubbery consistency. Microcystic lymphatic malformations tend to permeate the skin, whereas macrocystic lymphatic malformations manifest as smooth, translucent multiple masses under normal skin [17]. Unlike venous malformations, they are noncompressible. Diffuse soft-tissue thickening and surrounding lymphedema may occur locally with microcystic lesions [4].

8.1.1.1.3 Capillary Malformation

Capillary malformations are present at birth in 0.3% of children [4] and demonstrate cutaneous red discoloration. They are predominantly localized in the head and neck region [17]. Capillary malformations are areas of congenital ectasia of thin-walled small-caliber vessels of the skin. Although typically confined to the dermis or mucous membranes, they may also be the hallmark of more complex anomalies such as Sturge–Weber, Klippel–Trénaunay, and Parkes Weber syndromes [4,17]; symptoms in these patients are the result of deeper associated malformations [2,7].

The Sturge–Weber syndrome involves a unilateral capillary malformation in the distribution of the trigeminal nerve with ipsilateral leptomeningeal malformation, atrophy, and calcification of the subjacent cerebral cortex and malformation of the choroid [3,13]. The Klippel–Trénaunay syndrome involves a combined capillary-venous malformation of the trunk and extremities in association with limb overgrowth [3,13]. Parkes Weber syndrome involves a cutaneous capillary malformation with limb hypertrophy in combination with AVFs and congenital varicose veins [3].

8.1.1.1.4 Capillary-Venous Malformation

Capillary-venous malformations are combined low-flow malformations formed from dysplastic capillary vessels and enlarged postcapillary vascular spaces.

8.1.1.2 High-Flow Vascular Malformations

High-flow malformations make up approximately 10% of malformations in the extremities and include AVMs and AVFs. During the proliferating phase, infantile hemangiomas are also considered high-flow lesions.

AVFs are formed by a single vascular channel between an artery and a vein, whereas AVMs consist of feeding arteries, draining veins, and a nidus composed of multiple dysplastic vascular channels that connect the arteries and veins, with absence of a normal capillary bed [7,13,15].

8.1.1.2.1 Arteriovenous Malformation

AVMs are already present at birth in the early quiescent stage [17] but do not usually become evident until childhood or adulthood. Like other vascular malformations, they generally increase proportionally in size as the child grows, with growth being exacerbated due to hormonal changes during puberty or pregnancy [13] or as a result of thrombosis, infection, or trauma [7].

Owing to their high blood flow, they manifest as a red, pulsatile, warm mass with a thrill and may lead to bone overgrowth, arterial steal phenomenon, and cutaneous ischemia. Ulceration and hemorrhage may be seen in later stages [17].

8.1.1.2.2 Arteriovenous Fistula

Congenital AVFs, which usually occur in the head and neck, are different from the more common acquired AVFs, which are mostly the consequence of an iatrogenic or traumatic penetrating injury. MRI shows the arterial and venous components as large signal voids on spin-echo (SE) images or high-signal-intensity foci on gradient-echo (GRE) images, without a well-defined mass [16]. Chronic secondary AVFs can simulate AVMs because AVFs may have such strong flow that more proximal supplying arteries and distal draining veins also enlarge [18].

8.2 Imaging

Multiple imaging modalities should be used to evaluate characteristics of the lesion, such as size, flow velocity, flow direction, relation to the surrounding structures (vessels, muscle, nerve, bone, and the skin), and lesion contents.

8.2.1 Conventional Radiography

Conventional radiography plays only a small part in the diagnosis and classification of vascular lesions, but it provides useful information about bone and joint involvement. Bone erosion, sclerotic change, periosteal reaction, and pathologic fracture each suggest bone involvement.

8.2.2 Ultrasonography

Ultrasonography is an essential, noninvasive tool that is widely used to examine superficial vascular lesions. Color Doppler imaging permits real-time analyses of arterial and venous flow and measurement of flow velocities. It is an important method for monitoring patients who have undergone therapy, but it is limited to the assessment of deep lesions and lesions adjacent to interfering air or bone.

8.2.3 Computed Tomography

Computed tomography (CT) with intravenous contrast material is useful for assessment of vascular malformations. Multidetector row CT can be a valuable means to evaluate the effect of enhancement, existence of calcification or thrombosis, distal runoff (when lesions are located in the extremities), and concomitant lesions. The high temporal resolution of CT and the ease with which findings can be interpreted are advantageous for evaluating vascular lesions (Figures 8.1 and 8.2).

However, because CT involves considerable exposure to ionizing radiation and provides less information about blood flow, MRI has replaced CT in the evaluation of vascular malformations. However, phleboliths are more clearly depicted, and fatty components are sometimes demonstrated at CT. CT is also particularly useful in evaluating bone overgrowth or lysis that may accompany vascular anomalies and in patients who cannot be sedated for MRI or in patients in whom MRI is contraindicated.

8.2.4 Magnetic Resonance

MRI is the most valuable modality for classification of vascular anomalies [19]. It allows one to define the extension of vascular lesions and their anatomic relationship to adjacent structures [3], thus providing important information for therapy planning.

The selection of coils depends on the size and location of the lesion, but generally the smallest surface coil that covers the entire lesion should be chosen. If the lesion is palpable or visible, placement of a skin marker over the area of clinical concern is often useful.

The essential set of sequences is as follows: SE or fast SE T_1-weighted imaging, fat-suppressed fast SE T_2-weighted

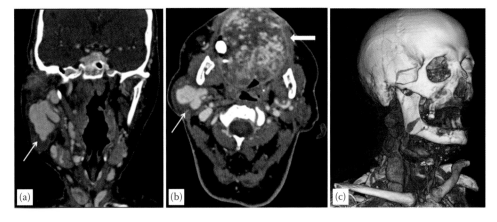

FIGURE 8.1
(a) Coronal plane, (b) axial plane, and (c) volume rendering of a 28-year-old male. CT scan, acquired on venous phase, shows a large AVM of the right malar region, which diffusely involves the parotid gland (thin arrow) and the tongue (thick arrow).

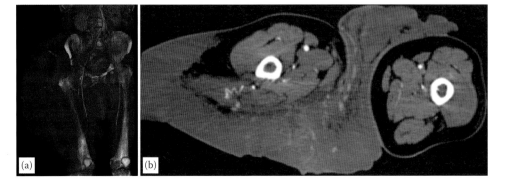

FIGURE 8.2
(a) Volume rendering and (b) axial plane of a 37-year-old male. CT scan on arterial phase shows a large AVM of the gluteal region with diffuse thickening of subcutaneous soft tissues and engorgement of regional small vessels.

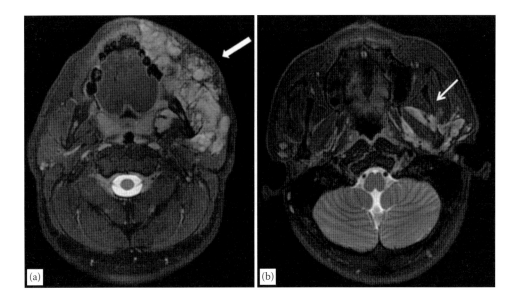

FIGURE 8.3

(a) T_2-weighted STIR sequence of a 23-year-old female acquired on axial plane shows a venous malformation of the left malar region (thick arrow). (b) T_2-weighted STIR sequence of an 18-year-old male on axial plane shows a venous malformation of the left parotid, also involving the pterygoid muscles (thin arrow).

or short τ inversion-recovery (STIR) imaging and 3D dynamic time-resolved MR angiography (MRA).

SE or fast SE T_1-weighted imaging permits basic anatomic evaluation and fat-suppressed fast SE T_2-weighted or STIR imaging is used to assess extension of the lesion (Figure 8.3).

A comprehensive assessment of vascular anomalies requires functional analysis of the involved vessels. For this purpose, dynamic time-resolved MRA has become an essential tool. This technique permits acquisition of images with high temporal and spatial resolution. These sequences allow acquisition of one 3D dataset every 2 s. This high temporal resolution enables (a) clear separation of arterial inflow from venous drainage and detection of early venous shunting, (b) acquisition of information about the contrast material arrival time (defined as the interval between the onset of enhancement and the maximal percentage of enhancement in the vessels) and flow direction, and (c) reduction of motion artifacts [4,13] (Figure 8.4).

Hence, excellent depiction of the architecture and hemodynamic properties of vascular malformations can be achieved [17], yielding clinically important data about feeding and draining vessels that is crucial for therapy planning [20]. The technique has been proved to allow discrimination between low-flow and high-flow malformations [17,21]. These sequences are started immediately after contrast agent administration in order to observe in real time the passage of the contrast agent bolus through the vascular bed of the lesion. Background subtraction and maximum intensity projection (MIP) reconstructions are used to generate fluoroscopic-like MRA datasets for functional image evaluation.

Finally, a modified 3D T_1-weighted high-resolution sequence is acquired during the equilibrium phase of contrast circulation at 3 min after contrast agent administration in order to obtain a detailed vascular map and to depict intralesional thrombosis (Figure 8.5).

GRE T_2*-weighted images can be used in some cases to demonstrate calcification or hemosiderin, as well as high-flow vessels. On GRE images, absence of signal in the blood vessel suggests a low-flow malformation [3,22], whereas high-flow vessels have high signal intensity.

To facilitate analysis of images from contrast-enhanced MRA, subtraction as well as 3D reformation techniques, including multiplanar reformation, MIP, and volume rendering are used as deemed necessary.

8.2.4.1 MRI Features of Low-Flow Vascular Malformations

In terms of treatment decisions, recognizing whether the lesion is a low-flow vascular malformation is more important than determining exactly whether the lesion is predominantly venous, lymphatic, or capillary [7,20,21,23]. Diagnosis of a low-flow malformation is based on the absence of flow voids on SE images. Occasionally, low-signal-intensity striations, septa, thrombosed vessels, or phleboliths may simulate flow voids on cross-sectional images. Contrast-enhanced and GRE images may be helpful in distinguishing these other causes of low signal intensity from flow-related signal voids. Phleboliths and calcifications typically appear as signal voids with

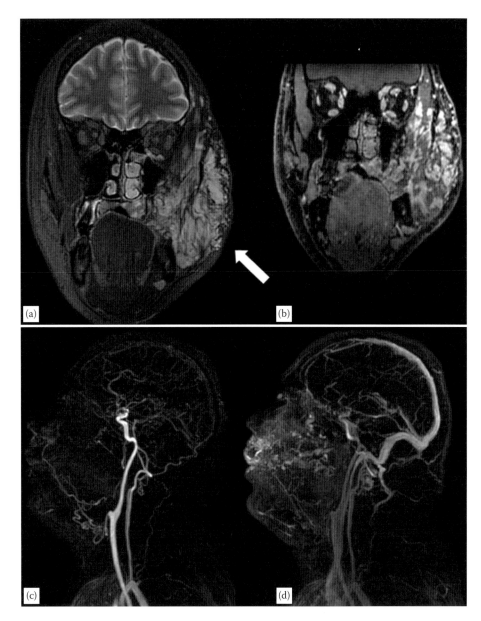

FIGURE 8.4

MR images show a large malformation involving the left malar region; dynamic time-resolved images allow functional evaluation of the involved vessels: the feeding arteries are mainly branches of the facial artery, whereas draining veins are tributary of the internal jugular vein. (a) T_2-weighted STIR sequence on coronal plane showing the malformation involving the left malar region (arrow). (b) 3D T_1-weighted high-resolution sequence after contrast media administration, on coronal plane. (c–d) Dynamic time-resolved MRA (arterial and venous phase) of a 23-year-old female.

all pulse sequences, whereas signal voids related to high flow demonstrate enhancement and appear as high-signal-intensity foci on GRE images.

Venous malformations are usually septated lesions with intermediate to decreased signal intensity on T_1-weighted images and increased signal intensity on T_2-weighted and STIR images. Occasionally, hemorrhage or high protein content may cause internal fluid-fluid levels.

In cases of thrombosis or hemorrhage, heterogeneous signal intensity can be observed on T_1-weighted images. The best clue for identification of a venous malformation

is the presence of phleboliths [17], which appear as small low-signal-intensity foci with all pulse sequences (Figures 8.6 and 8.7).

Fat-suppressed T_2-weighted and STIR images have been shown to exquisitely define the extent of venous malformations [3,13].

Venous malformations are characterized by lack of arterial and early venous enhancement and absence of enlarged feeding vessels or arteriovenous shunting. They typically show slow gradual filling with contrast material and may demonstrate characteristic nodular

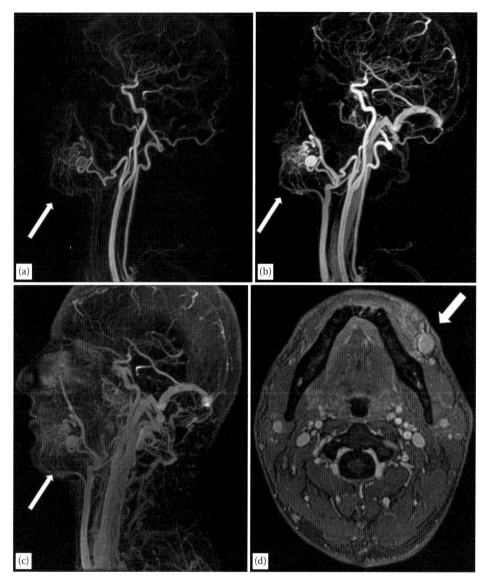

FIGURE 8.5

(a–c) Dynamic time-resolved MRA (arterial, venous, and delayed phase) and (d) 3D T_1-weighted high-resolution sequence after contrast media administration, on axial plane of a 22-year-old male. MR images depict an AVM of the left mental region (arrows). Dynamic time-resolved images permit a correct depiction of the architecture and hemodynamic properties of the malformation.

enhancement of tortuous vessels on delayed venous phase images [20]. Low-flow vascular malformations, mainly venous malformations, have a contrast material rise time of about 90 s, significantly higher than that of high-flow AVMs [23]. Delayed post-contrast T_1-weighted images of venous malformations usually show diffuse enhancement of the slow-flowing venous channels [4] (Figure 8.8).

Demonstration of a connection between a malformation and the deep venous system is useful for planning treatment, because such a finding increases the risk of deep venous thrombosis. The delayed contrast-enhanced sequence is perfectly suited for this purpose [4]. Together with the lack of evident venous drainage, well-defined lesion margins in MRI have been shown

to be a predictor of good outcome after percutaneous sclerotherapy [17,24,25].

Although venous malformations can be associated with surrounding edema or fibrofatty stroma, they rarely appear masslike [13,23]. Accordingly, lesions with unusual clinical and imaging features should be evaluated with biopsy [23].

Lymphatic malformations are usually seen as lobulated, septated masses with intermediate to decreased signal intensity on T_1-weighted images and increased signal intensity on T_2-weighted and STIR images. Internal fluid-fluid levels are common. Lymphatic malformations tend to be infiltrative, permeate across fat planes, and involve multiple tissues [4,13].

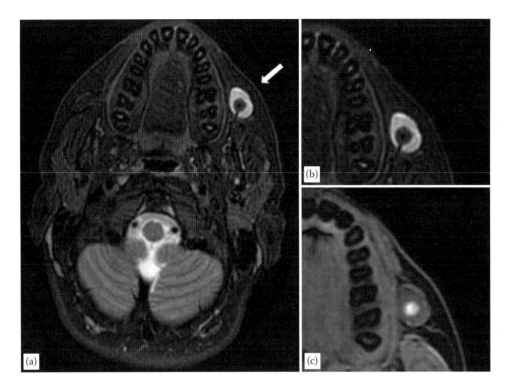

FIGURE 8.6
(a–b) T_2-weighted STIR sequence and (c) T_1-weighted FS sequence on the axial plane of a 29-year-old female showing a vascular malformation of the left malar region (arrow). MR images show an hemorrhagic focus as a hypointense on T_2 sequence and hyperintense on T_1 sequence.

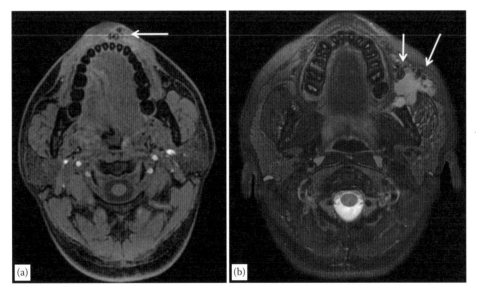

FIGURE 8.7
(a) T_1 FS-weighted sequence on axial plane of a 33-year-old female. (b) T_2-weighted STIR sequence on axial plane of a 24-year-old female. MR images shows phleboliths as small hypointense foci on both T_1 and T_2 images (arrows).

There is usually no significant enhancement of microcystic lymphatic malformations [4], whereas macrocystic lymphatic malformations exhibit only rim and septal enhancement, and no central filling of the cystic structures is expected after contrast material injection [4,7]. Occasionally, microcystic lymphatic malformations or combined lymphatic-venous malformations may show diffuse enhancement, which is due to septal enhancement of the small, nonperceptible cysts in microcystic lymphatic malformations or enhancement of the venous component in mixed malformations. This appearance may render them indistinguishable from venous malformations [7,10].

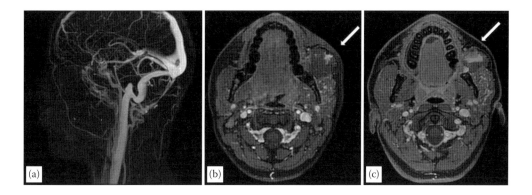

FIGURE 8.8

A 24-year-old female with a large venous malformation of the left malar region (arrows). (a) Dynamic time-resolved MRA (arterial phase) and (b–c) delayed 3D T_1-weighted high-resolution sequence after contrast media administration, on the axial plane. Dynamic time-resolved sequence shows the lack of arterial and early venous enhancement; delayed 3D T_1-weighted sequence reveal slow gradual filling with contrast material and the characteristic nodular enhancement of tortuous vessels.

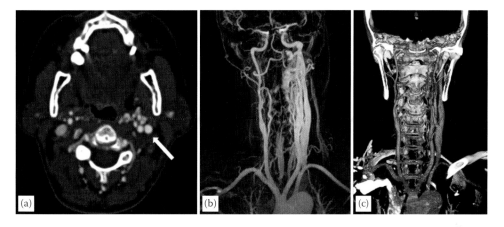

FIGURE 8.9

(a) CT on the axial plane, venous phase; (b) dynamic time-resolved MRA (MIP); and (c) Volume rendering of a 45-year-old male. CT and MR images show an AVM of the left cervical region. Images show earlier enhancement of the left internal jugular vein compared with the contralateral vessel (arrow).

In capillary malformations, MRI is not usually required because the diagnosis is made clinically. MRI findings of capillary malformations are subtle, with skin thickening and occasional increased subcutaneous thickness as the only findings [4,7,26]. MRI may be required to evaluate possible associated underlying disorders.

Imaging findings of capillary-venous malformation may be nondistinguishable from those of venous malformations. Dynamic contrast-enhanced MRI can be useful for this purpose, as capillary-venous malformations typically show early homogeneous enhancement, whereas only delayed enhancement is seen in venous malformations [4,27].

8.2.4.2 MRI Features of High-Flow Vascular Malformations

MRI findings of AVM include high-flow serpentine and enlarged feeding arteries and draining veins, which appear as large flow voids on SE images or high-signal-intensity foci on GRE images, with absence of a well-defined mass. Intraosseous extension of the lesion can be seen as decreased marrow signal intensity on T_1-weighted images [17]. Areas of high signal intensity on T_1-weighted images may represent areas of hemorrhage, intravascular thrombosis, or flow-related enhancement [28].

Gadolinium enhancement is useful in evaluating the feeding arteries and draining veins. The dynamic enhancement of the AVM is well assessed by using time-resolved dynamic 3D MRA, with a contrast material rise time of 5–10 s. Early venous filling is typically seen in AVMs (Figure 8.9).

8.3 Treatment

Infantile hemangiomas tend to spontaneously involute; thus, no treatment is usually required. However, vascular malformations should be treated to prevent permanent functional and aesthetic impairment. The treatment strategy, which often consists of multiple treatment sessions, depends on the type of malformation in terms of flow dynamics and includes both minimally invasive and surgical interventions.

Treatment of AVMs could be surgical, endovascular, and/or percutaneous.

Quite often, surgical approach (although permits radical excision) must be avoided due to localization of the lesion and for the high risk of bleeding.

For low-flow vascular malformations, the treatment of choice is percutaneous sclerotherapy; for high-flow lesions, it is transarterial embolization [3,4,15], with subsequent surgical resection occasionally being necessary. Percutaneous sclerotherapy is not effective for high-flow lesions because the infused agents are rapidly washed away from the lesion [4]. Usually more step treatment is needed because recruitment of new feeders (neoangiogenesis) from the nidus is often observed during follow-up.

Percutaneous treatment consists in the direct puncture of the lesion's nidus and injection of ethyl alcohol 95% or atoxysclerol 2% (either foam or liquid); treatment is performed with needles of different caliber (18–25 G). Before performing the sclerotization, digital subtraction angiography must be performed in order to study eventual venous drainage of the nidus; treatment must be performed only when direct injection of the nidus is possible, avoiding systemic spread of embolization agent.

Endovascular embolization, usually via the common femoral artery, is performed with selective or even superselective embolization of the nidus feeders by using Onyx®, polyvinyl alcohol particles or cyanoacrylate. Postprocedural complications are represented by pain, edema, swelling, skin ulceration, nontarget vessel embolization (Figure 8.10).

8.3.1 Posttreatment Appearances

MRI is an excellent tool for assessment of treatment results and establishment of the long-term management strategy [29]. Imaging modalities employed during follow-up do not differ in terms of technical aspects from the preoperative one.

8.3.1.1 Venous Malformations

Ethanol causes almost instantaneous denudation of endothelium, intense inflammatory reaction, and thrombosis of the malformation associated with significant swelling [17]. During the following weeks, fibrosis develops and progressive shrinking of the malformation is observed. A delay of up to several months is necessary to evaluate the therapeutic response after sclerotherapy, allowing time for the transient inflammatory response to resolve [9].

In MRI, venous malformations after sclerotherapy demonstrate heterogeneous signal intensity on both T_1- and T_2-weighted images [9]. Immediate posttreatment MRI shows high signal intensity in the treated areas as well as along the intermuscular septa on T_2-weighted and STIR images [30]. In our experience, the high signal intensity in the treated malformation persists up to 3 months after treatment, but it is no longer seen along the intermuscular septa [30].

At MRA, there is absence of enhancement in the central portion of the treated lesion with intense peripheral hyperenhancement secondary to reactive hyperemia [30]. This enhancement is already seen on arterial phase images. Beyond 3 months, the enhancement usually disappears and a scar is left, which appears dark on T_1-weighted images as well as on STIR images without gadolinium enhancement [30]. Progressive shrinkage of the lesion is often seen [30] (Figure 8.11).

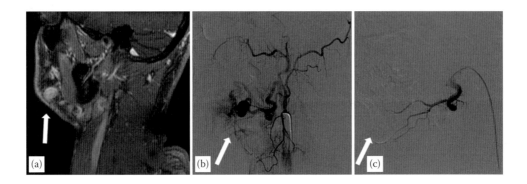

FIGURE 8.10
A 22-year-old male with an AVM of the left mental region (arrows). (a) 3D T_1-weighted sequence after contrast media administration, on the sagittal plane. (b–c) Angiographic images before and after embolization. Note that the malformation, whose feedings are branches of the facial artery, has almost completely disappeared at the end of the procedure.

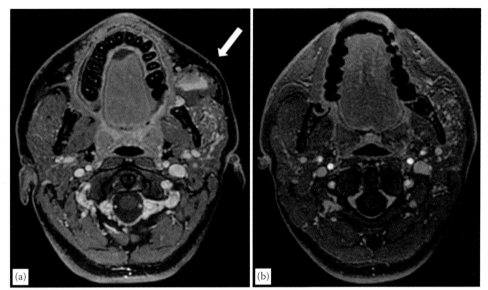

FIGURE 8.11
A 24-year-old female with an AVM of the left malar region (arrow). 3D T_1-weighted images after contrast media administration on the axial plane, respectively, before (a) and after (b) sclerotherapy. The malformation has almost completely disappeared with no enhancement on delayed images.

In the case of extensive malformations, it can be difficult to detect the effects of treatment despite multiple treatment sessions [30]. Gadolinium-enhanced imaging is therefore useful in demonstrating residual perfusion of the malformation and directing additional treatment [9].

8.3.1.2 Arterial Malformations and AVMs

The treatment strategy must be oriented toward achieving complete eradication of the nidus of a high-flow vascular malformation, since any incomplete treatment may stimulate more aggressive growth [17]. After transarterial embolization, thrombosis of the malformation is often seen; MRA may show reduced or absent shunting, with reduced or absent early opacification of the venous system. An early posttreatment study should be performer, and any remaining malformation must be treated in a second stage. In cases where ferromagnetic coils are used for embolization, susceptibility artifacts are present, potentially obscuring residual vascular malformation in their vicinity [30] (Figures 8.12 and 8.13).

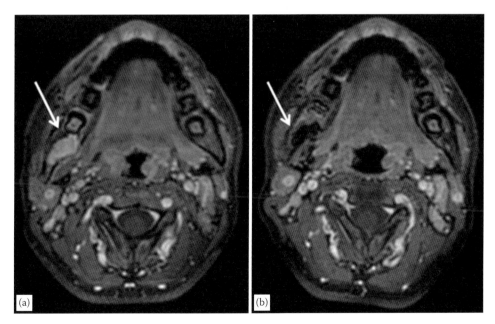

FIGURE 8.12
A 42-year-old female with an AVM of the right hemimandibula (arrow). 3D T_1-weighted images after contrast media administration on the axial plane, respectively, before (a) and after (b) embolization. Note that the lesion show a complete lack of enhancement after the treatment.

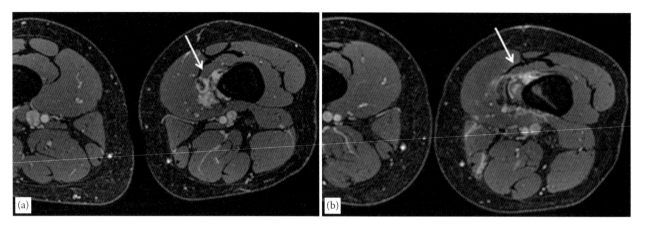

FIGURE 8.13
A 25-year-old female with a vascular malformation of the left knee (arrows). 3D T_1-weighted images after contrast media administration on the axial plane, respectively, before (a) and after (b) a procedure of embolization. Note that the lesion shows a smaller size and minor enhancement after the procedure.

8.4 Conclusions

Vascular malformations are rare but important pathologic conditions because they often require aggressive treatment, and imaging plays an important role in their diagnosis. MRI has emerged as the preeminent imaging modality for assessing these lesions. Contrast-enhanced MRA, especially dynamic time-resolved MRA, provides information about the hemodynamics of vascular anomalies and thus aids in the classification of the type of lesion. Furthermore, MRI is an excellent tool for the assessment of treatment results and establishment of a long-term management strategy.

References

1. Hand JL, Frieden IJ. Vascular birthmarks of infancy: Resolving nosologic confusion. *Am J Med Genet* 2002;108(4):257–264.
2. Mulliken JB, Glowacki J. Hemangiomas and vascular malformations in infants and children: A classification based on endothelial characteristics. *Plast Reconstr Surg* 1982;69(3):412–422.
3. Dubois J, Alison M. Vascular anomalies: What a radiologist needs to know. *Pediatr Radiol* 2010;40(6): 895–905.
4. Moukaddam H, Pollak J, Haims AH. MRI characteristics and classification of peripheral vascular malformations and tumors. *Skeletal Radiol* 2009;38(6):535–547.
5. Jackson IT, Carreño R, Potparic Z, Hussain K. Hemangiomas, vascular malformations, and lymphovenous malformations: Classification and methods of treatment. *Plast Reconstr Surg* 1993;91(7): 1216–1230.
6. Enjolras O. Classification and management of the various superficial vascular anomalies: Hemangiomas and vascular malformations. *J Dermatol* 1997;24(11):701–710.
7. Fayad LM, Hazirolan T, Bluemke D, Mitchell S. Vascular malformations in the extremities: Emphasis on MR imaging features that guide treatment options. *Skeletal Radiol* 2006;35(3):127–137.
8. Breugem CC, Maas M, Reekers JA, van der Horst CM. Use of magnetic resonance imaging for the evaluation of vascular malformations of the lower extremity. *Plast Reconstr Surg* 2001;108(4):870–877.
9. Dubois J, Soulez G, Oliva VL, Berthiaume MJ, Lapierre C, Therasse E. Soft-tissue venous malformations in adult patients: Imaging and therapeutic issues. *RadioGraphics* 2001;21(6):1519–1531.
10. Laor T, Burrows PE. Congenital anomalies and vascular birthmarks of the lower extremities. *Magn Reson Imaging Clin N Am* 1998;6(3):497–519.
11. Mulliken JB, Fishman SJ, Burrows PE. Vascular anomalies. *Curr Probl Surg* 2000;37(8):517–584.
12. Rak KM, Yakes WF, Ray RL et al. MR imaging of symptomatic peripheral vascular malformations. *AJR Am J Roentgenol* 1992;159(1):107–112.
13. Donnelly LF, Adams DM, Bisset GS 3rd. Vascular malformations and hemangiomas: A practical approach in a multidisciplinary clinic. *AJR Am J Roentgenol* 2000;174(3):597–608.
14. Marler JJ, Mulliken JB. Current management of hemangiomas and vascular malformations. *Clin Plast Surg* 2005;32(1):99–116.
15. Dubois J, Garel L. Imaging and therapeutic approach of hemangiomas and vascular malformations in the pediatric age group. *Pediatr Radiol* 1999;29 (12):879–893.
16. Navarro OM, Laffan EE, Ngan BY. Pediatric soft-tissue tumors and pseudotumors: MR imaging features with pathologic correlation. I. Imaging approach, pseudotumors, vascular lesions, and adipocytic tumors. *RadioGraphics* 2009;29(3):887–906.

17. Ernemann U, Kramer U, Miller S et al. Current concepts in the classification, diagnosis and treatment of vascular anomalies. *Eur J Radiol* 2010;75(1):2–11.

18. Lawdahl RB, Routh WD, Vitek JJ, McDowell HA, Gross GM, Keller FS. Chronic arteriovenous fistulas masquerading as arteriovenous malformations: Diagnostic considerations and therapeutic implications. *Radiology* 1989;170(3 Pt. 2):1011–1015.

19. Hyodoh H, Hori M, Akiba H, Tamakawa M, Hyodoh K, Hareyama M. Peripheral vascular malformations: Imaging, treatment approaches, and therapeutic issues. *RadioGraphics* 2005;25(suppl. 1): S159–S171.

20. Herborn CU, Goyen M, Lauenstein TC, Debatin JF, Ruehm SG, Kröger K. Comprehensive time-resolved MRI of peripheral vascular malformations. *AJR Am J Roentgenol* 2003;181(3):729–735.

21. Ohgiya Y, Hashimoto T, Gokan T et al. Dynamic MRI for distinguishing high-flow from low-flow peripheral vascular malformations. *AJR Am J Roentgenol* 2005;185(5):1131–1137.

22. Siegel MJ. Magnetic resonance imaging of musculoskeletal soft tissue masses. *Radiol Clin North Am* 2001;39(4):701–720.

23. Dobson MJ, Hartley RW, Ashleigh R, Watson Y, Hawnaur JM. MR angiography and MR imaging of symptomatic vascular malformations. *Clin Radiol* 1997;52(8):595–602.

24. Goyal M, Causer PA, Armstrong D. Venous vascular malformations in pediatric patients: Comparison of results of alcohol sclerotherapy with proposed MR imaging classification. *Radiology* 2002;223(3): 639–644.

25. Yun WS, Kim YW, Lee KB et al. Predictors of response to percutaneous ethanol sclerotherapy (PES) in patients with venous malformations: Analysis of patient self-assessment and imaging. *J Vasc Surg* 2009;50(3):581–589.

26. Breugem CC, Maas M, van der Horst CM. Magnetic resonance imaging findings of vascular malformations of the lower extremity. *Plast Reconstr Surg* 2001;108(4):878–884.

27. Van Rijswijk CS, van der Linden E, van der Woude HJ, van Baalen JM, Bloem JL. Value of dynamic contrast-enhanced MR imaging in diagnosing and classifying peripheral vascular malformations. *AJR Am J Roentgenol* 2002;178(5):1181–1187.

28. Abernethy LJ. Classification and imaging of vascular malformations in children. *Eur Radiol* 2003; 13(11):2483–2497.

29. Lee BB, Choe YH, Ahn JM et al. The new role of magnetic resonance imaging in the contemporary diagnosis of venous malformation: Can it replace angiography? *J Am Coll Surg* 2004;198(4):549–558.

30. Hagspiel K, Stevens P, Leung D et al. Vascular malformations of the body: Treatment follow-up using MRI and 3D gadolinium-enhanced MRA. In: CIRSE 2002. Abstracts of the annual meeting and postgraduate course of the Cardiovascular and Interventional Radiological Society of Europe and the 4th Joint Meeting with the European Society of Cardiac Radiology (ESCR). Lucern, Switzerland, October 5–9, 2002. *Cardiovasc Intervent Radiol* 2002;25(suppl. 2):S77–S265.

9

Four-Dimensional Flow Imaging

Sergio Uribe, Israel Valverde, and Pablo Bächler

CONTENTS

9.1 Introduction to Magnetic Resonance Flow Imaging

A phase contrast technique is usually performed to obtain velocity images of the heart and vessels. In this sequence, flow-encoding gradients are used to translate the velocity of the magnetization into a phase in the image [1]. The most common flow-encoding gradient is a bipolar velocity-encoding gradient (Figure 9.1).

The axis of the bipolar gradient determines the direction of flow sensitivity. This can be added to any of the three logical axes (slice selection, frequency encoding, or phase encoding) separately, but also simultaneously to two or more of the logical axes to achieve flow sensitivity along any arbitrary axis.

Because the net area under the bipolar gradient is zero, it produces no phase accumulation for stationary spins. However, for moving spins, such as blood flow, the phase introduced to the spins is proportional to the velocity and the area of each bipolar gradient.

A velocity-encoding gradient is often characterized by its aliasing velocity parameter, usually denoted as VENC. By definition, when the velocity component along the gradient direction is equal to ±VENC, the resulting phase difference is ±π as shown in Equation 9.1:

$$\text{VENC} = \frac{\pi}{\gamma A_g \tau} \tag{9.1}$$

Since MR images contain other contributions to their phase (e.g., B_0 inhomogeneities), phase contrast imaging acquires two complete sets of image data using the same imaging parameters, except for the first moment of the bipolar gradients. The phases or the complex values of the two resulting images are subtracted on a pixel-by-pixel basis. The subtraction process accentuates flow, while suppressing unwanted phase variation and the background stationary tissue.

Two different subtraction methods are commonly used: (1) complex difference reconstruction and (2) phase difference reconstruction.

Complex difference reconstruction is accomplished by subtracting the complex data from the two datasets. Result of this method is a magnitude value and therefore pixels intensity values are proportional to the speed of blood flow. This reconstruction method is also known as a *phase contrast angiography*, and it can be used to obtain angiograms or venograms of the brain vessels. The main problem with this type of reconstruction is that it does not provide directionality or quantitative flow information.

Phase difference reconstruction is also called *velocity mapping* or *phase velocity mapping* and is usually used to depict the direction of blood flow and to quantify the net flow rate through a vessel. Phase difference reconstruction is performed in the imaging domain. Each pair of datasets is Fourier transformed to yield two independent complex images. Then, the angle difference or phase difference (Δφ) is calculated from these two complex images, and flow velocity can be calculated as follows:

$$v = \left(\frac{\Delta\phi}{\pi}\right)\text{VENC} \tag{9.2}$$

where VENC is the *Velocity ENCoding* parameter.

Because the dynamic range of phase difference reconstruction is [−π,π], the flow direction can only be reliably represented when $|v| \leq$ VENC, unless a phase unwrapping algorithm is employed [2].

9.2 4D-Flow Magnetic Resonance Imaging

The axis of the flow-encoding gradient determines the direction of flow sensitivity. This information can be added to any of the three logical axes (slice selection, frequency encoding, or phase encoding) separately or simultaneously to achieve flow sensitivity along any arbitrary axis. When applying flow-encoding gradients consecutively along three logical axes, the technique can provide measurements of blood flow velocities in three orthogonal directions. This technique is commonly known as *three-dimensional phase contrast magnetic resonance imaging* (3D PC-MRI) or *four-dimensional (4D) flow MRI* [1,3,4] when 3D velocity encoding is applying to a 3D volume (Figure 9.2).

In the following sections, we will review different technical advances regarding acquisition techniques as well as quantification methods of 4D-flow data. Thereafter, we will show some potential clinical applications of 4D-flow MRI that has been reported in the literature. We will conclude this chapter with our vision on the future of 4D-flow MRI.

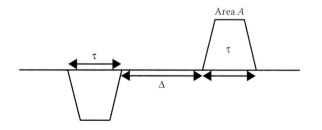

FIGURE 9.1
A bipolar gradient is used to encode velocity information. Stationary spins do not acquire any net phase; however, moving spins acquire a net phase proportional to the area of the gradients.

FIGURE 9.2
An example of 4D-flow MRI of the whole heart and great vessels.

9.2.1 Technical Aspects of Image Acquisition, Visualization Tools, and Quantification Methods of 4D-Flow Data

The amount of data that need to be collected in order to obtain 3D velocity information in a 3D volume is large. The scan time could range from 5 to 20 min, depending on the size of the field of view (FoV), and the spatial and temporal resolutions. This in turn implies that 4D-flow acquisitions cannot be performed in a breathhold. Therefore, methods for acquiring these datasets that minimize the scan time are needed. Furthermore, these acquisition techniques need to incorporate respiratory motion correction algorithms.

The standard method to acquire 4D-flow datasets consists of a segmented gradient eco sequence. The acquisition technique needs to collect $N_y \times N_z$ k-space lines during several heartbeats, where N_y and N_z are the number of k-space lines in the phase- and slice-encoding directions, respectively. A total of $N_y \times N_z/N_{LS}$ heartbeats are required to complete the 3D acquisition, where N_{LS} is the number of k-space lines on each segment. Each line in each segment needs to be acquired several times in order to achieve velocity encoding in three dimensions.

There are several approaches to obtain 3D velocity encoding [3]. One of the first approaches consisted of acquiring a pair of measurements for each direction, one flow-compensated and another flow-encoded, which lead to six measurements for obtaining flow encoding in three dimensions [1,3,5]. This method is known as *six-point method*. More efficient approaches are known as *four points* acquisition strategies. One of these strategies consists of four successive acquisitions, one flow-compensated, and then three acquisitions where the bipolar gradients are only applied in two axes (Figure 9.3). Some other approaches that use a *balanced* four-point method have been proposed [3,6,7]. Independent of the encoding strategy, 4D-flow acquisition obtains one anatomic image and three images with velocity encoding in each direction.

Nowadays, the four-point methods are the most efficient techniques to encode 4D-flow data. However, there is an inherent limitation of the temporal resolution equivalent to 4 TR times. This is the case when only one k-space line is acquired on each segment. The minimum TR depends on the velocity-encoding parameter VENC, which will impose the strength and duration of the bipolar gradients.

A lower VENC will require a strong gradient and/or a longer duration and therefore a lower temporal resolution could be achieved. On the other hand, setting a higher VENC will require a weak gradient allowing a higher temporal resolution. However, setting up the VENC too high might decrease the measurement accuracy for low velocities such as venous flow.

In the upcoming sections, we will review different sampling and acquisition strategies to obtain 4D-flow imaging and the respiratory motion compensation methods that can be used to control breathing motion artifacts.

9.2.2 Acquisition and Sampling Strategies

A 4D-flow data acquisition can be performed using a traditional 3D Cartesian sampling approach [4,6,8]. The advantage of this method is that the reconstruction of each dataset is straightforward; however, the

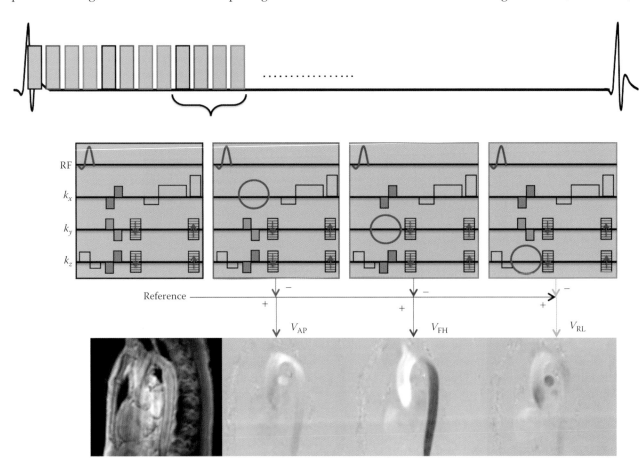

FIGURE 9.3

Schematic representation of a four-point acquisition strategy of a 4D-flow sequence. The acquisition consist of one flow-compensated image (reference image), and then three acquisitions, where the bipolar gradients are only applied in two axes. By subtracting each of these last three acquisitions from the reference image, it is possible to obtain three velocity images encoding in three different directions as shown in the last row. V_{AP}, V_{FH}, and V_{RL} represent the velocity in the anterior–posterior, foot–head, and right–left directions.

disadvantage is that the scan time acquisition tends to be long, limiting the achievable spatial resolution.

Another approach uses an undersampled 3D radial trajectory known as vastly undersampled isotropic projection reconstruction (PC-VIPR) [9]. This method can reduce the scan time and/or improve the spatial resolution of 4D-flow acquisitions. The *k*-space trajectory consists of multiple projections passing through the center of *k*-space. The advantage of this method is that it provides isotropic and high spatial resolution on a spherical and large FoV. This sequence has been applied for studying the hemodynamics in small vessels, such as brain aneurysm, in situations where a large FoV is required such as in acquisition of the whole heart and great vessels [10–13], and for studying hepatic flows [14] among other applications.

A stack of spiral trajectories [15] has also been used to efficiently sample *k*-space and to acquire 4D-flow dataset achieving twofold to threefold scan time reduction relative to Cartesian imaging methods. However, dedicated corrections for off-resonance effects and trajectory errors are necessary to maintain a high-quality image.

Parallel imaging has also been applied to speed up the acquisition of 4D-flow data. Results obtained using either GRAPPA [16] or SENSE [17] are in good agreement with fully sample images up to an acceleration factor of 3. However, larger undersampling factors deteriorate image quality and underestimate flow velocities [17]. Besides parallel imaging, alternative methods exploiting spatio-temporal correlations or exploring the sparsity of flow data have been devised, leading to higher undersampling factors. The following references can be reviewed [18–24] for the interested readers.

9.2.3 Respiratory Motion Compensation Techniques

Respiratory motion compensation techniques are needed to avoid inaccurate measurements of velocities, especially for thoracic and abdominal applications. Respiratory belts and respiratory navigator echoes are the most standard techniques that can be used for 4D-flow acquisitions. In both methods, the position of the diaphragm is measured and used to gate the acquisition, either indirectly (belt) or directly (navigator echoes). When gating the acquisition, a predefined expiratory window is usually used to accept or reject data. Typically, the efficiency of the scan is between 40% and 60%, and thus the use of this approach prolongs the scan time. *k*-space reordering algorithms have been applied to the acquisition of 4D-flow datasets in order to improve the efficiency of these gating strategies [25].

Although the respiratory belt and navigator echoes are the most available techniques across MR vendors, they have some disadvantages. The respiratory positions obtained from the belt could be inaccurate, have low reproducibility, and it is difficult to obtain the motion of the diaphragm in millimeters. On the other hand, respiratory navigator echoes have the main disadvantage of interrupting the acquisition, which disturbs the steady state and can result in artifacts. Moreover, the time needed for adding one or more navigators limits the temporal resolution of the 4D-flow acquisition, at the beginning and/or at the end of the cardiac cycle [25] with obvious disadvantages, for example, for assessment of diastolic function and retrograde flows [26].

To overcome these limitations, self-respiratory navigator techniques have been proposed. These methods acquire *k*-space data of the object being imaged, allow sampling 4D-flow data over the entire cardiac cycle and do not add or add little extra time to the acquisition. The two main strategies are the acquisition of the *k*-space center point or a *k*-space centerline. In both cases, the respiratory signal is used to gate the acquisition. The *k*-space center point provides the MR signal of the entire object, and therefore its amplitude is modulated by the cardiac cycle and respiratory motion. On the other hand, a 1D Fourier transform of the *k*-space center line provides a 1D projection of the entire volume, from which it is possible to obtain the motion of the object being imaged. An example of the *k*-space center profile to gate the acquisition of 4D-flow data can be found in [27]. One problem with both methods is that both techniques may fail in patients with high volume of adipose tissue, where high signal coming from fat may mask out the breathing motion, a situation that is more relevant for the *k*-space center point approach.

There are more advanced methods for retrospective motion correction for 2D and 3D dynamic and cine imaging, where multiple signals are obtained from navigators, belts, and/or from the same *k*-space data. However, the application of these methods for acquiring 4D-flow data has not yet been reported.

9.2.4 Visualization Methods

The acquired imaging data are a time-resolved 3D volume of velocity vectors (V_x, V_y, and V_z) and magnitude data. There are several programs to quantify and visualize the data: GTFlow (Gyrotools, Zurich, Switzerland) [28], EnSight (CEI Inc., Apex, NC) [29], and Mediframe (University of Karlsruhe, Karlsruhe, Germany) [30], among others. The velocity data can be visualized in one image for each velocity-encoding direction. For instance, Figure 9.4

Anatomy $\Delta\phi_x \approx V_{AP}Gt$ $\Delta\phi_y \approx V_{FH}Gt$ $\Delta\phi_z \approx V_{RL}Gt$

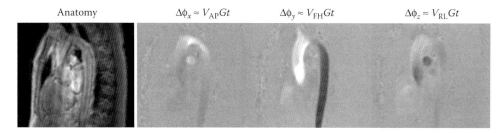

FIGURE 9.4
Examples of raw images obtained from a 4D-flow scan. An anatomic image and three flow-encoding images can be appreciated. The intensity on the flow-encoding images represents the flow velocity in each particular direction (AP, anterior–posterior; FH, foot–head; RL, right–left).

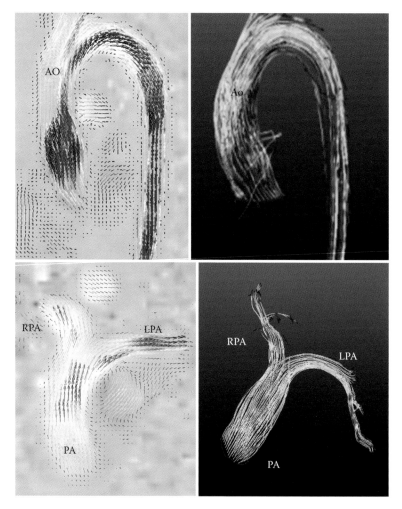

FIGURE 9.5
Left column shows a vector field representation of the velocity in the aorta and in the pulmonary artery (PA) and right (RPA) and left pulmonary (LPA) branches. Right column represents a streamline representation of the same vessels.

shows an anatomical image and three additional images with flow encoding in the anterior–posterior, right–left, and foot–head directions. In the last three images, the intensity represents the velocity of the blood in each particular direction.

The velocity data can also be pictured as a vector field in a single plane, as well as using streamline or particle traces.

By selecting a plane of interest, the velocities at the level of that plane can be represented as individual vectors (Figure 9.5). It is suitable for a detailed analysis of local flow patterns, such as helix direction [30], and to identify vortexes [31,32].

Streamlines are imaginary lines that align with the local velocity field in a specific phase within the cardiac

FIGURE 9.6
Representation of 4D-flow data using particle traces in the aorta and pulmonary artery in one particular cardiac phase.

cycle [33] (Figure 9.5). They provide a 3D perspective of the flow velocity vector field; however, streamlines are a static representation and do not necessarily show the path that blood may take along the cardiac cycle.

Particle traces are path-lines that an imaginary particle would take over time [34]. By selecting a plane of interest, imaginary particles can be released in a desired heart phase and can be followed through the cardiac cycle (Figure 9.6). They truly represent the pathways that blood flow follows.

9.2.5 Retrospective Quantification of Blood Flow

During the MRI offline review, it is common to find new diagnoses, which were not previously described or suspected during data acquisition. Hence, the required 2D PC-MRI sequences may have not been performed during the MRI scan, resulting in an essential loss of information. Nevertheless, 4D flow allows offline arbitrary selection of planes without being limited by the pre-acquired 2D PC-MRI. We will come back to this topic again in Section 9.4.

9.2.6 Advanced Quantification Methods of 4D-Flow Data

Several semiquantitative studies have described normal and abnormal blood flow patterns in the heart and large blood vessels by using 4D flow. The variety of postprocessing methods to visualize blood flow data and the ability of 4D flow to acquire 3D velocity vector fields within a single scan have allowed a detailed description of flow dynamics in different pathologies, such as aortic diseases, pulmonary hypertension, heart failure,

intracranial aneurysm, or in some congenital heart diseases. Some of the major findings in different diseases are discussed later in Section 9.3.

In addition to the semiquantitative analysis, 4D-flow MRI can be used to obtain a variety of quantitative hemodynamic indices. In the following sections, we will discuss some of these hemodynamic indices, highlighting their potential utility for clinical practice.

9.2.6.1 Wall Shear Stress

In non-pulsatile flow in a straight vessel, fluid flow is fastest at the center and slowest close to the wall. Therefore, the fluid velocities assume a parabolic profile referred to as the *laminar flow* [35]. This phenomenon is secondary to the friction within the fluid and between fluid and the vessel wall. This produces a frictional force exerted by the flowing fluid on the endothelium, which is known as *wall shear stress* (*WSS*). The magnitude of WSS depends on the velocity gradient when moving from the vessel wall to the center of the vessel. This velocity gradient near the wall is called the *wall shear rate*, and can be obtained by 4D flow to estimate WSS. However, because of limited spatiotemporal resolutions, segmentation errors, and partial volume effects, WSS is typically underestimated by MRI [36,37]. Nonetheless, although the absolute values of WSS might not be accurate, relative measurements may be useful for identifying regions of abnormal hemodynamic stress within a vessel. High WSS values drives cerebral aneurysm formation [38,39] and altered WSS have been shown to play an important role in atherosclerosis formation and plaque rupture [40]. In addition, it has been hypothesized that abnormal areas of WSS might promote aneurysm progression or rupture [41].

9.2.6.2 Pulse Wave Velocity

Pulse wave velocity (PWV) reflects vascular stiffness and is generally accepted as the most simple, noninvasive, robust, and reproducible method to determine arterial stiffness [42]. PWV refers to the velocity of propagation of a pulse wave along a vessel, usually an artery, over a known *vascular distance* [43]. This calculation may be performed using Doppler ultrasound, but MRI offers some advantages because the measurements are not limited by the geometric assumptions of ultrasound and regional analysis is possible [37,44]. It has been shown that quantification of PWV by using 4D-flow MRI is in good correlation with measurements obtained with conventional 2D PC-MRI [45,46], and provides better estimation in tortuous vascular anatomy. Carotid-femoral PWV and aortic PWV are strong predictors of adverse cardiovascular outcomes [42].

In addition, PWV might predict progression in specific disease states, such as aortic dilatation in patients with Marfan syndrome [47].

9.2.6.3 Flow Distribution

Particle traces or pathlines have been mainly used for dynamic flow visualization. However, they might provide valuable quantitative information related to flow distribution. Because particle traces emitted from a specific vessel can be traced both forward and backward in time, it is possible to identify where the flow goes or where it comes from. This can be used to determine the distribution of flow into branching vessels, especially in patients with congenital heart diseases after complex palliative procedures (such as Fontan procedure or other double inlet-double outlet connections), where standard methods cannot isolate the flow contribution of each vessel [48,49]. In addition, particle traces have been used to classify various subcomponents of flow within the ventricles [50], which might help in early identification and monitoring of cardiac dysfunction (see Section 9.3.1). It is important to note that streamlines should not be used for these purposes when there is pulsatile flow, as they reflect the flow field only at a given moment in time [37,51].

9.2.6.4 Energy Loss

Energy loss can also be estimated noninvasively by 4D-flow MRI. It has been defined as the difference in rate of the total energy transferred between vessels, for example, between the pulmonary artery branches and the main pulmonary artery [52]. This quantitative index might provide new insights to compare the performance of different palliative procedures in patients with congenital heart diseases.

9.2.6.5 Turbulence

Turbulent blood flow is characterized by fast random temporal and spatial velocities, in contrast to the ordered, typically laminar flow of the normal cardiovascular system. However, these irregular and rapid fluctuations are not represented in the mean intravoxel velocities calculated from 4D flow [43]. Nonetheless, the magnitude of the signal can be used as a measure of the standard deviation of the velocity distribution within the voxel [53,54], which is related to the turbulent kinetic energy, an independent measure of turbulence intensity. Validation studies have demonstrated good agreement with reference methods [55]. Since turbulent flow has been postulated to play a role in the pathogenesis of thrombus formation and atherosclerosis [56,57], turbulence imaging mapping might contribute to early identification of vascular regions at risk.

9.2.6.6 Pressure Gradients

Pressure gradients might be used to estimate the severity of a stenotic vessel or valve. Catheterization is currently the gold standard for pressure gradient estimation, but it is an invasive procedure. The 4D-flow MRI offers the possibility to obtain precise measurements of local pressure differences by using the Navier–Stokes equation. The velocity vector field obtained from 4D-flow data allows us to generate 3D pressure maps as well [58]. However, the main limitation of this approach is that assumes nonturbulent flow, which makes the technique of limited value for many stenotic areas, in which turbulence is present. Nonetheless, intracardiac flow has minimal turbulence, and thus 4D is a promising application for pressure mapping in this context [37].

9.2.6.7 Flow Eccentricity

As mentioned earlier, the fluid velocities assume a parabolic profile in non-pulsatile flow in a straight vessel, with the highest velocity in the center of the vessel. Although the flow at the ascending aorta is pulsatile, it exhibits normally an approximate parabolic velocity profile with the highest velocity in the center of the vessel. Interestingly, valvular and vascular disease may alter this velocity profile in many contexts, resulting in eccentric flow with high velocities at the periphery. This change in the normal velocity profile can be quantified by 4D-flow MRI in the ascending aorta with two different approaches: (1) measuring the displacement of systolic flow relative to the vessel centerline [59] and (2) measuring the angle of eccentric blood flow between the left ventricular outflow tract and aortic root [60]. It has been postulated that flow eccentricity might lead to aortic dilatation in patients with bicuspid aortic valve (BAV), although it might also be a consequence of aortic dilatation in this population [60].

9.2.6.8 Vorticity and Helicity

Vorticity is defined as a rotating or swirling motion around an orthogonal axis to the vessel centerline [43]. In contrast, helicity is defined as a rotational motion around the longitudinal axis of the vessel centerline, that is, the physiologic flow direction, creating a corkscrew-like flow pattern [43]. Although most of the previous works have analyzed vorticity and helicity semiquantitatively, it is possible to quantify both from 4D-flow data [32,61–63]. Helical flow is a normal physiological phenomenon in the aorta and pulmonary artery, and it is thought to protect against atherogenesis in the former [64].

9.3 Potential Clinical Applications of 4D Flow

9.3.1 Heart Failure

Heart failure is a complex clinical syndrome that results from any structural or functional impairment of ventricular filling or ejection of blood [65]. Intracardiac flow abnormalities are frequently seen in patients with heart failure, particularly in diastolic dysfunction. Some of these abnormalities can be quantified with 4D flow. For example, the blood that transits the left ventricle or right ventricle can be separated into four different functional components by using particle traces: (1) the direct flow, which refers to the blood that enters and exits during a single heartbeat; (2) the retained inflow, which is the blood that enters but does not exit; (3) the delayed ejection flow, which refers to the blood that starts inside and exits on the subsequent heartbeat; and (4) the residual volume, which remains in the ventricle for at least two cardiac cycles [50]. Isolation of these four flow components by 4D flow has shown that the direct flow percentage and the kinetic energy of direct flow at end diastole are reduced in patients with heart failure, leading to an increase in the workload of the left ventricle to eject the same stroke volume [50].

These unique markers of inefficient intracardiac flow may aid to identify patients early in the course of disease, to monitor therapy and to influence pacing strategies or target heart rates in cases in which diastolic–systolic coupling is diminished [37].

9.3.2 Valve Disease

The 4D-flow MRI can provide good accuracy for the assessment of valve abnormalities through all four-heart valves within a single MR acquisition [66]. Furthermore, because cardiac valves move considerably during the cardiac cycle, motion correction with retrospective 3D valve-tracking using a volumetric 4D-flow data can mitigate this effect, providing better assessment than 2D PC-MRI [67,68]. These advantages of 4D flow over standard 2D PC-MRI may facilitate the hemodynamic assessment of valve diseases in clinical practice.

9.3.3 Aortic Dissection

Stanford type B aortic dissection is usually managed medically. Current clinical practice is to treat patients with stent graft placement when complications of dissection are present, such as end-organ ischemia, aneurysm formation, or aortic rupture. Nonetheless, some patients develop aneurysm dilatation of the false lumen late in the course of disease, a critical complication [69]. Imaging plays a key role in the follow-up of these patients, because the aortic diameter and the amount of false lumen thrombosis are used to stratify patients according to risk of aortic rupture and to select patients for endovascular treatment [70,71].

It has been hypothesized that hemodynamic parameters such as flow patterns, volume, and velocity might have a role in false lumen expansion and aneurysm formation [72,73]. The 4D flow is able to quantify *in vivo* these hemodynamic parameters, information that could help to identify patients who might benefit for early endovascular treatment, before aortic expansion is already established [74].

9.3.4 Pulmonary Hypertension

Pulmonary hypertension has been defined as an increase in mean pulmonary arterial pressure greater than 25 mm Hg at rest [75]. Although the gold standard for diagnosing pulmonary hypertension is right heart catheterization, noninvasive predictors can be obtained from 4D-flow MRI. Vortices of blood flow in the main pulmonary artery detected by 4D flow enable the identification of manifest pulmonary hypertension with good sensitivity and specificity [76] (Figure 9.7). In addition, 4D-flow can provide information about average velocities and minimum pulmonary artery areas, the parameters that have shown the best diagnostic performance in phase contrast MRI [77]. Visualization and quantification of these flow features might be useful for guiding medical therapy and assessing treatment response in a way that is not possible with other modalities [37].

9.3.5 Chronic Liver Disease: Hepatic and Portal Venous Flow

Liver cirrhosis is a common and serious disease that induces significant abnormalities in the hepatic blood flow. Doppler ultrasonography is the current noninvasive standard evaluation for assessing hemodynamic changes in portal venous flow in patients with liver cirrhosis, portal hypertension, or liver transplant. However, Doppler ultrasound is limited by poor acoustic windows, observer dependency, and low inter-observer reproducibility. The 4D-flow MRI analysis of the hepatic and portal venous flow is feasible and might overcome these limitations [14,78]. Nonetheless, there are still some disadvantages when compared to Doppler ultrasound. The complexity of the hepatic and portal venous

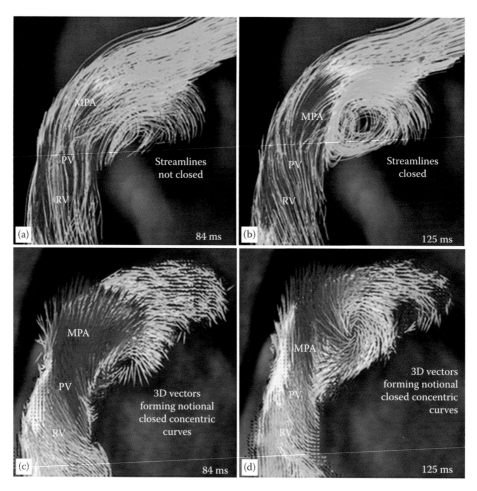

FIGURE 9.7
Typical flow patterns in the right ventricular outflow tract at different cardiac phases (a–d) for a patient with manifest PH. (Reproduced from Reiter, U. et al., *PLoS ONE*, 8, e82212, 2013. doi:10.1371/journal.pone.008221. With permission.)

flow—composed primarily by three vascular beds of interest (hepatic arterial portal venous inflow, and hepatic venous drainage) all with markedly different velocity profiles—makes 4D-flow acquisition challenging [79]. Dual-VENC encoding might be an alternative to obtain accurate measures of low portal venous velocities and arterial flows simultaneously avoiding aliasing formation [22] and achieving high velocity-to-noise ratio in all these vessels.

9.3.6 Neurovascular Applications

Perhaps the most exciting discovery revealed by 4D flow in this area is related to stroke. Plaques of the ascending aorta and aortic arch are a relevant cause of embolic stroke [80,81]. However, the incidence of complex aortic plaques is highest in the proximal descending aorta, but these plaques are only considered an embolic source of stroke in the unlikely coincidence of severe aortic valve insufficiency causing retrograde flow and embolization in case of plaque rupture [82].

Nonetheless, 4D flow have revealed that retrograde embolization from complex descending aortic plaques is frequent in patients with atherosclerosis and could reach all supra-aortic arteries, indicating that descending aortic plaques should be considered a new embolic source of stroke, even in the absence of aortic valve insufficiency [26,83,84] (Figure 9.8). This mechanism has been shown to constitute the only probable source of cerebral infarction in a subgroup of patients with cryptogenic stroke etiology.

Intracranial aneurysms have also been studied by using 4D-flow MRI [85,86]. Complex flow patterns with vortices formation have been described. Moreover, 4D flow can identify areas of altered WSS or quantified intra-aneurysmal pressure gradients, both variables that might contribute to aneurysm progression or rupture [10,41]. However, although eventually 4D flow might provide pieces of evidence to identify patients at risk of aneurysmal growth or rupture; large cohorts studies with long-term follow-up are needed to confirm this hypothesis.

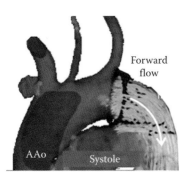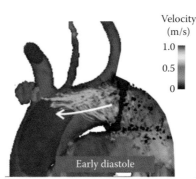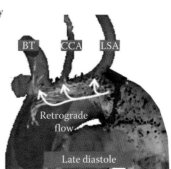

FIGURE 9.8

In vivo measurement and visualization of 3D blood flow in the proximal descending aorta using 4D-flow MRI. Increased aortic stiffness in a patient with aortic atherosclerosis results in substantial diastolic retrograde flow as demonstrated by time-resolved 3D pathlines originating from an emitter plane in the proximal descending aorta (DAo). Diastolic retrograde flow clearly reaches all three brain-feeding arteries and may provide a previously overlooked mechanism of retrograde embolization in stroke patients with high-risk plaques in the DAo. AAo, ascending aorta; BT, brachiocephalic trunk; CCA, left common carotid artery; DAo, descending aorta; LSA, left subclavian artery. (Reproduced from Markl, M. et al., *J. Magn. Reson. Imaging*, 36, 1015, 2012, doi: 10.1002/jmri.23632. With permission.)

9.3.7 Other Applications

Some studies have used 4D flow to analyze other vascular territories, such as carotid arteries [87,88], renal arteries [11,89], and arteries from lower extremities [90]. In carotid arteries, 4D flow could be a valuable technique for assessing the individual risk of flow-mediated atherosclerosis and plaque progression. In renal arteries, 4D flow can be used to provide morphological evaluation through unenhanced MR angiography, and functional information such as trans-stenotic pressure gradients.

9.4 4D Flow in Congenital Heart Diseases

In current clinical practice, cardiovascular MR is increasingly used in patients with congenital heart disease because it is a noninvasive technique, nonionizing, and is not limited by acoustic windows. The ability of combining precise depiction of the pathological anatomy plus accurate and reproducible quantification of blood flow using 4D flow makes cardiovascular MR a complete imaging technique in this patient group.

9.4.1 Advantages of 4D Flow in Congenital Heart Diseases

9.4.1.1 Time Efficiency

Quantification of 2D PC-MRI requires acquisition of a perpendicular plane in each one of the cardiovascular structures of interest. It can be particularly challenging in patients with congenital heart disease due to unusual cardiovascular structure arrangement and complex spatial orientations [91]. Quantification of the different blood flow sources requires planning and acquisition of one 2D PC-MRI scan for each structure of interest. Current methods for image planning such as real-time or interactive localizers [91,92] suffer from low spatial resolution and might fail to identify small or stenotic structures. Therefore, if multiple vessels need to be investigated, such as the case of multilevel shunting, it might result in a long scan time. In order to reduce the scan time, 2D PC-MRI can be performed during a single breath-hold of 10–15 s. However, free-breathing 2D PC-MRI, usually performed for 2–3 min, is a more physiological condition to study the flows. Moreover, ill and young patients might find difficult to hold their breaths. Hence, acquisition 2D PC-MRI is hard to plan and might result in no optimal through-plane cuts requiring multiple acquisitions, which in turn may lead to lengthen the scanning time. This is of relevance due to the elevated costs of MR scans and, even more, because studies in children are generally performed under general anesthesia.

The 4D-flow acquisition takes longer scanning time compared with a single 2D PC-MRI acquisition. Nevertheless, it has some time-efficiency advantages. For instance, 4D flow of the whole heart and great vessels [27] is an easy-to-plan sequence. The FoV geometry is a simple volume-box containing the heart and great vessels, and therefore, no expertise is required in case of complex cardiovascular anatomy. The analysis of the data is performed offline allowing assessment of all vessels of interest from a single MR scan (Figure 9.9). Furthermore, the assessment of flow of distorted and misaligned structures can be guided by other acquired high-resolution cardiovascular MR images such as 3D-steady-state free precession or black-blood sequences. An example of this time efficiency advantages has been shown in a case of multilevel shunting consisting of an atrial septal defect

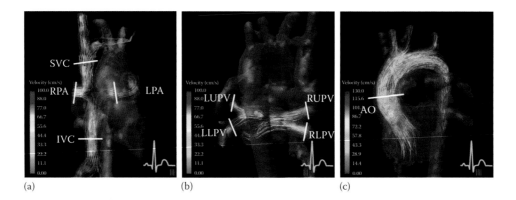

FIGURE 9.9
Result of a 4D-flow sequence obtained in a patient with palliated univentricular heart physiology shows that a single 4D-flow scan could provide multiple measurements of flow instead of acquiring multiple 2D PC-MRI images as it is shown in (a), (b) and (c). AO, aorta; IVC, inferior vena cava; LLPV, left lower pulmonary veins; LPA, left pulmonary artery; LUPV, left upper pulmonary veins; RLPV, right lower pulmonary veins; RPA, right pulmonary artery; RUPV, right upper pulmonary veins. (Reproduced from Valverde, I. et al., *J. Cardiovasc. Magn. Reson.*, 14, 25, 2012, doi: 10.1186/1532-429X-14-25. With permission.)

and partial anomalous pulmonary venous return. In this case, six 2D PC-MRI sequences under free-breathing were required, leading to a longer scan time compared to a single 4D-flow acquisition [93].

9.4.1.2 Retrospective Analysis

As we already mentioned, during the postprocessing of cardiovascular MR images, new findings can be discovered, which is especially common in the setting of congenital heart diseases. As the 2D PC-MRI acquisitions may have not been performed during the scanning, these clinically relevant flow investigations cannot be performed or the patient needs to be rebooked for an additional scan. This could be the case of systemic to pulmonary collaterals in the settings of hemi-Fontan circulation. These collateral vessels can be unnoticed during the MR scan and hence not investigated by conventional 2D PC-MRI. Unlike 2D PC-MRI, all the relevant flow information is stored within the 4D-flow volume as a backup. New vessels can be at anytime interrogated if new findings are discovered as shown in a recent report [94].

9.4.1.3 Individual Quantification of Blood Sources

Cardiovascular shunts are abnormal communication between heart chambers and vessels, and therefore, the blood flow is diverted from the normal circuit of the circulatory system. The relationship between the pulmonary (Q_p) and the systemic blood flow (Q_s), that is, the $Q_p{:}Q_s$ ratio describes the direction of the shunting blood flow (left-to-right or right-to-left) and its magnitude. This information is very helpful for follow-up and surgical management. It can be addressed by 2D PC-MRI by interrogating the main pulmonary artery and the aorta [95] On the other hand, in the case of multilevel

shunting, for example, anomalous pulmonary venous return and atrial septal defect, individual quantification of the shunting sources is important to decide whether the individual contribution to the total shunting flow justifies the risk of surgical repair. If this latter approach has to be performed by routine 2D PC-MRI investigations, it might suffer from the planning and time limitations described above, as several 2D PC-MRI scans need to be planned and acquired.

The 4D flow showed to be an accurate tool for depicted flow quantification of the blood sources contributing to the total shunting flow, such as in scimitar syndrome and atrial septal defect [96]. The depicted quantification of the different blood sources is also an excellent internal validator for flow measurement accuracy. For example, the pulmonary blood flow can be expressed as the pulmonary arteries flow or the pulmonary venous return [93].

9.4.2 Applications in CHD

9.4.2.1 Aorta

The 4D flow can help understand the aortic blood flow hemodynamics and the relationship between flow abnormalities and disease progression.

9.4.2.1.1 Aortic Valve

Congenital BAV occurs in 1.3% of the population and in 85% of patients with aortic coarctation [97]. About one-third of them will develop aortic regurgitation [98]. MRI with gradient-echo cine images has traditionally been used to visualize systolic signal void jets of aortic stenosis and diastolic jets of regurgitation, and 2D PC-MRI to quantify the net flow and regurgitant fraction. The association with aortic aneurysm in these patients is not well

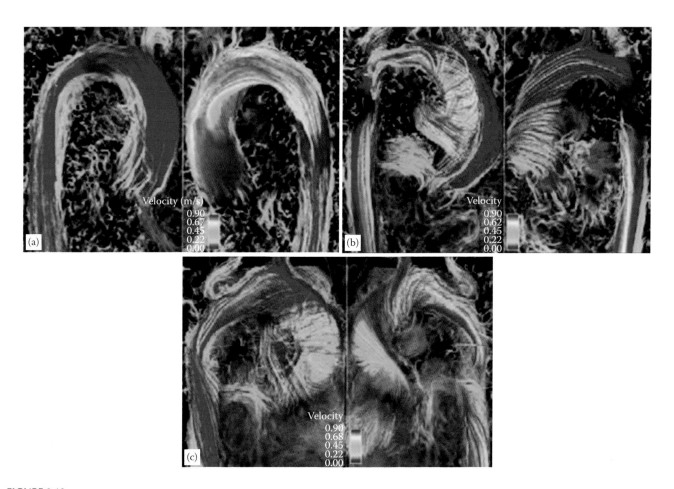

FIGURE 9.10
Typical flow patterns described in the ascending aorta of patients with BAV at different time during the cardiac cycle (a–c). (From Hope, M.D. et al., *J. Cardiovasc. Magn. Reson.*, 11, P184, 2009, doi: 10.1186/1532-429X-11-S1-P184.)

understood. Previous studies in the radial artery proximal to an arteriovenous dialysis fistula have shown that supra-physiological hemodynamics lead to an increase in the vessel size [99], supporting the theory that increased hemodynamic load in the proximal aorta of patients with BAV may result in progressive aortic dilatation [100]. Hope et al. [101] reported two discrete nested helices of mid-systolic blood flow in the ascending aorta of patients with BAV (Figure 9.10), demonstrating an increased hemodynamic burden, which may predispose to aneurysm formation [102]. The 4D-flow evaluation revealed altered helical systolic flow, with eccentric systolic flow jets, that result in focal WSS in specific regions of the ascending aorta. These locations coincide where aneurysms develop over time. Therefore, 4D flow may help to identify these patients at risk of developing aortic aneurysm and progression to rupture.

9.4.2.1.2 Aortic Coarctation

Coarctation of the aorta is a congenital stenosis almost invariably at the junction of the ductus arteriosus and the descending aorta. Assessment of the pressure

gradient across the coarctation can be carried out by PC-MRI. However, the previously explained pitfalls of plane align and plane placement in 2D PC-MRI can be solved by offline interrogation of the poststenotic jet performed in 4D-flow data. Far beyond quantification of collateral blood flow, 4D flow can also bring information about altered flow patterns not only downstream but also upstream the coarctation [103]. Aneurysm development following the repair of aortic coarctation have been related to operative procedure [104], patient's age at the initial repair [105], and altered hemodynamics [106]. From the literature, it is known that altered WSS forces can modify endothelial function and create areas at risk for cardiovascular remodeling, which may contribute to the development of aortic aneurysm and atherosclerosis [56]. Frydrychowicz et al. [107] demonstrated the diversity in hemodynamics associated with aneurysm formation after surgery in aortic coarctation. They hypothesized that flow pattern may predict the severity of the disease and might be related to aneurysm shapes (Figure 9.11).

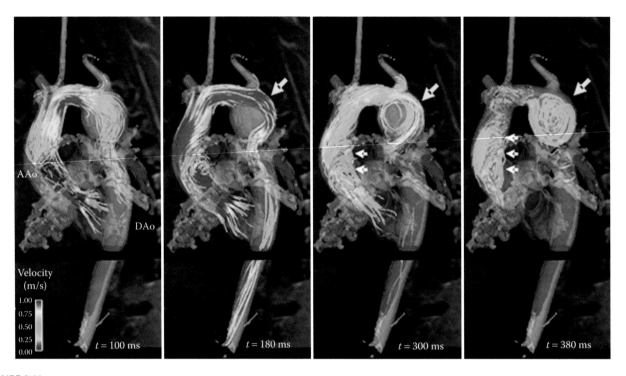

FIGURE 9.11
Flow pattern along the cardiac cycle in the thoracic aorta in a patient with an aneurysm of the proximal descending aorta. (Reproduced from Frydrychowicz, A. et al., *J. Cardiovasc. Magn. Reson.*, 10, 30, 2008, doi: 10.1186/1532-429X-10-30. With permission.)

9.4.2.2 Tetralogy of Fallot

Tetralogy of Fallot is the most common cyanotic congenital heart defect. Surgical repair results in an excellent overall prognosis; however, in the majority of patients, repair leads to chronic pulmonary regurgitation, which has been associated with the development of right ventricular dilatation and dysfunction, low exercise tolerance, propensity for significant cardiac arrhythmia, and sudden cardiac death [108].

As previously stated, 4D flow individual blood flow quantification within a single scan can be used to evaluate not only the degree of pulmonary regurgitation but also the flow ratio between both pulmonary arteries in the case of residual pulmonary stenosis. This valuable information is an additional parameter to anatomy that can be considered for endovascular intervention. Elevated peak systolic velocity and abnormal vortex flow in the main pulmonary artery have been demonstrated and may offer a new risk stratification parameter in this population [109].

9.4.2.3 Pulmonary Vein Stenosis

Pulmonary vein stenosis can be congenital, for example, total anomalous pulmonary venous return (TAPVR) [110], or acquired, for example, postcardiovascular surgical interventions [111], or radiofrequency ablation of atrial fibrillation [112]. Cardiac catheterization is currently considered the gold standard for the evaluation of

pulmonary venous anatomy but is invasive and X-ray dependent. Conventional cardiovascular MRI is limited by spatial resolution and therefore small pulmonary venous structures can be missed. Furthermore, turbulent flow in stenotic structures may be difficult to demonstrate with current 2D PC-MRI techniques. Transesophageal echocardiography [112] can provide this flow information, but probes need to be inserted in the esophagus during general anesthesia in children. Cardiovascular MR has been shown to be a complete approach for evaluation of this condition, as it can provide comprehensive anatomical and 4D-flow evaluation of postsurgical pulmonary vein stenosis, as reported by Valverde et al. [113]. The 4D-flow imaging correlated with transesophageal echocardiography images and was able to delineate the more distal hilar pulmonary veins, which could not be evaluated by conventional transesophageal echocardiography. Even more, 4D flow also allowed the quantification of blood flow in the pulmonary veins (Figure 9.12).

9.4.2.4 Cardiovascular Shunts

The quantification of shunt volumes in patients with congenital heart diseases and the flow ratio in the pulmonary (Q_p) and systemic circulation (Q_s) Q_p:Q_s are important parameters for clinical decisions and follow-up. As stated before, 2D PC-MRI can provide net flow ratios, but it is sometimes difficult to identify the different sources contributing to a shunt volume. The 4D flow

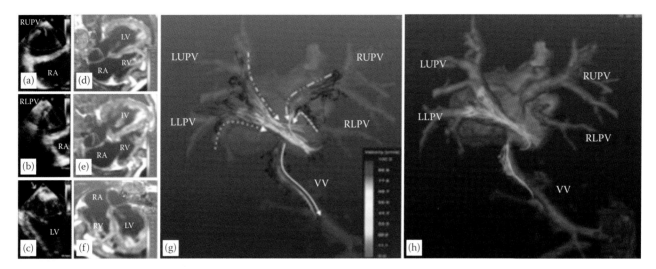

FIGURE 9.12
Patient diagnosed with right lower pulmonary vein (RLPV) stenosis and high-velocity Doppler (2.4 m/s) in the left pulmonary venous conflu-ence. However, transesophageal echocardiography (a–f) was unable to further delineate the more distal pulmonary veins (PVs) as shown by 4D flow (g and h). LA, left atrium; LLPV, left lower pulmonary vein; LUPV, left upper pulmonary veins; LV, left ventricle; RA, right atrium; RV, right ventricle; VV, vertical vein. (Reproduced from Valverde, I. et al., *J. Am. Coll. Cardiol.*, 58, e3, 2011. With permission.)

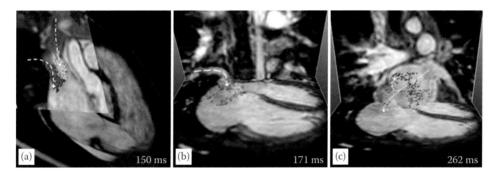

FIGURE 9.13
Particle released from different vessels allows the evaluation of the different contribution of shunting in this patient. Particles releases in (a) from the right upper pulmonary vein (RUPV, red) clearly showed the left to right shunting as the particles from the RUPV mix with the particles from the superior vena cava (turquoise) and drain into the right atrium. Particles releases from the right lower pulmonary vein (deep purplish red) drained into the right atrium, contributing as shown in (b). The particles released in the left pulmonary veins (pink) hit the atrial septum but some of them shunt through the ASD to the right atrium (c). (Reproduced from Valverde, I. et al., *Pediatr. Cardiol.*, 31, 1244, 2010. doi: 10.1007/s00246-010-9782-x. With permission.)

can depict and quantify shunts in atrial, ventricular, patent ductus arteriosus, aortopulmonary window, and partial anomalous pulmonary venous return.

9.4.2.4.1 Atrial Septal Defects and Anomalous Pulmonary Venous Return

Atrial septal defects are the most common shunt lesion detected de novo in adulthood and accounts for at least 30% of all congenital cardiac defects. Atrial septal defects may be associated with partial anomalous pulmonary venous return, which results in larger left-to-right shunt-ing. This may increase the risk of developing pulmonary artery hypertension and therefore planning for surgical correction should include not only precise delineation of the anatomy but also evaluation of the different blood sources contributing to the shunt. The evaluation of 4D

flow to accurately evaluate multilevel pre-tricuspid left-to-right shunting in congenital heart disease was shown by Valverde et al. [93]. They showed that 4D flow can accurately depict the individual quantification of the dif-ferent sources contributing to the left-to-right shunting from a single flow scan (Figure 9.13). They also showed the advantages over the current 2D PC-MRI in terms of internal validation of flow measurement accuracy and time efficiency.

9.4.2.4.2 Patent Ductus Arteriosus

The patent ductus arteriosus is the persistent distal left sixth aortic arch. It can be easily demonstrated using echocardiography and conventional MRI. Nevertheless, little is known about its hemodynamics. Recently, a comprehensive overview on blood flow patterns for the assessment of the temporal and spatial distributions of large patent ductus arteriosus with Eisenmenger's physiology was illustrated by Frydrychowicz et al. [114].

9.4.2.4.3 *Ventricular Septal Defects*

Methods based on net velocity measurements and planar imaging, such as Doppler ultrasound, have not been satisfactory to describe the time-varying 3D characteristics of blood flow within the cavities of the heart. Bolger et al. described and quantified the volume, distribution, and time changes of separate components of intraventricular flow in the normal left ventricle [115] using 4D flow. A 4D-flow study in a patient with a large ventricular septal defect revealed that in early systole the left ventricle particles crosses the ventricle septal defect and filled the main pulmonary artery, resulting in early pulmonary artery overflow and peak systolic pressure. During the late systole, because of the increase in the pulmonary artery pressure, streamlines from the left ventricle directs straight into the aorta [29].

9.4.2.4.4 *Aortopulmonary Window*

In a recent article, Wong et al. also showed the application of 4D flow in the presence of aortopulmonary window [116]. Retrospective analysis allowed accurate quantification of shunt flow within any plane of interest. This analysis was found to be difficult using traditional 2D PC-MRI due to abnormal flow patterns (please see Figure 1 in Wong et al. [116]).

9.4.2.5 Quantification of Collaterals

9.4.2.5.1 *Aortopulmonary Collaterals*

Following the staged Fontan palliation for univentricular heart physiology, development of systemic-to-pulmonary collateral circulation is frequently observed due to arterial hypoxemia. It has been associated with prolonged pleural effusions and increased mortality rate after the Fontan operation, but its effect on the long-term outcome has yet to be properly investigated [117]. It has been previously evaluated by conventional 2D PC-MRI at the expense of several individual flow investigations [117]. This small vessel network connects to the precapillary pulmonary vessels, resulting in an increased pulmonary overflow. In this situation, the pulmonary perfusion is greater than the pulmonary blood supplied by the pulmonary arteries. The 4D flow can assess the pulmonary perfusion by interrogating the pulmonary venous return and also provide information about the distribution between the right and the left lungs [118].

9.4.2.5.2 *Veno-Venous Collaterals*

The presence of veno-venous collaterals is also frequent in patients with palliated univentricular heart physiology. They usually connect the systemic neck veins and drains into the left atrium, increasing the premitral flow but with

no contribution to the pulmonary oxygenation. During the postprocessing analysis, these vessels can be identified and excluded for the pulmonary perfusion analysis [94].

9.4.2.5.3 *Collaterals in Aortic Coarctation*

In the case of hemodynamically significant aortic coarctation, there is a development of collateral circulation from the intercostal arteries and other collateral channels connecting the upper body to the descending aorta bypassing the aortic coarctation. Evaluation of these collaterals is important for surgical planning because it is an indirect sign of the severity of the coarctation. In the case of significant collateral blood flow, surgical repair can be performed by cross-clamping alone, as the collaterals may ensure proper blood flow to the lower limb. Otherwise, other surgical techniques are needed. Therefore, determining an increase in blood flow from the proximal to the distal descending aorta provides a direct measurement of the collateral blood flow to the lower part of the body [119]. Instead of prescribing several 2D PC-MRI planes, 4D flow can evaluate collateral blood flow in a single MR acquisition [103].

9.4.2.6 Univentricular Heart Physiology

9.4.2.6.1 *Stage II: Glenn/Bidirectional Cavopulmonary Connection*

Functional single ventricle palliation with a Glenn operation or superior cavopulmonary anastomosis is the preceding stage before the Fontan completion. As there is no effective pump to force the blood to enter into the lungs, optimal anastomosis and reconstructed anatomy should avoid any stenosis, which lead to energy loss and therefore upstream elevated pressure in the superior vena cava. In one-third of these patients, this may lead to raised trans-pulmonary pressure gradient and development of veno-venous collaterals. Such veno-venous collaterals channels may result in profound systemic desaturation. Unfortunately, because of its small size and anatomical variability, identification and more importantly quantification of the shunt is challenging.

In cases where the one-and-a-half ventricle approach is chosen as surgical repair, 4D-flow analysis of flow patterns allows us to evaluate the competition between the right ventricle and the Glenn shunt. Uribe et al. showed the asymmetric flow distribution and poor efficiency when severe pulmonary regurgitation is present (Figure 9.14), as most of the flow from the superior vena cava was suctioned toward the right ventricle instead of entering into the pulmonary arteries [49].

9.4.2.6.2 *Stage III: Fontan/Total Cavopulmonary Connection*

The Fontan operation completes the conversion of a single functional ventricle from a parallel circulation to an in-series circulation. Predictors of Fontan outcomes

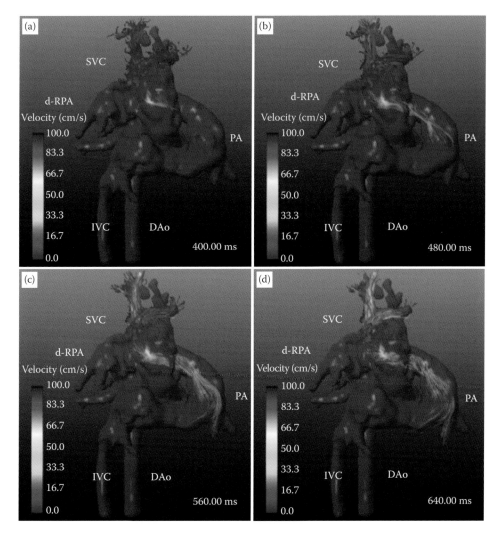

FIGURE 9.14
In this, patient with one-and-a-half ventricle repair 4D flow depicts that blood flow coming from the superior vena cava going into the right and main pulmonary artery instead of the right lung. Panels (a)–(d) show the hemodynamics at different time points during the cardiac cycle. (Reproduced from Uribe, S. et al., *Pediatr. Cardiol.*, 2012, doi: 10.1007/s00246-012-0288-6. With permission.)

involve several factors: anatomy (pulmonary artery size and reconstructed aortic arch), ventricular size and function, presence of collaterals and importantly blood flow hemodynamics, and efficiency. A 4D-flow evaluation showed that retrospective analysis allows quantification of any unexpected thoracic vessel such as the presence of a hemiazygos vein, resulting in steal hypoxic-syndrome (Figure 9.15).

The 4D-flow imaging can also provide new insights about caval flow contribution to both pulmonary arteries in patients with Fontan procedure. Estimation of flow contribution can be performed by using particle traces as recently validated by Bächler et al. [48,120] (Figures 9.16 and 9.17). Uneven flow distribution from the superior and inferior vena cava to the right and left pulmonary arteries might lead to the development of pulmonary artery venous malformations, and therefore this method might bring new insights in this patient population. Potential

clinical applications are evident: to identify patients at risk of developing pulmonary arteriovenous malformations as well as to aid in the follow-up of patients with Fontan circulation after cavopulmonary pathway modifications to restore a more balanced flow distribution.

9.4.3 4D-Flow and Cardiovascular Modeling

Computational fluid dynamics (CFD) is a technique of determining fluid flow characteristics by numerically solving equations of motion. In the field of cardiovascular disease, it has been applied to understand the blood flow dynamics and the effect of flow disturbances on atherosclerosis, aneurysm rupture, and several congenital heart diseases [121]. Solving the equations of fluid flow requires that certain initial conditions must be specified a priori. One great advantage of implementing numerical simulations with 4D flow is the possibility of using

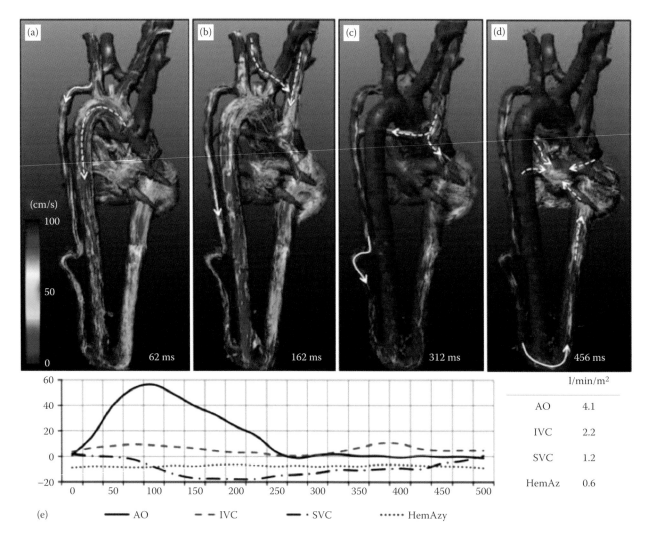

FIGURE 9.15
The 4D flow is able to identify and quantify downstream head-to-feet blood flow in the hemiazygos vein in this patient after after hemi-Fontan operation. Panels (a)–(d) show the nemodynamics along the cardiac cycle and panel (e) shows the flux curves and quantification of flow rate in the aorta (AO), inferior vena cava (IVC), superior vena cava (SVC) and in the HemAz. (Reproduced from Valverde, I. et al., *Cardiol. Young*, 1, 4, 2012. With permission.)

patient-specific hemodynamic information and therefore calculating subject specific relative pressure and other derived parameters. Noninvasive MR evaluation of cardiovascular geometry and velocity is of significant value to solve the equations using finite element computations. The combination of 4D-flow MRI and CFD can be used to refine both methodologies, which may help to enhance the assessment and understanding of blood flow *in vivo* [122].

9.5 Future Perspectives of 4D Flow

The 4D-flow imaging provides fabulous data to obtain new insight of the cardiovascular system. Solid steps in the development of this technique promise amazing perspectives. New advances on data acquisition allow reduction in the scanning time tremendously, and help to obtain 4D-flow data free of respiratory motion artifacts. Difficulties associated with both visualization and assessment of the data are still an issue; however, the amount of information that can be extracted from these datasets is worthwhile, at least for research purposes. Even though new software are available, there is still a need for more automatic processing methods, although this may be difficult to achieve in cases of complex cardiovascular anatomy, as is usually found in congenital heart diseases.

The 4D flow allows us to understand basic normal blood flow patterns, which is the first step to understand how different cardiovascular pathologies and diseases alter the normal hemodynamics. There has been a great effort for understanding flow patterns in a wide variety

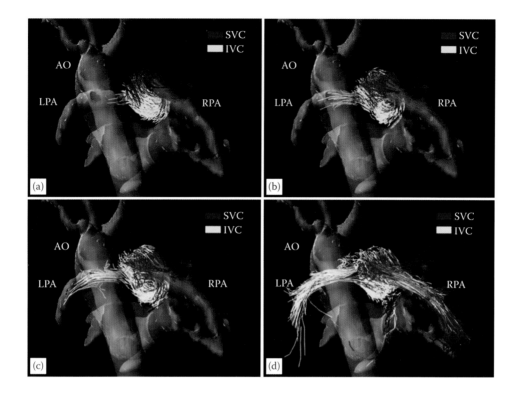

FIGURE 9.16
Emitting particle traces from the superior and inferior vena cava (SVC and IVC, red and yellow, respectively) it is possible to quantify the contribution of flow from each vein into the pulmonary arteries. Panels (a)–(d) show the flow distribution at different time points during the cardiac cycle. AO, aorta; IVC, inferior vena cava; LPA, left pulmonary artery; RPA, right pulmonary artery; SVC, superior vena cava. (Reproduced from Bachler, P. et al., *J. Thorac. Cardiovasc.*, 2012. With permission.)

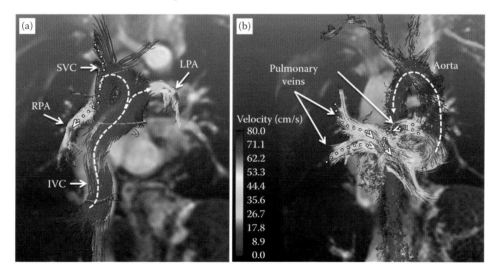

FIGURE 9.17
Particle traces in one patient with extracardiac total cavopulmonary connection show in (a) the Fontan circulation and in (b) the pulmonary venous return and aortic blood flow. Note that in (a), the flow split from the inferior vena cava (IVC) to the right and left pulmonary arteries (RPA and LPA) and the preferential stream flow from the superior vena cava (SVC) to the RPA. (Reproduced from Bächler, P. et al., *Radiology* [Internet], 1, 14, 2013. doi: 10.1148/radiol.12120778. With permission.)

of aortic and pulmonary diseases, as well as in healthy volunteers [32,76,109,123].

Not only flow patterns but also quantitative cardiovascular parameters provided by 4D flow promise new imaging biomarkers that may have a great

impact in clinical practice. For instance, altered WSS in poststenotic areas [124] may contribute to vascular remodeling as suggested in the recent literature [40]. Morbiducci et al. [31] proposed an analytical model for helical flow quantification, providing a better understanding of the role of pitch and torsion of the blood flow. Those investigations might allow the detection of anomalies in flow patterns, which could be useful in diagnostic [124–126], prognostic [107], and therapy [127–130]. Pressure mapping from 4D-flow data has been demonstrated in phantoms and in some patients; however, further studies are required to validate this application in patients with cardiovascular diseases [58], especially in patients with congenital heart disease.

Novel imaging techniques, such as 4D-flow MRI, are key to exploring new frontiers in the knowledge of flow dynamics, biomechanics, and cardiovascular remodeling, which might help to early detect pathophysiological adaptations in patients with several cardiovascular diseases.

References

1. Moran PR. A flow velocity zeugmatographic interlace for NMR imaging in humans. *Magn. Reson. Imaging* 1982;1:197–203. doi: 10.1016/0730-725X(82)90170-9.
2. Liang ZP. A model-based method for phase unwrapping. *IEEE Trans. Med. Imaging* 1996;15:893–897.
3. Pelc NJ, Shimakawa A, Bernstein MA. Encoding strategies for three-direction phase-contrast MR imaging of flow. *J. Magn. Reson. Imaging* 1991;1:405–413.
4. Markl M, Chan FP, Alley MT et al. Time-resolved three-dimensional phase-contrast MRI. *J. Magn. Reson. Imaging* 2003;17:499–506. doi: 10.1002/jmri.10272.
5. Dumoulin CL, Souza SP, Walker MF, Wagle W. Three-dimensional phase contrast angiography. *Magn. Reson. Med.* 1989;9:139–149. doi: 10.1002/mrm.1910090117.
6. Bernstein M, Shimakawa A, Pelc N. Minimizing TE in moment-nulled or flow-encoded two- and three-dimensional gradient-echo imaging. *J. Magn. Reson. Imaging* 1992;Sep–Oct:583–588.
7. Johnson KM, Markl M. Improved SNR in phase contrast velocimetry with five-point balanced flow encoding. *Magn. Reson. Med.* 2010;63:349–355. doi: 10.1002/mrm.22202.
8. Wigström L, Sjöqvist L, Wranne B. Temporally resolved 3D phase-contrast imaging. *Magn. Reson. Med.* 1996;36:800–803.
9. Gu T, Korosec FR, Block WF, Fain SB, Turk Q, Lum D, Zhou Y, Grist TM, Haughton V, Mistretta CA. PC VIPR: A high-speed 3D phase-contrast method for flow quantification and high-resolution angiography. *AJNR Am. J. Neuroradiol.* 2005;26:743–749. doi: 26/4/743 [pii].
10. Moftakhar R, Aagaard-Kienitz B, Johnson K, Turski PA, Turk AS, Niemann DB, Consigny D, Grinde J, Wieben O, Mistretta CA. Noninvasive measurement of intra-aneurysmal pressure and flow pattern using phase contrast with vastly undersampled isotropic projection imaging. *AJNR Am. J. Neuroradiol.* 2007;28:1710–1714. doi: 10.3174/ajnr.A0648.
11. François CJ, Lum DP, Johnson KM, Landgraf BR, Bley TA, Reeder SB, Schiebler ML, Grist TM, Wieben O. Renal arteries: Isotropic, high-spatial-resolution, unenhanced MR angiography with three-dimensional radial phase contrast. *Radiology* 2011;258:254–260. doi: 10.1148/radiol.10100443.
12. Turk AS, Johnson KM, Lum D, Niemann D, Aagaard-Kienitz B, Consigny D, Grinde J, Turski P, Haughton V, Mistretta C. Physiologic and anatomic assessment of a canine carotid artery stenosis model utilizing phase contrast with vastly undersampled isotropic projection imaging. *AJNR Am. J. Neuroradiol.* 2007;28:111–115.
13. Jiang J, Strother C, Johnson K, Baker S, Consigny D, Wieben O, Zagzebski J. Comparison of blood velocity measurements between ultrasound Doppler and accelerated phase-contrast MR angiography in small arteries with disturbed flow. *Phys Med Biol.* 2011 Mar 21;56(6):1755–1773. doi: 10.1088/0031-9155/56/6/015.
14. Frydrychowicz A, Landgraf BR, Niespodzany E, Verma RW, Roldán-Alzate A, Johnson KM, Wieben O, Reeder SB. Four-dimensional velocity mapping of the hepatic and splanchnic vasculature with radial sampling at 3 tesla: A feasibility study in portal hypertension. *J. Magn. Reson. Imaging* 2011;34:577–584. doi: 10.1002/jmri.22712.
15. Pike GB, Meyer CH, Brosnan TJ, Pelc NJ. Magnetic resonance velocity imaging using a fast spiral phase contrast sequence. *Magn. Reson. Med.* 1994;32:476–483.
16. Griswold MA, Jakob PM, Heidemann RM, Nittka M, Jellus V, Wang J, Kiefer B, Haase A. Generalized autocalibrating partially parallel acquisitions (GRAPPA). *Magn. Reson. Med.* 2002;47:1202–1210. doi: 10.1002/mrm.10171.
17. Peng H-H, Bauer S, Huang T-Y, Chung H-W, Hennig J, Jung B, Markl M. Optimized parallel imaging for dynamic PC-MRI with multidirectional velocity encoding. *Magn. Reson. Med.* 2010;64:472–480. doi: 10.1002/mrm.22432.
18. Giese D, Schaeffter T, Kozerke S. Highly undersampled phase-contrast flow measurements using compartment-based k-t principal component analysis. *Magn. Reson. Med.* 2013;69:434–443. doi: 10.1002/mrm.24273.
19. Knobloch V, Boesiger P, Kozerke S. Sparsity transform k-t principal component analysis for accelerating cine three-dimensional flow measurements. *Magn. Reson. Med.* 2012;63:53–63. doi: 10.1002/mrm.24431.
20. Kwak Y, Nam S, Akçakaya M, Basha TA, Goddu B, Manning WJ, Tarokh V, Nezafat R. Accelerated aortic flow assessment with compressed sensing with and without use of the sparsity of the complex difference image. *Magn. Reson. Med.* 2012;000:1–8. doi: 10.1002/mrm.24514.
21. Jung B, Stalder AF, Bauer S, Markl M. On the undersampling strategies to accelerate time-resolved 3D imaging using k-t-GRAPPA. *Magn. Reson. Med.* 2011;66:966–975. doi: 10.1002/mrm.22875.
22. Nett EJ, Johnson KM, Frydrychowicz A, Del Rio AM, Schrauben E, Francois CJ, Wieben O. Four-dimensional phase contrast MRI with accelerated dual velocity encoding. *J. Magn. Reson. Imaging* 2012;35:1462–1471. doi: 10.1002/jmri.23588.
23. Carlsson M, Töger J, Kanski M, Bloch K, Ståhlberg F, Heiberg E, Arheden H. Quantification and visualization of cardiovascular 4D velocity mapping accelerated with

parallel imaging or k-t BLAST: Head to head comparison and validation at 1.5 T and 3 T. *J. Cardiovasc. Magn. Reson.* 2011;13:55. doi: 10.1186/1532-429X-13-55.

24. Hsiao A, Lustig M, Alley MT, Murphy M, Chan FP, Herfkens RJ, Vasanawala SS. Rapid pediatric cardiac assessment of flow and ventricular volume with compressed sensing parallel imaging volumetric cine phase-contrast MRI. *AJR Am. J. Roentgenol.* 2012;198:W250–W259. doi: 10.2214/AJR.11.6969.

25. Markl M, Harloff A, Bley TA, Zaitsev M, Jung B, Weigang E, Langer M, Hennig J, Frydrychowicz A. Time-resolved 3D MR velocity mapping at 3T: Improved navigator-gated assessment of vascular anatomy and blood flow. *J. Magn. Reson. Imaging* 2007;25:824–831. doi: 10.1002/jmri.20871.

26. Harloff A, Strecker C, Frydrychowicz AP, Dudler P, Hetzel A, Geibel A, Kollum M, Weiller C, Hennig J, Markl M. Plaques in the descending aorta: A new risk factor for stroke? Visualization of potential embolization pathways by 4D MRI. *J. Magn. Reson. Imaging* 2007;26:1651–1655. doi: 10.1055/s-2007-987599.

27. Uribe S, Beerbaum P, Sørensen TS, Rasmusson A, Razavi R, Schaeffter T. Four-dimensional (4D) flow of the whole heart and great vessels using real-time respiratory self-gating. *Magn. Reson. Med.* 2009;62:984–992. doi: 10.1002/mrm.22090.

28. Stadlbauer A, van der Riet W, Crelier G, Salomonowitz E. Accelerated time-resolved three-dimensional MR velocity mapping of blood flow patterns in the aorta using SENSE and k-t BLAST. *Eur. J. Radiol.* 2010;75:e15–e21. doi: 10.1016/j.ejrad.2009.06.009.

29. Hope TA, Herfkens RJ. Imaging of the thoracic aorta with time-resolved three-dimensional phase-contrast MRI: A review. *Semin. Thorac. Cardiovasc. Surg.* 2008;20:358–364. doi: 10.1053/j.semtcvs.2008.11.013.

30. Unterhinninghofen R, Ley S, Ley-Zaporozhan J, von Tengg-Kobligk H, Bock M, Kauczor H-U, Szabó G, Dillmann R. Concepts for visualization of multidirectional phase-contrast MRI of the heart and large thoracic vessels. *Acad. Radiol.* 2008;15:361–369. doi: 10.1016/j.acra.2007.11.012.

31. Morbiducci U, Ponzini R, Rizzo G, Cadioli M, Esposito A, De Cobelli F, Del Maschio A, Montevecchi FM, Redaelli A. In vivo quantification of helical blood flow in human aorta by time-resolved three-dimensional cine phase contrast magnetic resonance imaging. *Ann. Biomed. Eng.* 2009;37:516–531. doi: 10.1007/s10439-008-9609-6.

32. Bächler P, Pinochet N, Sotelo J, Crelier G, Irarrazaval P, Tejos C, Uribe S. Assessment of normal flow patterns in the pulmonary circulation by using 4D magnetic resonance velocity mapping. *Magn. Reson. Imaging* 2013 Feb;31(2):178–188. doi: 10.1016/j.mri.2012.06.036.

33. Buonocore MH. Visualizing blood flow patterns using streamlines, arrows, and particle paths. *Magn. Reson. Med.* 1998;40:210–226. doi: 10.1002/mrm.1910400207.

34. Wigström L, Ebbers T, Fyrenius A, Karlsson M, Engvall J, Wranne B, Bolger AF. Particle trace visualization of intracardiac flow using time-resolved 3D phase contrast MRI. *Magn. Reson. Med.* 1999;41:793–799. doi: 10.1002/(SICI)1522-2594(199904)41:4<793::AID-MRM19>3.0.CO;2-2 [pii].

35. Shaaban A, Duerinckx A. Wall shear stress and early atherosclerosis: A review. *AJR Am. J. Roentgenol.* 2000;174:1657–1665.

36. Petersson S, Dyverfeldt P, Ebbers T. Assessment of the accuracy of MRI wall shear stress estimation using numerical simulations. *J. Magn. Reson. Imaging* 2012;36:128–38. doi: 10.1002/jmri.23610.

37. Hope MD, Sedlic T, Dyverfeldt P. Cardiothoracic magnetic resonance flow imaging. *J. Thorac. Imaging* 2013;28:217–230. doi: 10.1097/RTI.0b013e31829192a1.

38. Chalouhi N, Hoh B, Hasan D. Review of cerebral aneurysm formation, growth, and rupture. *Stroke* 44:3613–3622.

39. Meng H, Wang Z, Hoi Y, Gao L, Metaxa E, Swartz DD, Kolega J. Complex hemodynamics at the apex of an arterial bifurcation induces vascular remodeling resembling cerebral aneurysm initiation. *Stroke* 2007;38:1924–1931. doi: 10.1161/STROKEAHA.106.481234.

40. Cheng C, Tempel D, van Haperen R, van der Baan A, Grosveld F, Daemen MJAP, Krams R, de Crom R. Atherosclerotic lesion size and vulnerability are determined by patterns of fluid shear stress. *Circulation* 2006;113:2744–2753. doi: 10.1161/CIRCULATIONAHA.105.590018.

41. Boussel L, Rayz V, McCulloch C, Martin A, Acevedo-Bolton G, Lawton M, Higashida R, Smith WS, Young WL, Saloner D. Aneurysm growth occurs at region of low wall shear stress: Patient-specific correlation of hemodynamics and growth in a longitudinal study. *Stroke* 2008;39:2997–3002. doi: 10.1161/STROKEAHA.108.521617.

42. Laurent S, Cockcroft J, Van Bortel L, Boutouyrie P, Giannattasio C, Hayoz D, Pannier B, Vlachopoulos C, Wilkinson I, Struijker-Boudier H. Expert consensus document on arterial stiffness: Methodological issues and clinical applications. *Eur. Heart J.* 2006;27:2588–2605. doi: 10.1093/eurheartj/ehl254.

43. Markl M, Kilner PJ, Ebbers T. Comprehensive 4D velocity mapping of the heart and great vessels by cardiovascular magnetic resonance. *J. Cardiovasc. Magn. Reson.* 2011;13:7. doi: 10.1186/1532-429X-13-7.

44. Mohiaddin RH, Firmin DN, Longmore DB. Age-related changes of human aortic flow wave velocity measured noninvasively by magnetic resonance imaging. *J. Appl. Physiol.* 1993;74:492–497. doi: 10.1161/01.CIR.0000069826.36125.B4.

45. Wentland AL, Wieben O, François CJ, Boncyk C, Munoz Del Rio A, Johnson KM, Grist TM, Frydrychowicz A. Aortic pulse wave velocity measurements with under-sampled 4D flow-sensitive MRI: Comparison with 2D and algorithm determination. *J. Magn. Reson. Imaging* 2013;37:853–859. doi: 10.1002/jmri.23877.

46. Markl M, Wallis W, Brendecke S, Simon J, Frydrychowicz A, Harloff A. Estimation of global aortic pulse wave velocity by flow-sensitive 4D MRI. *Magn. Reson. Med.* 2010;63:1575–1582. doi: 10.1002/mrm.22353.

47. Nollen GJ, Groenink M, Tijssen JGP, Van Der Wall EE, Mulder BJM. Aortic stiffness and diameter predict progressive aortic dilatation in patients with Marfan syndrome. *Eur. Heart J.* 2004;25:1146–1152. doi: 10.1016/j.ehj.2004.04.033.

48. Bachler P, Valverde I, Uribe S. Quantification of caval flow contribution to the lungs in vivo after total cavopulmunary connection with 4-dimensional flow magnetic resonance imaging. *J. Thorac. Cardiovasc.* 2012 Mar;143(3):742–743. doi: 10.1016/j.jtcvs.2011.11.003.

49. Uribe S, Bächler P, Valverde I, Crelier GR, Beerbaum P, Tejos C, Irarrazaval P. Hemodynamic assessment in patients with one-and-a-half ventricle repair revealed by four-dimensional flow magnetic resonance imaging. *Pediatr. Cardiol.* 2013 Feb;34(2):447–451. doi: 10.1007/s00246-012-0288-6.

50. Eriksson J, Carlhäll CJ, Dyverfeldt P, Engvall J, Bolger AF, Ebbers T. Semi-automatic quantification of 4D left ventricular blood flow. *J. Cardiovasc. Magn. Reson.* 2010;12:9. doi: 10.1186/1532-429X-12-9.

51. Bogren H, Buonocore M. Helical-shaped streamlines do not represent helical flow. *Radiology* 2010 Dec;257(3):895–896; author reply 896. doi: 10.1148/radiol.101298.

52. Lee N, Taylor MD, Hor KN, Banerjee RK. Non-invasive evaluation of energy loss in the pulmonary arteries using 4D phase contrast MR measurement: A proof of concept. *Biomed. Eng. Online* 2013;12:93. doi: 10.1186/1475-925X-12-93.

53. Dyverfeldt P, Gårdhagen R, Sigfridsson A, Karlsson M, Ebbers T. On MRI turbulence quantification. *Magn. Reson. Imaging* 2009;27:913–922. doi: 10.1016/j.mri.2009.05.004.

54. Dyverfeldt P, Sigfridsson A, Kvitting J-PE, Ebbers T. Quantification of intravoxel velocity standard deviation and turbulence intensity by generalizing phase-contrast MRI. *Magn. Reson. Med.* 2006;56:850–858. doi: 10.1002/mrm.21022.

55. Petersson S, Dyverfeldt P, Gårdhagen R, Karlsson M, Ebbers T. Simulation of phase contrast MRI of turbulent flow. *Magn. Reson. Med.* 2010;64:1039–1046. doi: 10.1002/mrm.22494.

56. Malek AM, Alper SL, Izumo S. Hemodynamic shear stress and its role in atherosclerosis. *JAMA* 1999;282:2035–2042. doi: 10.1001/jama.282.21.2035.

57. Stein PD, Sabbah HN. Measured turbulence and its effect on thrombus formation. *Circ. Res.* 1974;35:608–614. doi: 10.1161/01.RES.35.4.608.

58. Tyszka JM, Laidlaw DH, Asa JW, Silverman JM. Three-dimensional, time-resolved (4D) relative pressure mapping using magnetic resonance imaging. *J. Magn. Reson. Imaging* 2000;12:321–329. doi: 10.1002/1522-2586(200008)12:2<321::AID-JMRI15>3.0.CO;2-2 [pii].

59. Sigovan M, Hope MD, Dyverfeldt P, Saloner D. Comparison of four-dimensional flow parameters for quantification of flow eccentricity in the ascending aorta. *J. Magn. Reson. Imaging* 2011;34:1226–30. doi: 10.1002/jmri.22800.

60. Den Reijer PM, Sallee D, van der Velden P et al. Hemodynamic predictors of aortic dilatation in bicuspid aortic valve by velocity-encoded cardiovascular magnetic resonance. *J. Cardiovasc. Magn. Reson.* 2010;12:4. doi: 10.1186/1532-429X-12-4.

61. Morbiducci U, Ponzini R, Rizzo G, Cadioli M, Esposito A, Montevecchi FM, Redaelli A. Mechanistic insight into the physiological relevance of helical blood flow in the human aorta: An in vivo study. *Biomech. Model. Mechanobiol.* 2011;10:339–355. doi: 10.1007/s10237-010-0238-2.

62. Lorenz R, Bock J, Barker AJ, von Knobelsdorff-Brenkenhoff F, Wallis W, Korvink JG, Bissell MM, Schulz-Menger J, Markl M. 4D flow magnetic resonance imaging in bicuspid aortic valve disease demonstrates altered distribution of aortic blood flow helicity. *Magn. Reson. Med.* 2013;00:1–12. doi: 10.1002/mrm.24802.

63. Wong KKL, Tu J, Kelso RM, Worthley SG, Sanders P, Mazumdar J, Abbott D. Cardiac flow component analysis. *Med. Eng. Phys.* 2010;32:174–188. doi: 10.1016/j.medengphy.2009.11.007.

64. Liu X, Pu F, Fan Y, Deng X, Li D, Li S. A numerical study on the flow of blood and the transport of LDL in the human aorta: The physiological significance of the helical flow in the aortic arch. *Am. J. Physiol. Heart Circ. Physiol.* 2009;297:H163–H170. doi: 10.1152/ajpheart.00266.2009.

65. Yancy CW, Jessup M, Bozkurt B et al. 2013 ACCF/AHA guideline for the management of heart failure: A report of the American College of Cardiology Foundation/American Heart Association Task Force on Practice Guidelines. *J. Am. Coll. Cardiol.* 2013;62:e147–e239. doi: 10.1016/j.jacc.2013.05.019.

66. Roes SD, Hammer S, van der Geest RJ, Marsan NA, Bax JJ, Lamb HJ, Reiber JHC, de Roos A, Westenberg JJM. Flow assessment through four heart valves simultaneously using 3-dimensional 3-directional velocity-encoded magnetic resonance imaging with retrospective valve tracking in healthy volunteers and patients with valvular regurgitation. *Invest. Radiol.* 2009;44:669–675. doi: 10.1097/RLI.0b013e3181ae99b5.

67. Van der Hulst AE, Westenberg JJM, Kroft LJM, Bax JJ, Blom NA, de Roos A, Roest AAW. Tetralogy of Fallot: 3D velocity-encoded MR imaging for evaluation of right ventricular valve flow and diastolic function in patients after correction. *Radiology* 2010;256:724–734. doi: 10.1148/radiol.10092269.

68. Westenberg JJM, Danilouchkine MG, Doornbos J, Bax JJ, van der Geest RJ, Labadie G, Lamb HJ, Versteegh MIM, de Roos A, Reiber JHC. Accurate and reproducible mitral valvular blood flow measurement with three-directional velocity-encoded magnetic resonance imaging. *J. Cardiovasc. Magn. Reson.* 2004;6:767–776. doi: 10.1081/JCMR-200036108.

69. Song J-M, Kim S-D, Kim J-H, Kim M-J, Kang D-H, Seo JB, Lim T-H, Lee JW, Song M-G, Song J-K. Long-term predictors of descending aorta aneurysmal change in patients with aortic dissection. *J. Am. Coll. Cardiol.* 2007;50:799–804. doi: 10.1016/j.jacc.2007.03.064.

70. Tsai TT, Evangelista A, Nienaber CA et al. Partial thrombosis of the false lumen in patients with acute type B aortic dissection. *N. Engl. J. Med.* 2007;357:349–359. doi: 10.1056/NEJMoa063232.

71. Svensson LG, Kouchoukos NT, Miller DC et al. Expert consensus document on the treatment of descending thoracic aortic disease using endovascular stent-grafts. *Ann. Thorac. Surg.* 2008; 85:S1–S41. doi: 10.1016/j.athoracsur.2007.10.099.

72. Tse KM, Chiu P, Lee HP, Ho P. Investigation of hemodynamics in the development of dissecting aneurysm within patient-specific dissecting aneurismal

aortas using computational fluid dynamics (CFD) simulations. *J. Biomech.* 2011;44:827–836. doi: 10.1016/j.jbiomech.2010.12.014.

73. Cheng Z, Tan FPP, Riga CV, Bicknell CD, Hamady MS, Gibbs RGJ, Wood NB, Xu XY. Analysis of flow patterns in a patient-specific aortic dissection model. *J. Biomech. Eng.* 2010;132:051007. doi: 10.1115/1.4000964.

74. Clough RE, Waltham M, Giese D, Taylor PR, Schaeffter T. A new imaging method for assessment of aortic dissection using four-dimensional phase contrast magnetic resonance imaging. *J. Vasc. Surg.* 2012;55:914–923. doi: 10.1016/j.jvs.2011.11.005.

75. McLaughlin V V, Archer SL, Badesch DB et al. ACCF/AHA 2009 expert consensus document on pulmonary hypertension: A report of the American College of Cardiology Foundation Task Force on Expert Consensus Documents and the American Heart Association: Developed in collaboration with the American College. *Circulation* 2009;119:2250–2294.

76. Reiter G, Reiter U, Kovacs G, Kainz B, Schmidt K, Maier R, Olschewski H, Rienmueller R. Magnetic resonance-derived 3-dimensional blood flow patterns in the main pulmonary artery as a marker of pulmonary hypertension and a measure of elevated mean pulmonary arterial pressure. *Circ. Cardiovasc. Imaging* 2008;1(1):23–30. doi: 10.1161/CIRCIMAGING.108.780247.

77. Sanz J, Kuschnir P, Rius T, Salguero R, Sulica R, Einstein AJ, Dellegrottaglie S, Fuster V, Rajagopalan S, Poon M. Pulmonary arterial hypertension: Noninvasive detection with phase-contrast MR imaging. *Radiology* 2007;243:70–79. doi: 10.1148/radiol.2431060477.

78. Stankovic Z, Csatari Z, Deibert P, Euringer W, Blanke P, Kreisel W, Abdullah Zadeh Z, Kallfass F, Langer M, Markl M. Normal and altered three-dimensional portal venous hemodynamics in patients with liver cirrhosis. *Radiology* 2012;262:862–873. doi: 10.1148/radiol.11110127.

79. Markl M, Frydrychowicz A, Kozerke S, Hope M, Wieben O. 4D flow MRI. *J. Magn. Reson. Imaging* 2012;36:1015–1036. doi: 10.1002/jmri.23632.

80. Sen S, Hinderliter A, Sen PK, Simmons J, Beck J, Offenbacher S, Ohman EM, Oppenheimer SM. Aortic arch atheroma progression and recurrent vascular events in patients with stroke or transient ischemic attack. *Circulation* 2007;116:928–935. doi: 10.1161/CIRCULATIONAHA.106.671727.

81. Amarenco P, Cohen A, Tzourio C, Bertrand B, Hommel M, Besson G, Chauvel C, Touboul PJ, Bousser MG. Atherosclerotic disease of the aortic arch and the risk of ischemic stroke. *N. Engl. J. Med.* 1994;331:1474–1479. doi: 10.1056/NEJM199412013312202.

82. Kronzon I. Aortic atherosclerotic disease and stroke. *Circulation* 2006;114:63–75. doi: 10.1161/CIRCULATIONAHA.105.593418.

83. Harloff A, Simon J, Brendecke S et al. Complex plaques in the proximal descending aorta: An underestimated embolic source of stroke. *Stroke* 2010;41:1145–1150. doi: 10.1161/STROKEAHA.109.577775.

84. Harloff A, Strecker C, Dudler P et al. Retrograde embolism from the descending aorta: Visualization by multidirectional 3D velocity mapping in cryptogenic stroke. *Stroke* 2009;40(4):1505–1508. doi: 10.1161/STROKEAHA.108.530030.

85. Hope TA, Hope MD, Purcell DD, von Morze C, Vigneron DB, Alley MT, Dillon WP. Evaluation of intracranial stenoses and aneurysms with accelerated 4D flow. *Magn. Reson. Imaging* 2010;28:41–46. doi: 10.1016/j.mri.2009.05.042.

86. Meckel S, Stalder AF, Santini F, Radü E-W, Rüfenacht DA, Markl M, Wetzel SG. In vivo visualization and analysis of 3-D hemodynamics in cerebral aneurysms with flow-sensitized 4-D MR imaging at 3 T. *Neuroradiol.* 2008;50(6):473–484. doi: 10.1007/s00234-008-0367-9.

87. Markl M, Wegent F, Zech T, Bauer S, Strecker C, Schumacher M, Weiller C, Hennig J, Harloff A. In vivo wall shear stress distribution in the carotid artery: Effect of bifurcation geometry, internal carotid artery stenosis, and recanalization therapy. *Circ. Cardiovasc. Imaging* 2010;3:647–655. doi: 10.1161/CIRCIMAGING.110.958504.

88. Harloff A, Albrecht F, Spreer J et al. 3D blood flow characteristics in the carotid artery bifurcation assessed by flow-sensitive 4D MRI at 3T. *Magn. Reson. Med.* 2009;61:65–74. doi: 10.1002/mrm.21774.

89. Bley TA, Johnson KM, Francois CJ, Reeder SB, Schiebler ML, R. Landgraf B, Consigny D, Grist TM, Wieben O. Noninvasive assessment of transstenotic pressure gradients in porcine renal artery stenoses by using vastly undersampled phase-contrast MR angiography. *Radiology* 2011;261:266–273. doi: 10.1148/radiol.11101175.

90. Frydrychowicz A, Winterer JT, Zaitsev M, Jung B, Hennig J, Langer M, Markl M. Visualization of iliac and proximal femoral artery hemodynamics using time-resolved 3D phase contrast MRI at 3T. *J. Magn. Reson. Imaging* 2007;25:1085–1092. doi: 10.1002/jmri.20900.

91. Boegaert J, Dymarkowski S, Taylor AM, Muthurangu V eds. Cardiovascular MRI planes and segmentations. In *Clinical Cardiac MRI*: With Interactive CD-ROM. Berlin: New York: Springer; 2005, p. 549.

92. Lee VS, Resnick D, Bundy JM, Simonetti OP, Lee P, Weinreb JC. Cardiac function: MR evaluation in one breath hold with real-time true fast imaging with steady-state precession. *Radiology* 2002;222:835–842. doi: 10.1148/radiol.2223011156.

93. Valverde I, Simpson J, Schaeffter T, Beerbaum P. 4D phase-contrast flow cardiovascular magnetic resonance: Comprehensive quantification and visualization of flow dynamics in atrial septal defect and partial anomalous pulmonary venous return. *Pediatr. Cardiol.* 2010;31:1244–1248. doi: 10.1007/s00246-010-9782-x.

94. Valverde I, Rachel C, Kuehne T, Beerbaum P. Comprehensive four-dimensional phase-contrast flow assessment in hemi-Fontan circulation: Systemic-to-pulmonary collateral flow quantification. *Cardiol. Young* 2011;21:116–119. doi: 10.1017/S1047951110001575.

95. Beerbaum P, Körperich H, Barth P, Esdorn H, Gieseke J, Meyer H. Noninvasive quantification of left-to-right shunt in pediatric patients: Phase-contrast cine

magnetic resonance imaging compared with invasive oximetry. *Circulation* 2001;103:2476–2482. doi: 10.1161/01.CIR.103.20.2476.

96. Frydrychowicz A, Landgraf B, Wieben O, François CJ. Images in Cardiovascular Medicine. Scimitar syndrome: Added value by isotropic flow-sensitive four-dimensional magnetic resonance imaging with PC-VIPR (phase-contrast vastly undersampled isotropic projection reconstruction). *Circulation* 2010;121:e434–e436.

97. Hoffman JIE, Kaplan S. The incidence of congenital heart disease. *J. Am. Coll. Cardiol.* 2002;39:1890–1900.

98. Fenoglio JJ, McAllister HA, DeCastro CM, Davia JE, Cheitlin MD. Congenital bicuspid aortic valve after age 20. *Am. J. Cardiol.* 1977;39:164–169. doi: 10.1016/S0002-9149(77)80186-0.

99. Girerd X, London G, Boutouyrie P, Mourad JJ, Safar M, Laurent S. Remodeling of the radial artery in response to a chronic increase in shear stress. *Hypertension* 1996;27:799–803. doi: 10.1161/01.HYP.27.3.799.

100. Davies RR, Kaple RK, Mandapati D, Gallo A, Botta DM, Elefteriades JA, Coady MA. Natural history of ascending aortic aneurysms in the setting of an unreplaced bicuspid aortic valve. *Ann. Thorac. Surg.* 2007;83:1338–1344. doi: 10.1016/j.athoracsur.2006.10.074.

101. Hope MD, Meadows AK, Hope TA, Ordovas KG, Saloner D, Reddy GP, Alley MT, Higgins CB. 4D flow evaluation of abnormal flow patterns with bicuspid aortic valve. *J. Cardiovasc. Magn. Reson.* 2009;11:P184. doi: 10.1186/1532-429X-11-S1-P184.

102. Hope MD, Meadows AK, Hope TA, Ordovas KG, Reddy GP, Alley MT, Higgins CB. Images in cardiovascular medicine. Evaluation of bicuspid aortic valve and aortic coarctation with 4D flow magnetic resonance imaging. *Circulation* 2008;117:2818–2819. doi: 10.1161/CIRCULATIONAHA.107.760124.

103. Hope MD, Meadows AK, Hope TA, Ordovas KG, Saloner D, Reddy GP, Alley MT, Higgins CB. Clinical evaluation of aortic coarctation with 4D flow MR imaging. *J. Magn. Reson. Imaging* 2010;31:711–718. doi: 10.1002/jmri.22083.

104. Martin RS 3rd, Edwards WH J, JM J, Edwards WH S, JL M. Ruptured abdominal aortic aneurysm: A 25-year experience and analysis of recent cases. *Am. Surg.* 1988;54:539–543.

105. De Divitiis M, Pilla C, Kattenhorn M, Zadinello M, Donald A, Leeson P, Wallace S, Redington A, Deanfield JE. Vascular dysfunction after repair of coarctation of the aorta: Impact of early surgery. *Circulation* 2001 Sep;104(12Suppl1):I165–I170. doi: 10.1161/hc37t1.094900.

106. Von Kodolitsch Y, Aydin MA, Koschyk DH et al. Predictors of aneurysmal formation after surgical correction of aortic coarctation. *J. Am. Coll. Cardiol.* 2002;39:617–624. doi: 10.1016/S1062-1458(02)00754-7.

107. Frydrychowicz A, Arnold R, Hirtler D, Schlensak C, Stalder AF, Hennig J, Langer M, Markl M. Multidirectional flow analysis by cardiovascular magnetic resonance in aneurysm development following repair of aortic coarctation. *J. Cardiovasc. Magn. Reson.* 2008;10:30. doi: 10.1186/1532-429X-10-30.

108. Gatzoulis MA, Balaji S, Webber SA et al. Risk factors for arrhythmia and sudden cardiac death late after repair of tetralogy of Fallot: A multicentre study. 2000;356(9234):975–981. doi: 10.1016/S0140-6736(00)02714-8.

109. Francois CJ, Srinivasan S, Schiebler ML, Reeder SB, Niespodzany E, Landgraf BR, Wieben O, Frydrychowicz A. 4D cardiovascular magnetic resonance velocity mapping of alterations of right heart flow patterns and main pulmonary artery hemodynamics in tetralogy of Fallot. *J. Cardiovasc. Magn. Reson.* 2012;14:16. doi: 10.1186/1532-429X-14-16.

110. Holt DB, Moller JH, Larson S, Johnson MC. Primary pulmonary vein stenosis. *Am. J. Cardiol.* 2007;99:568–572. doi: 10.1016/j.amjcard.2006.09.100.

111. Hancock Friesen CL, Zurakowski D, Thiagarajan RR, Forbess JM, del Nido PJ, Mayer JE, Jonas RA. Total anomalous pulmonary venous connection: An analysis of current management strategies in a single institution. *Ann. Thorac. Surg.* 2005;79:596–606; discussion 596–606. doi: 10.1016/j.athoracsur.2004.07.005.

112. Sigurdsson G, Troughton RW, Xu XF, Salazar HP, Wazni OM, Grimm RA, White RD, Natale A, Klein AL. Detection of pulmonary vein stenosis by transesophageal echocardiography: Comparison with multicletector computed tomography. *Am. Heart J.* 2007;153:800–806.

113. Valverde I, Miller O, Beerbaum P, Greil G. Imaging of pulmonary vein stenosis using multidimensional phase contrast magnetic resonance imaging (4 Dimensional flow). *J. Am. Coll. Cardiol.* 2011;58:e3.

114. Frydrychowicz A, Bley TA, Dittrich S, Hennig J, Langer M, Markl M. Visualization of vascular hemodynamics in a case of a large patent ductus arteriosus using flow sensitive 3D CMR at 3T. *J. Cardiovasc. Magn. Reson.* 2007;9:585–587. doi: 772835820.

115. Bolger AF, Heiberg E, Karlsson M, Wigström L, Engvall J, Sigfridsson A, Ebbers T, Kvitting J-PE, Carlhäll CJ, Wranne B. Transit of blood flow through the human left ventricle mapped by cardiovascular magnetic resonance. *J. Cardiovasc. Magn. Reson.* 2007;9:741–747. doi: 10.1080/10976640701544530.

116. Wong J, Marthur S, Giese D, Pushparajah K, Schaeffter T, Razavi R. Analysis of aortopulmonary window using cardiac magnetic resonance imaging. *Circulation* 2012;126:e228–e229.

117. Grosse-Wortmann L, Al-Otay A, Yoo S-J. Aortopulmonary collaterals after bidirectional cavopulmonary connection or Fontan completion: Quantification with MRI. *Circ. Cardiovasc. Imaging* 2009;2:219–225. doi: 10.1161/CIRCIMAGING.108.834192.

118. Valverde I, Nordmeyer S, Uribe S, Greil G, Berger F, Kuehne T, Beerbaum P. Systemic-to-pulmonary collateral flow in patients with palliated univentricular heart physiology: Measurement using cardiovascular magnetic resonance 4D velocity acquisition. *J. Cardiovasc. Magn. Reson.* 2012;14:25. doi: 10.1186/1532-429X-14-25.

119. Steffens JC, Bourne MW, Sakuma H, O'Sullivan M, Higgins CB. Quantification of collateral blood flow in coarctation of the aorta by velocity encoded cine magnetic resonance imaging. *Circulation* 1994;90:937–943. doi: 10.1161/01.CIR.90.2.937.

120. Bächler P, Valverde I, Pinochet N, Nordmeyer S, Kuehne T, Crelier G, Tejos C, Irarrazaval P, Beerbaum P, Uribe

S. Caval blood flow distribution in patients with Fontan circulation: Quantification by using particle traces from 4D flow MR imaging. *Radiology* 2013:267(1):67–75. doi: 10.1148/radiol.12120778.

121. DeCampli WM, Argueta-Morales IR, Divo E, Kassab AJ. Computational fluid dynamics in congenital heart disease. *Cardiol. Young* 2012;22:800–808.

122. Wittek A, Nielsen PMF, Miller K eds. Patient specific hemodynamics: Combined 4D flow-sensitive MRI and CFD. In *Computational Biomechanics for Medicine.* New York: Springer; 2011, pp. 27–38.

123. Hope TA, Markl M, Wigström L, Alley MT, Miller DC, Herfkens RJ. Comparison of flow patterns in ascending aortic aneurysms and volunteers using four-dimensional magnetic resonance velocity mapping. *J. Magn. Reson. Imaging* 2007;26:1471–1479. doi: 10.1002/jmri.21082.

124. Frydrychowicz A, Berger A, Russe MF et al. (2008). Time-resolved magnetic resonance angiography and flow-sensitive 4-dimensional magnetic resonance imaging at 3 Tesla for blood flow and wall shear stress analysis. *J. Thorac. Cardiovasc. Surg.* 2008;136:400–407. doi: 10.1016/j.jtcvs.2008.02.062.

125. Frydrychowicz A, Harloff A, Jung B, Zaitsev M, Weigang E, Bley TA, Langer M, Hennig J, Markl M. Time-resolved, 3-dimensional magnetic resonance flow analysis at 3 T: Visualization of normal and pathological aortic vascular hemodynamics. *J. Comput. Assist. Tomogr.* 31:9–15.

126. Frydrychowicz A, Weigang E, Langer M, Markl M. Flow-sensitive 3D magnetic resonance imaging reveals complex blood flow alterations in aortic Dacron graft repair. *Interact. Cardiovasc. Thorac. Surg.* 2006;5:340–2.

127. Fogel MA, Weinberg PM, Hoydu AK, Hubbard AM, Rychik J, Jacobs ML, Fellows KE, Haselgrove J. Effect of surgical reconstruction on flow profiles in the aorta using magnetic resonance blood tagging. *Ann. Thorac. Surg.* 1997;63(6):1691–1700. doi: 10.1016/S0003-4975(97)00330-5.

128. Frydrychowicz A, Arnold R, Harloff A, Schlensak C, Hennig J, Langer M, Markl M. Images in cardiovascular medicine. In vivo 3-dimensional flow connectivity mapping after extracardiac total cavopulmonary connection. *Circulation* 2008;118:e16–e17.

129. Reiter U, Reiter G, Kovacs G et al. Evaluation of elevated mean pulmonary arterial pressure based on magnetic resonance 4D velocity mapping: Comparison of visualization techniques. *PLoS ONE* 2013;8(12):e82212. doi:10.1371/journal.pone.008221.

130. Valverde I, Razishankar A, Miller O, Razavi R. Hemiazygos vein "steal hypoxic-sindrome" after hemi-Fontan operation: Comprehensive four-dimensional flow magnetic resonance imaging. *Cardiol. Young* 2012:22(4):481–484. doi: 10.1017/S1047951112000248.

10

Magnetic Resonance Imaging for Trachea, Bronchi, and Lung

Yoshiharu Ohno

CONTENTS

10.1 Introduction

When magnetic resonance imaging (MRI) was first implemented, many investigators were interested in this new technique for not only brain but also other areas including chest. As a result, from the 1980s to the early 1990s, MRI was tested to evaluate different lung diseases as well as mediastinal, pleural, and cardiac diseases by many physicists and radiologists. However, because the MR systems, sequences, and other applications at that time were very primitive and limited, adequate image quality within an appropriate examination time could not be realized, so that it could not be demonstrated that MR could be substituted for computed tomography (CT), pulmonary angiography, and/or nuclear medicine studies. Until 2000, MRI was used only for some minor clinical indications.

In the 2000s, however, technical advances in sequencing, scanners and coils, adaptation of parallel imaging techniques, utilization of contrast media, and development of post-processing tools were reported by many basic and clinical researchers, in particular for lung MRI, which has been one of the more challenging fields for MRI. State-of-the-art MRI of the lung may therefore be able to perform as a substitute and/or in a complimentary role in management of patients with pulmonary and/or cardiopulmonary diseases. In addition, continuing developments have made it possible to provide not only morphological but also pulmonary functional, physiological, and physiopathological information for 1.5 Tesla (T) as well as 3 T MR systems. The focus of this chapter is thus on recent advances in MRI for several important pulmonary diseases, which may constitute useful indications of the nature of current and/or future clinical practices. It is hoped the reader will gain further insights into the progress that has been made in lung MRI since the 1990s and be able to use this information for their own practice where applicable.

10.2 Problematic Factors for Lung MRI

Conventional MRI applied to routine clinical practice is based on proton MRI and mainly relies on the signal from the protons in the hydrogen nuclei of water molecules. For conventional proton MRI, the nuclear spins of the hydrogen atoms are brought into preferential alignment by the large static magnetic field of the MR scanner in an approximately 1 nuclei per million. This makes MRI of the lung difficult because of the low physical density of the lung and a weak signal due to the low proton density within the lung. In addition, the many air–tissue interfaces are responsible for the extremely heterogeneous magnetic susceptibility of the lung [1–6]. While essential for gas exchange, these air–soft-tissue interfaces produce large magnetic field gradients that dephase the MR signal and cause a rapid decay of the MR signal in the lung. Moreover, the T_2^* value of the lung parenchyma is very short, ranging from 0.9 to 2.2 ms [5,6]. As a result of these problematic factors and the need to overcome them, MRI of the lung is now considered one of the more challenging fields with reference to not only MRI but also radiology in general.

As the first step in this new academic and clinical research field, MRI of pulmonary parenchyma has recently proved feasible despite the short T_2^* of lung tissue. This was made possible by the use of several new techniques, which have been developed since the 1990s, such as hyperpolarized noble gas MRI, utilization of contrast media, application of turbo, or fast spin-echo (SE) or gradient-echo (GRE) sequences with ultra-short echo time (TE) or projection reconstruction techniques, in which the data are acquired during the free induction decay, and application of cardiac and/or respiratory triggering for fast MR sequences as well as utilization of new techniques for suppressing the signals from fat, blood, and so on [5–12].

10.3 MRI of Trachea

MRI of trachea is mainly used to examine tracheal invasion from thyroid cancer because tumor invasion of the trachea is a major cause of death for patients with thyroid carcinoma [13]. The incidence of tracheal invasion for thyroid cancer patients reportedly ranges from 1% to 13% [14,15], and surgical removal of the tumor is the primary treatment procedure for such patients with preserved healthy structures [16]. As a result, surgical planning for these patients is altered considerably by the presence of tumor invasion of the trachea [17,18], and preoperative radiologic assessment of the absence or presence and, in the latter case, of the extent of tumor invasion of the trachea is clinically advantageous for surgical planning and prediction of prognosis.

Currently, the presence of a mass lesion in the tracheal lumen was the only reliable radiologic sign of tumor invasion of the trachea, while invasion of the superficial tracheal layers could be identified by surgeons only

at the time of surgical exploration [17–19]. During the last decade, however, a few investigators have tried to determine the utility of MRI for the assessment of tracheal and/or surrounding organ invasions as well as laryngeal nerve recurrence from thyroid cancers [20–22].

According to the findings of one of the studies reported by these investigators [21], the trachea shows a horseshoe or circular configuration in many normal subjects, while the distance between the center of the trachea and the midline of the spinal body is normally less than 4 mm. Signal intensity on non-contrast-enhanced MR images and the thickness of the mucous membrane and posterior membranous portion of the trachea in normal subjects were found to be similar to the findings for the cadaveric trachea. The adventitia was too thin to be identified on MR images, but incomplete fat planes between the cartilage and the thyroid gland were sometimes observed. On T_2-weighted turbo SE images, over 80% of normal subjects had a horseshoe-shaped trachea of low signal intensity and less than 15% showed small areas of higher signal intensity in the cartilage. On contrast-enhanced (CE) T_1-weighted SE images, mucous membrane and the thyroid gland showed intermediate to fair enhancement, but no discernible enhancement, which could have made it easier to differentiate the tracheal rings from other structures, was found in the cartilage, intercartilaginous membranes, or posterior membranous portion.

On the other hand, the same study found that thyroid cancers could be identified as hyper- (48%), iso- (43%), and hypointensive (9%) on non-CE-T_1-weighted turbo SE images, with corresponding ratios of 85%, 9%, and 6% on T_2-weighted SE images. In addition, over 80% of thyroid cancers showed good enhancement (84%) on CE-T_1-weighted turbo SE images, while others were determined as moderate (11%) and mild (5%) enhancements. When radiological findings of thyroid cancer with and without tracheal invasion were compared, the maximum axial diameter and distance of displacement of the trachea, the incidence of soft tissue in the cartilage, tumor extension to the posterior membranous portion, and intraluminal mass showed significant differences between thyroid cancer patients with and without tracheal invasion. Moreover, these two types of patients displayed significant differences in the distribution of deformity and tumor circumference of the trachea. The soft tissue in the cartilage, intraluminal mass, and tumor circumference of the trachea of 180° or greater were also determined as significant factors for predicting tracheal invasion. Of the three factors, tracheal invasion was predicted with the following diagnostic performances: sensitivity: ranged from 43% to 100%, specificity: ranged from 84% to 100%, and accuracy: ranged from 81% to 90% (Figure 10.1). In addition, another study found that, when apparent intraluminal tumor extension, erosion of the walls, displacement of the structures by the tumors as

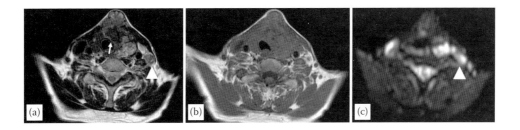

FIGURE 10.1

A 74-year-old male thyroid cancer patient with tracheal invasion and left cervical lymph node metastasis: (a) T_2-weighted turbo SE image, (b) T_1-weighted turbo SE image, and (c) DWI. Thyroid cancer is visualized as mass lesions from the left thyroid gland and thyroid isthmus. This tumor shows heterogeneous and intermediate and/or high signal intensities on T_2-weighted turbo SE image, as well as homogeneous and low signal intensity, and heterogeneous and low or high signal intensity on DWI. On the T_2-weighted turbo SE image, tracheal cartilage is missing (arrow), and the cervical lymph node (arrowhead) is represented as a high signal intensity area on T_2-weighted turbo SE image and DWI. Pathological examination of a surgical specimen from this patient led to a diagnosis of tracheal invasion and left cervical lymph node metastasis.

well as loss of the intervening fat plane, regardless of the size of the tumor in the affected organs, were used as morphological diagnosis criteria for carotid artery and cartilage invasion, diagnostic capability of the morphological diagnosis criteria was higher than that of determination of the degree of circumferential encroachment of tumors [20]. On the other hand, for tracheal and esophageal invasion, diagnostic capability was reversed. Therefore, tracheal MRI can be considered a good indicator for thyroid cancer in routine clinical practice.

10.4 MRI of Bronchus and Bronchi for Airway Disease

10.4.1 Cystic Fibrosis

Cystic fibrosis (CF) is an airway disease, which is caused by mutations in the *CFTR* gene, and remains one of the most frequent, lethal, inherited diseases among the populations of Europe and the United States. Because of progress in therapeutic options and the management of CF, the life expectancy of patients with CF has increased substantially, while radiologic examination is considered crucial for successful management of patients with CF. For this reason, CT is currently considered to be the gold standard for the assessment of morphologic changes of the airways and lung parenchyma [23–26], while during the last decade MRI has been proposed as a new technique for the assessment of CF.

Although the spatial resolution of MRI is lower than that of CT, it has the advantage of being able to distinguish different aspects of tissues on the basis of differences in contrast on T_1-weighted and T_2-weighted turbo SE images, as well as differences in enhancement after contrast media administration [27]. For the qualitative assessment of CF patients, the accuracy of MR assessment of bronchiectasis depends on the bronchial

level, bronchial diameter, wall thickness, wall signal, and the signal within the bronchial lumen. In general, the central bronchi and central bronchiectasis are well visualized. However, normal peripheral bronchi starting at the third to fourth generation are poorly visualized, and the depiction of bronchial wall thickening at these generations depends on bronchial size and signal [27]. A high signal of the bronchial wall on T_2-weighted turbo SE images represents edema, possibly caused by active inflammation. Moreover, enhancement of the thickened bronchial wall on CE-T_1-weighted MR images with and without fat-suppression pulse is related to inflammatory activity. Furthermore, mucus plugging is well visualized on T_2-weighted turbo SE image, even down to the small airways, and, since mucus plugs are not enhanced, they are easily differentiated from bronchial wall thickening [27]. Air-fluid levels are also indicative of active infection and can be visualized as high signal intensity. However, discriminating a bronchus with an air-fluid level from one with a partial mucus plug or a severely thickened wall can be difficult.

As well as bronchial findings obtained with MRI, pulmonary consolidation is an important finding for CF patients, and is mainly caused by alveoli filled with inflammatory material, leading to a high signal on T_2-weighted turbo SE imaging. Similar to CT, MR images visualize air bronchograms as low signal areas that follow the course of bronchi in consolidation [27,28]. With progression of the disease, complete destruction of lung segments or lobes can be assessed on T_1-weighted and T_2-weighted turbo SE images, as well as on conventional and/or thin-section CT.

In contrast to CT assessment of CF [29–33], MRI lacks a dedicated scoring system as well as quantitative readouts. In clinical practice, it is therefore important to be able to visual assessment of CF patients by means of MRI with a modified Brody [32] or Bhalla/Helbich score [29,31,34] to make the scores compatible with those of CT, and a few studies have convincingly validated these scoring systems for MRI and CT in the clinical setting [34].

As well as conventional MRI for CF patients, pulmonary functional MRI, such as CE-perfusion MRI, O_2-enhanced MRI, and hyperpolarized noble gas MRI, is capable of differentiating regional functional changes in lung parenchyma. Regional ventilatory defects in CF were found to cause regional lung perfusion changes due to the reflex of hypoxic vasoconstriction or tissue destruction. Perfusion defects assessed by CE-perfusion MRI, which is one of the proton MR techniques, showed good correlation with the degree of tissue destruction in patients assessed with CF. In addition, quantitatively analyzed CE-perfusion MRI visualized regional perfusion differences and could be used for therapeutic monitoring [27].

Another proton MR technique is oxygen-enhanced (O_2-enhanced) MRI, which is based on the paramagnetic properties of 100% molecular oxygen and can serve as a *contrast medium* to assess lung ventilation [35]. When this technique was used for patients with CF, an inhomogeneous signal change after 100% oxygen inhalation was observed, which was generated by the inhomogeneous combination of regional ventilation and perfusion [36]. The combination of morphological and functional assessment has demonstrated that proton MRI has the potential to evaluate disease severity in and therapeutic effects on CF patients and thus play a complementary role in CT and nuclear medicine study.

10.4.2 Asthma

Asthma is a chronic inflammatory disorder of the lung, which predominantly involves the small- and medium-sized airways. It has been suggested that the pathogenesis of asthma involves thickening of the airway walls, which were thickened due to increased smooth muscle mass, inflammatory cell infiltration, deposition of connective tissue (subepithelial fibrosis), vascular changes, and mucous gland hyperplasia. It has therefore been hypothesized that quantitative assessment of airway wall as well as lung density on CT may be useful for disease severity and therapeutic effect evaluation because the pathogenesis of asthma involves airway hyperresponsiveness, mucous hypersecretion, smooth muscle hypertrophy, and subepithelial fibrosis. Since the symptoms in asthmatics are usually associated with variable airflow obstruction caused by changes resulting from the aforementioned pathophysiology, hyperpolarized noble gas, the capability of MRI for visualization of these pathogenic symptoms has been extensively tested during the last decade.

On the other hand, conventional MRI based on proton MRI has not been found to be useful for asthmatics. Proton-based MRI had therefore not been employed for asthmatics until O_2-enhanced MRI was tested and its capability for asthmatics demonstrated in 2011 [37]. When O_2-enhanced MRI was directly compared with quantitatively assessed CT, it was found to be as effective as CT for pulmonary functional loss assessment and clinical stage classification of asthmatics [37]. With further clinical evaluation as well as improvements and software, O_2-enhanced MRI can be expected to function as an alternative MR method for management of asthmatics.

10.4.3 Chronic Obstructive Pulmonary Disease

According to the American Thoracic Society/European Respiratory Society consensus and the Global Initiative for Chronic Obstructive Lung Disease guidelines, the narrowing of small airways caused by inflammation and the scarring and blocking of the small airway lumens with mucinous secretions are thought to represent the primary pathology of airflow restriction, which is not fully reversible in patients with chronic obstructive pulmonary disease (COPD). Airflow restriction is caused by a mixture of abnormal inflammatory responses in the small airways and parenchymal destruction of the lung. Over the last few decades, many studies have suggested that the small airways constitute the most important sites that cause airflow restriction in COPD, and parenchymal destruction (i.e., emphysema) is a definite contributing factor, although the extent of this contribution varies. Thin-section CT is most commonly used for COPD severity assessment, and several commercially available or proprietary software and visual scoring systems have been adapted for CT-based assessment of COPD in clinical and academic practice [38–43]. In contrast to thin-section CT, the use of conventional proton MRI is limited to the assessment of morphological changes in COPD. It has been demonstrated on conventional proton MRI that the extent of pulmonary emphysema in patients with COPD is theoretically equal to the lung area with lower signal intensity than that in normal lung areas because of loss of tissue, reduced blood volume, hyperinflation, and hypoxic vasoconstriction on T_1- and T_2-weighted MRI [44]. However, such changes in emphysematous lung are not routinely evaluated or diagnosed on conventional proton MRI due to the greater difficulty of using this modality than CT.

Unlike with conventional proton MRI, COPD has been assessed by means of a few pulmonary functional MR techniques such as hyperpolarized noble gas MRI, dynamic CE-perfusion MRI, O_2-enhanced MRI, dynamic MRI for motion evaluation, and proton MRI with ultra-short TEs. Although studies using hyperpolarized noble gas MRI have produced positive results, this technique cannot be applied worldwide due to regulatory restrictions, so that every effort is being made to demonstrate the utility of proton MR-based techniques for COPD assessment.

As for the clinical feasibility of these proton MR techniques, dynamic CE-perfusion MRI or time-resolved

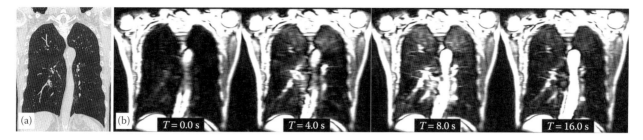

FIGURE 10.2
A 72-year-old male patient with chronic obstructive pulmonary disease. (a): Thin-section coronal multiplanar reconstructed (MPR) image shows heterogeneously distributed low attenuation area due to pulmonary emphysema. In addition, pulmonary emphysema is predominantly distributed in the bilateral lower lobe. (b) Time-resolved CE-MR angiographs obtained with 3 T system (left to right: time point after contrast media injection) demonstrate heterogeneous enhancement within lung parenchyma. Bilateral lower lung field, especially left lower lung field, shows less enhancement than bilateral upper lung field. The degree and distribution of enhancement correlate well with the degree of emphysema.

(or 4D) MR angiography has proven to be effective in routine clinical practice for quantitative and qualitative assessment of regional lung perfusion differences due to lung parenchyma destruction and disease severity of pulmonary emphysema [45–48]. COPD patients with emphysema generally show heterogeneously reduced lung parenchyma (Figure 10.2). In addition, it has been shown by means of quantitatively assessed dynamic CE-perfusion MRI that mean pulmonary blood flow (PBF), mean transit time (MTT), and mean pulmonary blood volume (PBV) were diffusely and significantly and heterogeneously reduced in COPD patients compared with healthy volunteers [45,48]. If this technique is used for postoperative lung function prediction, dynamic CE-perfusion MRI has the potential to function as a substitute for qualitatively and quantitatively assessed thin-section CT and nuclear medicine study, including perfusion scan, perfusion single-photon emission tomography (SPECT) and perfusion SPECT combined with CT (SPECT/CT) [49–51].

Although O_2-enhanced MRI is clinically less easily available than dynamic CE-perfusion MRI, it is a clinically applicable MR method without any need for modification. The underlying physiology for O_2-enhanced MRI may be different from that for hyperpolarized noble gas MRI as the former can provide regional information based on ventilation, perfusion, and oxygen transfer from alveoli to the capillary bed [52–58]. Since 2001, several studies have examined the capability of O_2-enhanced MRI for COPD as well as pulmonary emphysema assessment in comparison with the pulmonary function test quantitatively assessed thin-section CT [52–57]. On O_2-enhanced MRI, oxygen enhancement in COPD patients is heterogeneously reduced when compared with that in healthy volunteers. In these studies, it was determined that oxygen enhancement from O_2-enhanced MRI showed significant and good correlation with forced expiratory volume in one second (FEV_1), FEV_1/forced vital capacity (FEV_1/FVC), diffusing capacity of the lungs for carbon monoxide (DL_{CO}), and/or diffusing capacity divided by the alveolar volume (DL_{CO}/V_A). In addition, it had better potential for clinical staging of COPD, which was better than that of quantitatively assessed thin-section CT [52,53,56]. Therefore, quantitatively assessed O_2-enhanced MRI can be considered at least as effective as quantitatively assessed thin-section CT for pulmonary functional loss assessment and clinical stage classification in routine clinical practice [56]. Moreover, detailed assessment of the signal intensity–time course curve derived from dynamic O_2-enhanced MRI can provide air-flow limitation-dependent and oxygen transfer-dependent parameters as well as the quantitatively assessed degree of airway and lung parenchyma destruction on thin-section CT [55,57]. In addition, O_2-enhanced MRI has the potential for the evaluation of candidates for lung volume reduction surgery that is as good as that of thin-section CT and nuclear medicine study [58].

Dynamic MRI for respiratory motion analysis is a more easily applied, but clinically less useful proton MR-based technique for COPD. Several investigators have evaluated this technique for the assessment of diaphragm and chest wall motion, lung volume or morphologic changes in pulmonary emphysema patients, and lung volume in candidates for lung volume reduction surgery [59–65]. They found that the parameters derived from respiratory motion analyses correlated significantly with pulmonary functional parameters, and this technique also detected paradoxical diaphragm motion or targeted lesions for lung volume reduction surgery in routine clinical scanners [59–65]. Although respiration motion analysis using dynamic MRI can also provide opportunities for the assessment of changes in COPD patients' pulmonary function in terms of lung, chest wall, and diaphragm motion, further investigation may verify the real significance of this technique for pulmonary functional imaging and the value of the clinical evidence for patients with COPD.

Recently, body MR examination in routine clinical practice is gradually shifting from 1.5 to 3 T MR systems. Consequently, quantitatively assessed T_2 star (T_2^*) for the 3 T MR system has been advocated as a new proton MR-based technique with ultra-short TEs for assessing morphologic changes associated with lung parenchyma in smoking-related COPD patients [28,66–68]. T_2^* values of lung parenchyma are affected by reductions in lung tissues and PBF, heterogeneously increased and larger air–soft-tissue interface due to emphysema and air trapping. In one study, mean T_2^* value as well as CT-based pulmonary functional volume and ratio of wall area to total airway area (WA%) correlated significantly with pulmonary functional test results and clinical stage classification [67]. In addition, when compared with smokers without COPD, T_2^* within lung was found to be heterogeneously diminished in COPD patients according to pulmonary functional loss and clinical stages [67,68]. It has therefore been suggested that quantitative assessment of T_2^* using MRI with ultra-short TEs may be as useful as a quantitatively assessed thin-section CT for pulmonary functional loss assessment and clinical stage classification of COPD in smokers [67,68]. However, further investigations are needed to verify this new MR technique is as clinically useful as thin-section CT for management of COPD patients in routine clinical practice.

10.5 MRI of Lung

10.5.1 Pulmonary Nodules

Pulmonary nodules are a frequently detected finding on chest radiographs as well as during CT examinations. The vast majority of these nodules (90%) are incidental radiologic findings, detected accidentally on chest radiograms (CXRs) obtained for unrelated diagnostic workups [69,70]. In addition, results of nationwide lung cancer screening trials as well as previously reported results of CT-based lung cancer screenings have led to a growing need for management of pulmonary nodules in routine clinical practice [70]. In routine clinical practice, CT is the most frequently used modality for diagnosis, and positron emission tomography (PET) or PET combined with CT (PET/CT) scans with [18F]-fluorodeoxyglucose (FDG-PET/CT) is the second most frequently utilized modality for further radiological evaluation of patients with pulmonary nodules [71–74]. In addition, it has been suggested that conventional proton MRI may be useful for diagnosis of pulmonary nodules.

According to previous reports [75–83], T_2-weighted and/or pre- or post-CE-T_1-weighted SE or turbo SE images can be used for the diagnosis of bronchoceles, tuberculomas, mucin-containing tumors, hamartomas, and aspergillomas on the basis of their specific MR findings. However, many pulmonary nodules, including lung cancers, pulmonary metastases, and low-grade malignancies show low or intermediate signal intensity on T_1-weighted imaging and slightly high intensity on T_2-weighted imaging. This, as well as significant overlaps of relaxation time between benign and malignant nodules or masses, makes it difficult to differentiate malignant from benign nodules by means of these methods [81–83].

To overcome this limitation for relaxation time assessment, short inversion time (TI) inversion recovery (STIR) turbo SE imaging and diffusion-weighted MRI (DWI) were both introduced as more promising for non-CE MRI for nodule assessment than other SE or turbo SE imaging sequences [84–87]. The study of STIR turbo SE imaging demonstrated that the quantitative capability of this modality, with sensitivity, specificity, and accuracy of 83.3%, 60.6%, and 74.5%, respectively, for differentiating malignant solitary pulmonary nodules from benign solitary pulmonary nodules, was significantly better than that of non-CE-T_1-weighted and T_2-weighted turbo SE imaging [84]. DWI was introduced as another promising non-CE-MR method in this setting (Figures 10.4 and 10.5) [85–87]. Theoretically, DWI can assess the diffusivity of water molecules within tissue in terms of cellularity, perfusion, tissue disorganization, and extracellular space and other variables in the same manner as the apparent diffusion coefficient (ADC). For a maximum b value ranging from 500 to 1000 s/mm², quantitative and/or qualitative sensitivities and specificities of the ADC for differentiation of malignant from benign solitary pulmonary nodules were 70.0%–88.9% for sensitivity and 61.1%–97.0% for specificity [85–87], while specificity of DWI (97.0%) was significantly higher than that of FDG-PET/CT (79.0%) [85]. In addition, when the lesion-to-spinal cord signal intensity ratio rather than ADC was employed, one study found that sensitivity, specificity, and accuracy of DWI were 83.3%, 90.0%, and 85.7%, respectively, making the accuracy of this new parameter significantly higher than that of ADC (50.0%) [87]. Although the size of the study cohorts used in these studies was limited, the results can be considered promising. Since these non-CE MR techniques are regarded as at least as efficacious as FDG-PET or PET/CT, they can be expected to be used to differentiate malignant from benign pulmonary nodules in the near future.

During the last two decades, dynamic CE-MRI has been tested for its capability to distinguish malignant from benign nodules as well as to differentiate nodules requiring further intervention or treatment from those needing only follow-up examinations (Figures 10.3 and 10.4) [83,88–98]. The findings of these studies suggest that dynamic CE-MRI is at least as effective as dynamic CE-CT, FDG-PET, or PET/CT [83,97,98]. Various dynamic MR techniques were used in these studies to distinguish malignant from benign nodules with reported sensitivities ranging from 94% to 100%, specificities from 70% to 96%, and accuracies of more than 94% [88–98]. In addition, a meta-analysis reported that there were no significant differences in diagnostic performance among dynamic CE-CT, dynamic CE-MRI, and FDG-PET [98]. Dynamic MRI may thus be able to play a complementary role or function as a substitute in the characterization of pulmonary nodules assessed with dynamic CE-CT, FDG-PET, and/or PET/CT. However, it should be noted that MRI has several drawbacks such as being

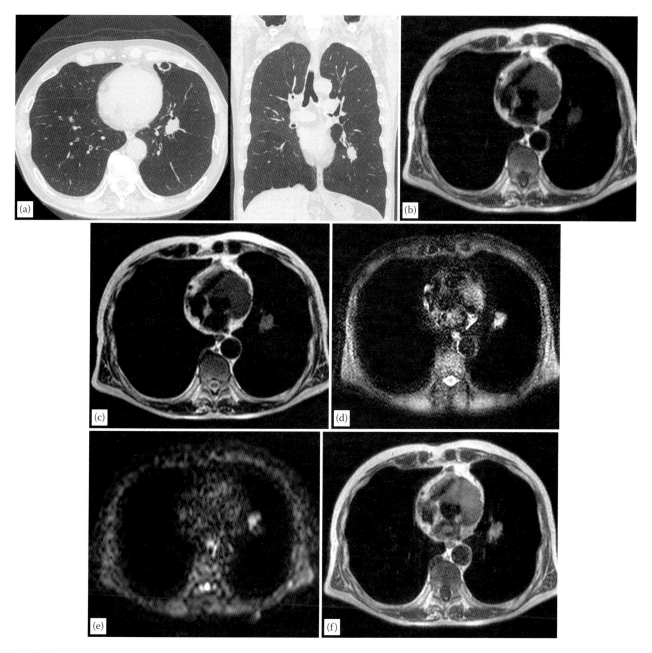

FIGURE 10.3

A 78-year-old male patient with invasive adenocarcinoma. (a) Thin-section CT and MPR images obtained with lung window setting show a nodule with a diameter of 20 mm in the left lower lobe. It was pathologically diagnosed as invasive adenocarcinoma. (b) The nodule seen on the non-CE-black-blood T_1-weighted turbo SE image shows homogeneous and low signal intensity. (c) On the non-CE-black-blood T_2-weighted turbo SE image, signal intensity of the nodule is also homogeneous but intermediate. (d) Black-blood STIR turbo SE image depicts the nodule as a high signal intensity area. The nodule-muscle contrast ratio was 1.7, and the nodule was diagnosed as malignant. This nodule was identified as true-positive on STIR turbo SE imaging. (e) DWI visualizes the nodule as a high signal intensity area. ADC was assessed as 0.9×10^{-3} mm^2/s. ADC evaluation assessed this nodule as true-positive. (f) On the CE-black-blood T_1-weighted turbo SE image, the nodule is seen to be homogeneously enhanced by gadolinium contrast media.

(Continued)

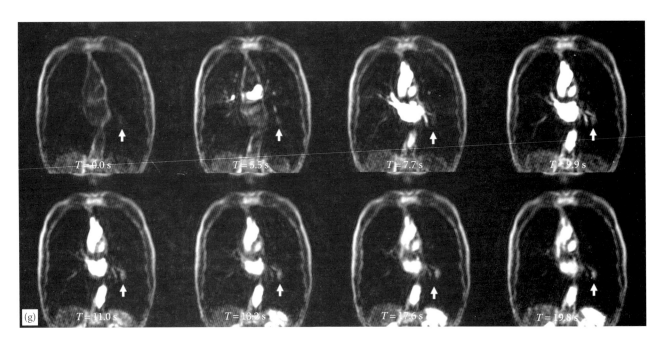

FIGURE 10.3 (Continued)
A 78-year-old male patient with invasive adenocarcinoma. (g) Dynamic CE-T_1-weighted GRE imaging with ultra-short TE (T: the time point after contrast media injection) shows a well-enhanced nodule after 9.9 s, and its blood supply was considered to be derived mainly from systemic circulation. Mean relative enhancement ratio and slope of enhancement were 0.6 and 0.2 s^{-1}. This nodule was assessed as true-positive case in terms of each index.

not viable for patients with many implanted devices, which render them unsuitable for MR examinations, claustrophobia, renal dysfunction, and so on.

10.5.2 Lung Cancer

Lung cancer is one of the most common cancers worldwide with the highest mortality for both men and women [99]. Currently, non-small cell lung cancer (NSCLC) accounts for 80% of all lung cancers and can be cured surgically if detected at an early stage [100]. However, most NSCLC patients present late in the course of their illness at an inoperable advanced stage [101–103]. Small cell carcinoma constitutes approximately 13%–20% of all lung cancers [104], and it is staged with a two-stage system developed by the Veterans' Administration Lung Cancer study group rather than with the tumor, lymph node, and metastasis staging system. With either of these two staging systems for lung cancer, radiological examinations including CT, PET, and PET/CT are recommended as important pretherapeutic assessment modalities. In contrast to CT, FDG-PET, and PET/CT, it was suggested in 1991 that MRI was of limited clinical relevance and would be more useful for identifying mediastinal and chest wall invasions because of its multiplanar capability and superior contrast resolution of tumor and mediastinum and/or tumor and chest wall to what could be attained with CT [105]. However, recent advances in MR systems, the introduction of improved or newly developed pulse

sequences and/or the more effective utilization of contrast media may well result in further improvements in tumor, lymph node, and metastasis staging accuracy for lung cancer patients in routine clinical practice [81–83].

10.5.2.1 T-Factor Assessment

Many surgeons consider minimal invasion of mediastinal fat as resectable [106], so that clinicians need to know whether minimal mediastinal invasion (T3 disease) or actual invasion (T4 disease) has occurred before considering surgical resection. Numerous studies [57–66,81–83,105–115] have reported sensitivity for the assessment of mediastinal invasion by CT with or without the use of helical scanning ranging from 38% to 84% and specificity from 40% to 94%. On the other hand, no major studies have investigated the capability of PET or PET/CT for T-factor assessment, even though these modalities are routinely used for lung cancer staging. The Radiologic Diagnostic Oncology Group (RDOG) study [105] used non-cardiac and respiratory-gated T_1- and T_2-weighted SE imaging for T-factor assessments and found that the diagnostic performance of this technique (sensitivity, 56.0%; specificity, 80.0%) was not significantly different from that of conventional CT (sensitivity, 63.0%; specificity, 84.0%) for differentiating T3–T4 from T1–T2 tumors [105]. However, non-cardiac and/or respiratory-gated T_1- and T_2-weighted SE imaging included motion artifacts, and pericardial or mediastinal invasion could therefore not be clearly identified. However, the use of cardiac and/or respiratory gating as well as fast MR sequences can result in better diagnostic performance by improving the depiction of tumor invasion of pericardium (T3) or heart (T4) on T_1- and T_2-weighted SE or turbo SE sequences. In addition to mediastinal invasion assessment, conventional non-CE MRI has also been advocated as effective for the assessment of chest wall invasion due to its multiplanar capability and better tissue contrast resolution than is obtainable with CT [81–83,105,115–120]. Sagittal and coronal multiplanar reconstructed images are superior to axial CT images for displaying the anatomical relationship between tumor and chest wall structures, while Padovani et al. have suggested that the diagnostic yield can be further improved by intravenous administration of contrast media [118]. In addition, when STIR turbo SE imaging is used, lung cancer can be identified as high signal intensity within the suppressed signal intensities of chest wall structures, thus enabling clinicians to determine the size of the tumor within the chest wall (Figure 10.5) [81–83].

Non-cardiac or cardiac-gated CE-MR angiography by fast GRE sequence has been recommended as useful for

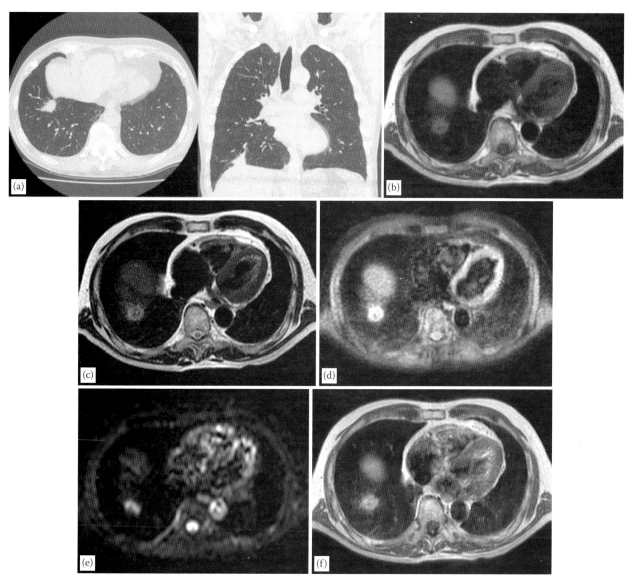

FIGURE 10.4

An 80-year-old male patient with organizing pneumonia. (a) Thin-section CT and MPR images obtained with lung window setting show a nodule with a diameter of 23 mm in the right lower lobe. It was pathologically diagnosed as organizing pneumonia. (b) On the non-CE-black-blood T_1-weighted turbo SE image, the nodule shows very low and low signal intensity. (c) On the non-CE-black-blood T_2-weighted turbo SE image, the signal intensity is low and intermediate. (d) Black-blood STIR turbo SE image depicts the nodule as a high signal intensity area. The nodule–muscle contrast ratio was 1.8. This nodule was assessed as false-positive on STIR turbo SE imaging. (e) DWI shows the nodule as a high signal intensity area. ADC value was 0.85×10^{-3} mm^2/s. ADC evaluation resulted in an assessment of this nodule as false positive. (f) On the CE-black-blood T_1-weighted turbo SE image, this nodule is relatively homogeneously enhanced by gadolinium contrast media. *(Continued)*

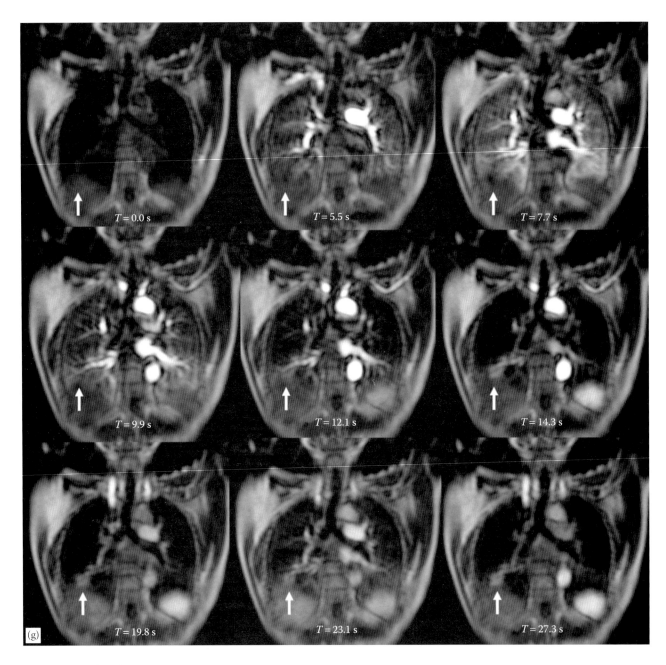

FIGURE 10.4 (Continued)
An 80-year-old male patient with organizing pneumonia. (g) Dynamic CE-T_1-weighted GRE imaging with ultra-short TE (*T*: the time point after contrast media injection) shows a well-enhanced nodule after 12.1 s, and its blood supply was considered to be derived from systemic circulation. Mean relative enhancement ratio and slope of enhancement were 0.4 and 0.03 s^{-1}. The mean relative enhancement ratio produced an assessment of this nodule as false positive, and the slope of enhancement identified it as true negative.

the assessment of cardiovascular or mediastinal invasion [121,122]. The use of cardiac gated CE-MR angiography [122] resulted in sensitivity, specificity, and accuracy for detection of mediastinal and hilar invasion ranging from 78% to 90%, 73% to 87%, and 75% to 88%, respectively, values that are higher than those attainable with CE-CT and conventional cardiac- and respiratory-gated T_1-weighted SE imaging [122]. CE-MR angiography is thus thought to improve the diagnostic capability

of MRI for T-factor assessment and therefore merits use for chest MR examination in this setting.

Another MR technique that can be beneficial is dynamic cine MRI, which can assess the movement of the tumor along the partial pleura during the respiratory cycle. In a study using this type of MRI, the tumor was seen to have invaded the chest wall, become affixed to the chest wall, and be moving freely along the pleura without invading it [123]. The sensitivity, specificity, and

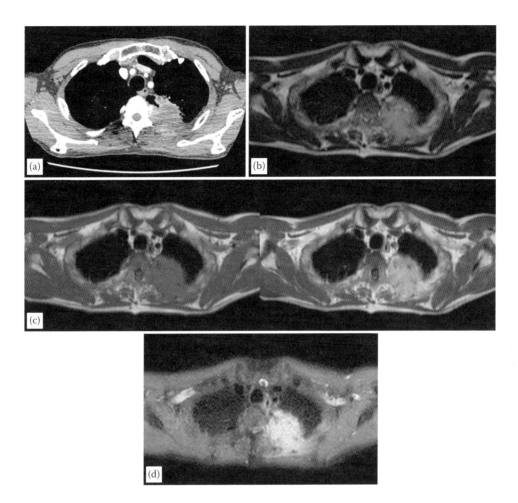

FIGURE 10.5
A 61-year-old female patient with chest wall invasion from squamous cell carcinoma. (a) Although lung cancer in the apex was suspected because chest wall and vertebra invasions were observed on CE-thin-section CT, it could not be clearly demonstrated due to severe artifacts and diminished contrast between tumor and chest wall. (b) On the black-blood T_2-weighted turbo SE image, the tumor is depicted as a high signal intensity area. Although chest wall invasion is clearly demonstrated, vertebral invasion could not be easily assessed. (c) Left to right: non-CE-to CE-black-blood T_1-weighted turbo SE image. The non-CE-black-blood T_1-weighted turbo SE image clearly shows chest wall and vertebral invasions and the primary tumor as a low signal intensity area. Although the CE-black-blood T_1-weighted turbo SE image shows good enhancement within the tumor, visualization of chest wall, and vertebral invasions is not improved. (d) The black-blood STIR turbo SE image shows the primary tumor, chest wall invasion, and vertebral invasion as markedly high signal intensity areas. In addition, this sequence clearly shows vertebral invasion only at the left pedicle.

accuracy of dynamic cine MRI for the detection of chest wall invasion attained in this study were 100%, 70%, and 76%, respectively, while those of conventional CT and MRI were 80%, 65%, and 68% [123]. Dynamic cine MRI in conjunction with static MRI can therefore be considered to be useful for chest wall invasion assessment.

10.5.2.2 N-Factor Assessment

CT is the standard noninvasive modality for staging of lung cancer, and the diagnosis of metastatic lymph nodes is based on the detection of enlarged lymph nodes with a short axis diameter. Two major papers from the RDOG and the Leuven Lung Cancer Group (LLCG) reported that the diagnostic performances of CT for N-factor assessment showed sensitivity from 52% to 69%

and specificity from 69% to 71% [105,124–130]. In view of these results, the authors of these papers currently recommend only traditional mediastinoscopy with biopsy for adequate assessment of N-factor in lung cancer.

Since the 1990s, FDG-PET has been used for distinguishing metastatic from non-metastatic lymph nodes in terms of the biochemical mechanism of increased glucose metabolism or tumor cell duplication [131–143]. However, glucose metabolism is also elevated within secondary tumors, infection, or inflammation. Moreover, spatial resolution of PET is inferior to that of CT and MR, so that during the last decade it was suggested that the diagnostic capability of the FDG-PET or PET/CT was limited because of some overlap between metastatic and non-metastatic lymph nodes in view of the aforementioned pathologic conditions [131–143].

It is thought that the diagnostic criteria of conventional T_1- or T_2-weighted SE or turbo SE imaging for N-factor assessment in lung cancer patients in order to differentiate metastatic from non-metastatic lymph nodes depend on lymph node size and are similar to CT criteria. The findings of a direct comparison of the diagnostic performances of MRI and CT from RDOG [105] have led to the notion that the direct multiplanar capability of MRI is its only advantage for the detection of lymph nodes in areas that are suboptimally imaged in the axial plane, such as in the aortopulmonary window and subcrainal regions. This report further suggested that there were no significant differences in diagnostic performance between conventional MRI and CT.

To improve the diagnostic performance of N-factor assessment on MRI, studies published since 2002 have examined the clinical efficacy of cardiac- and/or respiratory-triggered conventional or black-blood STIR turbo SE imaging and have demonstrated its superiority over that of CE-CT, FDG-PET, or PET/CT and other MR sequences [144–149]. These novel sequences have made it possible for the signal intensity of lymph nodes to be quantitatively assessed by means of comparison with a 0.9% normal saline phantom (lymph node–saline ratio) or muscle (lymph node–muscle ratio) [144–146,149]. In addition, they can visually differentiate metastatic from non-metastatic lymph nodes by comparing them with the signal intensity of muscle and/or primary tumor and have demonstrated that their diagnostic performance is superior to that of other modalities [144–149]. In these studies, the sensitivity, specificity, and accuracy of quantitatively and qualitatively assessed STIR turbo SE imaging ranged from 83.7% to 100.0%, from 75.0% to 93.1%, or from 86.0% to 92.2%, respectively, on a per-patient basis, and these values were equal to or higher than those for CE-CT, FDG-PET, or PET/CT. When STIR turbo SE imaging is used as part of chest MR examination for lung cancer patient, MRI becomes more effective than other modalities for differentiation of metastatic from non-metastatic lymph nodes. In addition, when this technique is combined with FDG-PET/CT in this setting, the combination was found to have a significantly higher capability on a per-patient basis (specificity: 96.9%, accuracy: 90.3%) than FDG-PET/CT alone (specificity: 65.6%, accuracy: 81.7%) [148].

More recently, DWI was introduced and promoted as another MR technique in this setting [149–152]. Sensitivity, specificity, and accuracy of quantitatively and qualitatively assessed DWI were found to range between 77.4% and 80.0%, 84.4% and 97.0%, or 89.0% and 95.0%, respectively, on a per-patient basis, and these results were reportedly equal to or higher than those for FDG-PET or PET/CT [149–152]. However, the capability of DWI as well as FDG-PET/CT to detect small metastatic foci or lymph nodes is limited because currently DWI for this purpose is acquired by using non-cardiac- and respiratory-gated

SE type echo-planar imaging, and the image quality becomes degraded due to cardiac and respiratory motion as well as susceptibility artifacts. A prospective and direct comparison of STIR turbo SE imaging with DWI and FDG-PET/CT for differentiation of metastatic from non-metastatic lymph nodes demonstrated that sensitivity and accuracy of STIR turbo SE imaging (quantitative sensitivity: 82.8%, qualitative sensitivity: 77.4%, quantitative accuracy: 86.8%) were significantly higher than those of DWI (74.2%, 71.0% and 84.4%, respectively) and FDG-PET/CT (quantitative sensitivity: 74.2%) [149].

STIR turbo SE imaging is a simple sequence, which can be easily incorporated into clinical protocols and perform like fat-suppressed MRI with additive T_1 and T_2 contrasts. In addition, another previously reported and successful STIR turbo SE imaging was cardiac- and/or respiratory-triggered conventional or black-blood T_1-weighted STIR turbo SE imaging. On STIR turbo SE images, metastatic lymph nodes are shown as areas of high signal intensity and non-metastatic lymph nodes as areas of low signal intensity (Figures 10.6 and 10.7). In view of these results, STIR turbo SE imaging may be a more efficacious MR technique than others in this setting for use before surgical treatment or lymph node sampling, during thoracotomy or mediastinoscopy for accurate pathologic tumor, lymph node, and metastasis staging after surgical treatment, or before either chemotherapy or radiation therapy or both [149]. In addition, STIR turbo SE imaging can be expected to perform as a substitute and in a complementary role for FDG-PET or PET/CT for diagnosis of N staging of NSCLC patients in routine clinical practice.

10.5.2.3 M-Factor Assessment

Detection of metastasis is one of the key issues in management and prognosis of lung cancer patients. And since extra-thoracic metastases are detected at presentation in approximately 40% of patients with newly diagnosed lung cancer [153,154], accurate diagnosis of not only intra-thoracic but also extra-thoracic metastases is frequently recommended by clinicians to help them provide the most appropriate treatment and/or management for lung cancer patients. For this type of detection, chest radiography, whole-body CT, brain MRI, and/or bone scintigraphy as well as clinical symptom and/or tumor markers have been recommended as surveillance modalities in preoperative guidelines [155–157]. Moreover, several investigators have suggested that whole-body FDG-PET or PET/CT, except for brain MRI, should be considered a more powerful tool for the assessment of extra-thoracic metastases in suspected NSCLC patients than conventional workups comprising CT and bone scintigraphy [135,136,138,140,141].

It has been suggested recently that in this setting, whole-body MRI can be performed by using the moving

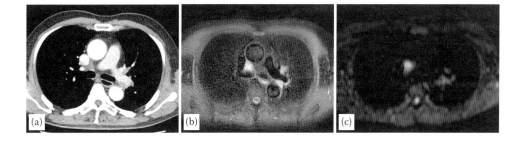

FIGURE 10.6

A 77-year-old male patient with invasive adenocarcinoma and mediastinal and hilar lymph node metastases (N3 disease). (a) CE-CT demonstrates left hilar and mediastinal lymph nodes with short axis diameters of less than 9 mm. All lymph nodes were assessed as non-metastatic sites, and this case was assessed as a false-negative node on a per-node basis. (b) The black-blood STIR turbo SE image clearly shows all lymph nodes as high signal intensity areas. All lymph nodes were diagnosed as metastatic lymph nodes and correctly assessed as N3 disease. (c) DWI also shows left hilar and right lower paratracheal lymph nodes as high signal intensity areas, although the presence of subcrainal lymph nodes could not be determined due to artifacts. All high signal intensity lymph nodes could be correctly diagnosed as metastatic lymph nodes, and the subcrainal lymph node was assessed as a false-negative node on a per-node basis. However, DWI could correctly evaluate this case as N3 disease, and this case was assessed as a true-positive node.

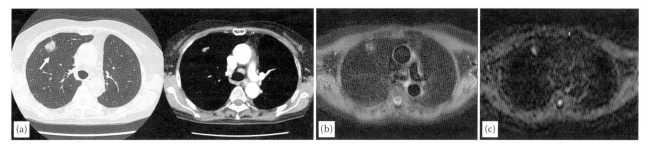

FIGURE 10.7

A 75-year-old female patient with invasive adenocarcinoma and non-metastatic lymph nodes at mediastinum (N0 disease). This case had been surgically treated for a left invasive adenocarcinoma 7 years previously. (a) Left to right: Thin-section CT with lung window setting and CE-CT. Thin-section CT shows invasive adenocarcinoma as a partly solid nodule in the right upper lobe. CE-CT shows right lower paratracheal lymph node with a short axis diameter of 11 mm. This case was diagnosed as N2 and assessed as false positive on a per-node basis. (b) Black-blood STIR turbo SE image shows the primary lesion as a high signal intensity area, and the right lower paratracheal node as a low signal intensity area. This case was diagnosed as N0 and assessed as true-negative node on a per-node basis. (c) DWI also shows the primary lesion with a very high signal intensity, although the right lower paratracheal node could not be visualized due to artifacts. This accidentally detected case was diagnosed as N0 and assessed as true-negative node on a per-node and per-patient basis. Limited visualization of lymph nodes on DWI is considered as one of the major limitations of the current DWI sequence.

table technique with a body coil or a combination of multiple array coils with parallel imaging capability and several newly developed fast MR sequences. This combined procedure is now seen as a single, cost-effective imaging test using 1.5 and 3 T systems for patients with not only lung cancer but other malignancies as well (Figure 10.8) [147,158–161]. Furthermore, whole-body DWI has been recommended as a promising new tool for whole-body MR examination of oncologic patients [159–161]. A comparison of whole-body MRI with FDG-PET or PET/CT for M-factor assessments showed that the diagnostic capability of the former with or without DWI (sensitivity: 52.0%–80.0%, specificity: 74.3%–94.0%, accuracy: 80.0%–87.7%) was equal to or significantly higher than that of FDG-PET or PET/CT (sensitivity: 48.0%–80.0%, specificity: 74.3%–96%, accuracy: 73.3%–88.2%) [147,158–161]. However, the following drawbacks of the use of whole-body DWI for whole-body MR examination in this setting should be taken into careful consideration. When only whole-body

DWI was used, specificity (87.7%) and accuracy (84.3%) of DWI were found to be significantly lower than those of FDG-PET/CT (specificity: 94.5%, accuracy: 90.4%) on a per-patient basis [159]. The same study showed that the diagnostic accuracy of whole-body MRI combined with DWI (87.8%) was not significantly different from that of FDG-PET/CT, while the accuracy of whole-body MRI without DWI (85.8%) was significantly lower than that of FDG-PET/CT. This indicates that the accuracy of whole-body MRI combined with DWI for M-stage assessment of patients with NSCLC is as good as that of FDG-PET/CT. Moreover, the use of whole-body DWI as part of whole-body MR examination would be advisable in order to improve the diagnostic accuracy of M-factor assessment of NSCLC patients [159].

In addition, when whole-body MRI was employed for postoperative recurrence assessment in NSCLC patients [162], it was found that a newly developed CE-T_1-weighted 3D high resolution GRE sequence combined

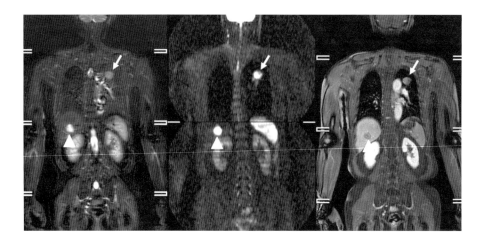

FIGURE 10.8
A 63-year-old male patient with invasive adenocarcinoma and liver metastasis (left to right: STIR turbo SE imaging, DWI, and CE-T_1-weighted high-resolution isotropic volume examination [THRIVE] imaging). On both the STIR turbo SE image and DWI, the primary lesion (arrow) and liver metastasis (arrow head) are shown as high or very high signal intensity areas, and this case is believed to be stage IV. On the CE-THRIVE image, the primary lesion appears as a well-enhanced mass (arrow). On the other hand, a mass within liver is shown as a ring-enhanced low intensity area (arrowhead). The enhancement pattern of this lesion made it easy to diagnose as liver metastasis, and this case was also easily diagnosed as stage IV.

with a double fat-suppression radiofrequency (RF) pulse technique may be more helpful than the previously applied CE-T_1-weighted GRE sequence and function in a complementary role in whole-body PET/CT in routine clinical practice. This study demonstrated that specificity (100%) and accuracy (95.5%) of whole-body MRI with CE-T_1-weighted 3D high resolution GRE sequence using the double fat-suppression RF pulse technique were significantly higher than those of FDG-PET/CT (specificity: 93.6%, $p = 0.02$; accuracy: 89.6%, $p = 0.01$) and conventional radiological examinations (specificity: 92.7%, $p = 0.01$; accuracy: 91.0%, $p = 0.03$), while specificity of whole-body MRI with the previously utilized CE-T_1-weighted GRE sequence (100%) was also significantly higher than that of FDG-PET/CT ($p = 0.02$) and conventional radiological examinations ($p = 0.01$). This indicates that whole-body MRI combined with DWI as well as newly developed post-CE-T_1-weighted GRE may be helpful for management of lung cancer patients as well as FDG-PET or PET/CT in routine clinical practice and merits worldwide use at many cancer centers or cancer research centers.

10.5.3 Interstitial Lung Disease

The official American Thoracic Society/European Respiratory Society statement is an updated version of the international multidisciplinary classification of interstitial lung disease (ILD) [163]. This statement asserts that clinical, sclerologic, thin-section CT, and histological findings may be helpful for distinguishing idiopathic interstitial pneumonias as well as other forms of ILD associated with connective tissue disease [163]. Thin-section CT as well as pulmonary function tests and serum markers are utilized for diagnosis, disease severity assessment,

and therapeutic effect evaluation of ILD in routine clinical practice. In addition to thin-section CT, MRI has been used to evaluate ILD, but its utility for ILD assessment could not be demonstrated because of its lesser capability for the depiction of lung parenchyma structures and pathological changes due to ILD compared with that of conventional T_1- and T_2-weighted and proton density SE sequences used in routine clinical practice.

Since 1999, advancements in MR systems, development of newer and faster MR sequences and applications of half-Fourier acquisition, and introduction of certain parallel imaging and RF pulse techniques for improving image contrast and signal-to-noise ratio within the lung have opened up a new scope for imaging lung parenchyma [164]. With applying above-mentioned issues, several investigators have reported that they used these advance and developments for the morphological assessment of ILD with short echo time three-dimensional (3D) breath-hold GRE imaging with shorter TE [165], T_2-weighted turbo SE imaging using spectrally selective attenuated inversion recovery pulse or triple inversion black-blood pulse [166,167], two-dimensional balanced steady-state free precession imaging [168], post-contrast T_1-weighted GRE imaging [169,170] and with dynamic CE-T_1-weighted turbo GRE imaging [167]. In addition and more recently, MRI with 3D radial ultra-short TE pulse sequence was found to be a more powerful tool than previously reported sequences for proton-based MRI to visualize lung parenchyma [171].

In contrast to the morphological approach using thin-section CT and MRI, O_2-enhanced MRI [172,173] and pulmonary MRI with ultra-short TEs [174] have been in use as proton-based pulmonary functional MRI since 2007. O_2-enhanced MRI reportedly has good potential for the assessment of regional ventilation, alveolo-capillary gas transfer of molecular oxygen, oxygen uptake per respiratory cycle, and airflow limitation [52–58]. In addition, findings based on regional oxygen enhancement from O_2-enhanced MRI and quantitatively assessed $T_2{}^*$ values obtained from pulmonary MRI with ultra-short TEs can be as effective as thin-section CT for pulmonary functional loss assessment and correlation with disease severity evaluated by serum markers in ILD patients with connective tissue disease [173,174]. Although further clinical assessments are warranted, morphological and functional assessment based on MRI with the above-mentioned techniques can be expected to become available to perform novel radiological role in the management of ILD in routine clinical practice.

10.5.4 Pulmonary Vascular Diseases Other Than Pulmonary Thromboembolism and Hypertension

During the last few decades, it has been suggested that MRI is useful for the assessment of acute or chronic pulmonary thromboembolisms and primary or secondary pulmonary hypertensions. However, since these diseases are dealt with in Chapter 13, this chapter will focus on pulmonary arteriovenous malformation and pulmonary sequestration.

10.5.5 Pulmonary Arteriovenous Malformation or Fistula (Pulmonary AVM or AVF)

Pulmonary arteriovenous malformation (AVM) or fistula (AVF) is an abnormal connection between a branch of a pulmonary artery and a pulmonary vein through a thin-walled aneurysmal sac [175]. Since the early 1980s, embolotherapy has been accepted as the treatment of choice for the majority of patients with pulmonary AVMs or AVFs since this procedure minimizes the risk of cerebral embolization and abscess formation without loss of pulmonary parenchyma [175–178]. A precise analysis of the outcome of the treatment of pulmonary AVMs and AVFs was not possible until the introduction of single- or multidetector row spiral CT, which made it possible to analyze morphologic changes at the level of the pulmonary vessels, while further improvements of this approach allowed for simultaneous evaluation of pulmonary and systemic vessels during the same acquisition [179–182].

A study published in 2002 demonstrated that all pulmonary AVMs with an aneurysmal sac and a feeding artery and draining vein with diameters of 3 mm or more were identified and measured with similar results by means of CT angiography, time-resolved CE-MR angiography, and pulmonary angiography [183]. Additional follow-up studies demonstrated that 7 of 12 (58.4%) treated pulmonary AVMs showed decrease in size and less residual contrast enhancement, and CE-perfusion MRI identified the latter as bronchial artery-to-pulmonary artery collateral

flow [183]. These proton-based MR techniques, such as time-resolved CE-MR angiography and CE-perfusion MRI, can therefore also be considered new techniques for management of pulmonary AVMs and AVFs.

10.5.6 Pulmonary Sequestration

Pulmonary sequestration is a relatively rare but clinically significant form of congenital bronchopulmonary foregut malformation and covers a spectrum of disorders involving the pulmonary airways, the arterial supply to the lungs, the lung parenchyma, and its venous drainage. Depending on the morphologic subtype, venous drainage is achieved via either the pulmonary or systemic veins. Although several imaging techniques, including routine chest radiography, CE-CT and MDCT angiography, nuclear medicine study, bronchography, sonography, and MRI, have also been used for the diagnosis of pulmonary sequestration [184], in routine clinical practice the definitive diagnosis of pulmonary sequestration is made with conventional or digital subtraction arterial angiography. Since the introduction of conventional or time-resolved CE-MR angiography [185–188], it was found that proton-based MRI can be also utilized as a radiological method for the assessment of pulmonary sequestration in routine clinical practice (Figure 10.9). Therefore, MRI can therefore be expected to come into widespread use in this setting. In addition, further improvements in the temporal and spatial resolutions for this technique can be expected to enable it to play an important role in definitive diagnosis, similar to that of conventional or digital subtraction arterial angiography.

10.6 Conclusion

This chapter has reviewed the state-of-the-art proton-based MR techniques for trachea, bronchi, and lung parenchyma assessment in various cervical and thoracic diseases. The current situation is that it has become possible to routinely use advanced MR systems and fast imaging and/or parallel imaging techniques to realize new contrast standards with recently developed MR sequences, superior contrast resolution with and without contrast media, and quantitative and qualitative analyses of MRI. In addition, the potential advantages and applications of MRI for trachea, bronchi, and lung parenchyma diseases are now being supported by recently published papers. Nevertheless, further development of protocols, more clinical trials, and the use of advanced analysis tools are needed to determine the real significance of MRI in this setting. In addition, previously and currently published results indicate that MRI used for these diseases can perform a complementary

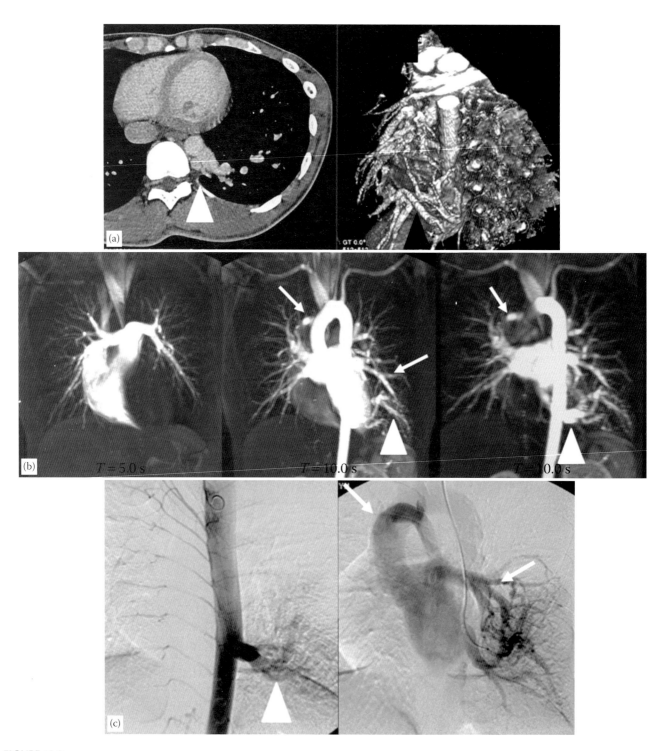

FIGURE 10.9

A 23-year-old male patient with pulmonary sequestration. (a) (Left to right: CE-CT to 3D CT angiography) CE-CT and 3D CT angiography depict an anomalous artery (white and black arrowheads) arising from the thoracic aorta. (b) (Left to right: Maximum intensity projections [MIPs] of time-resolved MR angiography 5 and 10 s after contrast media injection and partial MIP of time-resolved MR angiography 10 s after contrast media injection) MIP images show a perfusion defect in the pulmonary arterial phase (MIP image at 5.0 s), and an anomalous artery (arrowhead) arising from the thoracic aorta to sequestration in the pulmonary venous and systemic arterial phase (MIP image at 10.0 s). In addition, anomalous connections between the anomalous artery and pulmonary and azygos veins (arrows) are also seen on full and partial MIP images at 10.0 s. These findings coordinate well with those for aortography. (c) (Left to right: Aortography to selective angiography to anomalous artery) Aortography clearly shows the anomalous artery (arrowhead) arising from the thoracic aorta. Selective angiography shows anomalous connections between anomalous artery and pulmonary and azygos veins (arrows).

role for morphological procedures and/or function as a substitute for other modalities, and future study results can be expected to validate this use of MRI, similar to that of other modalities, in routine clinical practice.

References

1. Ailion DC, Case TA, Blatter DD et al. (1984) Application of NMR spin imaging to the study of lungs. *Bull Magn Reson* 6: 130–139.
2. Case TA, Durney CH, Ailion DC, Cutillo AG, Morris AH (1987) A mathematical model of diamagnetic line broadening in lung tissue and similar heterogenous systems: Calculations and measurements. *J Magn Reson* 73: 304–314.
3. Bergin CJ, Pauly JM, Macovski A (1991) Lung parenchyma: Projection reconstruction MR imaging. *Radiology* 179: 777–781.
4. Cutillo AG, Ganesan K, Ailion DC et al. (1991) Alveolar air–tissue interface and nuclear magnetic resonance behavior of lung. *J Appl Physiol* 70: 2145–2154.
5. Bergin CJ, Glover GH, Pauly JM (1991) Lung parenchyma: Magnetic susceptibility in MR imaging. *Radiology* 180: 845–848.
6. Hatabu H, Alsop DC, Listerud J, Bonnet M, Gefter WB (1999) T2* and proton density measurement of normal human lung parenchyma using submillisecond echo time gradient echo magnetic resonance imaging. *Eur J Radiol* 29: 245–252.
7. Mayo JR, MacKay A, Muller NL (1992) MR imaging of the lungs: Value of short TE spin-echo pulse sequences. *AJR Am J Roentgenol* 159: 951–956.
8. Alsop DC, Hatabu H, Bonnet M, Listerud J, Gefter W (1995) Multi-slice, breathhold imaging of the lung with submillisecond echo times. *Magn Reson Med* 33: 678–682.
9. Yamashita Y, Yokoyama T, Tomiguchi S, Takahashi M, Ando M (1999) MR imaging of focal lung lesions: Elimination of flow and motion artifact by breath-hold ECG-gated and black-blood techniques on T2-weighted turbo SE and STIR sequences. *J Magn Reson Imaging* 9: 691–698.
10. Biederer J, Reuter M, Both M et al. (2002) Analysis of artefacts and detail resolution of lung MRI with breath-hold T1-weighted gradient-echo and T2-weighted fast spin-echo sequences with respiratory triggering. *Eur Radiol* 12: 378–384.
11. Bruegel M, Gaa J, Woertler K et al. (2007) MRI of the lung: Value of different turbo spin-echo, single-shot turbo spin-echo, and 3D gradient-echo pulse sequences for the detection of pulmonary metastases. *J Magn Reson Imaging* 25: 73–81.
12. Yi CA, Jeon TY, Lee KS et al. (2007) 3-T MRI: Usefulness for evaluating primary lung cancer and small nodules in lobes not containing primary tumors. *AJR Am J Roentgenol* 189: 386–392.
13. Tollefsen HR, DeCosse JJ, Hutter RVP (1964) Papillary carcinoma of the thyroid: A clinical and pathological study of 70 fatal cases. *Cancer* 17: 1035–1044.
14. Cody HS 3rd, Shah JP (1981) Locally invasive, well-differentiated thyroid cancer. 22 years' experience at Memorial Sloan-Kettering Cancer Center. *Am J Surg* 142: 480–483.
15. Park CS, Suh KW, Min JS (1993) Cartilage-shaving procedure for the control of tracheal cartilage invasion by thyroid carcinoma. *Head Neck* 15: 289–291.
16. Clark OH, Levin K, Zeng QH, Greenspan FS, Siperstein A (1988) Thyroid cancer: The cases for total thyroidectomy. *Eur J Cancer Clin Oncol* 24: 305–313.
17. Ishihara T, Kikuchi K, Ikeda T et al. (1978) Resection of thyroid carcinoma infiltrating the trachea. *Thorax* 33: 378–386.
18. Nakao K, Miyata M, Izukura M, Monden Y, Maeda M, Kawashima Y (1984) Radical operation for thyroid carcinoma invading the trachea. *Arch Surg* 119: 1046–1049.
19. Ozaki O, Sugino K, Mimura T, Ito K (1995) Surgery for patients with thyroid carcinoma invading the trachea: Circumferential sleeve resection followed by end-to-end anastomosis. *Surgery* 117: 268–271.
20. Takashima S, Takayama F, Wang Q et al. (2000) Differentiated thyroid carcinomas. Prediction of tumor invasion with MR imaging. *Acta Radiol* 41: 377–383.
21. Wang JC, Takashima S, Takayama F et al. (2001) Tracheal invasion by thyroid carcinoma: Prediction using MR imaging. *AJR Am J Roentgenol* 177: 929–936.
22. Takashima S, Takayama F, Wang J, Kobayashi S, Kadoya M (2003) Using MR imaging to predict invasion of the recurrent laryngeal nerve by thyroid carcinoma. *AJR Am J Roentgenol* 180: 37–42.
23. Davis SD, Fordham LA, Brody AS et al. (2007) Computed tomography reflects lower airway inflammation and tracks changes in early cystic fibrosis. *Am J Respir Crit Care Med* 175: 943–950.
24. Davis SD, Brody AS, Emond MJ et al. (2007) Endpoints for clinical trials in young children with cystic fibrosis. *Proc Am Thorac Soc* 4: 418–430.
25. Sly PD, Brennan S, Gangell C et al. (2009) Lung disease at diagnosis in infants with cystic fibrosis detected by newborn screening. *Am J Respir Crit Care Med* 180: 146–152.
26. Tiddens HA (2006) Chest computed tomography scans should be considered as a routine investigation in cystic fibrosis. *Paediatr Respir Rev* 7: 202–208.
27. Eichinger M, Heussel CP, Kauczor HU, Tiddens H, Puderbach M (2010) Computed tomography and magnetic resonance imaging in cystic fibrosis lung disease. *J Magn Reson Imaging* 32: 1370–1378.
28. Ohno Y, Koyama H, Yoshikawa T et al. (2011) Pulmonary magnetic resonance imaging for airway diseases. *J Thorac Imaging* 26: 301–316.
29. Bhalla M, Turcios N, Aponte V et al. (1991) Cystic fibrosis: Scoring system with thin-section CT. *Radiology* 179: 783–788.
30. Santamaria F, Grillo G, Guidi G et al. (1998) Cystic fibrosis: When should high-resolution computed tomography of the chest be obtained? *Pediatrics* 101: 908–913.
31. Helbich TH, Heinz-Peer G, Eichler I et al. (1999) Cystic fibrosis: CT assessment of lung involvement in children and adults. *Radiology* 213: 537–544.
32. Brody AS, Kosorok MR, Li Z et al. (2006) Reproducibility of a scoring system for computed tomography scanning in cystic fibrosis. *J Thorac Imaging* 21: 14–21.
33. De Jong PA, Tiddens HA (2007) Cystic fibrosis specific computed tomography scoring. *Proc Am Thorac Soc* 4: 338–342.

34. Puderbach M, Eichinger M, Haeselbarth J et al. (2007) Assessment of morphological MRI for pulmonary changes in cystic fibrosis (CF) patients: Comparison to thin-section CT and chest x-ray. *Invest Radiol* 42: 715–725.

35. Edelman RR, Hatabu H, Tadamura E et al. (1996) Noninvasive assessment of regional ventilation in the human lung using oxygen-enhanced magnetic resonance imaging. *Nature Med* 2: 1236–1239.

36. Jakob PM, Wang T, Schultz G, Hebestreit H, Hebestreit A, Hahn D (2004) Assessment of human pulmonary function using oxygen-enhanced T(1) imaging in patients with cystic fibrosis. *Magn Reson Med* 51: 1009–1016.

37. Ohno Y, Koyama H, Matsumoto K et al. (2011) Oxygen-enhanced MRI vs. quantitatively assessed thin-section CT: Pulmonary functional loss assessment and clinical stage classification of asthmatics. *Eur J Radiol* 77: 85–91.

38. Mishima M, Oku Y, Kawakami K et al. (1997) Quantitative assessment of the spatial distribution of low attenuation areas on X-ray CT using texture analysis in patients with chronic pulmonary emphysema. *Front Med Biol Eng* 8: 19–34.

39. Madani A, Keyzer C, Gevenois PA (2001) Quantitative computed tomography assessment of lung structure and function in pulmonary emphysema. *Eur Respir J* 18: 720–730.

40. Goldin JG (2004) Quantitative CT of emphysema and the airways. *J Thorac Imaging* 19: 235–240.

41. Coxson HO, Rogers RM (2005) Quantitative computed tomography of chronic obstructive pulmonary disease. *Acad Radiol* 12: 1457–1463.

42. Hoffman EA, Simon BA, McLennan G (2006) State of the Art. A structural and functional assessment of the lung via multidetector-row computed tomography: Phenotyping chronic obstructive pulmonary disease. *Proc Am Thorac Soc* 3: 519–532.

43. Matsuoka S, Yamashiro T, Washko GR, Kurihara Y, Nakajima Y, Hatabu H (2010) Quantitative CT assessment of chronic obstructive pulmonary disease. *RadioGraphics* 30: 55–66.

44. Bankier AA, O'Donnell CR, Mai VM et al. (2004) Impact of lung volume on MR signal intensity changes of the lung parenchyma. *J Magn Reson Imaging* 20: 961–966.

45. Ohno Y, Hatabu H, Murase K et al. (2004) Quantitative assessment of regional pulmonary perfusion in the entire lung using three-dimensional ultrafast dynamic contrast-enhanced magnetic resonance imaging: Preliminary experience in 40 subjects. *J Magn Reson Imaging* 20: 353–365.

46. Morino S, Toba T, Araki M et al. (2006) Noninvasive assessment of pulmonary emphysema using dynamic contrast-enhanced magnetic resonance imaging. *Exp Lung Res* 32: 55–67.

47. Jang YM, Oh YM, Seo JB et al. (2008) Quantitatively assessed dynamic contrast-enhanced magnetic resonance imaging in patients with chronic obstructive pulmonary disease: Correlation of perfusion parameters with pulmonary function test and quantitative computed tomography. *Invest Radiol* 43: 403–410.

48. Sergiacomi G, Bolacchi F, Cadioli M et al. (2010) Combined pulmonary fibrosis and emphysema: 3D time-resolved MR angiographic evaluation of pulmonary arterial mean transit time and time to peak enhancement. *Radiology* 254: 601–608.

49. Ohno Y, Hatabu H, Higashino T et al. (2004) Dynamic perfusion MRI versus perfusion scintigraphy: Prediction of postoperative lung function in patients with lung cancer. *AJR Am J Roentgenol.* 182: 73–78.

50. Ohno Y, Koyama H, Nogami M et al. (2007) Postoperative lung function in lung cancer patients: Comparative analysis of predictive capability of MRI, CT, and SPECT. *AJR Am J Roentgenol.* 189: 400–408.

51. Ohno Y, Koyama H, Nogami M et al. (2011) State-of-the-art radiological techniques improve the assessment of postoperative lung function in patients with non-small cell lung cancer. *Eur J Radiol* 77: 97–104.

52. Ohno Y, Hatabu H, Takenaka D et al. (2001) Oxygen-enhanced MR ventilation imaging of the lung: Preliminary clinical experience in 25 subjects. *AJR Am J Roentgenol* 177: 185–194.

53. Ohno Y, Hatabu H, Takenaka D, Van Cauteren M, Fujii M, Sugimura K (2002) Dynamic oxygen-enhanced MRI reflects diffusing capacity of the lung. *Magn Reson Med* 47: 1139–1144.

54. Ohno Y, Hatabu H, Higashino T et al. (2005) Oxygen-enhanced MR imaging: Correlation with postsurgical lung function in patients with lung cancer. *Radiology* 236: 704–711.

55. Ohno Y, Koyama H, Nogami M et al. (2008) Dynamic oxygen-enhanced MRI versus quantitative CT: Pulmonary functional loss assessment and clinical stage classification of smoking-related COPD. *AJR Am J Roentgenol* 190: W93–W99.

56. Ohno Y, Iwasawa T, Seo JB et al. (2008) Oxygen-enhanced magnetic resonance imaging versus computed tomography: Multicenter study for clinical stage classification of smoking-related chronic obstructive pulmonary disease. *Am J Respir Crit Care Med.* 177: 1095–1102.

57. Ohno Y, Koyama H, Yoshikawa T et al. (2012) Comparison of capability of dynamic O$_2$-enhanced MRI and quantitative thin-section MDCT to assess COPD in smokers. *Eur J Radiol* 81: 1068–1075.

58. Ohno Y, Nishio M, Koyama H et al. (2012) Oxygen-enhanced MRI, thin-section MDCT, and perfusion SPECT/CT: Comparison of clinical implications to patient care for lung volume reduction surgery. *AJR Am J Roentgenol.* 199: 794–802.

59. Qanadli SD, Orvoen-Frija E, Lacombe P, Di Paola R, Bittoun J, Frija G (1999) Estimation of gas and tissue lung volumes by MRI: Functional approach of lung imaging. *J Comput Assist Tomogr* 23: 743–748.

60. Suga K, Tsukuda T, Awaya H et al. (1999) Impaired respiratory mechanics in pulmonary emphysema: Evaluation with dynamic breathing MRI. *J Magn Reson Imaging* 10: 510–520.

61. Suga K, Tsukuda T, Awaya H, Matsunaga N, Sugi K, Esato K (2000) Interactions of regional respiratory mechanics and pulmonary ventilatory impairment in pulmonary emphysema: Assessment with dynamic MRI and xenon-133 single-photon emission CT. *Chest* 117: 1646–1655.

62. Iwasawa T, Yoshiike Y, Saito K, Kagei S, Gotoh T, Matsubara S (2000) Paradoxical motion of the hemidiaphragm in patients with emphysema. *J Thorac Imaging* 15: 191–195.

63. Iwasawa T, Kagei S, Gotoh T et al. (2002) Magnetic resonance analysis of abnormal diaphragmatic motion in patients with emphysema. *Eur Respir J.* 19: 225–231.

64. Iwasawa T, Takahashi H, Ogura T et al. (2011) Influence of the distribution of emphysema on diaphragmatic motion in patients with chronic obstructive pulmonary disease. *Jpn J Radiol* 29: 256–264.

65. Shibata H, Iwasawa T, Gotoh T et al. (2012) Automatic tracking of the respiratory motion of lung parenchyma on dynamic magnetic resonance imaging: Comparison with pulmonary function tests in patients with chronic obstructive pulmonary disease. *J Thorac Imaging* 27: 387–392.

66. Takahashi M, Togao O, Obara M et al. (2010) Ultra-short echo time (UTE) MR imaging of the lung: Comparison between normal and emphysematous lungs in mutant mice. *J Magn Reson Imaging* 32: 326–333.

67. Ohno Y, Koyama H, Yoshikawa T et al. (2011) T2* measurements of 3-T MRI with ultrashort TEs: Capabilities of pulmonary function assessment and clinical stage classification in smokers. *AJR Am J Roentgenol* 197: W279–W285.

68. Ohno Y, Nishio M, Koyama H et al. (2014) Pulmonary 3 T MRI with ultrashort TEs: Influence of ultrashort echo time interval on pulmonary functional and clinical stage assessments of smokers. *J Magn Reson Imaging* 39: 988–997.

69. Ost D, Fein AM, Feinsilver SH (2003) Clinical practice. The solitary pulmonary nodule. *N Engl J Med* 348: 2535–2542.

70. National Lung Screening Trial Research Team, Aberle DR, Adams AM, Berg CD et al. (2011) Reduced lung-cancer mortality with low-dose computed tomographic screening. *N Engl J Med* 365: 395–409.

71. Patz EF Jr, Lowe VJ, Hoffman JM et al. (1993) Focal pulmonary abnormalities: Evaluation with F-18 fluorodeoxyglucose PET scanning. *Radiology* 188: 487–490.

72. Dewan NA, Gupta NC, Redepenning LS, Phalen JJ, Frick MP (1993) Diagnostic efficacy of PET-FDG imaging in solitary pulmonary nodules. Potential role in evaluation and management. *Chest* 104: 997–1002.

73. Croft DR, Trapp J, Kernstine K et al. (2002) FDG-PET imaging and the diagnosis of non-small cell lung cancer in a region of high histoplasmosis prevalence. *Lung Cancer* 36: 297–301.

74. Chun EJ, Lee HJ, Kang WJ et al. (2009) Differentiation between malignancy and inflammation in pulmonary ground-glass nodules: The feasibility of integrated (18) F-FDG PET/CT. *Lung Cancer* 65: 180–186.

75. Sakai F, Sone S, Maruyama A et al. (1992) Thin-rim enhancement in Gd-DTPA-enhanced magnetic resonance images of tuberculoma: A new finding of potential differential diagnostic importance. *J Thorac Imaging* 7: 64–69.

76. Sakai F, Sone S, Kiyono K et al. (1994) MR of pulmonary hamartoma: Pathologic correlation. *J Thorac Imaging* 9: 51–55.

77. Fujimoto K, Meno S, Nishimura H et al. (1994) Aspergilloma within cavitary lung cancer: MR imaging findings. *AJR Am J Roentgenol* 163: 565–567.

78. Blum U, Windfuhr M, Buitrago-Tellez C et al. (1994) Invasive pulmonary aspergillosis. MRI, CT, and plain radiographic findings and their contribution for early diagnosis. *Chest* 106: 1156–1161.

79. Chung MH, Lee HG, Kwon SS et al. (2000) MR imaging of solitary pulmonary lesion: Emphasis on tuberculomas and comparison with tumors. *J Magn Reson Imaging* 11: 629–637.

80. Gaeta M, Vinci S, Minutoli F et al. (2002) CT and MRI findings of mucin-containing tumors and pseudotumors of the thorax: Pictorial review. *Eur Radiol* 12: 181–189.

81. Ohno Y, Sugimura K, Hatabu H (2002) MR imaging of lung cancer. *Eur J Radiol* 44: 172–181.

82. Sieren JC, Ohno Y, Koyama H, Sugimura K, McLennan G (2010) Recent technological and application developments in computed tomography and magnetic resonance imaging for improved pulmonary nodule detection and lung cancer staging. *J Magn Reson Imaging* 32: 1353–1369.

83. Koyama H, Ohno Y, Seki S (2013) Magnetic resonance imaging for lung cancer. *J Thorac Imaging* 28: 138–150.

84. Koyama H, Ohno Y, Kono A et al. (2008) Quantitative and qualitative assessment of non-contrast-enhanced pulmonary MR imaging for management of pulmonary nodules in 161 subjects. *Eur Radiol* 18: 2120–2131.

85. Mori T, Nomori H, Ikeda K et al. (2008) Diffusion-weighted magnetic resonance imaging for diagnosing malignant pulmonary nodules/masses: Comparison with positron emission tomography. *J Thorac Oncol* 3: 358–364.

86. Satoh S, Kitazume Y, Ohdama S, Kimula Y, Taura S, Endo Y (2008) Can malignant and benign pulmonary nodules be differentiated with diffusion-weighted MRI? *AJR Am J Roentgenol* 191: 464–470.

87. Uto T, Takehara Y, Nakamura Y et al. (2009) Higher sensitivity and specificity for diffusion-weighted imaging of malignant lung lesions without apparent diffusion coefficient quantification. *Radiology* 252: 247–254.

88. Kono M, Adachi S, Kusumoto M, Sakai E (1993) Clinical utility of Gd-DTPA-enhanced magnetic resonance imaging in lung cancer. *J Thorac Imaging* 8: 18–26.

89. Kusumoto M, Kono M, Adachi S et al. (1994) Gadopentetate-dimeglumine-enhanced magnetic resonance imaging for lung nodules. Differentiation of lung cancer and tuberculoma. *Invest Radiol* 29: S255–S256.

90. Guckel C, Schnabel K, Deimling M, Steinbrich W (1996) Solitary pulmonary nodules: MR evaluation of enhancement patterns with contrast-enhanced dynamic snapshot gradient-echo imaging. *Radiology* 200: 681–686.

91. Ohno Y, Hatabu H, Takenaka D, Adachi S, Kono M, Sugimura K (2002) Solitary pulmonary nodules: Potential role of dynamic MR imaging in management initial experience. *Radiology* 224: 503–511.

92. Fujimoto K, Abe T, Muller NL et al. (2003) Small peripheral pulmonary carcinomas evaluated with dynamic MR imaging: Correlation with tumor vascularity and prognosis. *Radiology* 227: 786–793.

93. Ohno Y, Hatabu H, Takenaka D et al. (2004) Dynamic MR imaging: Value of differentiating subtypes of peripheral small adenocarcinoma of the lung. *Eur J Radiol* 52: 144–150.

94. Schaefer JF, Vollmar J, Schick F et al. (2004) Solitary pulmonary nodules: Dynamic contrast-enhanced MR imaging-perfusion differences in malignant and benign lesions. *Radiology* 232: 544–553.

95. Schaefer JF, Schneider V, Vollmar J et al. (2006) Solitary pulmonary nodules: Association between signal characteristics in dynamic contrast enhanced MRI and tumor angiogenesis. *Lung Cancer* 53: 39–49.

96. Kono R, Fujimoto K, Terasaki H et al. (2007) Dynamic MRI of solitary pulmonary nodules: Comparison of enhancement patterns of malignant and benign small peripheral lung lesions. *AJR Am J Roentgenol* 188: 26–36.

97. Ohno Y, Koyama H, Takenaka D et al. (2008) Dynamic MRI, dynamic multidetector-row computed tomography (MDCT), and coregistered 2-[fluorine-18]-fluoro-2-deoxy-d-glucose-positron emission tomography (FDG-PET)/CT: Comparative study of capability for management of pulmonary nodules. *J Magn Reson Imaging* 27: 1284–1295.

98. Cronin P, Dwamena BA, Kelly AM, Carlos RC (2008) Solitary pulmonary nodules: Meta-analytic comparison of cross-sectional imaging modalities for diagnosis of malignancy. *Radiology* 246: 772–782.

99. Parkin DM, Bray F, Ferlay J, Pisani P (2005) Global cancer statistics, 2002. *CA Cancer J Clin* 55: 74–108.

100. Melamed MR, Flehinger BJ, Zaman MB (1987) Impact of early detection on the clinical course of lung cancer. *Surg Clin North Am* 67: 909–924.

101. Geddes DM (1979) The natural history of lung cancer: A review based on rates of tumour growth. *Br J Dis Chest* 73: 1–17.

102. Spiro SG, Silvestri GA (2005) One hundred years of lung cancer. *Am J Respir Crit Care Med* 172: 523–529.

103. Nahmias C, Hanna WT, Wahl LM et al. (2007) Time course of early response to chemotherapy in non–small cell lung cancer patients with 18F-FDG PET/CT. *J Nucl Med* 48: 744–751.

104. Allen MS, Darling GE, Pechet TT et al.; ACOSOG Z0030 Study Group (2006) Morbidity and mortality of major pulmonary resections in patients with early-stage lung cancer: Initial results of the randomized, prospective ACOSOG Z0030 trial. *Ann Thorac Surg* 81: 1013–1019.

105. Webb WR, Gatsonis C, Zerhouni EA et al. (1991) CT and MR imaging in staging non-small cell bronchogenic carcinoma: Report of the Radiologic Diagnostic Oncology Group. *Radiology* 178: 705–713.

106. Baron RL, Levitt RG, Sagel SS, White MJ, Roper CL, Marbarger JP (1982) Computed tomography in the preoperative evaluation of bronchogenic carcinoma. *Radiology* 145: 727–732.

107. Martini N, Heelan R, Westcott J et al. (1985) Comparative merits of conventional, computed tomographic, and magnetic resonance imaging in assessing mediastinal involvement in surgically confirmed lung carcinoma. *J Thorac Cardiovasc Surg* 90: 639–648.

108. Rendina EA, Bognolo DA, Mineo TC et al. (1987) Computed tomography for the evaluation of intrathoracic invasion by lung cancer. *J Thorac Cardiovasc Surg* 94: 57–63.

109. Quint LE, Glazer GM, Orringer MB (1987) Central lung masses: Prediction with CT of need for pneumonectomy versus lobectomy. *Radiology* 165: 735–738.

110. Glazer HS, Kaiser LR, Anderson DJ et al. (1989) Indeterminate mediastinal invasion in bronchogenic carcinoma: CT evaluation. *Radiology* 173: 37–42.

111. Herman SJ, Winton TL, Weisbrod GL, Towers MJ, Mentzer SJ (1994) Mediastinal invasion by bronchogenic carcinoma: CT signs. *Radiology* 190: 841–846.

112. White PG, Adams H, Crane MD, Butchart EG (1994) Preoperative staging of carcinoma of the bronchus: Can computed tomographic scanning reliably identify stage III tumours? *Thorax* 49: 951–957.

113. Takahashi M, Shimoyama K, Murata K et al. (1997) Hilar and mediastinal invasion of bronchogenic carcinoma: Evaluation by thin-section electron-beam computed tomography. *J Thorac Imaging* 12: 195–199.

114. Quint LE, Francis IR (1999) Radiologic staging of lung cancer. *J Thorac Imaging* 14: 235–246.

115. Higashino T, Ohno Y, Takenaka D et al. (2005) Thin-section multiplanar reformats from multidetector-row CT data: Utility for assessment of regional tumor extent in non-small cell lung cancer. *Eur J Radiol* 56: 48–55.

116. Rapoport S, Blair DN, McCarthy SM, Desser TS, Hammers LW, Sostman HD (1988) Brachial plexus: Correlation of MR imaging with CT and pathologic findings. *Radiology* 167: 161–165.

117. Heelan RT, Demas BE, Caravelli JF et al. (1989) Superior sulcus tumors: CT and MR imaging. *Radiology* 170: 637–641.

118. Padovani B, Mouroux J, Seksik L et al. (1993) Chest wall invasion by bronchogenic carcinoma: Evaluation with MR imaging. *Radiology* 187: 33–38.

119. Bonomo L, Ciccotosto C, Guidotti A, Storto ML (1996) Lung cancer staging: The role of computed tomography and magnetic resonance imaging. *Eur J Radiol* 23: 35–45.

120. Freundlich IM, Chasen MH, Varma DG (1996) Magnetic resonance imaging of pulmonary apical tumors. *J Thorac Imaging* 11: 210–222.

121. Takahashi K, Furuse M, Hanaoka H et al. (2000) Pulmonary vein and left atrial invasion by lung cancer: Assessment by breath-hold gadolinium-enhanced three-dimensional MR angiography. *J Comput Assist Tomogr* 24: 557–561.

122. Ohno Y, Adachi S, Motoyama A et al. (2001) Multiphase ECG-triggered 3D contrast-enhanced MR angiography: Utility for evaluation of hilar and mediastinal invasion of bronchogenic carcinoma. *J Magn Reson Imaging* 13: 215–224.

123. Sakai S, Murayama S, Murakami J, Hashiguchi N, Masuda K (1997) Bronchogenic carcinoma invasion of the chest wall: Evaluation with dynamic cine MRI during breathing. *J Comput Assist Tomogr* 21: 595–600.

124. Glazer GM, Orringer MB, Gross BH, Quint LE (1984) The mediastinum in non-small cell lung cancer: CT-surgical correlation. *AJR Am J Roentgenol* 142: 1101–1105.

125. Glazer GM, Gross BH, Aisen AM, Quint LE, Francis IR, Orringer MB (1985) Imaging of the pulmonary hilum: A prospective comparative study in patients with lung cancer. *AJR Am J Roentgenol* 145: 245–248.

126. Musset D, Grenier P, Carette MF et al. (1986) Primary lung cancer staging: Prospective comparative study of MR imaging with CT. *Radiology* 160: 607–611.

127. Poon PY, Bronskill MJ, Henkelman RM et al. (1987) Mediastinal lymph node metastases from bronchogenic carcinoma: Detection with MR imaging and CT. *Radiology* 162: 651–656.

128. Laurent F, Drouillard J, Dorcier F et al. (1988) Bronchogenic carcinoma staging: CT versus MR imaging. Assessment with surgery. *Eur J Cardiothorac Surg* 2: 31–36.

129. McLoud TC, Bourgouin PM, Greenberg RW et al. (1992) Bronchogenic carcinoma: Analysis of staging in the mediastinum with CT by correlative lymph node mapping and sampling. *Radiology* 182: 319–323.

130. Dillemans B, Deneffe G, Verschakelen J, Decramer M (1994) Value of computed tomography and mediastinoscopy in preoperative evaluation of mediastinal nodes in non-small cell lung cancer. A study of 569 patients. *Eur J Cardiothorac Surg* 8: 37–42.

131. Wahl RL, Quint LE, Greenough RL, Meyer CR, White RI, Orringer MB (1994) Staging of mediastinal non-small cell lung cancer with FDG PET, CT, and fusion images: Preliminary prospective evaluation. *Radiology* 191: 371–377.

132. Patz EF, Jr., Lowe VJ, Goodman PC, Herndon J (1995) Thoracic nodal staging with PET imaging with 18FDG in patients with bronchogenic carcinoma. *Chest* 108: 1617–1621.

133. Boiselle PM, Patz EF, Jr., Vining DJ, Weissleder R, Shepard JA, McLoud TC (1998) Imaging of mediastinal lymph nodes: CT, MR, and FDG PET. *RadioGraphics* 18: 1061–1069.

134. Gupta NC, Graeber GM, Bishop HA (2000) Comparative efficacy of positron emission tomography with fluorodeoxyglucose in evaluation of small (<1 cm), intermediate (1 to 3 cm), and large (>3 cm) lymph node lesions. *Chest* 117: 773–778.

135. Marom EM, Erasmus JJ, Patz EF (2000) Lung cancer and positron emission tomography with fluorodeoxyglucose. *Lung Cancer* 28: 187–202.

136. Vansteenkiste JF (2002) Imaging in lung cancer: Positron emission tomography scan. *Eur Respir J Suppl* 35: S49–S60.

137. Antoch G, Stattaus J, Nemat AT et al. (2003) Non-small cell lung cancer: Dual-modality PET/CT in preoperative staging. *Radiology* 229: 526–533.

138. Acker MR, Burrell SC (2005) Utility of 18F-FDG PET in evaluating cancers of lung. *J Nucl Med Technol* 33: 69–74.

139. Kim BT, Lee KS, Shim SS et al. (2006) Stage T1 non-small cell lung cancer: Preoperative mediastinal nodal staging with integrated FDG PET/CT—A prospective study. *Radiology* 241: 501–509.

140. Bruzzi JF, Munden RF (2006) PET/CT imaging of lung cancer. *J Thorac Imaging* 21: 123–136.

141. Kligerman S, Digumarthy S (2009) Staging of non-small cell lung cancer using integrated PET/CT. *AJR Am J Roentgenol* 193: 1203–1211.

142. Billé A, Pelosi E, Skanjeti A et al. (2009) Preoperative intrathoracic lymph node staging in patients with non-small-cell lung cancer: Accuracy of integrated positron emission tomography and computed tomography. *Eur J Cardiothorac Surg* 36: 440–445.

143. Fischer B, Lassen U, Mortensen J et al. (2009) Preoperative staging of lung cancer with combined PET-CT. *N Engl J Med* 361: 32–39.

144. Takenaka D, Ohno Y, Hatabu H et al. (2002) Differentiation of metastatic versus non-metastatic mediastinal lymph nodes in patients with non-small cell lung cancer using respiratory-triggered short inversion time inversion recovery (STIR) turbo spin-echo MR imaging. *Eur J Radiol* 44: 216–224.

145. Ohno Y, Hatabu H, Takenaka D et al. (2004) Metastases in mediastinal and hilar lymph nodes in patients with non-small cell lung cancer: Quantitative and qualitative assessment with STIR turbo spin-echo MR imaging. *Radiology* 231: 872–879.

146. Ohno Y, Koyama H, Nogami M et al. (2007) STIR turbo SE MR imaging vs. coregistered FDG-PET/CT: Quantitative and qualitative assessment of N-stage in non-small-cell lung cancer patients. *J Magn Reson Imaging* 26: 1071–1080.

147. Yi CA, Shin KM, Lee KS et al. (2008) Non-small cell lung cancer staging: Efficacy comparison of integrated PET/CT versus 3.0-T whole-body MR imaging. *Radiology* 248: 632–642.

148. Morikawa M, Demura Y, Ishizaki T et al. (2009) The effectiveness of 18F-FDG PET/CT combined with STIR MRI for diagnosing nodal involvement in the thorax. *J Nucl Med* 50: 81–87.

149. Ohno Y, Koyama H, Yoshikawa T et al. (2011) N stage disease in patients with non-small cell lung cancer: Efficacy of quantitative and qualitative assessment with STIR turbo spin-echo imaging, diffusion-weighted MR imaging, and fluorodeoxyglucose PET/CT. *Radiology* 261: 605–615.

150. Nomori H, Mori T, Ikeda K et al. (2008) Diffusion-weighted magnetic resonance imaging can be used in place of positron emission tomography for N staging of non-small cell lung cancer with fewer false-positive results. *J Thorac Cardiovasc Surg* 135: 816–822.

151. Hasegawa I, Boiselle PM, Kuwabara K, Sawafuji M, Sugiura H (2008) Mediastinal lymph nodes in patients with non-small cell lung cancer: Preliminary experience with diffusion-weighted MR imaging. *J Thorac Imaging* 23: 157–161.

152. Pauls S, Schmidt SA, Juchems MS et al. (2012) Diffusion-weighted MR imaging in comparison to integrated [¹⁸F]-FDG PET/CT for N-staging in patients with lung cancer. *Eur J Radiol* 81: 178–182.

153. Quint LE, Tummala S, Brisson LJ et al. (1996) Distribution of distant metastases from newly diagnosed non-small cell lung cancer. *Ann Thorac Surg* 62: 246–250.

154. Pantel K, Izbicki J, Passlick B et al. (1996) Frequency and prognostic significance of isolated tumour cells in bone marrow of patients with non-small-cell lung cancer without overt metastases. *Lancet* 347: 649–653.

155. Silvestri GA, Gould MK, Margolis ML et al.; American College of Chest Physicians (2007) Noninvasive staging of non-small cell lung cancer: ACCP evidenced-based clinical practice guidelines (2nd edition). *Chest* 132(3 Suppl.): 178S–201S.

156. Simon GR, Turrisi A; American College of Chest Physicians (2007) Management of small cell lung cancer: ACCP evidence-based clinical practice guidelines (2nd edition). *Chest* 132: S324–S339.

157. Samson DJ, Seidenfeld J, Simon GR et al.; American College of Chest Physicians (2007) Evidence for management of small cell lung cancer: ACCP evidence-based clinical practice guidelines (2nd edition). *Chest* 132: S314–S323.

158. Ohno Y, Koyama H, Nogami M et al. (2007) Whole-body MR imaging vs. FDG-PET: Comparison of accuracy of M-stage diagnosis for lung cancer patients. *J Magn Reson Imaging* 26: 498–509.

159. Ohno Y, Koyama H, Onishi Y et al. (2008) Non-small cell lung cancer: Whole-body MR examination for M-stage assessment–utility for whole-body diffusion-weighted imaging compared with integrated FDG PET/CT. *Radiology* 248: 643–654.

160. Takenaka D, Ohno Y, Matsumoto K et al. (2009) Detection of bone metastases in non-small cell lung cancer patients: Comparison of whole-body diffusion-weighted imaging

(DWI), whole-body MR imaging without and with DWI, whole-body FDG-PET/CT, and bone scintigraphy. *J Magn Reson Imaging* 30: 298–308.

161. Sommer G, Wiese M, Winter L et al. (2012) Preoperative staging of non-small-cell lung cancer: Comparison of whole-body diffusion-weighted magnetic resonance imaging and 18F-fluorodeoxyglucose-positron emission tomography/computed tomography. *Eur Radiol* 22: 2859–2867.

162. Ohno Y, Nishio M, Koyama H et al. (2013) Comparison of the utility of whole-body MRI with and without contrast-enhanced Quick 3D and double RF fat suppression techniques, conventional whole-body MRI, PET/CT and conventional examination for assessment of recurrence in NSCLC patients. *Eur J Radiol* 82: 2018–2027.

163. Travis WD, Costabel U, Hansell DM et al.; ATS/ERS Committee on Idiopathic Interstitial Pneumonias (2013) An official American Thoracic Society/European Respiratory Society statement: Update of the international multidisciplinary classification of the idiopathic interstitial pneumonias. *Am J Respir Crit Care Med* 188: 733–748.

164. Puderbach M, Hintze C, Ley S, Eichinger M, Kauczor HU, Biederer J (2007) MR imaging of the chest: A practical approach at 1.5T. *Eur J Radiol* 64: 345–355.

165. Biederer J, Both M, Graessner J et al. (2003) Lung morphology: Fast MR imaging assessment with a volumetric interpolated breath-hold technique: Initial experience with patients. *Radiology* 226: 242–249.

166. Lutterbey G, Grohé C, Gieseke J (2007) Initial experience with lung-MRI at 3.0T: Comparison with CT and clinical data in the evaluation of interstitial lung disease activity. *Eur J Radiol* 61: 256–261.

167. Yi CA, Lee KS, Han J, Chung MP, Chung MJ, Shin KM (2008) 3-T MRI for differentiating inflammation- and fibrosis-predominant lesions of usual and nonspecific interstitial pneumonia: Comparison study with pathologic correlation. *AJR Am J Roentgenol* 190: 878–885.

168. Rajaram S, Swift AJ, Capener D et al. (2012) Lung morphology assessment with balanced steady-state free precession MR imaging compared with CT. *Radiology* 263: 569–577.

169. Semelka RC, Cem Balci N, Wilber KP et al. (2000) Breath-hold 3D gradient-echo MR imaging of the lung parenchyma: Evaluation of reproducibility of image quality in normals and preliminary observations in patients with disease. *J Magn Reson Imaging* 11: 195–200.

170. Karabulut N, Martin DR, Yang M, Tallaksen RJ (2002) MR imaging of the chest using a contrast-enhanced breath-hold modified three-dimensional gradient-echo technique: Comparison with two-dimensional gradient-echo technique and multidetector CT. *AJR Am J Roentgenol* 179: 1225–1233.

171. Johnson KM, Fain SB, Schiebler ML, Nagle S (2013) Optimized 3D ultrashort echo time pulmonary MRI. *Magn Reson Med* 70: 1241–1250.

172. Molinari F, Eichinger M, Risse F et al. Navigator-triggered oxygen-enhanced MRI with simultaneous cardiac and respiratory synchronization for the assessment of interstitial lung disease. *J Magn Reson Imaging* 26: 1523–1529.

173. Ohno Y, Nishio M, Koyama H et al. (2014) Oxygen-enhanced MRI for patients with connective tissue diseases: Comparison with thin-section CT of capability for pulmonary functional and disease severity assessment. *Eur J Radiol* 83: 391–397.

174. Ohno Y, Nishio M, Koyama H et al. (2013) Pulmonary MR imaging with ultra-short TEs: Utility for disease severity assessment of connective tissue disease patients. *Eur J Radiol* 82: 1359–1365.

175. White RI Jr, Pollak JS, Wirth JA (1996) Pulmonary arteriovenous malformations: Diagnosis and transcatheter embolotherapy. *J Vasc Interv Radiol* 7: 787–804.

176. Terry PB, Barth KH, Kaufman SL et al. (1980) Balloon embolization for treatment of pulmonary arteriovenous fistulas. *N Engl J Med* 302: 1189–1190.

177. White RI Jr., Lynch-Nyhan A, Terry P et al. (1988) Pulmonary arteriovenous malformations: Techniques and long-term outcome of embolotherapy. *Radiology* 69: 663–669.

178. Hartnell GG, Jackson JE, Allison DJ (1990) Coil embolization of pulmonary arteriovenous malformations. *Cardiovasc Intervent Radiol* 13: 347–350.

179. Remy J, Remy-Jardin M, Wattinne L, Deffontaines C (1992) Pulmonary arteriovenous malformations: Evaluation with CT of the chest before and after treatment. *Radiology* 182: 809–816.

180. Remy J, Remy-Jardin M, Giraud F, Wattinne L (1994) Angio architecture of pulmonary arteriovenous malformations: Clinical utility of three-dimensional helical CT. *Radiology* 191: 657–664.

181. Remy-Jardin M, Dumont P, Brillet PY, Dupuis P, Duhamel A, Remy J (2006) Pulmonary arteriovenous malformations treated with embolotherapy: Helical CT evaluation of long-term effectiveness after 2-21-year follow-up. *Radiology* 239: 576–585.

182. Brillet PY, Dumont P, Bouaziz N et al. (2007) Pulmonary arteriovenous malformation treated with embolotherapy: Systemic collateral supply at multidetector CT angiography after 2-20-year follow-up. *Radiology* 242: 267–276.

183. Ohno Y, Hatabu H, Takenaka D, Adachi S, Hirota S, Sugimura K (2002) Contrast-enhanced MR perfusion imaging and MR angiography: Utility for management of pulmonary arteriovenous malformations for embolotherapy. *Eur J Radiol* 41: 136–146.

184. Naidich DP, Rumancik WM, Lefleur RS, Estioko MR, Brown SM (1987) Intralobar pulmonary sequestration: MR evaluation. *J Comput Assist Tomogr* 11: 31–33.

185. Au VW, Chan JK, Chan FL (1999) Pulmonary sequestration diagnosed by contrast enhanced three-dimensional MR angiography. *Br J Radiol* 72: 709–711.

186. Lehnhardt S, Winterer JT, Uhrmeister P, Herget G, Laubenberger J (2002) Pulmonary sequestration: Demonstration of blood supply with 2D and 3D MR angiography. *Eur J Radiol* 44: 28–32.

187. Sancak T, Cangir AK, Atasoy C, Ozdemir N (2003) The role of contrast enhanced three-dimensional MR angiography in pulmonary sequestration. *Interact Cardiovasc Thorac Surg* 2: 480–482.

188. Epelman M, Kreiger PA, Servaes S, Victoria T, Hellinger JC (2010) Current imaging of prenatally diagnosed congenital lung lesions. *Semin Ultrasound CT MR* 31: 141–157.

11

Clinical Magnetic Resonance Imaging Applications for the Lung

Mark O. Wielpütz, Jim M. Wild, Grzegorz Bauman, Edwin J.R. van Beek, and Jürgen Biederer

CONTENTS

11.1 Introduction

Magnetic resonance imaging (MRI) has recently been introduced into clinical practice as a diagnostic lung imaging modality. After chest radiography and computed tomography (CT), which is regarded as the standard and most comprehensive three-dimensional (3D) imaging techniques of the lung, MRI is increasingly used as an alternative and supplementary method, albeit in selected cases. Once broadly available and sufficiently robust, it will likely become a modality of choice for cases in which exposure to ionizing radiation should be strictly avoided. This would comprise children, pregnant women, and disorders requiring repeated follow-up examinations over prolonged periods in which MRI could contribute significantly to lowering the cumulative radiation dose and therapy burden.

MRI combines structural and functional imaging aspects in a single examination and has the potential to provide metabolic molecular imaging information in the near future. The key technique for MRI of lung morphology is based on the detection of the magnetic resonance signal of the spins of protons in tissues and liquids, that is, proton- or ^{1}H-MRI. The basic pulse sequences can be further sensitized to physiological processes in the lungs and vasculature. It allows for multiple and repeated measurements, and can be used to assess the time course of motion, as well as lung expansion and blood flow in the thoracic organs and main vessels. These advantages are already well appreciated in MRI of the heart for example. Recent technical advances have enabled MRI to overcome its main limitations, due to low proton density and signal decay due to susceptibility artifacts at air–tissue interfaces. A range of pulse sequences has been introduced for specific clinical conditions (nodule detection, pneumonia, and pulmonary embolism), and tested for robustness and reproducibility of image quality and diagnostic accuracy. The structural information provided by proton MRI is supplemented by imaging alternative nuclei such as the inhaled gases hyperpolarized helium-3 and xenon-129 which also have a nuclear magnetic resonance signal.

Despite these benefits, generally MRI is currently underutilized. Among the reasons that have limited its broader application to date are the limited availability of MRI table time, the lack of standardized consistent protocols customized to clinical needs, and lack of experience of radiologists and referring clinicians. Since new users may find it difficult to implement appropriate protocols on their own platforms, it is the aim of this chapter (1) to explain the physical challenges of acquiring images from the lungs as seen through an MR scanner. The current technical aspects of MRI pulse sequences, radio-frequency (RF) coils, and MRI system requirements needed for imaging the pulmonary parenchyma and vasculature are outlined (see Section 11.2). (2) To present an affordable and effective lung protocol adapted to specific clinical conditions from the non-vendor–specific range of state-of-the-art MR sequences (see Section 11.3). (3) To explain the current scientific knowledge, advantages, and disadvantages of lung MRI in a clinical context (see Section 11.4). (4) To outline the future developments in lung MRI from a clinical perspective. This chapter is based on three previously published review articles and further extends to the more recently published literature in the field of lung MRI [1–3]. For pulmonary applications that demonstrate functional information on lung imaging and a dedicated description of the use of MRI for detection of pulmonary embolism and pulmonary perfusion, please refer to Chapter 13.

11.2 Lung Magnetic Resonance Physics—Seen through the MR Scanner

The physical properties of lung parenchyma are very different from those of tissue such as liver or brain [1]; from the perspective of MR imaging, the two properties of highest importance are the low proton density and the susceptibility differences between tissue and air. In healthy lungs the tissue density is 0.1–0.2 g/cm^3, which is about 10-fold lower than in other tissues. As the MR signal is directly proportional to the tissue proton density, even under perfect imaging conditions (i.e., neglecting relaxation effects), the MR signal of the lung is 5–10 times weaker than that of adjacent tissues. The low signal-to-noise ratio (SNR) makes structural proton MRI of the lung microstructure challenging: to increase the SNR, signal averaging can be employed, but this increases the image acquisition times which would make the protocols unsuitable for clinical routine. The SNR can be increased by using larger voxel sizes; however, smaller lesions such as peripheral lung metastases might not be visible due to partial volume effects. Oxygen in air is paramagnetic and tissue is diamagnetic, which leads to a bulk magnetic susceptibility difference ($\Delta\chi = 8$ ppm) at lung–air interfaces. At each tissue interface the susceptibility difference forms a static local field gradient. The multiple microscopic surfaces presented by the airways and alveoli in the lungs thus create highly inhomogeneous local magnetic field gradients on a spatial scale smaller than the size of a typical imaging voxel (2–5 mm). These microscopic field gradients lead to a rapid dephasing in gradient-echo (GRE) imaging—this signal decay is typically described by an apparent transverse relaxation time T_2^*, which can be as short as 2 ms or less at $B_0 = 1.5$ T. Thus, GRE MRI of lung parenchyma becomes highly challenging and requires pulse sequences with short echo times (TE < 1–2 ms). As the magnetic field inhomogeneity increases with B_0, even shorter T_2^* of about 0.5 ms are found at 3 T [4]. The expected SNR gain of 3 T over 1.5 T can often not be realized because it requires that TE is shortened accordingly. With identical pulse sequences a shorter TE can only be achieved if more powerful gradient systems are used, but current 3 T MR systems often utilize the same gradient units as high-end 1.5 T systems.

Low-field MRI at $B_0 = 0.5$ T or less has some potential advantages for lung MRI with respect to the magnetic inhomogeneity [4], and promising results have been demonstrated with low-field MRI as a non-ionizing alternative to chest X-ray [5]. Low-field strength does, however, require signal averaging to recover signal-to-noise, leading to longer imaging times, which is less attractive from a clinical perspective.

Thus, the methods shown in this chapter are focused on implementation of state-of-the-art 1.5 T scanners, as it represents a readily available and achievable optimal field strength for lung MRI.

Most of the methods outlined here involve imaging during a breath-hold of typically less than 20 s duration. Breath-hold is achieved either in end expiration or in full inspiration. The state of lung inflation plays a role, in that the signal intensity from the lung parenchyma is higher at expiration [6,7]. During expiration the relative parenchyma density of protons is increased, and the bulk volume magnetic susceptibility difference is reduced as air is expelled. If, on the other hand, high contrast-to-noise between the pulmonary vessels or nodules is sought, imaging at full inspiration can provide a darker background.

In order to reliably acquire images at defined stages during respiration (e.g., in full inspiration), respiratory or spirometric triggering and gating can be used. Synchronization of the image acquisition to the respiratory cycle requires the acquisition of a signal that is proportional to the respiratory state of the lung—in MRI this is achieved using either external hardware or the MR signal of the lung itself. The MR manufacturers often provide pneumatic bellows for this purpose; however, MR-compatible pneumotachographs have also been used [8].

Alternatively, the so-called navigator echoes can be applied to measure the breathing motion directly via MRI. Here, either a dedicated 90°–180° spin-echo excitation or a pencil beam readout is used to measure the MR signal orthogonal to the base of the diaphragm. The sub-second navigator scan is integrated into the imaging sequence to provide the motion information. Thus, motion of the diaphragm is detected and can be used for retrospective gating of scans during the free-breathing cycle [9]. A navigator signal can also be extracted from free-breathing non-gated two-dimensional (2D) lung images as long as the acquisition time is fast compared to the motion of the diaphragm. Motion information can also be acquired using self-navigated sequences [10] that were originally established for cardiac imaging. Here, the signal variations of an additional rapid signal acquisition without spatial encoding are utilized, which can be readily incorporated into nearly any imaging sequence. The self-navigator signal reflects the periodic changes in lung signal related to respiratory motion for retrospective sorting of acquired data.

11.2.1 Image Acquisition Techniques for the Lung

11.2.1.1 Parallel Imaging of the Lung

At a field strength of 1.5 T, the MR system's body coil provides a relatively homogeneous transmit field delivering a uniform flip angle over the fields of view (FOVs) of the lung. At 3 T, flip angles in the thorax are no longer homogeneous, which is a consequence of the shorter RF wavelength. In principle the body coil can be used as a receive coil for lung imaging, but a significantly higher SNR is achieved with a local receive coil array that is optimized for MRI of the thorax. Typically, thoracic MR coils consist of an anterior flexible part and a posterior part that is embedded in the patient table. Depending on the number of available receiver channels the coil arrays consist of 2–16 coil elements each. Elements are arranged in rows and columns to be able to accelerate the image acquisition in all directions using parallel imaging.

Parallel imaging methods [11,12] exploit the spatial sensitivity patterns of an array of coils to accelerate the image acquisition by a factor **R**. (see Chapter 1 in *Image Principles, Neck, and the Brain*.) Essentially, every coil array already provides an intrinsic spatial encoding of the MR signal, if the sensitivities of the coil elements are spatially discrete. Parallel imaging is useful in lung MRI where faster acquisition times can reduce breath-hold durations and increase temporal resolution in time-resolved contrast studies. Parallel imaging is particularly useful in single-shot fast spin-echo (SSFSE) imaging, as it shortens the echo train length [13] and the associated k-space filter thus reducing blurring. At 3 T, parallel imaging can help reduce the specific absorption rate (SAR) of an SSFSE sequence which can also allow a shorter echo time [14].

Parallel imaging should however be used carefully, because the acceleration achieved with parallel imaging comes at the price of a reduced SNR: an R-fold faster acquisition lowers the SNR by R and g-factor noise amplification further reduces SNR in a spatially varying manner. As lung parenchyma SNR is intrinsically low, high acceleration factors of $R > 4$ as used in cardiac imaging are not achievable in lung imaging. As such, coils with a large number of elements are not essential for the parallel imaging of the lung (even though they can help to increase the local SNR without parallel imaging) and acceleration factors of $R \approx 2$ in either phase-encode or slice selection/encode directions are appropriate. An important detail of parallel imaging is the method of coil sensitivity calibration, which is needed to determine the sensitivity profiles of the coils for image reconstruction. As the lungs are prone to movement, a method with integrated or autocalibration is preferable (e.g., generalized autocalibrating partially parallel acquisitions [GRAPPA], auto sensitivity encoding [SENSE], and volume acceleration flexible [FLEX]). A sequence that requires a separate sensitivity scan is potentially prone to artifacts due to spatial misregistration of calibration information with imaging data.

11.2.1.2 Pulse Sequences

Modern MR systems are equipped with gradient systems that offer a gradient strength of 40 mT/m or more with slew rates of more than 200 mT/m/ms—thus, for a typical spatial resolution of 1–3 mm very short echo times of TE < 1.5 ms can be realized with GRE sequences. At 1.5 T it is possible to acquire images that highlight regional fibrosis and parenchymal density. There are three basic sequences that were proven to be very effective for lung imaging. All share the common features of short TEs and short acquisition time needed when imaging the lungs and all can benefit from the judicious use of parallel imaging [11]. They can be used with and without respiratory gating, so that applications during free breathing or during breath-hold are both possible.

11.2.1.3 Spoiled GRE

The spoiled GRE sequence (fast low angle shot [FLASH], spoiled gradient recoiled echo imaging [SPGR], and fast field echo [FFE]) is conceptually the simplest imaging pulse sequence, and has been shown to be very robust in clinical practice. Essentially, the sequence acquires a GRE after RF excitation with a low flip angle RF pulse ($\alpha < 90°$), and destroys the remaining magnetization after the data acquisition (spoiling). The sequence is inherently T_2^*-weighted, and depending on TR and α, either an additional spin density weighting or a T_1 weighting can be achieved. As lung T_2^* values are short, a TE of 1 ms or less needs to be used to detect signal from lung parenchyma—furthermore, very low flip angles of $\alpha < 10°$ need to be applied to minimize T_1 weighting by the relatively long T_1 of lung parenchyma (1300 ms at 1.5 T). Use of an asymmetric echo (~30%) and partial Fourier reconstruction in the frequency-encoding direction with a high bandwidth >60 kHz ensure TEs of ~1 ms can be met.

The sequence has multiple applications in lung MRI; for anatomical imaging, it can be used for 2D and 3D acquisitions with or without fat suppression. Non-contrast–enhanced images are preferably acquired without fat suppression to facilitate the detection of lymph nodes surrounded by mediastinal fat. After injection of a T_1-shortening contrast agent (e.g., Gadolinium-DTPA [Gd-DTPA]) and with application of fat suppression, lymph nodes stand out within the suppressed background of the mediastinal fat [15]. Combination of the GRE sequence with parallel imaging methods, and/or slice interpolation techniques (e.g., volume interpolated breath-hold examination [VIBE]), enables full lung volume coverage with slices of 5 mm within short acquisition times of a breath-hold. Dynamic repetitive acquisitions at low resolution provide a robust means of assessment of lung wall motion for investigation of dynamic lung volumes [16], tumor motion for radiotherapy planning [17], and paradoxical diaphragmatic motion.

A heavily T_1-weighted 3D-spoiled GRE sequence with short TR and high α is the starting point for any contrast-enhanced structural imaging in the lung, as high flip angles provide excellent suppression of non-enhancing anatomical regions. Sequence parameters for pulmonary MR angiography and perfusion imaging will be dealt within Section 11.2.2.

11.2.1.4 Balanced Steady-State Free Precession

The balanced steady-state free precession (bSSFP) sequence (true fast imaging with steady state precession [TrueFISP], fast imaging employing steady-state acquisition [FIESTA], and breathhold fast encoding [BFE]) is structurally very similar to the spoiled GRE sequence; however, as it does not spoil but instead refocuses the transverse magnetization at the end of each TR interval, it makes more efficient use of the available magnetization. Within each TR all gradient dephasing is refocused (balanced), and by using RF pulses with alternating phases (i.e., the magnetization is flipped back and forth), a highly coherent steady state can be achieved. Compared to the spoiled GRE sequence, the propagation of transverse coherence leads to a more complex contrast behavior, and for short TR the contrast is proportional to the ratio T_2/T_1. These sequences use higher flip angles (about 70°), which lead not only to a stronger MR signal, but also to higher SAR values. For higher TR values the bSSFP sequences are susceptible to off-resonance and field inhomogeneities, which is manifested as the so-called banding artifacts where one to several dark bands occur. The fast bSSFP sequence has found many applications in particular in cardiac MRI, but it can also be advantageously applied to lung imaging. The similarity of the bSSFP sequence to the spoiled GRE sequence also means that short TE and TR can be realized so that it is well adapted for rapid imaging within the time window of a breath-hold. High readout bandwidths of 100 kHz are typically used to maximize magnetization recycling and to minimize TR, which reduce the size of the off-resonance banding artifacts; for example, ultra-fast SSFP [18] has recently been realized with TR < 1 ms. The inherent T_2/T_1 contrast of the sequence [19] makes it highly effective for imaging blood and mucus with their long T_2 when compared to tissue. Furthermore, a multislice or 3D bSSFP protocol can be used as a quick and contrast-free means of generating a pulmonary angiogram [20] and very promising results have recently been demonstrated with such an ultrashort radial 3D bSSFP sequence with diaphragmatic gating [21].

11.2.1.5 Single-Shot Fast Spin-Echo (RARE/ HASTE and Turbo FSE)

The SSFSE sequence or rapid acquisition with relaxation enhancement (RARE) sequence (RARE/half-Fourier acquisition single shot turbo spin echo [HASTE] and turbo fast spin echo [Turbo FSE]) [22] uses spin echoes (as compared to FLASH and bSSFP, which utilize GRE) and is thus conceptually different from the previous imaging sequences. To make the spin-echo acquisition suitable for rapid imaging, a train of spin echoes is created by repeated refocusing of the first spin echo. Thus, all lines in k-space are acquired in a single echo train. The SSFSE sequence is advantageous for lung imaging as the train of 180° pulses refocuses on any field inhomogeneity so that the images become T_2 weighted. The degree of T_2 weighting is determined by the echo time of the central k-space line (effective TE), which can be adjusted through the choice of different k-space sampling patterns. Echo times of the order of the T_2 of lung parenchyma (about 40 ms at 1.5 T) are still suitable for lung imaging. Compared to the sub-millisecond echo times needed for the gradient-echo sequences such a longer TE is easier to realize. However, with an inter-echo spacing of about 4 ms, a typical echo train for an image with 192 phase-encoding steps still amounts to 192 × 4 ms = 768 ms. In the later echoes of this echo train the T_2 decay will lead to significant signal reductions. The T_2 decay in the phase-encoding direction also leads to a filter effect in k-space that manifests as blurring in the images. Thus, it is advantageous to reduce the number of measured k-space lines, which can effectively be achieved by combinations of half Fourier encoding, rectangular FOV selection (e.g., in axial orientation, phase encoding is chosen in the anterior–posterior direction), and parallel imaging. The large number of 180° pulses in the SSFSE echo train also leads to high amounts of energy deposition (high SAR levels), which make the use of SSFSE at 3 T difficult, as SAR increases with the field strength.

Another option to reduce signal blurring in the phase-encoding direction is the segmentation of k-space.

Therefore, not all k-space lines are acquired in one echo train, but only a fraction N is sampled, and the acquisition is repeated N times to cover the complete k-space. Segmentation significantly prolongs the acquisition time, but is often required when a spatial resolution of 1 mm or better is needed or respiratory-triggered acquisition is needed for noncooperative patients. The SSFSE sequence can be combined with magnetization preparation modules such as fat saturation, inversion recovery preparation, or even diffusion weighting to impart additional contrast weighting to the images.

11.2.1.6 Ultrashort Echo Time Imaging

Originally proposed by Bergin et al. [23] for lung parenchyma imaging, ultrashort echo time (UTE) sequences have recently become available that use radial k-space sampling to shorten the echo times well below 1 ms. At a TE of 100 µs and less, the short T_2^* of lung parenchyma is less of a constraint, and MR images with parenchymal sensitivity can be acquired. Two-dimensional UTE sequences [23] require twice the number of RF pulses in order to achieve slice selection, which doubles the acquisition time. Furthermore, the T_2^* decay of the parenchymal signal introduces a k-space filter in the frequency-encoding direction (akin to that experienced in the phase-encoding direction in SSFSE imaging) which limits the achievable spatial resolution well below the nominal one. Nevertheless, UTE imaging is promising for lung MRI because a proton density contrast resembling that of CT can be achieved with that. Recently, a free-breathing, respiratory-gated 3D implementation of the UTE sequence allowed for substantially improved visualization of the lung parenchyma [24]. The potential application of the sequence is the visualization of destructive lung disease, for example, emphysema, which has so far been impossible with MRI.

11.2.2 Contrast Enhancement

11.2.2.1 IV Gadolinium

High-quality pulmonary angiograms can be obtained at breath-hold with injection of T_1-shortening contrast agents such as Gd-DTPA using heavily T_1-weighted 3D GRE sequences. The pulse sequence of choice for high-resolution 3D pulmonary MR angiography is a spoiled GRE (FLASH, SPGR, and FFE) with the shortest possible TR and TE and high readout bandwidth. Modest parallel imaging can improve the spatial resolution for a given breath-hold duration.

For an adult patient typical imaging parameters for a coronal 3D dataset are TR = 2.5–3 ms, TE = 1.0–1.5 ms, α = 30°–40°, matrix: 40 × 192 × 256, and FOV: 460 mm. Centric elliptic phase encoding [25], with the scan acquisition starting at peak enhancement, ensures maximum SNR and optimal separation of arterial and venous phases. Typical contrast doses used are for example 0.2 mmol/kg body weight Gd-DTPA followed by a 20 mL saline flush injected at 5 mL/s. The time from injection of the contrast agent to start of acquisition can be optimized with a time-resolved bolus-training scan following injection of a 1 mL bolus of contrast agent. Alternatively, a time-resolved 3D acquisition with view sharing such as a time resolved imaging of contrast kinetics (TRICKS)/ time-resolved angiography with interleaved stochastic trajectories (TWIST) sequence [26] can be used, which additionally delineates the arterial and venous regional hemodynamics, regional perfusion defects, and cardiac shunts (Figure 11.1). With TRICKS, *k*-space data are shared between successive datasets, which leads to minor temporal interpolation artifacts—thus, TRICKS data should be used with care when a rapid signal change is observed (e.g., during bolus arrival) (Figure 11.1).

Quantitative T_1-weighted perfusion images can be obtained by using a time-resolved low spatial resolution version of the sequences described above. Here, typical imaging parameters would be TR = 2.0–2.5 ms, TE = 0.8–1.0 ms, α = 30°–40°, matrix: 32 × 96 × 128, and FOV: 460 mm. With a temporal resolution of better than 1 s per 3D dataset this technique allows the generation of contrast passage kinetics curves and parametric maps of regional blood flow, volume, and transit times. Again, parallel imaging and view sharing can enhance the temporal resolution. The contrast agent dose should be less than needed for the pulmonary angiogram—for example, 0.05 mmol/kg Gd-BT-DO3A +20 mL saline flush at 5 mL/s provides robust results in patients with pulmonary vascular disease. If fully quantitative maps of pulmonary blood volume (PBV), blood flow, and transit time are sought, a deconvolution with an arterial input function (AIF) measured in the larger arteries is required. For this, weaker doses still may be needed to ensure a linear signal response between contrast concentration and the signal of the AIF due to T_1 weighting [27] (Figure 11.2).

11.2.3 Advanced Techniques

11.2.3.1 Oxygen-Enhanced ¹H-MRI

When a subject breathes pure molecular oxygen instead of room air, the paramagnetic oxygen shortens the T_1 of the blood, plasma, and tissue in the lungs. This T_1-shortening leads to a signal enhancement, which can be quantified with parametric maps with an inherent ventilation and perfusion weighting. Two strategies for ventilation and perfusion imaging have been proposed: The first technique is semiquantitative and utilizes an inversion-recovery HASTE sequence. The 2D HASTE readout uses a short echo time to visualize the lung parenchyma; however prior to the data acquisition a non-selective inversion

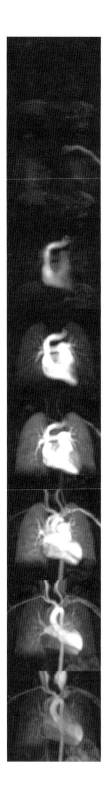

FIGURE 11.1

Time-resolved perfusion imaging with view sharing. Lung perfusion is ideally imaged with a time-resolved gradient-echo sequence with low spatial resolution tuned to a high temporal resolution of 1.5 s per whole chest volume by combining parallel imaging and view sharing. The lung can be thus imaged in a train of measurements during IV contrast injection. The image series depicts the first passage of an IV contrast bolus through the right heart, lung, left heart, and systemic circulation. Every second acquisition of the series is displayed as a subtracted maximum intensity projection.

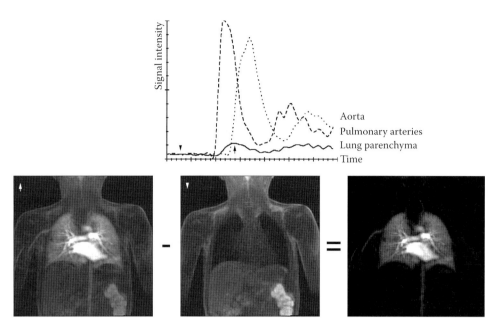

FIGURE 11.2
Generating perfusion maps. To facilitate an easy visual evaluation of extensive four-dimensional dynamic contrast-enhanced perfusion data-sets, the creation of subtracted perfusion maps is recommended. A time point before contrast inflow (white arrowhead) is subtracted from the time point with the highest enhancement of the pulmonary parenchyma (white arrow). Thus, a dataset displaying the pulmonary enhancement only is created. Note that this approach neglects the potential of a dynamic evaluation of the signal changes of the pulmonary parenchyma, which could help to distinguish areas of delayed perfusion from complete perfusion loss.

pulse is applied followed by an inversion delay TI to impart an additional T_1 contrast on the HASTE image. A global inversion pulse should be used rather than a slice selective one to avoid inflow effects of blood with a different magnetization history [28]. The TI is chosen so that the signal from the perfused pulmonary vascular bed is nulled when breathing room air (at the T_1 of the lung parenchyma 1100–1300 ms at 1.5 T [29], TI = 0.69 × T_1 = 700–900 ms). Pure oxygen is then administered via a tight-fitting face mask, which reduces the mean T_1 of the lung plasma and tissue, and a signal enhancement is seen in the IR-HASTE images

[30]. The technique is implicitly ventilation and perfusion weighted and thus represents an indirect way of measuring both aspects of lung function. The method can be made more quantitative with T_1 mapping which uses the inversion pulse as IR-HASTE, but a series of low flip angle FLASH images (Look–Locker sequence) [31]. This allows quantitative approximation of lung tissue/plasma partial pressure of oxygen (pO_2) to be made. Prolonged acquisition time and patient movement between pre-O_2 and post-O_2 measurements can be compensated by self-gating techniques and deformable registration [32] (Figure 11.3).

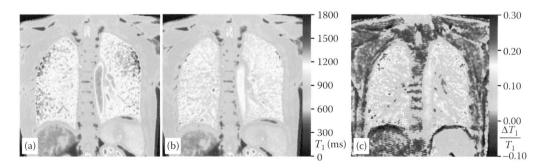

FIGURE 11.3
Oxygen-enhanced MRI for the evaluation of lung ventilation and diffusion. Tissue characterization by measuring T_1 with a radial self-gated inversion recovery multigradient-echo sequence in a healthy volunteer. Expiratory-weighted images are displayed. The subject first breathes room air (a). Supplementing pure O_2 by a mask leads to a subsequent drop of T_1 in ventilated lung areas after a phase of wash-in, as a result of alveolar and tissue dissolved O_2 (b). Exact deformable registration allows for the creation of a subtraction map displaying the gradual change in T_1 relaxation time by O_2 as a surrogate for ventilation (c). Note that venous pulmonary vessels do not show a difference in T_1 relaxation. (Courtesy of Dipl.-Phys. Simon Triphan, Heidelberg, Germany.)

11.2.3.2 Non-Contrast–Enhanced Ventilation and Perfusion Imaging

An alternative method to study lung function, the Fourier decomposition (FD) MRI technique, allows for simultaneous assessment of regional lung perfusion and ventilation-related information without exposure to ionizing radiation, administration of intravenous, or inhalational contrast agent [33]. The FD MRI method was developed on a 1.5 T clinical MR scanner and is based on free-breathing acquisitions of time-resolved datasets with sub-millisecond echo sampling using 2D bSSFP imaging. Subsequent non-rigid image registration is used to correct for the respiratory and cardiac motion, which allows tracing of signal intensity changes in corresponding lung areas caused by contraction of the lung parenchyma at the respiratory frequency and the blood flow at the cardiac frequency. Pixel-wise Fourier analysis can be used to separate the signal contribution from both frequencies and produce spatially resolved ventilation- and perfusion-weighted images. FD MRI allows for covering the whole lung volume using a multislice acquisition in a complete measurement time shorter than 10 min being applicable in clinical routine. It requires only minimal patient compliance and is not dependent on any triggering technique.

Previous studies demonstrated the feasibility of the FD MRI and tested reproducibility of the method in healthy volunteers [34]. The technique was successfully validated in animal experiments against other imaging modalities including the clinical gold standard of single photon emission computed tomography (SPECT)/CT, well-established dynamic contrast-enhanced (DCE) MRI, and ^{3}He-MRI [35,36]. First clinical data on FD MRI were obtained in a population of cystic fibrosis (CF) patients, where good correlation with DCE MRI technique was found [37,38]. The most recent study investigated the capability of FD MRI as a tool for preoperative assessment of regional lung perfusion in patients with

non-small cell lung cancer [39]. Moreover, an extension of image post-processing allowing for quantification of perfusion-weighted images was proposed [40]. There is also a perspective that FD MRI can contribute to the morphological and functional assessment of acute pulmonary embolism in the very near future with a non-contrast–enhanced free-breathing MR scan of 10–15 min (Figure 11.4).

11.2.4 Imaging the Lungs with Other Nuclei

MRI is sensitive to the nuclear magnetic resonance of other nuclei such as ^{3}He, ^{13}C, ^{19}F, ^{23}Na ^{31}P, and ^{129}Xe. Of particular interest in the lungs are the gaseous elements ^{3}He and ^{129}Xe, which can be used as inhaled contrast agents to directly image lung ventilation gas flow, diffusion, and gas exchange. The low spin density of the gaseous state and the inherently weaker MR signal of these nuclei when compared to ^{1}H mean that signal enhancement using a process of hyperpolarization is needed to boost the imaged signal. This is achieved using laser optical pumping techniques, which now allow polarization of both ^{3}He and ^{129}Xe to levels of between 20% and 60% in volumes large enough for routine clinical imaging with modest doses of inhaled gas (<1 L). In addition to the custom polarization systems needed for preparation of the gases, additional dedicated transmit/receive RF coils tuned to the ^{3}He or ^{129}Xe resonance frequency are also needed; this hardware is however commercially accessible and is less expensive than the MRI scanner itself.

The bulk of clinical work published to date with inhaled gas MRI has focused on the use of ^{3}He, and the technique has been shown to be an extremely sensitive method for visualizing and quantifying regional lung ventilation with 2D and 3D volumetric imaging with spoiled GRE, and SSFP sequences high SNR images of lung ventilation can be obtained in a single breath-hold with isotropic pixels size of 3–4 mm. ^{3}He ventilation MRI has been shown to have sensitivity to early signs

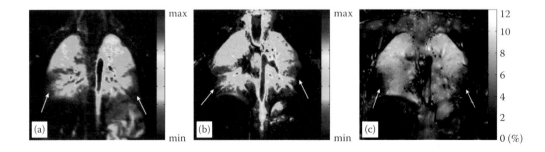

FIGURE 11.4
Fourier decomposition for simultaneous evaluation of ventilation and perfusion. Coronal FD-MRI images acquired in a 5-year-old male cystic fibrosis patient. Color-coded dynamic contrast-enhanced MRI shows perfusion abnormalities in the lower lobes of both lungs (white arrows) (a). An identical pattern of regions with reduced perfusion can be recognized in non-contrast perfusion-weighted FD-MRI (b). Furthermore, the ventilation-weighted FD-MRI shows decreased percental change of the pulmonary tissue signal during the respiratory cycle matching hypo-perfused lung areas, thus reflecting areas of reduced ventilation and subsequent hypoxic vasoconstriction (c).

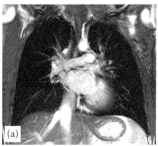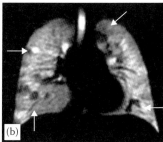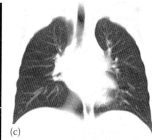

FIGURE 11.5
³He-MRI for ventilation imaging. Images of an 8-year-old child with Cystic Fibrosis with normal pulmonary function tests. Morphological ¹H-MRI 2D bSSFP (a) as well as computed tomography (c) did not reveal significant changes of airways or parenchyma. However, early signs of ventilation heterogeneity are evident on the ³He-MRI (arrows) (b).

of lung obstruction and ventilation heterogeneity before structural changes are manifested on lung function tests, CT, or ¹H-MRI in a variety of obstructive lung diseases such as CF (Figure 11.5) [41,42], asthma [43], and chronic obstructive pulmonary disease (COPD) [44,45]. With co-registered structural information from the ¹H signal in the same breath [46], the gas ventilation images can be interpreted quantitatively and assessed alongside lung contrast-enhanced perfusion images to give regional information on lung ventilation–perfusion [47]. ³He ventilation imaging has been shown to be useful in the assessment of regional changes in lung ventilation before and after therapy in CF [48,49], asthma [50,51], and COPD; the sensitivity, safety, and repeatability of the method make it well suited for use in evaluation of novel therapeutics in this respect.

Rapid imaging of dynamic gas flow in the airways can be performed to assess regional air trapping and collateral ventilation in COPD and asthma [52,53]. The diffusivity of the gases in the lungs can also be measured with apparent diffusion coefficient imaging which can provide non-invasive measurement of mean alveolar and acinar structural length scales. These techniques are of clinical interest in the detection of early emphysematous change and in understanding lung development, growth, and regeneration [54,55].

The limited availability of ³He has meant that clinical uptake has been limited although multicenter studies have been performed evaluating ³He ventilation MRI alongside CT and spirometry [56]. However, increases in levels of gas polarization and improved imaging methods mean that ¹²⁹Xe now provides a cheaper alternative for clinical lung ventilation imaging in years to come and is now ready for evaluation in larger clinical studies [57,58]. Of added interest is the solubility of xenon in the lung tissues and blood, making it sensitive to gas exchange pathways and changes in lung microstructure in interstitial lung diseases (ILDs) such as idiopathic pulmonary fibrosis as IPF [59–61].

11.3 MR Protocols for the Lung

11.3.1 Expectations of a Clinical Radiologist

For practicability, a basic protocol needs to cover the majority of expected clinical problems in lung MRI [15], which is now provided by all major vendors for contemporary standard scanners (1.5 T), which can be customized with additional sequences and IV contrast to a specific clinical situation such as staging of malignancy, evaluation of pulmonary vessels, and lung perfusion. Preferably, additional features complicating the procedure and prolonging acquisition time such as ECG triggering or respiratory gating should be avoided. However, sequence solutions for typical problems such as patients not being able to hold their breath or young children should be readily applicable. Given the high investment into state-of-the-art MRI equipment and the running costs for maintenance and personnel, the key parameter for cost-effectiveness of the method is in-room time. If the use of MRI is intended for emergency conditions such as acute pulmonary embolism, fast and efficient procedures are required with in-room times that make it possible to squeeze the examination into a full MR schedule [3,38]. Based on the currently available standards, in-room times of 15 min for the basic protocol or an emergency examination were considered desirable [3,15]. Protocol extensions, for example, contrast-enhanced series, dynamic contrast-enhanced MRI, and visualization of respiratory motion should take less than 15 min in addition to the basic protocol. This allows for appreciation of the specific features of lung MRI related to its excellent soft tissue contrast and functional imaging capacities. Previously published work gives an excellent overview of the different sequence techniques with vendor-specific acronyms, and most sequences have a counterpart on other vendors' platforms [2,3]. One step closer to routine clinical application in pulmonary imaging will be a consensus nomenclature for imaging findings as established for chest X-ray and CT by the Fleischner society [62]. For this purpose, Barreto and colleagues compared imaging findings in CT and MRI in a number of different pathologic condition, transferring some nomenclature from CT to MRI [63].

11.3.2 Protocol Components with Respect to Clinical Scenarios

11.3.2.1 Pathologic Conditions of the Lung

Pathologic conditions of the lung related to changes of lung structure result either in an increase of lung density (collapse of lung with atelectasis, fluid accumulation inside the alveolar space and/or the lung interstitium, fluid and cell accumulation in infiltrates, or growth of soft tissue resulting in nodules/masses or fibrosis) or in a decrease of lung density (over-inflation due to air trapping or enlargement of air spaces in emphysema) [6,7,64,65]. Given the well-known limitations of MRI in visualizing low proton density structures with high susceptibility artifacts at tissue–air interfaces, any pathology resulting in an increase of lung density (and consequently proton density) should be easy to detect while a decrease of lung density will remain more challenging to visualize with MRI. However, most conditions resulting in destruction and over-inflation of lung parenchyma such as in emphysema are associated with characteristic changes in the remaining lung parenchyma (fibrosis, septal thickening, parenchymal distortion, bronchial wall thickening, etc.). Hence morphologic MRI can still provide a correct diagnosis based on widened spaces of missing signal due to over-inflation and associated findings with increased signal intensity [66]. A direct visualization of the airspace remains the domain of sophisticated imaging technology in specialized centers (see Section 11.3.2.1.3). Furthermore, changes in lung perfusion (hypoxic vasoconstriction, pulmonary embolism, and congestion due to heart failure) result in changes of lung signal on MRI [33,35]. Non-enhanced MRI can directly visualize the vessel tree to detect macrovascular pathology [3].

11.3.2.1.1 Pathologic Conditions with Increased Signal Intensity

Significant experience exists for detection of pathologic conditions that result in an increase of lung proton density. Experimental work has shown the high sensitivity of MRI with T_2- and proton-density–weighted sequences for fluid accumulation inside the lung [67–70]. One can conclude that the sensitivity of MRI for pulmonary infiltrates is at least as good as with chest radiography and CT [71,72]. The majority of the applied protocols are based on T_2- or proton-density–weighted FSE sequences, either with respiration gating or triggering or in breath-hold acquisition modes.

Another key clinical demand is the detection of small solid or soft tissue lesions (*nodules*). The sensitivity of MRI for lung nodules larger than 4 mm ranges between 80% and 90% and reaches 100% for lesions larger than 8 mm [73]. Depending on the sequence technique and the signal intensity of the lesions and given that conditions are optimal (i.e., the patient can keep his breath for 20 s or perfect gating/triggering), a threshold size of 4 mm can be assumed for lung nodule detection with MRI [74–76]. In comparison to CT, lung MRI readout for pulmonary nodules is simpler, since they appear with bright signal against the dark background of the healthy lung tissue [76]. Calcified nodules tend to disappear in the background, as they have no inherent signal, whereas contrast-filled vascular lesions will be highly visible on T_1-weighted images [77]. So far, a variety of sequence types has been evaluated for lung nodule detection with MRI. The spectrum comprises T_2-weighted FSE imaging with and without fat saturation [78–81], inversion recovery techniques [82], T_1-weighted SE [80,83], and GRE sequences [84,85]. The implications for protocol design are to include at least one T_2-weighted or proton-density–weighted or short tau inversion recovery (STIR) sequence to cover infiltrates and nodular lesions with high fluid content. This should be combined with a second sequence to cover nodules with high signal on T_1-weighted images as well. In particular for the detection of malignant lesions with hyperperfusion and intense enhancement, the IV administration of paramagnetic contrast material might be helpful and even increase detection rates. However, this has not yet been confirmed by appropriate studies so far.

11.3.2.1.2 Vascular and Lung Perfusion Disorders

MR angiography and MR perfusion have mainly been applied to study acute pulmonary embolism, a key clinical entity requiring high diagnostic accuracy. Many approaches include direct visualization of the thrombus inside the pulmonary artery either with positive or with negative contrast against the signal of the flowing blood. Presently, fast steady-state free precession GRE sequences (SSFP-GRE) appear to be one of the most effective techniques. Lung vessels are filled with bright signal against which thrombotic material is contrasted with low signal intensity. A sensitivity of 90% and a specificity of 97% have been reported for acute central to segmental pulmonary embolism [86–88]. Another approach has used double inversion recovery, demonstrating stagnant blood clot as bright signal, but insufficient data exist as to the diagnostic accuracy at this time [89,90].

At present, highest spatial resolution is warranted by MR angiography with T_1-weighted contrast-enhanced 3D GRE acquisitions with k-space–centered contrast bolus in breath-hold [91]. For optimum contrast, it is recommended to use a power injector and selected injection protocols [92]. Several studies have shown very encouraging results for MR angiography in the work-up of patients with suspected pulmonary embolism [93–96]. However, more recently, the PIOPED III study involving 7 centers and 371 patients with suspected pulmonary embolism demonstrated sufficient image quality in only 75% of the patients. Overall sensitivity and specificity for acute pulmonary embolism were 78% and 99%, respectively [97]. The most frequent reason for insufficient image quality was dyspnea, coughing, or insufficient timing of contrast material injection.

Aside from the pulmonary artery tree, this technique can also be utilized to detect dilated bronchial arteries and abnormalities of the pulmonary venous system, depending on bolus timing. To overcome the problem of bolus timing, other approaches favored time-resolved MR angiography with multiple acquisitions of lung volumes using very fast 3D GRE techniques. Extremely short acquisition times of 1.5 s or less were achieved by combining parallel imaging and data sharing [98,99]. With the so-called 4D MR angiogram or dynamic first pass perfusion MRI (also known as DCE-MRI), Ersoy et al. realized a sensitivity of 98% for lobar and 92% for segmental embolism [100]. Due to lower spatial resolution, this technique can only detect thrombus in large vessels, but it is unable to demonstrate thrombus-related perfusion defects (Figure 11.1).

The development of dynamic first pass perfusion MRI has opened further perspectives in studying lung parenchyma perfusion in other disease conditions than pulmonary embolism as well. Limited to one plane, the temporally resolved so-called 2D dynamic (2D+t) perfusion MRI reaches excellent temporal resolution of up to 10 images/s with reasonable spatial resolution [101]. However, volume coverage needs multiple series and contrast injections, which are usually not feasible. Thus, 4D approaches are preferred. Other techniques such as arterial spin labeling, well known in brain imaging, based on intrinsic contrast of magnetized, inflowing blood into the imaging plane, or volume have been investigated and applied for scientific studies, but have not yet found their way into clinical applications [102,103].

The visual evaluation of 4D MRI image sets is facilitated by subtraction of the non-enhanced from the contrast-enhanced image with highest lung parenchyma signal, which results in a bright display of the contrast-enhanced lung vessels and parenchyma (Figure 11.1). The clinical value of lung perfusion studies with visual and semiquantitative evaluation is being tested in the assessment of lung perfusion deficits in CF patients who suffer from mucus retention and hypoxic vasoconstriction (Figures 11.6 and 11.7) [99,104–106]. The perfusion series allow indirect visualization of lung parenchyma abnormalities due to emphysema or conditions such

as pneumothorax, infiltrates, or abscesses due to the absence of perfused lung tissue. Please note that the sole evaluation of a perfusion map generated from a perfusion series of around 20–30 acquisitions does not exploit the full potential of such datasets to detect areas of delayed perfusion with increased system arterial supply, shunt volumes, or pulmonary venous abnormalities. Thus, various software-based post-processing tools have been developed, which have not made their way into clinical routine yet. Taking the time component into account, other parameters such as mean transit time, time-to-peak, peak enhancement, PBV, or pulmonary blood flow (PBF) can be calculated for each lung region and displayed as 3D color-coded parameter maps. These allow for differentiation of areas with reduced from those with delayed perfusion [103,107–109]. Due to the non-linearity between the MR signal and the concentration of applied contrast media as stated above, quantitative parameters such as PBV or PBF have to be interpreted with caution [27]. So far, these parameters have been evaluated in the context of COPD and pulmonary hypertension [110,111].

11.3.2.1.3 Airways and Lung Ventilation Disorders

Direct visualization of the airways with MRI is limited to airways in excess of 3 mm diameter unless filled with bright materials such as retained mucus in CF patients [112,113]. In young healthy subjects, lung MRI depicted airways down to the first subsegmental level, however, close to the heart detection rates were lower due to cardiac pulsation [114]. As expected from the higher spatial resolution, high resolution CT remains superior to MRI in the depiction of small peripheral airways.

However, MRI is capable of studying respiratory motion of the lung, central airways, and diaphragm. The technical advances described above for 2D+t MRI or 3D+t MRI with different variants of T_1-weighted GRE or bSSFP sequences have been applied for this purpose [115], for example, to study tracheobronchomalacia in children [116,117] as well as dynamic airway collapse in chronic bronchitis, COPD, or CF [118]. Apart from airways disease, one of the most important clinical applications will be radiotherapy planning for organs with respiration-correlated motion [119,120].

Lung ventilation can be estimated from regional volume changes and signal changes in dynamic acquisitions, or paired inspiratory and expiratory sequences with good parenchymal signal [6,121]. Since air has no signal on MRI, ventilation can only be visualized directly with hyperpolarized noble gases [122], or inhaled contrast materials such as sulphur hexafluoride (SF6) [123] and aerosolized paramagnetic contrast agent [124], while indirect assessment can take place using oxygen-enhanced imaging [125]. See Chapter 13 for further details about these methods. Presently, the use of dynamic image acquisitions

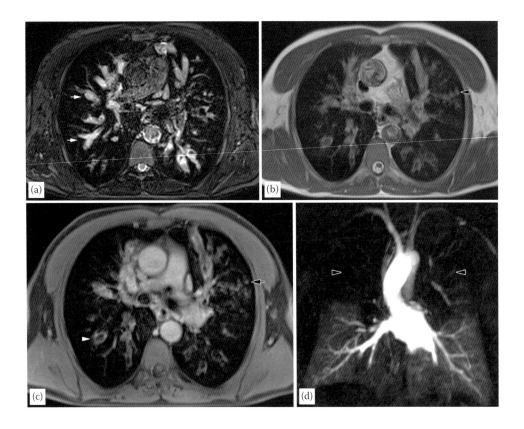

FIGURE 11.6

Advanced CF lung disease. Typical stigmata of advanced cystic fibrosis in a 10-year-old boy. Bronchiectasis with wall thickening and filled with mucus can be appreciated with high signal intensity on T_2-weighted sequences (white arrows) (a, b). A tree-in-bud pattern reflecting small airways disease (black arrow) is usually best appreciated on a high spatial resolution T_1-weighted sequence (c). Inflammatory wall thickening can be distinguished from mucus filling by contrast uptake of the airway wall (white arrowhead) (c). The subtracted perfusion maps show severe perfusion abnormalities of the superior lobes (black arrowheads) and also inhomogeneous perfusion of the lower lobes (5 mm maximum intensity projection) (d).

in single 2D planes (2D+t) to study respiratory motion appears to be closest to routine use.

11.3.2.2 Imaging Pediatric Patients

The lack of radiation exposure makes MRI of the lung particularly attractive for pediatric radiology. Depending on experience with the technology and the availability of scanner time, MRI has become the first line cross-sectional imaging technique for pulmonary disease in many departments [126]. The available sequence techniques are the same as for adult patients, but characteristic features of the pediatric setting influence the choice of protocols. First of all, breath-hold imaging is frequently not possible in young children up to 6–8 years. Depending on the size of the patients and their ability to comply with the procedure and breathing instructions, it appears useful to prepare a separate protocol not only with motion compensated protocols, but also with adjusted FOV, slice thickness, and in-plane resolution with optimized SNRs for smaller children [127].

Two basic strategies have been pursued for motion compensation: bSSFP or partial Fourier single-shot sequences (e.g., HASTE) have been successfully implemented [128]. The bSSFP sequences allow for a rapid acquisition of 10 slices with breath-hold times below 10 s even on low-field MRI systems. Alternatively, they can be performed during free breathing [129,130]. Typically, these fast and robust sequences would be used for an initial overview with further high resolution acquisitions in the second part of the examination. The second approach, gated or triggered acquisition, increases imaging time but provides better spatial resolution and soft tissue contrast [127,131]. The higher respiration frequencies of young children are of a certain advantage since they help speed up the acquisition. In most cases, the difference to non-gated acquisitions is much less than in adult subjects with low respiration rates. T_2-weighted fast spin-echo sequences can be applied with repetition times of 2000 ms or less, usually triggered to the expiratory phase which is around 2 s depending on the individual respiration frequency. This time frame allows for excellent T_2-weighted images without relevant motion artifacts [132]. Depending on the available hardware and specific experience of the team, both mechanical (respiratory belt or cushion technique [132])

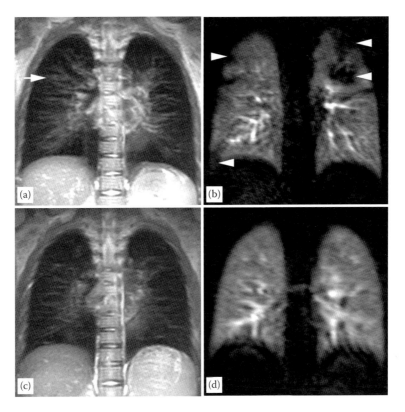

FIGURE 11.7
Response to therapy in cystic fibrosis lung disease. A T_2-weighted navigated as well as a dynamic contrast-enhanced perfusion sequence was acquired in free breathing in a 10-year-old female cystic fibrosis patient. Pulmonary exacerbations in CF can be associated with increased pulmonary mucus content (white arrows, a) and extensive perfusion abnormalities (white arrowheads, b). Repeat MRI one month after intravenous antibiotic treatment demonstrates a reduction in pulmonary mucus (c) and perfusion abnormalities (d).

or image-based (e.g., navigator) devices for the detection of respiratory motion can be applied with good results. A radial readout scheme of the k-space further improves the robustness against motion artifacts. The application of additional cardiac triggering may be helpful in specific cases, but will result in a significant increase of acquisition time [133].

A second characteristic feature of the pediatric setting is the need for sedation in many cases—of course demanding a non–breath-hold protocol, or general anesthesia, allowing respirator-controlled breath-holding. In low-field lung MRI it remains a matter of debate [130,134]. Anesthesia will result in dependent lung atelectasis; for propofol sedation, an incidence of up to 42% of dorsal atelectasis has been described [135]. Other groups use chloral hydrate or phenobarbital and report less frequent atelectasis [136]. However, it is important to know this condition, since it may mask or even simulate relevant pathology. Therefore, additional scans in prone position may be warranted. In case exact correlation with CT is needed, for example, when switching follow-up examinations from CT to MRI, it may be useful to acquire the MRI examination with elevated arms, but this is limited as children usually do not tolerate this for more than 15–20 min.

11.3.3 Protocol Suggestions

11.3.3.1 Basic Protocol

More than in X-ray or CT, image quality in MRI depends on patient compliance over a longer time period. Excellent results will be obtained in young, compliant subjects with good breath-holding capability. In contrast to this, it is extremely challenging to obtain diagnostic image quality examinations in non-compliant patient, obese patients, or those who are unable to hold their breath or follow breathing instructions. Therefore, the use of fast imaging protocols is advocated, with additional sequences to allow for with a certain degree of redundancy to compensate for any failed acquisitions [1–3]. Parts of the protocol can be acquired in free breathing, either due to their fast acquisition scheme or with respiration gated or triggered acquisition modes, at the cost of prolonged acquisition time. Respiration-triggered versions of the T_2-weighted FSE sequences are available, increasing the total in-room time by 10 min. Technicians should test patient compliance early on in the study and can adjust the sequences accordingly. However, it needs to be emphasized that coaching of patients is very important to gain most information from any MRI scan. Respiration belts may be helpful, since this allows for monitoring patient compliance even during image acquisition.

Though protocol suggestions were developed and evaluated mostly on 1.5 T systems, they can be transferred to 3 T scanners [1,2]. While most sequences, in particular 3D GRE techniques, profit from a higher lesion-to-background contrast to a certain degree, artifacts may deteriorate the image quality of steady-state GRE sequences [1,2]. Overall, the changes of image quality with transfer of the concept to 3 T are acceptable or even positive for most sequence types.

Adjusted to the size of the patient, the typical FOV are 450–500 mm in coronal and approximately 400 mm in transverse acquisitions with matrices of 256–384 pixel (for triggered fast spin-echo series up to 512), resulting in pixel sizes smaller than 1.8 × 1.8 mm. Slice thicknesses for the 2D acquisitions ranges from 4 mm to 6 mm. Three-dimensional acquisitions for imaging lung morphology in transverse orientation uses slice thicknesses of 4 mm or less, for pulmonary angiography in coronal orientation this is 2 mm or less [137].

A basic study combines T_1- and T_2-weighted images based on GRE and FSE sequences, respectively [2,15]. T_1–GRE sequences are available as 3D acquisitions and may be limited to a volume acquisition of the chest in one breath-hold. T_2-weighted FSE sequences should cover at least two planes, for example, with a half Fourier breath-hold acquisition in coronal and transverse orientation. This results in a high sensitivity of the protocol for infiltrates and small nodular lesions. To improve the sensitivity of the protocol for mediastinal lymph nodes and bone lesions (e.g., metastases), an STIR or fat-saturated T_2-weighted fast spin-echo sequence should be added. Rib metastases are easier to detect on transverse slices, spinal lesions on coronal orientation. The list of obligatory sequences is concluded with a coronal steady-state free precession sequence in free breathing, which contributes to a high sensitivity for central pulmonary embolism and gross cardiac or respiratory dysfunction if performed in free breathing. With this selection of non-contrast-enhanced sequences, the study covers some of the most common pathologies, including pneumonia, atelectasis, pulmonary nodules or masses, mediastinal masses (lymphoma, goiter, cyst, and thymoma), and acute central pulmonary embolism. The following protocol suggestions cover all expected clinical questions with different branches of the protocol tree for specific problems. They are a good point to start from and allow users to expand the protocol according to their own experience.

11.3.3.2 Extension I: Tumor

Depending on the initial or further expected findings, additional contrast-enhanced acquisitions are required, which would use the same type of volume interpolated 3D GRE sequence, but now with fat saturation to improve the visibility of contrast-enhanced tissues and mediastinal lymph nodes. Although the 3D sequences cover the whole chest, in-plane resolution is optimized in either transverse or coronal section (Figure 11.8). Since

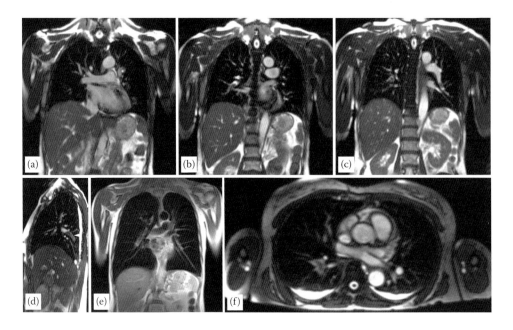

FIGURE 11.8
Pulmonary embolism. The 59-year-old immunocompromised female presented with fever and dyspnea. She initially received a bSSFP in three planes in free breathing and a T_2-weighted HASTE in two planes. The bSSFP allowed a detection of multiple segmental and subsegment emboli (white arrows) (a–e). Pleural effusion was evident on both sides. Pulmonary infiltrates could be ruled out and the examination was concluded at this point after less than 10 min in-room time (f).

acquisition times are just one breath-hold, it appears to be feasible to acquire the 3D GRE studies in at least transverse and coronal planes. In particular for staging of lung cancer, it might be helpful for the detection of small lymph nodes metastases to add a diffusion-weighted sequence as well. Both sequences extend the total imaging time by approximately 5 min [15]. As an option (but at the cost of additional imaging time exceeding the intended limitation to 15 min), motion-compensated T_2-weighted FSE sequences can be added to improve the depiction of masses with chest wall invasion [2,38]. An alternative approach uses bSSFP sequences acquired in free breathing in coronary orientation focusing on the tumor (2D+t or cine MRI). A sliding motion of the tumor can rule out chest wall invasion with high accuracy [138].

11.3.3.3 Extension II: Vasculature and Perfusion

The available options for imaging disorders of lung vasculature comprise three components: First of all, the free-breathing bSSFP study which is also part of the general protocol. Then two variations of 3D GRE-based contrast-enhanced MR angiography: (a) A time-resolved, low spatial resolution acquisition for first pass perfusion imaging and (b) a high spatial resolution acquisition for a breath-hold angiogram. Depending on the performance of the MR scanner, the dynamic study produces a comprehensive lung perfusion study with excellent temporal resolution, and at the same time serves as the determinant for optimal contrast bolus timing for the acquisition of the high-resolution angiogram. These protocol extensions optimize the sensitivity for acute and chronic pulmonary embolism, arterio–venous malformation (e.g., Osler-Weber-Rendu's disease), dilated bronchial arteries, lung sequestration, pulmonary arterial aneurysm, abnormalities of pulmonary venous drainage, and any other pathology of lung vasculature [15]. It is also recommended in airways disease (COPD and CF). As in other parts of the protocol recommendations, it is suggested to combine different fast imaging sequences to increase sensitivity and specificity of the examination [139]. Therefore, Kluge et al. suggested combinations of different available MRI techniques for the detection of pulmonary embolism [88]. The lung vessel imaging branch of the protocol tree can be used just for a study of vascular pathology, for example, in suspected acute pulmonary embolism or in combination with the tumor protocol for the comprehensive evaluation of a central mass with vessel invasion.

11.3.4 Expanding the Scope—One-Stop-Shop MRI?

Any suggested standard protocol for lung diseases should offer sufficient coverage of the surrounding structures potentially involved, which may then be further examined with dedicated protocols, for example, the assessment of the adjacent upper abdomen (e.g., adrenal and liver metastases in lung cancer). A wide field, for instance, is the interaction between the heart and the lung. Most obviously, patients with pulmonary (arterial) hypertension will benefit from a comprehensive cardiopulmonary protocol, covering a range of potential causes of pulmonary arterial pressure increase (e.g., COPD and ILD) and the pathophysiological consequences with right heart strain [140]. In this context, MRI showed highest reproducibility in measuring right ventricular output and hypertrophy, the latter being an important predictor of mortality [141]. Phase-contrast MRI may be utilized to determine the pulmonary arterial flow in the large vessels. Importantly, chronic thromboembolic pulmonary hypertension can be detected by typical webs or bands with a sensitivity of 98% and specificity of 94%, and determining the central thrombus load by MRI is essential for curative surgery [142].

11.4 Clinical Application—State of the Art

In comparison to other modalities, lung MRI has specific advantages and disadvantages that predetermine its potential role in a routine setting. In first line, MRI can be used to replace CT and X-ray when radiation exposure needs to be minimized. Typical indications are acute and chronic lung diseases in children (pneumonia and CF) or pulmonary embolism in young or pregnant patients (see Section 11.4.1). In some situations lung MRI can serve as a valuable adjunct to standard methods (e.g., lung cancer) like an *ace in the sleeve* for difficult clinical questions (see Section 11.4.2). Finally, in other pathological conditions lung MRI is feasible, but does not offer specific advantages over other modalities at present (e.g., ILDs) (see Section 11.4.3). In these cases, it is still an interesting option for research or short-term follow-up. Since technical developments continue, it can be anticipated that the spectrum of indications for lung MRI will grow constantly.

11.4.1 MRI First Line Indications

11.4.1.1 Pneumonia in Young Patients

The potential to replace chest radiography with low-field MRI and SSFP sequences was already shown several years ago [129,143–146]. Since then the feasibility of lung MRI in disease entities encompassing community-acquired pneumonia, empyema, fungal infections, and chronic bronchitis has been demonstrated by various studies with different protocols in vivo and ex vivo [106,132]. The sequences and sequence types that were successfully tested and approved with these investigations were incorporated into different parts of the suggested basic protocol. As expected, experience with 3 T systems is limited. For example, a recent comparison between 3 T lung MRI and high resolution computed tomography (HRCT) as gold standard showed an excellent correlation for non-CF chronic lung disease in children [147]. Breath-hold T_2- as well as T_1-weighted sequences with ECG triggering were acquired. In summary, lung MRI, as available today, can be considered a valuable tool for detection as well as characterization of inflammatory lung disease in children (Figure 11.9).

11.4.1.2 Cystic Fibrosis

CF remains the most frequent lethal inherited disease in the Caucasian population, in which the pulmonary phenotype is determining mortality [148]. Due to improvements in therapy and management, life expectancy of CF patients has increased substantially, with a current median survival of approximately 40 years [149,150]. Imaging plays a significant role in the management of the disease, since clinical parameters, including pulmonary function testing, provide only global information

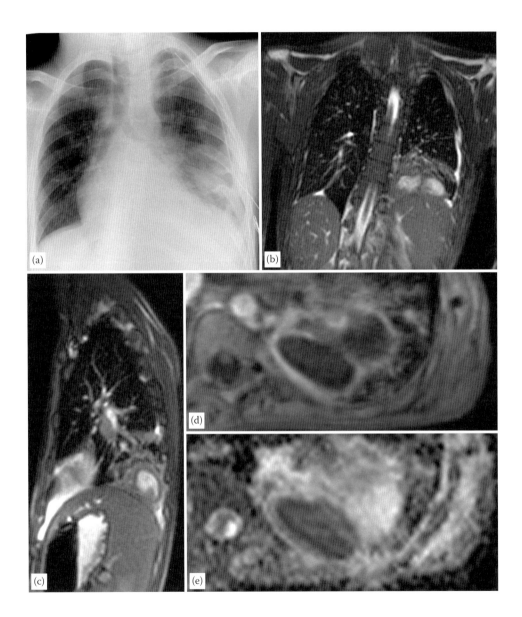

FIGURE 11.9
Detection of complications in pneumonia in pediatric patients. The 7-year-old male was treated with i.v. antibiotics for clinically and radiographically diagnosed pneumonia over one week. A chest tube was inserted due to pleural effusion and suspected empyema. Marked infiltrates of the left lower lobe did not resolve while the clinical symptoms and fever persisted (a). The bSSFP showed several areas of hyperintensity within the infiltrates of the left lower lobe, which showed smooth margins and a hypointense capsule (b, c). The capsule exhibited enhancement after i.v. contrast application (d). Diffusion-weighted imaging revealed a reduced apparent diffusion coefficient for some of the lesions (e). The final diagnosis was lung abscess secondary to pneumonia (e).

while the observation of regional changes of lung structure and function are increasingly regarded as important markers for the activity and course of CF lung disease. With respect to the young age of the patients (from newborns detected by screening programs) and prolonged median survival, radiation exposure becomes a dominant issue and makes a compelling case for use of MRI [151,152].

MRI is reported to be comparable to CT with regard to the detection of morphological changes in the CF lung of newborn to adult patients, and clinical experience with MRI is rapidly growing [104,106,112,113,128,153]. Moreover, MRI is superior to CT when it comes to the assessment of functional changes such as altered pulmonary perfusion [99,108]. Using the basic MRI protocols, it is possible to visualize bronchiectasis, bronchial wall thickening, mucus plugging, small airways disease, air–fluid levels, consolidation, and lung destruction [104,106,113]. The accuracy of MRI in detecting bronchiectasis is dependent on a number of factors, including bronchial level and diameter, wall thickness, and the signal from within the bronchial wall and lumen. Central bronchi and peripheral bronchiectasis are well visualized on MRI, whereas normal peripheral bronchi starting at the third to fourth generation are poorly visualized. The depiction of bronchial wall thickening depends on bronchial size and signal [113]. A high signal of the bronchial wall on T_2-weighted images represents increased fluid, that is, edema, possibly caused by active inflammation. Enhancement of the thickened bronchial wall on post-contrast, fat suppressed T_1-weighted images is thought to be related to inflammatory activity. It is important to note that compared to MRI, CT can only detect wall thickening and is not able to comment on the cause [104,113].

Mucus plugging is well visualized on MRI even down to the small airways due to the high T_2 signal of its fluid content. It is recognized as a high T_2 signal filling of the bronchus along its course, with branching in the periphery giving a grape-like or tree-in-bud appearance. As mucus plugs do not enhance, they are easily differentiated from bronchial wall thickening [104]. Bronchial air–fluid levels are indicative of active infection, occurring in saccular or varicose bronchiectasis, and can be visualized by their high T_2 signal. However, discriminating a bronchus with an air–fluid level or with partial mucus plugging from a severely thickened wall can be difficult. When evaluating the signal characteristics on T_2- and T_1-weighted images with and without contrast enhancement, air–fluid levels can usually be differentiated.

Pulmonary consolidation in CF is mainly caused by alveolar filling with inflammatory material leading to a high signal on T_2-weighted images, often accompanied by volume loss. Similar to CT, MRI is able to visualize air bronchograms as low signal areas following the course of the bronchi within the consolidation [71,144]. With progression of the disease, complete destruction of lung segments or lobes can occur with similar appearances on MRI and CT.

In CF, regional ventilatory defects cause changes in regional lung perfusion due to the hypoxic vasoconstriction response or tissue destruction. Using contrast-enhanced 3D MRI, perfusion defects in 11 children with CF were reported to correlate well with the degree of tissue destruction [99,154]. More recently it could be shown that approximately 20% of the perfusion abnormalities detected in 50 infants and preschool children with CF were induced by bronchiectasis, wall thickening, and mucus plugging [106]. It was also shown that at the age of 0–6 years lung perfusion changes were often more prominent than morphological changes. It is conceivable that perfusion imaging may be more sensitive to pathologic conditions in CF, because perfusion alterations may be induced by peripheral mucus plugging, which can otherwise not be visualized by MRI or even CT [106].

Due to the non-linearity between the MR signal and the concentration of applied contrast media, quantification of pulmonary perfusion is challenging [27]. Risse et al. described the mere importance of the qualitative assessment of the contrast time-course component when analyzing contrast-enhanced 3D MRI to categorize perfusion changes as normal, delayed, reduced, reduced and delayed, as well as perfusion loss [108]. Using dedicated post-processing tools, these data can be displayed in 3D [107,109]. Further research is warranted to extract clinically relevant information from the perfusion datasets. Thus, clinical practice relies on visual assessment. A recently presented morphofunctional MRI score is easily applicable and reproducible for the semiquantitative morphological and functional evaluation of a large severity spectrum of CF lung disease [105], and is comparable to common CT scoring systems [155]. Based on the current state of affairs, morphofunctional MRI can be applied to virtually all CF patients for assessment of disease severity to monitor therapy, and may be capable to differentiate between regions with reversible and irreversible disease [106].

11.4.1.3 Acute Pulmonary Embolism in Young or Pregnant Patients

The current imaging reference technique in evaluation of acute pulmonary embolism is CT pulmonary angiography [156]. Its major advantages over ventilation and perfusion scintigraphy and SPECT are the availability, the comparably short acquisition time of the study, and the diagnostic accuracy with ability to detect alternative diagnoses. However, even with recent protocol modifications, radiation exposure by CT remains an issue, particularly for young patients and pregnant women. To be competitive with CT, an abbreviated MR protocol focusing on lung vessel imaging and lung perfusion may be accomplished within 15 min in-room time.

Although MR angiography has been demonstrated as an excellent tool in dedicated centers, more recent data from a large multicenter study suggest that the technique in isolation produced unsatisfactory results [95,97]. Therefore, combinations of different available MRI techniques for the detection of pulmonary embolism may be of better value [88]. This protocol was further modified and extended into a two-step algorithm. As a first step, a steady-state GRE sequence acquired in two or three planes during free breathing would allow early detection of large central emboli within the first 5 min of the examination—according to the literature with a sensitivity of 90% and a specificity of close to 100% [87,88,157]. Any massive, central embolism detected at this point could be directly referred to intensive care and treatment; the time to diagnosis would be at least as short as with CT pulmonary angiography. If this first step of the examination produces a negative or unclear result, the protocol would be continued with the contrast-enhanced steps including first pass perfusion imaging, multiphasic high spatial resolution contrast-enhanced MRA, and a final acquisition with a volumetric interpolated 3D FLASH sequence in transverse orientation. Despite its composition of multiple sequences, the two-step examination could be completed within 15 min in-room time, which makes it feasible as a quick test for daily clinical routine (11.9). In many cases, such as

in pregnant woman, when administration of contrast material is relatively contraindicated, the examination can be limited to the first step, the free breathing, or breath-hold acquisition of steady-state free precession sequences alone in three orientations. Furthermore, since these steps are partially redundant, at least one acquisition would be expected to be diagnostic even in non-compliant patients. Most recently, Schiebler and colleagues presented their experience with routine MRI for suspected pulmonary embolism in 190 consecutive symptomatic patients, resulting in diagnostic quality in 97.4% of patients [158]. With a follow-up of 3 months and 1 year, the negative predictive values of 97% and 96%, respectively, were similar to previously published results of multidetector CT [158].

11.4.2 MRI as a Supplemental Modality

11.4.2.1 Non-Small Cell Lung Cancer

In detection and staging of thoracic malignancies MRI is an alternative when CT is contraindicated, for example, when the application of iodinated contrast media is not possible. For this purpose, MRI with dedicated standard protocols can provide a comprehensive morphologic tumour node metastasis staging principle (TNM) stage evaluation [65]. A contrast-enhanced examination can be achieved within 25 min in-room time. Intra-pulmonary masses larger than the clinically relevant size of 4–5 mm in diameter can be easily detected (Figure 11.10). The

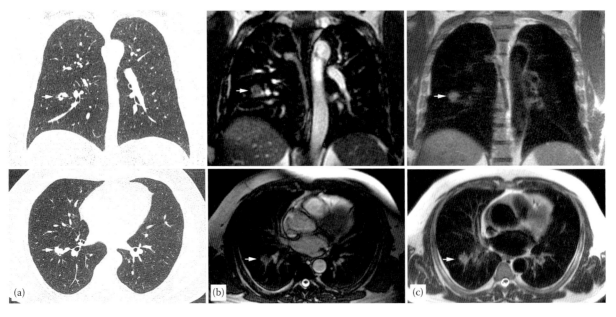

FIGURE 11.10
Detection of pulmonary nodules. The coronary and axial low-dose CT (1 mm slice thickness) detected an adenocarcinoma of the right inferior lobe in a 52-year-old heavy smoker with a maximum diameter of approx. 15 mm (white arrows) (a). The nodule tends to stand out against the dark lung background in a bSSFP (slice thickness 4 mm, free breathing) (b) and T_2 (HASTE) sequence (6 mm slice thickness, breath-hold) (c) in coronary and axial plane.

extent of mediastinal, hilar, and supraclavicular lymph node enlargement can be assessed with excellent soft tissue contrast. Additional phenomena such as the *dark lymph node sign* and contrast enhancement may increase its specificity for malignant involvement (Figure 11.11) [159]. Metastatic disease involving the liver, the adrenal glands, and the skeleton of the thorax are fully covered. The feasibility of MRI-based whole body lung cancer staging with comparable results to positron emission tomography (PET)/CT has been demonstrated [160–163]. Initial results in a screening setting of a high-risk population indicate that non-enhanced MRI may be more sensitive to malignant nodules than to benign lesions, thus making MRI an interesting modality for lung cancer screening (Figures 11.10 and 11.11) [76].

The role of diffusion-weighted imaging (DWI) of lung lesions still requires further evaluation. DWI is recommended for whole body staging of lung cancer including mediastinal metastases [160,164]; however, thus far DWI has not shown a clear advantage over other MRI protocols in the chest [165,166]. STIR sequences might be even more sensitive for the detection and classification of lung cancer and mediastinal metastases than DWI [167–169]. One potential role of DWI might be to predict tumor invasiveness for clinical stage IA non–small cell lung cancer and to separate the mass from atelectasis [170,171]. The role of DWI sequences for the differentiation of malignant or benign lung lesions or for the discrimination of subtypes of lung cancer remains controversial [169,172–174]. In the authors' own experience, DWI helps to demarcate lesions adjacent to the pleura, assess mediastinal extension, and may serve as a *second reader* for detection of small nodules. Thus, it makes sense to include a fast DWI acquisition in the protocol recommendations for MRI of the lung (Figure 11.12) [64,175].

In large pulmonary masses, the soft tissue contrast of MRI contributes to distinguish tumor from atelectasis and pleural effusion better than CT, for example, for image-guided radiotherapy planning. Administration of T_1-shortening contrast material specifically contributes to detect tumor necrosis, chest wall or mediastinal invasion, and pleural reaction/carcinomatosis. An excellent example for the clinical value of lung MRI is the examination of superior sulcus tumors (Pancoast tumor). Tumor invasion of the brachial plexus above the level of the T_1 roots is currently regarded as the main contraindication for resection. MRI is superior to CT in delineation of involvement of neurovertebral foramina, spinal canal, and the brachial plexus, because of its high soft tissue contrast and multiplanar capabilities [176]. Thus, MRI is mandatory in patients with Pancoast tumor (Figure 11.13).

Moreover, MRI contributes comprehensive functional information on respiratory mechanics and tumor displacement [138,177], and lung perfusion [178,179]. The clinical value of these functional imaging capabilities is subject to ongoing investigation, for example, the prediction of postoperative lung function with various techniques, taking it one step further toward a one-stop-shop modality [39].

The role of hybrid systems, that is, PET/MRI, for lung cancer staging is not yet defined. Principally, PET/MRI should profit from combining the advantages of lung MRI with respect to soft tissue contrast and functional imaging capacities with metabolic information from PET. However, the integration of the two systems is technically challenging [180]. In particular the preference of 3 T MR components for the hybrid scanners may not be favorable for lung MRI, as outlined above. In TNM-staging, the advantages of PET/MRI on local T-staging of lung cancer might be significant in cases of tumor inside atelectatic or consolidated lung, or local recurrence inside residual scar or pneumonitis after radiotherapy [181]. In N-staging, DWI and STIR sequences might contribute to improve the sensitivity and specificity of staging with fluoro-2-deoxy-D-glucose-PET (FDG-PET), as outlined above [167,182]. However, the most significant impact of hybrid PET/MRI would be expected for M-staging due to the already superior accuracy of MRI for brain, liver, and bone metastases [183]. Last but not the least, the replacement of the CT component by MRI

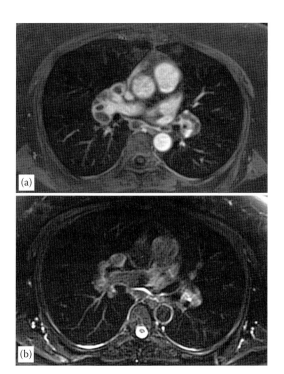

FIGURE 11.11
Evaluation of mediastinal lymph nodes. Contrast-enhanced T_1-weighted (VIBE) (a) and fat-suppressed T_2-weighted (BLADE) (b) images of a 46-year-old female with a history of sarcoidosis. Bihilar and mediastinal lymphadenopathy can easily be appreciated. The lymph nodes display a typical rim enhancement together with a homogeneously low signal intensity of their center on both sequences. (Courtesy of Prof. Jonathan H. Chung MD, Denver, CO.)

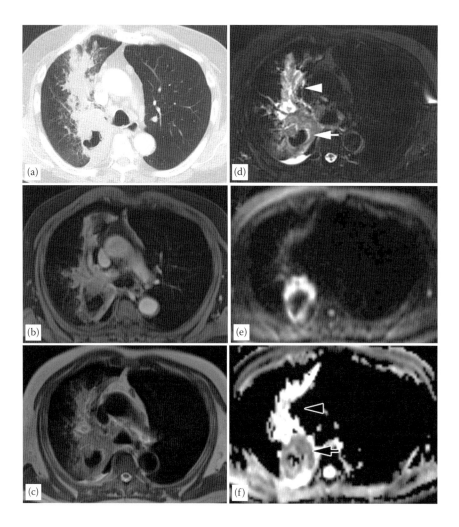

FIGURE 11.12

Differentiating tumor from atelectasis. (a) CT detected a large mass of the right lung with central necrosis. (b) 3D GRE sequences (VIBE) after IV contrast administration clearly depicts the extent of mediastinal invasion to the tracheal bifurcation. (c, d) Especially T_2-weighted sequences (HASTE and BLADE) are well suited for differentiating the underlying bronchogenic carcinoma (white arrow) from the adjacent atelectasis (white arrowhead). Some pleural effusion can be identified as well. (e, f) DWI with $b = 1000$ demonstrates a very high signal of the tumor and a relatively lower signal for atelectasis against the dark background of the lung. Only the tumor shows a reduced apparent diffusion coefficient value (black arrow), as opposed to the atelectatic lung (black arrowhead). (Courtesy of Prof. Dr. Claus Peter Heussel, Thorax klinik at the University of Heidelberg, Heidelberg, Germany.)

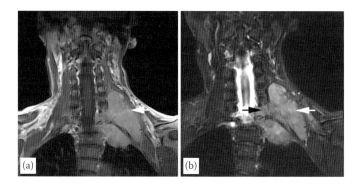

FIGURE 11.13

Tumors of the superior sulcus. (a) In tumors of the superior sulcus, contrast-enhanced 3D GRE and (b) T_2-weighted fast spin-echo sequences with fat suppression in coronary orientation are necessary to define the tumor margins. In this case, a large mass (white arrow) extends from the left apex into the cervical soft tissue and penetrates through the neural foramina into the spinal canal (black arrow). (Courtesy of Prof. Dr. Claus Peter Heussel, Thorax klinik at the University of Heidelberg, Heidelberg, Germany.)

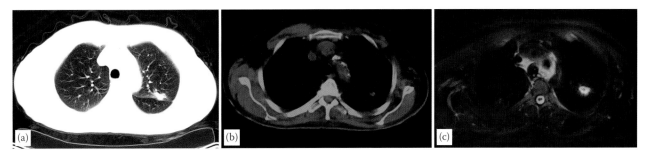

FIGURE 11.14
Potential value of hybrid imaging with PET/MRI. The 71-year-old male patient with a smoking history was routinely referred to FDG—PET/CT because of a part-solid nodule of the left superior lobe, which did not show growth over an observation period of 6 months (a, b). The PET/CT was inconclusive with regard to the FDG uptake of the nodule (b). PET/MRI with motion correction for the PET component was performed directly after PET/CT, and resulted in a remarkably increased uptake of the nodule compared to PET/CT (c). The histopathological diagnosis confirmed a bronchogenic adenocarcinoma. (Images courtesy of Prof. Dr. H. P. Schlemmer and Dr. O. Sedlaczek, German Cancer Research Center (DKFZ), Heidelberg, Germany.)

in PET hybrid imaging would be appreciated for reduction of radiation exposure, which might be of interest for research subjects and young patients. Despite these potential advantages it is not yet determined, if the additional technical efforts and the higher examination costs justify the broad application of lung PET/MRI in clinical routine. Further, very promising approaches such as using 4D MRI for 3D-motion correction of PET are subject to current investigations (Figure 11.14).

11.4.3 Difficult Terrain for Lung MRI— Potential Future Indications

11.4.3.1 Emphysema and COPD

COPD is one of the leading causes of morbidity and mortality worldwide. At present it is the fourth most common cause of death among adults, but its prevalence is increasing [184]. COPD is characterized by incompletely reversible airflow obstruction due to inter-individually variable combination of airway obstruction (obstructive bronchiolitis) and parenchymal destruction (emphysema) [184]. Predominance of one component usually correlates with distinct clinical features, the so-called phenotypes, with potential implications for differential therapy. Several studies have shown that airway obstruction in patients with COPD tends to be located in airways smaller than 2 mm internal diameter [185]. These airways are located between the 4th and the 14th generation of the tracheobronchial tree, and can mostly not be visualized by clinical imaging.

Severity of COPD is assessed by lung function tests and diffusion capacity for carbon monoxide. CT is considered the reference standard for imaging COPD-related morphological changes with emphasis on emphysema and quantitative imaging. Lung hyperinflation, reduced blood volume due to hypoxic pulmonary vasoconstriction, and progressive tissue loss to emphysema (i.e., *minus pathologies*) deteriorate the conditions

for lung ^{1}H-MRI, because all of these result in a marked reduction of lung parenchymal signal (Figure 11.15) [6]. One study demonstrated a correlation between the change of parenchymal signal intensity measured by MRI at inspiration and expiration and FEV_1 ($r = 0.508$) as a predictor of airflow obstruction [7].

The depiction of airway dimensions and size of the airway walls by MRI in physiological condition is limited to the central bronchial tree. For the depiction of bronchiectasis, sequences with a high spatial resolution are essential (see Section 11.4.1.2). The previously described 3D volume interpolated GRE sequence offers sufficient spatial resolution with a sensitivity of 79% and a specificity of 98% regarding visual depiction of bronchiectasis compared to CT (Figure 11.16) [85].

The strength of MRI for imaging COPD lies with the assessment of functional parameters like perfusion and respiratory dynamics [186]. Hyperinflation severely affects diaphragmatic geometry with subsequent reduction of the mechanical properties, while the accessory neck and rib muscles become less effective [187]. The common clinical measurements of COPD do not provide insights into how structural alterations in the lung lead to dysfunction in the breathing mechanics, although treatments such as lung volume reduction by surgery or endoscopy are thought to improve lung function by facilitating breathing mechanics and increasing elastic recoil [188,189]. In contrast to normal subjects, patients with emphysema frequently have reduced, irregular, or asynchronous motion, with a significant decrease in the maximum amplitude and the length of apposition of the diaphragm [190,191]. In some patients the diaphragmatic movement is not coordinated (e.g., the ventral portion of the hemidiaphragm moves inferiorly while the dorsal part moves cranially) [192], while paradoxical diaphragmatic motion correlated with mild and moderate hyperinflation [193]. Severe

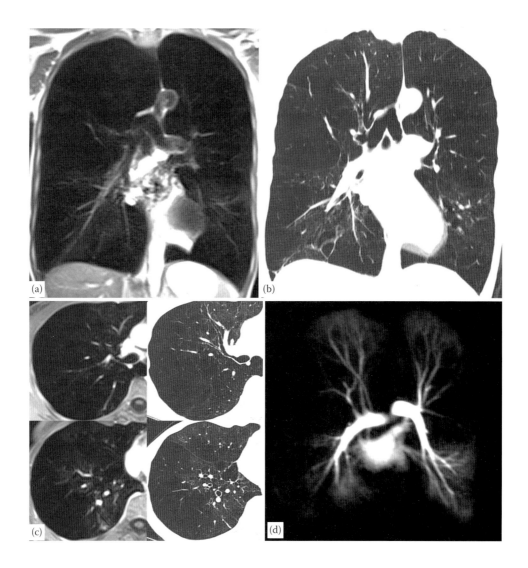

FIGURE 11.15
Emphysema phenotype of COPD. The coronary T_2-weighted (HASTE) sequence depicts extensive emphysema characterized by a lack of signal in this heavy smoker (a). These areas correspond with the low-dose CT scan acquired the same day (b). In relation to the tissue loss, airways show a wide lumen compared to the accompanying artery. Because of the missing wall-thickening bronchi are hard to identify on axial T_1-weighted (VIBE) images after contrast injection compared to corresponding CT sections (1 mm slice thickness) (c). Additional MR perfusion imaging (TWIST) reveals the functional impairment with perfusion deficits roughly matching emphysema (10 mm coronary maximum intensity projection of subtracted perfusion map) (d).

peripheral airflow obstruction also affects the proximal airways from subsegmental bronchi to trachea. MR cine acquisitions during continuous respiration or forced expiration can be recommended for the assessment of tracheal instability, such as seen in tracheobronchial malacia or excessive dynamic airway collapse, which may mimic the clinical appearance of small airways obstruction [118].

Gas exchange in the lungs is optimally maintained by matching of ventilation and perfusion. In regions with reduced ventilation, hypoxic vasoconstriction occurs [194,195] causing a reduction of local PBF with redistribution to better ventilated lung regions [196,197]. A reduction of the pulmonary vascular bed is related to the severity of parenchymal destruction; however, the distribution of perfusion does not necessarily match parenchymal destruction [198,199]. MR perfusion allows for a high diagnostic accuracy in detecting perfusion abnormalities of the lung in COPD [98,110,200]. Additionally, MR perfusion ratios correlate well with radionuclide perfusion scintigraphy ratios [201,202]. Lobar and segmental analysis of the perfusion defects can be performed [199]. Perfusion abnormalities in COPD clearly differ from those caused by vascular obstruction. While wedge-shaped perfusion defects occur in embolic obstruction, a generally low degree of inhomogeneous contrast enhancement is found in COPD with emphysema (Figure 11.15) [65,203]. Furthermore, the peak signal intensity is reduced. These

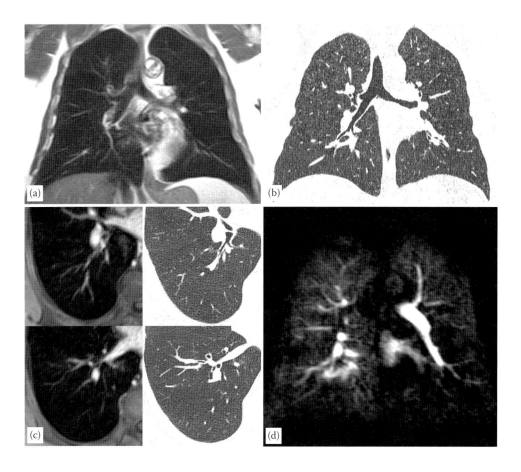

FIGURE 11.16
Airway phenotype of COPD. Lung parenchyma of this heavy smoker shows a relatively homogeneous signal on the coronary T_2-weighted (HASTE) sequence (a). The low-dose CT scan acquired the same day also does not show relevant emphysema (b). The lobar and segmental airways show thickened walls and a narrow lumen reflecting remodeling, and can thus easily be identified on the axial T_1-weighted (VIBE) images after contrast injection. Corresponding CT sections (1 mm slice thickness) are shown also (c). MR perfusion imaging (TWIST) shows a heterogeneously distributed patchy perfusion impairment (10 mm coronary maximum intensity projection of subtracted perfusion map) (d).

features allow for easy visual differentiation and compare well with work done using CT perfusion experiments [204]. In patients with COPD, these changes can be quantified with dedicated software: the mean PBF, PBV, and mean transit time are diffusely decreased and the changes are heterogeneous [109,110].

11.4.3.2 Interstitial Lung Disease

ILD encompasses numerous pathologic disorders of different etiologies, generally manifesting with an inflammatory reaction known as *alveolitis*, which may progress toward fibrosis. Because the nature of these disorders is highly heterogeneous, imaging findings alone are often insufficient for making the final diagnosis, and integration of morphologic aspects with clinical and functional data is required. CT has clarified the elementary alterations and morphologic patterns characterizing the infiltrative changes of ILD. With the recent revision of the guidelines for diagnosing IPF,

imaging features on CT have been assigned a key role [205]. In contrast, there are a relatively limited number of MRI studies that have been clinically performed in ILD patients. Nonetheless, published data suggest at least three possible applications for lung MRI in ILD: (1) visualization and recognition of morphological changes and their patterns, (2) assessment of the inflammatory activity of the disease, and (3) effects of lung morphologic changes on functional parameters such as contrast enhancement and perfusion.

The essential morphologic findings in ILD include air–space disease, interstitial abnormalities, or a combination of the two. Because MR signal increases proportionally to proton density, air–space infiltrates appear on the T_2-weighted images as hyperintense areas against the dark background of the normal lung parenchyma. When pulmonary vascular markings are not obscured, these areas can be considered similar to the ground glass opacities detected by CT (Figure 11.17) [71,206]. More dense opacities appear as consolidations, which

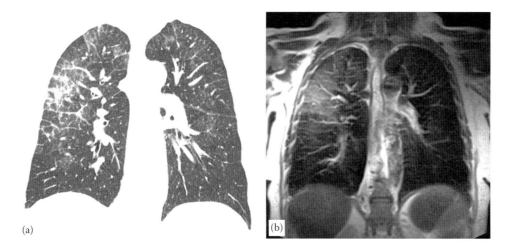

(a)   (b)

FIGURE 11.17
Detection of ground glass opacities. The 53-year-old female received chemotherapy for acute myeloicleukemia and now suffers from fever during leukopenia. She underwent low-dose CT (a) and non-contrast-enhanced MRI (b) the same day for the detection of infiltrates. Widespread areas of ground-glass opacities on CT nicely correspond with a signal increase on T_2-weighted (HASTE) MRI in both lungs, markedly of the right superior lobe. Note that vascular structures are not obscured with both techniques. She was later diagnosed with viral pneumonia.

can be easily assessed by MRI [72]. Similar to consolidations, interstitial abnormalities and fibrosis increase signal intensity presenting with curvilinear bands, nodules, and reticulations, which can be associated to a variable degree of parenchymal distortion [207–209]. Fibrotic changes that extensively involve both peripheral and perihilar portions of the lung are generally well demonstrated on T_2-weighted images, albeit that one needs to consider extracellular interstitial water as a potential differential diagnosis in patients with suspected congestive heart failure. T_1-weighted VIBE images offer higher spatial resolution, and post-contrast acquisition with fat suppression is recommended to increase signal of altered subpleural lung tissue against background represented by chest-wall muscles, ribs, and normal lung parenchyma. Honeycombing, which manifests with reticular changes and irregular cystic transformation of the lung, can also be assessed using this technique [209].

Differentiation of active inflammation from fibrosis is of significant clinical importance both for the prediction of therapy response and for clinical outcome of ILD. Both MR signal and contrast-enhancement characteristics of inflammation and fibrosis have been investigated. Although initial studies performed on 1.5 T lacked sufficient image quality [210–212], the feasibility of the assessment of disease activity in ILD was demonstrated. Only recently, Yi et al. [160] reported that MR signal of inflammatory and fibrotic lesions on T_2-weighted images is hyperintense and isointense, respectively, compared to the signal from chest wall muscle [209]. Dynamic MRI using with contrast administration also indicated that early enhancement and washout with discernible peak enhancement can predict disease activity [209]. The earlier enhancement and rapid washout would be in agreement with higher permeability

of capillaries in the areas of inflammation compared to those of fibrosis. In conclusion, these data are encouraging and support potential future applications of MRI in ILD both in research and in clinical settings.

References

1. Wild JM, Marshall H, Bock M, Schad LR, Jakob PM et al. (2012) MRI of the lung (1/3): Methods. *Insights Imaging* 3: 345–353.
2. Biederer J, Beer M, Hirsch W, Wild J, Fabel M et al. (2012) MRI of the lung (2/3). Why ... when ... how? *Insights Imaging* 3: 355–371.
3. Biederer J, Mirsadraee S, Beer M, Molinari F, Hintze C et al. (2012) MRI of the lung (3/3)-current applications and future perspectives. *Insights Imaging* 3: 373–386.
4. Müller CJ, Löffler R, Deimling M, Peller M, Reiser M (2001) MR lung imaging at 0.2 T with T1-weighted true FISP: Native and oxygen-enhanced. *J Magn Reson Imaging* 14: 164–168.
5. Anjorin A, Schmidt H, Posselt H-G, Smaczny C, Ackermann H et al. (2008) Comparative evaluation of chest radiography, low-field MRI, the Shwachman-Kulczycki score and pulmonary function tests in patients with cystic fibrosis. *Eur Radiol* 18: 1153–1161.
6. Bankier AA, O'Donnell CR, Mai VM, Storey P, De Maertelaer V et al. (2004) Impact of lung volume on MR signal intensity changes of the lung parenchyma. *J Magn Reson Imaging* 20: 961–966.
7. Iwasawa T, Takahashi H, Ogura T, Asakura A, Gotoh T et al. (2007) Correlation of lung parenchymal MR signal intensity with pulmonary function tests and quantitative computed tomography (CT) evaluation: A pilot study. *J Magn Reson Imaging* 26: 1530–1536.

8. Arnold JF, Morchel P, Glaser E, Pracht ED, Jakob PM (2007) Lung MRI using an MR-compatible active breathing control (MR-ABC). *Magn Reson Med* 58: 1092–1098.

9. Oechsner M, Pracht ED, Staeb D, Arnold JF, Kostler H et al. (2009) Lung imaging under free-breathing conditions. *Magn Reson Med* 61: 723–727.

10. Lin W, Guo J, Rosen MA, Song HK (2008) Respiratory motion-compensated radial dynamic contrast-enhanced (DCE)-MRI of chest and abdominal lesions. *Magn Reson Med* 60: 1135–1146.

11. Pruessmann KP, Weiger M, Scheidegger MB, Boesiger P (1999) SENSE: Sensitivity encoding for fast MRI. *Magn Reson Med* 42: 952–962.

12. Griswold MA, Blaimer M, Breuer F, Heidemann RM, Mueller M et al. (2005) Parallel magnetic resonance imaging using the GRAPPA operator formalism. *Magn Reson Med* 54: 1553–1556.

13. Heidemann RM, Griswold MA, Kiefer B, Nittka M, Wang J et al. (2003) Resolution enhancement in lung 1H imaging using parallel imaging methods. *Magn Reson Med* 49: 391–394.

14. Henzler T, Dietrich O, Krissak R, Wichmann T, Lanz T et al. (2009) Half-Fourier-acquisition single-shot turbo spin-echo (HASTE) MRI of the lung at 3 Tesla using parallel imaging with 32-receiver channel technology. *J Magn Reson Imaging* 30: 541–546.

15. Puderbach M, Hintze C, Ley S, Eichinger M, Kauczor HU et al. (2007) MR imaging of the chest: A practical approach at 1.5T. *Eur J Radiol* 64: 345–355.

16. Swift AJ, Woodhouse N, Fichele S, Siedel J, Mills GH et al. (2007) Rapid lung volumetry using ultrafast dynamic magnetic resonance imaging during forced vital capacity maneuver: Correlation with spirometry. *Invest Radiol* 42: 37–41.

17. Biederer J, Hintze C, Fabel M, Dinkel J (2010) Magnetic resonance imaging and computed tomography of respiratory mechanics. *J Magn Reson Imaging* 32: 1388–1397.

18. Bieri O (2013) Ultra-fast steady state free precession and its application to in vivo H morphological and functional lung imaging at 1.5 tesla. *Magn Reson Med* 70(3): 657–663.

19. Huang TY, Huang IJ, Chen CY, Scheffler K, Chung HW et al. (2002) Are TrueFISP images T2/T1-weighted? *Magn Reson Med* 48: 684–688.

20. Hui BK, Noga ML, Gan KD, Wilman AH (2005) Navigator-gated three-dimensional MR angiography of the pulmonary arteries using steady-state free precession. *J Magn Reson Imaging* 21: 831–835.

21. Miller GW, Mugler JP, Sá RC, Altes TA, Prisk GK et al. (2014) Advances in functional and structural imaging of the human lung using proton MRI. *NMR Biomed* 27(12): 1542–1556.

22. Hennig J, Nauerth A, Friedburg H (1986) RARE imaging: A fast imaging method for clinical MR. *Magn Reson Med* 3: 823–833.

23. Bergin CJ, Pauly JM, Macovski A (1991) Lung parenchyma: Projection reconstruction MR imaging. *Radiology* 179: 777–781.

24. Johnson KM, Fain SB, Schiebler ML, Nagle S (2013) Optimized 3D ultrashort echo time pulmonary MRI. *Magn Reson Med* 70: 1241–1250.

25. Wilman AH, Riederer SJ (1997) Performance of an elliptical centric view order for signal enhancement and motion artifact suppression in breath-hold three-dimensional gradient echo imaging. *Magn Reson Med* 38: 793–802.

26. Korosec FR, Frayne R, Grist TM, Mistretta CA (1996) Time-resolved contrast-enhanced 3D MR angiography. *Magn Reson Med* 36: 345–351.

27. Puderbach M, Risse F, Biederer J, Ley-Zaporozhan J, Ley S et al. (2008) In vivo Gd-DTPA concentration for MR lung perfusion measurements: Assessment with computed tomography in a porcine model. *Eur Radiol* 18: 2102–2107.

28. Wang T, Schultz G, Hebestreit H, Hebestreit A, Hahn D et al. (2003) Quantitative perfusion mapping of the human lung using 1H spin labeling. *J Magn Reson Imaging* 18: 260–265.

29. Stadler A, Jakob PM, Griswold M, Barth M, Bankier AA (2005) T1 mapping of the entire lung parenchyma: Influence of the respiratory phase in healthy individuals. *J Magn Reson Imaging* 21: 759–764.

30. Edelman RR, Hatabu H, Tadamura E, Li W, Prasad PV (1996) Noninvasive assessment of regional ventilation in the human lung using oxygen-enhanced magnetic resonance imaging. *Nat Med* 2: 1236–1239.

31. Jakob PM, Hillenbrand CM, Wang T, Schultz G, Hahn D et al. (2001) Rapid quantitative lung (1)H T(1) mapping. *J Magn Reson Imaging* 14: 795–799.

32. Triphan SM, Breuer FA, Gensler D, Kauczor HU, Jakob PM (2015) Oxygen enhanced lung MRI by simultaneous measurement of T1 and T2* during free breathing using ultrashort TE. *J Magn Reson Imaging* 41: 1708–1714.

33. Bauman G, Puderbach M, Deimling M, Jellus V, Chefd'hotel C et al. (2009) Non-contrast-enhanced perfusion and ventilation assessment of the human lung by means of Fourier decomposition in proton MRI. *Magn Reson Med* 62: 656–664.

34. Lederlin M, Bauman G, Eichinger M, Dinkel J, Brault M et al. (2013) Functional MRI using Fourier decomposition of lung signal: Reproducibility of ventilation- and perfusion-weighted imaging in healthy volunteers. *Eur J Radiol* 82: 1015–1022.

35. Bauman G, Lutzen U, Ullrich M, Gaass T, Dinkel J et al. (2011) Pulmonary functional imaging: Qualitative comparison of Fourier decomposition MR imaging with SPECT/CT in porcine lung. *Radiology* 260: 551–559.

36. Bauman G, Scholz A, Rivoire J, Terekhov M, Friedrich J et al. (2013) Lung ventilation- and perfusion-weighted Fourier decomposition magnetic resonance imaging: In vivo validation with hyperpolarized 3He and dynamic contrast-enhanced MRI. *Magn Reson Med* 69: 229–237.

37. Bauman G, Puderbach M, Heimann T, Kopp-Schneider A, Fritzsching E et al. (2013) Validation of Fourier decomposition MRI with dynamic contrast-enhanced MRI using visual and automated scoring of pulmonary perfusion in young cystic fibrosis patients. *Eur J Radiol* 82: 2371–2377.

38. Biederer J, Heussel CP, Puderbach M, Wielpuetz MO (2014) Functional magnetic resonance imaging of the lung. *Semin Respir Crit Care Med* 35: 74–82.

39. Sommer G, Bauman G, Koenigkam-Santos M, Draenkow C, Heussel CP et al. (2013) Non-contrast-enhanced preoperative assessment of lung perfusion in patients with non-small-cell lung cancer using Fourier decomposition magnetic resonance imaging. *Eur J Radiol* 82: e879–e887.

40. Kjorstad A, Corteville DM, Fischer A, Henzler T, Schmid-Bindert G et al. (2014) Quantitative lung perfusion evaluation using Fourier decomposition perfusion MRI. *Magn Reson Med* 72(2): 558–562.

41. van Beek EJ, Hill C, Woodhouse N, Fichele S, Fleming S et al. (2007) Assessment of lung disease in children with cystic fibrosis using hyperpolarized 3-Helium MRI: Comparison with Shwachman score, Chrispin-Norman score and spirometry. *Eur Radiol* 17: 1018–1024.

42. Donnelly LF, MacFall JR, McAdams HP, Majure JM, Smith J et al. (1999) Cystic fibrosis: Combined hyperpolarized 3He-enhanced and conventional proton MR imaging in the lung—Preliminary observations. *Radiology* 212: 885–889.

43. de Lange EE, Altes TA, Patrie JT, Gaare JD, Knake JJ et al. (2006) Evaluation of asthma with hyperpolarized helium-3 MRI: Correlation with clinical severity and spirometry. *Chest* 130: 1055–1062.

44. Woodhouse N, Wild JM, Paley MN, Fichele S, Said Z et al. (2005) Combined helium-3/proton magnetic resonance imaging measurement of ventilated lung volumes in smokers compared to never-smokers. *J Magn Reson Imaging* 21: 365–369.

45. Kirby M, Pike D, Coxson HO, McCormack DG, Parraga G (2014) Hyperpolarized He ventilation defects used to predict pulmonary exacerbations in mild to moderate chronic obstructive pulmonary disease. *Radiology*: 140161.

46. Wild JM, Marshall H, Xu X, Norquay G, Parnell SR et al. (2013) Simultaneous imaging of lung structure and function with triple-nuclear hybrid MR imaging. *Radiology* 267: 251–255.

47. Marshall H, Kiely DG, Parra-Robles J, Capener D, Deppe MH et al. (2014) Magnetic resonance imaging of ventilation and perfusion changes in response to pulmonary endarterectomy in chronic thromboembolic pulmonary hypertension. *Am J Respir Crit Care Med* 190: e18–e19.

48. Mentore K, Froh DK, de Lange EE, Brookeman JR, Paget-Brown AO et al. (2005) Hyperpolarized HHe 3 MRI of the lung in cystic fibrosis: Assessment at baseline and after bronchodilator and airway clearance treatment. *Acad Radiol* 12: 1423–1429.

49. Woodhouse N, Wild JM, van Beek EJ, Hoggard N, Barker N et al. (2009) Assessment of hyperpolarized 3He lung MRI for regional evaluation of interventional therapy: A pilot study in pediatric cystic fibrosis. *J Magn Reson Imaging* 30: 981–988.

50. Thomen RP, Sheshadri A, Quirk JD, Kozlowski J, Ellison HD et al. (2014) Regional ventilation changes in severe asthma after bronchial thermoplasty with He MR imaging and CT. *Radiology*: 140080.

51. Johansson MW, Kruger SJ, Schiebler ML, Evans MD, Sorkness RL et al. (2013) Markers of vascular perturbation correlate with airway structural change in asthma. *Am J Respir Crit Care Med* 188: 167–178.

52. Wild JM, Paley MN, Kasuboski L, Swift A, Fichele S et al. (2003) Dynamic radial projection MRI of inhaled hyperpolarized 3He gas. *Magn Reson Med* 49: 991–997.

53. Marshall H, Deppe MH, Parra-Robles J, Hillis S, Billings CG et al. (2012) Direct visualisation of collateral ventilation in COPD with hyperpolarised gas MRI. *Thorax* 67: 613–617.

54. Swift AJ, Wild JM, Fichele S, Woodhouse N, Fleming S et al. (2005) Emphysematous changes and normal variation in smokers and COPD patients using diffusion 3He MRI. *Eur J Radiol* 54: 352–358.

55. Altes TA, Mata J, de Lange EE, Brookeman JR, Mugler JP, 3rd (2006) Assessment of lung development using hyperpolarized helium-3 diffusion MR imaging. *J Magn Reson Imaging* 24: 1277–1283.

56. van Beek EJ, Dahmen AM, Stavngaard T, Gast KK, Heussel CP et al. (2009) Hyperpolarised 3He MRI versus HRCT in COPD and normal volunteers: PHIL trial. *Eur Respir J* 34: 1311–1321.

57. Kirby M, Svenningsen S, Owrangi A, Wheatley A, Farag A et al. (2012) Hyperpolarized 3He and 129Xe MR imaging in healthy volunteers and patients with chronic obstructive pulmonary disease. *Radiology* 265: 600–610.

58. Driehuys B, Martinez-Jimenez S, Cleveland ZI, Metz GM, Beaver DM et al. (2012) Chronic obstructive pulmonary disease: Safety and tolerability of hyperpolarized 129Xe MR imaging in healthy volunteers and patients. *Radiology* 262: 279–289.

59. Cleveland ZI, Cofer GP, Metz G, Beaver D, Nouls J et al. (2010) Hyperpolarized Xe MR imaging of alveolar gas uptake in humans. *PLoS One* 5: e12192.

60. Mugler JP, 3rd, Altes TA, Ruset IC, Dregely IM, Mata JF et al. (2010) Simultaneous magnetic resonance imaging of ventilation distribution and gas uptake in the human lung using hyperpolarized xenon-129. *Proc Natl Acad Sci USA* 107: 21707–21712.

61. Stewart NJ, Leung G, Norquay G, Marshall H, Parra-Robles J et al. (2014) Experimental validation of the hyperpolarized Xe chemical shift saturation recovery technique in healthy volunteers and subjects with interstitial lung disease. *Magn Reson Med*. doi:10.1002/mrm.25400.

62. Hansell DM, Bankier AA, MacMahon H, McLoud TC, Muller NL et al. (2008) Fleischner Society: Glossary of terms for thoracic imaging. *Radiology* 246: 697–722.

63. Barreto MM, Rafful PP, Rodrigues RS, Zanetti G, Hochhegger B et al. (2013) Correlation between computed tomographic and magnetic resonance imaging findings of parenchymal lung diseases. *Eur J Radiol* 82: e492–e501.

64. Biederer J, Hintze C, Fabel M (2008) MRI of pulmonary nodules: Technique and diagnostic value. *Cancer Imaging* 8: 125–130.

65. Wielputz M, Kauczor HU (2012) MRI of the lung: State of the art. *Diagn Interv Radiol* 18: 344–353.

66. Ley-Zaporozhan J, Ley S, Eberhardt R, Kauczor HU, Heussel CP (2010) Visualization of morphological parenchymal changes in emphysema: Comparison of different MRI sequences to 3D-HRCT. *Eur J Radiol* 73: 43–49.

67. Biederer J, Busse I, Grimm J, Reuter M, Muhle C et al. (2002) Sensitivity of MRI in detecting alveolar Infiltrates: Experimental studies. *Rofo* 174: 1033–1039.

68. Kersjes W, Hildebrandt G, Cagil H, Schunk K, von Zitzewitz H et al. (1999) Differentiation of alveolitis and pulmonary fibrosis in rabbits with magnetic resonance imaging after intrabronchial administration of bleomycin. *Invest Radiol* 34: 13–21.

69. Fink C, Puderbach M, Biederer J, Fabel M, Dietrich O et al. (2007) Lung MRI at 1.5 and 3 Tesla: Observer preference study and lesion contrast using five different pulse sequences. *Invest Radiol* 42: 377–383.

70. Jacob RE, Amidan BG, Soelberg J, Minard KR (2010) In vivo MRI of altered proton signal intensity and T2 relaxation in a bleomycin model of pulmonary inflammation and fibrosis. *J Magn Reson Imaging* 31: 1091–1099.

71. Eibel R, Herzog P, Dietrich O, Rieger CT, Ostermann H et al. (2006) Pulmonary abnormalities in immunocompromised patients: Comparative detection with parallel acquisition MR imaging and thin-section helical CT. *Radiology* 241: 880–891.

72. Rieger C, Herzog P, Eibel R, Fiegl M, Ostermann H (2008) Pulmonary MRI—A new approach for the evaluation of febrile neutropenic patients with malignancies. *Support Care Cancer* 16: 599–606.

73. Biederer J, Schoene A, Freitag S, Reuter M, Heller M (2003) Simulated pulmonary nodules implanted in a dedicated porcine chest phantom: Sensitivity of MR imaging for detection. *Radiology* 227: 475–483.

74. Both M, Schultze J, Reuter M, Bewig B, Hubner R et al. (2005) Fast T1- and T2-weighted pulmonary MR-imaging in patients with bronchial carcinoma. *Eur J Radiol* 53: 478–488.

75. Bruegel M, Gaa J, Woertler K, Ganter C, Waldt S et al. (2007) MRI of the lung: Value of different turbo spin-echo, single-shot turbo spin-echo, and 3D gradient-echo pulse sequences for the detection of pulmonary metastases. *J Magn Reson Imaging* 25: 73–81.

76. Sommer G, Tremper J, Koenigkam-Santos M, Delorme S, Becker N et al. (2014) Lung nodule detection in a high-risk population: Comparison of magnetic resonance imaging and low-dose computed tomography. *Eur J Radiol* 83: 600–605.

77. Gamsu G, de Geer G, Cann C, Muller N, Brito A (1987) A preliminary study of MRI quantification of simulated calcified pulmonary nodules. *Invest Radiol* 22: 853–858.

78. Kersjes W, Mayer E, Buchenroth M, Schunk K, Fouda N et al. (1997) Diagnosis of pulmonary metastases with turbo-SE MR imaging. *Eur Radiol* 7: 1190–1194.

79. Chung MH, Lee HG, Kwon SS, Park SH (2000) MR imaging of solitary pulmonary lesion: Emphasis on tuberculomas and comparison with tumors. *J Magn Reson Imaging* 11: 629–637.

80. Kirchner J, Kirchner EM (2001) Melanoptysis: Findings on CT and MRI. *Br J Radiol* 74: 1003–1006.

81. Regier M, Kandel S, Kaul MG, Hoffmann B, Ittrich H et al. (2007) Detection of small pulmonary nodules in high-field MR at 3 T: Evaluation of different pulse sequences using porcine lung explants. *Eur Radiol* 17: 1341–1351.

82. Baumann T, Ludwig U, Pache G, Gall C, Saueressig U et al. (2008) Detection of pulmonary nodules with move-during-scan magnetic resonance imaging using a free-breathing turbo inversion recovery magnitude sequence. *Invest Radiol* 43: 359–367.

83. Khalil AM, Carette MF, Cadranel JL, Mayaud CM, Akoun GM et al. (1994) Magnetic resonance imaging findings in pulmonary Kaposi's sarcoma: A series of 10 cases. *Eur Respir J* 7: 1285–1289.

84. Semelka RC, Cem Balci N, Wilber KP, Fisher LL, Brown MA et al. (2000) Breath-hold 3D gradient-echo MR imaging of the lung parenchyma: Evaluation of reproducibility of image quality in normals and preliminary observations in patients with disease. *J Magn Reson Imaging* 11: 195–200.

85. Biederer J, Both M, Graessner J, Liess C, Jakob P et al. (2003) Lung morphology: Fast MR imaging assessment with a volumetric interpolated breath-hold technique: Initial experience with patients. *Radiology* 226: 242–249.

86. Kluge A, Muller C, Hansel J, Gerriets T, Bachmann G (2004) Real-time MR with TrueFISP for the detection of acute pulmonary embolism: Initial clinical experience. *Eur Radiol* 14: 709–718.

87. Kluge A, Gerriets T, Stolz E, Dill T, Mueller KD et al. (2006) Pulmonary perfusion in acute pulmonary embolism: Agreement of MRI and SPECT for lobar, segmental and subsegmental perfusion defects. *Acta Radiol* 47: 933–940.

88. Kluge A, Luboldt W, Bachmann G (2006) Acute pulmonary embolism to the subsegmental level: Diagnostic accuracy of three MRI techniques compared with 16-MDCT. *AJR Am J Roentgenol* 187: W7–W14.

89. Moody AR, Liddicoat A, Krarup K (1997) Magnetic resonance pulmonary angiography and direct imaging of embolus for the detection of pulmonary emboli. *Invest Radiol* 32: 431–440.

90. Moody AR (2003) Magnetic resonance direct thrombus imaging. *J Thromb Haemost* 1: 1403–1409.

91. Biederer J, Liess C, Charalambous N, Heller M (2004) Volumetric interpolated contrast-enhanced MRA for the diagnosis of pulmonary embolism in an ex vivo system. *J Magn Reson Imaging* 19: 428–437.

92. Matsuoka S, Uchiyama K, Shima H, Terakoshi H, Oishi S et al. (2002) Effect of the rate of gadolinium injection on magnetic resonance pulmonary perfusion imaging. *J Magn Reson Imaging* 15: 108–113.

93. Meaney JF, Weg JG, Chenevert TL, Stafford-Johnson D, Hamilton BH et al. (1997) Diagnosis of pulmonary embolism with magnetic resonance angiography. *N Engl J Med* 336: 1422–1427.

94. Gupta A, Frazer CK, Ferguson JM, Kumar AB, Davis SJ et al. (1999) Acute pulmonary embolism: Diagnosis with MR angiography. *Radiology* 210: 353–359.

95. Oudkerk M, van Beek EJ, Wielopolski P, van Ooijen PM, Brouwers-Kuyper EM et al. (2002) Comparison of contrast-enhanced magnetic resonance angiography and conventional pulmonary angiography for the diagnosis of pulmonary embolism: A prospective study. *Lancet* 359: 1643–1647.

96. Goyen M, Laub G, Ladd ME, Debatin JF, Barkhausen J et al. (2001) Dynamic 3D MR angiography of the pulmonary arteries in under four seconds. *J Magn Reson Imaging* 13: 372–377.

97. Stein PD, Chenevert TL, Fowler SE, Goodman LR, Gottschalk A et al. (2010) Gadolinium-enhanced magnetic resonance angiography for pulmonary embolism: A multicenter prospective study (PIOPED III). *Ann Intern Med* 152: 434–443, W142–W433.

98. Fink C, Puderbach M, Bock M, Lodemann KP, Zuna I et al. (2004) Regional lung perfusion: Assessment with partially parallel three-dimensional MR imaging. *Radiology* 231: 175–184.

99. Eichinger M, Puderbach M, Fink C, Gahr J, Ley S et al. (2006) Contrast-enhanced 3D MRI of lung perfusion in children with cystic fibrosis—Initial results. *Eur Radiol* 16: 2147–2152.

100. Ersoy H, Goldhaber SZ, Cai T, Luu T, Rosebrook J et al. (2007) Time-resolved MR angiography: A primary screening examination of patients with suspected pulmonary embolism and contraindications to administration of iodinated contrast material. *AJR Am J Roentgenol* 188: 1246–1254.

101. Levin DL, Chen Q, Zhang M, Edelman RR, Hatabu H (2001) Evaluation of regional pulmonary perfusion using ultrafast magnetic resonance imaging. *Magn Reson Med* 46: 166–171.

102. Arai TJ, Henderson AC, Dubowitz DJ, Levin DL, Friedman PJ et al. (2009) Hypoxic pulmonary vasoconstriction does not contribute to pulmonary blood flow heterogeneity in normoxia in normal supine humans. *J Appl Physiol (1985)* 106: 1057–1064.

103. Hopkins SR, Wielputz MO, Kauczor HU (2012) Imaging lung perfusion. *J Appl Physiol* 113: 328–339.

104. Wielputz MO, Eichinger M, Puderbach M (2013) Magnetic resonance imaging of cystic fibrosis lung disease. *J Thorac Imaging* 28: 151–159.

105. Eichinger M, Optazaite DE, Kopp-Schneider A, Hintze C, Biederer J et al. (2012) Morphologic and functional scoring of cystic fibrosis lung disease using MRI. *Eur J Radiol* 81: 1321–1329.

106. Wielputz MO, Puderbach M, Kopp-Schneider A, Stahl M, Fritzsching E et al. (2014) Magnetic resonance imaging detects changes in structure and perfusion, and response to therapy in early cystic fibrosis lung disease. *Am J Respir Crit Care Med* 189: 956–965.

107. Kuder TA, Risse F, Eichinger M, Ley S, Puderbach M et al. (2008) New method for 3D parametric visualization of contrast-enhanced pulmonary perfusion MRI data. *Eur Radiol* 18: 291 297.

108. Risse F, Eichinger M, Kauczor HU, Semmler W, Puderbach M (2011) Improved visualization of delayed perfusion in lung MRI. *Eur J Radiol* 77: 105–110.

109. Kohlmann P, Strehlow J, Jobst B, Krass S, Kuhnigk J-M et al. (2014) Automatic lung segmentation method for MRI-based lung perfusion studies of patients with chronic obstructive pulmonary disease. *Int J Comput Assist Radiol Surg* 10(4): 403–417.

110. Ohno Y, Hatabu H, Murase K, Higashino T, Kawamitsu H et al. (2004) Quantitative assessment of regional pulmonary perfusion in the entire lung using three-dimensional ultrafast dynamic contrast-enhanced magnetic resonance imaging: Preliminary experience in 40 subjects. *J Magn Reson Imaging* 20: 353–365.

111. Ley S, Mereles D, Risse F, Grunig E, Ley-Zaporozhan J et al. (2007) Quantitative 3D pulmonary MR-perfusion in patients with pulmonary arterial hypertension: Correlation with invasive pressure measurements. *Eur J Radiol* 61: 251–255.

112. Puderbach M, Eichinger M, Haeselbarth J, Ley S, Kopp-Schneider A et al. (2007) Assessment of morphological MRI for pulmonary changes in cystic fibrosis (CF) patients: Comparison to thin-section CT and chest x-ray. *Invest Radiol* 42: 715–725.

113. Puderbach M, Eichinger M, Gahr J, Ley S, Tuengerthal S et al. (2007) Proton MRI appearance of cystic fibrosis: Comparison to CT. *Eur Radiol* 17: 716–724.

114. Biederer J, Reuter M, Both M, Muhle C, Grimm J et al. (2002) Analysis of artifacts and detail resolution of lung MRI with breath-hold T1-weighted gradient-echo and T2-weighted fast spin-echo sequences with respiratory triggering. *Eur Radiol* 12: 378–384.

115. Fabel M, Wintersperger BJ, Dietrich O, Eichinger M, Fink C et al. (2009) MRI of respiratory dynamics with 2D steady-state free-precession and 2D gradient echo sequences at 1.5 and 3 Tesla: An observer preference study. *Eur Radiol* 19: 391–399.

116. Faust RA, Rimell FL, Remley KB (2002) Cine magnetic resonance imaging for evaluation of focal tracheomalacia: Innominate artery compression syndrome. *Int J Pediatr Otorhinolaryngol* 65: 27–33.

117. Ley S, Loukanov T, Ley-Zaporozhan J, Springer W, Sebening C et al. (2010) Long-term outcome after external tracheal stabilization due to congenital tracheal instability. *Ann Thorac Surg* 89: 918–925.

118. Heussel CP, Ley S, Biedermann A, Rist A, Gast KK et al. (2004) Respiratory lumenal change of the pharynx and trachea in normal subjects and COPD patients: Assessment by cine-MRI. *Eur Radiol* 14: 2188–2197.

119. Cai J, Read PW, Altes TA, Molloy JA, Brookeman JR et al. (2007) Evaluation of the reproducibility of lung motion probability distribution function (PDF) using dynamic MRI. *Phys Med Biol* 52: 365–373.

120. Adamson J, Chang Z, Wang Z, Yin FF, Cai J (2010) Maximum intensity projection (MIP) imaging using slice-stacking MRI. *Med Phys* 37: 5914–5920.

121. Tetzlaff R, Schwarz T, Kauczor HU, Meinzer HP, Puderbach M et al. (2010) Lung function measurement of single lungs by lung area segmentation on 2D dynamic MRI. *Acad Radiol* 17: 496–503.

122. Wild JM, Schmiedeskamp J, Paley MN, Filbir F, Fichele S et al. (2002) MR imaging of the lungs with hyperpolarized helium-3 gas transported by air. *Phys Med Biol* 47: N185–N190.

123. Scholz AW, Wolf U, Fabel M, Weiler N, Heussel CP et al. (2009) Comparison of magnetic resonance imaging of inhaled SF6 with respiratory gas analysis. *Magn Reson Imaging* 27: 549–556.

124. Suga K, Ogasawara N, Tsukuda T, Matsunaga N (2002) Assessment of regional lung ventilation in dog lungs with Gd-DTPA aerosol ventilation MR imaging. *Acta Radiol* 43: 282–291.

125. Molinari F, Puderbach M, Eichinger M, Ley S, Fink C et al. (2008) Oxygen-enhanced magnetic resonance imaging: Influence of different gas delivery methods on the T1-changes of the lungs. *Invest Radiol* 43: 427–432.

126. Peltola V, Ruuskanen O, Svedstrom E (2008) Magnetic resonance imaging of lung infections in children. *Pediatr Radiol* 38: 1225–1231.

127. Ley-Zaporozhan J, Ley S, Sommerburg O, Komm N, Muller FM et al. (2009) Clinical application of MRI in children for the assessment of pulmonary diseases. *Rofo* 181: 419–432.

128. Failo R, Wielopolski PA, Tiddens HA, Hop WC, Mucelli RP et al. (2009) Lung morphology assessment using MRI: A robust ultra-short TR/TE 2D steady state free precession sequence used in cystic fibrosis patients. *Magn Reson Med* 61: 299–306.

129. Wagner M, Bowing B, Kuth R, Deimling M, Rascher W et al. (2001) Low field thoracic MRI—A fast and radiation free routine imaging modality in children. *Magn Reson Imaging* 19: 975–983.

130. Rupprecht T, Kuth R, Bowing B, Gerling S, Wagner M et al. (2000) Sedation and monitoring of paediatric patients undergoing open low-field MRI. *Acta Paediatr* 89: 1077–1081.

131. Serra G, Milito C, Mitrevski M, Granata G, Martini H et al. (2011) Lung MRI as a possible alternative to CT scan for patients with primary immune deficiencies and increased radiosensitivity MRI for lung evaluation in immunodeficiencies. *CHEST J* 140: 1581–1589.

132. Hirsch W, Sorge I, Krohmer S, Weber D, Meier K et al. (2008) MRI of the lungs in children. *Eur J Radiol* 68: 278–288.

133. Schaefer JF, Kramer U (2011) Whole-body MRI in children and juveniles. *Rofo* 183: 24–36.

134. Sanborn PA, Michna E, Zurakowski D, Burrows PE, Fontaine PJ et al. (2005) Adverse cardiovascular and respiratory events during sedation of pediatric patients for imaging examinations. *Radiology* 237: 288–294.

135. Lutterbey G, Wattjes MP, Doerr D, Fischer NJ, Gieseke J, Jr. et al. (2007) Atelectasis in children undergoing either propofol infusion or positive pressure ventilation anesthesia for magnetic resonance imaging. *Paediatr Anaesth* 17: 121–125.

136. Blitman NM, Lee HK, Jain VR, Vicencio AG, Girshin M et al. (2007) Pulmonary atelectasis in children anesthetized for cardiothoracic MR: Evaluation of risk factors. *J Comput Assist Tomogr* 31: 789–794.

137. Biederer J (2009) General requirements of MRI of the lung and suggested standard protocol. In: Kauczor H-U, ed., *MRI of the Lung*. Berlin, Germany: Springer, pp. 3–16.

138. Seo JS, Kim YJ, Choi BW, Choe KO (2005) Usefulness of magnetic resonance imaging for evaluation of cardiovascular invasion: Evaluation of sliding motion between thoracic mass and adjacent structures on cine MR images. *J Magn Reson Imaging* 22: 234–241.

139. Stein PD, Gottschalk A, Sostman HD, Chenevert TL, Fowler SE et al. (2008) Methods of Prospective Investigation of Pulmonary Embolism Diagnosis III (PIOPED III). *Semin Nucl Med* 38: 462–470.

140. Schiebler ML, Bhalla S, Runo J, Jarjour N, Roldan A et al. (2013) Magnetic resonance and computed tomography imaging of the structural and functional changes of pulmonary arterial hypertension. *J Thorac Imaging* 28: 178–193.

141. Vonk Noordegraaf A, Galie N (2011) The role of the right ventricle in pulmonary arterial hypertension. *Eur Respir Rev* 20: 243–253.

142. Rajaram S, Swift AJ, Capener D, Telfer A, Davies C et al. (2012) Diagnostic accuracy of contrast-enhanced MR angiography and unenhanced proton MR imaging compared with CT pulmonary angiography in chronic thromboembolic pulmonary hypertension. *Eur Radiol* 22: 310–317.

143. Cohen MD, Eigen H, Scott PH, Tepper R, Cory DA et al. (1986) Magnetic resonance imaging of inflammatory lung disorders: Preliminary studies in children. *Pediatr Pulmonol* 2: 211–217.

144. Rupprecht T, Bowing B, Kuth R, Deimling M, Rascher W et al. (2002) Steady-state free precession projection MRI as a potential alternative to the conventional chest X-ray in pediatric patients with suspected pneumonia. *Eur Radiol* 12: 2752–2756.

145. Hebestreit A, Schultz G, Trusen A, Hebestreit H (2004) Follow-up of acute pulmonary complications in cystic fibrosis by magnetic resonance imaging: A pilot study. *Acta Paediatr* 93: 414–416.

146. Abolmaali ND, Schmitt J, Krauss S, Bretz F, Deimling M et al. (2004) MR imaging of lung parenchyma at 0.2 T: Evaluation of imaging techniques, comparative study with chest radiography and interobserver analysis. *Eur Radiol* 14: 703–708.

147. Montella S, Santamaria F, Salvatore M, Pignata C, Maglione M et al. (2009) Assessment of chest high-field magnetic resonance imaging in children and young adults with noncystic fibrosis chronic lung disease: Comparison to high-resolution computed tomography and correlation with pulmonary function. *Invest Radiol* 44: 532–538.

148. Gibson RL, Burns JL, Ramsey BW (2003) Pathophysiology and management of pulmonary infections in cystic fibrosis. *Am J Respir Crit Care Med* 168: 918–951.

149. Dodge JA, Lewis PA, Stanton M, Wilsher J (2007) Cystic fibrosis mortality and survival in the UK: 1947–2003. *Eur Respir J* 29: 522–526.

150. Stern M, Wiedemann B, Wenzlaff P (2008) From registry to quality management: The German Cystic Fibrosis Quality Assessment project 1995–2006. *Eur Respir J* 31: 29–35.

151. de Jong PA, Mayo JR, Golmohammadi K, Nakano Y, Lequin MH et al. (2006) Estimation of cancer mortality associated with repetitive computed tomography scanning. *Am J Respir Crit Care Med* 173: 199–203.

152. O'Connell OJ, McWilliams S, McGarrigle A, O'Connor OJ, Shanahan F et al. (2012) Radiologic imaging in cystic fibrosis: Cumulative effective dose and changing trends over 2 decades. *Chest* 141: 1575–1583.

153. Sileo C, Corvol H, Boelle PY, Blondiaux E, Clement A et al. (2014) HRCT and MRI of the lung in children with cystic fibrosis: Comparison of different scoring systems. *J Cyst Fibros* 13: 198–204.

154. Hatabu H, Gaa J, Kim D, Li W, Prasad PV et al. (1996) Pulmonary perfusion: Qualitative assessment with dynamic contrast-enhanced MRI using ultra-short TE and inversion recovery turbo FLASH. *Magn Reson Med* 36: 503–508.

155. de Jong PA, Tiddens HA (2007) Cystic fibrosis specific computed tomography scoring. *Proc Am Thorac Soc* 4: 338–342.

156. Stein PD, Fowler SE, Goodman LR, Gottschalk A, Hales CA et al. (2006) Multidetector computed tomography for acute pulmonary embolism. *N Engl J Med* 354: 2317–2327.

157. Kluge A, Gerriets T, Muller C, Ekinci O, Neumann T et al. (2005) Thoracic real-time MRI: Experience from 2200 examinations in acute and ill-defined thoracic diseases. *Rofo* 177: 1513–1521.

158. Schiebler ML, Nagle SK, Francois CJ, Repplinger MD, Hamedani AG et al. (2013) Effectiveness of MR angiography for the primary diagnosis of acute pulmonary embolism: Clinical outcomes at 3 months and 1 year. *J Magn Reson Imaging* 38: 914–925.

159. Chung JH, Cox CW, Forssen AV, Biederer J, Puderbach M et al. (2014) The dark lymph node sign on magnetic resonance imaging: A novel finding in patients with sarcoidosis. *J Thorac Imaging* 29: 125–129.

160. Yi CA, Shin KM, Lee KS, Kim BT, Kim H et al. (2008) Non-small cell lung cancer staging: Efficacy comparison of integrated PET/CT versus 3.0-T whole-body MR imaging. *Radiology* 248: 632–642.

161. Ohba Y, Nomori H, Mori T, Ikeda K, Shibata H et al. (2009) Is diffusion-weighted magnetic resonance imaging superior to positron emission tomography with fludeoxyglucose F 18 in imaging non-small cell lung cancer? *J Thorac Cardiovasc Surg* 138: 439–445.

162. Ohno Y, Hatabu H, Takenaka D, Higashino T, Watanabe H et al. (2004) Metastases in mediastinal and hilar lymph nodes in patients with non-small cell lung cancer: Quantitative and qualitative assessment with STIR turbo spin-echo MR imaging. *Radiology* 231: 872–879.

163. Ohno Y, Koyama H, Nogami M, Takenaka D, Yoshikawa T et al. (2007) Whole-body MR imaging vs. FDG-PET: Comparison of accuracy of M-stage diagnosis for lung cancer patients. *J Magn Reson Imaging* 26: 498–509.

164. Chen W, Jian W, Li HT, Li C, Zhang YK et al. (2010) Whole-body diffusion-weighted imaging vs. FDG-PET for the detection of non-small-cell lung cancer. How do they measure up? *Magn Reson Imaging* 28: 613–620.

165. Hasegawa I, Boiselle PM, Kuwabara K, Sawafuji M, Sugiura H (2008) Mediastinal lymph nodes in patients with non-small cell lung cancer: Preliminary experience with diffusion-weighted MR imaging. *J Thorac Imaging* 23: 157–161.

166. Pauls S, Schmidt SA, Juchems MS, Klass O, Luster M et al. (2012) Diffusion-weighted MR imaging in comparison to integrated [(18)F]-FDG PET/CT for N-staging in patients with lung cancer. *Eur J Radiol* 81(1): 178–182.

167. Koyama H, Ohno Y, Aoyama N, Onishi Y, Matsumoto K et al. (2010) Comparison of STIR turbo SE imaging and diffusion-weighted imaging of the lung: Capability for detection and subtype classification of pulmonary adenocarcinomas. *Eur Radiol* 20: 790–800.

168. Liu H, Liu Y, Yu T, Ye N (2010) Usefulness of diffusion-weighted MR imaging in the evaluation of pulmonary lesions. *Eur Radiol* 20: 807–815.

169. Tondo F, Saponaro A, Stecco A, Lombardi M, Casadio C et al. (2011) Role of diffusion-weighted imaging in the differential diagnosis of benign and malignant lesions of the chest-mediastinum. *Radiol Med* 116: 720–733.

170. Kanauchi N, Oizumi H, Honma T, Kato H, Endo M et al. (2009) Role of diffusion-weighted magnetic resonance imaging for predicting of tumor invasiveness for clinical stage IA non-small cell lung cancer. *Eur J Cardiothorac Surg* 35: 706–710; discussion 710–701.

171. Qi LP, Zhang XP, Tang L, Li J, Sun YS et al. (2009) Using diffusion-weighted MR imaging for tumor detection in the collapsed lung: A preliminary study. *Eur Radiol* 19: 333–341.

172. Uto T, Takehara Y, Nakamura Y, Naito T, Hashimoto D et al. (2009) Higher sensitivity and specificity for diffusion-weighted imaging of malignant lung lesions without apparent diffusion coefficient quantification. *Radiology* 252: 247–254.

173. Karabulut N (2009) Accuracy of diffusion-weighted MR imaging for differentiation of pulmonary lesions. *Radiology* 253: 899; author reply 899–900.

174. Henzler T, Schmid-Bindert G, Schoenberg SO, Fink C (2010) Diffusion and perfusion MRI of the lung and mediastinum. *Eur J Radiol* 76: 329–336.

175. Regier M, Schwarz D, Henes FO, Groth M, Kooijman H et al. (2011) Diffusion-weighted MR-imaging for the detection of pulmonary nodules at 1.5 Tesla: Intraindividual comparison with multidetector computed tomography. *J Med Imaging Radiat Oncol* 55: 266–274.

176. Bruzzi JF, Komaki R, Walsh GL, Truong MT, Gladish GW et al. (2008) Imaging of non-small cell lung cancer of the superior sulcus: Part 2: Initial staging and assessment of resectability and therapeutic response. *RadioGraphics* 28: 561–572.

177. Akata S, Kajiwara N, Park J, Yoshimura M, Kakizaki D et al. (2008) Evaluation of chest wall invasion by lung cancer using respiratory dynamic MRI. *J Med Imaging Radiat Oncol* 52: 36–39.

178. Ohno Y, Koyama H, Takenaka D, Nogami M, Maniwa Y et al. (2008) Dynamic MRI, dynamic multidetector-row computed tomography (MDCT), and coregistered 2-[fluorine-18]-fluoro-2-deoxy-d-glucose-positron emission tomography (FDG-PET)/CT: Comparative study of capability for management of pulmonary nodules. *J Magn Reson Imaging* 27: 1284–1295.

179. Pauls S, Mottaghy FM, Schmidt SA, Kruger S, Moller P et al. (2008) Evaluation of lung tumor perfusion by dynamic contrast-enhanced MRI. *Magn Reson Imaging* 26: 1334–1341.

180. Pichler BJ, Kolb A, Nagele T, Schlemmer HP (2010) PET/MRI: Paving the way for the next generation of clinical multimodality imaging applications. *J Nucl Med* 51: 333–336.

181. Hintze C, Dimitrakopoulou-Strauss A, Strauss LG, Risse F, Thieke C et al. (2007) Fusion of FDG-PET and proton-MRI of the lung in patients with lung cancer: Initial results in differentiating tumor from atelectasis. *Fortschr Röntgenstr* 179 – VO_315_3.

182. Pauls S, Schmidt SA, Juchems MS, Klass O, Luster M et al. (2012) Diffusion-weighted MR imaging in comparison to integrated [(1)(8)F]-FDG PET/CT for N-staging in patients with lung cancer. *Eur J Radiol* 81: 178–182.

183. Antoch G, Vogt FM, Freudenberg LS, Nazaradeh F, Goehde SC et al. (2003) Whole-body dual-modality PET/CT and whole-body MRI for tumor staging in oncology. *JAMA* 290: 3199–3206.

184. Rabe KF, Hurd S, Anzueto A, Barnes PJ, Buist SA et al. (2007) Global strategy for the diagnosis, management, and prevention of chronic obstructive pulmonary disease: GOLD executive summary. *Am J Respir Crit Care Med* 176: 532–555.

185. Hogg JC, Chu F, Utokaparch S, Woods R, Elliott WM et al. (2004) The nature of small-airway obstruction in chronic obstructive pulmonary disease. *N Engl J Med* 350: 2645–2653.

186. Ley-Zaporozhan J, Ley S, Kauczor HU (2008) Morphological and functional imaging in COPD with CT and MRI: Present and future. *Eur Radiol* 18: 510–521.

187. Decramer M, Gosselink R, Troosters T, Verschueren M, Evers G (1997) Muscle weakness is related to utilization of health care resources in COPD patients. *Eur Respir J* 10: 417–423.

188. Henderson AC, Ingenito EP, Salcedo ES, Moy ML, Reilly JJ et al. (2007) Dynamic lung mechanics in late-stage emphysema before and after lung volume reduction surgery. *Respir Physiol Neurobiol* 155: 234–242.

189. Sciurba FC, Ernst A, Herth FJ, Strange C, Criner GJ et al. (2010) A randomized study of endobronchial valves for advanced emphysema. *N Engl J Med* 363: 1233–1244.

190. Suga K, Tsukuda T, Awaya H, Takano K, Koike S et al. (1999) Impaired respiratory mechanics in pulmonary emphysema: Evaluation with dynamic breathing MRI. *J Magn Reson Imaging* 10: 510–520.

191. Wielputz MO, Eberhardt R, Puderbach M, Weinheimer O, Kauczor HU et al. (2014) Simultaneous assessment of airway instability and respiratory dynamics with low-dose 4D-CT in chronic obstructive pulmonary disease: A technical note. *Respiration* 87: 294–300.

192. Iwasawa T, Yoshiike Y, Saito K, Kagei S, Gotoh T et al. (2000) Paradoxical motion of the hemidiaphragm in patients with emphysema. *J Thorac Imaging* 15: 191–195.

193. Iwasawa T, Kagei S, Gotoh T, Yoshiike Y, Matsushita K et al. (2002) Magnetic resonance analysis of abnormal diaphragmatic motion in patients with emphysema. *Eur Respir J* 19: 225–231.

194. Euler U, Liljestrand G (1946) Observations on the pulmonary arterial blood pressure in the cat. *Acta Physiol Scand* 12: 301–320.

195. Theissen IL, Meissner A (1996) Hypoxic pulmonary vasoconstriction. *Anaesthesist* 45: 643–652.

196. Morrison NJ, Abboud RT, Muller NL, Miller RR, Gibson NN et al. (1990) Pulmonary capillary blood volume in emphysema. *Am Rev Respir Dis* 141: 53–61.

197. Cederlund K, Hogberg S, Jorfeldt L, Larsen F, Norman M et al. (2003) Lung perfusion scintigraphy prior to lung volume reduction surgery. *Acta Radiol* 44: 246–251.

198. Sandek K, Bratel T, Lagerstrand L, Rosell H (2002) Relationship between lung function, ventilation-perfusion inequality and extent of emphysema as assessed by high-resolution computed tomography. *Respir Med* 96: 934–943.

199. Ley-Zaporozhan J, Ley S, Eberhardt R, Weinheimer O, Fink C et al. (2007) Assessment of the relationship between lung parenchymal destruction and impaired pulmonary perfusion on a lobar level in patients with emphysema. *Eur J Radiol* 63: 76–83.

200. Sergiacomi G, Sodani G, Fabiano S, Manenti G, Spinelli A et al. (2003) MRI lung perfusion 2D dynamic breath-hold technique in patients with severe emphysema. *In Vivo* 17: 319–324.

201. Ohno Y, Koyama H, Nogami M, Takenaka D, Matsumoto S et al. (2007) Postoperative lung function in lung cancer patients: Comparative analysis of predictive capability of MRI, CT, and SPECT. *AJR Am J Roentgenol* 189: 400–408.

202. Molinari F, Fink C, Risse F, Tuengerthal S, Bonomo L et al. (2006) Assessment of differential pulmonary blood flow using perfusion magnetic resonance imaging: Comparison with radionuclide perfusion scintigraphy. *Invest Radiol* 41: 624–630.

203. Amundsen T, Torheim G, Kvistad KA, Waage A, Bjermer L et al. (2002) Perfusion abnormalities in pulmonary embolism studied with perfusion MRI and ventilation-perfusion scintigraphy: An intra-modality and inter-modality agreement study. *J Magn Reson Imaging* 15: 386–394.

204. Alford SK, van Beek EJ, McLennan G, Hoffman EA (2010) Heterogeneity of pulmonary perfusion as a mechanistic image-based phenotype in emphysema susceptible smokers. *Proc Natl Acad Sci USA* 107: 7485–7490.

205. Raghu G, Collard HR, Egan JJ, Martinez FJ, Behr J et al. (2011) An official ATS/ERS/JRS/ALAT statement: Idiopathic pulmonary fibrosis: Evidence-based guidelines for diagnosis and management. *Am J Respir Crit Care Med* 183: 788–824.

206. Muller NL, Mayo JR, Zwirewich CV (1992) Value of MR imaging in the evaluation of chronic infiltrative lung diseases: Comparison with CT. *AJR Am J Roentgenol* 158: 1205–1209.

208. Lutterbey G, Grohe C, Gieseke J, von Falkenhausen M, Morakkabati N et al. (2007) Initial experience with lung-MRI at 3.0T: Comparison with CT and clinical data in the evaluation of interstitial lung disease activity. *Eur J Radiol* 61: 256–261.

209. Yi CA, Lee KS, Han J, Chung MP, Chung MJ et al. (2008) 3-T MRI for differentiating inflammation- and fibrosis-predominant lesions of usual and nonspecific interstitial pneumonia: Comparison study with pathologic correlation. *AJR Am J Roentgenol* 190: 878–885.

210. McFadden RG, Carr TJ, Wood TE (1987) Proton magnetic resonance imaging to stage activity of interstitial lung disease. *Chest* 92: 31–39.

211. Berthezene Y, Vexler V, Kuwatsuru R, Rosenau W, Muhler A et al. (1992) Differentiation of alveolitis and pulmonary fibrosis with a macromolecular MR imaging contrast agent. *Radiology* 185: 97–103.

212. Gaeta M, Blandino A, Scribano E, Minutoli F, Barone M et al. (2000) Chronic infiltrative lung diseases: Value of gadolinium-enhanced MRI in the evaluation of disease activity—Early report. *Chest* 117: 1173–1178.

12

Pleura and Diaphragm

Francesco Molinari

CONTENTS

12.1 Introduction

Magnetic resonance imaging (MRI) is rarely considered the initial imaging examination of choice in the evaluation of the pleura and diaphragm. However, current research provides evidence that MRI is a robust and safe alternative for assessing thoracic lesions. In particular, the array of MR techniques clinically available for the advanced characterization of thoracic lesions is continuously evolving. Moreover, the possibility to perform imaging studies without ionizing radiation exposure is attractive to clinicians dealing with young patients who may require short-term longitudinal examinations. Therefore, the number of potential clinical indications to perform chest MRI is recently increased, and MRI has become an excellent problem-solving tool for the assessment of pleural and diaphragmatic abnormalities.

12.2 Pleura

12.2.1 Anatomy and Function

The term *pleura* refers to the two layers of mesothelial cells that covers (1) the inner surface of the chest wall, the mediastinum, and the diaphragm (i.e., the parietal pleura) and (2) the surfaces of the pulmonary lobes (i.e., the visceral pleura). The pleural space, which is not visible in normal conditions, is bound by the parietal and visceral pleura at the lung–chest wall interface, and by two layers of visceral pleura at the level of the interlobar fissures. The parietal pleura adheres externally to the thoracic rib cage by interposition of extrapleural fat and the endothoracic fascia. The intercostal space between the ribs includes the intercostal muscles, fat, and vessels. The pleural layers and cavity participate in the lymphatic drainage of the lung and chest wall.

12.2.2 Pleural Disease

Pleural disease includes different clinical entities that usually require specific assessment by thoracic imaging. The following paragraphs describe the pleural diseases for which MRI may have a role as a diagnostic tool.

12.2.2.1 Pleural Effusions

The normal pleural space contains 5 mL of fluid. When the fluid exceeds that amount, the resulting condition is compatible with a pleural effusion. Pleural effusions are classified according to their mechanism of formation, their location, morphology, and inner content.

Pleural effusions that result from imbalanced intra-capillary pressures (increased hydrostatic or decreased osmotic pressure) are named transudates. Transudative causes typically include left ventricular failure, constrictive pericarditis, hypoalbuminemia, cirrhosis, and nephrotic syndrome. Transudates are most commonly mobile and accumulate in the most dependent part of the pleural cavity. In the upright position, small pleural transudate locates in the subpulmonic space, between the inferior surface of the lower lobes and the diaphragmatic leaflets. In the supine position, transudate fills the posterior costophrenic angles. As it accumulates, the mobile effusion tends to fill the rest of the cavity surrounding the lung.

Pleural effusions that result from an altered pleural membrane and capillary permeability are named exudates. Exudates are typically those of sterile parapneumonic effusions, empyemas, and pleural effusions due to malignant tumors. Those pleural effusions may be associated with pleural thickening and internal adhesions. As a result, the exudate may be restricted into loculations (Figure 12.1). Loculations are most commonly found in hemothorax, pyothorax, chylothorax, or infectious pleuritis. Multiple loculations are also frequent depending on the number and sites of adhesions (Figure 12.2). Loculated pleural effusions may appear as *mass-like* lesions at the pleuro-pulmonary interface or within the fissure (Figure 12.3).

Conventional chest radiography, computed tomography (CT), and ultrasound are the first-line imaging modalities for detecting and characterizing pleural effusions. MRI has a role in selected clinical circumstances, and in particular, it can be useful in characterizing pleural fluid of unknown origin, either transudates or exudates, or in assessing the presence of tissular lesions associated with the effusion.

Exudates typically show a higher degree of enhancement after IV gadolinium-based contrast administration than transudative pleural effusions (Frola et al. 1997). Diffusion-weighted imaging may be helpful in differentiating exudates from transudates based on the apparent diffusion coefficient value (Baysal et al. 2004). In addition, MRI can directly contribute in the definition of the type of effusion when the pleural fluid has a bloody or fat component. The signal characteristics on the T_1- and T_2-weighted images can help determine the type of hemoglobin derivate in the effusion and differentiate a subacute from a chronic pleural hemorrhage, according to its content (oxyhemoglobin, deoxyhemoglobin, methemoglobin, and hemosiderin) (Mitchell 2003). This information is valuable for orienting the diagnosis in the correct clinical setting (trauma, metastatic disease, anticoagulant therapy, etc.) and may obviate the need for invasive procedures, especially in high-risk patients (Figure 12.3). The high signal on the T_1-weighted sequences in a non-loculated pleural effusion can indicate a high fat content, which is most likely the result of chylothorax associated with a mediastinal tumor (McLoud and Flower 1991). Gas–liquid levels in an empyema, indicating the presence of a broncho-pleural fistula, can be detected by MRI. Irregular or nodular pleural thickening in the effusion that suggests a malignant pleural effusion can also be visualized by MRI (see the following paragraphs).

12.2.2.2 Tissular Lesions

12.2.2.2.1 Pleural Thickening and Plaques

Diffuse pleural thickening usually manifests as a response to injuries that produce an intense and/or chronic pleural inflammation. Inflammation most commonly results from pleural infections (i.e., empyema and tuberculosis infection), but it is also seen in noninfectious conditions (pleural hemorrhage, trauma, radiation, pulmonary embolism, medication, collagen vascular disease, neoplasm, and occupational exposure). Regardless of its cause, pleural thickening can be associated with pleural calcifications. Calcifications in diffuse pleural thickening are most common in patients with healed tuberculosis, bacterial empyema, hemothorax, or asbestos exposure. Because of the diffuse character of the thickening, the pleural fibrosis may lead to a fibrothorax with a progressive limitation to expand the hemithorax. Therefore, patients that present diffuse pleural thickening should be followed with serial assessment by physical examination, chest imaging, and pulmonary function testing.

Pleural plaques are deposits of hyalinized collagen fibers in the parietal pleura and are presumed to be the result of pleural inflammation caused by asbestos fibers that are transported to the pleural surface

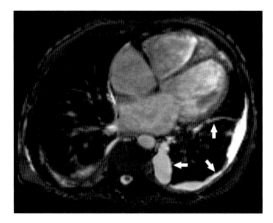

FIGURE 12.1
Pleural effusion. Balanced gradient-echo imaging obtained during free breathing. The MR image shows a pleural effusion with a loculated appearance and interlobar extension on the left side (white arrows). A small pleural effusion is also present on the right side (not indicated).

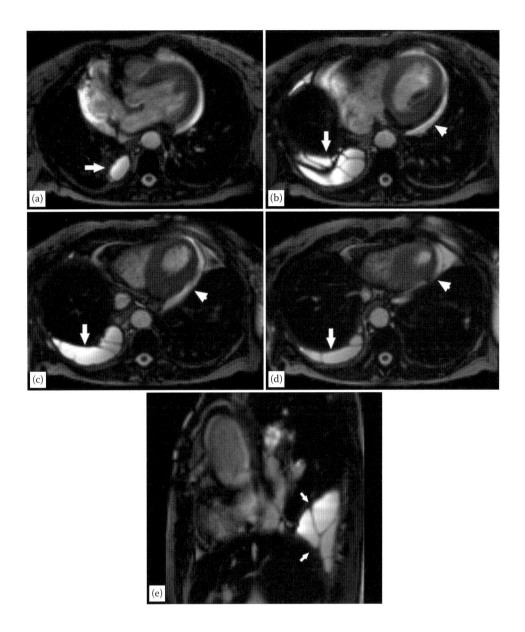

FIGURE 12.2
(a–e) Pleural effusion. Balanced gradient-echo MR images obtained during free breathing in the axial (a–d) and sagittal (e) planes. A right-sided loculated pleural effusion (white arrows) is seen in this patient with coexistent pericardial effusion (white arrowheads). The loculated effusion appears sharply demarcated from the adjacent lung. The pleural layers and internal septa are moderately thickened. The adjacent lung appears displaced and moderately compressed. The balanced gradient-echo MR technique provides an excellent contrast between the hyperintense fluid of the loculations and the hypointense septa.

along lymphatic channels. Bilateral scattered calcified pleural plaques are pathognomonic signs for prolonged asbestos exposure (Peacock et al. 2000). They are observed in 60%–70% of exposed workers (Falaschi et al. 1995) and typically become visible with an average latency period of about 15 years for uncalcified and at least of 20 years for calcified plaques (Muller 1993), after the inhalation of asbestos fibers. Although pleural plaques are not premalignant and do not have further disease potential, they are a marker of a significant asbestos exposure. Since asbestos exposure is a risk factor for higher incidence of malignancy, such as

mesothelioma (approximately threefold higher risk) and lung cancer (Mossman and Gee 1989; Schwartz 1991), the detection of pleural plaques by imaging is important.

The gold standard imaging modality for detecting and characterizing diffuse pleural thickening and pleural plaques is CT (Lynch et al. 1989). MRI is rarely used for the diagnosis of benign changes of the pleura. However, the detection of these pleural changes is important for the correct assessment of the patient. Therefore, these benign pleural alterations should not be overlooked when occasionally found at chest MRI.

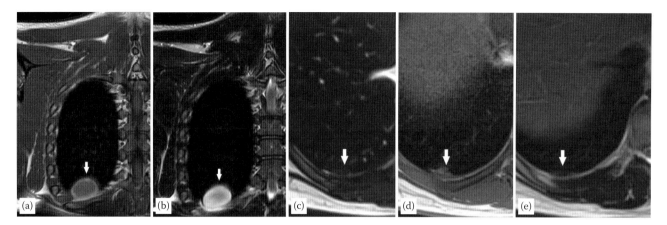

FIGURE 12.3

(a–e) Posttraumatic pleural hematoma. Turbo spin-echo MR images showing an elliptical-shaped lesion of the right posterior costal pleura (white arrows in a and b), which is sharply demarcated from the adjacent lung, in this patient with a recent history of high-impact trauma. The MRI was performed for suspicion of spinal injuries. The lesion manifests with a thick peripheral rim, hyperintense on both T_1- and T_2-weighted images, and with an inner component that appears slightly hyperintense on the T_1-weighted image (a) and moderately hyperintense on the T_2-weighted image (b), relative to the muscle, consistently with a recent hematoma. A follow-up MRI scan was performed 6 months after the first MRI examination (c–d). The lesion was dramatically reduced in size and appears as a slightly lobulated pleural nodule (white arrows), moderately hyperintense on the T_1-weighted fast gradient-echo image (d), and hypointense on the balanced gradient-echo and T_2-weighted fast spin-echo images (c, e). No additional follow-up was required.

Diffuse pleural thickening is defined by the following morphologic criteria: thickening of more than 3 mm in thickness that extends for more than 8 cm in the craniocaudal direction and along more than 5 cm of the chest wall in the lateral direction. It is usually bilateral. On the T_1- and T_2-weighted MR images, it appears as a subcostal linear band of diffusely hypointense signal. Fatty tissue deposition external to the parietal pleura may occur as a result of pleural retraction of subpleural fat. Subpleural fat appears at MRI as a T_1 and T_2 signal linear hyperintensity that becomes hypointense using fat-saturation techniques.

Pleural plaques appear as focal areas of pleural thickening that involve preferentially the parietal pleura adjacent to ribs, particularly the sixth through ninth ribs, and the diaphragmatic pleura (Figure 12.4). Pleural plaques are less extensive in the intercostal spaces, only rarely occur in the visceral pleura. They are generally absent in the region of the costophrenic sulci and the lung apices. They can have a linear, band-like, or nodular appearance and may impinge on the adjacent lung parenchyma. This impingement may occasionally cause a pulmonary subpleural curvilinear line adjacent to the plaque, which is indicative of focal lung fibrosis. The MR signal intensity of noncalcified pleural plaques varies from moderate to low on the T_1- and T_2-weighted images. Calcifications in the plaques are easily detected as focal or band-like structures of very low signal intensity on both T_1- and T_2-weighted images, which is due to the rapid signal decay of calcium aggregates on the standard fast spin-echo sequences. Blooming artifact may be also found around the calcified foci on the T_2^*-weighted images (gradient-echo sequence). Advanced MRI techniques using high-resolution ultra-short TE sequence with radial k-space sampling have been also proposed for the assessment of pleural plaques in a research setting (Weber et al. 2004), which could be useful for characterizing the calcium content of the plaque.

12.3 Pleural Tumors

According to the WHO histological classification of 2006, pleural tumors are divided into mesothelial tumors, which essentially include diffuse malignant mesothelioma and localized malignant mesothelioma, mesenchymal tumors, which include solitary fibrous tumor, and lymphoproliferative disorders. However, most of the pleural tumors are due to metastatic disease, whereas primary pleural malignancy is rare.

The role of imaging in the assessment of patients with suspected pleural malignancy is first to identify the radiological features that distinguish malignant from benign pleural disease and second to narrow the differential diagnosis of the different pleural malignancies. The mainstay for imaging of pleural tumors remains multi-detector row CT with intravenous, iodinated contrast-material enhancement. MRI with gadolinium-based contrast material has a high sensitivity and specificity for pleural malignancy, but is usually not performed due to general availability of this modality for thoracic examinations (Downer et al. 2013).

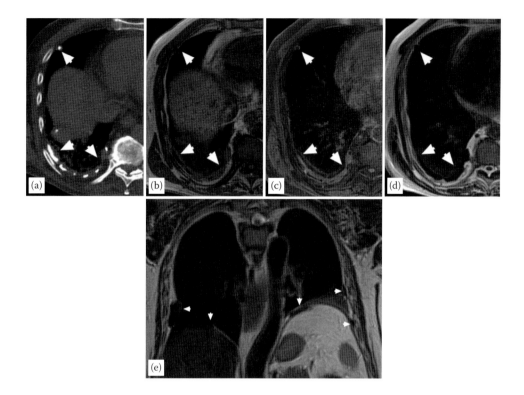

FIGURE 12.4

(a–e) Pleural plaques and calcifications. The CT scan (a) shows multiple plaques and calcifications (white arrowheads) of the costal and diaphragmatic pleura at the level of the posterolateral midportion of the chest wall in this patient with known asbestos exposure. The noncalcified part of the pleural plaques appears slightly hyperintense relative to the muscle on the T_1-weighted gradient-echo (b) and fat-saturated T_1-weighted gradient-echo (c) images and show remarkable signal loss on the T_2-weighted single-shot fast spin-echo image (d), due to the long TE and TR used of the single-shot fast spin-echo measurements. The calcified component of the plaques appears markedly hypointense on both T_1-weighted and T_2-weighted images, due to the extremely fast signal decay of calcium deposits. The morphologic aspect of the pleural plaques and their distribution is well depicted on the coronal T_2-weighted single-shot fast spin-echo image (e). The markedly hypointense thick lesions are located on both the costal and diaphragmatic pleura (white arrowheads).

Regardless of the preferred cross-sectional imaging modalities (CT or MRI) used for the assessment of patients with suspected pleural tumor, the features that almost invariably indicate malignant disease include (1) nodular and irregular thickening, (2) involvement of the mediastinal pleura, (3) circumferential thickening, and (4) thickness greater than 1 cm. Malignant pleural disease tends to involve the entire pleural surface, whereas reactive pleurisy usually does not affect the mediastinal pleura except in cases of tuberculous empyema (Hierholzer et al. 2000).

12.3.1 Metastases

Pleural metastases are the most common malignancies involving the pleura. The primary tumors that metastasize to pleura are usually cancer originating from the lung (40%) and the breast (20%), and lymphoma (10%). Other cancers that may metastasize to the pleura include colon, pancreas, kidney, or ovary (30%). Tumors can spread from a distant primary malignancy to the pleura by hematogenous dissemination, direct invasion from an adjacent carcinoma or subpleural tumor

(lymphoma), and pleural seeding (thymoma or bronchogenic carcinoma).

CT is usually the imaging modality of choice for assessing patients with suspected pleural metastases. Imaging features indicating the presence of pleura metastases can also be detected at chest MRI. Those features vary from a normal pleura with pleural effusion, smooth thickening, localized masses, and extensive nodular thickening. The extensive circumferential pleural thickening may be associated with fissure nodularity and volume loss in the hemithorax of the affected side. As previously mentioned, all these features should be regarded as signs of malignant disease but may not be specific for pleural metastasis. In particular, in the presence of a circumferential nodular thickening of the pleura associated with volume loss of the hemithorax, both metastatic disease and mesothelioma should be suspected.

12.3.2 Primary Tumors

Primary pleural lesions arise directly from the pleural layers. The following paragraphs focus on the most frequent primary tumors of the pleura.

12.3.2.1 Malignant Pleural Mesothelioma

Malignant mesothelioma is a rare, insidious, and highly aggressive neoplasm with a very poor prognosis. It may arise from all mesothelial layers of the body, namely, from the pleura, pericardium, peritoneum, or tunica vaginalis. Eighty percent of all cases are pleural in origin.

Malignant pleural mesothelioma (MPM) is divided into different histological subtypes, with the sarcomatoid being the subtype with the worst prognosis. MPM is the most common cancer in persons with a history of prolonged occupational exposure to asbestos. MPM is 1000 times more common in workers exposed to asbestos (occupational exposure is found in 80% of MPM). Therefore, MPM is relatively more frequent in referral hospitals that deal with occupational disease. The latency period found between the first asbestos exposure and clinical manifestation of the tumor typically ranges from 20 to 40 years (Kishimoto et al. 2003). The peak incidence of this tumor in asbestos-exposed patients is 50–70 years. The cancer incidence is expected to peak between 2010 and 2030 in industrialized countries.

Most patients with MPM present with nonspecific pulmonary symptoms for months before the diagnosis is made. At the time of the diagnosis, the tumor is already at an advanced stage. Morbidity and mortality of this tumor are primarily related to its local invasion. As the tumor spreads, it gradually obliterates the pleural space, encases the lung, invades crucial thoracic structures such as the heart or pericardium, and extends through the diaphragm into the abdominal cavity. Metastases may occur to the opposite lung, brain, and other extrathoracic sites. The majority of affected patients die from local extension and respiratory failure.

Chest imaging is generally performed as part of the initial assessment of patients with nonspecific pulmonary symptoms that may ultimately lead to the diagnosis of MPM. In these patients, conventional chest radiography typically shows a unilateral pleural effusion that will recur after drainage and will eventually prompt further evaluation by cross-sectional imaging. At CT or MRI, the most common tumor-related findings include (1) nodular concentric pleural thickening with more than 1 cm of thickness and with involvement of the fissures; (2) ipsilateral pleural effusion; (3) encasement of the lung due to the thick rind of tumor, which may lead to ipsilateral mediastinal shift; (4) significant unilateral loss of lung volume; (5) chest wall invasion and rib destruction; and (6) metastatic spread to the lung, cardiac, and pericardium (Figures 12.5 through 12.7).

Concerning the choice of the cross-sectional imaging modality for assessing MPM, chest CT is currently used as primary imaging tool for assessing and for staging the tumor (Layer et al. 1999; Marom et al. 2002; Wang et al. 2004). Since CT is generally superior than

MRI in the evaluation of calcifications, CT can suggest a benign cause for extensive or multi-focal calcified pleural thickening. However, MPM does not present as a calcified tumor and does not develop as malignant transformation of calcified plaques. Since the soft tissue component is predominant in MPM and because MRI provides high soft-tissue contrast, MRI can provide significant advantages for the morphologic assessment of the tumor, particularly in regard of its local invasion. MPM appears inhomogeneously hypointense to isointense on T_1-weighted and hyperintense on T_2-weighted MR images and shows contrast media enhancement (Figures 12.5 through 12.7) (Bonomo et al. 2000). High signal intensity in relation to the intercostal muscles on T_2-weighted and contrast-enhanced T_1-weighted MR images is suggestive for malignant disease with a sensitivity of 100% and a specificity of 93% for MRI in the detection of pleural malignancy (Hierholzer et al. 2000).

In recent studies, it has been shown that MRI can be used to address equivocal CT findings concerning the local extent of tumor, particularly in patients with MPM being considered for surgical resection (Patz et al. 1992; Wang et al. 2004). MRI is particularly suitable for the evaluation of local invasion of the endothoracic fascia, chest wall, and the diaphragm, as well as transdiaphragmatic tumor growth. Additional approaches for assessing MPM via dynamic MRI have been recently investigated. In previous studies, MRI was able to measure lung volumes in steady state (Qanadli et al. 1999; Ley et al. 2004) and during different phases of the breathing cycle (Whitelaw 1987; Plathow et al. 2004; Haage et al. 2005) as a functional parameter of therapy response (Plathow et al. 2006). Dynamic MRI techniques that use FLASH (fast low angle shot) sequences and trueFISP (true fast imaging with steady-state precession) sequences can be used to visualize and quantify volumetric changes of the thorax in patients with MPM during therapy (Plathow et al. 2006) and evaluate the effect of treatment on the breathing cycle. Dynamic contrast-enhanced (DCE) MRI can also be a useful tool to evaluate noninvasively the kinetics of contrast media arrival and clearance through the tumor microcirculation and adjacent tissues. It was shown that the gadolinium-contrast enhancement pattern can be used to characterize tumor micro vascular properties and to demonstrate tumor heterogeneity for therapy monitoring (Giesel et al. 2006, 2008).

Despite CT and MRI features are highly indicative of pleural malignancy, they are not specific of MPM as they can be equally mimicked by metastatic pleural disease. The diagnosis of MPM requires histopathologic examination of tissue and cannot be based exclusively on clinical or imaging findings even when these are combined with a history of asbestos exposure. When a patient presents with a significant pleural effusion, thoracentesis for cytology and closed pleural biopsy is generally the initial procedure and may be sufficient to establish the diagnosis. However, this approach may not provide enough tissue to establish the diagnosis of malignancy and to distinguish mesothelioma from other tumors. Negative results from thoracentesis and/or pleural biopsy do not exclude the diagnosis of mesothelioma. If the initial thoracentesis and pleural biopsy are nondiagnostic, surgical intervention (via video thoracoscopic biopsy or open thoracotomy) has a higher diagnostic yield.

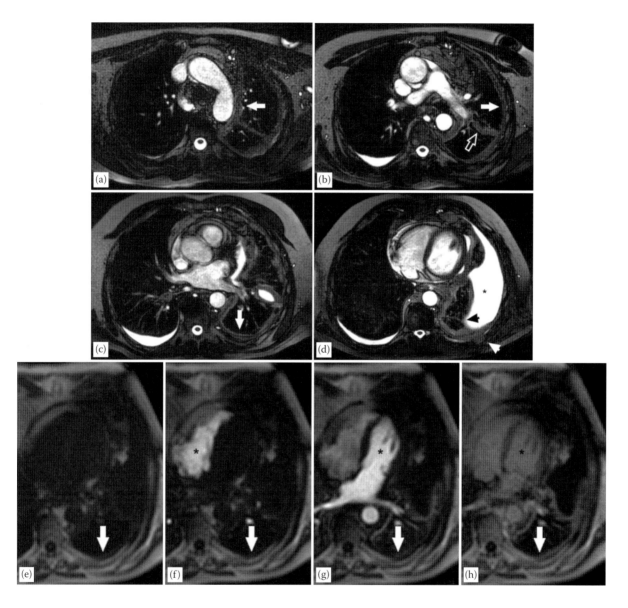

FIGURE 12.5

(a–m) Pleural mesothelioma. The balanced gradient-echo MR images (a–d) show multiple lobulated pleural-based coalescing masses that appear as a diffuse irregular thickening of the left pleural surfaces (white arrows). The thickening extends to the costal, mediastinal, and diaphragmatic pleura (white arrows) as well as to the interlobar fissure (open arrow) and causes a circumferential encasement of the left lung with contraction of the affected hemithorax and slight ipsilateral mediastinal shift. The pleural process infiltrates the underlying lung parenchyma primarily extending into interlobular septa (black arrowhead). Note also the infiltration of the posterior chest wall through the intercostal space (white arrowhead). The infiltrative pleural thickening is accompanied by a left-sided pleural effusion (black asterisk), without contralateral deviation of the mediastinum, confirming the fixation by pleural rind of neoplastic tissue (frozen hemithorax). The four perfusion images (e–h) are extracted from a series of 60 dynamic measurements obtained using a fast gradient-echo MR sequence during IV administration of a single dose (0.1 mmoL/kg) of a gadolinium-based contrast agent. The unenhanced T_1-weighted image (e) shows the pleural thickening, minimally hyperintense relative to the muscle (white arrow). During the early phase of the dynamic acquisition (f, note the contrast medium in the right cardiac cavities, black asterisk), the pleural lesions do not show substantial enhancement (white arrow in f). During the passage of contrast medium in the left cardiac cavities (g, black asterisk) and in the late phase of contrast administration (black asterisk in h), the pleural thickening shows a progressive enhancement. *(Continued)*

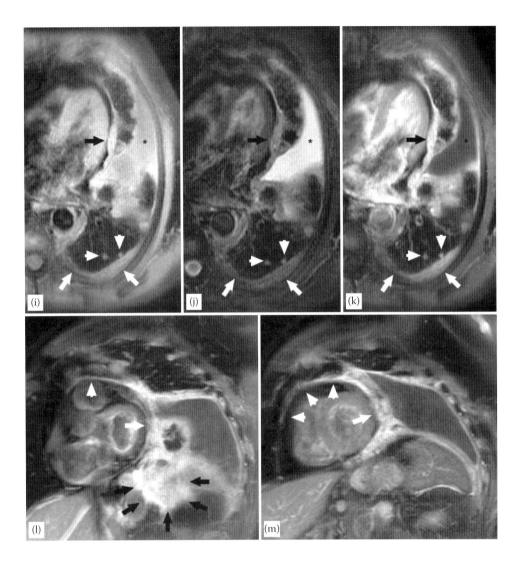

FIGURE 12.5 (Continued)
Pleural mesothelioma. Images (i–k) indicates characterization by T_1-weighted, T_2-weighted, and post-contrast T_1-weighted imaging. The pleural-based coalescing masses are moderately hyperintense relative to muscle on the T_1-weighted fat-saturated turbo spin-echo image (i, white arrows) and moderately hyperintense relative to muscle on the T_2-weighted STIR turbo spin-echo image (j, white arrows). The post-contrast T_1-weighted fat-saturated turbo spin-echo image (k) shows intense enhancement of the pleural tumor (white arrows). The mediastinal and pericardial pleura appear infiltrated (black arrows). Note the two pulmonary nodules in the left lower lobe (white arrowheads) indicating metastatic hematogenous spread to the lung. The pleural effusion (black asterisk) appears hyperintense on both T_1-weighted (i) and T_2-weighted (j) pre-contrast images, as from a hemorrhagic content. Images (l, m) represent evaluation of the diaphragmatic and pericardial spread on the post-contrast T_1-weighted fat-saturated turbo spin-echo images. The pleural tumor infiltrates the inferior pleural surface and extends through the diaphragm to the upper abdomen (black arrows in l). The pericardium (white arrowheads) and epicardial fat (white arrows) are also infiltrated.

12.3.3 Other Primary Pleural Tumors

Other relatively frequent primary tumors of the pleura include lipomas and fibrous tumors.

12.3.3.1 Lipomas of the Pleura

Lipomas are rare benign tumors of typically uniform and homogeneous fatty content. They tend to occur in the upper chest along the second or third ribs, but may be found everywhere in the thorax. The fatty content is easily shown on both CT (fat's low attenuation coefficient of −50 HU to −100 HU) and MRI (fat's high signal intensity on both T_1- and T_2-weighted images, and low signal intensity on the fat-saturated sequences) and allows for a relatively easy diagnosis of this tumor.

12.3.3.2 Solitary Fibrous Tumor of the Pleura

Solitary fibrous tumors of the pleura (SFTP, also known as localized fibrous tumors) are mesenchymal tumors of fibroblastic origin (Travis et al. 2004; Cardillo et al. 2012). SFTP represent less than 5% of primary pleural tumors and have not been found to be related to mesothelioma

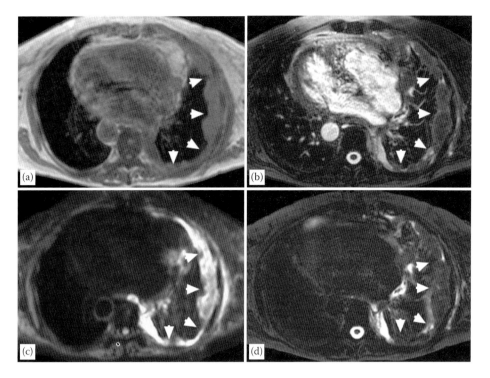

FIGURE 12.6

(a–d) Pleural mesothelioma. The MR images show multiple lobulated pleural-based coalescing masses (white arrowheads) that appear as a diffuse irregular thickening of the right pleural surfaces (white arrowheads). The pleural thickening appears moderately hyperintense on the T_1-weighted fast gradient-echo image (b), slightly hyperintense on the T_2-weighted fat-saturated turbo spin-echo and on the balanced gradient-echo images (c and a, respectively), and hyperintense on the DW image (d).

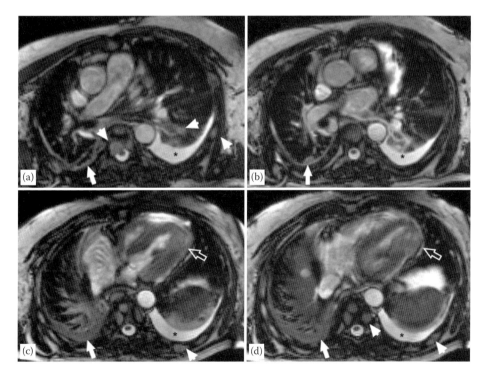

FIGURE 12.7

(a–g) Pleural mesothelioma with cardiac invasion. The balanced gradient-echo images (a–d) show bilateral lobulated pleural-based masses and pleural thickening (white arrows) in this patient with advanced mesothelioma, with hematogenous spread to the ribs, vertebral bodies, and lung (white arrowheads). A left-sided pleural effusion is also present (black asterisks). The cardiac apex and the apico-lateral segments of the left ventricle show irregular margins and heterogeneous signal intensity (white open arrows). *(Continued)*

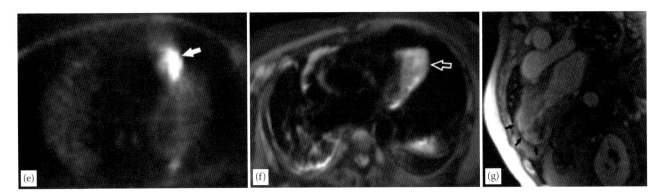

FIGURE 12.7 (Continued)
Pleural mesothelioma with cardiac invasion. At the FDG-PET scan (e), the cardiac lesion appears as a hypermetabolic mass (white arrow). The DW image (f) shows a hyperintense area in the same cardiac region (white open arrow). The post-contrast fast gradient-echo image (g) confirms an irregular infiltration of the left ventricle with heterogeneous enhancement (black arrows).

nor asbestosis. Histologically identical tumors have been reported at other extrathoracic sites (Wignall et al. 2010). Although most SFTP are benign, some may have histologic features of malignancy such as the number of mitoses, the presence of necrosis, hypercellularity, or the presence of nuclear atypia. Malignant SFTP typically have a higher tendency toward a rapid growth compared to benign SFTP. Complete surgical resection is the standard treatment approach for SFTP (Lahon et al. 2012).

The mean age at presentation of SFTP is between 55 and 65 years (Magdeleinat 2002). Most patients with SFPT are asymptomatic at presentation or may show nonspecific pulmonary symptoms due to the presence of an intrathoracic mass (Lahon et al. 2012; Lococo et al. 2012). An intrathoracic mass may be also detected in asymptomatic individuals when chest imaging is performed for an unrelated reason. Hypertrophic pulmonary osteoarthropathy (Pierre Marie–Bamberger syndrome) has been described in 20% of cases, and refractory hypoglycemia (Doege–Potter syndrome) has been reported in up to 5% of cases. The diagnosis

of SFTP requires pathologic examination. Transthoracic needle biopsies are often inadequate for a definitive diagnosis. In most cases, the diagnosis of SFTP is made after complete resection of the lesion.

Chest imaging is generally performed in the assessment of patients with SFTP. At conventional chest radiography and cross-sectional imaging (CT or MRI), SFTP typically manifest a well-delineated and occasionally lobulated thoracic mass of 5–30 cm (Figures 12.8 and 12.9). The lesion is frequently located in the paravertebral area; however, other locations are not uncommon, and the tumor may also arise in the fissures, mimicking intrapulmonary mass. The site of origin (i.e., the pleura) is usually not evident until surgery. A pedicle is present in 40% of the cases. Due to its pedicle, tumor mobility can present and may be detected by changing patient's position during the imaging examination. This sign can be helpful for narrowing the differential diagnosis (Desser and Stark 1998). SFTP are typically characterized by heterogeneous structures. Intralesional cystic or necrotic components of myxoid degeneration have been described, which appear of high signal

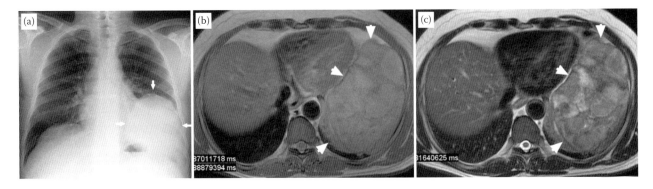

FIGURE 12.8
(a–c) Solitary fibrous tumors of the pleura. Radiographic appearance of solitary fibrous tumors of the pleura (SFTP, a). A rounded, slightly lobulated mass of more than 10 cm is detected at the base of the left hemithorax (white arrows). The MRI turbo spin-echo T_1-weighted (b) and T_2-weighted (c) images show a highly heterogeneous mass without signs of pleural origin (white arrowheads).

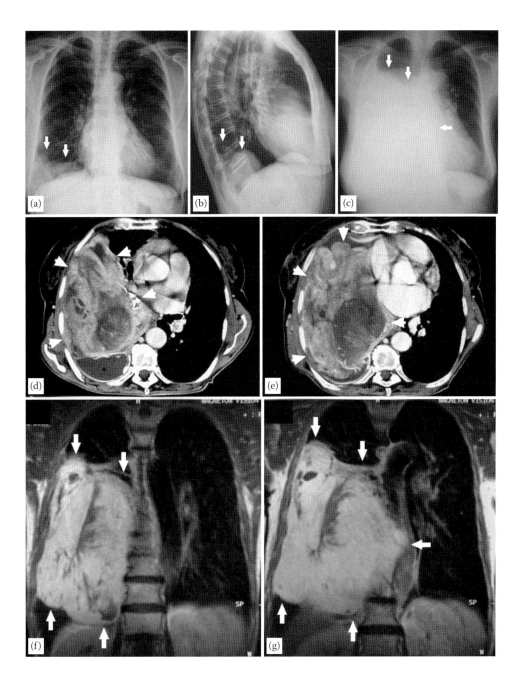

FIGURE 12.9

(a–g) Solitary fibrous tumors of the pleura. Chest radiography of SFTP (a). A lobulated mass of the right posterior costophrenic angle in an asymptomatic patient (white arrows) who refused surgical treatment. Follow-up chest radiography 9 months after the initial radiographic detection of the lesion (a). The right hemithorax is almost completely filled by the mass and the mediastinum is slightly dislocated toward the left (white arrows). The rapid growth of the tumor indicates its malignancy. At chest CT, the mass presents as a highly heterogeneous enhancing tissular lesion with hypodense intralesional necrotic areas (white arrows). Hypertrophic vessels probably originating from the diaphragmatic arteries are also present at the periphery of the mass. At chest MRI, the heterogeneity of the mass and its enhancement are well depicted on these T_1-weighted turbo spin-echo images (white arrows).

intensity on T_2-weighted MR images. Calcifications within the mass are rare (7%). Due to its vascularity, the mass enhances after intravenous contrast material administration and the diaphragmatic arteries and their branches may be visible around or within the mass (Figure 12.9). Rarely, pleural effusion is associated with solitary fibrous tumors of the pleura (De Perrot et al. 2002). Although MRI has limited use in the assessment of such pleural diseases, the morphology and relationship of large lesions to adjacent mediastinal and major vascular structures may be better delineated using MRI compared with CT.

References

Baysal T, Bulut T, Gokirmak M, Kalkan S, Dusak A, Dogan M. Diffusion-weighted MR imaging of pleural fluid: Differentiation of transudative vs exudative pleural effusions. *Eur Radiol* 2004;14(5):890–896.

Bonomo L, Feragalli B, Sacco R, Merlino B, Storto ML. Malignant pleural disease. *Eur J Radiol* 2000;34(2): 98–118.

Cardillo G, Lococo F, Carleo F, Martelli M. Solitary fibrous tumors of the pleura. *Curr Opin Pulm Med* 2012;18:339.

De Perrot M, Fischer S, Brundler MA, Sekine Y, Keshavjee S. Solitary fibrous tumors of the pleura. *Ann Thorac Surg* 2002;74(1):285–293.

Desser TS, Stark P. Pictorial essay: Solitary fibrous tumor of the pleura. *J Thorac Imaging* 1998;13(1):27

Downer NJ, Ali NJ, Au-Yong IT. Investigating pleural thickening. *BMJ* 2013;346:e8376. doi: 10.1136/bmj.e8376.

Falaschi F, Battolla L, Paolicchi A et al. High-resolution computed tomography compared with the thoracic radiogram and respiratory function tests in assessing workers exposed to silica. *Radiol Med* 1995;89(4):424–429.

Frola C, Cantoni S, Turtulici I et al. Transudative vs exudative pleural effusions: Differentiation using Gd-DTPA-enhanced MRI. *Eur Radiol* 1997;7(6):860–864.

Giesel FL, Bischoff H, von Tengg-Kobligk H et al. Dynamic contrast-enhanced MRI of malignant pleural mesothelioma: A feasibility study of noninvasive assessment, therapeutic follow-up, and possible predictor of improved outcome. *Chest* 2006;129(6):1570–1576.

Giesel FL, Choyke PL, Mehndiratta A et al. Pharmacokinetic analysis of malignant pleural mesothelioma-initial results of tumor microcirculation and its correlation to microvessel density (CD-34). *Acad Radiol* 2008;15(5): 563–570.

Haage P, Karaagac S, Spuntrup E et al. Feasibility of pulmonary ventilation visualization with aerosolized magnet resonacne contrast media. *Invest Radiol* 2005;40:85–88.

Hierholzer J, Luo L, Bittner RC et al. MRI and CT in the differential diagnosis of pleural disease. *Chest* 2000;118:604–609.

Kishimoto T, Ohnishi K, Saito Y. Clinical study of asbestos-related lung cancer. *Ind Health* 2003;41(2):94–100.

Lahon B, Mercier O, Fadel E et al. Solitary fibrous tumor of the pleura: Outcomes of 157 complete resections in a single center. *Ann Thorac Surg* 2012;94(2):394–400.

Layer G, Schmitteckert H, Steudel A et al. MRT, CT, and sonography in the preoperative assessment of the primary tumor spread in malignant pleural mesothelioma. *Rofo* 1999;170(4):365–370.

Ley S, Zaporozhan J, Morbach A et al. Functional evaluation of emphysema using diffusion-weighted 3Helium-magnetic resonance imaging, high-resolution computed tomography, and lung function tests. *Invest Radiol* 2004;39:427–434.

Lococo F, Cesario A, Cardillo G et al. Malignant solitary fibrous tumors of the pleura: Retrospective review of a multicenter series. *J Thorac Oncol* 2012;7:1698.

Lynch DA, Gamsu G, Aberle DR. Conventional and high resolution computed tomography in the diagnosis of asbestos-related diseases. *RadioGraphics* 1989;9(3): 523–551.

Magdeleinat P, Alifano M, Petino A et al. Solitary fibrous tumors of the pleura: Clinical characteristics, surgical treatment and outcome. *Eur J Cardiothorac Surg* 2002;21(6): 1087–1093.

Marom EM, Erasmus JJ, Pass HI, Patz EF, Jr. The role of imaging in malignant pleural mesothelioma. *Semin Oncol* 2002;29(1):26–35.

McLoud TC, Flower CD. Imaging the pleura: Sonography, CT, and MR imaging. *AJR Am J Roentgenol* 1991;156(6):1145–1153.

Mitchell JD. Solitary fibrous tumor of the pleura. *Semin Thorac Cardiovasc Surg* 2003;15(3):305–309.

Mossman BT, Gee JB. Asbestos-related diseases. *N Engl J Med* 1989;320(26):1721–1730.

Muller NL. Imaging of the pleura. *Radiology* 1993;186(2): 297–309.

Patz EF Jr, Shaffer K, Piwnica-Worms DR et al. Malignant pleural mesothelioma: Value of CT and MR imaging in predicting resectability. *AJR* 1992;159(5):961–966.

Peacock C, Copley SJ, Hansell DM. Asbestos-related benign pleural disease. *Clin Radiol* 2000;55(6):422–432.

Plathow C, Klopp M, Schoebinger M et al. Monitoring of lung motion in patients with malignant pleural mesothelioma using two-dimensional and three-dimensional dynamic magnetic resonance imaging: Comparison with spirometry. *Invest Radiol* 2006;41(5):443–448.

Plathow C, Ley S, Fink C et al. Evaluation of chest motion and volumetry during the breathing cycle by dynamic MRI in healthy subjects—comparison with pulmonary function tests. *Invest Radiol* 2004;39:202–209.

Qanadli SD, Orvoen-Frija E, Lacombe P, Di Paola R, Bittoun J, Frija G. Estimation of gas and tissue lung volumes by MRI: Functional approach of lung imaging. *J Comput Assist Tomogr* 1999;23(5):743–748.

Schwartz DA. New developments in asbestos-induced pleural disease. *Chest* 1991;99(1):191–198

Travis WD, Churg A, Aubry MC et al. Mesenchymal tumours. In: *Tumours of the LUng, Pleura, Thus and Heart*, Travis WD, Brambilla E, Muller-Hermelink HK, and Harris CC. (Eds), IARC Press, Lyon, France, 2004, p.142.

Wang ZJ, Reddy GP, Gotway MB et al. Malignant pleural mesothelioma: Evaluation with CT, MR imaging, and PET. *RadioGraphics* 2004;24(1):105–119.

Weber MA, Bock M, Plathow C et al. Asbestos-related pleural disease: Value of dedicated magnetic resonance imaging techniques. *Invest Radiol* 2004;39(9):554–564.

Whitelaw WA. Shape and size of the human diaphragm in vivo. *J Appl Physiol (1985)* 1987;62(1):180–186.

Wignall OJ, Moskovic EC, Thway K, Thomas JM. Solitary fibrous tumors of the soft tissues: Review of the imaging and clinical features with histopathologic correlation. *AJR Am J Roentgenol* 2010;195:W55.

13

Functional Magnetic Resonance Imaging of the Lung and Pulmonary Vasculature

Saeed Mirsadraee, Andrew J. Swift, Jim M. Wild, and Edwin J.R. van Beek

CONTENTS

13.1 Introduction

The standard imaging techniques of the lungs, such as high-resolution computed tomography (CT), have many limitations, do not always diagnose the pathology at early treatable stage, and cannot provide physiological information. Although additional methods have since been developed, they require an increase in radiation dose, inspiratory and expiratory volumetric CT, and advanced software tools, which usually aren't available immediately in the course of reporting workflow.

Functional lung imaging is increasingly recognized as an important addendum to the traditional morphologic assessment. Function can be derived from CT (using inspiration, expiration, and contrast-enhanced methods), but an overall assessment is difficult due to restrictions in radiation dose. Hence, magnetic resonance imaging (MRI) has the ideal components to allow functional lung imaging to thrive: it is fast, and it allows multiple assessments of the subsystems of the lungs and the integrated evaluation of the heart without the need for ionizing radiation.

MRI employs contrast and non-contrast imaging approaches to study lungs and airways function by evaluating pulmonary perfusion and ventilation, dynamics of the flow in the pulmonary artery, and motion. In this chapter, the role of novel imaging techniques in the functional assessment of the lungs is discussed. Finally, the role of MRI in the assessment of pulmonary hypertension is discussed.

13.2 MR Techniques

Imaging of the lung is challenging due to low proton concentration, accelerated decay of MR signal, and the effect of motion (breathing) (Wielpütz and Kauczor 2012; Wild et al. 2012). Recent advances in MR techniques have made lung imaging possible with adequate temporal and spatial resolution in a single or multiple breath holds, or during free breathing with respiratory gating. Application of various MR techniques has enabled functional lung imaging that is discussed in this section.

13.2.1 Perfusion Imaging

Blood flow at microvascular level in the tissue parenchyma can be depicted and quantified by MR perfusion imaging (Harris et al. 2009). This technique has been used to investigate various benign and malignant lung disorders such as pulmonary embolism (Sebastian and Julia 2012) and lung cancer (Ohno et al. 2014a). Imaging techniques require total or partial lung imaging at high temporal resolution (e.g., 1.5 s) and adequate spatial resolution. Perfusion imaging can be performed with and without the administration of intravenous contrast.

13.2.1.1 Contrast-Enhanced Pulmonary Perfusion Imaging

The most common approach in quantitative perfusion imaging is time-resolved (repetitive) imaging while administering paramagnetic contrast agent (gadolinium–diethylenetriamine pentaacetic acid [DPTA]). Heavily T_1-weighted 3D spoiled gradient-echo sequence (short TR and high flip angle to suppress non-enhancing tissues) is the most commonly used pulse sequence in angiography and perfusion imaging (Wild et al. 2012). The most common alternative imaging sequence is time-resolved dynamic MR angiography with k-space manipulation (Table 13.1). Parallel imaging can be applied within the time limits of a breath hold. The aim of perfusion imaging technique is to achieve an optimal combination of temporal resolution and contrast-to-noise ratio. It is reported that a temporal resolution of approximately 2 s is necessary for correct application of the deconvolution algorithms. High image noise leads to a substantial underestimation of pulmonary blood flow (PBF) estimates. Lower acceleration factor (e.g., 2) and increased voxel sizes are employed to improve contrast-to-noise ratio (Ingrisch et al. 2010).

Gadolinium–DTPA shortens T_1, which results in increased signal from enhancing structures on heavily T_1-weighted imaging. The injection of the contrast agent is commonly commenced with the start of breath hold and imaging, or after a short delay (3–5 s). The time-resolved signal data would define the time–signal curve of the arterial input function (AIF; pulmonary artery), which will be used in the mathematical process of perfusion quantification (Figure 13.1). A single breath hold (20–30 s) is usually sufficient to obtain the time–signal curve (either in part or completely) from the normal lung parenchyma, but longer image acquisition times, with or without respiratory gating, may be necessary to adequately assess enhancement and wash out patterns in areas with delayed contrast transit time (e.g., tumors). Imaging beyond 20–30 s is usually performed during shallow abdominal breathing, and post-processing applications (e.g., non-rigid registration) will be necessary to reduce the effect of motion.

The aim of an optimal injection protocol is to deliver a dense contrast bolus followed by saline flush to achieve a short lasting first pass peak enhancement followed by rapid wash out within the pulmonary artery and the lung parenchyma. To achieve this, the contrast bolus is injected at a fast rate (4–5 mL/s) using a large bore venous line in a medium caliber vein (e.g., antecubital veins). Viscosity of the contrast can be reduced by 50% by pre-warming it to body temperature. Bilateral venous lines may be used to improve delivery of the contrast bolus. In comparison to iodinated contrast agents, where there is a linear relationship between the contrast dose and CT density, the relationship between the MR signal and gadolinium is nonlinear resulting in signal saturation at higher blood concentrations. Signal saturation makes perfusion quantification more complicated, and to avoid the effect on MRI, the contrast concentration has to be kept minimum (Puderbach et al. 2008). The effect is higher at higher magnetic field (i.e., 3 T vs. 1.5 T) and a lower concentration of gadolinium–DTPA should be used (0.05 mmol/kg). A dual bolus approach has been applied in pulmonary perfusion imaging to overcome the problem (Risse et al. 2006). In this approach, a low dose bolus (e.g., 0.05 mmol/kg at 1.5 T) is given to achieve a linear signal–time curve in the AIF, followed by standard dose injection (e.g., 0.1 mmol/kg at 1.5 T) (Hueper et al. 2013) to investigate enhancement of the lung tissue. While this technique avoids signal saturation in the input function, it will result in underestimation of perfusion values as a result of AIF underestimation. Previous studies have reported techniques to correct for such effect.

Pulmonary perfusion is defined as capillary supply to the lung parenchyma (PBF) measured in mL/100 g/min. For the quantification of the perfusion, an AIF has to be defined by placing a region of interest in the main pulmonary artery (Sebastian and Julia 2012) or the right ventricle (Hueper et al. 2013). Tissue response function of the lung parenchyma is determined by the region of interests within the lung parenchyma (Sebastian and

TABLE 13.1
MRI Sequence Acronyms Used for Contrast-Enhanced Perfusion Imaging

Type of Sequence	Philips	Siemens	GE	Hitachi	Toshiba
Spoiled gradient-echo sequences	T1-FFE	FLASH	SPGR MPSPGR	RSSG	RF-spoiled FE
Dynamic MR angiography with k-space manipulation	Keyhole (4D-TRAK)	TWIST	TRICKS	—	—

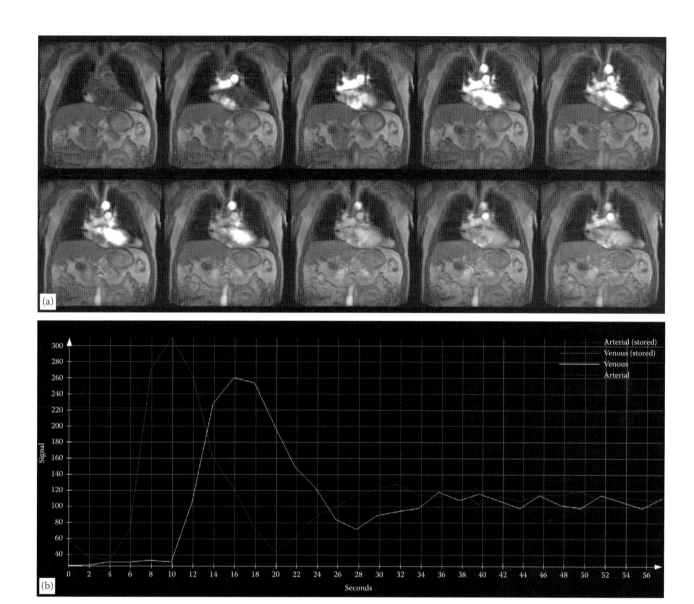

FIGURE 13.1

Dynamic contrast-enhanced MR perfusion imaging. Coronal images (a) show the passage of dense contrast from the right to the left heart and later recirculation of diluted contrast (temporal resolution: 1.5 s). Image (b) shows time–signal intensity curves from regions of interests in the pulmonary artery (red) and left atrium (yellow).

Julia 2012). Analysis of the data is performed using validated software. Software provides parametric colored map of its pixel-by-pixel analysis. Various models are used to quantify perfusion from the time–signal intensity curves. Assuming a linear relation between the signal and the concentration of contrast agent, signal–time curves are converted to concentration–time curves (Risse et al. 2006), and various perfusion values are calculated. Pulmonary blood volume (PBV; mL/100 mL lung tissue) is calculated by normalizing the area under the tissue concentration–time curve to the integral of the AIF (Ley 2012).

By plotting the changes in signal intensity in a region of interest, a time–signal intensity curve is constructed

in the input function (e.g., pulmonary artery) and the lung parenchyma. The perfusion parameters are calculated from the time–signal intensity curves. Time to peak is defined as the time from the start of the contrast injection to maximal tissue enhancement and is measured in seconds. The pulmonary tissue blood flow (PBV) is a measure of the total volume of blood within an imaging voxel (tissues and blood vessels; mL/100 g) and is determined by mathematic integration of the area under the tissue time–signal curve as described in previous paragraph (Aksoy and Lev 2000; Allmendinger et al. 2012). PBF is defined as the volume of blood moving through a given region per unit time (mL/100 mg/min), representing the capillary flow in the lung tissue.

The mean transit time is the average transit time of all the contrast medium molecules through a given volume of tissue and is measured in seconds and can be approximated according to the central volume principle: mean transit time = PBV/PBF (Allmendinger et al. 2012). Various models are used to calculate perfusion values from the time–density curves including the deconvolution techniques. Absolute measurement of perfusion depends on the cardiac output at the time of measurement. To compare perfusion values of a patient, it has been proposed that the values are normalized to the cardiac output (Miles 2001).

To avoid overestimation of perfusion values as a result of the inclusion of pulmonary artery branches in the perfusion analysis, correlation analysis techniques may be used to suppress the pulmonary vasculature in lung perfusion MRI. In this technique, pixels with high correlation coefficients to the pulmonary artery and left atrium are considered as arterial or venous and are excluded from further analysis. It was reported that this correction could result in a reduction of 15% and 25% for PBF and PBV values, respectively, compared to inclusion of all vessel structures (Risse et al. 2009).

13.2.1.2 Non-Contrast MR Pulmonary Perfusion Techniques

Arterial spin labeling has been used as an alternative to contrast-enhanced perfusion imaging. In this technique, blood protons in the supplying arteries are magnetically tagged (labeling), usually with an inversion recovery pulse, and used as endogenous contrast (Rizi et al. 2003). Tissue magnetization changes according to its perfusion status. Since no extraneous contrast is used, the technique can be repeated without restriction. The technique is slower than contrast-enhanced perfusion MRI and more prone to motion artifacts, but recently it has been shown to approximate contrast-enhanced perfusion techniques' outcomes (Martirosian et al. 2010).

Other non-contrast perfusion techniques use a single-slice time-resolved balanced steady-state free procession sequence. Fourier decomposition is then applied to obtain perfusion data from the time–signal within the parenchyma (see Chapter 11 for more details) (Kjørstad et al. 2014).

13.2.2 Ventilation Imaging

Imaging of ventilation can be achieved through different contrast mechanisms. Ventilation imaging includes proton-based methods with administration of paramagnetic contrast agents (oxygen and nebulized gadolinium solutions) and non-proton-based methods (hyperpolarized noble gases) (Ohno et al. 2008a; Mugler et al. 2010; Wild et al. 2012; Mirsadraee and van Beek 2015).

Clearly, by adding ventilation to the arsenal of lung imaging methods, it completes the overall ability to assess the main function of the lungs: the exchange of oxygen and carbon dioxide through matched ventilation and perfusion.

Hyperpolarized gas MRI has been under development for over two decades and has made significant progress in spite of several limitations and pressures, including the need for advanced hardware (including dedicated tuned RF transmit/receive coils, polarizers for the production of hyperpolarized gas, and the competition with other strategic needs for 3-Helium in particular) (Van Beek et al. 2004; Mugler et al. 2010). In spite of this, the technique has reached a level of development that now allows the application in dedicated clinical research studies, and multicenter studies are feasible (Van Beek et al. 2009; Kirby et al. 2014).

In parallel with 3-Helium, hardware development in the polarizer domain has led to improvements of polarization levels and volume production (Becker et al. 1994), allowing the reintroduction of 129-Xenon as a potential contrast agent (Hersman et al. 2008). Other novel gases are being piloted, such as 19-Fluorine, which may also be useful in the future (Couch et al. 2013).

All the above described gases need to undergo energy input (hyperpolarization), which renders them detectable in specifically tuned RF systems. With the use of RF pulses, the polarization is destroyed, and therefore, the methods require different approaches to imaging sequences with reduced flip angles to avoid depolarization.

Although the high signal introduced with hyperpolarized gas makes this technique relatively less dependent on field strength, most research has been performed on 1.5-T systems that are fairly routinely available in most hospitals. Non-standard imaging sequences have been developed to allow maximal use of the hyperpolarization. The utility of hyperpolarized noble gases has elicited interesting insights into pathophysiological relationships and its potential to serve as a quantifiable biomarker for disease and treatment response in a variety of clinical conditions (Kirby et al. 1985; Samee et al. 2003; Woodhouse et al. 2009; Ireland et al. 2010; Kirby et al. 2011; Couch et al. 2013).

Normal subjects will have relatively homogeneous distribution of ventilation with a slight hyperventilation of dependent lungs (which in fact matches perfusion) (De Lange et al. 1999). Patients with a range of pulmonary and pulmonary vascular diseases will develop ventilation defects (De Lange et al. 1999; Marshall et al. 2014), which tend to correlate with extent of disease and offer the study of ventilation distribution. What is particularly of interest is that the technique makes no use of ionizing radiation and therefore can be repeated both for short-term and long-term follow-up. This allows for direct assessment of challenges of ventilation (e.g., allergic challenge) and response to therapies (e.g., bronchodilator or physiotherapy measures) in patients as young as 5–6 years of age. In addition, ventilation MRI has been used to study the size of small airways (down to alveolar level), the functional distribution of ventilation (Saam et al. 2000; Salerno et al. 2002; Fichele et al. 2004), the dynamics of gas flow into the lungs (Salerno et al. 2001; Wild et al. 2003), and the oxygen uptake from the lungs into the blood (Wild et al. 2005; Hamedani et al. 2015). More recently, 129-Xenon has been shown to enable the study of alveolar–blood barrier (interstitium) thickness and the overall surface area of available gas exchange membrane (Mugler et al. 2010; Patz et al. 2007; Muradyan et al. 2013).

The technique has been applied to the entire range of lung diseases, ranging from asthma, cystic fibrosis, chronic obstructive pulmonary diseases (COPD), and lung cancer to lung transplant rejection assessment.

13.2.3 Oxygen-Enhanced ¹H MRI

Oxygen-enhanced MRI is able to obtain images of pulmonary ventilation through the application of the paramagnetic properties of oxygen. The principle is a respiratory gated centrically reordered inversion-recovery single-shot fast spin-echo pulse sequence at end expiration, which obtains a mask image of the lungs, followed by the same imaging sequence after the subject has been allowed to breath 100% oxygen for 5 min (Ohno et al. 2014a). The images are subsequently subtracted, bringing out the difference in signal due to the dephasing of the proton MR images through the paramagnetic effects of oxygen.

The method has been piloted in a few centers, but has thus far not made a full entry into clinical applications. Nevertheless, the technique is relatively straightforward and does not require additional hardware (oxygen is normally available within any MRI suite), and translation into clinical practice should be feasible. More recently, it was shown that using a segmented inversion-recovery Look-Locker multiecho sequence based on a multiecho 2D ultrashort TE (UTE), it was feasible to detangle T_1 and T_2^* quantification, correlated to dissolved molecular oxygen in lung tissue version oxygen concentration in alveolar gas, respectively (Triphan et al. 2015). These developments make it likely that this method will enter clinical routine in the near future.

13.2.4 Dynamic Imaging of Motion

MRI can be used to assess mobility of structures within the thoracic cavity (e.g., diaphragm, airways, and peripheral tumors). Time-resolved imaging techniques such as spoiled gradient-echo (e.g., FLASH) or balanced steady-state free precession pulse sequences are used to assess the mobility during inspiration and expiration (Figure 13.2). Thus, motion of the diaphragm has been tracked and shown of clinical interest in patients with diaphragm pathologies and also correlates well with spirometric measures (Cluzel et al. 2000; Voorhees et al. 2005; Swift et al. 2007).

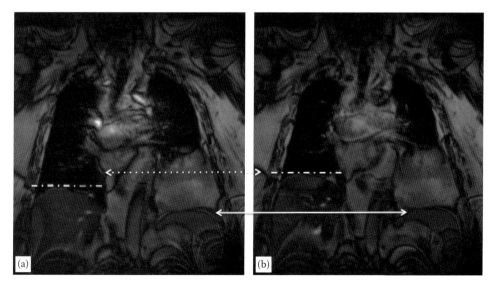

FIGURE 13.2
Dynamic MRI of the lung. Sixty-two-year-old male presented with a lobulated mass on CT. Real-time steady-state free precession imaging showed a normal movement of right hemidiaphragm, but a paralyzed left hemidiaphragm, indicating possibility of phrenic nerve palsy. This precluded suitability for surgical resection of the mass. Image (a): inspiration; image (b): expiration. Note the changed levels of the right hemidiaphragm (dotted line) and the unchanged levels of the left hemidiaphragm.

13.3 Clinical Applications

13.3.1 Clinical Applications of Pulmonary Perfusion Imaging

Pulmonary perfusion is assessed by dynamic contrast-enhanced MRI. Areas with altered perfusion would demonstrate slower and less intense enhancement, or no perfusion (infarcted tissue).

Pulmonary perfusion imaging has been used in the assessment of pulmonary thromboembolism (Figure 13.3). Some studies reported the dynamic MR perfusion imaging to be superior to MR and CT angiography in the prediction of disease severity and outcome after acute pulmonary thromboembolism (Ohno et al. 2010). A sensitivity of 98% for lobar and 92% for the detection of segmental pulmonary embolism has been reported (Ersoy et al. 2007). MR is most extensively used in the assessment of patients with chronic thromboembolic disorders that will be discussed later in this chapter (Tsai et al. 2011; Pena et al. 2012).

More recently, contrast media-free Fourier decomposition MRI is reported to produce qualitative assessment of regional lung ventilation and perfusion that was comparable to that of conventional single-photon emission computed tomography/computed tomography (Bauman et al. 2011). This technique is discussed in Chapter 11.

There has been interest in perfusion imaging in patients with COPD (Figures 13.4 and 13.5). This is based on the knowledge that endothelial dysfunction and abnormal pulmonary vascular response play a key role in the pathogenesis of this disease (Kanazawa et al. 2003; Barr et al. 2007). When emphysema is established, alveolar surface reduction is associated with reduction in capillary volume (Morrison et al. 1990; Barberà et al. 1994) and hence further reduction in perfusion (Pansini et al. 2009). MRI of emphysematous lungs is challenging due to the T_2^* effect from the additional air volume and further reduction in proton density. Multiple quantitative and semi-quantitative MR perfusion studies demonstrated reduced perfusion in emphysematous lungs when compared to normal (Amundsen et al. 2002; Sergiacomi et al. 2003; Ohno et al. 2004a,b; Hueper et al. 2013).

A small-scale study reported enhancement patterns using dynamic contrast-enhanced MRI and ventilation–perfusion (V–Q) scintigraphy in various pulmonary conditions (20 with PE, 11 with acute pneumonia, and 13 with exacerbation of COPD) (Amundsen et al. 2002). Perfusion MRI was performed during free breathing and without cardiac gating. The contrast enhancement pattern was qualitatively assessed. The potential effect of motion in the measurements was not investigated. The inter-modality agreement (kappa value) in the evaluation of perfusion abnormalities ranged from 0.52 to 0.57. In pneumonia, distinct high signal intensity was present in the consolidation compared to the surrounding normal lung tissue, pre- and post-contrast. In COPD with emphysema, a generally low-degree contrast enhancement compared to normal lung tissue was observed. In acute *pulmonary embolism*, distinct lung regions without visible contrast enhancement (perfusion defects) were observed. The time–signal intensity curves showed no peak in signal in early phase and only a delayed slight increase in signal. A higher proportion of lungs with perfusion defects were observed on MR perfusion compared to V–Q scintigraphy.

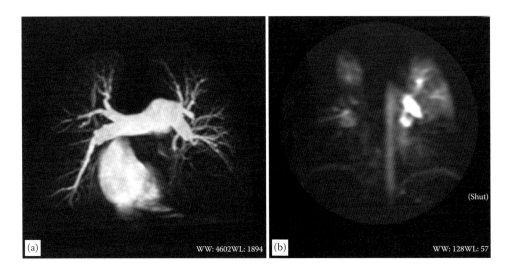

FIGURE 13.3
A patient with a history of previous pulmonary emboli and progressive shortness of breath. There are extensive chronic thromboembolic disease shown on the MR angiogram (a) and a matched pattern of segmental perfusion defects on MR perfusion (b). This patient was found to have chronic thromboembolic pulmonary hypertension (note the size of the proximal pulmonary artery branches).

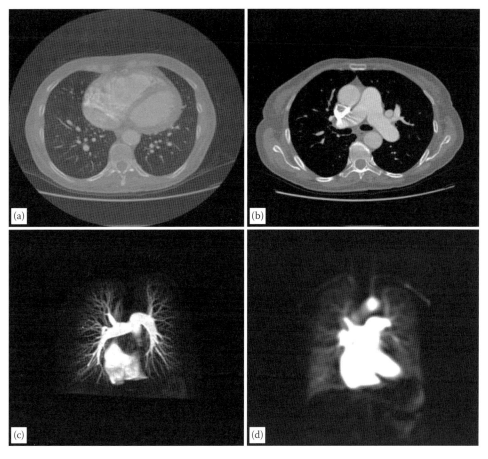

FIGURE 13.4
Fifty-five-year-old male patient with known chronic bronchitis. (a) Axial CT image at the mid-ventricular level. The RV is hypertrophied as it is commonly seen in patients with COPD. (b) Axial CT image at the level of the main pulmonary artery. The pulmonary artery is dilated, measuring 3.5 cm (normal range up to 3.3 cm). (c and d) 3D MR angiogram coronal projection image and coronal perfusion image show normal pulmonary artery branching and perfusion.

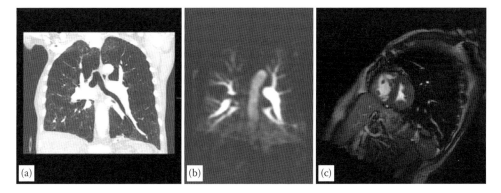

FIGURE 13.5
MR perfusion in a patient with emphysema. (a) Coronal CT of a patient with centrilobular emphysema with a typical upper zone predominance. Image (b) shows matched coronal MR perfusion image illustrating poor perfusion in the areas of lung destruction. Image (c) shows the effect in this patient of emphysema on the pulmonary vasculature with resultant elevated afterload on the right ventricle with RV dilatation and hypertrophy and paradoxical deviation of the interventricular septum.

Another semi-quantitative assessment of pulmonary perfusion in 30 patients with severe emphysema demonstrated high sensitivity (86.7%) and good specificity (80.0%) in detecting perfusion abnormalities by MRI, compared to V–Q scintigraphy. Similarly, lower peak signal intensity in emphysematous regions was observed (Sergiacomi et al. 2003). A larger scale study in 143 patients (80 with COPD) applied dual bolus MR perfusion. The study demonstrated close correlation between quantitative and semi-quantitative MRI parameters ($r = 0.86$). The study reported that MRI-derived perfusion measures correlated with global lung perfusion (cardiac output divided by total lung capacity) and lung diffusing capacity (Hueper et al. 2013).

Similar to emphysema, microvascular abnormalities such as microvascular thrombosis and injury play a role in the pathogenesis of pulmonary fibrosis (Magro et al. 2003), and hence, imaging techniques have been used to map perfusion in these patients. A study reported prolonged mean transit time and time to peak enhancement in patients with combined pulmonary fibrosis and emphysema (Sergiacomi et al. 2010). Compared to standard MR techniques that can not differentiate inflammation from fibrosis (Jacob et al. 2010), a 3-T dynamic MRI with a T_1-weighted sequence before and 1, 3, 5, and 10 min after IV contrast injection demonstrated different enhancement patterns with sensitivity, specificity, and accuracy of 82%, 92%, and 88%, respectively, in differentiating inflammation- and fibrosis-predominant changes. Early enhancement was identified in majority of inflammation predominant lesions and late onset/persistent enhancement in fibrosis-predominant lesions (Yi et al. 2008). Figure 13.6 shows perfusion changes in areas with apparent fibrosis. Figure 13.7 shows an example of a patient with pulmonary hypertension secondary to pulmonary fibrosis.

Dynamic MR perfusion is reported to characterize pulmonary nodules. A sensitivity of 52%–100%, specificities of 17%–100%, and accuracies of 58%–96% are reported by various studies (Hittmair et al. 1995; Gückel et al. 1996; Ohno et al. 2002, 2008b; Kim et al. 2004; Schaefer et al. 2004; Kono et al. 2007; Zou et al. 2008).

Differences in perfusion patterns can be explained by the knowledge of the pathological changes. Previous studies demonstrated abnormal vascular architecture and interstitial spaces in tumors that would ultimately affect tumor's enhancement characteristics (Less et al. 1991). When comparing malignant to benign nodules, there are differences in perfusion, extracellular space volume, and the permeability of capillaries that could be investigated by perfusion imaging (Ohno et al. 2014b).

Compared to normal lungs where the majority of perfusion is from pulmonary arteries, there is a significant contribution from bronchial arteries in pulmonary nodules (Wright 1967; Milne 1976, 1987). This results in reduced wash-in and impeded flow during the pulmonary arterial phase of the imaging (Ohno et al. 2004b). Similarly, compared to normal lung, the wash-out of intravascular contrast is mainly by bronchial veins and not the pulmonary vein branches. Moreover, the normal lymphatic wash-out from the interstitial spaces is not present. All these would result in delayed wash-out and retention of contrast from lung nodules (Littleton et al. 1990).

Inflammatory nodules are mainly supplied by the increase in size and number of bronchial arteries. The pulmonary arterial supply is reduced in most cases due to diffuse thrombosis at the arteriolar level. This results

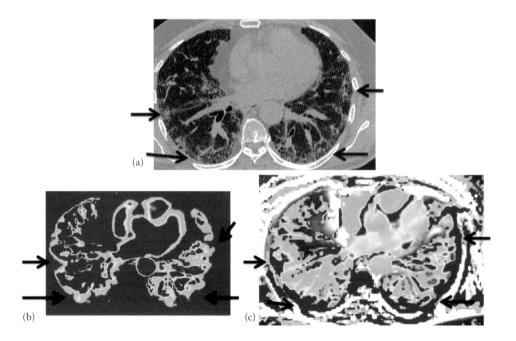

FIGURE 13.6
MR perfusion imaging in a 67-year-old patient with idiopathic pulmonary fibrosis. (a) High-resolution computed tomography of the lung demonstrating sub-pleural interstitial changes (arrow heads) indicative of pulmonary fibrosis. Parametric perfusion maps show reduction in pulmonary blood flow (PBF) (b) and prolonged mean transit time (c) in the sub-pleural region that corresponds to the fibrotic changes. Pixels are color-coded in the parametric maps according to the degree of perfusion (from high to low PBF values: red, yellow, green, blue; from high to low mean transit time values: red, yellow, green).

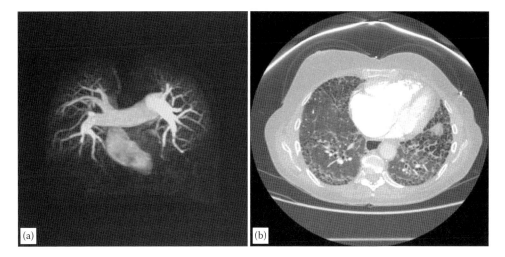

FIGURE 13.7

An elderly patient with connective tissue disease and related fibrosis. There is marked pulmonary artery dilatation and tortuosity of the vasculature on MR angiography (a). Image (b) shows the fibrotic changes ate the lung bases and the resultant effect on the heart with right ventricular and atrial dilatation and RV hypertrophy.

in diminished and delayed blood supply. In comparison, the wash-out of contrast medium from the interstitial space is enhanced by increased convection and diffusion to dilated lymphatic vessels (Ohno et al. 2014b). A strong relationship is also reported between thin peripheral enhancement and benign nodules, which is thought to be an inflammatory response (Schaefer et al. 2004).

13.3.2 Ventilation Imaging in Lung Disorders

The application of pulmonary ventilation MRI in a true routine clinical setting is still in its infancy (Figures 13.8

and 13.9). In spite of the many (mainly small) studies, there are no currently accepted clinical indications for use of the method. Nevertheless, there are some advances that need to be seen as precursors to clinical use (whether routine or part of clinical trials).

Oxygen-enhanced imaging is starting to get used in clinical settings. Clearly, this method is most easy to introduce as no particular hardware or software is required, while it is normally performed at 1.5-T scanners. (Renne et al. 2015b). The technique described in the paragraph above is able to predict the outcome of patients who are to undergo lung resection for lung cancer, in a study in 30 patients who underwent oxygen-enhanced MRI in

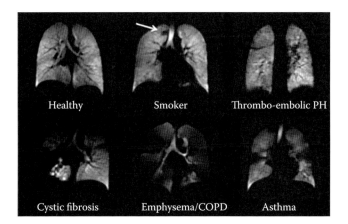

FIGURE 13.8

Hyperpolarized 3-Helium MRI ventilation imaging. Image in a normal volunteer demonstrate homogeneous signal (normal ventilation). In a smoker, a small ventilation defect was found in the right upper lobe (arrow). Slightly inhomogeneous ventilation is demonstrated in a patient with chronic thrombo-embolic pulmonary hypertension. In a patient with cyctic fibrosis, large ventilation defects and reduced ventilation is demonstrated mainly in the right lung. Image of a patient with emphysema shows large segmental ventilation defects. In an asthmatic patient, ventilation defects are noted particularly in the periphery (subsegmental).

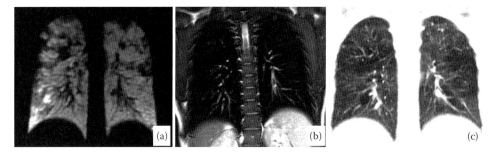

FIGURE 13.9

MRI of cystic fibrosis. Hyperpolarized 3-Helium image demonstrates upper zones dominant ventilation defects (a) with normal proton MRI (b). The CT scan (c) shows corresponding areas of low attenuation consistent with air trapping and some centrilobular nodules due to mucus plugging.

comparison to pre- and postoperative pulmonary function tests (Ohno et al. 2005). There was an excellent correlation between the two methods, suggesting that the method can predict postoperative lung function and also may offer functional and regional information that cannot be obtained by pulmonary function tests.

In another study, 30 patients with asthma, dynamic oxygen-enhanced MRI was compared with quantitative CT measurements and pulmonary function test (Ohno et al. 2014a). The wash-in time of patients with asthma showed a better correlation with pulmonary function tests than quantitative CT measures.

A small study in 9 asthmatics and 4 healthy individuals compared oxygen-enhanced MRI before and after an allergen challenge and demonstrated that it was feasible to visualize and quantify regional allergic reaction effects (Renne et al. 2015a).

Hyperpolarized noble gas imaging is currently merely being used in relatively small-scale clinical trials. Its use is largely limited to centers with the appropriate hardware and knowledge base. In addition, the gas is relatively expensive and not easily available (e.g., 3He) or not simple to use in a routine clinical setting (e.g., 129Xe). Although this method is being used to demonstrate effectiveness of interventions, such as through bronchial hyperresponsiveness challenge/relief tests, and has shown its propensity to perform detailed physiological analysis of small airways morphology to complex ventilation/perfusion interactions, it is unlikely that it will make it into routine clinical practice.

13.3.3 MRI of Pulmonary Hypertension

Pulmonary hypertension is a disease characterized by increased blood pressure within the pulmonary vasculature (Kiely et al. 2013) (Figures 13.7, 13.10 through 13.13). This leads to progressive right ventricular (RV)

failure and ultimately death. Pulmonary arterial hypertension has a poor prognosis if untreated, with a median survival time of just 2.8 years following diagnosis (D'Alonzo et al. 1991). Recent advances in treatment, however, have improved the outlook for patients suffering with the disease. With this, there has been a drive to develop biomarkers that can track the response to therapies and stratify patient risk.

Cardiac MRI is recognized as an increasingly important imaging modality for the assessment and management of pulmonary hypertension (Swift et al. 2014b). MRI has the advantage that it is noninvasive, uses no ionizing radiation, and provides excellent contrast resolution, temporal resolution, and freedom to image in any plane. Furthermore, MRI has sensitivity to blood flow, allowing measurement of blood velocity and flow, and post-contrast imaging allows assessment of the myocardium and lung vasculature and perfusion (Bradlow et al. 2012).

13.3.3.1 Imaging of the Heart

MRI is routinely used to assess cardiac morphology, chamber volumes, myocardial mass, and cardiac function in cardiovascular diseases (Figures 13.11 through 13.13). The advent of late gadolinium imaging has revolutionized cardiac imaging allowing the characterization of myocardial tissue changes in a variety of ischemic and inflammatory pathologies.

Cardiac and respiratory motion, claustrophobia, relatively long scan acquisition times, and specific contraindications to *in vivo* metallic devices limit the application of MRI. Methods to improve scan time include development of parallel imaging techniques (Wieben et al. 2008). In addition, new sequences such as k-t blast (Kozerke et al. 2004) and compressed sensing (Gamper et al. 2008) have lead to reduced acquisition time while maintaining a high signal-to-noise ratio. The solution

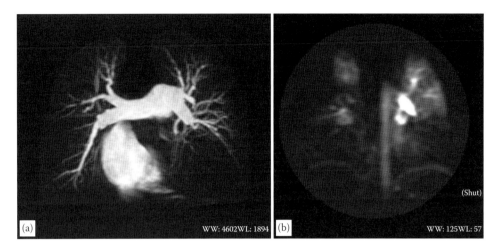

FIGURE 13.10
A patient with a history of previous pulmonary emboli and progressive shortness of breath. There are extensive chronic thromboembolic disease shown on the MR angiogram (a) and a matched pattern of segmental perfusion defects on MR perfusion (b). This patient was found to have chronic thromboembolic pulmonary hypertension (note the size of the proximal pulmonary artery branches).

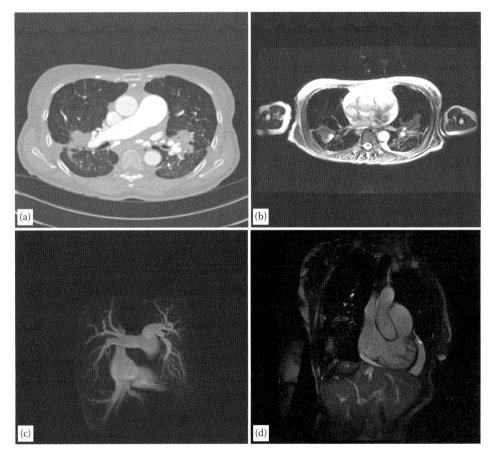

FIGURE 13.11
Pulmonary hypertension in a patient with sarcoidosis. CT (a) and MRI (b) demonstrate interstitial thickening, conglomerate nodules, and lymphadenopathy in a patient with pulmonary sarcoidosis. Angiography (c) demonstrated dilated pulmonary artery and tortuous branches as a result of pulmonary hypertension. Coronal steady-state free precession imaging (d) shows dilated right atrium and hypertrophied ventricle with pericardial effusion.

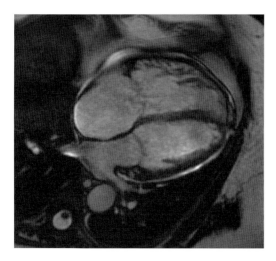

FIGURE 13.12
Four-chamber CINE cardiac MR image in a patient with pulmonary arterial hypertension. The image displays poor prognostic features, RV dilatation, and a small pericardial effusion.

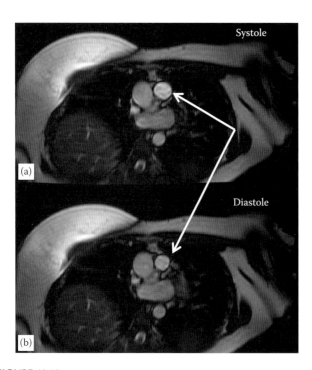

FIGURE 13.13
CINE MRI orthogonal to the main pulmonary artery shows good pulsatility of the pulmonary artery with change in caliber of the vessel, increasing in size in systole. Pulsatility of the pulmonary artery is a surrogate marker of compliance of the pulmonary vasculature and is a good marker of long-term outcome. (a) Systole and (b) diastole.

to cardiac motion is electrocardiogram (ECG) gating, through simultaneous acquisition of ECG often with imaging data acquired over repeated heartbeats, in some circumstance real-time imaging is advantageous. Further challenges relate to implanted metal devices in terms of patient safety and impact on image quality;

however, the majority of implanted metallic devices are now produced with materials that are MRI compatible.

13.3.3.1.1 Right Ventricle

Cardiac MRI is a highly reproducible method for quantifying ventricular volumes, mass, and function (Grothues et al. 2002; Grothues et al. 2004) (Figures 13.11 and 13.12). Balanced steady-state free precession MRI is considered the gold standard method for ventricular volumetry, achieved by multi-slice image acquisition through the heart. Typical measurements of each chamber include the end-diastolic volume, which is the maximal chamber volume prior to contraction timed to the R wave at ECG, and the end-systolic volume or minimal chamber volume on each short axis slice.

Multiple MRI parameters are of potential benefit in the work-up of patients with suspected pulmonary hypertension. MR ventricular mass index, defined as right ventricular mass divided by left ventricular mass, closely relates to pulmonary arterial pressure (Saba et al. 2002; Hagger et al. 2009) and is a useful diagnostic marker. In addition, the presence of late gadolinium enhancement at the inter-ventricular hinge points and pulmonary artery flow artifact seen on black blood images help to predict the presence of pulmonary hypertension in unselected patients with suspected pulmonary vascular disease.

Inter-ventricular septal position can be assessed by its angle or by the measurement of its curvature (Roeleveld et al. 2005; Dellegrottaglie et al. 2007); septal angle is quick and easy to measure and is highly accurate as a diagnostic marker. In addition, the added value of combining septal angle with ventricular mass index has been demonstrated. The composite index of septal angle and ventricular mass index improved the correlation with mPAP and strengthened the diagnostic accuracy.

Furthermore, pulmonary vascular resistance can be estimated noninvasively with MRI with the addition of surrogate markers of pulmonary capillary wedge pressure and cardiac output substituting left atrial volume and phase contrast MRI, respectively, indicating that all major hemodynamic measurements can be estimated noninvasively with MRI.

Measurement of the RV at baseline allows for an assessment as to the level of compensatory dilatation relative to the increase in RV afterload. Progressive dilatation due to failure of RV adaptation of increasing RV afterload leads to right ventricular failure and death (van Wolferen et al. 2007). A single-center prognostic study in 64 patients with idiopathic pulmonary arterial hypertension found that indexed right ventricular end-diastolic volume (RVEDV) >84 mL/m^2 is associated with poor survival ($P = 0.011$). In addition, at 1-year follow-up, change in indexed RVEDV was independently associated with prognosis (van Wolferen et al. 2007). In a cohort of 110 patients with pulmonary arterial hypertension, at baseline, RV ejection fraction was found to independently confer a poor prognosis, and a change in RV ejection fraction at follow-up was independently related to survival hazard ratio of 0.93 (95% CI 0.88–0.99). RV end-diastolic and end-systolic volumes are predictors of adverse outcome in patients with pulmonary arterial hypertension at baseline and at follow-up. Of the two volume states, RV end-systolic volume has been shown to be of proportionately greater prognostic value, likely because it represents both dilatation of the chamber and low systolic function (Swift et al. 2014a).

13.3.3.1.2 Left Ventricle

Stroke volume has been suggested as a more important prognostic factor in pulmonary hypertension than cardiac output, as patients can maintain cardiac output by an increase in hazard ratio when stroke volume can no longer be augmented (van Wolferen et al. 2007). Holverda et al. (2006) studied a very small cohort of 10 patients with idiopathic pulmonary arterial hypertension in functional class III. These patients were compared to 10 age- and gender-matched controls. Patients with pulmonary hypertension were unable to augment (stroke volume) in response to exertion in order to produce an adequate cardiac output. Left ventricular (stroke volume) has been shown to be a consistent indicator of poor prognosis in several studies (van Wolferen et al. 2007; Swift et al. 2014a).

Low LV end-diastolic volume has been associated with a worse prognosis in pulmonary hypertension; this has been hypothesized to be due the effect of RV volume overload, compression of the LV, and poor LV filling. Underfilling of the LV leads to poor LV stroke volume, which is also a strong indicator of adverse outcome in patients with pulmonary arterial hypertension. Furthermore, changes in LV end-diastolic volume have been found to be markers of treatment response in pulmonary hypertension.

13.3.3.2 Imaging of Pulmonary Vasculature

Previous imaging studies have focused on the clinical utility of pulmonary arterial size in patients with suspected pulmonary hypertension (Devaraj and Hansell 2009; Devaraj et al. 2010) (Figure 13.13). Pulmonary artery size is of value in the screening for pulmonary hypertension, however is limited in the assessment of disease severity. This is because the degree of pulmonary arterial dilatation is related to the duration of the condition rather than directly related to mean pulmonary artery pressure (mPAP) (Boerrigter et al. 2010).

Loss of pulmonary artery relative area change (RAC) is a marker of mild pulmonary vascular disease, explained by an inverse relationship with pulmonary vascular resistance, with small increases in pulmonary vascular resistance leading to larger proportional reductions in RAC. This corroborates a previous study (Sanz et al. 2009) that proposed RAC as a marker sensitive to the detection of preclinical disease by identifying significantly reduced RAC in patients with exercise-induced pulmonary hypertension. In patients with mild pulmonary vascular disease, RAC is reduced, whereas RV volume and function were not significantly altered, in keeping with the theory that the pulmonary artery becomes stiff prior to the loss of RV function.

Chronic thromboembolic pulmonary hypertension (CTEPH) is a serious complication of pulmonary thromboembolic disease and is associated with considerable morbidity and mortality (Riedel et al. 1982). It is estimated that up to 3.8% of patients suffering a symptomatic acute pulmonary embolism will develop CTEPH at 2 years (Pengo et al. 2004), and even among those who receive appropriate treatment during their acute episode, the thrombus may not resolve completely, resulting in CTEPH (Remy-Jardin et al. 1997; Ribeiro et al. 1999; Wartski and Collignon 2000).

The diagnosis of CTEPH is principally made by imaging. Traditionally, invasive pulmonary angiography has been considered the definitive investigation for the diagnosis of chronic thromboembolism, especially in the context of surgical planning (Daily et al. 1980; Jamieson et al. 1993). With the advent of multi-detector CT scanners, CTPA has replaced pulmonary angiography (Kauczor et al. 1994; Remy-Jardin et al. 1992; Wittram et al. 2006).

High spatial resolution MR angiography is capable of visualizing the central parts of the pulmonary vasculature. The breath hold for high spatial resolution pulmonary angiography can be as long as 25 s, which is too long for patients with marked dyspnea as in pulmonary hypertension. Parallel imaging techniques allow spatial resolution to be increased, and whole-lung volume coverage can be achieved with a shorter breath hold.

Time-resolved MRA techniques are available, which can easily differentiate patients with idiopathic pulmonary arterial hypertension from patients with CTEPH, allowing visualization of the pulmonary arteries down to the segmental level. Furthermore, the addition of 3D MR perfusion imaging can aid the characteristic pattern of loss of perfusion. Non-ionizing 3D MR lung perfusion imaging has high diagnostic accuracy, comparable to that of perfusion scintigraphy, in patients with suspected CTEPH, with a normal MR lung perfusion study excluding operable CTEPH (Rajaram et al. 2013).

13.3.3.3 Imaging of Pulmonary Flow Characteristics

Phase contrast is a powerful method for estimation of blood flow with MRI (Lotz et al. 2002; Gatehouse et al. 2005). The basic principle underpinning phase contrast

MRI is that proton spins moving in the direction of a magnetic field gradient undergo phase shift that is proportional to velocity. Hemodynamic measurements such as forward flow, retrograde flow, average velocity, and peak velocity can be determined.

Cardiac output has been identified in several studies as a key independent marker of adverse outcome in patients with PAH (D'Alonzo et al. 1991; Sandoval et al. 1994; McLaughlin et al. 2002; Sitbon et al. 2002; Humbert et al. 2010). There are several noninvasive MRI methodologies that are well suited to the evaluation of cardiac output and stroke volume. MR volumetry can accurately and reproducibly evaluate the change in volume of the RV and LV chambers in healthy and diseased states (Heusch et al. 1999; Benza et al. 2008; Mooij et al. 2008; Bradlow et al. 2010; Marrone et al. 2010), and phase contrast imaging is a robust method of evaluating pulmonary arterial and aortic blood flow (Kondo et al. 1991; Gatehouse et al. 2005; Ley et al. 2007; Sanz et al. 2007; Mauritz et al. 2008). Phase contrast MRI-derived flow measurements have also been shown to correlate with invasive measurements of pressure and resistance. For example, Sanz et al. have demonstrated pulmonary pressures are negatively correlated with the average velocity of blood flow in the main pulmonary artery (Sanz et al. 2007), and pulmonary vascular resistance has been estimated by calculating the ratio of the maximal change in flow rate during ejection by the acceleration volume (Mousseaux et al. 1999). Moreover, a recent study has identified that early retrograde flow is a characteristic feature in patients with pulmonary hypertension (Helderman et al. 2011).

References

Aksoy FG, Lev MH. Dynamic contrast-enhanced brain perfusion imaging: Technique and clinical applications. *Semin Ultrasound CT MR*. 2000 Dec;21(6):462–477.

Allmendinger AM, Tang ER, Lui YW et al. Imaging of stroke: Part 1, Perfusion CT—Overview of imaging technique, interpretation pearls, and common pitfalls. *AJR Am J Roentgenol*. 2012;198:1, 52–62.

Amundsen T, Torheim G, Kvistad KA, Waage A, Bjermer L, Nordlid KK, Johnsen H, Asberg A, Haraldseth O. Perfusion abnormalities in pulmonary embolism studied with perfusion MRI and ventilation-perfusion scintigraphy: An intra-modality and inter-modality agreement study. *J Magn Reson Imaging*. Apr 2002;15(4):386–394.

Barberà JA, Riverola A, Roca J et al. Pulmonary vascular abnormalities and ventilation-perfusion relationships in mild chronic obstructive pulmonary disease. *Am J Respir Crit Care Med*. 1994;149(2 Pt 1):423–429.

Barr RG, Mesia-Vela S, Austin JH et al. Impaired flow-mediated dilation is associated with low pulmonary function and emphysema in ex-smokers: The Emphysema and Cancer Action Project (EMCAP) Study. *Am J Respir Crit Care Med* 2007;176:1200–1207.

Bauman G, Lützen U, Ullrich M et al. Pulmonary functional imaging: Qualitative comparison of Fourier decomposition MR imaging with SPECT/CT in porcine lung. *Radiology*. 2011;260(2):551–559.

Becker J, Heil W, Krug B et al. Study of mechanical compression of spin-polarized 3He gas. *Nucl Instrum Meth A* 1994;346:45–51.

Benza R, Biederman R, Murali S, Gupta H. Role of cardiac magnetic resonance imaging in the management of patients with pulmonary arterial hypertension. *J Am Coll Cardiol*. 2008;52:1683–1692.

Boerrigter B, Mauritz GJ, Marcus JT, Helderman F, Postmus PE, Westerhof N, Vonk-Noordegraaf A. Progressive dilatation of the main pulmonary artery is a characteristic of pulmonary arterial hypertension and is not related to changes in pressure. *Chest*. 2010;138(6):1395–1401.

Bradlow WM, Gibbs JS, Mohiaddin RH. Cardiovascular magnetic resonance in pulmonary hypertension. *J Cardiovasc Magn Reson*. 2012;14:6.

Bradlow WM, Hughes ML, Keenan NG, Bucciarelli-Ducci C, Assomull R, Gibbs JS, Mohiaddin RH. Measuring the heart in pulmonary arterial hypertension (pah): Implications for trial study size. *J Magn Reson Imaging*. 2010;31:117–124.

Cluzel P, Similowski T, Chartrand-Lefebvre C et al. Diaphragm and chest wall: Assessment of the inspiratory pump with MR imaging preliminary observations. *Radiology*. 2000;215: 574–583.

Couch MJ, Ball IK, Li T et al. Pulmonary ultrashort echo time 19F MR imaging with inhaled fluorinated gas mixtures in healthy volunteers: Feasibility. *Radiology*. 2013;269:903–909.

D'Alonzo GE, Barst RJ, Ayres SM et al. Survival in patients with primary pulmonary hypertension. Results from a national prospective registry. *Ann Inter Med*. 1991;115:343–349.

Daily PO, Johnston GG, Simmons CJ, Moser KM. Surgical management of chronic pulmonary embolism: Surgical treatment and late results. *J Thorac Cardiovasc Surg*. 1980;79:523–531.

De Lange EE, Mugler III JP, Brookeman JR et al. Lung air spaces: MR imaging evaluation with hyperpolarized 3He gas. *Radiology*. 1999;210:851–857.

Dellegrottaglie S, Sanz J, Poon M, Viles-Gonzalez JF, Sulica R, Goyenechea M, Macaluso F, Fuster V, Rajagopalan S. Pulmonary hypertension: Accuracy of detection with left ventricular septal-to-free wall curvature ratio measured at cardiac MR. *Radiology*. 2007;243:63–69.

Devaraj A, Hansell DM. Computed tomography signs of pulmonary hypertension: Old and new observations. *Clin Radiol*. 2009;64:751–760.

Devaraj A, Wells AU, Meister MG, Corte TJ, Wort SJ, Hansell DM. Detection of pulmonary hypertension with multidetector ct and echocardiography alone and in combination. *Radiology*. 2010;254:609–616.

Ersoy H, Goldhaber SZ, Cai T et al. Time-resolved MR angiography: A primary screening examination of patients with suspected pulmonary embolism and contraindications to administration of iodinated contrast material. *AJR Am J Roentgenol.* 2007;188:1246–1254.

Fichele S, Woodhouse N, Swift AJ et al. MRI of Helium-3 gas in healthy lungs: Posture related variation of alveolar size. *J MRI.* 2004;20:331–335.

Gamper U, Boesiger P, Kozerke S. Compressed sensing in dynamic MRI. *Magn Reson Med.* 2008;59:365–373.

Gatehouse PD, Keegan J, Crowe LA, Masood S, Mohiaddin RH, Kreitner KF, Firmin DN. Applications of phase-contrast flow and velocity imaging in cardiovascular MRI. *Eur Radiol.* 2005;15:2172–2184.

Grothues F, Moon JC, Bellenger NG, Smith GS, Klein HU, Pennell DJ. Interstudy reproducibility of right ventricular volumes, function, and mass with cardiovascular magnetic resonance. *Am Heart J.* 2004;147:218–223.

Grothues F, Smith GC, Moon JC, Bellenger NG, Collins P, Klein HU, Pennell DJ. Comparison of interstudy reproducibility of cardiovascular magnetic resonance with two-dimensional echocardiography in normal subjects and in patients with heart failure or left ventricular hypertrophy. *Am J Cardiol.* 2002;90:29–34.

Gückel C, Schnabel K, Deimling M, Steinbrich W. Solitary pulmonary nodules: MR evaluation of enhancement patterns with contrast-enhanced dynamic snapshot gradient-echo imaging. *Radiology.* 1996;200:681–686.

Hagger D, Condliffe R, Woodhouse N, Elliot CA, Armstrong IJ, Davies C, Hill C, Akil M, Wild JM, Kiely DG. Ventricular mass index correlates with pulmonary artery pressure and predicts survival in suspected systemic sclerosis-associated pulmonary arterial hypertension. *Rheumatology.* 2009;48:1137–1142.

Hamedani H, Kadlecek SJ, Ishii M et al. Alterations of regional alveolar oxygen tension in asymptomatic current smokers: Assessment with hyperpolarized 3He MR imaging. *Radiology.* 2015;274:585–596.

Harris AD, Coutts SB, Frayne R. Diffusion and perfusion MR imaging of acute ischemic stroke. *Magn Reson Imaging Clin N Am.* 2009;17(2):291–313.

Helderman F, Mauritz GJ, Andringa KE, Vonk-Noordegraaf A, Marcus JT. Early onset of retrograde flow in the main pulmonary artery is a characteristic of pulmonary arterial hypertension. *J Magn Reson Imaging.* 2011;33:1362–1368.

Hersman FW, Ruset IC, Ketel S et al. Large production system for hyperpolarized 129Xe for human lung imaging studies. *Acad Radiol.* 2008;15:683–692.

Heusch A, Koch JA, Krogmann ON, Korbmacher B, Bourgeois M. Volumetric analysis of the right and left ventricle in a porcine heart model: Comparison of three-dimensional echocardiography, magnetic resonance imaging and angiocardiography. *Eur J Ultrasound.* 1999;9:245–255.

Hittmair K, Eckersberger F, Klepetko W, Helbich T, Herold CJ. Evaluation of solitary pulmonary nodules with dynamic contrast-enhanced MR imaging: A promising technique. *Magn Reson Imaging.* 1995;13:923–933.

Hueper K, Parikh MA, Prince MR et al. Quantitative and semiquantitative measures of regional pulmonary microvascular perfusion by magnetic resonance imaging and their relationships to global lung perfusion and lung diffusing capacity: The multiethnic study of atherosclerosis chronic obstructive pulmonary disease study. *Invest Radiol.* 2013 Apr;48(4):223–230.

Humbert M, Sitbon O, Chaouat A et al. Survival in patients with idiopathic, familial, and anorexigen-associated pulmonary arterial hypertension in the modern management era. *Circulation.* 2010;122:156–163.

Ingrisch M, Dietrich O, Attenberger UI, Nikolaou K, Sourbron S, Reiser MF, Fink C (2010) Quantitative pulmonary perfusion magnetic resonance imaging: Influence of temporal resolution and signal-to-noise ratio. *Invest Radiol.* 45:7–14.

Ireland RH, Din O, Swinscoe JA et al. Detection of radiation-induced lung injury in non-small cell lung cancer patients using hyperpolarized helium-3 magnetic resonance imaging. *Radiother Oncol* 2010;97: 244–248.

Jacob RE, Amidan BG, Soelberg J, Minard KR. In vivo MRI of altered proton signal intensity and T2 relaxation in a bleomycin model of pulmonary inflammation and fibrosis. *J Magn Reson Imaging.* 2010;31(5):1091–1099.

Jamieson SW, Auger WR, Fedullo PF, Channick RN, Kriett JM, Tarazi RY, Moser KM. Experience and results with 150 pulmonary thromboendarterectomy operations over a 29-month period. *J Thorac Cardiovasc Surg.* 1993;106:116–126; discussion 126–117.

Kanazawa H, Asai K, Hirata K et al. Possible effects of vascular endothelial growth factor in the pathogenesis of chronic obstructive pulmonary disease. *Am J Med.* 2003;114: 354–358.

Kauczor HU, Schwickert HC, Mayer E, Schweden F, Schild HH, Thelen M. Spiral ct of bronchial arteries in chronic thromboembolism. *J Comput Assist Tomogr.* 1994;18:855–861.

Kiely DG, Elliot CA, Sabroe I, Condliffe R. Pulmonary hypertension: Diagnosis and management. *BMJ.* 2013;346:f2028.

Kim JH, Kim HJ, Lee KH, Kim KH, Lee HL. Solitary pulmonary nodules: A comparative study evaluated with contrast-enhanced dynamic MR imaging and CT. *J Comput Assist Tomogr.* 2004;28:766–775.

Kirby M, Heydarian M, Wheatley A et al. Evaluating bronchodilator effects in chronic obstructive pulmonary disease using diffusion-weighted hyperpolarized helium-3 magnetic resonance imaging. *J Appl Physiol (1985).* 2012;112:651–657.

Kirby M, Mathew L, Heydarian M et al. Chronic obstructive pulmonary disease: Quantification of bronchodilator effects using hyperpolarized 3He MR imaging. *Radiology.* 2011;261:283–292.

Kirby M, Pike D, Coxson HO et al. Hyperpolarized 3He ventilation defects used to predict pulmonary exacerbations in mild to moderate chronic obstructive pulmonary disease. *Radiology.* 2014;273:887–896.

Kjørstad Å, Corteville DM, Fischer A, Henzler T, Schmid-Bindert G, Zöllner FG, Schad LR. Quantitative lung perfusion evaluation using Fourier decomposition perfusion MRI. *Magn Reson Med.* 2014 Aug;72(2):558–562.

Kondo C, Caputo GR, Semelka R, Foster E, Shimakawa A, Higgins CB. Right and left ventricular stroke volume measurements with velocity-encoded cine MR imaging: In vitro and in vivo validation. *AJR*. 1991;157:9–16.

Kono R, Fujimoto K, Terasaki H et al. Dynamic MRI of solitary pulmonary nodules: Comparison of enhancement patterns of malignant and benign small peripheral lung lesions. *AJR*. 2007;188:26–36.

Kozerke S, Tsao J, Razavi R, Boesiger P. Accelerating cardiac cine 3d imaging using k-t blast. *Magn Reson Med*. 2004;52:19–26.

Less JR, Skalak TC, Sevick EM, Jain RK. Microvascular architecture in a mammary carcinoma: Branching patterns and vessel dimensions. *Cancer Res*. 1991;51:265–273.

Ley S, Mereles D, Puderbach M, Gruenig E, Schock H, Eichinger M, Ley-Zaporozhan J, Fink C, Kauczor HU. Value of MR phase-contrast flow measurements for functional assessment of pulmonary arterial hypertension. *Eur Radiol*. 2007;17:1892–1897.

Ley S, Ley-Zaporozhan J. Pulmonary perfusion imaging using MRI: Clinical application. *Insights Imaging*. 2012;3(1):61–71.

Littleton JT, Durizch ML, Moeller G, Herbert DE. Pulmonary masses: Contrast enhancement. *Radiology*. 1990;177:861–871

Lotz J, Meier C, Leppert A, Galanski M. Cardiovascular flow measurement with phase-contrast MR imaging: Basic facts and implementation. *RadioGraphics*. 2002;22:651–671.

Magro CM, Allen J, Pope-Harman A et al. The role of microvascular injury in the evolution of idiopathic pulmonary fibrosis. *Am J Clin Pathol*. Apr 2003;119(4):556–567.

Marrone G, Mamone G, Luca A, Vitulo P, Bertani A, Pilato M, Gridelli B. The role of 1.5t cardiac MRI in the diagnosis, prognosis and management of pulmonary arterial hypertension. *Int J Cardiovasc Imaging*. 2010;26:665–681.

Marshall H, Kiely DG, Parra-Robles J et al. Magnetic resonance imaging of ventilation and perfusion changes in response to pulmonary endarterectomy in chronic thromboembolic pulmonary hypertension. *Am J Resp Crit Care Med*. 2014;190:e18–e19.

Martirosian P1, Boss A, Schraml C, Schwenzer NF, Graf H, Claussen CD, Schick F. Magnetic resonance perfusion imaging without contrast media. *Eur J Nucl Med Mol Imaging*. 2010 Aug;37(Suppl 1):S52–S64.

Mauritz GJ, Marcus JT, Boonstra A, Postmus PE, Westerhof N, Vonk-Noordegraaf A. Non-invasive stroke volume assessment in patients with pulmonary arterial hypertension: Left-sided data mandatory. *J Cardiovasc Magn Reson*. 2008;10:51.

McLaughlin VV, Shillington A, Rich S. Survival in primary pulmonary hypertension: The impact of epoprostenol therapy. *Circulation*. 2002;106:1477–1482.

Miles KA, Griffiths MR, Fuentes MA. Standardized perfusion value: Universal CT contrast enhancement scale that correlates with FDG PET in lung nodules. *Radiology*. 2001;220:548–53.

Milne EN. Circulation of primary and metastatic pulmonary neoplasms: a postmortem micro-arteriographic study. *AJR*. 1967;100:603–619.

Milne EN. Pulmonary metastases: vascular supply and diagnosis. *Int J Radiat Oncol Biol Phys*. 1976;1:739–742.

Milne EN, Zerhouni EA. Blood supply of pulmonary metastases. *J Thorac Imaging*. 1987;2:15–23.

Mirsadraee S, van Beek EJR. Functional imaging: CT and MRI. *Clin Chest Med*. 2015;36(2):349–363.

Mooij CF, de Wit CJ, Graham DA, Powell AJ, Geva T. Reproducibility of MRI measurements of right ventricular size and function in patients with normal and dilated ventricles. *J Magn Reson Imaging*. 2008;28:67–73.

Morrison NJ, Abboud RT, Müller NL et al. Pulmonary capillary blood volume in emphysema. *Am Rev Respir Dis*. 1990;141(1):53–61.

Mousseaux E, Tasu JP, Jolivet O, Simonneau G, Bittoun J, Gaux JC. Pulmonary arterial resistance: Noninvasive measurement with indexes of pulmonary flow estimated at velocity-encoded MR imaging—preliminary experience. *Radiology*. 1999;212:896–902.

Mugler JP 3rd, Altes TA, Ruset IC et al. Simultaneous magnetic resonance imaging of ventilation distribution and gas uptake in the human lung using hyperpolarized xenon-129. *Proc Natl Acad Sci USA*. 2010;107:21707–21712.

Muradyan I, Butler JP, Dabaghyan M et al. Single-breath xenon polarization transfer contrast (SB-XTC): Implementation and initial results in healthy humans. *J MRI*. 2013;37:457–470.

Ohno Y, Hatabu H, Higashino T et al. Oxygen-enhanced MR imaging: Correlation with postsurgical lung function in patients with lung cancer. *Radiology*. 2005;236:704–711.

Ohno Y, Hatabu H, Murase K et al. Quantitative assessment of regional pulmonary perfusion in the entire lung using three-dimensional ultrafast dynamic contrast-enhanced magnetic resonance imaging: Preliminary experience in 40 subjects. *J Magn Reson Imaging*. 2004a;20:353–365.

Ohno Y, Hatabu H, Takenaka D et al. Dynamic MR imaging: Value of differentiating subtypes of peripheral small adenocarcinoma of the lung. *Eur J Radiol*. 2004b;52:144–150.

Ohno Y, Hatabu H, Takenaka D, Adachi S, Kono M, Sugimura K. Solitary pulmonary nodules: Potential role of dynamic MR imaging in management—Initial experience. *Radiology*. 2002;224:503–511.

Ohno Y, Iwasawa T, Seo JB et al. Oxygen-enhanced magnetic resonance imaging versus computed tomography: Multicentre study for clinical stage classification of smoking-related chronic obstructive pulmonary disease. *Am J Respir Crit Care Med*. 2008a;177:1095–1102.

Ohno Y, Koyama H, Matsumoto K et al. Dynamic MR perfusion imaging: Capability for quantitative assessment of disease extent and prediction of outcome for patients with acute pulmonary thromboembolism. *J Magn Reson Imaging*. May 2010;31(5):1081–1090.

Ohno Y, Koyama H, Takenaka D et al. Dynamic MRI, dynamic multidetector-row computed tomography (MDCT), and coregistered 2-[fluorine-18]-fluoro-2-deoxy-d-glucose-positron emission tomography (FDG-PET)/CT: Comparative study of capability for management of pulmonary nodules. *J Magn Reson Imaging* 2008b;27:1284–1295.

Ohno Y, Nishio M, Koyama H et al. Asthma: Comparison of dynamic oxygen-enhanced MR imaging and quantitative thin-section CT for evaluation of clinical treatment. *Radiology*. 2014a;273:907–916.

Ohno Y, Nishio M, Koyama H, Miura S, Yoshikawa T, Matsumoto S, Sugimura K. Dynamic contrast-enhanced CT and MRI for pulmonary nodule assessment. *AJR Am J Roentgenol*. Mar 2014b;202(3):515–529.

Pansini V, Remy-Jardin M, Faivre JB et al. Assessment of lobar perfusion in smokers according to the presence and severity of emphysema: Preliminary experience with dual-energy CT angiography. *Eur Radiol*. Dec 2009;19(12):2834–2843.

Patz S, Hersman PW, Muradyan I et al. Hyperpolarized (129) Xe MRI: A viable functional lung imaging modality? *Eur J Radiol* 2007;64:335–344.

Pena E, Dennie C, Veinot J et al. Pulmonary hypertension: How the radiologist can help. *RadioGraphics*. 2012;32:9–32.

Pengo V, Lensing AW, Prins MH et al. Incidence of chronic thromboembolic pulmonary hypertension after pulmonary embolism. *N Engl J Med*. 2004;350:2257–2264.

Puderbach M, Risse F, Biederer J, Ley-Zaporozhan J, Ley S, Szabo G, Semmler W, Kauczor HU. In vivo Gd-DTPA concentration for MR lung perfusion measurements: Assessment with computed tomography in a porcine model. *Eur Radiol*. 2008 Oct;18(10):2102–2107.

Rajaram S, Swift AJ, Telfer A et al. 3d contrast-enhanced lung perfusion MRI is an effective screening tool for chronic thromboembolic pulmonary hypertension: Results from the aspire registry. *Thorax*. 2013;68(7):677–678.

Remy-Jardin M, Louvegny S, Remy J, Artaud D, Deschildre F, Bauchart JJ, Thery C, Duhamel A. Acute central thromboembolic disease: Posttherapeutic follow-up with spiral ct angiography. *Radiology*. 1997;203:173–180.

Remy-Jardin M, Remy J, Wattinne L, Giraud F. Central pulmonary thromboembolism: Diagnosis with spiral volumetric CT with the single-breath-hold technique–comparison with pulmonary angiography. *Radiology*. 1992;185:381–387.

Renne J, Hinrichs J, Schonfeld C et al. Noninvasive quantification of airway inflammation following segmental allergen challenge with functional MR imaging: A proof of concept study. *Radiology*. 2015a;274:267–275.

Renne J, Lauermann P, Hinrichs J et al. Clinical use of oxygen-enhanced T1 mapping MRI of the lung: Reproducibility and impact of closed versus loose fit oxygen delivery system. *J MRI*. 2015b;41:60–66.

Ribeiro A, Lindmarker P, Johnsson H, Juhlin-Dannfelt A, Jorfeldt L. Pulmonary embolism: One-year follow-up with echocardiography Doppler and five-year survival analysis. *Circulation*. 1999;99:1325–1330.

Riedel M, Stanek V, Widimsky J, Prerovsky I. Longterm follow-up of patients with pulmonary thromboembolism. Late prognosis and evolution of hemodynamic and respiratory data. *Chest*. 1982;81:151–158.

Risse F, Kuder TA, Kauczor HU, Semmler W, Fink C. Suppression of pulmonary vasculature in lung perfusion MRI using correlation analysis. *Eur Radiol*. 2009;19:2569–2575.

Risse F, Semmler W, Kauczor HU, Fink C. Dual-bolus approach to quantitative measurement of pulmonary perfusion by contrast-enhanced MRI. *J Magn Reson Imaging*. 2006 Dec;24(6):1284–1290.

Rizi RR, Lipson DA, Dimitrov IE, Ishii M, Roberts DA. Operating characteristics of hyperpolarized 3He and arterial spin tagging in MR imaging of ventilation and perfusion in healthy subjects. *Acad Radiol*. 2003 May;10(5):502–508.

Roeleveld RJ, Marcus JT, Faes TJ, Gan TJ, Boonstra A, Postmus PE, Vonk-Noordegraaf A. Interventricular septal configuration at MR imaging and pulmonary arterial pressure in pulmonary hypertension. *Radiology*. 2005;234:710–717.

Saam BT, Yablonskiy DA, Kodibagkar VD et al. MR imaging of diffusion of 3He gas in healthy and diseased lungs. *Magn Reson Med*. 2000;44:174–179.

Saba TS, Foster J, Cockburn M, Cowan M, Peacock AJ. Ventricular mass index using magnetic resonance imaging accurately estimates pulmonary artery pressure. *Eur Respir J*. 2002;20:1519–1524.

Salerno M, Altes TA., Brookeman JR, de Lange EE, Mugler JP 3rd. Dynamic spiral MR imaging of pulmonary gas flow using hyperpolarized 3He: Preliminary studies in healthy and diseased lungs. *Magn Reson Med*. 2001;46:667–677.

Salerno M, de Lange EE, Altes TA, Truwit JD, Brookemann JR, Mugler III JP. Emphysema: Hyperpolarized helium 3 diffusion MR imaging of the lungs compared with spirometric indexes-initial experience. *Radiology*. 2002;222:252–260.

Samee S, Altes T, Powers P et al. Imaging the lungs in asthmatic patients by using hyperpolarized helium-3 magnetic resonance: Assessment of response to methocholine and exercise challenge. *J Allergy Clin Immunol*. 2003;111:1205–1211.

Sandoval J, Bauerle O, Palomar A, Gomez A, Martinez-Guerra ML, Beltran M, Guerrero ML. Survival in primary pulmonary hypertension. Validation of a prognostic equation. *Circulation*. 1994;89:1733–1744.

Sanz J, Kariisa M, Dellegrottaglie S, Prat-Gonzalez S, Garcia MJ, Fuster V, Rajagopalan S. Evaluation of pulmonary artery stiffness in pulmonary hypertension with cardiac magnetic resonance. *JACC Cardiovasc Imaging*. 2009;2:286–295.

Sanz J, Kuschnir P, Rius T, Salguero R, Sulica R, Einstein AJ, Dellegrottaglie S, Fuster V, Rajagopalan S, Poon M. Pulmonary arterial hypertension: Noninvasive detection with phase-contrast MR imaging. *Radiology*. 2007;243:70–79.

Schaefer JF, Vollmar J, Schick F et al. Solitary pulmonary nodules: Dynamic contrast-enhanced MR imaging–perfusion differences in malignant and benign lesions. *Radiology*. 2004;232:544–553.

Sebastian L, Julia L-Z. Pulmonary perfusion imaging using MRI: Clinical application. *Insights Imaging*. Feb 2012;3(1):61–71.

Sergiacomi G, Bolacchi F, Cadioli M et al. Combined pulmonary fibrosis and emphysema: 3D time-resolved MR angiographic evaluation of pulmonary arterial mean transit time and time to peak enhancement. *Radiology*. Feb 2010;254(2):601–608.

Sergiacomi G, Sodani G, Fabiano S, Manenti G, Spinelli A, Konda D, Di Roma M, Schillaci O, Simonetti G. MRI lung perfusion 2D dynamic breath-hold technique in patients with severe emphysema. *In Vivo*. Jul–Aug 2003;17(4):319–324.

Sitbon O, Humbert M, Nunes H, Parent F, Garcia G, Herve P, Rainisio M, Simonneau G. Long-term intravenous epoprostenol infusion in primary pulmonary hypertension: Prognostic factors and survival. *J Am Coll Cardiol*. 2002;40:780–788.

Swift AJ, Rajaram S, Campbell MJ, Hurdman J, Thomas S, Capener D, Elliot C, Condliffe R, Wild JM, Kiely DG. Prognostic value of cardiovascular magnetic resonance imaging measurements corrected for age and sex in idiopathic pulmonary arterial hypertension. *Circ Cardiovas Imaging*. 2014a;7:100–106.

Swift AJ, Wild JM, Nagle SK et al. Quantitative magnetic resonance imaging of pulmonary hypertension: A practical approach to the current state of the art. *J Thorac Imaging*. 2014b;29:68–79.

Swift AJ, Woodhouse N, Fichele S, Siedel J, Mills GH, van Beek EJR, Wild JM. Rapid lung volumetry using ultrafast dynamic magnetic resonance imaging during forced vital capacity maneuver. Correlation with spirometry. *Invest Radiol*. 2007;42:37–41.

Triphan SMF, Breuer FA, Gensler D, Kauczor HU, Jakob PM. Oxygen enhanced lung MRI by simultaneous measurement of T_1 and T_2^* during free breathing using ultrashort TE. *J MRI*. Jul 7 2015;41(6):1708–1714.

Tsai IC, Tsai WL, Wang KY et al. Comprehensive MDCT evaluation of patients with pulmonary hypertension: Diagnosing underlying causes with the updated Dana Point 2008 classification. *Amer J Roentgenol*. 2011;197:W471–W481.

Van Beek EJR, Dahmen AM, Stavngaard T et al. Comparison of hyperpolarised 3-He MRI and HRCT in normal volunteers, patients with COPD and patients with alpha-1-antitrypsin deficiency—PHIL trial. *Eur Resp J*. 2009;34:1–11.

Van Beek EJR, Wild JM, Kauczor HU, Schreiber W, Mugler JP 3rd, De Lange EE. Functional MRI of the lung using hyperpolarized 3-Helium gas. *J MRI*. 2004;20:540–554.

van Wolferen SA, Marcus JT, Boonstra A, Marques KM, Bronzwaer JG, Spreeuwenberg MD, Postmus PE, Vonk-Noordegraaf A. Prognostic value of right ventricular mass, volume, and function in idiopathic pulmonary arterial hypertension. *Eur Heart J*. 2007;28:1250–1257.

Voorhees A, An J, Berger KI et al. Magnetic resonance imaging-based spirometry for regional assessment of pulmonary function. *Magn Reson Med*. 2005;54:1146–1154.

Wartski M, Collignon MA. Incomplete recovery of lung perfusion after 3 months in patients with acute pulmonary embolism treated with antithrombotic agents. Thesee study group. Tinzaparin ou heparin standard: Evaluation dans l'embolie pulmonaire study. *J Nucl Med*. 2000;41:1043–1048.

Wieben O, Francois C, Reeder SB. Cardiac MRI of ischemic heart disease at 3 t: Potential and challenges. *Eur J Radiol*. 2008;65:15–28.

Wielpütz M, Kauczor HU. MRI of the lung: State of the art. *Diagn Interv Radiol*. 2012 Jul–Aug;18(4):344–353.

Wild JM, Fichele S, Woodhouse N, Paley MNJ, Kasuboski L, van Beek EJR. 3D Volume-localized pO2 measurement in the human lung with 3He MRI. *Magn Reson Med*. 2005;53:1055–1064.

Wild JM, Marshall H, Bock M, Schad LR, Jakob PM, Puderbach M, Molinari F, Van Beek EJ, Biederer J. MRI of the lung (1/3): Methods. *Insights Imaging*. 2012 Aug;3(4):345–353.

Wild JM, Paley MNJ, Kasuboski L et al. Dynamic radial projection MRI of inhaled hyperpolarized 3He. *Magn Reson Med*. 2003;49:991–997.

Wittram C, Kalra MK, Maher MM, Greenfield A, McLoud TC, Shepard JA. Acute and chronic pulmonary emboli: Angiography-CT correlation. *AJR Am J Roentgenol*. 2006;186:S421–S429.

Woodhouse N, Wild JM, van Beek EJR et al. Hyperpolarized 3He-MRI for the evaluation of CF therapies. *J MRI* 2009;30:981–988.

Wright RD. The blood supply of abnormal tissues in the lungs. *J Pathol Bacteriol*. 1938;47:489–499.

Yi CA, Lee KS, Han J, Chung MP, Chung MJ, Shin KM. 3-T MRI for differentiating inflammation- and fibrosis-predominant lesions of usual and nonspecific interstitial pneumonia: Comparison study with pathologic correlation. *AJR Am J Roentgenol*. 2008;190(4):878–885.

Zou Y, Zhang M, Wang Q, Shang D, Wang L, Yu G. Quantitative investigation of solitary pulmonary nodules: Dynamic contrast-enhanced MRI and histopathologic analysis. *AJR*. 2008;191:252–259.

14

Breast MRI

Diana L. Lam and Habib Rahbar

CONTENTS

14.1 Introduction

MRI was first proposed for use as an imaging tool for cancer detection in the early 1970s [1]. Subsequently, it was discovered that normal and abnormal breast tissue can be differentiated using tissue variations in longitudinal (T_1) and transverse (T_2) relaxation times in vitro [2]. In the early 1980s, MRI was performed successfully on ex vivo mastectomy specimens to identify breast cancers, followed by the first reported in vivo study that identified characteristics between normal and abnormal breast tissue [3,4]. It was not until it was determined that breast cancers were reliably visible only by administering gadolinium-based contrast material that breast MRI blossomed into a dependable clinical tool for in vivo detection of breast malignancies [5]. Once the optimal clinical indications for breast MRI were determined, a standardized reporting system (i.e., the American College of Radiology Breast Imaging Reporting and Data System, or BI-RADS) was created, and more uniform technical approaches across sites were established, breast MRI became widely recognized to be the most sensitive imaging tool for the detection of breast cancer and the most accurate imaging modality for assessment of local extent of disease [6].

14.2 Indications for Breast MRI

There are six primary indications for breast MRI, and data to support each are summarized below.

14.2.1 Screening

Currently, the American Cancer Society recommends the use of MRI to supplement annual mammographic breast cancer screening for women who have a >20% lifetime risk of breast cancer, which includes women with a strong family history of breast or ovarian cancer, personal history of radiation therapy to the chest between 10 and 30 years of age, BRCA mutation carriers (or first-degree relatives who are untested), and women affected with a genetic syndrome (or first-degree relatives who are untested) that portends a high lifetime risk of breast cancer (Table 14.1) [7]. While breast MRI's sensitivity for breast cancer detection is unrivaled by any other current imaging modality (Figure 14.1), its specificity remains modest; as a result, the risk of additional false-positive abnormalities and associated patient anxiety along with the high cost of MRI relative to mammography has prevented application of breast MRI to the general screening population. However, there is growing evidence to support the use of MRI in the intermediate risk population (lifetime risk of 15%–20%), including patients with a personal history of treated breast cancer or diagnosis of high-risk pathologies, such as lobular neoplasia, that are associated with an elevated lifetime risk of developing breast cancer [8–11].

TABLE 14.1

Indications for Annual Breast MRI Based on the American Cancer Society 2007 Guidelines

Indication for Annual Breast MRI Screening	Insufficient Evidence for Annual Breast MRI Screening
• BRCA mutation (BRCA1 or BRCA2)[a] • First-degree relative of BRCA carrier, but untested[a] • Lifetime risk >20%, as defined by BRCAPRO or other models that are largely dependent on family history[a] • Radiation to the chest between 10 and 30 years old • Li–Fraumeni syndrome and first-degree relatives • Cowden syndrome and Bannayan–Riley–Ruvalcaba syndrome and first-degree relatives	• Lifetime risk <15%–20%, as defined by BRCAPRO or other models that are largely dependent on family history • Lobular carcinoma in situ (LCIS) or atypical lobular hyperplasia (ALH) • Atypical ductal hyperplasia (ADH) • Heterogeneously or extremely dense breast on mammography • Women with a personal history of breast cancer, including ductal carcinoma in situ (DCIS)

[a] Based on evidence from nonrandomized clinical trials and observational studies.

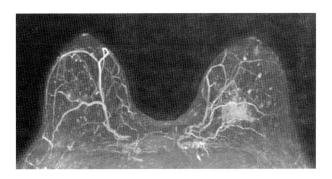

FIGURE 14.1
Mammographically occult cancer. A 48-year-old woman with >20% lifetime risk of breast cancer and personal history of right excisional biopsy showing lobular carcinoma in situ. Axial maximum intensity projection (post-contrast) demonstrates an irregular mass with spiculated margins within the left breast, upper inner quadrant. Biopsy showed invasive lobular carcinoma.

14.2.2 Evaluation of Women Newly Diagnosed with Breast Cancer

Current National Comprehensive Cancer Network (NCCN) practice guidelines recommend that breast MRI be considered in patients with newly diagnosed breast cancer to evaluate the extent of newly diagnosed cancer in the affected breast and to screen the contralateral breast [12]. There are multiple potential advantages to the routine use of breast MRI in this clinical setting. First, MRI provides the most accurate imaging assessment of disease extent, which allows clinicians to more confidently preoperatively assess for the presence or absence of multifocal (more than one area of malignancy within a breast quadrant) or multicentric (multiple quadrants) disease in the ipsilateral breast (Figures 14.2 and 14.3) [6]. In a review of published

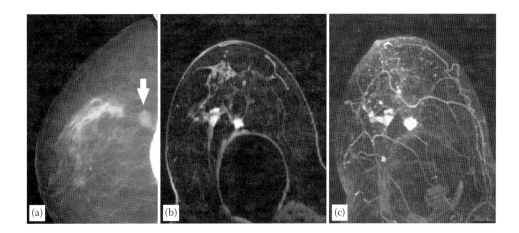

FIGURE 14.2
Multifocal disease. A 61-year-old woman with known invasive ductal carcinoma. MRI was performed for extent of disease workup. Implant displaced mammogram (a) shows an irregular mass with spiculated margins within the posterior breast (arrow). Post-contrast axial T_1-weighted fat-saturated (b) and maximum intensity projection (c) image demonstrates multiple similar appearing irregular masses that occupy one quadrant. Biopsy revealed invasive ductal carcinoma compatible with multifocal disease.

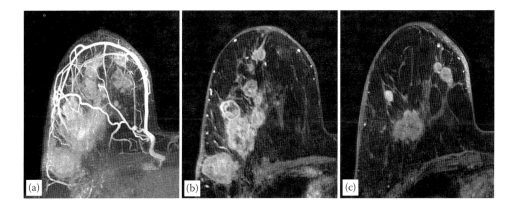

FIGURE 14.3
Multicentric disease. A 61-year-old woman's recent diagnosis of invasive ductal carcinoma. MRI for extent of disease workup demonstrates NME in the area of known cancer (not shown). Axial maximum intensity projection image (a) and post-contrast axial T_1-weighted images different levels (b and c) demonstrate multiple additional masses in different quadrants compatible with multicentric disease.

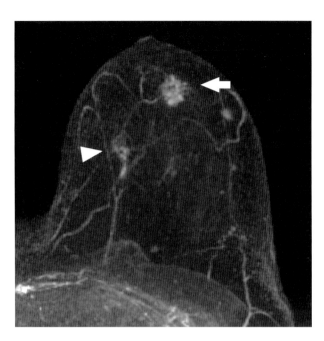

FIGURE 14.4
Mammographically occult ipsilateral disease. A 52-year-old woman with a mammographically occult, clinically palpable left breast invasive ductal carcinoma. Axial maximum intensity projection shows the biopsy-proven invasive ductal carcinoma at the 12 o'clock position (arrow). Additional focal area of NME posteromedial to the known cancer was biopsy-proven low-grade DICS (arrowhead).

studies, approximately 10%–34% of patients who had a preoperative breast MRI for extent of disease had additional ipsilateral disease that was clinically and mammographically occult (Figure 14.4) [13]. Furthermore, breast MRI can also help evaluate for the presence of direct extension of disease to the adjacent structures, including the pectoralis musculature, chest wall (including the ribs and intercostal muscles), and overlying skin (Figures 14.5 and 14.6). Breast MRI can also assess for the presence of axillary and intramammary lymphadenopathy suspicious for lymph node metastasis, which can facilitate preoperative ultrasound-guided sampling of worrisome nodes (Figure 14.7) [14]. Finally, breast MRI is useful for detecting clinically and mammographically occult contralateral breast cancers in women newly diagnosed with a breast malignancy. A large multicenter trial demonstrated that MRI identified synchronous contralateral breast cancers that were otherwise clinically and mammographically occult in 3% of women newly diagnosed with breast cancer (Figure 14.8) [12]. Although large trials have not yet been performed to evaluate the long-term impact of breast MRI on patients who have breast conservation surgery, such as positive margins, recurrence rates, and mortality [13], it has been shown that women with early breast cancer who undergo breast conservation therapy have a lower reexcision rate if they had a preoperative breast MRI than those who did not [15].

14.2.3 Problem Solving for Equivocal Clinical or Imaging Findings

The use of MRI to further evaluate equivocally suspicious clinical or imaging findings remains controversial at this time. This breast MRI application, sometimes referred to as "problem solving," is not currently well supported by the literature [13]. Barriers to cost-effective implementation of MRI to reduce unnecessary biopsies prompted by mammography include the relatively low cost associated with image-guided biopsies, which are highly accurate and safe, compared with serial imaging [16], in addition to the high negative predictive value (~98%) needed to obviate the biopsy of a suspicious finding on the basis of BI-RADS recommendations [13]. Recent studies have demonstrated that the negative predictive value of MRI for this clinical indication ranges from 76% for further evaluation of suspicious mammographic calcifications [17] to 85% in the setting of any suspicious mammographic or clinical finding [18]. Studies from our institution found that while breast MRI had high sensitivity and high negative predictive value for this clinical indication, the added breast cancer yield was too low to routinely recommend breast MRI for problem solving [19]. Interestingly, two-thirds of cancers that were identified in this cohort were in the setting of suspicious nipple discharge with negative mammographic and sonographic evaluation (Figure 14.9), and the use of MRI for this specific clinical scenario is actively being investigated [20].

14.2.4 Evaluation of Patients with Metastatic Axillary Adenocarcinoma with Unknown Primary

Although the clinical presentation of metastatic adenocarcinoma to the axillary lymph nodes without an identified primary is rare, these metastases occur frequently from ipsilateral breast cancer in women (Figure 14.10). A meta-analysis of all available studies shows that breast MRI can identify the mammographically and clinically occult primary malignancy presenting with metastatic axillary lymphadenopathy in approximately 61% of patients [13].

14.2.5 Evaluation of Breast Cancer Response to Neoadjuvant Chemotherapy

In the National Surgical Adjuvant Breast and Bowel Project B-18 (NSABP B-18), comparison was made between pre- (neoadjuvant) and postoperative (adjuvant) chemotherapy in patients with operable breast cancer. No statistical significance was found in either overall or disease-free survival between the two groups; however, more patients in the neoadjuvant chemotherapy group

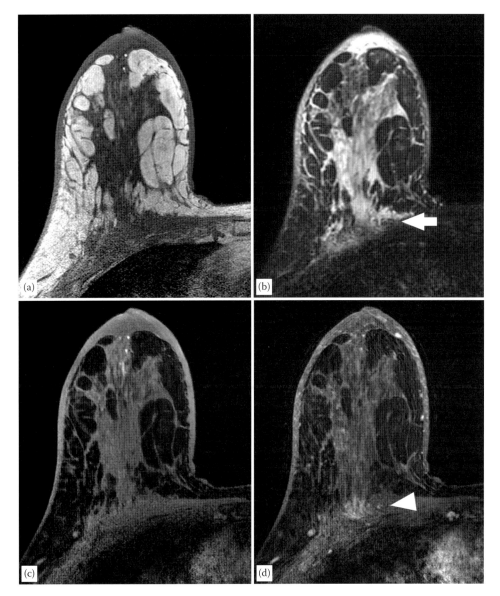

FIGURE 14.5
Pectoralis muscle invasion. A 64-year-old woman with personal history of treated right-sided breast cancer presenting with a recurrent breast cancer invading the pectoralis muscle. Pre-contrast axial T_1-weighted (a), T_2-weighted SPAIR (b), T_1-weighted fat-saturated (c), and post-contrast T_1-weighted fat-saturated (d) images show enhancement that involves the pectoralis muscle (arrow), consistent with pectoralis muscle invasion. Note the diffuse skin thickening compatible with previous radiation treatment.

were able to successfully undergo breast-conserving surgery as opposed to mastectomy. Furthermore, primary tumors' responses to chemotherapy were evaluated clinically prior to surgery, and response (as determined by change in size on clinical examination) to neoadjuvant chemotherapy correlated with patient outcomes. This led way to interest for examining the use of MRI to improve both assessment of early response to therapy and posttherapeutic assessment of extent of disease prior to surgery [21,22]. Initial studies examining the use of MRI for early treatment response found that MRI-detected changes in size or volume and enhancement

kinetic profiles were associated with response to therapy [23,24], suggesting that MRI could facilitate changes in neoadjuvant chemotherapy regimens in women whose tumors were responding poorly (Figures 14.11 and 14.12). Other single-center studies also demonstrated that MRI was superior to clinical breast examination, mammography, and ultrasound to assess for residual disease upon completion of neoadjuvant chemotherapy [25,26], but could not be relied on to confirm the complete absence of residual disease [25]. Recent results from the multisite trial sponsored by the American College of Radiology Imaging Network (ACRIN 6657) confirmed that MRI

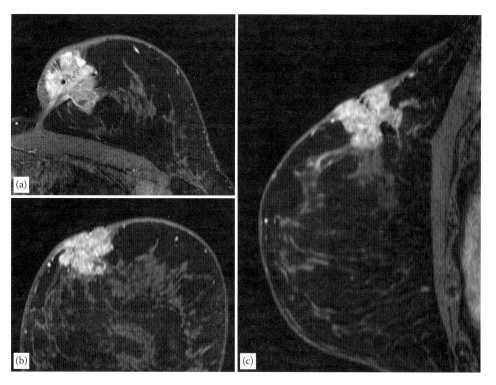

FIGURE 14.6
Skin involvement. A 31-year-old woman with newly diagnosed recurrent left breast invasive ductal carcinoma. MRI was performed for extent of disease. Post-contrast axial (a), coronal (b), and sagittal (c) T_1-weighted fat-saturated images demonstrate an irregular mass with spiculated margins invading the adjacent skin at 11 o'clock at the site of previous lumpectomy.

findings were a stronger predictor of pathologic response to neoadjuvant chemotherapy than clinical assessment [27]. Additional ACRIN-sponsored investigations are ongoing, studying whether advanced MRI techniques, such as diffusion-weighted imaging (DWI) and MR spectroscopy, can further improve MRI's ability to assess for early response to neoadjuvant therapies.

14.2.6 Evaluation of Silicone Implant Integrity

Breast augmentation has been the most commonly performed cosmetic surgical procedure in the United States annually, since 2006. Implants may be composed of a single or double lumen that contains saline, silicone, or a combination of both. Implant complications are common, with over 50% of patients found to have a silicone implant rupture within 12 years after placement [28]. Saline implant complications are usually apparent on clinical examination with rapid loss of breast volume and saline, which is typically reabsorbed by the body in 5–14 days [29]; as a result, imaging to evaluate integrity of saline implants is usually not required. However, the evaluation of implants containing silicone for complications is more challenging on clinical examination.

Although mammography alone can provide findings of silicone implant failure, particularly when higher density silicone material can be seen outside of the fibrous capsule and inside the breast parenchyma, mammography has a low sensitivity for detecting silicone implant rupture (range: 11%–69%) [13], particularly in the setting of intracapsular ruptures. MRI has been shown to be the most sensitive (range 46%–98%) imaging study to detect silicone implant complications, with good to excellent specificity (range: 55%–92%) [30–34], and is the preferred modality for detection of silicone implant ruptures that are not apparent on clinical examination.

14.3 Technique Considerations

Although breast MRI protocols can vary somewhat between institutions, high-quality MRIs share several basic technical principles. These include adequate breast and axillary coverage, sufficient spatial and temporal resolution to allow lesion characterization, and an approach that prevents and minimizes artifacts.

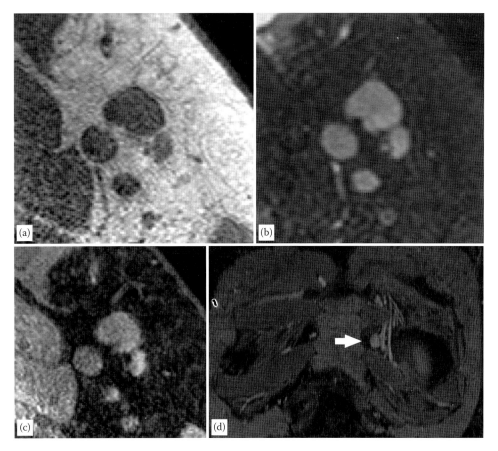

FIGURE 14.7
Abnormal lymph node identification. A 27-year-old woman with recently diagnosed invasive ductal carcinoma. MRI was performed for extent of disease. Pre-contrast axial T_1-weighted (a), T_2-weighted SPAIR (b), and post-contrast T_1-weighted fat-saturated (c) images show morphologically abnormal level I axillary lymph nodes, which are rounded in appearance with loss of the fatty hilum. Post-contrast coronal T_1-weighted fat-saturated image (d) demonstrates an abnormally enlarged internal mammary lymph node (arrow), also suspicious for metastatic involvement.

14.3.1 Patient Positioning and Comfort

Breast MRIs should be performed with patients prone so that their breasts are positioned pendently within the breast coil [35]. This results in less cardiac and respiratory motion, and also serves to extend the breast for optimal breast tissue coverage. It is also essential that scan times be kept as short as possible and that clear patient instructions are given prior to the dynamic-contrast-enhanced (DCE) sequences in order to limit significant interscan patient motion.

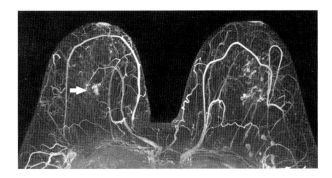

FIGURE 14.8
Synchronous contralateral cancer. A 67-year-old woman with recently diagnosed right breast cancer. MRI was performed for extent of disease. The axial maximum intensity projection image shows an irregular-shaped mass with spiculated margins in the right breast (arrow). There is asymmetric stippled NME in the left breast, biopsy-proven DCIS.

14.3.2 Bilateral Imaging

Bilateral imaging is recommended for breast MRI for nearly every clinical indication because it provides important technical and clinical advantages. First, by employing a bilateral technique, one can minimize

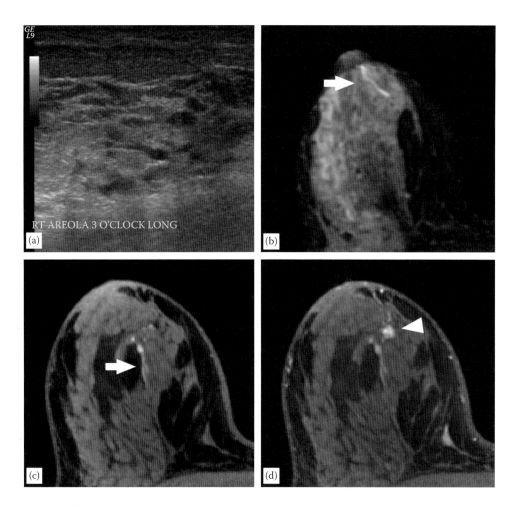

FIGURE 14.9

Evaluation of suspicious nipple discharge. A 46-year-old woman with spontaneous bloody right nipple discharge. Mammogram (not shown) and targeted ultrasound (a) were negative. A bilateral breast MRI was performed for further evaluation. Axial T_2-weighted SPAIR (b) and T_1-weighted fat-saturated (c) images prior to contrast administration demonstrated dilated ducts with a high signal (arrows) consistent with fluid and debris within the ducts. Post-contrast T_1-weighted fat-saturated (d) images revealed a round intraductal mass with circumscribed margins and heterogeneous internal enhancement (arrowhead). Pathology revealed an intraductal papilloma with usual ductal hyperplasia.

the potential for wrap-around artifacts in plane of the phase-encoding gradient, which is typically applied in the left–right direction to minimize the effects of cardiac motion [36]. Second, bilateral imaging allows assessment of both breasts and axillae, which is particularly important for women undergoing MRI for high-risk screening or for extent of disease for a newly diagnosed breast cancer [37]. Furthermore, much like conventional mammography, imaging both breasts allows for assessment of symmetry, which can improve diagnostic accuracy [38]. For example, symmetry is useful for establishing patterns and degree of background parenchymal enhancement (BPE, a physiologic and benign phenomenon described in more detail below) and for identifying areas of unique, suspicious enhancement distinct from BPE. Finally, bilateral imaging can aid in the assessment of multiple, bilateral, benign-appearing masses, such as intramammary lymph nodes and fibroadenomas.

14.3.3 Coils

It is imperative that breast MRI be performed using only dedicated breast surface coils because the built-in body coils of MRI systems provide an insufficient signal to optimally image the breasts. It is most ideal to use breast coils with a high number of coil elements (currently ranging from 7 to 32 simultaneous RF channels) so that parallel imaging can be used. Parallel imaging is particularly efficient for breast imaging [39] and can facilitate reduced scan times or higher spatial resolution.

14.3.4 Contrast Agent

Although some noncontrast techniques, such as DWI and MR spectroscopy, are being investigated for breast cancer detection, clinical breast MRI currently requires the administration of a gadolinium contrast agent for breast cancer detection or assessment. Typically, the gadolinium chelate delivered intravenously at a dose of 0.1 mmol/kg followed by a 20 mL saline flush using a power injector, all at a rate of approximately 2 mL/s, which ensures that contrast quickly reaches the intravascular space with consistent contrast-enhancement timing across examinations. However, it should be noted that fluid-sensitive imaging, such as T_2-weighted series, should be acquired prior to the administration of contrast because injection of chelated gadolinium results in decreased T_2 and T_2^* relaxation times [40].

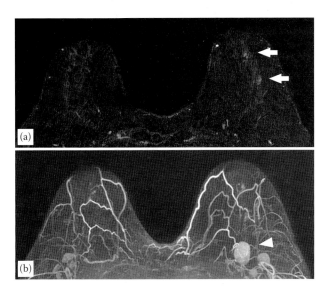

FIGURE 14.10
Axillary nodal metastatic carcinoma with unknown primary malignancy. A 63-year-old woman with axillary nodal metastatic carcinoma with unknown breast primary malignancy. MRI demonstrates discontinuous NME in the upper outer quadrant of the left breast (arrows) on the axial post-contrast subtraction image (a). Axial maximum intensity projection image demonstrates multiple enlarged left axillary lymph nodes (arrowhead) (b). Pathology revealed invasive lobular carcinoma.

14.3.5 Spatial and Temporal Resolution

High spatial and temporal resolution scans are essential to facilitate identification and characterization of lesions on breast MRI. However, these two factors have technological demands that must be balanced—achieving high spatial resolution requires a longer imaging time, and thus a relative trade-off in temporal resolution. As a result, an optimally balanced breast MRI requires understanding of the technical demands and clinical value of these intertwined acquisition factors.

In the authors' opinion, for most clinical purposes, breast MRIs should emphasize spatial resolution over temporal resolution. This is due to the fact that high spatial resolution allows greater anatomic detail, enabling the interpreting radiologist to assess lesion morphology characteristics as well as potential disease involvement of the nipple–areola complex, skin, and chest wall. In general, morphological features have been shown to have greater diagnostic value than kinetic enhancement features [18], which are more precisely defined with scans that emphasize temporal resolution. In order to observe some of the most specific MRI morphologic details, such as the presence of spiculations (typically malignant) or dark internal septations (typically benign), a spatial resolution of at ≤1 mm in plane pixel size should be achieved.

Temporal resolution, or the speed with which dynamic post-contrast scans are obtained, dictates the imaging sampling points that provide information on kinetic enhancement patterns over time. Typically, invasive breast cancers demonstrate early-phase rapid enhancement with subsequent delayed-phase washout. Prior studies have demonstrated that this initial rapid enhancement occurs within 60–120 s after injection [41]. Although breast cancers may exhibit initial rapid and

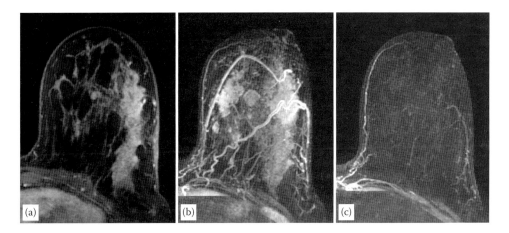

FIGURE 14.11
Tumor response to neoadjuvant chemotherapy. A 39-year-old woman with extensive biopsy-proven IDC. The post-contrast axial T_1-weighted fat-saturated image demonstrates NME predominantly along the lateral breast (a), and the axial maximum intensity projection image demonstrates multicentric disease (b). The patient received neoadjuvant chemotherapy with complete resolution of the multiple masses and NME on the axial maximum intensity projection image (c), compatible with favorable response to neoadjuvant chemotherapy.

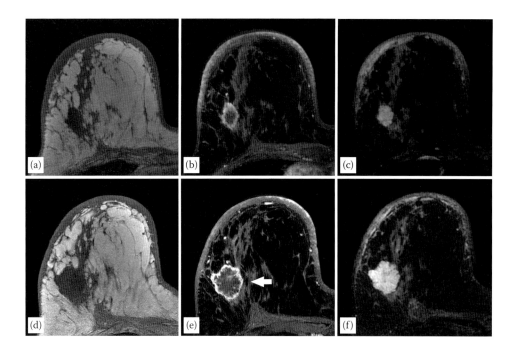

FIGURE 14.12
Tumor progression on neoadjuvant chemotherapy. A 51-year-old woman with invasive ductal carcinoma and inflammatory breast cancer currently undergoing neoadjuvant chemotherapy. Breast MRI was obtained before (top row) and after (bottom row) neoadjuvant chemotherapy to evaluate for treatment response. Axial T_1-weighted (a and d), post-contrast T_1-weighted fat-saturated (b and e), and T_2-weighted SPAIR (c and f) images show increase in size of an irregular mass with irregular margins, rim enhancement (arrow), and central necrosis as noted by the increased signal intensity on the T_2-weighted images. There is associated diffuse skin thickening and enhancement consistent with clinical presentation of inflammatory breast cancer.

delayed washout enhancement more frequently than benign breast lesions do, there remains substantial overlap in the kinetics of malignant and nonmalignant lesions of the breast [42]. While high temporal resolution techniques have shown promise to provide highly quantitative pharmacokinetic information, which may prove to have high value for lesion characterization and monitoring of treatment response [43,44], this area requires additional research prior to routine implementation in clinical practice. Recently, hybrid techniques have been described that employ novel k-space sampling methods, which allow the acquisition of very high temporal resolution images prior to and immediately following a high spatial resolution series.

14.3.6 Key Pulse Sequences and Reformatted Images

Essential acquired sequences for breast MRIs include a T_1-weighted DCE series with pre-contrast, initial post-contrast, and at least two delayed post-contrast acquisitions, as well as a pre-contrast fluid-sensitive series. If the DCE series is performed with active fat suppression, it is recommended that a pre-contrast T_1-weighted sequence without fat suppression be also obtained, which can be useful for confirming the presence of fat in benign lesions such as fat-necrosis and intramammary lymph nodes. Gradient-recalled echo (GRE) sequences

should be used for the T_1-weighted portions of the examination in order to achieve high-quality imaging at acceptable speeds. The GRE pulse sequence should be spoiled in order to avoid any T_2-like weighting effects or streaking artifacts [45]. Because GRE sequences are not capable of providing true T_2-weighted images [45], pre-contrast fluid-sensitive sequences require the use of spin-echo, fast spin-echo, or short tau inversion recovery (STIR) techniques.

Useful reconstructed and reformatted images include the axial subtraction series (pre-contrast DCE sequence subtracted from initial post-contrast DCE sequence), maximum intensity projection (MIP) images from the subtraction series, and orthogonal plane (sagittal and coronal) reformatted images from the initial post-contrast series. Subtraction images are particularly important in protocols that do not employ fat saturation, as described below.

14.3.7 Fat Suppression: Saturation and Subtraction

Because enhancing lesions within the breast can be obscured by a high signal related to fat within the breast, it is important to employ a fat-suppression technique that results in homogeneous fat suppression free from major artifacts. This can be achieved during image acquisition through fat saturation techniques or through postprocessing by utilizing subtraction techniques to exclude fat (and other nonenhancing signal on T_1-weighting) [36].

Fat saturation, also known as active fat suppression, is typically achieved either by applying additional RF pulses to cancel out the signal from fat or through selective water excitation [46]. The major drawback of fat-saturation techniques when compared with subtraction techniques is that it incurs an acquisition time penalty, and thus, it is often favored in techniques that emphasize spatial resolution over temporal resolution. Furthermore, a homogeneous magnetic field (B_0) over the entire field of view is required for successful fat suppression to be obtained (Figure 14.13) [46]. Because there is greater separation of the spectral peaks of fat and water at higher field strengths, fat-suppression techniques may be more effective at high B_0 strengths such as 3T [47].

Subtraction techniques subtract data from the post-contrast T_1-weighted acquisition from the pre-contrast

T_1-weighted acquisition to remove nonenhancing signals from fat and other structures with intrinsic high signals on T_1-weighting. However, the major drawback of relying solely on such an approach for fat suppression is that even slight interscan patient motion can create artifacts and that can simulate suspicious enhancement. In particular, this issue is even more problematic when spatial resolution is emphasized, and thus, it is generally favored in protocols that prioritize temporal resolution.

It is the authors' preference to combine both fat-saturated and subtraction techniques, which allows the interpreting radiologist to quickly evaluate for areas of enhancement on the subtraction images by eliminating the potentially confounding issue of superimposed high pre-contrast T_1 signal on the post-contrast images while also allowing reference to the source fat saturated DCE images in cases of significant "pseudoenhancement" on the subtraction series created by interscan patient motion (Figure 14.14).

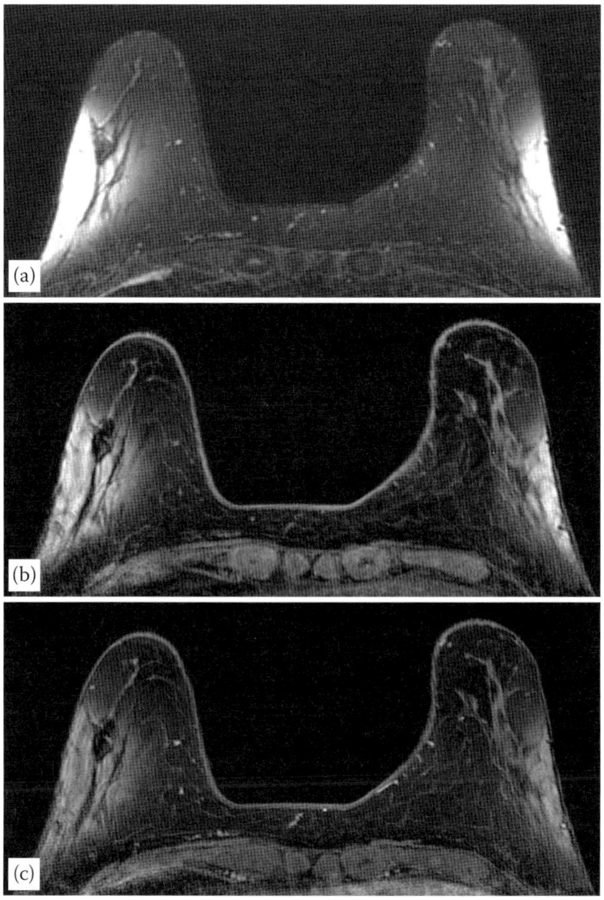

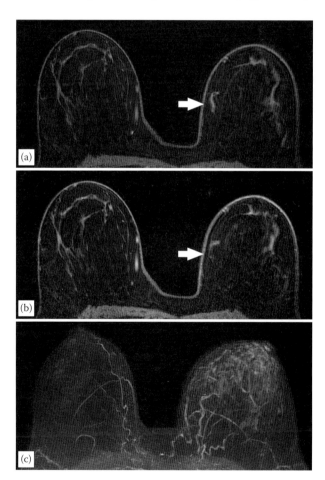

FIGURE 14.13
Incomplete fat saturation. A 35-year-old woman. MRI obtained for high-risk screening. Axial T_2-weighted SPAIR (a), pre- (b), and post-contrast (c) T_1-weighted fat-saturation images demonstrate poor fat saturation along the lateral aspects of both breasts due to magnetic field (B_0) inhomogeneity.

FIGURE 14.14
Interscan motion causing pseudoenhancement on subtraction images. A 58-year-old woman; MRI obtained for high-risk screening. There is significant motion within the left between the pre- (a) and post-contrast (b) axial T_1-weighted fat-saturated images as can be seen by the differences in fibroglandular tissue when comparing images at the same slice level (arrows). This causes the illusion of asymmetric BPE between the right and left breasts seen on the subtraction maximum intensity projection images (c).

14.3.8 Magnetic Strength (B_0) and Homogeneity

The signal-to-noise ratio (SNR) of breast MRI needs to be maximized in order to balance spatiotemporal resolution demands. Theoretically, SNR is directly proportional to B_0 strength, and thus, lower magnetic strengths (typically those less than 1–1.5 T) are not recommended for breast imaging [46]. Higher field strengths also improve B_0 homogeneity across both breasts, which can allow more consistent fat saturation [35,36]. Inhomogeneous magnetic fields can also cause signal dropout such as at the air–soft tissue interfaces (Figure 14.15). Breast MRI is increasingly being used at 3 T, offering a higher SNR, which can be translated into higher spatiotemporal resolution and improved fat suppression. However, it should be noted that, to date, there are few published data supporting a direct clinical benefit of performing breast MRI at 3 T over 1.5 T, with only one published study demonstrating improved diagnostic accuracy with 3 T technique compared with 1.5 T [48].

14.3.9 Phase-Encoding Gradient

Motion artifacts can be problematic for breast imaging due to the relatively close position of the heart and lungs. As a result, the direction of the phase-encoding gradient should be selected to minimize the effects of motion artifacts. For bilateral axial imaging, the phase-encoding gradient should be in the left–right plane to prevent the propagation of cardiac and respiratory motion artifacts into the breasts. Similarly, when performing coronal or sagittal acquisitions, the phase-encoding gradient should be in the superior–inferior direction.

14.3.10 Technical Considerations for Silicone Implant Evaluation

A variety of noncontrast sequences can be used for evaluation of silicone implant integrity. The most useful sequence in the authors' experience is the use of a silicone-only inversion recovery technique where both fat and water are dark, which allows for highly sensitive assessment for intra- and extracapsular rupture. As with other breast MRI sequences, the study is performed with the patient in the prone position, and is best performed at high field strengths (at least 1 T) since intracapsular silicone ruptures may have only subtle MRI findings.

14.4 Common Artifacts

There are a number of common MRI artifacts, some of which may mimic pathology, that one must be aware of when reading breast MRIs. A detailed discussion of

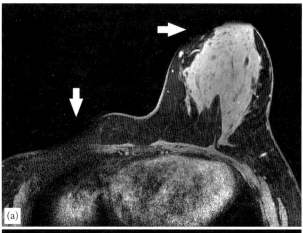

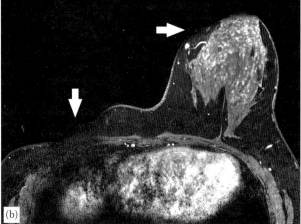

FIGURE 14.15
Inhomogeneous magnetic field. A 33-year-old woman with a personal history of right breast DCIS status postmastectomy. MRI obtained for high-risk screening. Pre- (a) and post-contrast (b) axial T_1-weighted fat-saturated images show loss of signal near the right anterior chest wall and left anterior breast (arrows), due to artifacts from an inhomogeneous magnetic field (B_0).

artifacts is beyond the scope of this chapter; however, as discussed above, it is important to recognize such artifacts, particularly those related to patient motion, inhomogeneous fat suppression, magnetic susceptibility, and ghosting (Figures 14.15 through 14.20).

14.5 Normal Breast Anatomy on MRI

The adult female breast is composed of three primary structures: skin, subcutaneous tissue, and breast tissue [49]. The upper outer quadrant of the breast contains the greatest portion of breast tissue, which is further composed of functional elements (parenchyma) comprised of glandular tissue and milk ducts and supporting elements (stroma) consisting of fat, connective tissue, blood vessels, nerves, and lymphatics. A network of

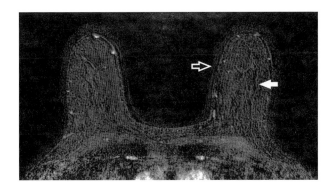

FIGURE 14.16

Interscan motion. A 60-year-old woman presenting for high-risk screening breast MRI. Axial post-contrast subtraction images demonstrate multiple linear areas of signal void (white arrow) with signal also outside the breast (hollow arrow). This represents misregistration on the subtraction images due to the motion between the pre- and post-contrast T_1-weighted images. This can potentially obscure lesions or mimic linear ductal enhancement. Gentle compression can be applied around the breast to avoid motion in addition to maximizing patient comfort and providing clear instructions (e.g., shallow breathing).

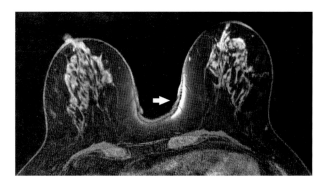

FIGURE 14.17

Incomplete fat saturation. A 71-year-old woman presents for high-risk screening breast MRI. The pre-contrast T_1-weighted image with fat saturation demonstrates incomplete fat saturation within both medial breasts (arrow). This is particularly common near the edges of an image and should not be mistaken for skin thickening or enhancement on post-contrast images.

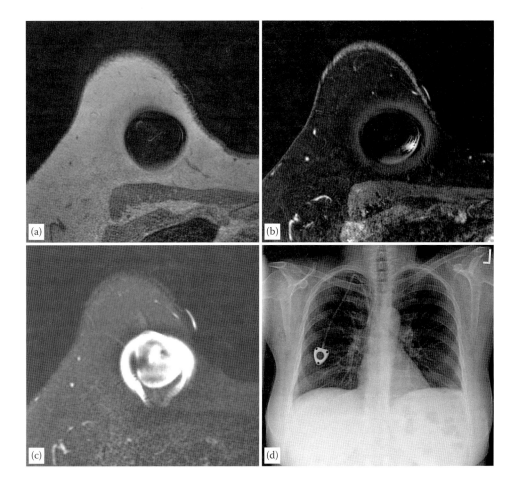

FIGURE 14.18

Magnetic susceptibility artifact. A 66-year-old woman with invasive ductal carcinoma of the left breast. The pre-contrast T_1-weighted image (a) in addition to the post-contrast T_1-weighted image with fat saturation (b) demonstrates a large round area of signal void within the right upper inner breast. This demonstrates round, heterogeneous but predominantly high signal on T_2-weighted images (c). This corresponds to the patient's port catheter as seen on chest radiograph (d). These artifacts can potentially obscure findings on MRI.

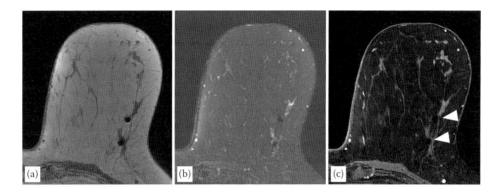

FIGURE 14.19

Magnetic susceptibility artifact. A 66-year-old woman with invasive ductal carcinoma of the left breast. Pre-contrast T_1-weighted image (a) and T_2-weighted SPAIR (b) sequences show two foci of signal void within the left upper outer quadrant. These correspond to MRI biopsy markers. Note that these are faintly visualized on the post-contrast T_1 fat-saturated images (arrowheads) (c). The non-fat-saturated images are the best sequence to evaluate for biopsy clips and surgical staples.

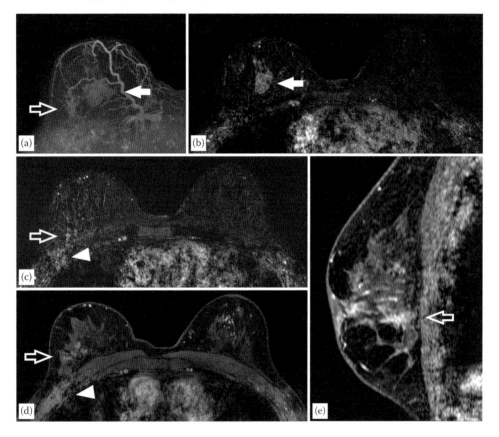

FIGURE 14.20

Ghosting. A 61-year-old woman with biopsy-proven IDC in the right breast (solid arrow) in addition to nonmass enhancement (NME) at the 7 o'clock position posteriorly (open arrow) on post-contrast maximum intensity projection (a) and axially subtraction (b) images. Ghosting represents a replication of interfaces or bright areas, which occurs in the phase-encoding direction, and in this case mimics invasion of the pectoralis musculature (arrow head) on post-contrast axial subtraction (c) and T_1-weighted fat-suppressed images (d). When breast MRI is acquired axially, the phase-encoding gradient should be selected in the left to right direction to avoid artifacts within the breast parenchyma. The reconstructed sagittal post-contrast T_1-weighted image (e) confirms the NME (open arrow) without true pectoralis invasion.

Cooper suspensory ligaments, which connect underlying layers of fascia, supports the breast. The skin of the breast is thin (less than 3 mm on imaging) and contains hair follicles and sebaceous and sweat glands. The nipple typically contains five to nine ductal openings, and the surrounding areola contains Morgagni tubercles at its periphery, which represent openings of the ducts of the Montgomery glands. The lymphatics, which are present in the subcutaneous tissue and stroma of the breast, drain predominantly to the axillary nodes (97%), with the remaining lymphatic flow to the internal mammary chain. There are varying numbers of intramammary lymph nodes, the majority of which are located in the upper outer quadrant of the breast. The axillary lymph nodes are organized by anatomic levels on the basis of relationship to the pectoralis minor: level I lymph nodes are lateral to the lateral margin of the pectoralis minor, level II lymph nodes are posterior to the body of the pectoralis minor, and level III lymph nodes are medial to the medial margin of the pectoralis minor (Figures 14.21 and 14.22). The majority of blood supply to the breasts arises from the internal mammary arteries and lateral thoracic arteries via the axillary artery. These vessels are readily apparent on post-contrast images, particularly MIPs.

Representative images from a normal MRI with relevant anatomy are shown in Figure 14.23. On noncontrast T_1-weighted images (without fat saturation), breast (fibroglandular) tissue, lymph nodes, muscle, and skin exhibit intermediate signal intensity and can be readily differentiated from adipose tissue containing higher signal intensity. Ducts within the breast parenchyma leading to the nipple–areolar complex may contain fluid with variable amounts of fat and protein and, as a result, may exhibit high signal on T_1 and/or T_2-weighted images. On the noncontrast T_2-weighted images, the fibroglandular tissue (FGT) usually has higher signal intensity than that of muscle but less than that of the adjacent subcutaneous blood vessels. On the post-contrast images, the normal breast tissue can enhance to a variable degree, a phenomenon called background parenchymal enhancement (BPE). The nipple–areolar complex typically enhances intensely on post-contrast images because of its rich vascularity (Figures 14.24 and 14.25). Other normal enhancing structures include blood vessels and the nipple–areolar complex. Normal skin and fat do not enhance.

Kinetic enhancement curves (described in more detail below) with the DCE technique can also help distinguish normal from abnormal tissue on MRI. In general, FGT has mild slow initial enhancement, while muscle has mild rapid initial enhancement. Breast malignancies generally have greater enhancement than the normal surrounding breast parenchyma and usually exhibit unique kinetic enhancement profiles, described in more detail below.

14.6 Interpretation and Reporting of Breast MRI

The development of the ACR BI-RADS atlas for MRI [50] has helped standardize interpretation and reporting and has allowed for increased consistency in recommendations among radiologists and institutions. As with mammography and ultrasound, a consistent approach to interpreting MRIs that incorporates the BI-RADS

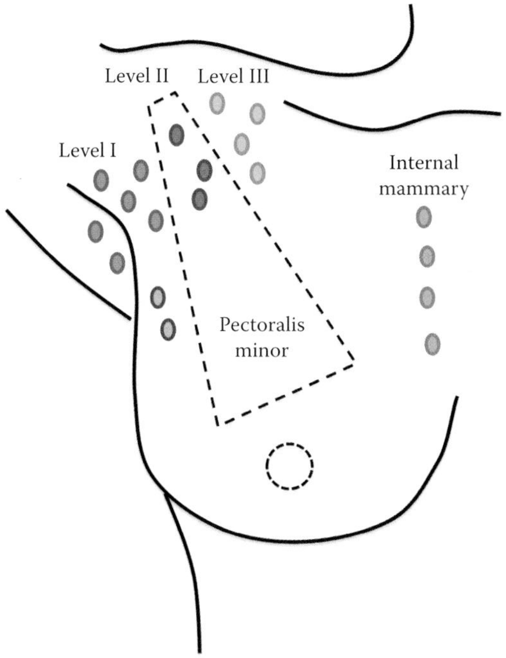

FIGURE 14.21
Axillary lymph node levels. Classification of the three levels of axillary lymph nodes is defined by the locations of the lymph nodes relative to the pectoralis minor. Level I lymph nodes (blue) are lateral to the lateral margin of the pectoralis minor, level II lymph nodes (purple) are posterior to the body of the pectoralis minor, and level III lymph nodes (green) are medial to the medial margin of the pectoralis minor. Internal mammary lymph nodes (orange) and intramammary lymph nodes (red) are also demonstrated. Interpectoral nodes (or Rotter's nodes, not shown) are located between the pectoralis major and minor muscles.

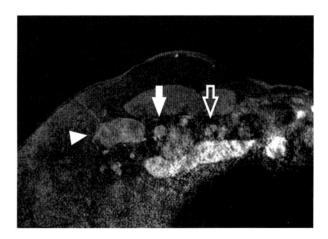

FIGURE 14.22
Axillary lymphadenopathy. A 61-year-old woman with screen-detected invasive ductal carcinoma and ductal carcinoma in situ and biopsy-proven metastatic right axillary lymphadenopathy. Axial T_1-post-contrast T_1 images demonstrate abnormal level 1 (arrowhead), level 3 (hollow arrow), and interpectoral (Rotter's) lymph nodes (solid arrow).

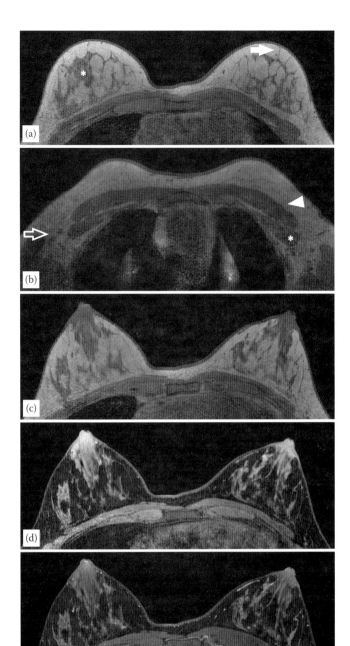

FIGURE 14.23
Normal breast MRI. A 28-year-old woman with family history of breast cancer. MRI obtained for high-risk screening. The axial noncontrast T_1-weighted image at a level just above the nipple (a) demonstrates intermediate signal intensity of the fibroglandular tissue composed of stromal and parenchymal elements (asterisk), and Cooper's ligaments (solid arrow) are seen as thin, intermediate signal intensity lines extending to the skin from the fibroglandular tissue, which are connected in a reticular manner. The axial noncontrast T_1-weighted image more superiorly (b) shows the intermediate signal intensity of normal axillary lymph nodes (hollow arrow), skin, and pectoralis major (arrowhead) and minor (asterisk) musculature. Axial noncontrast T_1-weighted (c), T_2-weighted SPAIR (d), and post-contrast T_1-weighted with fat saturation (e) images at the same level show the normal nipple–areolar complex. Note the normal high-signal intensity of the fibroglandular tissue on the T_2-weighted image.

lexicon is essential to providing clinicians accurate, focused, and clinically relevant breast MRI reports and recommendations [51]. Box 14.1 outlines the authors' suggested step-by-step approach to interpreting breast MRIs.

14.6.1 Amount of FGT

Similar to breast density on a mammogram, the amount of FGT can also be characterized on MRI. This is best assessed on the T_1-weighted pre-contrast images and is divided into four groups on the basis of qualitative assessments using the ACR BI-RADS lexicon: almost entirely fat, scattered FGT, heterogeneous FGT, and extreme FGT (Box 14.2, Figure 14.26).

14.6.2 Background Parenchymal Enhancement

Background parenchymal enhancement (BPE) refers to benign, hormone-related enhancement of normal breast tissue. The amount of BPE varies among and within patients, depending on pre- or postmenopausal status, stage of the menstrual cycle in premenopausal patients, and exogenous hormone or hormone-modulating therapies. The assessment of BPE is based on the first post-contrast image with the center of k-space acquired at approximately 90 s. In general, BPE demonstrates gradually increasing enhancement over time and may be more apparent on delayed imaging. According to the 5th edition of the ACR BI-RADS atlas (2nd edition for MRI), BPE should be categorized as one of four categories with respect to the amount of FGT: minimal, mild, moderate, or marked (Box 14.3, Figures 14.27 and 14.28).

It was previously hypothesized that the amount of BPE on breast MRI may either mask or be mistaken for an enhancing breast cancer, thereby adversely affecting breast MRI performance. However, increased BPE is not related to significant differences in the positive biopsy rate, cancer yield of MRI, sensitivity, or specificity [52]. Nonetheless, increased background enhancement is associated with a higher abnormal interpretation rate, which may lead to additional imaging [52,53]. Therefore, in premenopausal patients with more elective examinations such as high-risk screening, it may be preferable to image patients 7–10 days after the onset of their menstrual cycle when the background enhancement is at its lowest (Figure 14.29). However, such timing should never delay obtaining imaging for patients with a recent diagnosis of breast cancer in which the breast MRI is used for evaluation of disease extent.

While most commonly symmetric, BPE can occasionally be asymmetric. This most commonly occurs in women who have undergone prior therapy for breast

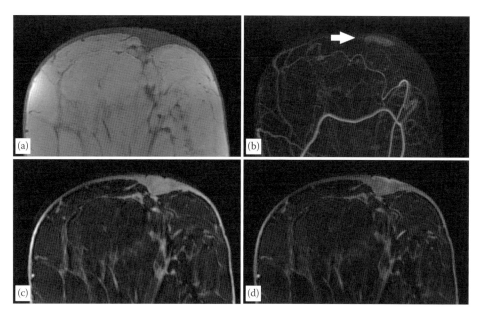

FIGURE 14.24

Nipple–areolar complex. A 52-year-old woman. High-risk screening breast MRI. Pre-contrast axial T_1-weighted (a), maximum intensity projection (b), T_1-weighted fat-saturated (c), and post-contrast T_1-weighted fat-saturated (d) images demonstrate normal enhancement of the nipple–areolar complex (arrow).

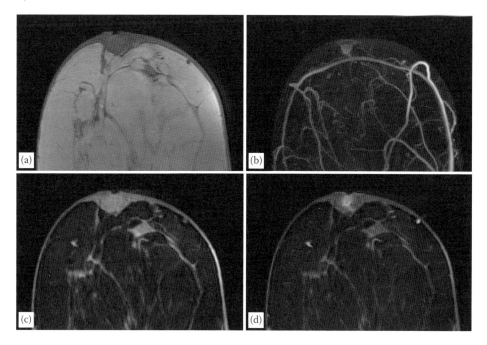

FIGURE 14.25

Inverted nipple. A 52-year-old woman. High-risk screening breast MRI. Pre-contrast axial T_1-weighted (a), maximum intensity projection (b), T_1-weighted fat-saturated (c), and post-contrast T_1-weighted fat-saturated (d) images demonstrate an inverted nipple, which mimics an enhancing subareolar enhancing mass.

cancer (Figure 14.30). Typically, BPE initially increases in the affected breast relative to the unaffected breast after radiation therapy and subsequently decreases over time. Asymmetrical benign patterns can also occur in women with substantially asymmetrical amounts of breast tissue and in lactating women who are only nursing from one breast. Asymmetrical BPE without an anatomic or a physiologic cause should raise suspicion for mastitis or malignancy, and close clinical correlation is required.

14.6.3 Lesion Evaluation

A lesion on MRI is defined as an enhancing finding on post-contrast T_1-weighted images, which is unique to the BPE. Each enhancing lesion can be described further with specific morphological and kinetic curve enhancement features that are defined by the ACR BI-RADS atlas. Furthermore, depending on the lesion type, additional features on the pre-contrast T_1-weighted and T_2-weighted (or other fluid-sensitive sequence) can be helpful for lesion characterization.

14.6.3.1 Morphology

The ACR BI-RADS atlas describes three general lesion types: foci, masses, and nonmass enhancement (NME). Each lesion type, with the exception of foci, can be characterized further with additional morphological descriptors that are useful for assessing the appropriate level of suspicion for malignancy. In general, morphology is the single most important factor on MRI for distinguishing benign from malignant lesions [18].

14.6.3.1.1 Foci

On the post-contrast image, a focus of enhancement is used to describe a lesion typically less than 5 mm in size that cannot be characterized morphologically and does not have a corresponding finding on pre-contrast T_1-weighted images (Figure 14.31). These lesions are typically nonspecific in nature and generally are the least suspicious breast MRI lesion. Although foci can represent small malignancies, the majority of such findings represent BPE (particularly when bilateral), intramammary lymph nodes, or benign pathology. If a focus is deemed to be unique (i.e., stands out when compared with BPE or is larger and conspicuous compared with other foci in the breasts), it is useful to follow a checklist to determine the appropriate level of suspicion for malignancy and management. Table 14.2 summarizes features to consider when evaluating unique foci. In general, foci that exhibit high signal on T_2-weighted images and medium initial with plateau or persistent delayed enhancement kinetic features on DCE sequences that are either stable compared with prior examinations or identified on a baseline examination are benign and can either be safely ignored or followed with serial imaging. Foci that represent small intramammary lymph nodes can often be resolved by identifying high signal on T_2-weighting with a small invagination of fat (or "fatty notch") on T_1-weighted images without fat suppression. However, if a unique focus is new compared with prior studies and does not exhibit most of these typically benign features, biopsy should be considered. Finally, it should be noted that not all unique lesions measuring less than 5 mm are obligatorily classified as foci. As MRI technologies improve, particularly with higher magnetic field strength and increased emphasis on high spatial resolution imaging, it is anticipated that an increasing number of small lesions will meet the criteria for description as masses.

14.6.3.1.2 Masses

A mass on breast MRI is a space-occupying, three-dimensional lesion with distinct shapes, margins, and internal enhancement features. While this definition based on the ACR BI-RADS atlas is fairly specific, overlap exists between foci and masses, and at times, they may be difficult to distinguish. A useful rule of thumb for distinguishing small masses from foci is assessing

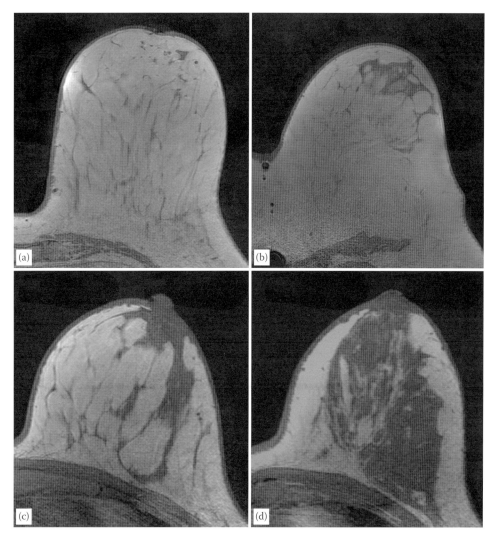

FIGURE 14.26
Breast composition. Examples of the amount of fibroglandular tissue on T_1-weighted axial images: almost entirely fat (a), scattered fibroglandular tissue (b), heterogeneous fibroglandular (c), and extreme fibroglandular (d).

BOX 14.3   BACKGROUND PARENCHYMAL ENHANCEMENT

- Minimal (1%–25%)
- Mild (25%–50%)
- Moderate (50%–75%)
- Marked (>75%)

whether suspicious shape and margin descriptors can be described for the lesion: if a small enhancing lesion has an irregular shape and/or noncircumscribed margins, it should be described as a mass.

Compared with foci, masses should be approached with a higher index of suspicion and a lower threshold to recommend biopsy. In general, because most breast cancers enhance rapidly because of associated neovascularity,

it is most useful to evaluate the shape and margins of a mass on the first T_1 post-contrast image. According to the 5th edition of the BI-RADS atlas, the shape of a mass can be described using three terms: oval (which includes lobulated, meaning 2–3 gentle undulations), round, or irregular. The margin of a mass may be circumscribed (well demarcated from the surrounding breast tissue) or not circumscribed (irregular or spiculated) (Figures 14.32 and 14.33). As with mammography, masses with circumscribed margins are more likely to be benign than those with spiculated or irregular margins; however, it is important to recognize that depending on spatial resolution of the scan, volume averaging can cause spiculated masses to appear circumscribed on MRI.

Internal architecture and enhancement kinetics can also help distinguish benign from malignant lesions. Masses that exhibit homogeneous (confluent and uniform) enhancement are more suggestive of benign processes,

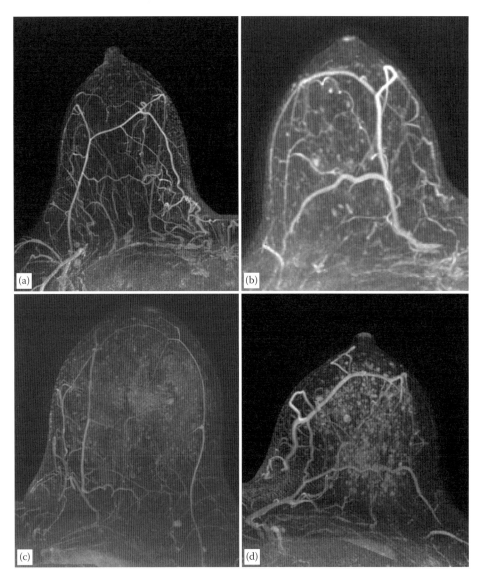

FIGURE 14.27
Background parenchymal enhancement. Representative examples of background parenchymal enhancement on maximum intensity projection images with minimal (a), mild (b), moderate (c), and marked (d) background parenchymal enhancement.

whereas those with heterogeneous enhancement are suggestive of malignant pathology. There are also two special enhancement patterns: nonenhancing dark internal septations and rim enhancement. Nonenhancing dark internal septations have been reported to be specific for fibroadenomas when identified in conjunction with other benign morphologic and kinetic characteristics. However, caution should be exercised when using this enhancement characteristic while determining lesion suspicion, as a prior study found that nonenhancing septations alone were not predictive of benignancy [54]. The second is rim enhancement, which is considered particularly suspicious for malignancy (Figure 14.34) except for the cases where the enhancement is clearly related to an inflamed cyst or evolving fat necrosis (Figure 14.35). Usually, benign cysts and fat necrosis can be easily

distinguished from necrotic or mucinous malignancies exhibiting rim enhancement on the basis of pre-contrast intrinsic signal properties. Inflamed cysts presenting as rim-enhancing masses typically exhibit high signal on T_2 and/or T_1-weighted images depending on cyst content. In cases of fat necrosis, internal signal characteristics should follow fat signal on T_1-weighted images without (high signal) and with fat suppression (dark).

14.6.3.1.3 Nonmass Enhancement

NME describes areas with a distinct pattern of enhancement different from normal breast tissue enhancement (BPE) but not fitting the definition of a focus or a mass. NME may extend over small or large areas and are classified on the basis of their distribution (Table 14.3).

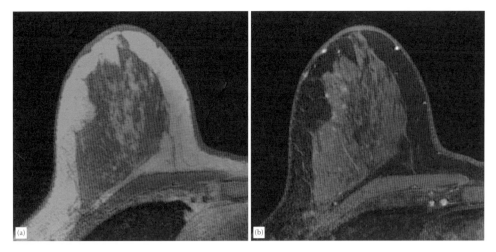

FIGURE 14.28
Breast composition and background parenchymal enhancement. A 48-year-old woman with an extreme amount of fibroglandular tissue on the axial T_1-weighted image (a). The axial post-contrast T_1-weighted fat-saturated image (b) demonstrates minimal background parenchymal enhancement (BPE). Note that the amount of fibroglandular tissue does not directly correlate with quantity of BPE.

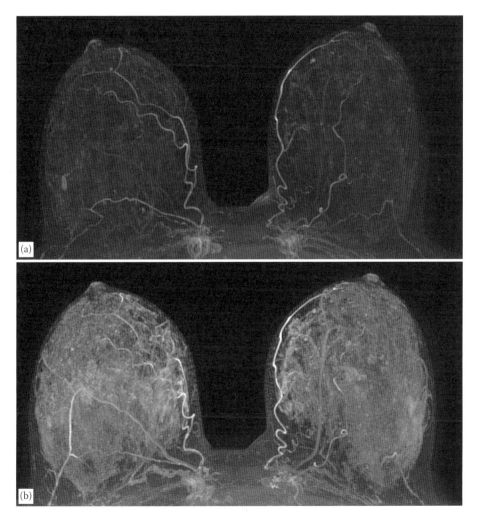

FIGURE 14.29
Differences in BPE due to stages in the menstrual cycle. A 31-year-old woman presenting for a high-risk screening breast MRI. The patient had minimal BPE on the recent examination (a); however, she had marked BPE on examination performed 1 year ago (b). This variation is secondary to changes in hormonal influence due to different stages in the menstrual cycle.

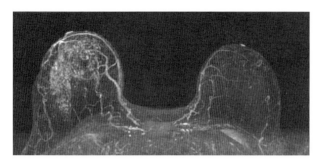

FIGURE 14.30
Asymmetric BPE due to radiation therapy. A 48-year-old woman with history of treated left breast cancer. MRI obtained for high-risk screening. Axial post-contrast maximum intensity projection shows asymmetric BPE with minimal BPE on the left and moderate on the right, due to history of radiation therapy to the left breast.

Linear and segmental distributions of NME are concerning for carcinoma (Figures 14.36 through 14.38), particularly DCIS, whereas a regional or diffuse distribution suggests a benign process. Multicentric carcinoma may also have these appearances, although kinetic curves can help distinguish benign from malignant lesions. The internal enhancement characteristics of NME can also be described using the BI-RADS

descriptors of homogeneous, heterogeneous, clumped, and clustered ring enhancement.

14.6.3.2 Kinetic Enhancement Features on DCE-MRI

The abnormal kinetic enhancement characteristics of breast cancer are based on the theory that tumor angiogenesis is associated with abnormal, increased periductal or stromal vascularity, which allows for increased perfusion and leaky vessels. There is no consensus on the number of scans needed to accurately determine kinetic enhancement features, but most protocols include at least three post-contrast time points, typically performed with *k*-space centered at approximately 90, 270, and 450 s after contrast injection. Obtaining DCE information allows evaluation of two major phases of enhancement: initial phase enhancement and late phase enhancement (Figure 14.39). Initial phase reflects enhancement features until peak enhancement is reached (which typically occurs within the first 2 min after intravenous contrast injection). The second is the late phase of enhancement, which occurs after peak enhancement. The late phase of enhancement can be described using three major curve types: persistent, plateau, and washout.

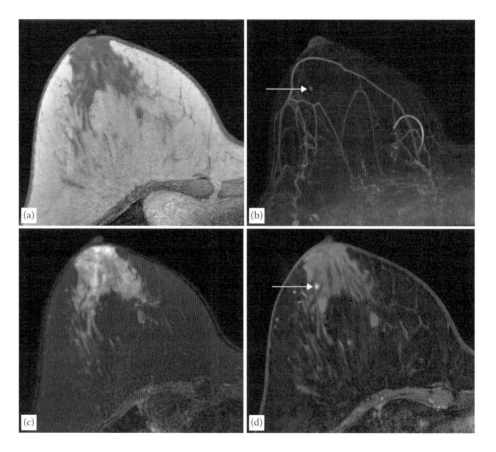

FIGURE 14.31
Focus. A 54-year-old woman with an MRI obtained for high-risk screening. Pre-contrast axial T_1-weighted (a), maximum intensity projection (b), T_2-weighted SPAIR and (c) post-contrast T_1-weighted fat-saturated (d) images show a 3 mm focus at the 9 o'clock position (arrow). Note that there is no correlate on the T_1-weighted image, and this does not exhibit high signal on the T_2-weighted image.

TABLE 14.2

Features of Foci Useful for Determining Need for Pathological Sampling

Likely Benign	Suspicious
Not unique to BPE	Conspicuous accounting for BPE
High signal on fluid-sensitive sequences	Low signal on fluid-sensitive sequences
Medium initial enhancement with delayed persistent or plateau kinetics	Rapid initial enhancement with delayed washout kinetics
Round or oval with circumscribed margins	Irregular shape with noncircumscribed margins (note: if such features are present, finding should be described as a mass)
Morphology suggestive of a lymph node (presence of "fatty notch")	
Stable or identified on baseline study	New when compared with prior examinations

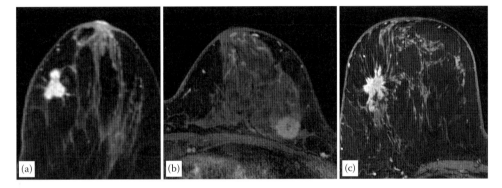

FIGURE 14.32

Suspicious shape and margins. Post-contrast axial T_1-weighted fat-saturated images in three different women show a lobular mass with circumscribed borders in a 49-year-old woman with invasive ductal carcinoma (a), an irregular mass with irregular margins and heterogeneous internal enhancement in a 55-year-old woman with invasive ductal carcinoma (b), and an irregular mass with spiculated margins and heterogeneous internal enhancement in a 40-year-old woman with invasive lobular carcinoma (c).

In the initial phase of enhancement, classification of "slow," "medium," and "rapid" are defined by the percent increase in intensity being <50%, 50%–100%, and >100% respectively, where signal intensity increase is defined by the equation

$$SI_{\%increase} = [(SI_{post} - SI_{pre})/SI_{pre}] \times 100\%$$

where SI_{pre} is the baseline signal intensity of a region of interest, and SI_{post} is the signal intensity of the same region of interest after contrast injection. In the delayed phase, there are three main curve types: persistent, plateau, and washout. Persistent delayed enhancement is defined as continuous increase enhancement ≥10% initial enhancement and is considered to be the most benign delayed enhancement curve type. Plateau delayed enhancement refers to constant signal intensity once peak is reached (±10% initial enhancement) and is of intermediate suspicion. Delayed washout indicates decreasing signal intensity after peak enhancement ≥10% initial enhancement and is considered the curve type most suggestive of malignancy.

Although the most classic curve type for a malignant breast lesion is rapid initial enhancement followed by early washout (Figure 14.40), there is significant overlap of semiquantitative kinetic curve types among benign and malignant lesions. As a result, kinetic enhancement features should be interpreted in the context of other important clinical and imaging features, such as patient history, lesion morphology, and comparison studies. In our experience, kinetic evaluation is most helpful in confirming benignancy in lesions that demonstrate mostly, but not unequivocally, benign morphological features. Lesions with suspicious morphologic features should be biopsied regardless of enhancement characteristics. Computer-aided analysis tools are used to evaluate lesion kinetics, and different enhancement thresholds can be set to show which tissues enhance more than a certain range (such as 50%–100% of increased signal intensity).

In addition to the standard semiquantitative kinetic parameters, there is great interest in developing advanced MRI techniques that can employ pharmacokinetic models to obtain highly quantitative information that may serve as biomarkers of malignancy (e.g., rate constant for passive contrast reagent transfer between plasma and interstitium, K^{trans}). These parameters, while promising, are technically demanding and fraught with limited reproducibility at this time and, therefore, are not recommended for routine clinical applications.

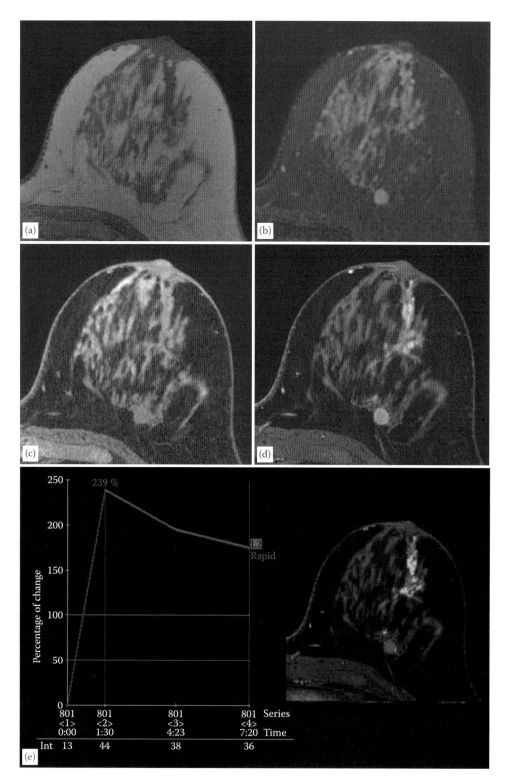

FIGURE 14.33
Round circumscribed mass. A 57-year-old woman with newly diagnosed IDC and DCIS. MRI performed for extent of disease workup. Pre-contrast axial T_1-weighted (a), T_2-weighted SPAIR (b), T_1-weighted fat-saturated (c), and post-contrast T_1-weighted fat-saturated (d) images show a round mass with circumscribed margins in the central breast, posterior depth with rim enhancement. The kinetic enhancement curve (e) demonstrates rapid initial enhancement with late-phase washout. Biopsy revealed fibrocystic change, usual ductal hyperplasia, and sclerosing adenosis.

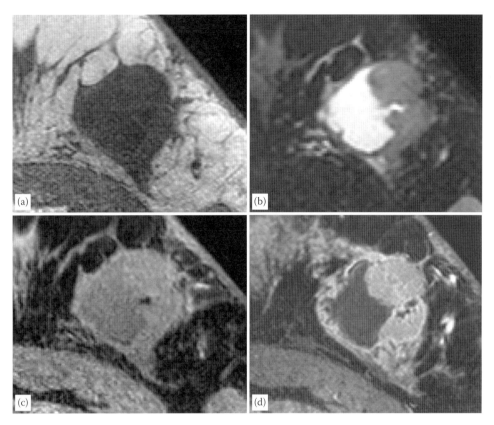

FIGURE 14.34

Rim enhancement. A 27-year-old woman with invasive ductal carcinoma. Pre-contrast T_1-weighted (a), T_2-weighted SPAIR (b), T_1-weighted fat-saturated pre- (c), and post-contrast (d) images show an irregular mass with irregular margins and rim internal enhancement. There is a high signal on T_2 within portions of the mass, which could represent necrosis.

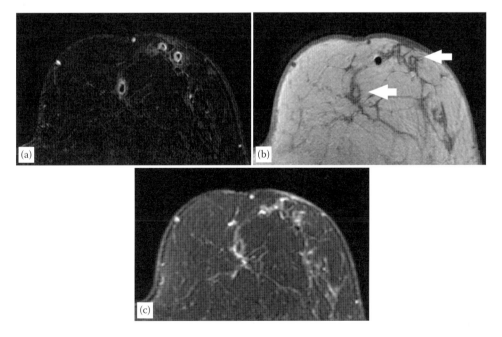

FIGURE 14.35

Rim enhancement from fat necrosis. A 69-year-old woman's status postreduction mammoplasty. Axial post-contrast subtraction (a), pre-contrast T_1-weighted (b) and T_2-weighted SPAIR (c) images show multiple scattered rim-enhancing masses with central fat intensity (arrows) and no increased signal on T_2-weighted images within the masses consistent with fat necrosis.

TABLE 14.3

Distribution of NME

Distribution of Enhancement	Extent of Enhancement
Focal	Less than one quadrant of the breast within a single ductal system
Linear	Single duct
Segmental	Triangular or cone shaped with the apex at the nipple
Regional	Spans at least 1 quadrant, which encompasses a larger area than a single duct system
	Lacks a convex-outward contour for this to be defined as a mass
Multiple regions	Enhancement over at least 2 broad areas, which are separated by normal glandular tissue or fat
Diffuse	Evenly distributed, wide area of enhancement, which is similar appearing throughout the entire breast tissue

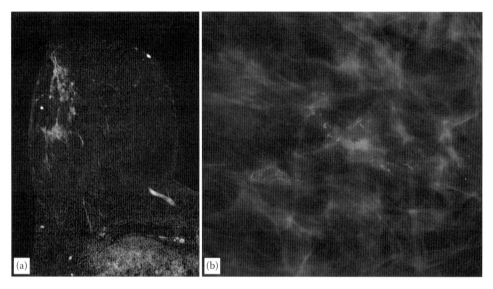

FIGURE 14.36

Segmental nonmass enhancement. A 50-year-old woman with newly diagnosed ductal carcinoma in situ (DCIS). Post-contrast subtraction images (a) demonstrate segmental nonmass enhancement extending to the nipple compatible with biopsy-proven DCIS. Corresponding spot magnification views from patient's mammogram (b) demonstrate segmentally distributed fine pleomorphic calcifications.

14.6.3.3 T_2-Weighted Imaging

In general, nonenhancing findings with homogeneously high intrinsic signal on T_2-weighted images relative to normal breast tissue represent benign pathologies (Box 14.4). These include lesions such as cysts (which may also exhibit peripheral enhancement) and fluid-filled ectatic ducts. T_2-weighted images are also useful for confirming the benignity of small (typically less than 10 mm in size) circumscribed masses with homogeneous enhancement: those with high intrinsic signal on T_2-weighting are most likely fibroadenomas or intramammary lymph nodes and thus can be safely ignored or followed at short interval. The presence of high signal on T_2-weighted images should never be used to obviate biopsy of any mass or other suspicious finding that demonstrates suspicious morphological features. In fact, several malignant subtypes are known to often exhibit high signals on T_2-weighting, such as mucinous and metaplastic carcinoma.

14.6.3.4 Advanced and Emerging Breast MRI Technique Features

While DCE MRI techniques provide a remarkably high sensitivity for the detection of breast cancer (~90%), its moderate specificity (~72%) [55], need for contrast agent, and relatively high cost limit access to this tool. Recently, multiple studies have demonstrated that DWI, a noncontrast technique that measures the ability of water to freely diffuse within tissue, holds much potential to improve the specificity and positive predictive value of DCE MRI by reducing false-positive MRI findings and unnecessary biopsies [56]. Furthermore, a recent study demonstrated potential of DWI as a stand-alone noncontrast screening technique [57], which could reduce costs and increase accessibility of MRI to patients with renal insufficiency or contrast-material allergies who cannot undergo a contrast-enhanced examination. MR spectroscopy (MRS), a technique that can measure chemical information in vivo and is widely used for brain and prostate applications, has

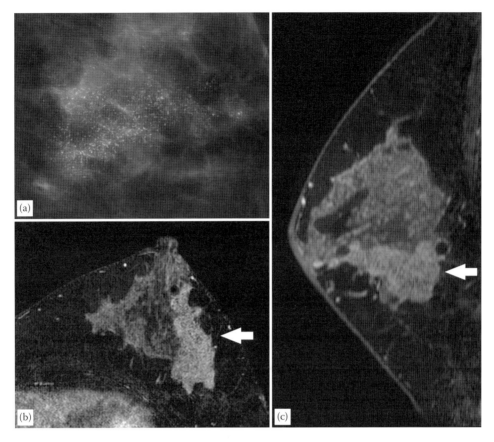

FIGURE 14.37
Segmental nonmass enhancement. A 36-year-old woman with newly diagnosed ductal carcinoma in situ with segmental pleomorphic calcifications on mammogram (a). Pre-contrast T_1-weighted fat-saturated axial (b) and sagittal (c) images show segmental nonmass enhancement (arrows) extending from the posterior edge of the breast parenchyma to the nipple with homogeneous internal enhancement.

also shown promise for differentiating between benign and malignant breast lesions [58]. However, despite these techniques' potential to improve breast MRI performance and/or increase access to breast MRI, significant unresolved technical and standardization issues remain at this time, and thus, they cannot be relied upon for routine clinical use at this time. Several multicenter studies examining DWI and MRS of the breast are underway, sponsored by the American College of Radiology Imaging Network (ACRIN trials 6698, 6657, and 6702), which should shed additional light on the clinical utility of these techniques.

14.6.4 Location of Lesion

It is important to describe the location of a lesion in a consistent manner that clearly communicates the position of the finding so that it can be easily identified on correlative imaging with other modalities, on follow-up MRIs, and during localization for surgical planning and guidance. As with mammography, the breast is viewed as a clock face while the patient is facing the physician, and the use of quadrants in addition to clock face will minimize confusion (Figure 14.41). In the report description, the side of the lesion is given first, followed by

location and then depth. Depth of a lesion should also be reported and is determined by dividing the breast tissue in thirds (anterior, middle, and posterior; Figure 14.42). A distance from the nipple, which is drawn from the anterior margin of the lesion to the nipple, can also be used to help define depth and location. The subareolar region can also be used if the lesion is just deep to the nipple. Central, subareolar, and axillary tail descriptions can also be used in lieu of quadrants when they are more appropriate.

14.7 Benign MRI Findings

In additon to malignancies, MRI can identify a range of benign findings, and it is important for interpreting radiologists to be familiar with their appearances.

14.7.1 Fluid-Filled Cysts

Cysts are fairly common benign lesions and have distinguishing features on MRI. Like elsewhere within the body, simple cysts are round or oval in shape with

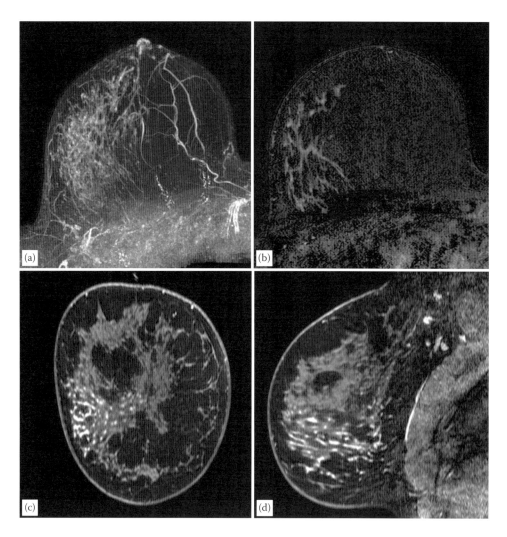

FIGURE 14.38

Segmental nonmass enhancement. A 31-year-old woman with newly diagnosed recurrent left breast invasive carcinoma. MRI was performed for extent of disease. Axial maximum intensity projection (a), axial post-contrast subtracted image (b), and post-contrast T_1-weighted fat-saturated axial (c) and sagittal (d) images show segmental linear and branching nonmass enhancement in the lower outer quadrant with dendritic internal enhancement. This was biopsied and shown to be ductal carcinoma in situ.

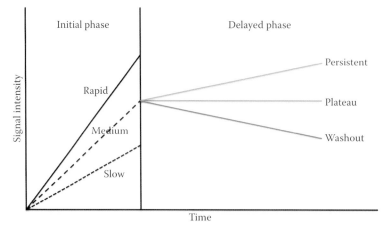

FIGURE 14.39

Dynamic contrast-enhancement kinetic curves. A representative image showing the different initial- and delayed-phase kinetic enhancement curves during dynamic contrast-enhancement MRI. There can be slow, medium, or rapid initial-phase enhancement with persistent, plateau, or washout type late-phase enhancement.

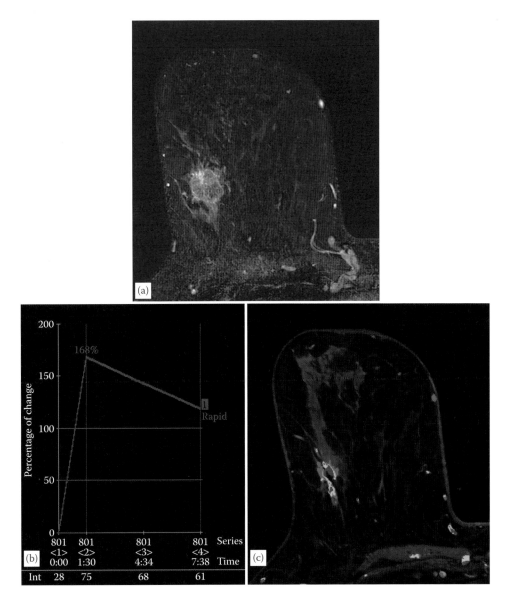

FIGURE 14.40
Washout kinetics. A 50-year-old woman with newly diagnosed invasive ductal carcinoma presented for breast MRI to evaluate extent of disease. The post-contrast axial T_1-weighted image (a) shows an irregular mass with irregular margins and heterogeneous internal enhancement. The kinetic curve (b) in a portion of the biopsy-proven malignancy (c) shows rapid initial enhancement and delayed washout (red overlay).

BOX 14.4   MASSES THAT ARE HYPERINTENSE ON T_2-WEIGHTED IMAGES

- Cysts
- Ectatic ducts
- Intramammary lymph nodes
- Fibroadenomas
- Mucinous or metaplastic carcinoma

circumscribed margins. In general, simple cysts have high signal on fluid-sensitive sequences with no internal enhancement (Figure 14.43) and may have thin septations.

Cysts that contain high protein content or may contain blood products, resulting in lower signal on fluid-sensitive imaging and high signal on pre-contrast T_1-weighted images (Figure 14.44). Cysts never enhance internally; on occasion, however, the wall of a cyst can enhance on post-contrast T_1-weighted images when inflamed (Figure 14.45). This is a benign finding as long as the enhancement is uniform, thin, and free from nodularity or associated internal enhancement.

14.7.2 Duct Ectasia

Duct ectasia, or dilated, fluid-filled ducts, tend to radiate from the nipple and may show a branching-type

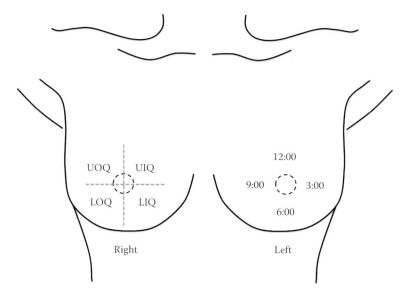

FIGURE 14.41
Lesion location—quadrant and clock-face method. UOQ, upper outer quadrant; LOQ, lower inner quadrant; UIQ, upper inner quadrant; LIQ, lower inner quadrant.

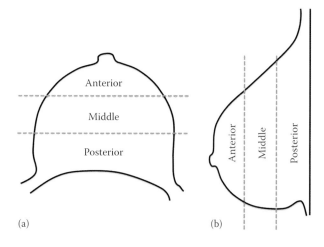

FIGURE 14.42
Lesion depth. Axial (a) and sagittal (b) views of the breast showing how depth can be defined by dividing the breast into anterior, middle, and posterior thirds.

pattern with signal intensity characteristics similar to those of cysts. These tubular structures can exhibit high intrinsic signal on pre-contrast T_1-weighted images when filled with proteinaceous fluid (Figure 14.46). This is a normal finding and has no known associated increased risk for breast cancer; however, one must be careful to verify that there is no ductal enhancement or intraductal mass present that would explain ductal ectasia, particularly in the setting of a solitary dilated duct. Clustered ring enhancement surrounding ducts or enhancement within a duct raises concern for malignancy.

14.7.3 Fibrocystic Change

Fibrocystic changes refer to a heterogeneous group of benign, nonproliferative epithelial breast lesions that are not associated with an increased risk of breast cancer. This includes entities such as simple breast cysts, papillary apocrine change, epithelial-related calcifications, and mild hyperplasia of the usual type [49]. There have been few studies evaluating MRI features of fibrocystic changes of the breast, as this pathological entity is usually diagnosed on the basis of clinical findings. Fibrocystic changes may appear as segmental NME that is asymmetric compared with the contralateral breast (Figure 14.47), and usually includes associated high signal intensity on T_2-weighted images.

14.7.4 Lymph Nodes

The visualization of intramammary lymph nodes is common, particularly in the upper outer quadrant of the breast, and these benign masses must be distinguished from suspicious masses and foci. Intramammary lymph nodes are typically small, ≤5 mm masses that are oval (often with gentle lobulations) in shape with circumscribed margins and uniform cortical high T_2 signal intensity. Secondary findings to help confirm a mass in question is a benign lymph node in the presence of a central fatty hilum, which can be seen on the T_1-weighted non-fat-suppressed images and close association with blood vessels.

DCE kinetic features typically are not helpful in distinguishing normal lymph nodes from malignant lesions, as both entities frequently exhibit rapid initial

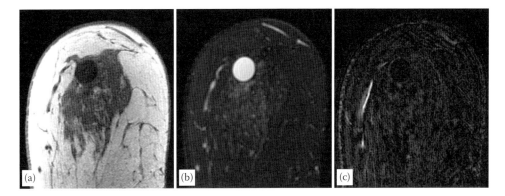

FIGURE 14.43
Simple cyst. A 54-year-old woman with right breast cancer—extent of disease workup with preoperative MRI. Pre-contrast axial T_1-weighted (a), T_2-weighted SPAIR (b), and T_1 post-contrast subtracted (c) images demonstrate a round mass with well-circumscribed borders and homogeneous high T_2 signal intensity without enhancement consistent with a simple cyst.

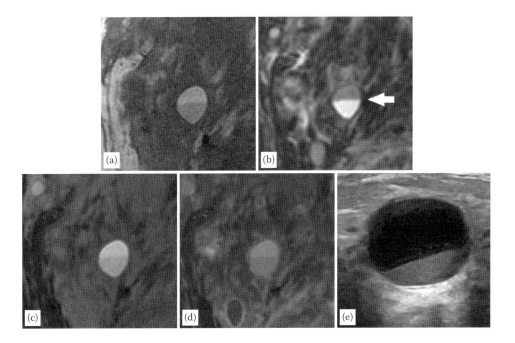

FIGURE 14.44
Proteinaceous/hemorrhagic cyst. A 63-year-old woman with history of cysts in both breast presents for high-risk screening breast MRI. Precontrast T_1-weighted (a), T_2-weighted SPAIR (b), T_1-weighted fat-saturated (c), and post-contrast T_1-weighted fat-saturated (d) images show an oval mass with circumscribed borders with internal fluid–fluid level (arrow) and no evidence of enhancement compatible with a hemorrhagic or proteinaceous component. Note that the patient is lying prone on the MRI scanner and the images are oriented so that anterior is at the superior aspect of the image. The high signal intensity on T_1-weighted and low signal intensity on T_2-weighted images demonstrate the hemorrhagic or proteinaceous component, while the low signal on T_1-weighted images and high signal on the T_2-weighted images are the simple fluid component. Targeted ultrasound with the patient lying supine (e) shows an oval mass with circumscribed borders and a fluid–fluid level with the hemorrhagic or proteinaceous component layering dependently.

phase enhancement with late-phase plateau or washout. As a result, one must rely heavily on morphologic features and signal on T_2-weighted images (Figure 14.48). Occasionally, small masses that are likely intramammary lymph nodes exhibit indeterminate morphology and signal on breast MRI. In such cases, targeted ultrasound can be helpful to clarify whether these masses are in fact intramammary lymph nodes. If such an approach

is taken, it is recommended that a definitive BI-RADS assessment (e.g., BI-RADS category 3 or 4) be given on the basis of MRI features in case the finding either is not identified or remains equivocal on ultrasound.

Abnormal lymph nodes, whether involved by metastatic disease from a breast primary or another malignant process such as lymphoma, can demonstrate loss of the fatty hilum, enlarged and rounded shape (opposed

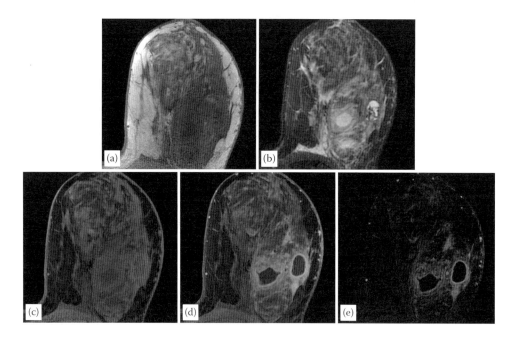

FIGURE 14.45

Inflamed cysts. A 63-year-old woman with history of left breast pain and mastitis symptoms with multiple inflamed cysts. Pre-contrast T_1-weighted (a), T_2-weighted SPAIR (b), T_1-weighted fat-saturated (c), post-contrast T_1-weighted fat-saturated (d) and T_1 post-contrast subtracted images (e) show at least two rim-enhancing masses at the 3 o'clock position, with the more lateral mass demonstrating an oval shape with circumscribed margins and containing increased signal intensity on the T_2- and T_1-weighted images indicating the presence of hemorrhagic or proteinaceous contents.

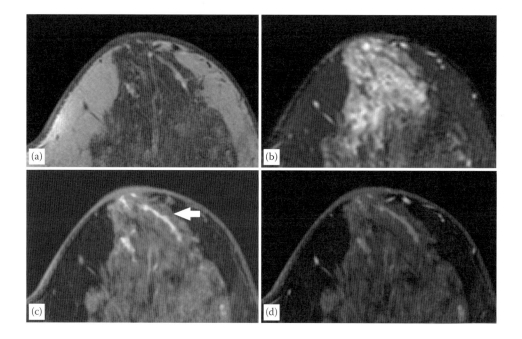

FIGURE 14.46

Duct ectasia. A 55-year-old woman with newly diagnosed invasive carcinoma. Breast MRI obtained for extent of disease workup. Pre-contrast axial T_1-weighted (a), T_2-weighted SPAIR (b), T_1-weighted fat-saturated (c), and post-contrast T_1-weighted fat-saturated (d) images show a normal dilated duct filled with proteinaceous fluid (arrow). There is no intraductal enhancement.

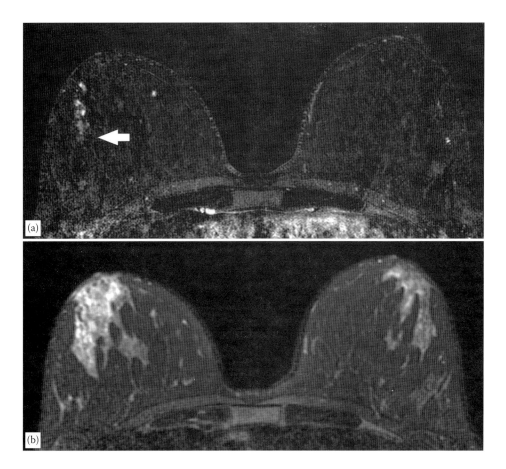

FIGURE 14.47
Fibrocystic changes and usual ductal hyperplasia. A 37-year-old woman with strong family history of breast cancer. The post-contrast sub-tracted image (a) demonstrates segmental nonmass enhancement with clumped internal enhancement in the right breast (arrow) that is asymmetric compared with the contralateral breast. The nonmass enhancement exhibits high signal intensity on the T_2-weighted image (b); however, a fluid signal on MRI cannot be relied on as an indicator of benign pathology in the cases of otherwise suspicious morphology. Biopsy showed fibrocystic changes and usual ductal hyperplasia, which was concordant with imaging findings.

to kidney–bean shaped), or focal eccentric cortical thickening (Figure 14.49). Unfortunately, morphological and kinetic features do not have sufficient ability to dis-tinguish malignant lymph nodes from benign/reactive ones to replace traditional surgical staging in patients newly diagnosed with breast cancer [37].

14.7.5 Fibroadenoma

Fibroadenomas are benign fibroepithelial tumors that account for the most common solid mass in women of all ages, though they are most often diagnosed at 30 years of age. Depending on the stage of maturation, fibroadeno-mas can exhibit a varied appearance on breast MRI. Most commonly, they present on breast MRI as oval or lobular enhancing masses with circumscribed margins with high signal on T_2-weighted images. However, as some fibroad-enomas involute and become fibrotic, they often become dark on T_2-weighted images due to sclerosis (Figure 14.50). Nonenhancing internal septations can be seen in approxi-mately 20% of fibroadenomas (Figure 14.51). Phyllodes

tumors commonly share morphological features of fibro-adenomas, and it is often difficult to distinguish between these two entities [59]. As a result, any enlarging mass (not explained by a physiological hormonal state, such as preg-nancy) with imaging features of a fibroadenoma should be biopsied and/or excised.

In general, fibroadenomas have slow to medium ini-tial phase enhancement with delayed phase persistent kinetics. However, not all fibroadenomas demonstrate benign enhancement kinetic curves, with myxoid fibro-adenomas in particular exhibiting initial-phase rapid enhancement [60]. Finally, sclerotic fibroadenomas typi-cally do not exhibit any internal enhancement and may exhibit irregular internal signal voids that correspond to the popcorn calcifications visible on mammography.

14.7.6 Intraductal Papilloma

Intraductal papillomas are a proliferation of epithelial and myoepithelial cells, which may have a fibrovas-cular stalk. Most commonly, these are small (<5 mm)

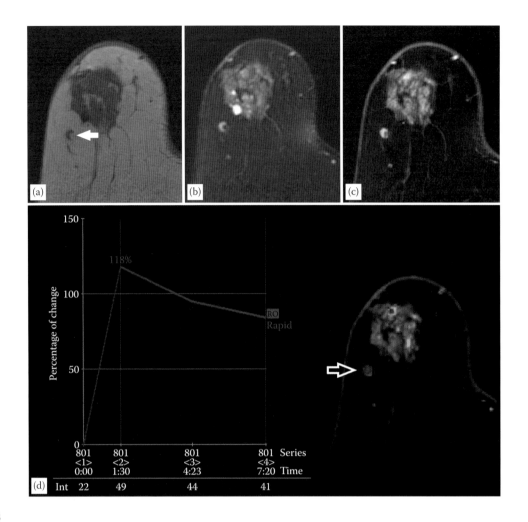

FIGURE 14.48

Normal intramammary lymph node. Pre-contrast axial T_1-weighted (a), T_2-weighted SPAIR (b), and post-contrast T_1-weighted fat-saturated (c) images show a normal appearing intramammary lymph node with a fatty hilum (arrow) and high signal intensity on the T_2-weighted image. Note that the dynamic contrast-enhancement kinetic curve (d) of this benign lymph node (open arrow) demonstrates rapid initial-phase enhancement with late-phase washout, which is not helpful for distinguishing normal intramammary lymph nodes from malignant lesions, and thus, it is most important to evaluate the morphology in this case.

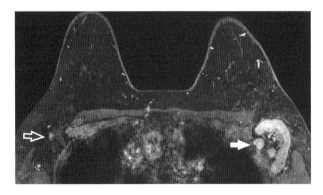

FIGURE 14.49

Abnormal axillary lymph node. A 66-year-old woman with IDC. The post-contrast T_1-weighted fat-saturated axial image demonstrates multiple enlarged and morphologically abnormal left axillary lymph nodes with thickened cortices and one with a rounded configuration and loss of the fatty hilum (solid arrow). A normal right axillary lymph node is shown (hollow arrow).

intraductal masses found in the subareolar region that present clinically with bloody nipple discharge. Differing data have been reported in regard to the rate of intraductal papillomas or associated high-risk lesions upgrading to malignancy on surgical excision; thus, the decision for surgical excision is often institution dependent [61–63]. On MRI, intraductal papillomas can be seen as avidly enhancing masses at the distal end of a fluid-filled duct. On pre-contrast imaging, a hyperintense duct on T_1- or T_2-weighted series may be identified containing an intraductal hypointense mass, analogous to a filling defect on ductography (Figure 14.52). Time intensity curves most commonly demonstrate rapid initial enhancement with washout on the delayed images [64].

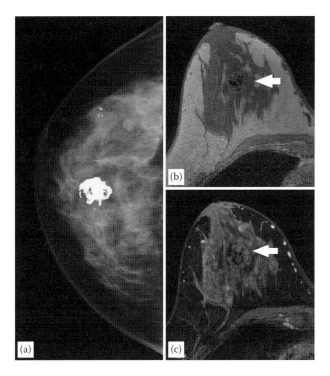

FIGURE 14.50

Fibrotic fibroadenoma. A 61-year-old woman with newly diagnosed invasive carcinoma presented for a breast MRI for extent of disease workup. Craniocaudal mammogram (a) demonstrates coarse calcifications within the central breast, compatible with a fibrotic fibroadenoma. This has dark signal intensity on the pre-contrast T_1-weighted image (b) and does not demonstrate contrast enhancement (arrow) on the post-contrast T_1-weighted fat-saturated images (c).

14.7.7 Postoperative Changes

Postsurgical scar tissue may enhance for up to 2 years after surgery. In general, this resolves in approximately 6 months but can vary based on wound-healing capability. Most commonly, a postsurgical scar can be differentiated from recurrent or residual cancer by the degree of enhancement. In general, scar tissue has mild enhancement that is uniform along the incision site, whereas a recurrent or residual tumor often exhibits rapid enhancement with a more nodular appearance and may cause mass effect. Fat necrosis, which is also common in the postsurgical breast, can appear as a mass with irregular, intense rim-like enhancement. The hallmark of fat necrosis, though, is the presence of a signal consistent with fat in the center of the mass. This can be seen on the pre-contrast T_1-weighted sequences as high fat signal intensity and should demonstrate suppression on the fat-suppressed images (Figure 14.53).

Postoperative or postbiopsy seromas and hematomas are fluid-filled collections with variable internal signal intensity (Figure 14.54). Hematomas typically exhibit high signal on T_1-weighted images because of the presence of blood products, whereas seromas more commonly are bright on T_2-weighting. Both seromas and hematomas may exhibit uniform, thin peripheral enhancement. The presence of more nodular areas of enhancement should raise suspicion for the presence of residual or recurrent malignancy.

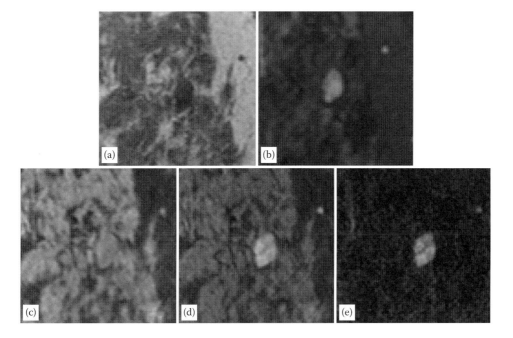

FIGURE 14.51

Fibroadenoma. A 60-year-old woman presented for high-risk screening breast MRI for personal history of treated breast cancer. Pre-contrast axial T_1-weighted (a), T_2-weighted SPAIR (b), T_1-weighted fat-saturated (c), post-contrast T_1-weighted fat-saturated (d), and T_1-weighted post-contrast subtracted (e) images show an enhancing oval mass with circumscribed margins, high T_2 signals, and dark internal septations consistent with a fibroadenoma.

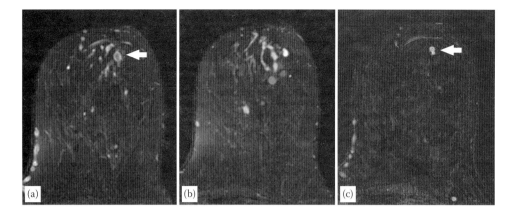

FIGURE 14.52

Duct ectasia due to an intraductal mass. A 54-year-old woman with newly diagnosed invasive carcinoma of the right breast. MRI was obtained to evaluate for extent of disease. In the contralateral breast, the subareolar region is an intraductal oval mass (arrow) with surrounding high signal on T_2-weighting related to fluid within a dilated duct on the axial T_2-weighted SPAIR image (a). The T_2-weighted SPAIR image at another level demonstrates several dilated fluid-filled ducts (b). This oval, circumscribed mass demonstrates avid heterogeneous enhancement on the T_1 post-contrast subtracted image (c). Biopsy showed sclerotic papilloma with focal atypical lobular hyperplasia.

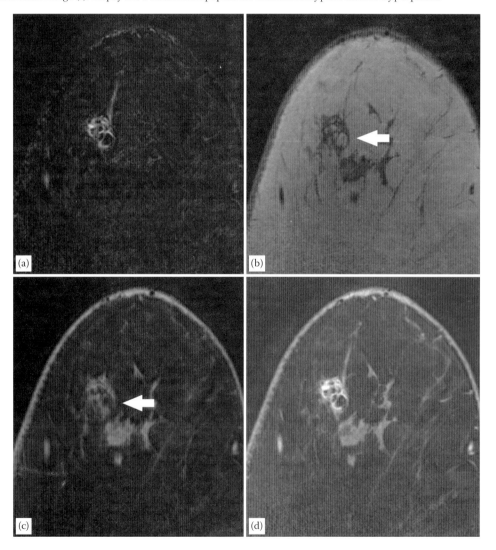

FIGURE 14.53

Fat necrosis. A 52-year-old woman with recent left mastopexy. Axial T_1 post-contrast subtracted images (a), pre-contrast T_1 (b), T_2-weighted SPAIR (c), and post-contrast T_1-weighted fat-saturated images (d) demonstrate a lobular mass with rim enhancement within the left breast upper inner quadrant with high central T_1 signal intensity (arrow) and low T_2 signal intensity (arrow), which demonstrates fat saturation. These findings are consistent with fat necrosis.

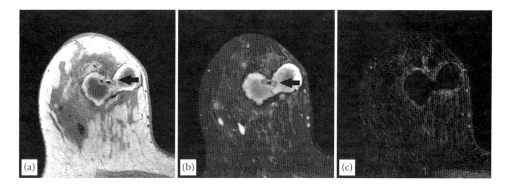

FIGURE 14.54
Hematoma. A 54-year-old woman's status post recent percutaneous needle biopsy. Pre-contrast axial T_1-weighted (a), T_2-weighted SPAIR (b), and post-contrast subtraction images (c) demonstrate a bi-lobed heterogeneous signal intensity collection with mild thin rim enhancement consistent with a hematoma. The biopsy marker is in the center of the collection (arrow).

14.8 Malignant MRI Findings

There are specific imaging features that have been shown to be highly specific for malignancy, which are described in greater detail below. In addition to assessment of a primary lesion, it is important to evaluate other associated or secondary features of breast cancer such as nipple retraction or invasion (Figures 14.55 and 14.56). These are listed in Box 14.5. It is also important to pay close attention to common sites of breast metastases that are included in the field of view of breast MRIs, such as the sternum and ribs (Figure 14.57), the pleural spaces (malignant pleural effusions), and liver, particularly in patients with evidence of locally advanced breast cancer.

14.8.1 Invasive Carcinoma

Most commonly, invasive breast cancers (regardless of histopathologic subtype) present as masses on MRI. Multiple prior studies have demonstrated that the specific morphological features of spiculated margins (positive predictive value = 76%–88%) and rim enhancement (positive predictive value 79%–92%) are highly predictive for the presence of invasive breast cancer [65]. It has also been shown that masses at least 10 mm in size that exhibit heterogeneous internal enhancement and irregular or spiculated margins are highly likely to be malignant [66,67], while those that demonstrate smooth (now referred to as circumscribed in the 5th edition of BI-RADS) margins with homogeneous enhancement have a very low probability of malignancy [67]. Kinetic

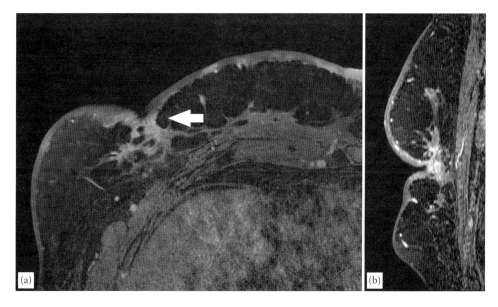

FIGURE 14.55
Nipple retraction. A 71-year-old woman with history of infiltrating ductal carcinoma with evidence of nipple retraction (arrow) secondary to the known malignancy evident on the axial (a) and sagittal (b) T_1-weighted post-contrast fat-saturated images.

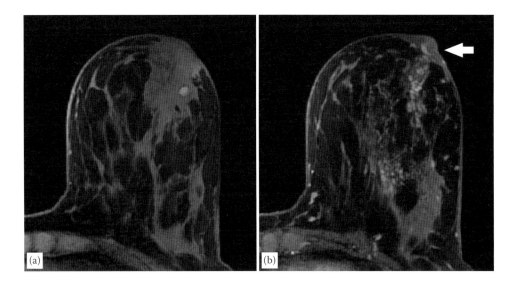

FIGURE 14.56
Nipple invasion. A 39-year-old woman presenting for MRI for high-risk screening. Pre- (a) and post-contrast (b) axial T_1-weighted fat-saturated images demonstrate nonmass enhancement within the lateral breast with enhancement extending to the nipple, indicating nipple invasion (arrow). This was biopsy-proven infiltrating ductal carcinoma.

BOX 14.5 ASSOCIATED FEATURES AND SECONDARY SIGNS OF BREAST CANCER

- Skin thickening, retraction or invasion (through direct invasion or inflammatory breast cancer)
- Nipple retraction or invasion
- Muscle invasion
- Growth through Cooper ligaments
- Architectural distortion
- Round, axillary lymph nodes without fatty hilum

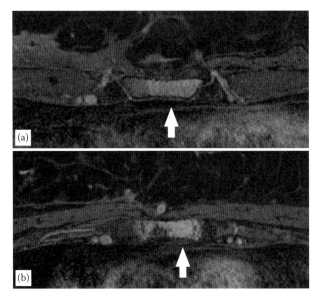

FIGURE 14.57
Sternal metastasis. A 70-year-old woman with history of stage IV right breast cancer. Breast MRI was performed to evaluate for treatment response. Post-contrast axial T_1-weighted images demonstrate enhancing sternal metastases (arrows) at two different levels of the sternum.

features, such as the presence of initial rapid enhancement [66] and delayed washout [68], have been shown to be predictive of the presence of malignancy but are less predictive overall than morphological features [54].

Invasive ductal carcinoma accounts for the vast majority (70%–90%) of invasive breast cancers and typically exhibits the MRI features described above (Figure 14.58). Invasive lobular carcinoma, the second most common subtype of invasive breast cancer, is widely recognized to account for a relatively high proportion of mammographically occult breast malignancies due to its infiltrative growth pattern. However, detection sensitivity of invasive lobular carcinoma on MRI is approximately 93%, which is not significantly lower than invasive ductal phenotype [69]. Furthermore, while the majority of invasive lobular carcinomas present as masses on MRI, those that present as NME are more likely to represent mammographically occult invasive lobular carcinomas that exhibit a diffuse, infiltrating growth pattern (Figure 14.59).

Several special subtypes of invasive breast cancer, including invasive papillary, mucinous, metaplastic,

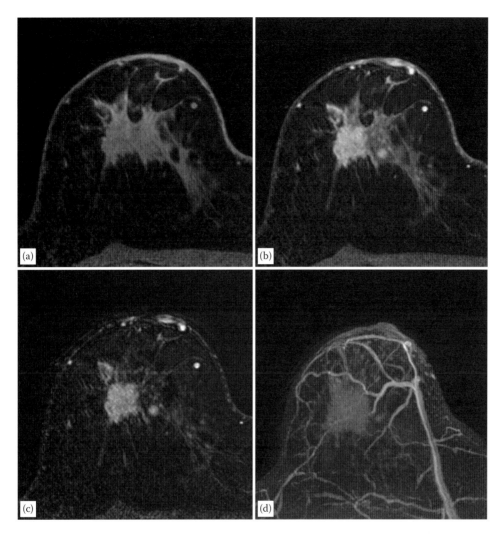

FIGURE 14.58
Invasive ductal carcinoma. A 42-year-old woman with newly diagnosed breast cancer, breast MRI obtained for extent of disease. Pre- (a) and post-contrast (b) axial T_1-weighted fat-saturated, post-contrast T_1-weighted subtracted image (c) and maximum intensity projection image (d) show an irregular mass with spiculated margins and heterogeneous enhancement, biopsy-proven invasive ductal carcinoma.

and medullary carcinomas, more often present with relatively benign imaging features on MRI, such as masses with round or oval shapes and circumscribed margins [70]. Invasive papillary carcinomas carry a favorable prognosis and are more frequently encountered in postmenopausal women and exhibit homogeneous internal enhancement. Both mucinous and metaplastic carcinomas are also more common in older women, and both can exhibit areas of high signal on T_2-weighted images, owing to the presence of mucin (for mucinous carcinomas) or cystic degeneration (for metaplastic carcinomas). Mucinous carcinomas also often exhibit variable amounts of signals on T_1-weighting depending on the amount of protein present in the lesion (Figure 14.60). In both metaplastic and mucinous tumors, viable tumor cells in the periphery of the lesion often account for the presence of irregular rim enhancement with variable

kinetic features on MRI. Medullary carcinomas are usually seen in relatively younger patients, with approximately 60% of patients less than 50 years old. In addition to oval or round morphology, these cancers also often exhibit rim enhancement with or without enhancing internal septations [71] and at times can be difficult to distinguish from fibroadenomas.

Malignant phyllodes tumors most commonly present in women in their fifth decade of life. These tumors more closely resemble a sarcomatous lesion on histopathology than other primary breast cancers, and they also typically demonstrate round or oval shapes with circumscribed margins on MRI, making them difficult to distinguish from fibroadenomas on the basis of MRI features alone. While less useful for primary diagnosis, MRI may be useful for delineating the full extent of disease for surgical planning when a known malignant

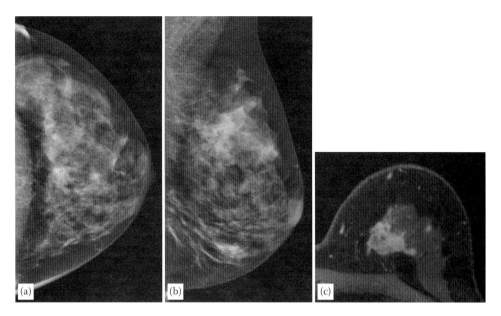

FIGURE 14.59
Invasive lobular carcinoma. A 48-year-old woman with personal history of LCIS. Craniocaudal and medial–lateral oblique conventional mammographic views of the left breast (a and b) were negative. MRI was obtained for high-risk screening, which demonstrated an irregular mass with spiculated borders. This was biopsied and shown to be invasive lobular carcinoma.

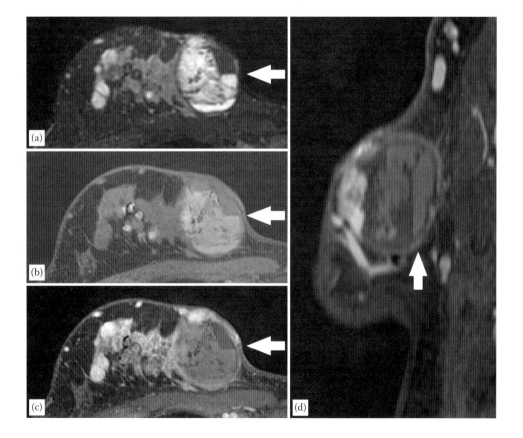

FIGURE 14.60
Mucinous carcinoma. A 44-year-old woman with newly diagnosed mucinous carcinoma. Pre-contrast axial T_2-weighted SPAIR (a), axial pre-contrast T_1-weighted fat-saturated (b), and axial (c) and sagittal (d) T_1-weighted fat-saturated post-contrast images show an oval mass with circumscribed margins and nodular enhancement with heterogeneous but mainly high signal intensity on the T_2-weighted image with evidence of fluid–fluid levels (arrows). Note the intrinsic high signal intensity on the pre-contrast T_1-weighted image. This is consistent with a mucinous-type tumor, confirmed on pathology.

phyllodes tumor is challenging to define by clinical assessment and/or standard imaging alone [49].

Inflammatory breast cancer is a form of locally advanced breast cancer that can be associated with any pathological subtype of invasive adenocarcinoma of the breast, most commonly invasive ductal carcinoma. Inflammatory breast cancer is diagnosed clinically by identifying the presence of erythema, edema (peau d'orange of the skin), and enlargement. Pathological identification of dermal lymphatic involvement can confirm the diagnosis, but it is not necessary for diagnosis since it is only present in 75% of skin biopsies. Similar to mammography, findings suggestive of inflammatory breast cancer on MRI include a unilaterally enlarged breast with skin thickening, skin enhancement, and diffuse breast edema (Figure 14.61) [72]. MRI is particularly useful for identifying the primary cancer (sensitivity of 98% compared to 68%), which can present as a mass or NME and for determining whether the

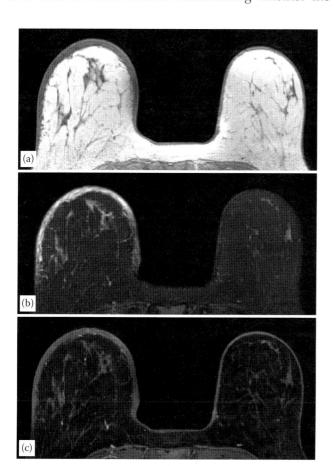

FIGURE 14.61
Inflammatory breast cancer. A 46-year-old woman with newly diagnosed inflammatory breast cancer. MRI was performed to evaluate extent of disease. Pre-contrast axial T_1-weighted (a), T_2-weighted SPAIR (b), and post-contrast T_1-weighted subtraction images (c) show skin thickening, edema, and enhancement with mild enlargement of the right breast compared with the left compatible with inflammatory breast cancer.

pectoralis musculature and regional lymph node chains are involved when compared with mammography [73]. Furthermore, MRI is superior to all other modalities for determining response to neoadjuvant chemotherapy, which has become the mainstay for inflammatory breast cancer therapy due to its ability to reduce the rate of metastatic spread and optimize locoregional control.

14.8.2 Ductal Carcinoma In Situ

Ductal carcinoma in situ (DCIS) is a nonobligatory precursor to invasive breast cancer that has been diagnosed with greater frequency due to more prevalent screening. Breast MRI was initially considered to be relatively poor for DCIS evaluation, with high false-negative rates [74]. As MRI techniques shifted from an emphasis on high temporal resolution to high spatial resolution, morphologic features such as NME that commonly represent DCIS became increasingly recognized. Multiple studies have since shown the superiority of MRI over mammography for both sensitivity of detection of DCIS (92% vs. 56%) [12] and accuracy of determination of recently diagnosed DCIS extent of disease [75]. Interestingly, MRI also identifies a greater fraction of high-grade lesions when compared with mammography [38]. Because it has been estimated that up to half of diagnosed DCIS lesions would never adversely impact a woman's life if left untreated, the realization that MRI may preferentially identify biologically aggressive forms of DCIS has prompted investigation of potential MRI biomarkers of DCIS risk [76].

DCIS most commonly presents on breast MRI as segmental or ductal NME with clumped internal enhancement morphology, accounting for 60%–80% of DCIS lesions visible on MRI. Less common presentations of DCIS on MRI include enhancing masses (14%–34%) and foci of enhancement (1%–12%) [77–79]. DCIS can have variable kinetic features, ranging from rapid initial enhancement with washout to no enhancement in approximately 5%–10% of cases (Figure 14.62) [80].

14.9 High-Risk Lesions on MRI

High-risk lesions encompass a diverse group of breast pathologies that, when diagnosed by core needle biopsy, have a significant chance to be upgraded to malignancy upon surgical excision due to undersampling of the lesion or variations in pathological interpretation from the core sample. The rate of upgrade of high-risk lesions diagnosed initially from MRI findings ranges from

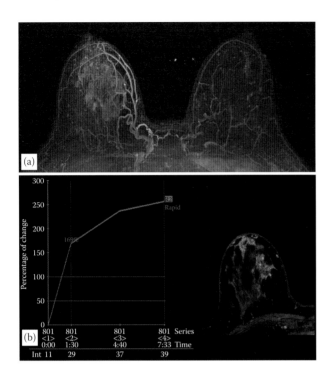

FIGURE 14.62
Ductal carcinoma in situ. A 53-year-old woman with right breast ductal carcinoma in situ (DCIS). MRI obtained for extent of disease. On the axial maximum intensity projection image (a), there is asymmetric segmental nonmass enhancement with heterogeneous internal enhancement extending from the nipple–areolar complex to 10 mm from the chest wall within the right breast. Computer-aided evaluation of the enhancement kinetics (b) demonstrates initial rapid enhancement (percent increased intensity >100%). However, there are persistent late-phase features, which are more typical of benign pathologies, but are relatively more common in DCIS. This illustrates the overlap in kinetic features among benign and malignant pathologies, and that kinetic curve evaluation should be performed only in the context of morphological features.

13% to 57% [81]. These lesions include atypical ductal hyperplasia, lobular neoplasia (including lobular carcinoma in situ [LCIS] and atypical lobular hyperplasia), radial scars/complex sclerosing lesions, papillomas, and flat epithelial atypia. Management (surgical excision versus observation) of some of these entities remains controversial and varies by institution, particularly for radial scars/complex sclerosing lesions and papillomas without atypia. Unfortunately, there are no specific morphologic or kinetic features that can be used to discriminate among high-risk lesions on MRI or determine suitability for follow-up rather than surgical excision at this time [82]. Regardless, as with benign pathologies, any finding on MRI that is deemed highly suspicious for malignancy (e.g., irregularly shaped mass with spiculated margins exhibiting heterogeneous or rim enhancement, with rapid initial phase and delayed washout kinetic curve type) is not concordant with core needle biopsy diagnosis of high-risk pathology, and additional pathological sampling should be pursued.

14.10 Evaluation of Silicone Breast Implants

MRI is the most sensitive and specific imaging modality to evaluate for silicone breast implant complications [30,32,83]. Single lumen-type silicone implants are composed of an elastomer shell filled with liquid silicone. On silicone-only inversion recovery images, the silicone should be bright and the elastomer shell itself should have no signal intensity and appear as a thin black line. To allow for the natural curvature of the breast, the implant itself is not taught, and thus, on imaging there may be normal undulations or infolding of the implant, called "radial folds," which can cause confusion when evaluating for rupture (Figure 14.63). Over time, the body creates a fibrous capsule surrounding the implant, which should also appear as a thin black line and, given its location adjacent to the shell, usually cannot be distinguished as a unique structure from the elastomer shell in a normal-appearing implant.

Implants are described on the basis of composition and location. As previously mentioned, MRI may be used in patients with suspected silicone implant complications, as saline implant rupture is typically clinically apparent. Implants may be located in one of two locations: prepectoral (also known as subglandular, retroglandular, or retromammary) or subpectoral. Prepectoral implants are located anterior to the pectoralis musculature but posterior to the FGT, while subpectoral implants are located posterior to the pectoralis major (but usually anterior to the pectoralis minor) muscle.

14.10.1 Silicone Implant Complications

The age of the silicone implant has been shown to be the most important predisposing factor for implant rupture, with studies showing that the number of patients with intact implants decrease with the age of the implant [31].

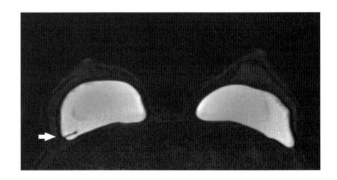

FIGURE 14.63
Radial fold. A 50-year-old woman with bilateral prepectoral silicone implants. High-risk screening. Axial silicone-only sequence demonstrates a thin hypointense line that can be traced to the surrounding implant shell, consistent with a radial fold, which is a normal finding.

One study estimated that the survival curve of an implant decreases from 89% after 8 years of implantation to 29% after 14 years to 5% after 20 years of implantation [28].

There are two major types of silicone implant ruptures—intracapsular and extracapsular. In intracapsular ruptures, there is a disruption of the implant shell; however, the outer fibrous capsule is still intact, and thus, the silicone is outside the shell but still contained by the fibrous capsule. In extracapsular ruptures, there is disruption of both the elastomer shell and the fibrous capsule, which allows for silicone to diffuse into the normal breast tissue. In these cases, there may be drainage of the silicone through the lymphatics and into axillary and intramammary lymph nodes (Figure 14.64). Over time, granulomatous inflammation will occur, and the signal related to extracapsular silicone may decrease.

Trauma is the most common cause of silicone implant rupture, which can start at the time of implantation when a small rent can be created that allows for leakage of silicone outside of the shell. Silicone oils can dissociate from the gel, diffuse through the elastomer shell, and over time can cause adherence of the elastomer shell to the fibrous capsule. This is the cause for more focal intracapsular ruptures, as the silicone can no longer freely travel between the implant shell and the fibrous capsule. Over time, the silicone instead causes focal invagination of the implant shell itself and appears as an "inverted-loop" or "keyhole" sign of an uncollapsed intracapsular rupture (Figure 14.65a). As the leakage progresses, silicone gel may interdigitate between the shell and the fibrous capsule, causing the "subcapsular line sign" (Figure 14.65b). As leakage continues, the implant shell can then collapse and fold upon itself, creating a stack of low-signal intensity lines. The leaked silicone will slowly surround the implant capsule, which will appear as a dark or wavy

"floating" line, sometimes called the "linguine sign" of an intracapsular rupture. This is the most specific sign of implant rupture (Figure 14.66).

Long radial folds can be confused with the linguine sign of intracapsular rupture; however, this can be resolved by tracing the line back to an intact elastomer shell. Radial folds should always trace back to the implant shell, whereas the lines from a collapsed intracapsular rupture do not. Nonetheless, this at times can be difficult to resolve, and complex radial folds can be mistakenly interpreted as evidence of intracapsular rupture.

14.11 Histopathologic Sampling of Suspicious Lesions Identified on MRI

When a suspicious lesion is identified on breast MRI, it requires histopathologic evaluation for definitive diagnosis. In order to achieve this, there are two approaches that are typically pursued: targeted ultrasound to evaluate for a sonographic correlate of the suspicious lesion followed by ultrasound-guided core needle biopsy or MRI-guided core needle biopsy. There are several advantages to performing a targeted ultrasound of suspicious MRI findings. First, it allows further characterization of the MRI finding, at times resolving the finding as benign (e.g., intramammary lymph node). Second, ultrasound-guided biopsy is less expensive and is more comfortable for the patient compared with MRI-guided biopsy. Finally, ultrasound allows greater exposure to the posterior and axillary tissue and can provide access to lesions intimately associated with prepectoral implants.

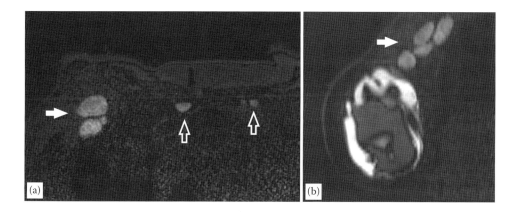

FIGURE 14.64
Extracapsular silicone rupture with silicone deposits in the axillary lymph nodes. Axial (a) and sagittal (b) silicone-only sequences in a patient with bilateral double-lumen implants with a saline inner and silicone outer lumen. There is extracapsular rupture with silicone deposits in the axillary (solid arrow) and internal mammary lymph nodes (hollow arrow).

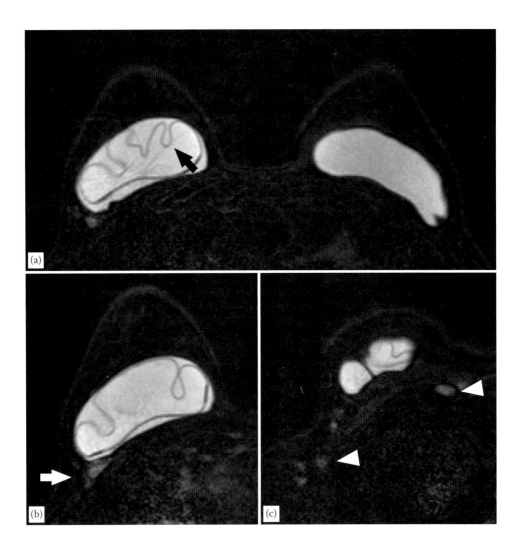

FIGURE 14.65
Uncollapsed intracapsular rupture. A 60-year-old woman with bilateral prepectoral silicone implants, MRI performed to evaluate for rupture. Axial silicone-only sequence of the bilateral breasts (a), and the right breast at different levels (b and c) demonstrates the "keyhole sign" (black arrow) of intracapsular rupture in addition to the high silicone weighted signal at the posterolateral aspect of the right silicone implant (white arrow) consistent with extracapsular rupture. In addition, there is a high silicone weighted signal in level 1 axillary lymph nodes and in a single internal mammary lymph node, consistent with silicone deposits, thus further evidence of extracapsular rupture (arrowheads).

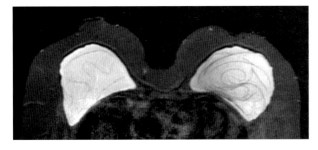

FIGURE 14.66
Linguine sign of intracapsular silicone implant rupture. A 76-year-old woman's status post skin-sparing bilateral mastectomies with bilateral subpectoral silicone implant reconstruction. The axial silicone-only sequence shows the "linguine sign" of bilateral collapsed intracapsular rupture with multiple thin hypointense lines that do not track to the implant shell.

There are also significant risks in routinely performing a targeted ultrasound to further evaluate MRI-suspicious lesions. At times, it can be challenging to definitively identify a sonographic correlate for a suspicious MRI finding, particularly in the cases of NME, small lesions less than 5–10 mm, and in women with heterogeneous breast tissue on ultrasound. In addition, performing a targeted ultrasound routinely to evaluate MRI findings can lead to unnecessary delays in a patient's workup and definitive diagnosis of a lesion. Finally, a negative sonographic evaluation cannot obviate the need for histopathologic sampling of MRI-detected suspicious findings since 43%–53% of MRI-detected malignancies do not have a sonographic correlate [84,85]. Because of these factors, the authors advocate only performing targeted ultrasound after breast MRI when it is felt that there is a high likelihood that the lesion will be found with ultrasound guidance; typically, this is reserved for masses that are larger in size (typically at least 10 mm in size), in a location that is felt highly conducive for sonographic identification, or located in the axilla or near a prepectoral implant. We do not routinely evaluate suspicious NME or foci by ultrasound because of the lower likelihood of identification with ultrasound [85]. When a targeted ultrasound is recommended, it is important that the interpreting radiologist provide clear instructions that sampling of the lesion, when possible, should be pursued with MRI guidance in the event that the lesion cannot be identified sonographically.

While multiple acceptable approaches exist, the basic requirements and steps in performing an MRI-guided breast biopsy are similar, regardless of vendor or platform. Detailed guidance on how to perform MRI-guided biopsies for each system is beyond the scope of this chapter. To summarize the basic steps, first the patient is imaged in prone position using a dedicated breast coil with the breast containing the lesion of interest compressed in a grid that contains a fiducial marker, which is visible on T_1-weighted sequences. Pre-contrast and post-contrast images are obtained (typically in sagittal orientation as most techniques only allow for either a medial or lateral approach) and the location of the lesion is then determined (using either a manual method or with the assistance of software), allowing the appropriate box on the grid to enter on the skin and the correct needle depth to be identified. After sterilizing the skin with a cleaning agent, local anesthesia is administered, and the outer component of an MRI biopsy coaxial system is introduced through the skin (use of a scalpel is optional and in the authors' opinion rarely needed) to the appropriate depth in the breast using the inner cutting stylet. The coaxial sheath is held into placed by a plastic needle guide (a square plastic box typically with 9 tunnels) that fits into the appropriate grid space, and the inner stylet is removed and replaced with a plastic obturator. A scan of the breast is then performed to confirm the location of the sheath, and the obturator is replaced with an MRI-compatible vacuum-assisted breast biopsy device. Typically 8–12 samples are obtained, and a biopsy marker clip is placed through the sheath. A final postbiopsy MRI scan and two-view conventional mammogram are then performed to confirm clip placement and location. Placement of a clip is essential, as it facilitates mammographic wire localization if needed, which is easier to perform and much less expensive than MRI-guided wire localization.

14.12 Summary

Breast MRI is a powerful imaging tool that offers the greatest sensitivity for the detection of breast cancer and silicone implant complications and provides unparalleled imaging-based lesion characterization useful for evaluation of disease extent and response of breast cancers to neoadjuvant chemotherapy. Due to its relatively high cost compared with ultrasound and conventional mammography and its imperfect specificity, breast MRI must be used judiciously and for appropriate indications so that its clinical value can be maximized. Furthermore, it is essential to have a thorough understanding of basic technical requirements for high-quality MRI acquisition in order to avoid suboptimal images and artifacts. Finally, the use of standardized reporting, as outlined by the ACR BI-RADS atlas, is essential to provide consistent and clinically relevant reports and recommendations.

FOCUS POINTS

- Indications for breast MRI include high-risk screening for patients who have a greater than 20% lifetime risk of developing cancer, evaluation of patients with newly diagnosed breast cancer for extent of disease, evaluation of response to neoadjuvant chemotherapy, identification of an occult breast cancer in women with axillary metastasis with unknown primary, and evaluation of silicone implant integrity.

- Breast MRI should be performed with a high field-strength magnet and a dedicated breast coil to allow high spatial and temporal resolution dynamic-contrast-enhanced images.

- Thorough understanding of standardized reporting with BI-RADS is essential to providing focused and clinically relevant MRI assessments and recommendations.

- The most suspicious characteristic of an MRI finding should be used to guide BI-RADS assessments, and suspicious morphologic features trump reassuring kinetic enhancement curves.

- Because of the imperfect specificity of breast MRI, suspicious findings identified on breast MRI require histopathologic sampling prior to affecting clinical decision making, which can be performed either under MRI guidance or by ultrasound in appropriately selected cases.

References

1. Damadian R. Tumor detection by nuclear magnetic resonance. *Science (New York)*. 1971;171(3976):1151–1153.

2. Bovee WM, Getreuer KW, Smidt J, Lindeman J. Nuclear magnetic resonance and detection of human breast tumor. *Journal of the National Cancer Institute*. 1978;61(1):53–55.

3. Mansfield P, Morris PG, Ordidge RJ, Pykett IL, Bangert V, Coupland RE. Human whole body imaging and detection of breast tumours by n.m.r. *Philosophical Transactions of the Royal Society of London Series B, Biological Sciences*. 1980;289(1037):503–510.

4. Ross RJ, Thompson JS, Kim K, Bailey RA. Nuclear magnetic resonance imaging and evaluation of human breast tissue: Preliminary clinical trials. *Radiology*. 1982;143(1):195–205.

5. Kaiser WA, Zeitler E. MR imaging of the breast: Fast imaging sequences with and without Gd-DTPA. Preliminary observations. *Radiology*. 1989;170(3 Pt 1):681–686.

6. Orel SG, Schnall MD. MR imaging of the breast for the detection, diagnosis, and staging of breast cancer. *Radiology*. 2001;220(1):13–30.

7. Saslow D, Boetes C, Burke W et al. American Cancer Society guidelines for breast screening with MRI as an adjunct to mammography. *CA: A Cancer Journal for Clinicians*. 2007;57(2):75–89.

8. Mainiero MB, Lourenco A, Mahoney MC et al. ACR appropriateness criteria breast cancer screening. *Journal of the American College of Radiology: JACR*. 2013;10(1):11–14.

9. Brennan S, Liberman L, Dershaw DD, Morris E. Breast MRI screening of women with a personal history of breast cancer. *AJR American Journal of Roentgenology*. 2010;195(2):510–516.

10. Sung JS, Malak SF, Bajaj P, Alis R, Dershaw DD, Morris EA. Screening breast MR imaging in women with a history of lobular carcinoma in situ. *Radiology*. 2011;261(2):414–420.

11. Schacht DV, Yamaguchi K, Lai J, Kulkarni K, Sennett CA, Abe H. Importance of a personal history of breast cancer as a risk factor for the development of subsequent breast cancer: Results from screening breast MRI. *AJR American Journal of Roentgenology*. 2014;202(2):289–292.

12. Lehman CD, Gatsonis C, Kuhl CK et al. MRI evaluation of the contralateral breast in women with recently diagnosed breast cancer. *The New England Journal of Medicine*. 2007;356(13):1295–1303.

13. DeMartini W, Lehman C. A review of current evidence-based clinical applications for breast magnetic resonance imaging. *Topics in Magnetic Resonance Imaging: TMRI*. 2008;19(3):143–150.

14. Liberman L. Breast MR imaging in assessing extent of disease. *Magnetic Resonance Imaging Clinics of North America*. 2006;14(3):339–349, vi.

15. Sung JS, Li J, Costa GD et al. Preoperative breast MRI for early-stage breast cancer: Effect on surgical and long-term outcomes. *AJR American Journal of Roentgenology*. 2014;202(6):1376–1382.

16. Lee CI, Bensink ME, Berry K et al. Performance goals for an adjunct diagnostic test to reduce unnecessary biopsies after screening mammography: Analysis of costs, benefits, and consequences. *Journal of the American College of Radiology: JACR*. 2013;10(12):924–930.

17. Cilotti A, Iacconi C, Marini C et al. Contrast-enhanced MR imaging in patients with BI-RADS 3–5 microcalcifications. *La Radiologia Medica*. 2007;112(2):272–286.

18. Bluemke DA, Gatsonis CA, Chen MH et al. Magnetic resonance imaging of the breast prior to biopsy. *JAMA*. 2004;292(22):2735–2742.

19. Yau EJ, Gutierrez RL, DeMartini WB, Eby PR, Peacock S, Lehman CD. The utility of breast MRI as a problem-solving tool. *The Breast Journal*. 2011;17(3):273–280.

20. Lorenzon M, Zuiani C, Linda A, Londero V, Girometti R, Bazzocchi M. Magnetic resonance imaging in patients with nipple discharge: Should we recommend it? *European Radiology*. 2011;21(5):899–907.

21. Fisher B, Bryant J, Wolmark N et al. Effect of preoperative chemotherapy on the outcome of women with operable breast cancer. *Journal of Clinical Oncology: Official Journal of the American Society of Clinical Oncology*. 1998;16(8):2672–2685.

22. Wolmark N, Wang J, Mamounas E, Bryant J, Fisher B. Preoperative chemotherapy in patients with operable breast cancer: Nine-year results from National Surgical Adjuvant Breast and Bowel Project B-18. *Journal of the National Cancer Institute Monographs*. 2001(30):96–102.

23. Martincich L, Montemurro F, De Rosa G et al. Monitoring response to primary chemotherapy in breast cancer using dynamic contrast-enhanced magnetic resonance imaging. *Breast Cancer Research and Treatment*. 2004;83(1):67–76.

24. Pickles MD, Lowry M, Manton DJ, Gibbs P, Turnbull LW. Role of dynamic contrast enhanced MRI in monitoring early response of locally advanced breast cancer to neoadjuvant chemotherapy. *Breast Cancer Research and Treatment*. 2005;91(1):1–10.

25. Rosen EL, Blackwell KL, Baker JA et al. Accuracy of MRI in the detection of residual breast cancer after neoadjuvant chemotherapy. *AJR American Journal of Roentgenology*. 2003;181(5):1275–1282.

26. Weatherall PT, Evans GF, Metzger GJ, Saborrian MH, Leitch AM. MRI vs. histologic measurement of breast cancer following chemotherapy: Comparison with x-ray mammography and palpation. *Journal of Magnetic Resonance Imaging*. 2001;13(6):868–875.

27. Hylton NM, Blume JD, Bernreuter WK et al. Locally advanced breast cancer: MR imaging for prediction of response to neoadjuvant chemotherapy—results from ACRIN 6657/I-SPY TRIAL. *Radiology*. 2012;263(3):663–672.

28. Robinson OG, Jr., Bradley EL, Wilson DS. Analysis of explanted silicone implants: A report of 300 patients. *Annals of Plastic Surgery*. 1995;34(1):1–6; discussion 7.

29. Brenner RJ. Evaluation of breast silicone implants. *Magnetic Resonance Imaging Clinics of North America*. 2013;21(3):547–560.

30. Ahn CY, DeBruhl ND, Gorczyca DP, Shaw WW, Bassett LW. Comparative silicone breast implant evaluation using mammography, sonography, and magnetic resonance imaging: Experience with 59 implants. *Plastic and Reconstructive Surgery.* 1994;94(5):620–627.

31. Berg WA, Caskey CI, Hamper UM et al. Diagnosing breast implant rupture with MR imaging, US, and mammography. *RadioGraphics: A Review Publication of the Radiological Society of North America, Inc.* 1993;13(6):1323–1336.

32. Everson LI, Parantainen H, Detlie T et al. Diagnosis of breast implant rupture: Imaging findings and relative efficacies of imaging techniques. *AJR American Journal of Roentgenology.* 1994;163(1):57–60.

33. Reynolds HE, Buckwalter KA, Jackson VP, Siwy BK, Alexander SG. Comparison of mammography, sonography, and magnetic resonance imaging in the detection of silicone-gel breast implant rupture. *Annals of Plastic Surgery.* 1994;33(3):247–255; discussion 56–57.

34. Weizer G, Malone RS, Netscher DT, Walker LE, Thornby J. Utility of magnetic resonance imaging and ultrasonography in diagnosing breast implant rupture. *Annals of Plastic Surgery.* 1995;34(4):352–361.

35. Hendrick RE. High-quality breast MRI. *Radiologic Clinics of North America.* 2014;52(3):547–562.

36. Rausch DR, Hendrick RE. How to optimize clinical breast MR imaging practices and techniques on Your 1.5-T system. *RadioGraphics: A Review Publication of the Radiological Society of North America, Inc.* 2006;26(5): 1469–1484.

37. Rahbar H, Partridge SC, Javid SH, Lehman CD. Imaging axillary lymph nodes in patients with newly diagnosed breast cancer. *Current Problems in Diagnostic Radiology.* 2012;41(5):149–158.

38. Kuhl CK, Schrading S, Bieling HB et al. MRI for diagnosis of pure ductal carcinoma in situ: A prospective observational study. *Lancet.* 2007;370(9586):485–492.

39. Nnewihe AN, Grafendorfer T, Daniel BL et al. Custom-fitted 16-channel bilateral breast coil for bidirectional parallel imaging. *Magnetic Resonance in Medicine.* 2011;66(1):281–289.

40. Hendrick RE, Haacke EM. Basic physics of MR contrast agents and maximization of image contrast. *Journal of Magnetic Resonance Imaging.* 1993;3(1):137–148.

41. Kuhl CK. Breast MR imaging at 3T. *Magnetic Resonance Imaging Clinics of North America.* 2007;15(3):315–320, vi.

42. Wang LC, DeMartini WB, Partridge SC, Peacock S, Lehman CD. MRI-detected suspicious breast lesions: Predictive values of kinetic features measured by computer-aided evaluation. *AJR American Journal of Roentgenology.* 2009;193(3):826–831.

43. Huang W, Tudorica LA, Li X et al. Discrimination of benign and malignant breast lesions by using shutter-speed dynamic contrast-enhanced MR imaging. *Radiology.* 2011;261(2):394–403.

44. Li X, Huang W, Yankeelov TE, Tudorica A, Rooney WD, Springer CS, Jr. Shutter-speed analysis of contrast reagent bolus-tracking data: Preliminary observations in benign and malignant breast disease. *Magnetic Resonance in Medicine.* 2005;53(3):724–729.

45. Bitar R, Leung G, Perng R et al. MR pulse sequences: What every radiologist wants to know but is afraid to ask. *RadioGraphics.* 2006;26(2):513–537.

46. Kuhl CK. Current status of breast MR imaging. Part 2. Clinical applications. *Radiology.* 2007;244(3):672–691.

47. Rahbar H, Partridge SC, DeMartini WB, Thursten B, Lehman CD. Clinical and technical considerations for high quality breast MRI at 3 Tesla. *Journal of Magnetic Resonance Imaging: JMRI.* 2013;37(4):778–790.

48. Kuhl CK, Jost P, Morakkabati N, Zivanovic O, Schild HH, Gieseke J. Contrast-enhanced MR imaging of the breast at 3.0 and 1.5 T in the same patients: Initial experience. *Radiology.* 2006;239(3):666–676.

49. Calhoun KE, Allison KA, Kim JN, Rahbar H, Anderson BO. Phyllodes tumors. In *Diseases of the Breast*, 5th edn, Harris JR, Lippman ME, Morrow M, Osborne CK (eds), 2014. Wolters Kluwer Health, Philadelphia, PA.

50. Morris EA, Comstock CE, Lee CH et al. *ACR BI-RADS Atlas, Breast Imaging Reporting and Data System.* Reston, VA, American College of Radiology; 2013.

51. Burnside ES, Sickles EA, Bassett LW et al. The ACR BI-RADS experience: Learning from history. *Journal of the American College of Radiology: JACR.* 2009;6(12):851–860.

52. DeMartini WB, Liu F, Peacock S, Eby PR, Gutierrez RL, Lehman CD. Background parenchymal enhancement on breast MRI: Impact on diagnostic performance. *AJR American Journal of Roentgenology.* 2012;198(4):W373–W380.

53. Hambly NM, Liberman L, Dershaw DD, Brennan S, Morris EA. Background parenchymal enhancement on baseline screening breast MRI: Impact on biopsy rate and short-interval follow-up. *AJR American Journal of Roentgenology.* 2011;196(1):218–224.

54. Schnall MD, Blume J, Bluemke DA et al. Diagnostic architectural and dynamic features at breast MR imaging: Multicenter study. *Radiology.* 2006;238(1):42–53.

55. Peters NH, Borel Rinkes IH, Zuithoff NP, Mali WP, Moons KG, Peeters PH. Meta-analysis of MR imaging in the diagnosis of breast lesions. *Radiology.* 2008;246(1):116–124.

56. Partridge SC, McDonald ES. Diffusion weighted magnetic resonance imaging of the breast: Protocol optimization, interpretation, and clinical applications. *Magnetic Resonance Imaging Clinics of North America.* 2013;21(3):601–624.

57. Kuroki-Suzuki S, Kuroki Y, Nasu K, Nawano S, Moriyama N, Okazaki M. Detecting breast cancer with non-contrast MR imaging: Combining diffusion-weighted and STIR imaging. *Magnetic Resonance in Medical Sciences: MRMS: An Official Journal of Japan Society of Magnetic Resonance in Medicine.* 2007;6(1):21–27.

58. Bolan PJ. Magnetic resonance spectroscopy of the breast: Current status. *Magnetic Resonance Imaging Clinics of North America.* 2013;21(3):625–639.

59. Wurdinger S, Herzog AB, Fischer DR et al. Differentiation of phyllodes breast tumors from fibroadenomas on MRI. *AJR American Journal of Roentgenology.* 2005;185(5):1317–1321.

60. Brinck U, Fischer U, Korabiowska M, Jutrowski M, Schauer A, Grabbe E. The variability of fibroadenoma in contrast-enhanced dynamic MR mammography. *AJR American Journal of Roentgenology.* 1997;168(5):1331–1334.

61. Lewis JT, Hartmann LC, Vierkant RA et al. An analysis of breast cancer risk in women with single, multiple, and atypical papilloma. *The American Journal of Surgical Pathology.* 2006;30(6):665–672.

62. Liberman L, Tornos C, Huzjan R, Bartella L, Morris EA, Dershaw DD. Is surgical excision warranted after benign, concordant diagnosis of papilloma at percutaneous breast biopsy? *AJR American Journal of Roentgenology.* 2006;186(5):1328–1334.

63. Jaffer S, Bleiweiss IJ, Nagi C. Incidental intraductal papillomas (<2 mm) of the breast diagnosed on needle core biopsy do not need to be excised. *The Breast Journal.* 2013;19(2):130–133.

64. Daniel BL, Gardner RW, Birdwell RL, Nowels KW, Johnson D. Magnetic resonance imaging of intraductal papilloma of the breast. *Magnetic Resonance Imaging.* 2003;21(8):887–892.

65. Heller SL, Hernandez O, Moy L. Radiologic-pathologic correlation at breast MR imaging: What is the appropriate management for high-risk lesions? *Magnetic Resonance Imaging Clinics of North America.* 2013;21(3):583–599.

66. Mahoney MC, Gatsonis C, Hanna L, DeMartini WB, Lehman C. Positive predictive value of BI-RADS MR imaging. *Radiology.* 2012;264(1):51–58.

67. Gutierrez RL, DeMartini WB, Eby PR, Kurland BF, Peacock S, Lehman CD. BI-RADS lesion characteristics predict likelihood of malignancy in breast MRI for masses but not for nonmasslike enhancement. *AJR American Journal of Roentgenology.* 2009;193(4):994–1000.

68. Kuhl CK, Mielcareck P, Klaschik S et al. Dynamic breast MR imaging: Are signal intensity time course data useful for differential diagnosis of enhancing lesions? *Radiology.* 1999;211(1):101–110.

69. Mann RM, Hoogeveen YL, Blickman JG, Boetes C. MRI compared to conventional diagnostic work-up in the detection and evaluation of invasive lobular carcinoma of the breast: A review of existing literature. *Breast Cancer Research and Treatment.* 2008;107(1):1–14.

70. Yoo JL, Woo OH, Kim YK et al. Can MR Imaging contribute in characterizing well-circumscribed breast carcinomas? *RadioGraphics.* 2010;30(6):1689–1702.

71. Jeong SJ, Lim HS, Lee JS et al. Medullary carcinoma of the breast: MRI findings. *AJR American Journal of Roentgenology.* 2012;198(5):W482–W487.

72. Chow CK. Imaging in inflammatory breast carcinoma. *Breast Disease.* 2005;22:45–54.

73. Le-Petross HT, Cristofanilli M, Carkaci S et al. MRI features of inflammatory breast cancer. *AJR American Journal of Roentgenology.* 2011;197(4):W769–W776.

74. Boetes C, Strijk SP, Holland R, Barentsz JO, Van Der Sluis RF, Ruijs JH. False-negative MR imaging of malignant breast tumors. *European Radiology.* 1997;7(8):1231–1234.

75. Berg WA, Gutierrez L, NessAiver MS et al. Diagnostic accuracy of mammography, clinical examination, US, and MR imaging in preoperative assessment of breast cancer. *Radiology.* 2004;233(3):830–849.

76. Rahbar H, Partridge SC, Demartini WB et al. In vivo assessment of ductal carcinoma in situ grade: A model incorporating dynamic contrast-enhanced and diffusion-weighted breast MR imaging parameters. *Radiology.* 2012;263(2):374–382.

77. Jansen SA, Newstead GM, Abe H, Shimauchi A, Schmidt RA, Karczmar GS. Pure ductal carcinoma in situ: Kinetic and morphologic MR characteristics compared with mammographic appearance and nuclear grade. *Radiology.* 2007;245(3):684–691.

78. Rosen EL, Smith-Foley SA, DeMartini WB, Eby PR, Peacock S, Lehman CD. BI-RADS MRI enhancement characteristics of ductal carcinoma in situ. *The Breast Journal.* 2007;13(6):545–550.

79. Menell JH, Morris EA, Dershaw DD, Abramson AF, Brogi E, Liberman L. Determination of the presence and extent of pure ductal carcinoma in situ by mammography and magnetic resonance imaging. *The Breast Journal.* 2005;11(6):382–390.

80. Kuhl CK. Concepts for differential diagnosis in breast MR imaging. *Magnetic Resonance Imaging Clinics of North America.* 2006;14(3):305–328, v.

81. Heller SL, Moy L. Imaging features and management of high-risk lesions on contrast-enhanced dynamic breast MRI. *AJR American Journal of Roentgenology.* 2012;198(2):249–255.

82. Strigel RM, Eby PR, Demartini WB et al. Frequency, upgrade rates, and characteristics of high-risk lesions initially identified with breast MRI. *AJR American Journal of Roentgenology.* 2010;195(3):792–798.

83. Gorczyca DP, DeBruhl ND, Ahn CY et al. Silicone breast implant ruptures in an animal model: Comparison of mammography, MR imaging, US, and CT. *Radiology.* 1994;190(1):227–232.

84. LaTrenta LR, Menell JH, Morris EA, Abramson AF, Dershaw DD, Liberman L. Breast lesions detected with MR imaging: Utility and histopathologic importance of identification with US. *Radiology.* 2003;227(3):856–861.

85. Demartini WB, Eby PR, Peacock S, Lehman CD. Utility of targeted sonography for breast lesions that were suspicious on MRI. *AJR American Journal of Roentgenology.* 2009;192(4):1128–1134.

15

Liver: Technique and Diffuse Pathology

Michele Di Martino and Carlo Catalano

CONTENTS

15.1 Introduction

In the past few decades, with the development of new hardware and software which lead to the acquisition of faster sequences, magnetic resonance (MR) overcomes the main problem in the study of the liver which is the motion artifacts. Moreover, the better contrast resolution compared to computed tomography (CT) results in the characterization of different components (i.e., fat, water, and blood) within the same parenchyma. However in most cases, these features are not enough to correctly detect and characterize liver diseases [1]. It is important to use appropriate technical equipment to obtain good image quality. MR liver examination should be performed with a high field magnet (≥1.5 T) with fast gradients; a phase-array surface body coil increases the signal-to-noise ratio (SNR) providing a better spatial resolution, and the parallel imaging reduces the scan time acquisition [2–4]. The introduction of gadolinium-based contrast agents has also improved the diagnostic accuracy of MR since it has similar values to that of CT. Moreover, the introduction in clinical practice of liver-specific contrast agents has provided functional information in adjunction to morphological appearance. New MR techniques have also been developed such as diffusion, perfusion, spectroscopy, and elastography that are now assuming an increasing role in clinical practice.

15.2 MR Technique

The magnetic field strength of choice currently employed for body MR is 1.5 T. At the current state of development, this field strength provides an optimal combination of SNR and speed, allowing optimization of rapid acquisition techniques while staying within government institution-determined energy deposition-rate limits. These systems also provide a good balance between T_1 values, which are dependent on field strength, and achievable contrast effects. In addition, field distortion and paramagnetic effects that increase with increasing field strength possibly resulting in undesirable image artifacts remain within tolerable limits at 1.5 T. There are theoretical considerations favoring development of higher field systems for body MR and efforts are well underway to transfer techniques used at 1.5 T to 3.0 T. However, it has become apparent that the approaches used previously to migrate from lower to higher (1.5 T) field systems have not proven successful in migrating from 1.5 T to 3 T (see below). MR scan protocol for the study of the liver disease includes T_1-weighted and T_2-weighted sequences with and without suppression of adipose tissue and T_1-weighted sequences after the intravenous injection on contrast agent.

15.2.1 T_2-Weighted Sequence

Currently, T_2-weighted turbo spin echo (TSE) or fast spin echo (FSE) is used for the study of liver parenchyma (Figure 15.1). The half-Fourier single-shot turbo spin echo (HASTE) sequence is helpful in uncooperative, medically unstable or claustrophobic patients; another acronym is single-shot FSE (SSFSE) [5]. With HASTE only half the K-space is filled, with a few additional lines to correct the imperfections, the remaining part being reconstructed by software. The time saving, however, has its price as the SNR is decreased [6]. Sequences that are sensitive to magnetic susceptibility, such as T_2^*-weighted sequences, are useful for the study of hemochromatosis or for focal liver lesions following the injection of superparamagnetic iron oxide (SPIO) contrast agents. T_2^*-weighted gradient-recalled-echo (GRE) images are more sensitive to the effects of SPIO in comparison to T_2-weighted FSE, because they lack a 180° refocusing pulse, so the effect of susceptibility from local field inhomogeneities is increased [7]. T_2-weighted sequences are extremely helpful in the differentiation between fluid and solid lesions: the fluid lesions show high signal intensity on T_2, by contrast solid lesions appear as isointense or slightly hyperintense to the surrounding liver parenchyma (Figure 15.2).

15.2.2 T_1-Weighted Sequence

T_1-weighted sequences, as well as T_2-weighted, are routinely acquired in the evaluation of liver disease. Precontrast sequences are useful in the depiction of fibrous tissue or fluid collection that generally shows a low signal intensity; by contrast, elements such as fat, protein, and blood collection (hemorrhage/hemorrhagic cysts) reveal a high signal intensity. The possibility of suppressing the adipose tissue signal intensity helps in the characterization of fat-containing lesions. After contrast medium administration, it is recommended

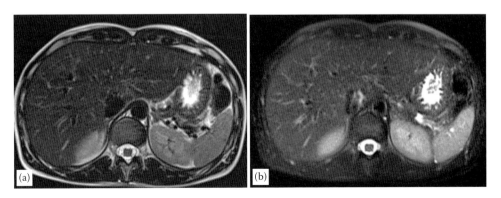

FIGURE 15.1
T_2-weighted images for the study of the liver. (a and b) Turbo spin echo T_2-weighted MR images with and without fat saturation respectively.

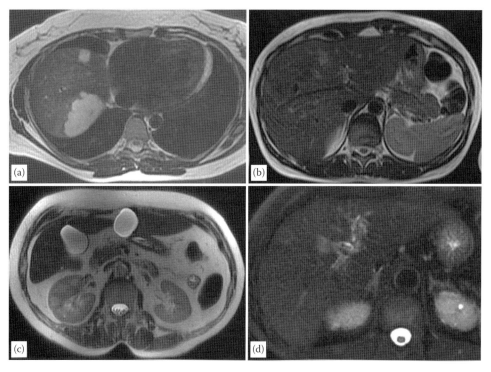

FIGURE 15.2
T_2-weighted images of focal liver lesions. (a) Hemangioma, (b) focal nodular hyperplasia, (c) hepatic cyst, and (d) liver metastases. The signal intensity of the cyst is similar to that of gallbladder.

to apply the fat suppression to increase the contrast between the lesion and the adjacent structures. T_1-weighted spoiled gradient echo sequences (SGE) are the best in the study of the upper abdomen. The relatively long repetition time (TR) and the short echo time (TE), with the adjunction of parallel imaging, enable to scan the region of interest (ROI) in a single breath-hold of 20 s. Among the MR techniques that allow the suppression of fat tissue, in-phase/out-of-phase imaging is useful for demonstrating disease in which fat and water are present within the same voxel. Another use of dual-echo imaging is to identify paramagnetic effects associated with iron [8]. On the longer second echo (4.4 ms at 1.5 T) image, iron will lead to the loss of tissue signal due to T_2^* effect [9,10]. Fat-suppressed images are acquired before contrast injection to detect protein (mucin, melanin) or blood elements [11]. Fat suppression is achieved by using a tuned radio-frequency (RF) pulse for selectively stimulating the fat protons, prior to acquiring the sequence. The image parameters are similar to those for standard SGE sequences. On current MRI machines, fat-suppressed SGE permits the acquisition of 22 sections in a 20 s breath-hold with reproducible uniform fat suppression (Figure 15.3). In addition to its use in pre-contrast T_1-weighted imaging, SGE sequences are routinely used for multi-phase image acquisition after intravenous administration of gadolinium contrast agents. The short acquisition time

allows to study the liver parenchyma according to its vascular supply (arterial phase, venous phase, and equilibrium phase) (see below).

15.2.3 Contrast Agents

Contrast agents in liver imaging are used to delineate blood vessels and hepatic parenchyma and detect focal and diffuse abnormalities. To date, there is a general consensus that use of contrast media is mandatory in liver MRI. Different compounds have been developed and according to their distribution and target may be classified as extracellular, hepatocytes specific, reticuloendothelial, and blood. Extracellular contrast agents are hydrophilic, small molecular-weight gadolinium chelates. After intravenous administration, these substances are rapidly cleared from the intravascular space to the interstitial space. They do not penetrate into intact cells and are eliminated through the urinary system. The most common gadolinium compounds used for MR examination of the liver are gadopentetate dimeglumine (Gd-DTPA, Magnevist, Bayer), gadoterate meglumine (Gd-DOTA, Dotarem, Laboratoires Guerbet), gadodiamide (Gd-DTPA-BMA, Omniscan, Amersham Health), and gadoteridol (Gd-HP-DO3A, ProHance, Bracco). As paramagnetic compounds, their principal effect is to shorten the T_1 relaxation time resulting in an increase in tissue signal intensity. This effect is best captured on

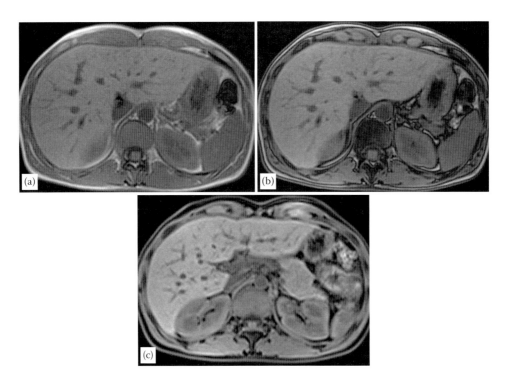

FIGURE 15.3

T_1-weighted images. Gradient echo T_1-weighted image for the study of the liver: (a) in-phase, (b) out-of-phase, and (c) with fat selective pre-pulse saturation.

heavily T_1-weighted images [11–14]. Due to rapid redistribution of gadolinium chelates from the intravascular compartment to the extracellular space, the contrast agents must be administered as a rapid bolus, typically at 2 mL/s, followed by a bolus of 20 mL of saline flush (0.9% sodium chloride solution) at the same injection rate. Thereafter, imaging of the entire liver is performed in a single breath-hold during the dynamic phase of contrast enhancement. Using fast sequence such as 2D or 3D T_1-weighted spoiled GRE sequence, it is possible to study liver parenchyma at different vascular phases according to its vascularization: liver presents a dual blood supply, which originates from hepatic artery (25%) and from portal vein (75%). Following an intravenous bolus of contrast material, the hepatic artery enhances first approximately 15 s and reaches peak attenuation at 30 s. The contrast agent returns to the liver from the intraperitoneal organs through the portal system after 30 s. Liver parenchyma peak enhancement ranges from 60 to 70 s. The equilibrium phase, when the contrast agent is equally divided between intra- and extracellular space, occurs after almost 3 min.

The *hepatic arterial phase* is the most important phase dataset [15]. It allows to highlight lesions with strong arterial vascularization, whereas the liver parenchyma remains relatively unenhanced (Figure 15.4). To get this topic, it is crucial to fill the central of the κ-space at the right time. The landmarks of an optimal arterial phase are the contrast enhancement in the hepatic artery and portal vein, the absence of contrast agent in hepatic veins, the strong enhancement of the renal cortex, and the mottled appearance of the spleen (Figure 15.5). Different techniques should be used in the acquisition of the optimal arterial phase: (*a*) fixed delay time, which is approximately ≈25 s after contrast medium administration [16,17]; (*b*) test bolus technique, which evaluates the contrast medium transit time by a pre-bolus (1–2 mL) before its injection; and (*c*) automated bolus detection technique (CARE-bolus, SmartPrep), the optimal arterial phase is after 8 s after the peak enhancement at D9 level (Figure 15.6) [16,18,19].

The *venous phase* is characterized by the maximum enhancement of the liver parenchyma, and it is approximately 60 s after contrast injection. In this phase, the portal and hepatic veins are well opacificated, so thrombus within the lumen vessels is well appreciable (Figure 15.7). Hypovascular metastases are better depicted in this phase as well as thrombus within vein vessels (Figure 15.8).

The *delayed phase* is the time interval between 90 s and 5 min in which the contrast agent is equally diffused between the vascular and interstitial space (Figure 15.9). It is generally acquired after 180 s after contrast medium injection. This phase is extremely important to assess enhancement of hypovascular lesions (cholangiocarcinoma and hemangioma), homogeneous *fading* of hypervascular lesions (focal nodular hyperplasia and adenoma), or capsule enhancement and/or wash-out sign in hepatocellular carcinoma (Figure 15.10) [20–24].

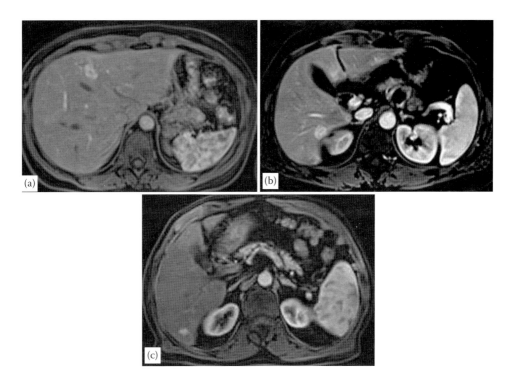

FIGURE 15.4
Hepatic arterial phase. Examples of liver lesions with high arterial vascular supply: (a) focal nodular hyperplasia, (b) hemangioma, and (c) hepatocellular carcinoma.

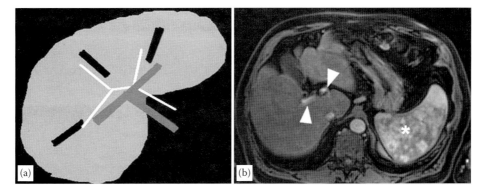

FIGURE 15.5
Hepatic arterial phase. (a and b) Drawing and MR image of hepatic arterial phase: the main characteristics are the mottled appearance of the spleen (asterisk), the slight opacification of the portal vein (arrowheads), and the high enhancement of the renal cortex. The hepatic veins must not be enhanced.

15.2.4 Liver-Specific Contrast Agents

15.2.4.1 Reticuloendothelial Contrast Agents

Reticuloendothelial contrast agents are iron oxide particulate agents that are selectively taken up by reticuloendothelial system (RES) cells in the liver, spleen, and bone marrow and result in signal loss at T_2-weighted imaging owing to the susceptibility effects of iron [25]. SPIOs have a mean iron oxide particulate size of 50 nm, whereas ultra-small SPIOs have a mean size of less than 50 nm. RES-specific contrast agents are superparamagnetic causing shortening of both T_2 and T_1 relaxivity. T_1- and

T_2^*-weighted gradient echo and T_2-weighted echo train spin echo imaging sequences are used for image acquisition. Ferumoxides are administered by intravenous infusion [26,27]. Before the infusion, the patient undergoes a noncontrast MRI of the liver with the use of T_1-weighted gradient echo and T_2-weighted sequences. Approximately 30 min following the completion of contrast administration, the patient undergoes a repeat MRI with the same imaging sequences. The particles are cleared from the plasma by the RES of the liver (80%) and the spleen (12%); minimal uptake occurs in the lymph nodes and bone marrow. Ferucarbotran is administered by direct bolus

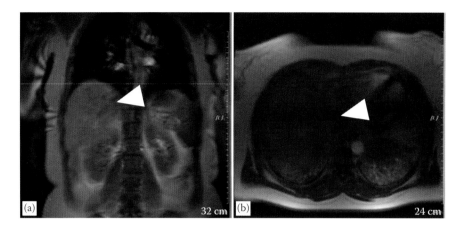

FIGURE 15.6
CARE bolus technique in the acquisition of the arterial phase. Axial trigger is placed at 9th dorsal vertebra: the sequence starts after 8 s from the arrival of the contrast agent.

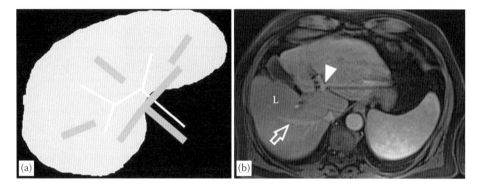

FIGURE 15.7
Venous phase. (a and b) Drawing and MR image of the venous phase illustrate the main imaging features: high enhancement of the liver parenchyma (L), portal vein (arrowhead), and hepatic veins (open arrow).

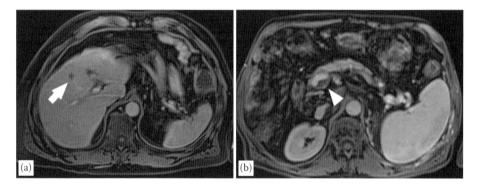

FIGURE 15.8
Venous phase. (a) In the venous phase, there is higher contrast-to-noise ratio between the liver parenchyma and hypovascular metastases (arrow). (b) Venous phase is also useful to detect thrombus within the vessels (arrowhead).

injection of a small volume (<2 mL) of contrast. Serial contrast-enhanced liver imaging is performed with the use of T_1-weighted gradient echo images. During the late-phase imaging (20 min after injection), there is increased uptake of the iron particles in the Kupffer cells, and liver parenchyma is rendered hypointense, especially on T_2-weighted images [28].

15.2.4.2 Hepatocytes-Selective Agents

Hepatobiliary agents are paramagnetic compounds that are taken up by functioning hepatocytes and excreted in bile. Due to the paramagnetic properties, agents of this class increase the signal intensity of the liver, bile ducts, and some hepatocyte-containing lesions at T_1-weighted imaging. They may be up taken only by the

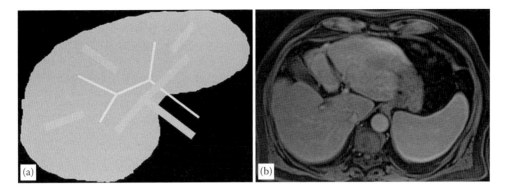

FIGURE 15.9
Delayed phase. (a and b) Drawing and MR image illustrate a similar distribution of contrast agent between vessels and liver parenchyma.

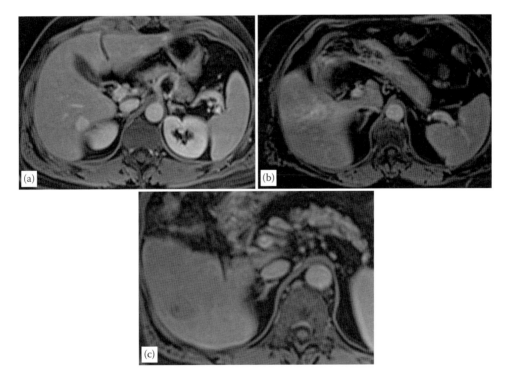

FIGURE 15.10
Delayed phase. Delayed phase is important to demonstrate late enhancement such as typical of hemangioma or cholangiocarcinoma (a and b) or wash-out which is the main feature of hepatocellular carcinoma (c).

hepatocytes or can be distributed both into the extracellular space and into the hepatocytes. Both contrasts are T_1 agents, which shorten T_1 time and result in increased signal on T_1-weighted images; however, the last ones are the most commonly used. In clinical practice are available two compounds that combine vascular and hepatocyte-selective properties: gadobenate dimeglumine (Gd-BOPTA, Bracco, Milan, Italy) and gadoxetic acid (Gd-EOB-DTPA, Bayer, Berlin Germany). Gd-BOPTA has a higher relaxivity but an uptake from the functional hepatocytes of about 5%; by contrast, Gd-EOB-DTPA has an uptake of approximately 50%, which determines a higher lesion-to-liver contrast [29–31]. The higher percentage of contrast uptake by the hepatocytes led to acquire the hepatobiliary phase after 20 min with Gd-EOB-DTPA compared with 45 min of GD-BOPTA: some authors also suggest the acquisition of gadoxetic acid hepatobiliary phase after 10 min. [32,33]. Hepatobiliary-specific agents are particularly useful in determining whether a lesion contains functional hepatocytes (e.g., focal nodular hyperplasia) or not (e.g., cyst, hemangioma, adenoma, metastasis) (Figure 15.11). In cirrhotic liver, detection of cirrhosis-associated nodules (regenerative nodules, dysplastic nodules, and hepatocellular carcinoma) remains

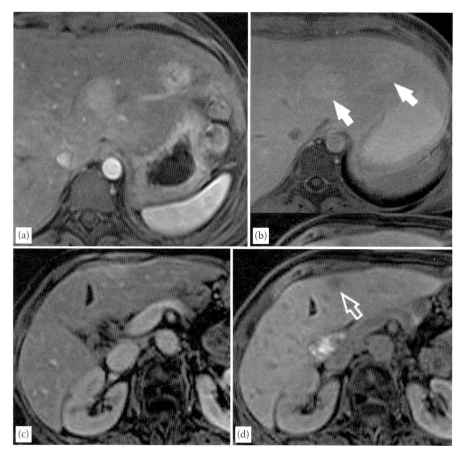

FIGURE 15.11
Liver-specific contrast agent. (a and b) Focal lesions such as focal nodular hyperplasia, which contains functional hepatocytes, retain the hepatobiliary contrast agent (arrows). (c and d) focal lesion, which does not origin from hepatocellular line, does not up take the contrast agent (i.e., metastases, open arrow)

of challenge also with hepatobiliary contrast agent, both because overlap among these nodules at histologic analysis and at imaging and by the inhomogeneous appearance of the cirrhotic liver caused by fibrosis and hemodynamic alterations. Due to the strong biliary excretion of Gd-EOB-DTPA, it should also be used for the study of the biliary tree giving additional information to MR cholangiography. For visualization of the contrast-enhanced biliary system, a T_1-weighted 3D GRE pulse sequence in the coronal/oblique plane provides the best spatial resolution (Figure 15.12).

15.2.5 Diffusion-Weighted Imaging

The main field of application of diffusion-weighted imaging (DWI) is neuroradiology; it plays a crucial role in the detection of tissue ischemia in hyper-acute phase (0–6 h); it is useful in the differentiation between solid and cystic lesions (i.e., abscess and tumor) and recently in the assessment of demyelinating diseases. The inherent sensitivity to motion and the magnetic susceptibility of DW sequences nonetheless still create

problems in the study of the abdomen due to artifacts caused by the heartbeat and intestinal peristalsis, as well as the presence of various parenchymal gas interfaces. The development of ultrafast sequences (echoplanar sequences, EPI) reduces the artifacts motion to be acceptable with minimal influence on the spatial resolution [34]. The sequence should be performed with a high-field superconductive magnet (1.5 T or more) equipped with 23–30 mT/m intensity gradients, a slew rate of 150 mT/m per second, and phased-array surface coils [35]. DW images are acquired using a variation of the spin echo sequence (Stejskal and Tanner sequence) performed with the single-shot echoplanar technique (SE-EPI-SSh) with the following parameters: (TR = 2883, TE = 61, flip angle = 90°, FOV variable, matrix = 128 × 256), with b values of 0 and 500 s/mm². (More recently, multi b value sequence is generally used with the following b values: 0, 50, 400, 800 s/mm².) In addition, where possible the use of parallel imaging can be an advantage, a technique capable of reducing acquisition time or increasing the matrix and therefore the spatial resolution without altering the acquisition time

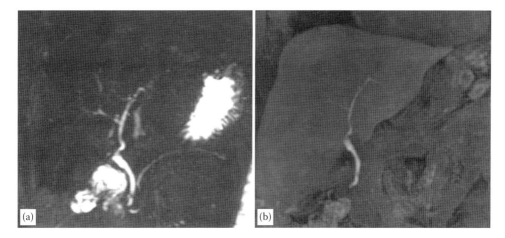

FIGURE 15.12
MR-cholangiogram with hepatobiliary contrast agent. (a and b) Coronal T_1-weighted image during the hepatobiliary phase shows similar delineation of biliary tree as T_2-weighted thick slab image (a).

and without significant loss in image quality (sensitivity encoding, SENSE) [35,36]. In the post-processing phase, the various images are used to obtain the respective ADC maps on which quantitative analysis of the signal is performed by positioning a ROI on the structure being studied. As cited above, the choice of the *b* value and therefore the degree of weighting in diffusion is still an unsolved problem. A lesion is generally considered malignant if it is mild-to-moderate hyperintensity compared with liver parenchyma on DW images at low *b* values and remains (restricted diffusion) at high *b* values: at ADC map, it should be hypointense to the surrounding liver parenchyma (Figure 15.13).

15.2.6 Perfusion Imaging

Perfusion MR refers to imaging of tissue blood flow (i.e., tissue microcirculation), which is beyond the resolution of the MR scanner to directly visualize. This technique allows the quantitative characterization of parenchymal and tumor microcirculatory alterations in the liver [37]. Currently, characterization of focal liver lesions is based on rate and pattern of contrast enhancement during the three different vascular phases. Perfusion MRI could extend the currently used qualitative assessment applied for the differential diagnosis of focal liver lesions, by applying quantitative metrics to describe their vascular behavior. MR perfusion is generally acquired with a 3D T_1-weighted sequence with parallel imaging to reduce the scan time and improve temporal resolution. MR parameters should be the following: TR 2.7 mms, TE 1 ms, slice thickness 8 mm, matrix 256 × 159, FA = 14°, bandwidth 490 Hz, and temperature resolution 1.98 per slab of 10 slices [38].

15.2.7 MR Spectroscopy

For diffusion imaging, MR spectroscopy (MRS) had been previously developed for the characterization of different brain disease. In the past few decades, MRS has been also tested in the evaluation of diffuse and focal liver disease. MRS reveals signals from chemicals in tissue or metabolites, which are identified primarily by their frequency and are expressed as a shift in frequency relative to a standard [39]. In 1H MRS, the frequency location of a metabolite or chemical compound depends on the configuration of the protons within the chemical. Water is abundant in tissues, and its frequency location is used as the conventional standard for *in vivo* 1H MRS, meaning that all other chemicals are identified by comparing their frequency location (frequency shift) to that of water. High field strengths (1.5 T or higher) and a torso phased-array coil for signal reception are desirable for adequate SNR. Single-voxel liver spectroscopy is performed in a 10–20 mm³ voxel placed to avoid large intrahepatic vessels and at least 10 mm from the edge of the liver. Spectra are acquired with the use of CHESS for water suppression and with a PRESS technique (time series of 128 acquisitions at a TR of 2000–3000 ms and TE of 20–30 ms) [40,41]. Liver MRS often is performed with free breathing to achieve greater SNR. Commercial software with various algorithms is available for post-processing of MR spectroscopic data. Post-processing includes motion correction (correction for frequency and phase shifts), automated water suppression, low-frequency filtration of residual water signal, Fourier transform, and Lorentzian-to-Gaussian transformation, and it may be fully or partially automated. MRS is the gold standard for the noninvasive quantification of liver steatosis due to its ability in the quantification of triglycerides within the voxel (Figure 15.14). At present, MRS in the detection of focal liver is limited

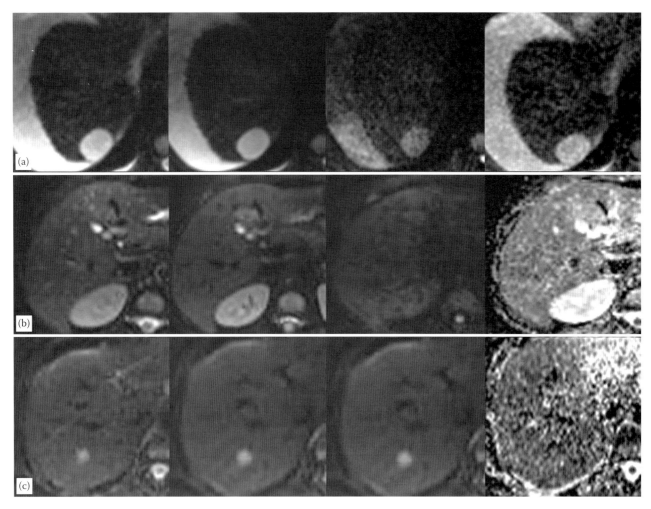

FIGURE 15.13
Diffusion-weighted imaging. (a through c) Examples of focal liver lesions with progressive restriction of molecular diffusion assessable at ADC map: (a) hepatic cyst, (b) focal nodular hyperplasia, and (c) metastases.

because motion artifacts and voxel size are significant drawbacks when small lesions are evaluated. Choline is the metabolite that is generally used in the evaluation of malignant tumor, such as hepatocellular carcinoma, because it reflects the high cell turnover [42]. However, the ability to reliably distinguish benign and malignant tumors from normal liver parenchyma has yet to be established [43–45].

15.2.8 MR Elastography

The elasticity of liver parenchyma shows a strong correlation with degree of hepatic fibrosis and an association with increased vascular resistance, as seen in elevated portal venous pressure [46]. In a technique originally described as ultrasonography (US)-based elastography, external compressional impulses are visualized with Doppler US as shear-wave transients propagating through liver parenchyma and directly depending

on the elasticity of liver parenchyma itself [47]. When limited-organ Doppler US is substituted with whole-organ MR, this relatively new application allows quantification of the viscoelastic properties of the liver, in particular for the assessment of hepatic fibrosis [48]. MR elastography combines an external sinusoidal vibration pattern applied to the liver parenchyma with a 2D gradient-echo MR elastography sequence that applies motion-encoding gradients along the section-selection direction to detect cyclic motion in the through-plane direction; the external acoustic driver and motion-encoding gradients have to be synchronized in frequency and phase [49].

15.2.9 High Field MR (3 T)

Three-T MR substantially provides a better SNR and contrast-to-noise ratio than 1.5 T, which can be used to improve image quality and reduce the scan time. In MR of the liver, better SNR and higher resolution

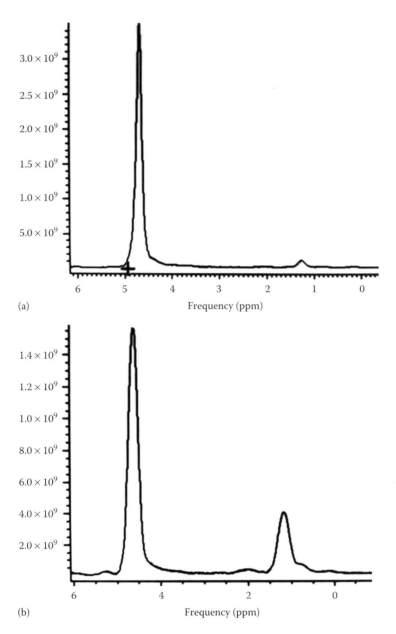

FIGURE 15.14
MR spectroscopy. (a) Liver spectrum from a healthy volunteer does not show lipid peak relative to that of water. (b) Liver spectrum from a man with nonalcoholic fatty liver disease shows a small lipid peak relative to that of water due to mild steatosis.

achieved on T_2-weighted images from a 3.0-T system result in better detection of malignant lesions. T_1-weighted 2D GRE sequences are still difficult to optimize on a 3.0-T system; however, use of a 3D GRE sequence provides comparable lesion contrast and greater SNR (Figure 15.15). The improved image contrast at higher field strengths is largely the result of exogenous contrast media such as gadolinium, a paramagnetic substance that disrupts the local magnetic field and leads to T_1 shortening [50]. T_1 is lengthened at 3.0-T imaging, even with the use of a paramagnetic agent such as gadolinium. However, because the

T_1 of gadolinium is shorter than that of the soft tissues, gadolinium-enhanced tissues stand out markedly against the background. Diagnostic sensitivity is improved by the enhanced contrast (Figure 15.16), and this improvement in turn provides an opportunity for gadolinium dose reduction. Substantial modification of examination protocols with optimized imaging parameters and sequence designs along with ongoing development of hardware systems should increase the role of 3.0-T systems for abdominal MR examinations. DWI substantially benefits from increased SNR at 3.0 T, with improved sensitivity to areas of restricted

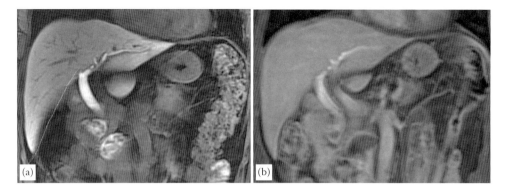

FIGURE 15.15
High field magnetic resonance. Coronal hepatobiliary phase T_1-weighted image acquired in the same patient at 3 T (a) and 1.5 T (b). The SNR is significantly better with high field magnet.

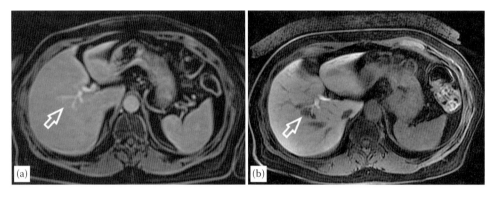

FIGURE 15.16
High field magnetic resonance. Liver metastases in the same patient evaluated at 1.5 T (a) and 3 T (b). Lesion conspicuity is better with high field magnet (open arrow).

diffusion. However, image distortion from increased magnetic susceptibility often greatly contributes to a loss of image quality at 3.0 T. Magnetic susceptibility artifacts may be limited by using parallel imaging techniques. However, the difference with 1.5 T may not be clinically important. Increased chemical shift effect at 3.0 T results in better spectral resolution, with improved separation of metabolite peaks [51]. This allows for more accurate quantification of individual metabolites during MRS [52]. A concomitant increase in SNR also allows for a shorter overall imaging time. As with other applications, an increase in SNR may be translated into improved spatial resolution with the use of smaller voxels. Along with direct increases in SNR and CNR come substantial increases in power deposition, RF field inhomogeneity, magnetic susceptibility, chemical shift artifacts, and concerns regarding MR device compatibility. At present, 3.0-T MRI technology does not yield an image quality improvement sufficient to justify its routine use for abdominal and liver evaluation. In addition, cost and design issues weigh heavily in decisions about the use of clinical 3.0-T MRI.

15.3 Liver Anatomy

The liver parenchyma is radiologically divided into eight segments according to the system proposed by Couinaud and later modified by Bismuth, each with specific vascular and biliary connections [53,54]. It provides a reproducible and clinically meaningful description of the detail needed for surgery. Except the caudate lobe and medial segment of the left lobe, the segments are defined not only by the three vertical fissures described by the three major hepatic veins, but also by a transverse fissure described by the left and right portal venous branches. The segment I is the caudate lobe. The segments II-III-IV are placed in the left lobe, while segments V-VI-VII-VIII are located in the right lobe (Figure 15.17). Each segment has an independent blood supply (arterious and venous) and biliary drainage. The amendment made by Bismuth IV divides the segment into subsegments higher (IVa) and lower (IVb). The most common anatomic variation is the Riedel lobe, which is an extended, tongue-like, right lobe of the liver. It is not pathological; it is

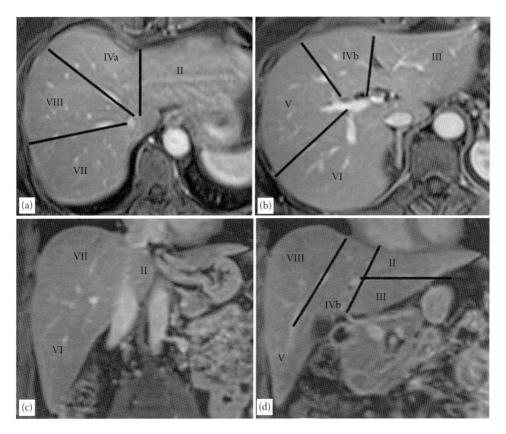

FIGURE 15.17
Liver anatomy. (a and b) Axial T_1-weighted images show liver segment location according to Couinaud system. (c and d) Coronal T_1-weighted images demonstrate liver segment's placement.

a normal anatomical variant and may extend into the pelvis. It is often mistaken for a distended gallbladder or liver tumor.

Other anatomic variations include liver lobe agenesia. Complex congenital heart defects may be associated with a symmetrical, centralized liver. A liver located on the left is found in a full or partial situs viscerum inversus.

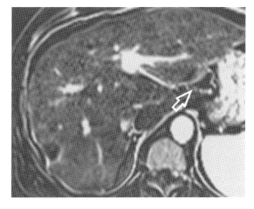

FIGURE 15.18
Variant hepatic artery anatomy. Subtracted image from arterial phase shows the origin of the left hepatic artery from the left gastric artery (open arrow).

15.4 Vascular Anatomy and Variants

The hepatic artery carries only 25%–30% of the blood, which is related to the liver. The common hepatic artery usually originates from the celiac trunk. After gastroduodenal artery originated behind the pylorus, it becomes the proper hepatic artery, which passes into the hepatoduodenal ligament to the anteromedial portal vein and anterior to the bile duct. At hepatic hilum, it divides into left and right branch. The middle hepatic artery that supplies the segment IV with equal frequency originates from the left or right hepatic artery. The classical distribution of the hepatic arteries is only slightly more than half of the subjects, while 45% have one or more

variants. The most common variants include the left hepatic branch that originates from the left gastric artery (10%–25%) and the right hepatic branch that originates from the superior mesenteric artery (11%–17%) (Figure 15.18) [55]. The portal vein divides into two branches, right and left, which run next to the hepatic arteries right and left, and to the bile ducts. In the parenchyma of the

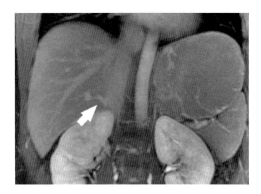

FIGURE 15.19
Variant hepatic veins anatomy. Maximum intensity projection (MIP) from MR post-contrast image shows the presence of an accessory inferior right hepatic vein draining segment VI directly into the inferior vena cava.

right lobe, the right portal vein divides into anterior and posterior branches, which go to the corresponding liver segments. The portal system can be affected by anatomical variants, such as the trifurcation of the portal vein hilum (8%–10%); other anomalies can be also recognized including agenesis or atrophy of one of the two main branches, with atrophy of the liver parenchyma. In the classic anatomy, three main hepatic veins drain into the inferior vena cava: the left hepatic vein drains segments II and III; the middle hepatic vein drains segments IV, V, and VIII; and the right hepatic vein drains segments V, VI, and VII. In approximately 60% of the population, the middle and left hepatic veins join to form a common trunk. The most common variant seen is an accessory right hepatic vein draining the segment VI directly into the inferior vena cava (Figure 15.19) [56].

15.5 Diffuse Liver Disease

Diffuse liver disease, including all causes of chronic liver disease, affects millions of people worldwide. There is a growing need for diagnostic evaluation as treatments become more readily available. MRI provides unique capabilities for noninvasive characterization of liver tissue that overcome the diagnostic utility of liver biopsy. There has been incremental improvement in the use of standardized MRI sequences, acquired before and after administration of contrast agent for the evaluation of diffuse liver disease, and this includes study of the liver parenchyma and blood supply. More recent developments have led to methods for quantifying important liver metabolites, including fat and iron, and liver fibrosis, which is the hallmark for chronic liver disease. Diffuse hepatic diseases can be usually classified as storage, vascular, and inflammatory.

15.5.1 Storage

15.5.1.1 Fat Storage

Hepatic steatosis is the storage of triglycerides within the hepatocytes; it should be focal or diffuse. Fatty liver disease can cover a severity spectrum ranging from dominant noninflammatory to steatohepatitis, fibrosis, and cirrhosis. The main cause of liver steatosis is related to the alcohol abuse; non-alcoholic fatty liver disease (NAFLD) is associated with obesity, type 2 diabetes, and dyslipidemia, all of these are manifestations of metabolic syndrome [57,58]. Recently, an interaction between genetic and environmental factors in the genesis of NAFLD has also been demonstrated; in particular, it has been demonstrated that PNAPLA3 polymorphism rs738409 is strongly correlated with NAFLD [59]. Hepatic steatosis can be either focal or diffuse. Focal steatosis is often easily recognized on the basis of the typical periligamentous or peri-hilar location and the presence of non-distorted, traversing blood vessels. Multinodular or patchy focal fat deposition can be also visible, and they may be mistaken for an infiltrative neoplasm (Figure 15.20) [60]. MRI represents the best noninvasive diagnostic method for the detection of liver steatosis. Chemical shift MRI is certainly the easiest method for the depiction of liver steatosis: the signal drop-off in out-of-phase T_1-weighted sequence is a very accurate tool. Liver fat quantification using MRS is considered the only noninvasive reference standard for detecting and quantifying the biochemical fat content in the liver due to its ability in the quantification of triglycerides within the voxel (Figure 15.21) [61]. It has been used in several research studies, but it has limited clinical applicability or availability because it requires sophisticated post-processing methods and not every MRI is routinely equipped with MRS capabilities. At the same time, some MR sequences have been also developed in the past years as alternative and faster methods for the quantification of liver steatosis: (a) the two-point Dixon modified by Fishbein, (b) the three-point TE reconstructed with IDEAL, and (c) the three- and six-echo GRE chemical shift sequence [62–65]. Debate still persists in literature over the best FAST-MR approach in quantification liver steatosis.

15.5.1.2 Iron Storage

Hepatic iron overload is the abnormal accumulation of iron within the hepatocytes, the Kupffer cells, or both. Storage of iron in the liver derives from intestinal abnormal absorption and from damaged erythrocytes (hereditary hemochromatosis, sideroblastic anemia, and sickle cell disease). Transfusional iron overload results from the parenteral administration of red blood cells and results in iron deposits of the RE cells of the liver,

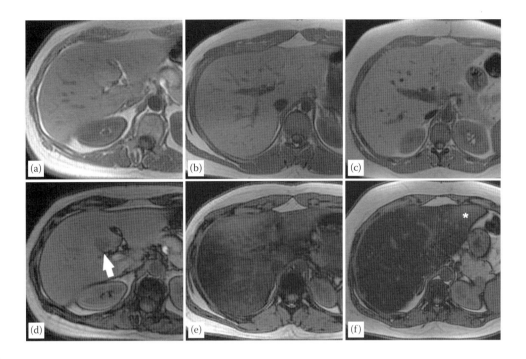

FIGURE 15.20

Fat storage. Imaging pattern at dual-echo gradient-echo T_1-weighted images. (a and d) Focal area of hepatic steatosis hyperintense on in-phase T_1-weighted image and hypointense on out-of-phase T_1-weighted image. (b and e) Patchy distribution of steatosis within the liver parenchyma. (c and f) Diffuse and homogeneous fat storage within the liver parenchyma. Asterisk is used to note the presence of an area of fatty sparing in liver segment II.

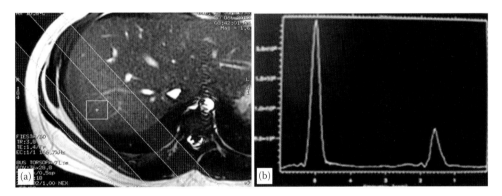

FIGURE 15.21

Fat storage with MR spectroscopy. (a) Axial T_2-weighted MR image shows the correct placement of the MR spectroscopy voxel in liver segment VII. (b) Liver spectrum from a child shows a smaller lipid peak relative to that of water.

spleen, pancreas, and bone marrow [66]. Iron overload within hepatocytes creates oxidative stress that can lead to end-stage cirrhosis, liver failure, and the development of hepatocellular carcinoma [67]. MRI does not image the iron directly but instead detects the effect of iron on water protons in the tissue of interest. The basis of using MRI for iron detection is that iron, a superparamagnetic element, accelerates T_2 relaxation and T_2^* signal decay, thereby causing signal loss on T_2-weighted spin echo/fast spin echo and T_2^* gradient echo MR images: the latest have greater sensitivity for detecting iron and delineating its distribution (Figures 15.22 and 15.23) [68].

The signal intensity of the skeletal muscle is used as the internal reference to compare the change in the signal intensities of other organs and for quantitative analysis [69]. Iron and fat may also occur together and quantitative analysis may result more complicated.

15.5.1.3 Wilson's Disease

Hepatolenticular degeneration, more commonly known as Wilson's disease, is an autosomal recessive disease characterized by increased intestinal uptake of copper and subsequent deposition, predominantly in the

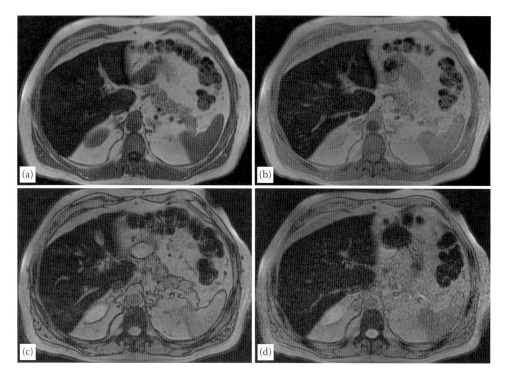

FIGURE 15.22
Iron storage. Genetic hemochromatosis (GE). (a through d) Different T_2*-weighted sequences for the quantification of iron storage.

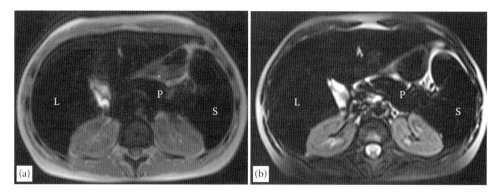

FIGURE 15.23
Iron storage. (a and b) Transfusional T_2* GRE sequence shows low signal intensity in the liver (L), spleen (S), and pancreas (P). The abnormal low pancreatic signal intensity suggests that the amount of transfused iron has exceeded the storage capacity of the RE cells.

liver, brain, and cornea. Whereas malignant transformation into hepatocellular carcinoma may occur only very rarely, Wilson's disease can manifest as acute and even fulminant hepatitis with rapid progression into mostly macronodular type cirrhosis. Clinically, patients present with very low levels of ceruloplasmin. At present, to our knowledge, MRI of Wilson's disease does not show characteristic features that allow distinction from other forms of chronic hepatic disease (Figure 15.24). In clinical scenarios of Wilson's disease-induced hepatitis or liver failure, early nonspecific and patchy enhancement patterns with significantly delayed stromal gadolinium uptake have been observed [70]. Some authors have described unusual imaging findings in patients with Wilson's disease such as perihepatic fat layer, multiple hypervascular nodules, and absence of caudate lobe hypertrophy [71].

15.5.1.4 Glycogen Storage Disease

Patients with glycogen storage disease are characterized by an inability to convert glucose-6-phosfate dehydrogenase to glucose; however, they can produce glucose endogenously. The storage of glycogen in the hepatocytes causes an increased signal intensity on T_1-weighted images compared with bone marrow. Patients with this disease may also have hepatocellular adenoma [72].

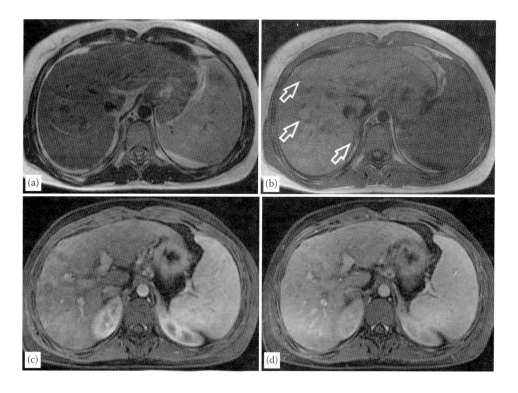

FIGURE 15.24
Wilson disease. (a and b) Axial T_1- and T_2-weighted sequence show irregular margins of liver surface with multiple parenchymal nodules (open arrows). (c and d) Axial T_1-weighted post-contrast images do not reveal enhancement of these nodule. MR findings are not specific and are similar to that of liver cirrhosis.

15.5.2 Vascular

15.5.2.1 Hereditary Hemorrhagic Telangiectasia (Osler–Weber–Rendu Syndrome)

Hereditary hemorrhagic telangiectasia, also called Osler–Weber–Rendu disease, is a vascular, hereditary, autosomic dominant disorder that occurs with a prevalence of approximately 1–2 cases/100,000 and hepatic involvement occurs in 8%–31% of cases [73]. Hereditary hemorrhagic telangiectasia is characterized by the presence of mucocutaneous or visceral angiodysplastic lesions, the latter most frequently seen in the liver, lung, brain, and gastrointestinal tract. Hepatic involvement is characterized by the presence of vascular abnormalities such as dilatation of the hepatic artery, intrahepatic telangiectases, arteriosystemic venous shunting, arterio-portal shunting, vascular mass-forming lesions, and aneurysms of the hepatic artery [74,75]. Regenerative nodular hyperplasia was identified in 17 patients (74%) and ischemic cholangitis in 9 patients (39%) [76]. Arteriovenous malformations are usually distributed diffusely throughout the liver and may be associated with enlargement of the hepatic artery and increased tortuosity of vessels in the liver hilum and in the central portions of the liver lobes. In Osler's disease, increased arterial perfusion of the liver tissue frequently leads to secondary nodular hypertrophy, which may be misinterpreted as a malignant hepatic tumor. These pseudotumors, as in focal nodular hyperplasia, represent a localized overgrowth of hepatocellular tissue and are not real liver tumors. Patients with hepatic involvement can be asymptomatic, but heart failure, portosystemic encephalopathy, cholangitis, portal hypertension, and cirrhosis have been reported [75]. On MRI, telangiectases appear as small hypo- to isointense lesions on unenhanced T_1-weighted images and as hyper- or isointense lesions on T_2-weighted images. Dynamic MRI reveals strong arterial phase enhancement and subsequent isointensity with the surrounding liver tissue in the portal venous and equilibrium phases. Normal enhancement of the affected tissue in the hepatobiliary phase can be noted with the use of hepatobiliary contrast agents (Figure 15.25). Whereas arteriovenous shunts are poorly detected on unenhanced T_1- and T_2-weighted images, dilated and tortuous vessels can usually be seen near the arteriovenous shunts (Figure 15.26).

15.5.2.2 Passive Hepatic Congestion

Passive hepatic congestion is caused by stasis of blood within the liver parenchyma due to compromise of hepatic venous drainage. It is a common complication of congestive heart failure and constrictive pericarditis, wherein elevated central venous pressure is directly transmitted from the right atrium to the hepatic veins because of their close anatomic relationship.

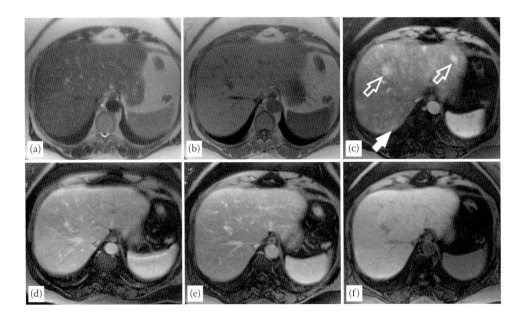

FIGURE 15.25
Hereditary hemorrhagic teleangiectasia. (a and b) T_2- and T_1-weighted MR images do not reveal focal lesions in liver parenchyma. (c) Post-contrast T_1-weighted sequence during the arterial phase shows image intrahepatic (open arrow) and subcapsular (open arrow) angioplastic lesions, including arteriovenous malformations and angiectasia. (d through f) On the following acquired sequences, the liver shows homogeneous enhancement.

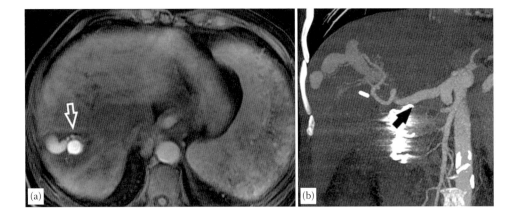

FIGURE 15.26
Hereditary hemorrhagic teleangiectasia. (a) Post-contrast axial T_1-weighted image shows abnormal ectatic vessels into the liver parenchyma. (b) Coronal MIP images obtained in the arterial phase show arterial-portal shunting, with the dilated hepatic artery (black arrow) communicating with the portal vein.

Hepatocytes are highly sensitive to even short periods of ischemic trauma and can be injured by a variety of cardiac-induced circulatory derangements, including arterial hypoxia, acute left-sided heart failure with a fall in cardiac output, and central venous hypertension [77]. In acute right-sided heart failure, the hepatic veins distend and the sinusoids become congested because of stasis of blood. Compression, atrophy, and necrosis of centrolobular hepatocytes occur to a variable degree, and mild centrolobular fatty infiltration may ensue. In chronic passive congestion, persistent hypoxia prevents hepatocellular regeneration and causes fibrous tissue bands from adjoining centrolobular areas to join and encircle relatively normal portal tracts, eventually leading to the development of cardiac cirrhosis (nutmeg liver). Distension of the inferior vena cava and hepatic veins is the main feature of hepatic congestion: diameter of the main trunk of the right hepatic more than 8.8 mm is considered pathologic [78]. Moreover, when the bolus of contrast material injected IV reaches the failing right atrium, it may flow directly into the inferior vena cava and hepatic veins rather than flow normally into the right ventricle, *retrograde opacification* (Figure 15.27). On T_2-weighted images, a slightly high signal intensity may

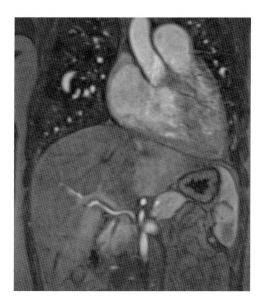

FIGURE 15.27

Hepatic congestion. Post-contrast T_1-weighted image shows enlargement of the liver parenchyma and ectatic hepatic veins with retrograde opacification.

be appreciable along the periportal space due to perivascular lymphedema. After contrast medium administration, a reticulated mosaic pattern of low signal intensity linear markings is detectable: after 1–2 min, the liver parenchyma becomes more homogeneous.

15.5.2.3 Budd–Chiari Syndrome

Budd–Chiari syndrome is defined as hepatic venous outflow obstruction at any level from the small hepatic veins to the junction of the inferior vena cava and the right atrium, regardless of the cause of the obstruction [79]. It is generally classified as *primary* Budd–Chiari syndrome when it is caused by an intrinsic luminal web or thrombus, and as *secondary* when there is an extraluminal compression or tumoral invasion [80–82].

Membranous obstruction of suprahepatic veins, which is also known as primary Budd–Chiari syndrome, is due to fibromuscular membrane (web) or an acquired lesion. Membrane or web arises from the wall of the vessel and may obliterate the lumen completely or partially [83]. This type of lesion is believed to be a sequela of long-standing thrombosis [84]. The etiology of the venous thrombosis is obviously an important consideration. Hematologic abnormalities such as myeloproliferative disorders; paroxymal nocturnal hemoglobinuria; the antiphospholipid syndrome; inherited deficiencies of protein C, protein S, and antithrombin III; factor V Leiden mutation; prothrombin–gene mutation; and methylene tetrahydrofolate reductase mutation are responsible for the majority of cases of Budd–Chiari syndrome [83,85]. The other factors that contribute to the development of Budd–Chiari syndrome include pregnancy, immediate postpartum period, and use of oral contraceptives. The secondary Budd–Chiari syndrome is caused by an extraluminal compression of a space occupying lesion or luminal invasion of malignant tumor (renal cell carcinoma, hepatocellular carcinoma, adrenal carcinoma, hepatic metastasis, and primary leiomyosarcoma of inferior vena cava) [82–85]. Main alterations in hepatic morphology include atrophy of peripheral regions, hypertrophy of the caudate lobe and central portions of the liver, and nodular transformation of liver parenchyma (Figure 15.28) [86,87]. Liver parenchyma enhancement during the acute phase is higher in the central zone than that of peripheral areas (flip-flop sign). In the chronic phase, the contrast enhancement of the liver parenchyma is more homogeneous [87].

Benign regenerative nodules have been described in the literature in association with chronic Budd–Chiari syndrome [86,88–90]. The pathogenesis of benign regenerative nodules in Budd–Chiari syndrome is unclear. Hepatocellular growth factors and the disturbance of hepatic microcirculation may play role in their

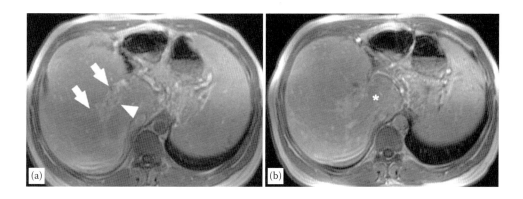

FIGURE 15.28

Budd–Chiari syndrome. (a and b) Post-contrast T_1-weighted images acquired during the venous phase show the enlargement of the liver parenchyma and mainly the hypertrophy of the caudate lobe (asterisk). The hepatic veins are not visible as the portal vein (arrowhead), which shows ectatic collateral vessels (arrow).

development [90,92]. Hepatic nodules in Budd–Chiari syndrome are typically multiple (>10), small (<4 cm), and hypervascular. On MRI, the benign hepatic nodules are often noted as hyperintense on T_1-weighted images and isointense/hypointense on T_2-weighted images [90–93]. The nodules have an increased arterial supply corresponding to their dense enhancement on dynamic imaging. A number of such hepatic nodules progressively increase during the consecutive phases of MR angiography and their enhancement persists until the late venous phase. The patients with chronic end-stage liver disease have increased risk for development of hepatocellular carcinoma. Thus, it is important to distinguish benign regenerative nodules from hepatocellular carcinoma (Figure 15.29). In general, the pre-contrast signal intensity characteristics and enhancement patterns of these two lesions permit their distinction. However, in some cases, the differentiation cannot be made on the basis of MR signal intensity.

15.5.2.4 Hepatic Infarction

Hepatic infarction is defined as areas of coagulation necrosis from hepatocyte death cells caused by local ischemia because of the obstruction of the circulation of the affected area by a thrombus or embolus. It is rare because of the dual vascular supply from the hepatic artery and portal vein. Hepatic infarction may be iatrogenic (transarterial chemoembolization) or posttraumatic (laceration of hepatic artery or portal vein); it can occur as a complication of liver transplantation and may be secondary to vasculitis or infection [33]. Hepatic artery thrombosis leading to infarction

most often occurs after liver transplantation has been reported in almost 3% of adult transplant recipients. On T_1-weighted images, hepatic infarction reveals a well-defined, hypointense area. On T_2-weighted images, the liver parenchyma shows a heterogeneous pattern with high signal intensity. After contrast medium administration, hepatic infarction shows an area of perfusion defect predominantly hypointense compared with the surrounding liver parenchyma (Figure 15.30). Although the diagnosis of liver infarction is evident in most cases, it may be confused with focal fatty infiltration, abscess, or even a tumor [37].

15.5.3 Inflammatory

15.5.3.1 Cirrhosis

Liver cirrhosis is a liver disease characterized by an irreversible remodeling process of hepatic parenchyma with fibrous septa and hepatocellular nodules [94]. Cirrhosis is most commonly the result of hepatitis B and C virus infection or chronic alcoholism; other causes are biliary, cryptogenic, and metabolic. Usual clinical manifestations result from portal hypertension, portosystemic shunting, and hepatic insufficiency. Common complications are ascites, gastrointestinal bleeding, encephalopathy, and coagulopathy. At an early stage of cirrhosis, liver appears normal in size with regular margins. In advanced stage of disease, some typical features may be present on diagnostic imaging such as nodular contours, enlargement of the caudate lobe and lateral segment of the left lobe and atrophy of the right lobe, the

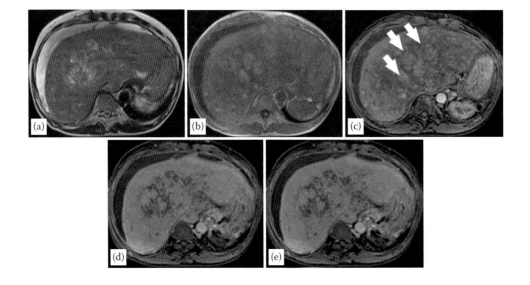

FIGURE 15.29

Budd–Chiari syndrome. (a and b) T_1-weighted and T_2-weighted sequences reveal a heterogeneous appearance of liver parenchyma with necrotic area and multinodular pattern. After contrast agent administration, during the arterial phase (c) the nodules slightly enhance but do not show wash-out during the venous (d) or delayed phases (e). These nodules are sometimes very difficult to distinguish from hepatocellular carcinoma.

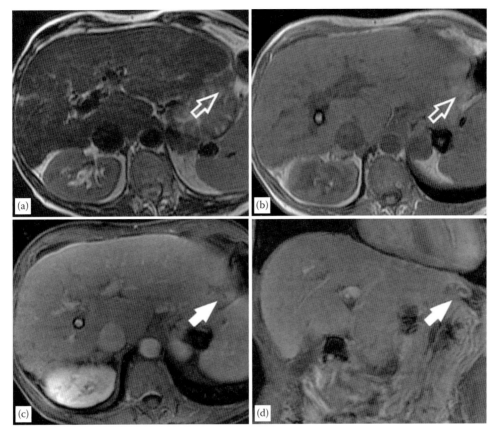

FIGURE 15.30
Hepatic infarction. (a and b) T_2- and T_1-weighted images show a peripheral wedge-shaped, non-homogeneous area with retraction of liver surface (open arrows). (c and d) On axial and coronal post-contrast images during the venous phase, the area is hypointense to the surrounding liver parenchyma (arrows).

notch sign, and enlargement of hilar periportal space (Figure 15.31) [95–99]. A ratio of transverse caudate lobe width to right lobe width greater than or equal to 0.65 constitutes a positive indicator for the diagnosis of cirrhosis with high level of accuracy (Figure 15.32). A modified caudate lobe width to right lobe width ratio, using the right portal vein instead of the main portal vein to set the lateral boundary, has been also proposed [100]. Changes in architecture tissue and fibrous septa in chronic liver disease lead to increased vascular resistance at the level of the hepatic sinusoids. The increased pressure gradient is defined as portal hypertension and causes complications such as ascites, splenomegaly, and the development of engorged and tortuous collateral vessels that may develop at the lower end of the esophagus and at the gastric fundus (Figure 15.33) [101]. The paraumbilical veins and the left gastric vein, both draining into the portal vein, also reopen to form portosystemic shunts. Other shunts between the portal and the systemic circulation include splenorenal collaterals, hemorrhoidal veins, abdominal wall, and retroperitoneal collaterals. As cited above, one of the main features

of cirrhosis is liver fibrosis: it is typically detected as patchy fibrosis, as a lacelike pattern, or as a confluent mass. The lacelike type of fibrosis is best described as thin or thick bands that surround regenerative nodules. Focal confluent fibrosis is observed in end-stage liver disease and is usually a wedge-shaped lesion located in the subcapsular portion of segment IV, V, or VIII, with associated capsular retraction (Figure 15.34) [102]. Sometimes focal confluent fibrosis shows enhancement during the arterial phase and should be misdiagnosed with hepatocellular carcinoma: persistent delayed enhancement, wedge-shaped retraction, and capsule retraction, which are typical of fibrosis, may facilitate the differential diagnosis. The second most important characteristic of cirrhosis is hepatocellular nodules: most of them are benign (regenerative nodules); however, regenerative nodules may progress along a well-described carcinogenic pathway to become dysplastic nodules (low and high grade) or hepatocellular carcinomas [103]. Regenerative nodules are the most common cirrhosis-associated hepatocellular nodules: they may be classified according to size as either micronodules (<3 mm) or macronodules

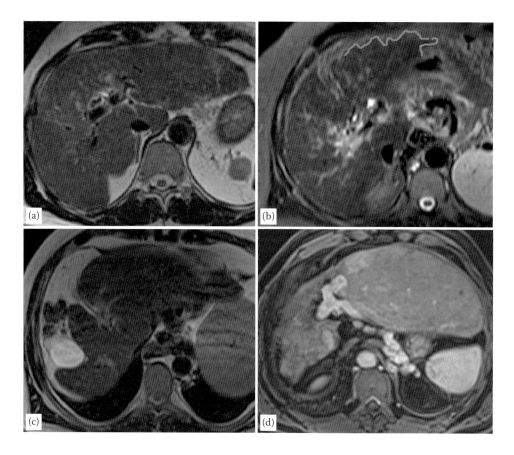

FIGURE 15.31

Liver cirrhosis. Morphologic pattern. (a through c) T_2-weighted images show right posterior hepatic notch: this finding is more frequent in alcoholic cirrhosis, (b) irregular margin of hepatic surface (line), and (c) an expanded gallbladder fossa (open arrow). (d) Post-contrast T_1-weighted image reveals an atrophy of the right lobe and hypertrophy of the left.

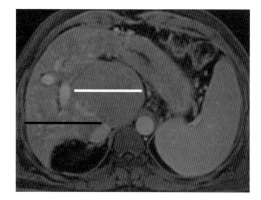

FIGURE 15.32

Liver cirrhosis. Caudate ratio. Post-contrast T_1-weighted image shows a huge enlargement of the caudate lobe with hypotrophy of the right lobe with a ratio of 1.15.

(≥3 mm). AT MRI when they are visible, regenerative nodules appear sharply circumscribed within the liver parenchyma. On unenhanced T_2- and T_2*-weighted images, the nodules typically display low signal intensity; their signal intensity on T_1-weighted images is variable [104]. Lipid-containing regenerative nodules display signal loss on out-of-phase GRE images and unenhanced asymmetric spin echo images in comparison with in-phase images. Contrast-enhanced imaging features are diagnostically more specific than findings at unenhanced imaging. After extracellular contrast agent, most regenerative nodules enhance to the same degree as the adjacent liver or show slightly more enhancement during the arterial and portal-venous phase. Uptake and excretion of liver-specific contrast agent by these nodules is usually preserved. Consequently, on images acquired during the hepatocellular, all regenerative nodules have a similar signal intensity of the surrounding liver parenchyma (Figure 15.35). Occasionally, regenerative nodules may have sufficient hepatocellular function to take up the hepatocellular agent but not to excrete it; such nodules show hyperintense signal [105]. Finally, because most regenerative nodules have a preserved phagocytic function, they are SPIO avid and appear hypointense on SPIO-enhanced T_2- and T_2*-weighted images.

Siderotic nodules are regenerative or dysplastic nodules with high levels of endogenous iron. Although the term *siderotic nodule* is not included in the International Working Party lexicon, it was coined by radiologists to

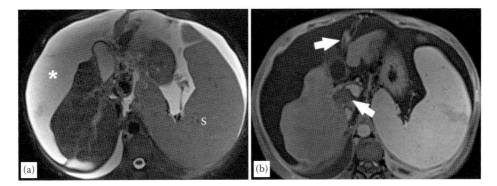

FIGURE 15.33
Liver cirrhosis. Extrahepatic findings. (a) T_2-weighted image shows a huge enlargement of the spleen (S) due to portal hypertension and large amount of ascites (asterisk). (b) Post-contrast T_1-weighted image demonstrates blood flow into the paraumbilical vein and thrombus within the portal vein (arrows).

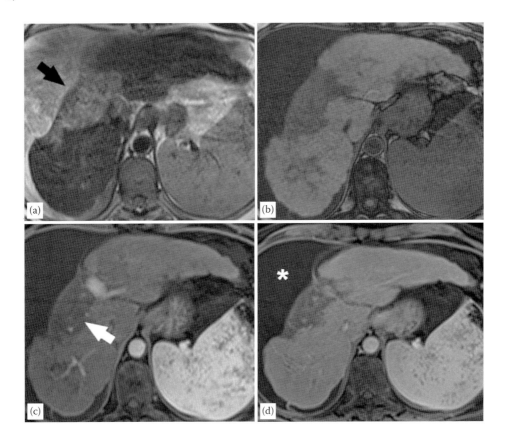

FIGURE 15.34
Liver cirrhosis. Confluent hepatic fibrosis. (a and b) T_2- and T_1-weighted images show, respectively, a hyperintense area and a hypointense area in the central segments of the liver with irregular margins and deep retraction of liver capsule (black arrow). (c and d) Post-contrast MR scan during the arterial and delayed phase shows wedge-shaped lesion (white arrow) of lower attenuation than adjacent liver parenchyma in the anterior segment of the right lobe. Asterisk is to note the large amount of intra-abdominal fluid.

describe cirrhosis-associated nodules with high levels of iron. Siderotic nodules have low signal intensity on T_1- and T_2^*-weighted unenhanced MR images [106].

Dysplastic nodules have variable appearances on MR images, and their signal intensity characteristics overlap with those of regenerative nodules and well-differentiated hepatocellular carcinomas. On T_2-weighted images, dysplastic nodules tend to have low signal intensity relative to adjacent liver [107]. T_1-weighted images are not helpful because they display variable (low, intermediate, or high) signal intensity. On gadolinium- and SPIO-enhanced images, low-grade dysplastic nodules generally do not show enhancement during the arterial phase; moreover, several dysplastic nodules show wash-out

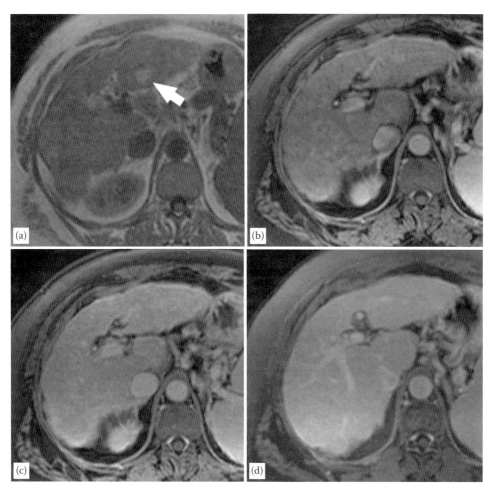

FIGURE 15.35
Liver cirrhosis. Regenerative nodule. (a) T_1-weighted image shows a slightly hyperintense focal lesion in the liver segment III. (b through d) After contrast medium injection during the arterial, delayed, and hepatobiliary phase, no lesions are detectable into the liver parenchyma. This nodule was confirmed at liver transplantation.

sign on delayed phase (Figure 15.36) [108]. Arterial phase enhancement should suggest development of a focus of hepatocellular carcinoma within high-grade dysplastic nodule, the so-called nodule within a nodule appearance on MRI [109]. Dysplastic nodules are detected and characterized better by MR than by CT; however, accurate diagnosis may be made in only about 15% of cases.

15.5.3.2 Sclerosing Cholangitis

Primary sclerosing cholangitis (PSC) is a chronic cholestatic disease characterized by inflammatory fibrosis and destruction of intra- and extrahepatic biliary ducts. The PSC has been more commonly described in middle-aged males with mean age of 40 years at the time of diagnosis. Approximately 75% of the patients are associated with inflammatory bowel disease such as ulcerative colitis and Crohn's disease [110,111]. PSC is also considered a risk factor for cholangiocarcinoma, which has been described in 10%–15% of patients [112,113]. Although several factors have been

implicated, no single mechanism has been identified. The PSC is thought to be related to an autoimmune process because it has been associated with abnormalities in humoral and cell-mediated immunologic processes, and the aberrant expression of class II HLA antigens on biliary epithelial cells. The association of PSC with other disease entities such as retroperitoneal fibrosis, mediastinal fibrosis, and Sjogren's syndrome also suggests an autoimmune process [114,115]. Classic findings of PSC are related to progressive inflammation and fibrosis of the biliary tree, resulting in segmental dilatation of bile ducts alternated with stenotic or obliterated segments [116]. The MR-cholangiography (MRC) clearly shows cholangiographic features of the PSC. The most common findings are diffuse, multifocal annular short biliary strictures involving intrabiliary and extrabiliary duct alternated with normal or slightly dilated segments producing *beaded* appearance. Other cholangiographic findings include webs, diverticula, and stones. Webs are described as focal,

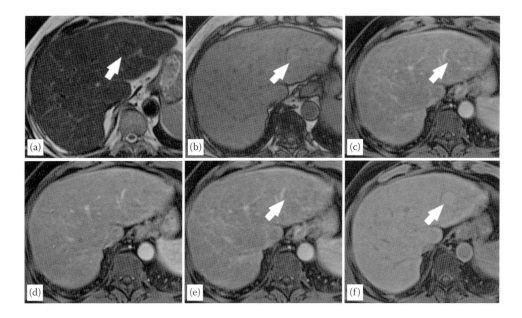

FIGURE 15.36
Liver cirrhosis. Dysplastic nodule. (a) T_2-weighted image shows a slightly hypointense liver nodule in segment II. (b) On T_1-weighted image, the lesion is hyperintense, probably due to fat or glycol-protein storage. After contrast medium administration, the lesion is hypovascular and shows wash-out on delayed phase (c through e). On T_1-weighted hepatobiliary phase, the lesion is hyperintense to the surrounding liver parenchyma (f).

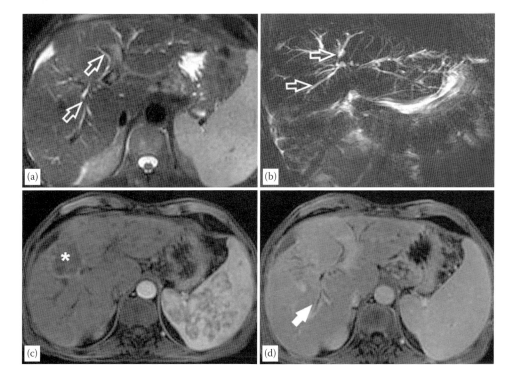

FIGURE 15.37
Primary sclerosing cholangitis. (a and b) Axial T_2-weighted and coronal MIP images show irregular boundaries and dilated bile ducts in the periphery of the liver (open arrows). (c and d) After contrast medium administration, some bile ducts show enhancements due to inflammation and fibrosis (arrow). Asterisk is used to note the presence of an area of confluent fibrosis in the liver segment VIII.

1–2 mm thick areas of incomplete circumferential narrowing. Increase in intraductal pressure during conventional cholangiography may lead to focal dilatation of the bile duct, with consequent conversion of the web into diverticulum. Also liver parenchyma undergoes morphologic changes, which consist in hypertrophy of caudate lobe and hypotrophy of the lateral-posterior segments. After contrast medium injection, it has been described a slightly greater enhancement during the arterial phase [117,118] (Figure 15.37).

15.5.3.3 Hepatitis

Hepatitis is defined as a nonspecific inflammatory response of liver to various agents. The most common cause is due to viral infections (HBV, HCV, EBV, HAV) diagnosed primarily by clinical or serological examination; other causes of hepatitis can be bacterial or fungal infection, autoimmune reaction, drug-induced injury, and radiation therapy.

Cross-sectional imaging is not normally part of the primary diagnostic approach. Typical MR findings in acute viral hepatitis are hepatomegaly combined with edema of the liver capsule. In fulminant forms of acute viral hepatitis, diffuse or focal necrosis may be detected on MR images (Figure 15.38). In patients suffering from chronic hepatitis, cross-sectional imaging, especially MRI, is performed to determine the presence of cirrhosis or ascites and to screen for the presence of hepatocellular carcinoma. A region of high signal intensity surrounding the portal vein branches can frequently be found on T_2-weighted images in patients suffering from acute or chronic active hepatitis, but is considered a nonspecific sign. In addition, diffuse or regional high signal areas can be identified on T_2-weighted images [119]. Patients with viral hepatitis typically have enlarged lymph nodes at the liver hilum presenting as solitary or confluent. However, in contrast to lymph node metastasis, the portal veins and other structures are not compressed and maintain their caliber.

Due to its size and anatomical location in the abdomen, the liver is frequently affected secondarily by radiation therapy of extrahepatic malignancies. Within six months of radiation injury, diffuse edema of the liver can be seen, appearing as increased signal intensity on T_2-weighted images and decreased signal intensity on T_1-weighted images [119]. Portal flow is generally reduced in patients with radiation injured regions of the liver, and, as in the case of patients with concomitant fatty infiltration, the deposition of fat in these areas is usually reduced (Figure 15.39) [120].

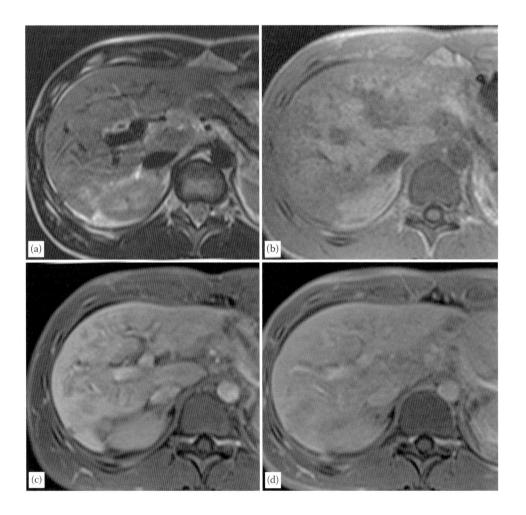

FIGURE 15.38

Acute hepatitis from drug abuse. (a and b) T_2- and T_1-weighted images show non-homogeneous signal intensity in the liver periphery. (c and d) After contrast medium injection, the liver parenchyma shows heterogeneous enhancement during the arterial phase that becomes more homogeneous on delayed phase.

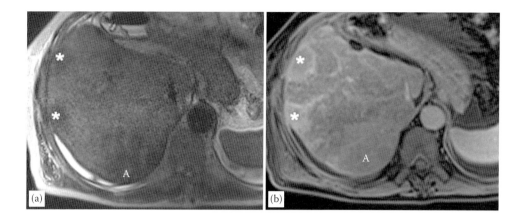

FIGURE 15.39
Chronic hepatitis from radiation exposure during radioembolization. (a) T_2-weighted image reveals a diffuse high signal intensity of liver parenchyma, due to tissue edema. (b) Post-contrast T_1-weighted image shows a diffuse low signal intensity of liver parenchyma with an area of healthy parenchyma in segment VI (A). Asterisks are used to note the presence of two liver metastases.

References

1. Bartolozzi C, Cioni D, Donati F et al. (2001) Focal liver lesions: MR imaging-pathologic correlation. *Eur Radiol* 11:1374–1388.
2. Keogan MT, Edelman RR (2001) Technologic advances in abdominal MR imaging. *Radiology* 220:310–320.
3. Morrin MM, Rofsky NM (2001) Techniques for liver MR imaging. *Magn Reson Imaging Clin N Am* 9:675–696.
4. Schwartz LH, Panicek DM, Thomson E et al. (1997) Comparison of phased-array and body coils for MR imaging of liver. *Clin Radiol* 52:745–749.
5. Helmberger TK, Schroder J, Holzknecht N et al. (1999) T2-weighted breathhold imaging of the liver: A quantitative and qualitative comparison of fast spin echo and half Fourier single shot fast spin echo imaging. *MAGMA* 9:42–51.
6. Kim TK, Lee HJ, Jang HJ et al. (1998) T2-weighted breathhold MRI of the liver at 1.0 T: Comparison of turbo spin-echo and HASTE sequences with and without fat suppression. *J Magn Reson Imaging* 8:1213–1218.
7. Ward J, Robinson PJ, Guthrie JA et al. (2005) Liver metastases in candidates for hepatic resection: Comparison of helical CT and gadolinium- and SPIO-enhanced MR imaging. *Radiology* 237:170–180.
8. Siegelman ES (1997) MR imaging of diffuse liver disease: Hepatic fat and iron. *Magn Reson Imaging Clin N Am* 5:347–365.
9. Engelhardt R, Langkowski JH, Fischer R et al. (1994) Liver iron quantification: Studies in aqueous iron solutions, iron overloaded rats, and patients with hereditary hemochromatosis. *Magn Reson Imaging* 12:999–1007.
10. Ernst O, Sergent G, Bonvarlet P et al. (1997) Hepatic iron overload: Diagnosis and quantification with MR imaging. *AJR Am J Roentgenol* 168:1205–1208.
11. Ferrucci JT (1998) Advances in abdominal MR imaging. *RadioGraphics* 18:1569–1586.
12. Schneider G, GrazioliL, Saini S (2006) *MR Imaging of the Liver*, 2nd edition. Springer, New York.
13. Larson RE, Semelka RC, Bagley AS, Molina PL, Brown ED, Lee JK (1994) Hypervascular malignant liver lesions: Comparison of various MR imaging pulse sequences and dynamic CT. *Radiology* 192:393–399.
14. Low RN, Francis IR, Sigeti JS, Foo TK (1993) Abdominal MR imaging: Comparison of T2-weighted fast conventional spin-echo, and contrast-enhanced fast multiplanar spoiled gradient-recalled imaging. *Radiology* 186:803–811.
15. Yamashita Y, Mitsuzaki K, Yi T et al. (1996) Small hepatocellular carcinoma in patients with chronic liver damage: Prospective comparison of detection with dynamic MR imaging and helical CT of the whole liver. *Radiology* 200:79–84.
16. Hussain HK, Londy FJ, Francis IR et al. (2003) Hepatic arterial phase MR imaging with automated bolus-detection three-dimensional fast gradient-recalled-echo sequence: Comparison with test-bolus method. *Radiology* 226:558–566.
17. Kanematsu M, Semelka RC, Matsuo M et al. (2002) Gadolinium-enhanced MR imaging of the liver: Optimizing imaging delay for hepatic arterial and portal venous phases—A prospective randomized study in patients with chronic liver damage. *Radiology* 225:407–415.
18. Mori K, Yoshioka H, Takahashi N et al. (2005) Triple arterial phase dynamic MRI with sensitivity encoding for hypervascular hepatocellular carcinoma: Comparison of the diagnostic accuracy among the early, middle, late, and whole triple arterial phase imaging. *AJR Am J Roentgenol* 184:63–69.
19. Goshima S, Kanematsu M, Kondo H et al. (2009) Optimal acquisition delay for dynamic contrast-enhanced MRI of hypervascular hepatocellular carcinoma. *AJR Am J Roentgenol* 192:686–692.
20. Hussain HK, Londy FJ, Francis IR et al. (2003) Hepatic arterial phase MR imaging with automated bolus-detection three-dimensional fast gradient-recalled-echo sequence: Comparison with test-bolus method. *Radiology* 226:558–566.

21. Whitney WS, Herfkens RJ, Jeffrey RB et al. (1993) Dynamic breath-hold multiplanar spoiled gradient-recalled MR imaging with gadolinium enhancement for differentiating hepatic hemangiomas from malignancies at 1.5 T. *Radiology* 189:863–870.

22. Mathie D, Rahmouni A, Anglade MC et al. (1991) Focal nodular hyperplasia of the liver: Assessment with contrast-enhanced Turbo-FLASH MR imaging. *Radiology* 180:25–30.

23. Mahfouz AE, Hamm B, Wolf KJ (1994) Peripheral washout: A sign of malignancy on dynamic gadolinium-enhanced MR images of focal liver lesions. *Radiology* 190:49–52.

24. Low RN, Sigeti JS, Francis IR et al. (1994) Evaluation of malignant biliary obstruction: Efficacy of fast multiplanar spoiled gradient-recalled MR imaging vs. spin-echo MR imaging, CT and cholangiography *AJR Am J Roentgenol* 162:315–323.

25. Ros PR, Freeny PC, Harms SE et al. (1995) Hepatic MR imaging with ferumoxides: A multicenter clinical trial of the safety and efficacy in the detection of focal hepatic lesions. *Radiology* 196:481–488.

26. Balci NC, Semelka RC (2005) Contrast agents for MR imaging of the liver. *Radiol Clin North Am* 43:887–898.

27. Semelka RC, Helmberger T.K.G (2001) Contrast agent for MR imaging of the liver. *Radiology* 218:27–38.

28. Bellin MF, Zaim S, Auberton E et al. (1994) Liver metastases: Safety and efficacy of detection with superparamagnetic iron oxide in MR imaging. *Radiology* 193:657–663.

29. Winter III TC, Freeny PC, Nghiem HV et al. (1993) MR imaging with i.v. superparamagnetic iron oxide: Efficacy in the detection of focal hepatic lesions. *AJR Am J Roentgenol* 161:1191–1198.

30. Kirchin MA, Pirovano GP, Spinazzi A (1998) Gadobenate dimeglumine (Gd-BOPTA). An overview. *Invest Radiol* 33:798–809.

31. Huppertz A, Haraida S, Kraus A et al. (2005) Enhancement of focal liver lesions at gadoxetic acid-enhanced MR imaging: Correlation with histopathologic findings and spiral CT—Initial observations. *Radiology* 234:468–478.

32. Bashir MR, Breault SR, Braun R, Do RK, Nelson RC, Reeder SB (2014) Optimal timing and diagnostic adequacy of hepatocyte phase imaging with gadoxetate-enhanced liver MRI. *Acad Radiol* 21:726–732.

33. Van Kessel CS, Veldhuis WB, van den Bosch MA, van Leeuwen MS (2012) MR liver imaging with Gd-EOB-DTPA: A delay time of 10 minutes is sufficient for lesion characterisation. *Eur Radiol* 22:2153–2160.

34. Colagrande S1, Carbone SF, Carusi LM, Cova M, Villari N (2006) Magnetic resonance diffusion-weighted imaging: Extraneurological applications. *Radiol Med* 111:392–419.

35. Taouli B, Martin AJ, Qayyum A et al (2004) Parallel imaging and diffusion tensor imaging for diffusion-weighted MRI of the liver: Preliminary experience in healthy volunteers. *AJR Am J Roentgenol* 183:677–680.

36. Yoshikawa T, Kawamitsu H, Mitchell DG et al. (2006) ADC measurement of abdominal organs and lesions using parallel imaging technique. *AJR Am J Roentgenol* 187:1521–1530.

37. Thng CH, Koh TS, Collins DJ, Koh DM. Perfusion magnetic resonance imaging of the liver. *World J Gastroenterol* 16:1598–1609.

38. Jackson A, Haroon H, Zhu XP et al. (2002) Breath-hold perfusion and permeability mapping of hepatic malignancies using magnetic resonance imaging and a first-pass leakage profile model. *NMR Biomed* 15:164–173.

39. Qayyum A (2009) MR Spectroscopy of the liver: Principles and clinical applications *RadioGraphics* 29:1653–1664.

40. Noworolski SM, Tien PC, Merriman R, Vigneron DB, Qayyum A (2009) Respiratory motion-corrected proton magnetic resonance spectroscopy of the liver. *Magn Reson Imaging* 27:570–576.

41. Cowin GJ, Jonsson JR, Bauer JD et al. (2008) Magnetic resonance imaging and spectroscopy for monitoring liver steatosis. *J Magn Reson Imaging* 28:937–945.

42. Soper R, Himmelreich U, Painter D et al. (2002) Pathology of hepatocellular carcinoma and its precursors using proton magnetic resonance spectroscopy and a statistical classification strategy. *Pathology* 34:417–422.

43. Li CW, Kuo YC, Chen CY et al. (2005) Quantification of choline compounds in human hepatic tumors by proton MR spectroscopy at 3 T. *Magn Reson Med* 53:770–776.

44. Wu B, Peng WJ, Wang PJ et al. (2006) In vivo 1H magnetic resonance spectroscopy in evaluation of hepatocellular carcinoma and its early response to transcatheter arterial chemoembolization. *Chin Med Sci J* 21:258–264.

45. Kuo YT, Li CW, Chen CY, Jao J, Wu D, Liu GC (2004) In vivo proton magnetic resonance spectroscopy of large focal hepatic lesions and metabolite change of hepatocellular carcinoma before and after transcatheter arterial chemoembolization using a 3.0 T MR scanner. *J Magn Reson Imaging* 19:598–604.

46. Hernandez-Guerra M, Garcia-Pagan JC, Bosch J (2005) Increased hepatic resistance: A new target in the pharmacologic therapy of portal hypertension. *J Clin Gastroenterol* 39:S131–S137.

47. Ganne-Carrié N, Ziol M, de Ledinghen V et al. (2006) Accuracy of liver stiffness measurement for the diagnosis of cirrhosis in patients with chronic liver diseases. *Hepatology* 44:1511–1517.

48. Talwalkar JA, Yin M, Fidler JL, Sanderson SO, Kamath PS, Ehman RL (2008) Magnetic resonance imaging of hepatic fibrosis: Emerging clinical applications. *Hepatology* 47:332–342.

49. Yin M, Talwalkar JA, Glaser KJ et al. (2007) Assessment of hepatic fibrosis with magnetic resonance elastography. *Clin Gastroenterol Hepatol* 5:1207–1213.

50. Elster AD (1997) How much contrast is enough? Dependence of enhancement on field strength and MR pulse sequence. *Eur Radiol* 7(suppl 5):276–280.

51. Chang KJ, Kamel IR, Macura KJ, Bluemke DA (2008) 3.0-T MR imaging of the abdomen: Comparison with 1.5 T. *RadioGraphics* 28:1983–1998.

52. Barth MM, Smith MP, Pedrosa I, Lenkinski RE, Rofsky NM (2007) Body MR imaging at 3.0 T: Understanding the opportunities and challenges. *RadioGraphics* 27:1445–1462.

53. Couinaud C (1957) *Le foie: Études anatomiques et chirurgicales.* Masson, Paris, France, pp. 9–12.

54. Bismuth H (1982) Surgical anatomy and anatomical surgery of the liver. *World J Surg* 6:3–8.

55. Catalano OA, Singh AH, Uppot RN et al. (2008) Vascular and biliary variants in the liver: Implications for liver surgery *RadioGraphics* 28:359–378.

56. Makuuchi M, Hasegawa H, Yamazaki S et al. (1983) The inferior right hepatic vein ultrasonic demonstration. *Radiology* 148:213–217.

57. Charlton M (2004) Non-Alcoholic fatty liver disease: A review of current understanding and future impact. *Clin Gastroenterol Hepatol* 2:1048–1058.

58. Fabbrini E, Sullivan S, Klein S (2010) Obesity and noalcoholic fatty liver disease: Biochemical, metabolic, and clinical implications. *Hepatology* 51:679–689.

59. Romeo S, Kozlitina J, Xing C, Pertsemlidis A, Cox D, Pennacchio LA, Boerwinkle E, Cohen JC, Hobbs HH (2008) Genetic variation in PNPLA3 confers susceptibility to nonalcoholic fatty liver disease. *Nat Genet* 40:1461–1465.

60. Prasad SR, Wang H, Rosas H et al. (2005) Fat-containing lesions of the liver: Radiologic-pathologic correlation. *RadioGraphics* 25:321–331.

61. Qayyum A (2009) MR spectroscopy of the liver: Principles and clinical applications. *RadioGraphics* 29:1653–1664.

62. Kim H, Taksali SE, Dufour S et al. (2008) Comparative MR study of hepatic fat quantification using single-voxel proton spectroscopy, two-point dixon and three-point IDEAL. *Magn Reson Med* 59:521–527.

63. Hussain HK, Chenevert TL, Londy FJ et al. (2005) Hepatic fat fraction: MR imaging for quantitative measurement and display—Early experience. *Radiology* 237:1048–1055.

64. Guiu B, Petit JM, Loffroy R et al. (2009) Quantification of liver fat: Comparison of triple-echo chemical shift gradient-echo imaging and in vivo proton MR Spectroscopy. *Radiology* 1:95–102.

65. Shwartz LH, Panicek DM, Koutcher JA et al. (1995) Adrenal masses in patients with malignancy: Prospective comparison of echo-planar, fast spin-echo, and chemical shift MR Imaging. *Radiology* 197:421–425.

66. Siegelman ES, Mitchell DG, Semelka RC (1996) Abdominal iron deposition: Metabolism, MR findings, and clinical importance. *Radiology* 199:13–22.

67. Sirlin CB, Reeder SB (2010) Magnetic resonance imaging quantification of liver iron. *Magn Reson Imaging Clin N Am* 18:359–381.

68. Alústiza JM, Artetxe J, Castiella A et al. (2004) MR quantification of hepatic iron concentration. *Radiology* 230:479–484.

69. Li TQ, Aisen AM, Hindmarsh T (2004)Assessment of hepatic iron content using magnetic resonance imaging. *Acta Radiol* 45:119–129.

70. Vogl TJ, Steiner S, Hammerstingl R et al. (1994) MRT of the liver in Wilson's disease. *Rofo* 160:40–45.

71. Akhan O, Akpinar E, Karcaanticaba M et al. (2009) Imaging findings of liver involvement of Wilson's disease. *Eur Radiol* 69:147–155.

72. Tani I, Kurihara Y, Kawaguchi A et al. (2000) MR imaging of diffuse liver disease. *AJR Am J Roentgenol* 174:965–971.

73. Buscarini E, Buscarini L, Civardi G, Arruzzoli S, Bossalini G, Piantanida M (1994) Hepatic vascular malformations in hereditary hemorrhagic telangiectasia: Imaging findings. *AJR Am J Roentgenol* 163:1105–1110.

74. Dakeishi M, Shioya T, Wada Y et al. (2002) Genetic epidemiology of hereditary hemorrhagic teleangiectasia in a local community in the northern part of Japan. *Hum Mutat* 19:140–148.

75. Milot L, Kamaoui I, Gautier G, Pilleul F (2008) Hereditary-hemorrhagic telangiectasia: One-step magnetic resonance examination in evaluation of liver involvement. *Gastroenterol Clin Biol* 32:677–685.

76. Memeo M, Stabile Ianora AA, Scardapane A et al. (2004) Hepatic involvement in hereditary hemorrhagic teleangiectasia: CT findings. *Abdom Imaging* 29:211–220.

77. Hennksson L, Hedman A, Johansson A, Lindstrom K (1982) Ultrasound assessment of liver veins in congestive heart failure. *Acta Radiol* 23:361–363.

78. Moulton JS, Miller BL, Dodd GD, Vu DN (1988) Passive hepatic congestion in heart failure: CT abnormalities. *AJR* 51:939–942.

79. Janssen HL, Garcia-Pagan JC, Elias E et al. (2003) Budd-Chiari syndrome: A review by an expert panel. *J Hepatol* 38:364–71.

80. Dilawari JB, Bambery P, Chawla Y et al. (1994) Hepatic outflow obstruction (Budd-Chiari syndrome). Experience with 177 patients and a review of the literature. *Medicine* 73:21–36.

81. Ludwig J, Hashimoto E, McGill DB, van Heerden JA (1990) Classification of hepatic venous outflow obstruction: Ambiguous terminology of the Budd-Chiari syndrome. *Mayo Clin Proc* 65:51–55.

82. Noone TC, Semelka RC, Siegelman ES et al. (2000) Budd-Chiari syndrome: Spectrum of appearances of acute, subacute, and chronic disease with magnetic resonance imaging. *J Magn Reson Imaging* 11:44–50.

83. Bogin V, Marcos A, Shaw-Stiffel T (2005) Budd-Chiari syndrome: In evolution. *Eur J Gastroenterol Hepatol* 17:33–35.

84. Lim JH, Park JH, Auh YH (1992) Membranous obstruction of the inferior vena cava: Comparison of findings at sonography, CT, and venography. *AJR Am J Roentgenol* 159:515–520.

85. Kimura C, Matsuda S, Koie H, Hirooka M (1972) Membranous obstruction of the hepatic portion of the inferior vena cava: Clinical study of nine cases. *Surgery* 72:551–559.

86. Menon KV, Shah V, Kamath PS (2004) The Budd-Chiari syndrome. *N Engl J Med* 350:578–585.

87. Stark DD, Hahn PF, Trey C, Clouse ME, Ferrucci Jr JT (1986) MRI of the Budd-Chiari syndrome. *AJR Am J Roentgenol* 146:1141–1148.

88. Soyer P, Lacheheb D, Caudron C, Levesque M (1993) MRI of adenomatous hyperplastic nodules of the liver in Budd-Chiari syndrome. *J Comput Assist Tomogr* 17:86–89.

89. Federle MP (2004) *Diagnostic Imaging. Abdomen.* Amirsys, Salt Lake City, Utah.

90. Vilgrain V, Lewin M, Vons C et al. (1999) Hepatic nodules in Budd-Chiari syndrome: Imaging features. *Radiology* 210:443–450.

91. Brancatelli G, Federle MP, Grazioli L, Golfieri R, Lencioni R (2002) Large regenerative nodules in Budd-Chiari syndrome and other vascular disorders of the liver: CT and MR imaging findings with clinicopathologic correlation. *AJR Am J Roentgenol* 178:877–883.

92. Maetani Y, Itoh K, Egawa H et al. (2002) Benign hepatic nodules in Budd-Chiari syndrome: Radiologic–pathologic correlation with emphasis on the central scar. *AJR Am J Roentgenol* 178:869–875.

93. Boll DT, Merkle EM (2009) Diffuse liver disease: Strategies for hepatic CT and MR imaging. *RadioGraphics* 29:1591–1614.

94. Hanna RF, Aguirre DA, Kased N et al. (2008) Cirrhosis-associated hepatocellular nodules: Correlation of histopathologic and MR Imaging Features. *RadioGraphics* 28:747–769.

95. Giorgio A, Amoroso P, Lettieri G et al. (1986) Cirrhosis: Value of caudate to right lobe ratio in diagnosis with US. *Radiology* 161:443–445.

96. Stark DD, Goldberg HI, Moss AA, Bass NM (1984) Chronic liver disease: Evaluation by magnetic resonance. *Radiology* 150:149–151.

97. Okazaki H, Ito K, Fujita T, Koike S, Takano K, Matsunaga N (2000) Discrimination of alcoholic from virus-induced cirrhosis on MR imaging. *AJR Am J Roentgenol* 175:1677–1681.

98. Ito K, Mitchell DG, Kim MJ, Awaya H, Koike S, Matsunaga N (2003) Right posterior hepatic notch sign: A simple diagnostic MR sign of cirrhosis. *J Magn Reson Imag* 18:561–6.

99. Ito K, Mitchell DG, Gabata T, Hussain SM (1999) Expanded gallbladder fossa: Simple MR imaging sign of cirrhosis. *Radiology* 211:723–6.

100. Brancatelli G, Federle MP, Ambrosini R et al. (2007) Cirrhosis: CT and MR imaging evaluation. *Eur J Radiol* 61:57–69.

101. Vilgrain V (2001) Ultrasound of diffuse liver disease and portal hypertension. *Eur Radiol* 11:1563–77.

102. Ohtomo K, Baron RL, Dodd III GD et al. (1993) Confluent hepatic fibrosis in advanced cirrhosis: Evaluation with MR imaging. *Radiology* 189:871–874.

103. Coleman WB (2003) Mechanisms of human hepatocarcinogenesis. *Curr Mol Med* 3:573–588.

104. Hussain SM, Zondervan PE, IJzermans JN, Schalm SW, de Man RA, Krestin GP (2002) Benign versus malignant hepatic nodules: MR imaging findings with pathologic correlation. *RadioGraphics* 22:1023–1036; discussion 1037–1039.

105. Manfredi R, Maresca G, Baron RL et al. (1998) Gadobenate dimeglumine (BOPTA) enhanced MR imaging: Patterns of enhancement in normal liver and cirrhosis. *J Magn Reson Imaging* 8:862–867.

106. Krinsky GA, Lee VS, Nguyen MT et al. (2000) Siderotic nodules at MR imaging: Regenerative or dysplastic? *J Comput Assist Tomogr* 24:773–776.

107. Krinsky GA, Lee VS, Theise ND et al. (2001) Hepatocellular carcinoma and dysplastic nodules in patients with cirrhosis: Prospective diagnosis with MR imaging and explanation correlation. *Radiology* 219:445–454.

108. Di Martino M, Anzidei, M, Zaccagna F, Saba L, Catalano C (2014) Gadoxetic Acid MR Imaging in the characterization of the "grey zone" of the hepatocarcinogenesis. In: *Scientific Assembly and Annual Meeting*, Chicago, IL, Radiological Society of North America.

109. Mitchell DG, Rubin R, Siegelman ES, Burk Jr DL, Rifkin MD (1991) Hepatocellular carcinoma within siderotic regenerative nodules: Appearance as a nodule within a nodule on MR images. *Radiology* 78:101–103.

110. Cohen SA, Siegel JH, Kasmin FE (1996) Complication of diagnostic and therapeutic ERCP. *Abdom Imaging* 21:385–394.

111. Silverman WB, Kaw M, Rabinovitz M et al. (1994) Complication rate of endoscopic retrograde cholangiopancreatography (ERCP) in patients with primary sclerosing cholangitis: Is it safe? *Gastroenterology* 106:359.

112. Lee YM, Kaplan MM (1995) Primary sclerosing cholangitis. *N England J Med* 332:924–933.

113. Boberg KM, Schrumpf E (2004) Diagnosis and treatment of cholangiocarcinoma. *Curr Gastroenterol Rep* 6:52–59.

114. Chapman RW, Jewell DP (1985) Primary sclerosing cholangitis: An immunologically mediated disease? *West J Med* 143:193–195.

115. Crippin JS, Lindor KD (1992) Primary sclerosing cholangitis: Etiology and immunology. *Eur J Gastroenterol Hepato* 4:261–265.

116. McCarty RC, LaRusso NF, Wiesner RH et al. (1983) Primary sclerosing cholangitis: Findings on cholangiography and pancreatography. *Radiology* 149:39–44.

117. Bader TR, Beavers KL, Semelka RC (2003) MR imaging features of primary sclerosing cholangitis: Patterns of cirrhosis in relationship to clinical severity of disease. *Radiology* 226:675–685.

118. Ito K, Mitchell DG, Outwater EK et al. (1999) Primary sclerosing cholangitis: MR imaging features. *AJR* 172:1527–1533.

119. Matsui O, Kadoya M, Takashima T, Kameyama T, Yoshikawa J, Tamura S (1989) Intrahepatic periportal abnormal intensity on MR images: An indication of various hepatobiliary diseases. *Radiology* 171:335–338.

120. Unger EC, Lee JK, Weyman PJ (1987) CT and MR imaging of radiation hepatitis. *J Comput Assist Tomogr* 11:264–268.

Michele Di Martino and Carlo Catalano

CONTENTS

16.1 Introduction

Focal liver lesions are a very common disease, and its prevalence in the general population is more than 20%. In the past few decades, with the significant developments in technology, magnetic resonance (MR) has acquired a leading role in the evaluation of focal liver pathologies, since it has overcome computed tomography (CT) in sensitivity and specificity.

16.2 Cystic Focal Lesions

Cystic focal lesions of the liver represent a comprehensive heterogeneous cluster with regard to pathogenesis, clinical presentation, diagnostic findings, and therapeutic management. They can be classified as congenital, inflammatory, and neoplastic [1]. Congenital cysts include different entities (hepatic cyst, biliary hamartomas, polycystic disease, and Caroli's disease) that can be grouped under the "fibropolycystic liver disease:" this term identifies lesions that stem from a derangement of embryonic ductal plate development at various stages [2]. Inflammatory cysts may be abscesses on an intrahepatic hydatid cyst, whereas neoplastic cystic lesions may be cystadenoma cystadenocarcinoma, undifferentiated sarcoma, or cystic metastases.

16.2.1 Hepatic Cysts

Hepatic cysts are common findings, which may occur in up to 2.5% of the general population [3]. They can be solitary or multiple, variable in size (a few millimeters to 5 cm and over), and do not communicate with the biliary tree and have often serous content; sometimes they may be complicated by hemorrhage or inflammation. At MR imaging, hepatic cysts appear as homogeneous very low hypointense lesions on T_1-weighted sequences and very high hyperintense lesions on T_2-weighted sequences. They do not show enhancement after contrast-medium injection, and the wall is never seen (Figure 16.1).

16.2.2 Biliary Hamartomas

Biliary hamartomas, also called the von Meyenburg complex, are composed of one or more dilated duct-like structures lined by biliary epithelium and accompanied by a variable amount of fibrous stromata. They occur very often as multiple scattered lesions, less than 1.5 cm in size, which do not communicate with the biliary tree. On the basis of hamartomas consistency and biliary dilatation, they should be classified as follows: class 1 solid lesions with narrow bile channels, class 2 mild or focal dilatation of channels, and class 3 prominent dilated bile channels [4]. At MR imaging, biliary hamartomas have been reported as hypointense lesions on T_1-weighted images and very strong hyperintense

lesions on T_2-weighted sequences (Figure 16.2). After contrast-medium injection, some researchers have observed homogeneous enhancement or rim enhancement of these lesions [5–8]. In oncologic patients, they can be misdiagnosed as liver metastases [9].

16.2.3 Polycystic Liver Disease

Autosomal-dominant polycystic liver disease is characterized by multiple, sometimes innumerable, hepatic cysts, often associated with renal polycystic disease [3]. It is thought that they result from progressive dilation of the abnormal ducts in biliary hamartomas as part of a ductal plate malformation at the level of small intrahepatic bile ducts. As biliary hamartomas, they lose continuity with the biliary tree. Patients are generally asymptomatic or have abdominal distension, discomfort, and dyspnea. At imaging, the liver parenchyma appears enlarged and replaced by diffuse cysts ranging from few millimeters in diameter to 12 cm or more (Figure 16.3) [10]. The features of the cysts are the same as that of simple cysts: very low signal intensity on the T_1-weighted sequence and very high signal intensity on the T_2-weighted sequence; some of them may be complicated by hemorrhage and/or infections and show nonhomogeneous signal intensity. Wall calcifications may be present. Malignant degeneration is extremely rare, and liver transplantation should be performed for symptomatic cases.

16.2.4 Caroli's Disease

Caroli's disease is the result of ductal plate malformation of the large intrahepatic bile ducts. In the literature, two forms of the disease have been described: Caroli's disease, which involves the larger intrahepatic ducts, and Caroli's syndrome, which involves both peripheral and main bile ducts. At MR imaging, Caroli's disease

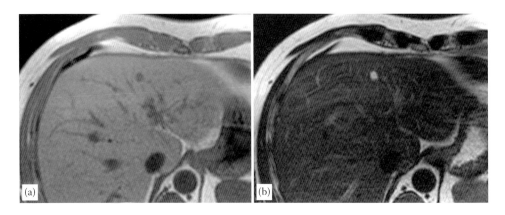

FIGURE 16.1
Hepatic cyst. (a) Axial T_1-weighted image shows a homogeneously, rounded, well-defined, hypointense lesion to the surrounding liver. (b) On T_2-weighted image the lesion appears markedly hyperintense.

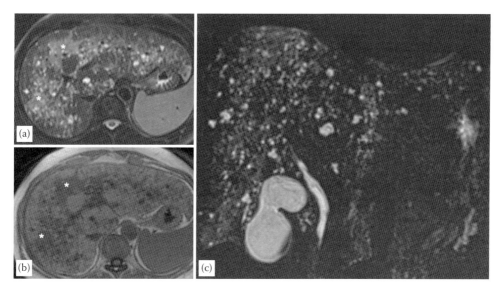

FIGURE 16.2

Biliary hamartomas. (a and b) T_1- and T_2-weighted images show multiple small (1.5 cm diameter), nodules consistent with biliary hamartomas. (c) Coronal projection MR cholangiogram shows that all of the lesions are smaller than 1.5 cm in diameter and do not communicate with the biliary tree. Note extended liver fibrosis (asterisks).

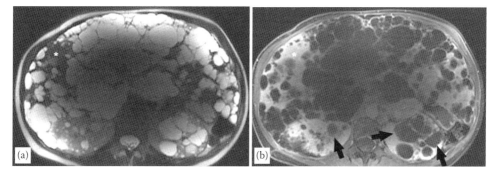

FIGURE 16.3

Polycystic liver disease. (a and b) T_1- and T_2-weighted images show multiple typical MR imaging appearance of hepatic cysts. Note also the involvement of the kidneys (arrows).

appears as sacciform-dilated bile ducts with very low signal intensity on T_1 and very high on T_2. After contrast-medium administration, a portal vein branch enhances within the lumen of the bile duct dilatation—"dot sign" (Figure 16.4) [11]. MR-cholangiography is extremely useful to demonstrate saccular or fusiform dilatations of bile ducts that communicate with the biliary tree, sometimes containing filling defects representing intrahepatic calculi (Figure 16.4) [12]. The main complications of Caroli's disease are cholangitis, stones, and abscesses; in Caroli's syndrome in addition to the complications cited above, secondary biliary cirrhosis and portal hypertension may occur. Prevalence of cholangiocarcinoma in patients with Caroli's disease is higher than in the general population [13].

16.2.5 Abscesses

Abscesses can be classified as pyogenic, amebic, or fungal. Pyogenic hepatic abscesses are most commonly caused by *Clostridium* species and gram-negative bacteria, such as *Escherichia coli*, *Klebsiella*, *Enterococcus*, which enter the liver via the portal venous system or biliary tree [14].

At an early stage, the *pyogenic* abscess shows a multiple clusters pattern, and after that, it coalesces into a single larger cavity. It may also appear as a large multiloculated lesion with septa and rim enhancement; peripheral enhancement is caused by granulation tissue. At MR imaging, the characteristic features of these abscesses include intense enhancement of the rim and septa on early gadolinium-enhanced images, which

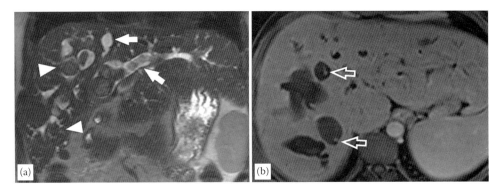

FIGURE 16.4
Caroli's disease. (a) Coronal T_2-weighted image shows multiple hyperintense cystic ectasias (arrows) and calculi (arrowheads). (b) After contrast-medium injection, the central fibrovascular bundle (central dot sign) is also seen (open arrows).

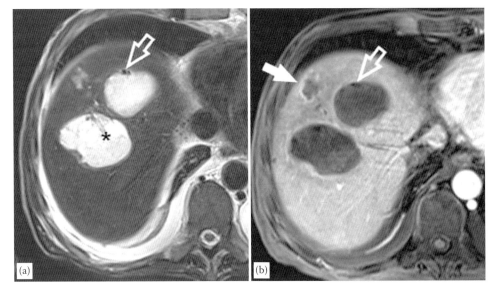

FIGURE 16.5
Pyogenic abscess. (a) Axial T_2-weighted image shows a multiloculated lesion surrounded by another smaller one. It is also appreciable bubble area within the biggest lesion (open arrow). (b) Axial venous phase T_1-weighted image shows rim enhancement of the smaller lesion (arrow). Note the presence of drainage catheter (asterisk).

persists with a negligible change in thickness and intensity on later postgadolinium images (Figure 16.5) [15].

The *amebic* liver abscess is the most common extraintestinal complication of amebiasis, occurring in 3%–9% of the cases. It is characterized by a large uniloculated cavity with thickened wall and peripheral edema. The central cystic area is composed of hemorrhage, granular necrotic material, and known cellular debris known as "anchovy paste" [16]. At MR imaging, T_1 and T_2 signal intensity is variable and depends on the cyst content; after gadolinium injection a rapid enhancement of the wall is usually seen (Figure 16.6).

16.2.6 Intrahepatic Hydatid Cyst

The intrahepatic hydatid cyst is a severe and common parasitic disease; it is generally caused by larval stages of taeniid cestodes (tapeworms) belonging to the *Echinococcus* species. MR imaging clearly demonstrates the pericyst, the matrix, and daughter cysts. The pericyst is seen as a hypointense rim on both T_1- and T_2-weighted images because of its fibrous composition and the presence of calcifications. The hydatid matrix (hydatid "sand") appears hypointense on T_1-weighted images and markedly hyperintense on T_2-weighted images; when present, daughter cysts are more hypointense than the matrix on

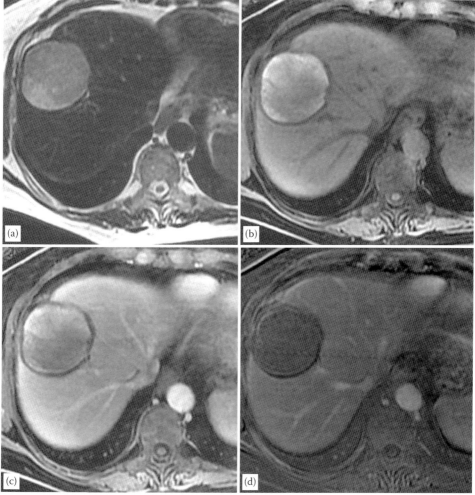

FIGURE 16.6
Amebic abscess. (a and b) T_1- and T_2-weighted images demonstrate a large, rounded, well-defined cystic mass in the right hepatic lobe with nonhomogeneous and hemorrhagic fluid content. (c and d) Venous phase T_1-weighted image and the corresponding subtracted image do not reveal enhancement within the lesion.

T_2-weighted images, and after contrast-medium administration, no enhancement is seen (Figure 16.7) [17].

16.2.7 Biliary Cystadenoma/Cystadenocarcinoma

Biliary cystadenoma is a rare, slow-growing, multilocular cystic tumor that represents less than 5% of intrahepatic cystic masses of the biliary tract. It occurs predominantly in middle-aged women (mean age, 38 years) and is considered a premalignant lesion because it can progress to cystadenocarcinoma; therefore, surgical resection is the treatment of choice. It is generally intrahepatic (85%), even though some extrahepatic lesions have been reported. Biliary cystadenoma ranges in diameter from 1.5 to 35 cm. Proteinaceous, mucinous, and occasionally gelatinous, purulent, or hemorrhagic fluid may be present inside the tumor [18]. The appearance at MR imaging is typical of multilocular fluid mass, with generally homogeneous low signal intensity on T_1-weighted images and homogeneous high signal intensity on T_2-weighted images [19]. Variable signal intensities on both T_1- and T_2-weighted images depend on the presence of solid components, hemorrhage, and protein content (Figure 16.8) [20]. The presence of irregular wall thickening, mural solid nodules, thick calcification, and papillary projections is suggestive of a cystadenocarcinoma. A recent tumor marker called tumor-associated glycoprotein (TAG) 72 has recently been proposed over carcinoembryonic antigen (CEA), carbohydrate antigen (CA) 19-9, for the differentiation between a simple cyst and cystadenoma [21].

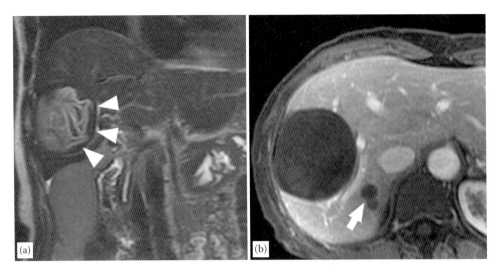

FIGURE 16.7
Hidatid cyst. (a) Coronal T_2-weighted image shows a large nonhomogeneous mass with hypointense pseudocapsule (arrowheads). (b) Axial venous phase T_1-weighted image reveals the main lesion with the daughter cysts (arrow).

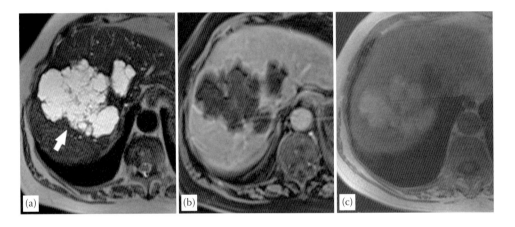

FIGURE 16.8
Biliary cystadenoma. (a) Axial T_2-weighted MR image shows a multilocular, septate mass (arrow) in the right lobe of the liver. (b) Corresponding venous-phase T_1-weighted MR image shows enhancement of the capsule and septa. (c) At follow-up imaging, pre-contrast T_1-weighted image demonstrates the presence of mucin content within the lesion.

16.2.8 Cystic Metastases

Cystic metastases are less common than solid metastases (both hypovascular and hypervascular). They are defined as large intratumoral areas of liquid content with faint enhancement in the periphery. Two interpretations have been proposed for the origin of cystic metastases: (a) necrosis of hypervascular metastases secondary to rapid growth beyond the vascular supply (neuroendocrine tumor, GIST) and (b) mucin production by mucinous adenocarcinoma (colon and ovarian) [22]. Ovarian metastases commonly spread by means of peritoneal seeding; therefore, they appear as cystic serosal implants on the liver surface rather than intraparenchymal lesions. Cystic metastases appear as single or multiloculated cysts. The wall tends to have irregular thickness, and the inner surface can be nonhomogeneous, ragged with multiple mural nodules. After contrast-medium administration, the peripheral rim or enhancement of the nodular component is appreciable (Figure 16.9) [15].

16.3 Benign Focal Lesions

With the widespread use of imaging, benign liver lesions are commonly encountered on MR examination. The most commonly diffuse in the general population

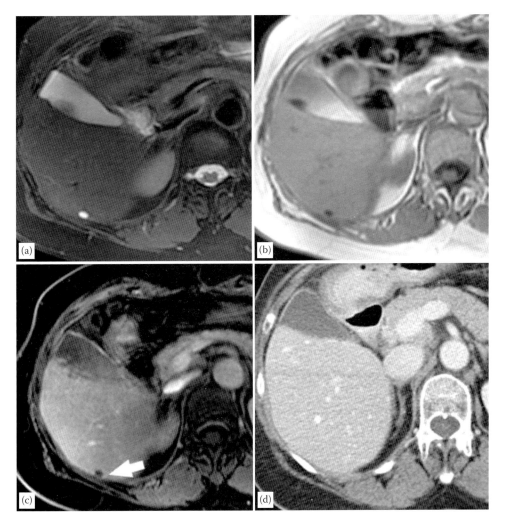

FIGURE 16.9
Cystic metastases. (a and b) T_2- and T_1-weighted sequences show a cystic-like lesion on segment six under the liver surface. (c) Axial T_1-weighted venous-phase image shows a peripheral faint rim enhancement. (d) CT image 3 months before does not reveal any focal lesion.

are haemangioma (4%), focal nodular hyperplasia (FNH, 0.4%) and hepatic adenomas (0.004%).

16.3.1 Hemangioma

Hepatic hemangioma is the most common benign liver lesion, and its prevalence ranges from 2% to 20% in the general population. It is generally asymptomatic, and only large masses may produce complications such as compression to adjacent structures and/or hemorrhage after rupture [23,24]. Hemangioma can be multiple in up to 50% of cases and should also be noted with other focal lesions such as focal nodular hyperplasia (FNH) [25]. Histologically, it is a tumor composed of multiple vascular channels lined by a single layer of endothelial cells supported by a thin, fibrous stroma. Large lesions almost always have a heterogeneous composition with areas of fibrosis, necrosis, cystic changes, and intratumoral calcifications.

At MR imaging, it appears as a well-defined lesion, with rounded margins, hypointense on T_1 and homogeneously marked hyperintense on T_2; the last sequence plays an important role in differential diagnosis. After gadolinium injection, hemangioma can show two different patterns: (a) small lesion (<1.5 cm) often homogeneously enhanced during the arterial phase with retention of the contrast agent in the venous and equilibrium phase—these features can also be observed in other hypervascular tumors such as FNH, hepatocellular carcinoma (HCC), and hypervascular metastases (Figure 16.10) and (b) typical and giant hemangiomas peripherally enhance during the arterial phase with a pseudonodular pattern and progressive centripetal filling in venous and delayed phases (Figures 16.11 and 16.12). With T_2-weighted spin–echo and dynamic gadolinium-enhanced T_1-weighted gradient-echo sequences, the sensitivity and specificity of MR imaging are 98%

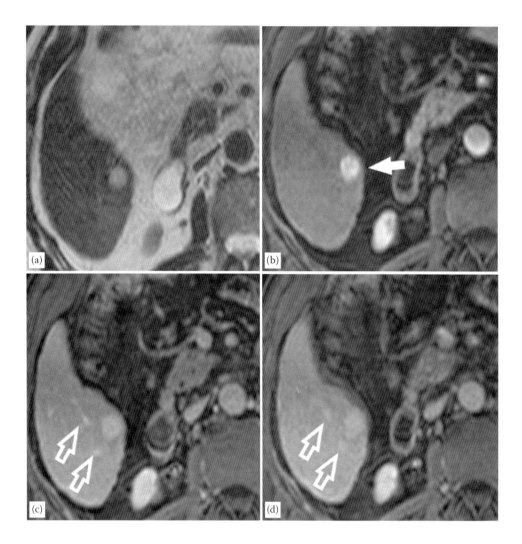

FIGURE 16.10
"Flash filling" hemangioma. (a) T_2-weighted image shows a homogeneous, brightly rounded lesion in liver segment five. (b) Arterial phase T_1-weighted MR image shows immediate homogeneous enhancement of a small lesion (arrow). (c and d) Venous and delayed-phases T_1-weighted MR images show persistent enhancement of the lesion similar to that of liver vessels (open arrows).

and the accuracy is 99% [26]. Hepatic hemangiomas rarely demonstrate calcifications [27,28].

16.3.2 Focal Nodular Hyperplasia

FNH is the second most common benign tumor, and its prevalence ranges between 3% and 8%. It is most frequently found in young and middle-aged women (ratio 8:1) during the third or fifth decade [29]. FNH is defined as a nodule of functional hepatocytes surrounded by normal or nearly normal liver parenchyma [30]. Its pathogenesis is not well known: vascular malformation and vascular injury have been proposed as underlying mechanisms [31]. FNH is divided into two types: classic and nonclassic. The nonclassic type contains three subtypes: (a) telangiectatic FNH, (b) FNH with cytologic atypia, and (c) mixed hyperplastic and adenomatous FNH [29]. Approximately 20% of the patients have

multiple FNH lesions. The combination of multiple FNH lesions and one or more of the following lesions— liver hemangiomas, central nervous system vascular malformation, meningioma, astrocytoma—is considered to be the multiple FNH syndrome [32,33]. The gross appearance of classic FNH consists of lobulated contours and parenchyma that is composed of nodules surrounded by radiating fibrous septa originating from a central scar. FNH does not have a tumor capsule, although the pseudocapsule surrounding some FNH lesions may be quite prominent. On histologic analysis, classic FNH shows nodular hyperplastic parenchyma. These nodules are completely or incompletely surrounded by circular or short fibrous septa. The hepatic plates may be moderately thickened hepatocytes. The central scar contains fibrous connective tissue, cholangiolar proliferation with surrounding inflammatory infiltrates, and malformed vessels of various calibers.

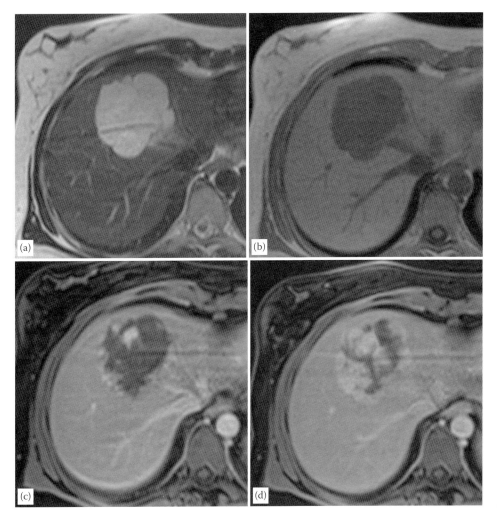

FIGURE 16.11
Cavernous hemangioma. (a and b) T_1- and T_2-weighted images show a homogeneous, rounded, well-defined lesion between segments four and eight. Arterial phase (c) and delayed-phase (d) contrast-material-enhanced MR sequences show progressive, peripheral, globular enhancement, which is highly suggestive of hemangioma.

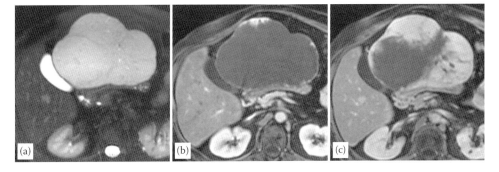

FIGURE 16.12
Giant hemangioma. (a through c) T_2-weighted image and post-contrast T_1-weighted images during the arterial and delayed phase show a focal lesion of the whole left lobe of the liver with the typical imaging pattern of hemangioma.

MR imaging has higher sensitivity (70%) and specificity (98%) for FNH than ultrasound and CT [34]. At MR imaging, typical FNH is isointense or slightly hypointense on T_1-weighted images and slightly hyperintense or isointense on T_2-weighted images [35]. After contrast-medium injection, FNH shows homogeneous enhancement during the arterial phase, and becomes isointense during the portal venous and equilibrium phase. The central scar, when present, is slightly hyperintense on T_2 and hypointense on T_1 and shows delayed

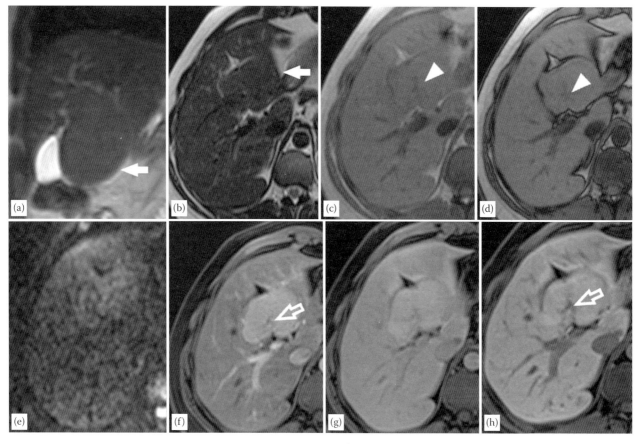

FIGURE 16.13
Focal nodular hyperplasia with scar. (a and b) On coronal and axial T_2-weighted images FNH is isointense to the surrounding liver but visible due to the mass effect. (c and d) On axial in-phase and out-of-phase T_1-weighted images FNH is isointense to the liver with a faintly darker central scar (arrowheads). (e) On b-400 EPI DW image the signal intensity of the lesion is similar to that of liver parenchyma. (f) Axial arterial phase image: FNH shows a very intense and homogeneous enhancement with central scar sparing (open arrow). (g and h) On venous and delayed-phase FNH reveals similar contrast enhancement to the surrounding liver.

enhancement after contrast-medium administration (Figures 16.13 and 16.14).

With the introduction of MR liver-specific contrast agents (hepatobiliary and reticuloendothelial) in clinical practice, the diagnosis of FNH has significantly improved. After the administration of paramagnetic hepatobiliary contrast agents, FNH appears hyper- or isointense compared with the surrounding liver parenchyma because the contrast agents tend to remain in the non-well-formed bile duct system [36,37]. This feature has a sensitivity ranging from 92% to 96.9% (Figure 16.15). Atypical FNH hypointense signal intensity in the hepatobiliary phase may be explained by a large central scar, an abundant fat component within the lesion, or the telangiectatic variant (Figure 16.16) [38]. Using superparamagnetic iron oxide (SPIO) based media, FNH nodules show a signal loss on T_2- and T_2^*-weighted images related to the uptake of iron oxide particles by Kupffer cells within the lesion. Therefore, the central scar appears

more conspicuous when these contrast agents are used (Figure 16.15) [39]. Liver-specific contrast agents are a very useful tool in the differential diagnosis of FNH from other hypervascular tumors such as hepatocellular adenoma, HCC and hypervascular metastases, and "flash filling" hemangioma; all these lesions do not uptake liver-specific contrast agents because of the absence of bile duct system in hepatocellular adenoma and the lack of functional hepatocytes in HCC and hypervascular metastases.

16.3.3 Hepatic Adenoma

Hepatic adenoma (HA) is an uncommon benign neoplasm that usually develops in young middle-aged women. It is usually solitary (70%–80%), but it is not unusual to encounter two or three adenomas; multiple (>10) adenomas are considered as "liver adenomatosis" [40]. The pathogenesis of the tumor is not yet clear, but it is likely caused by multiple factors. The two most common complications of

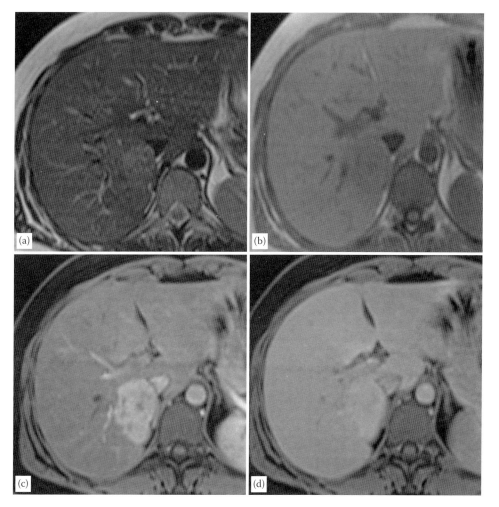

FIGURE 16.14
Focal nodular hyperplasia without scar. (a and b) Pre-contrast sequences show a slightly hyperintense lesion on T_2-weighted image in liver segment, VII which is faintly hypointense on T_1. (c and d) After contrast-medium administration, the lesion enhances during the arterial phase, and it is homogeneous at liver parenchyma in the delayed phase.

hepatocellular adenoma is bleeding with associated rupture and the development of HCC.

Accordingly, hepatocellular adenomas are currently categorized into three distinct histopathologic subtypes: (a) inflammatory hepatocellular adenomas, (b) hepatocyte nuclear factor 1 alpha (HNF-1a)-mutated hepatocellular adenomas, and (c) b-catenin-mutated hepatocellular adenomas [41].

Inflammatory hepatocellular adenoma is the most common subtype and accounts for about 40%–50% of all hepatocellular adenomas. It occurs frequently in women with women using oral contraceptive and in obese patients [42]. On gross pathologic examination, inflammatory hepatocellular adenomas are heterogeneous in appearance, with areas of congestion and frank hemorrhage. On histopathologic examination, inflammatory hepatocellular adenomas show intense polymorphous inflammatory infiltrates, marked sinusoidal dilatation or congestion, and thick-walled arteries [43]. On T_2-weighted

images, it often appears as a hyperintense lesion; on T_1 it is usually iso- or hyperintense without signal drop-off on the opposed-phase sequence. After contrast-medium administration, inflammatory HA shows homogeneous enhancement, which persists during the venous and delayed phases [44,45].

At MR imaging, HNF-1a-mutated hepatocellular adenoma is predominantly hyper- or isointense on T_1-weighted images, with diffuse signal drop-off on the opposed-phase T_1-weighted sequence; by contrast T_2-weighted images, it is isointense to slightly hyperintense. After contrast-agent administration, HNF-1a-mutated hepatocellular adenoma is slightly hypervascular in the arterial phase with no persistent enhancement in the portal venous and delayed phases. No specific MR imaging patterns have yet been proposed to identify b-catenin-mutated hepatocellular adenomas, and on T_1- and T_2-weighted images, these tumors may show homogeneous or heterogeneous hyperintense signal intensity, depending on the presence

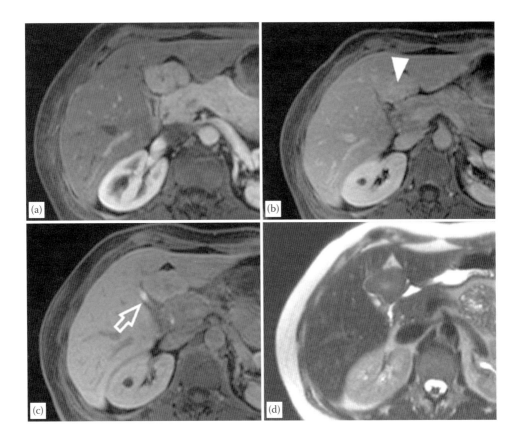

FIGURE 16.15
Focal nodular hyperplasia: liver-specific contrast-agent pattern. (a and b) Axial arterial and delayed phase of a hypervascular focal liver lesion in liver segment four, which becomes isointense to the liver parenchyma. Note the enhancement of the central scar in the delayed phase (arrowhead). (c) On T_1-weighted hepatobiliary phase after paramagnetic contrast-agent injection, the lesion is iso- to hyperintense, which demonstrates the presence of functional hepatocytes. Note the contrast-agent output through the biliary tree (open arrow). (d) On T_2-weighted hepatobiliary phase superparamagnetic contrast-agent injection, the same lesion is isointense to the surrounding liver parenchyma.

of hemorrhage and/or necrosis. b-Catenin-mutated hepatocellular adenomas commonly demonstrate strong arterial enhancement that may or may not persist in the portal venous and delayed phases, and these tumors may mimic hepatocellular carcinomas at imaging.

Since HA lack bile ducts, there is altered hepatocellular transport compared with normal hepatocytes. Thus, while HA may contain functioning hepatocytes to take up paramagnetic liver-specific contrast agent, the absence of an intracellular transport gradient due to the lack of active transport across the sinusoid membrane results in these lesions appearing hypointense on delayed hepatobiliary phase images against normal enhanced liver parenchyma. This has proven a highly accurate means of differentiating HA from FNH (Figures 16.17 and 16.18) [35,36].

Adenomas usually do not show uptake of SPIO particles, resulting in decreased signal intensity on T_2-weighted images. After injection of a hepatocellular-specific contrast agent, there is usually no substantial uptake. Almost 4% of hepatic adenomas may show uptake of liver-specific contrast agents during the hepatobiliary

phase; unfortunately, the explanation for this behavior is not well known (Figure 16.19) [36].

In liver adenomatosis, which is characterized by the presence of multiple adenomas in the same patient, the lesions may appear small or large, noncomplicated or complicated. Moreover, some nodules may appear with fatty metamorphosis, while others may have a homogeneous appearance. For these reasons, the enhancement behavior typically varies for different lesions (Figure 16.20).

16.4 Other Benign Lesions

16.4.1 Leiomyoma

Leiomyoma is a benign tumor composed of interlacing bundles of smooth muscle fibers [46]. Though this type of lesion is frequently discovered in the genitourinary and gastrointestinal tracts, a few cases of primary leiomyoma of the liver have been reported (2–9). Clinical presentation can range from small, incidentally discovered asymptomatic lesions to large, palpable upper abdominal masses (up to 15 cm in maximum diameter) often accompanied by abdominal pain. Although primary leiomyoma of the liver rarely degenerates into a malignancy, liver resection is often required to yield a definite diagnosis. At MR imaging, small hepatic leiomyoma is hypointense on T_1 and iso- to hyperintense on T_2. After contrast-medium administration, leiomyoma strongly enhances with persistent evidence of contrast medium in venous and delayed phases without evidence of washout [47]. During the hepatobiliary phase, it is hypointense to the surrounding live parenchyma due to the absence of functional hepatocytes (Figure 16.21) [48].

16.4.2 Angiomyolipoma

Hepatic angiomyolipoma (HAML) is a rare, benign mesenchymal neoplasm composed of variable amounts of smooth cells, fat, and blood vessel components [49].

Its diagnosis is sometimes difficult due to the presence of different components; it generally appears as an undeveloped and nonencapsulated nodule with intratumoral fat and a prominent central vessel. T_1- and T_2-weighted sequences with fat suppression are able to detect intratumoral fat [50]. After contrast-medium injection, it shows intense heterogeneous enhancement in the arterial phase. The enhancement may be somewhat decreased but grossly persists in the delayed phase (Figure 16.22) [51]. In hepatobiliary phase, the absence of functional hepatocytes within the lesion is shown as a loss of signal intensity to the surrounding liver parenchyma. Differential diagnoses of HAML are HCC, adenoma, FNH, lipoma, and focal steatosis; this underlines how HAML is difficult to diagnosis, and sometimes histologic analysis is necessary.

16.5 Malignant Focal Lesions

Malignant tumors can arise in healthy liver or in parenchyma affected by chronic liver disease. The most commonly encountered malignant lesions in the normal liver are metastases from dissemination of a primary tumor outside the liver. Hepatocellular carcinomas (HCC), and to a lesser extent intrahepatic cholangiocarcinomas (IHC),

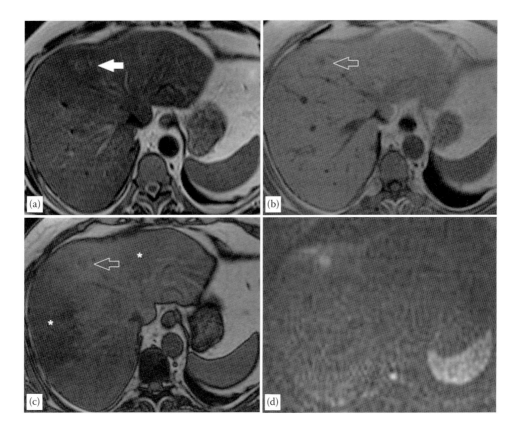

FIGURE 16.16
FNH: hypointense on hepatobiliary phase (teleangiectatic). (a) Axial T_2-weighted image shows a faintly hyperintense lesion in liver segment four. (b and c) Axial T_1-weighted in-phase and out-of-phase images reveal hypointense lesion (open arrow). Note the presence of patchy areas of liver steatosis (asterisks). (d) Diffusion-weighted image at b-500 shows faintly high signal intensity of the lesion to the liver parenchyma.
(Continued)

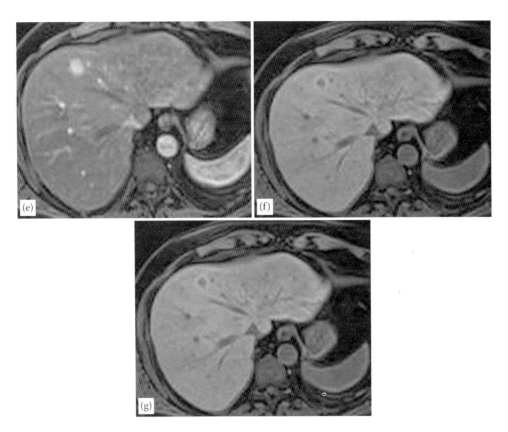

FIGURE 16.16 (Continued)
FNH: hypointense on hepatobiliary phase (telangiectatic). After contrast-agent administration, the lesion is hypervascular during the late arterial phase (e) and becomes isointense in the delayed phase (f) with a thin rim. (g) On T_1-weighted hepatobiliary phase the lesion is hypointense to the surrounding liver parenchyma.

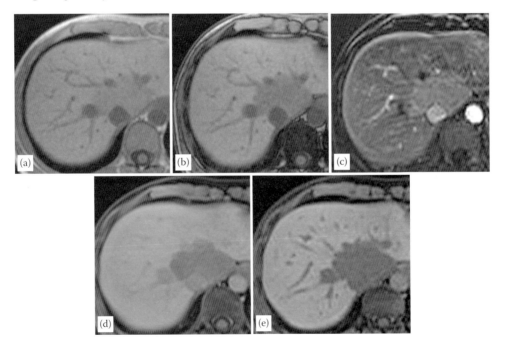

FIGURE 16.17
Hepatic adenoma. (a and b) Axial T_1-weighted in-phase and out-of-phase images show a lesion in liver segment eight with drop-off signal intensity, which demonstrates the presence of fat content. (c and d) Axial post-contrast T_1-weighted images during the arterial and delayed phases: the lesion shows faint enhancement with low signal intensity. (e) During the hepatobiliary phase, the lesion is hypointense to the surrounding liver parenchyma due to the absence of bile structures.

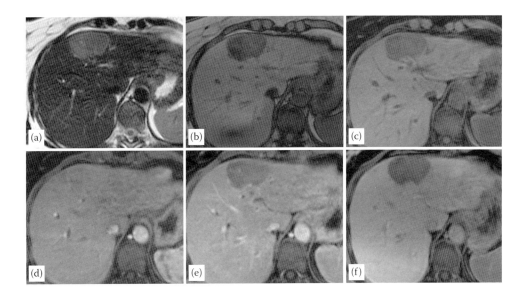

FIGURE 16.18
Hepatic adenoma. (a) On T_2-weighted image the lesion is slightly hyperintense to the surrounding liver parenchyma. (b and c) Out-of-phase and fat-saturated T_1-weighted images demonstrate the presence of fat component. (d and e) After contrast-medium administration, the lesion is isointense to the liver parenchyma on the arterial phase and becomes hypointense during the delayed phase. (f) On hepatobiliary phase the lesion is hypointense due to the absence of bile duct structures.

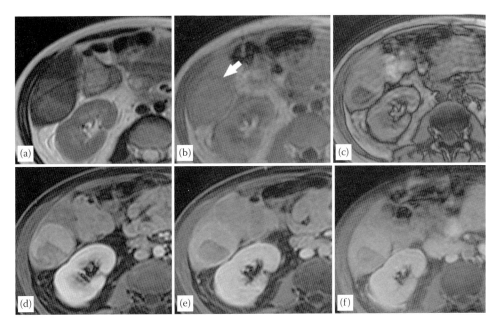

FIGURE 16.19
Hepatic adenoma. (a) Axial T_2-weighted image shows a rounded, well-defined, slightly hyperintense lesion in liver segment VI. (b and c) Axial T_1-weighted in-phase and out-of-phase images show drop-off signal intensity, which demonstrates the presence of fat content: a thin lesion capsule is also visible (arrow). (d) Axial T_1-weighted image in the late arterial phase shows heterogeneous enhancement. (e and f) On delayed and hepatobiliary phase, the lesion is isointense to the surrounding liver parenchyma.

occur mainly in the setting of chronic liver disease, and represent the most common primary liver malignancies.

16.5.1 Hepatocellular Carcinoma

HCC is the most common primary tumor of the liver and it is the fifth tumor in men and the seventh in women. The very poor prognosis associated with this disease makes it the third leading cause of cancer-related mortality worldwide, with an estimated 694,000 deaths in 2008 [52]. It occurs primarily in subjects who have chronic liver disease or liver cirrhosis and is the primary cause of death among this group. HCC incidence is increasing and is much more common among males than females. In high-incidence countries, the male-to-female ratio may be as high as 7:1 or 8:1 [53–57].

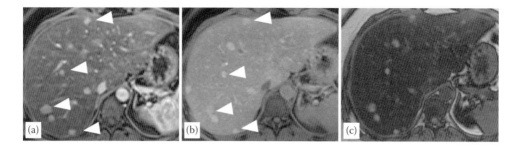

FIGURE 16.20
Hepatic adenomatosis. (a and b) Post-contrast T_1-weighted images in arterial and delayed phases show multiple, rounded, well-defined adenomas (arrowheads). The lesions appear hyperattenuating because of the underlying liver steatosis (c).

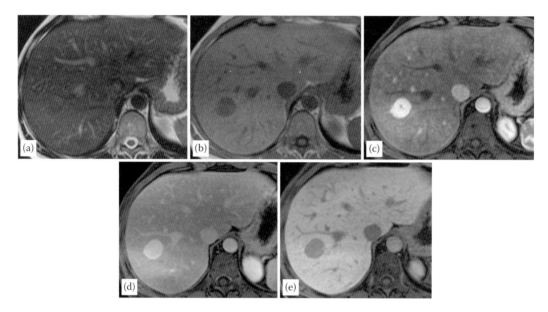

FIGURE 16.21
Leiomyoma. (a) Axial T_2-weighted image with fat saturation shows a slightly hyperintense area in liver segment VIII. (b) Axial T_1-weighted pre-contrast scan illustrates a well-defined hypointense lesion, with smooth margin. (c and d) Axial T_1-weighted image during the late arterial phase shows a brightly hypervascular lesion with prolonged enhancement during the delayed phase. (e) On T_1-weighted post-contrast image during the hepatobiliary phase the lesion shows low signal intensity, which confirms the absence of hepatocytes.

The development of HCC may arise from de novo hepatocarcinogenesis or by means of a multistep progression from regenerative nodules, through low-grade and high-grade dysplastic nodules to HCC [58,59]. HCC may show different growth patterns. The most common is a solitary small or large mass with or without capsules. The second most common pattern is multifocal HCC, which is characterized by multiple separate nodules. At least diffuse/infiltrative HCC shows multiple small tumoral foci distributed throughout the liver, mimicking nodules of cirrhosis (Figure 16.23) [60].

The recently published guidelines report a surveillance with ultrasound and α-fetoprotein (AFP) every 6 months as the best cost-effective strategy [61]. Although AFP assessment is the most common blood serum test for screening patients with cirrhosis, several reports show that false-negative interpretations are relatively common, particularly for small lesions [62–64]. Although the sensitivity of ultrasound for HCC detection is good, this technique suffers in terms of specificity, primarily because of a relatively high number of false positive interpretations [61,65,66]. At MR imaging, small HCC have variable signal intensity on T_1-weighted pre-contrast imaging: generally they are hypointense, but high signal intensity has been reported with a frequency ranging between 34% and 61% [67,68]. On T_2-weighted images, HCC is iso- to hyperintense to the surrounding liver parenchyma [69]. Generally, hyperintense lesions on T_1 and isointense in T_2 are well differentiated, due to the presence of fat and/or glycoprotein (Figure 16.24); by contrast, lesions hypointense on T_1 and hyperintense on T_2 are moderately/poorly differentiated [70,71]. The tumor arterial recruitment is the most important feature in the differentiation of preneoplastic lesions and HCC, so arterial enhancement after contrast-medium injection is the main imaging finding of HCC.

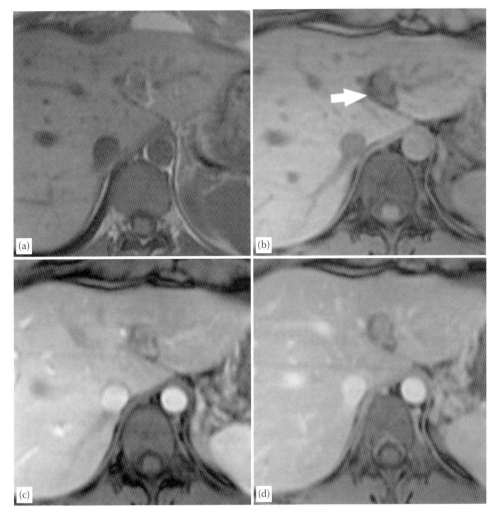

FIGURE 16.22
Angiomyolipoma. (a and b) Axial T_1-weighted in-phase and out-of-phase images show a hyperintense lesion in liver segment two with signal drop-off in the out-of-phase image (arrow). After contrast-medium administration, during the late arterial phase (c) the lesion shows heterogeneous enhancement with prolonged contrast-medium retaining in the delayed phase (d).

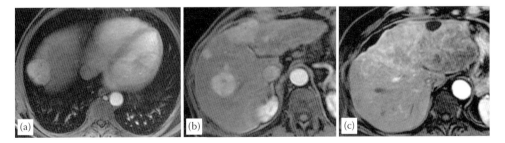

FIGURE 16.23
Hepatocellular carcinoma: morphologic pattern. (a) Solitary nodule, (b) multiple nodules, and (c) infiltrative.

Small HCC have homogeneous enhancement, and by contrast, large HCC show heterogeneous contrast-medium uptake; some lesions may present corona enhancement, and almost 7% are hypovascular (Figure 16.25) [72].

Unfortunately, not all arterially enhancing masses in patients with cirrhosis are due to HCC; other focal lesions such as flash filling hemangioma, arterio-portal shunt, transient hepatic intensity difference, regenerative nodules may be encountered and sometimes may be the cause of misinterpretation [73,74]. Loss of signal intensity during the delayed-phase "wash-out" sign is the second main feature; it significantly improves sensitivity and specificity in the detection of HCC and

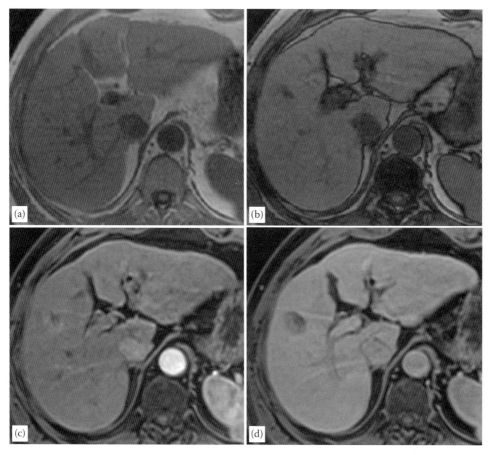

FIGURE 16.24
Hepatocellular carcinoma: well-differentiated. (a and b) Axial T_1-weighted in-phase and out-of-phase images reveal fat content within a typical HCC nodule (c and d), which is associated with well-differentiated neoplasm.

probably is the strongest independent predictor of HCC (Figures 16.24 and 16.26) [74–76]. The presence of a tumor capsule is another important sign of malignancy: it appears as a slightly hypointense peripheral rim on T_1- and slightly hyperintense on T_2-weighted pre-contrast sequences. During the delayed phase, the tumor capsule enhances to the surrounding liver parenchyma because of the presence of fibrous stroma (Figure 16.26). The introduction of liver-specific contrast agents in clinical practice, superparamagnetic as well as paramagnetic, significantly improves the detection and characterization of HCC in particular for lesions between 1 and 2 cm. With paramagnetic contrast agents, the absence of functional hepatocytes, which is considered a sign of malignancy, is represented as a loss of signal intensity during the hepatobiliary phase (Figure 16.27). Nevertheless, fewer than 20% of well-differentiated and moderately differentiated HCC appear iso- or hyperintense on hepatobiliary phase images [77–79]. SPIO agents are helpful for detecting small HCC in cirrhotic livers. A recent report [80] investigated the relationship between the number of Kupffer cells in HCC and DN and the degree of SPIO uptake. This study showed that the ratio between the number of Kupffer cells in tumorous versus nontumorous tissue decreased with the degree of cellular differentiation. Diffusion-weighted MRI (DWI) has recently gained interest in liver imaging, showing improved detection of liver lesions and enabling lesion characterization using the apparent diffusion coefficient (ADC) [81]. With regard to HCC, several studies suggest that DW-MR images may be able to improve detection of HCC and may be useful to better characterize small HCC with atypical findings on dynamic-phase MRI [82,83]. However, debate still persists regarding the role of quantitative and qualitative analysis of DWI in the evaluation of HCC; some authors suggest that it is able to just slightly improve the sensitivity (Figure 16.28) [84–86]. HCC can develop in the absence of cirrhosis and even without identifiable risk factors. In this setting, HCC is more likely to manifest as a symptomatic mass in a middle-aged man. The hepatic mass is likely to be a large solitary or dominant mass with lobulated and possibly encapsulated margins.

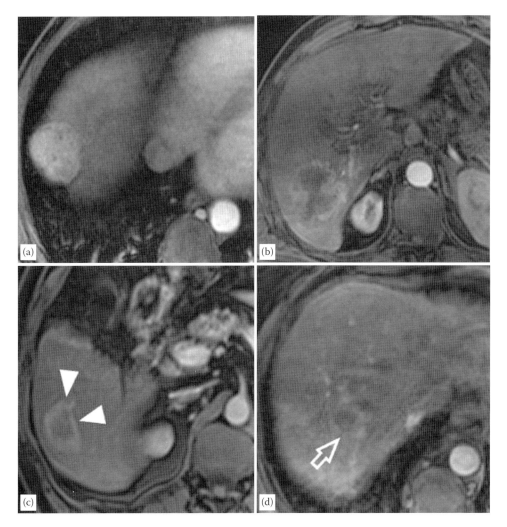

FIGURE 16.25
Hepatocellular carcinoma: arterial phase enhancement. (a) Homogeneous, (b) heterogeneous, (c) corona enhancement (arrowheads), and (d) hypovascular (open arrow).

MR imaging shows similar patterns to those of HCC in cirrhotic patients (Figure 16.29) [87–89].

Fibrolamellar carcinoma is a rare variant of HCC with quite unique features [90]. It is seen in noncirrhotic livers of young population. Hemorrhage or necrosis are not typical. Macroscopically, it resembles FNH. On MR images, it is usually seen as hypointense lesions with T_1-weighted sequences and hyperintense lesions with T_2-weighted techniques. They have a central scar, which is of low signal intensity on both T_1-weighted and T_2-weighted images in contrast to the central scar seen in FNH, which is bright on T_2-weighted images [91]. On dynamic gadolinium-enhanced images, fibrolamellar carcinoma demonstrates early diffuse heterogeneous enhancement, and promptly thereafter the enhancement is homogeneous and isointense signal intensity is noted [92].

16.5.2 Intrahepatic Cholangiocarcinoma

Intrahepatic cholangiocarcinoma (ICC) is the adenocarcinoma that arises from the epithelium of the bile ducts. It is the second most common primary tumor of the liver. On the basis of the site of origin, it can be classified as peripheral and hilar (Klatskin tumor) [93]. The first one is considered to originate from the second-order branches, while the second one from the first-order branches, right and left bile ducts, and their bifurcation. The radiological pattern and management of these two types of neoplasm significantly differ.

Peripheral cholangiocarcinoma is almost 10% of cholangiocarcinoma; it is associated with intrahepatic cholelithiasis, primary sclerosing cholangitis, Caroli's disease, *Clonorchis* infection, and exposure to thorium oxide suspension (Thorotrast) [94]. The most common symptoms are abdominal pain, fatigue, and loss of

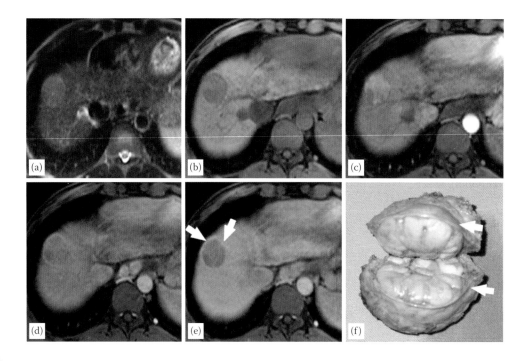

FIGURE 16.26
Hepatocellular carcinoma: typical pattern with capsule. (a) Axial T_2-weighted image shows a hyperintense lesion in liver segment VIII, which is hypointense on the corresponding T_1-weighted image (b). After gadolinium injection the lesion enhances during the arterial phase (c) with the wash-out sign in venous and delayed phases (d and e). The tumor capsule is better visible in delayed phase (arrows). (f) Photograph of the gross specimen shows an encapsulated lesion (arrows).

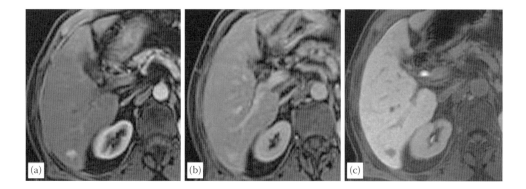

FIGURE 16.27
Hepatocellular carcinoma: role of liver-specific contrast agent. (a through c) Small hypervascular HCC without wash-out during the delayed phase. The hepatobiliary phase (c) shows loss of signal intensity within the lesion, which means the absence of functional hepatocytes and thus considered as a sign of malignancy. Data from the literature report that liver-specific contrast agents significantly improve the diagnostic accuracy of MR in the detection of small HCC (1–2 cm).

weight. On the basis of macroscopic appearance and growth pattern, there are three types of peripheral cholangiocarcinoma: (a) mass-forming, (b) intraductal, and (c) periductal infiltrating type [95]. At MR imaging, cholangiocarcinoma is homogeneously hypointense relative to the normal liver on T_1-weighted MR images but may range from markedly to mildly hyperintense on T_2-weighted images. The signal intensity of the tumor is variable and depends on the amount of mucinous material, fibrous tissue, hemorrhage, and necrosis within the tumor [96]. Although a central scar is not pathognomonic, this finding that reflects severe fibrosis appears to be a characteristic marker. On dynamic T_1-weighted MR images acquired after the intravenous administration of gadolinium, minimal or moderate incomplete enhancement is seen at the tumor periphery in the arterial phase, whereas progressive central contrast enhancement is seen in the delayed phase acquired after 3 and 5 minutes [97,98]. This pattern is consistent with the fibrous nature of the tumor, and it is typical of peripheral ICC (Figure 16.30). On DWI, the presence of high neoplastic cells determines the restriction of water

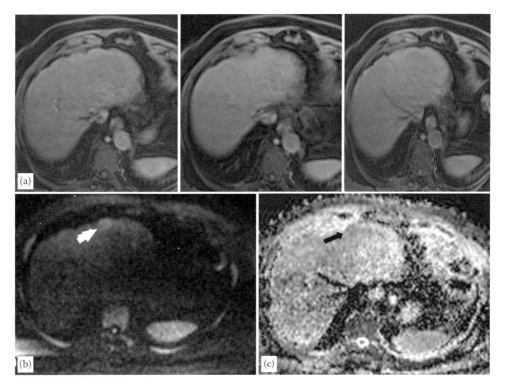

FIGURE 16.28
Hepatocellular carcinoma: DWI imaging. (a through c) Post-contrast T_1-weighted images acquired during the arterial, venous, and delayed phases do not reveal any focal lesion in liver segment II. DW image at b-400 shows a slightly focal area of restriction of molecular diffusion under the liver surface, which is confirmed as ADC map.

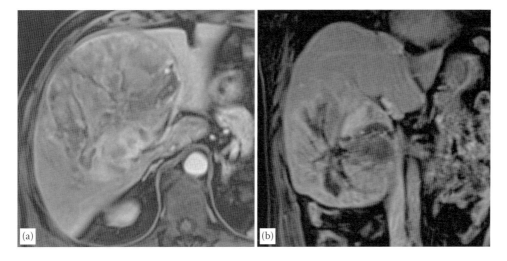

FIGURE 16.29
Hepatocellular carcinoma in not-at-risk patient. (a and b) Axial T_1-weighted during the arterial phase and coronal delayed-phase images demonstrate a large, solitary, encapsulated, heterogeneous, hypervascular mass with the wash-out sign.

diffusion with an increase of signal intensity at high b values: some authors suggest the target sign on DWI to differentiate small ICC from small HCC [99].

Imaging with liver-specific contrast agents is similar to imaging with conventional nonspecific Gd-based contrast agents during the dynamic phase of contrast enhancement. However, delayed imaging with paramagnetic contrast agents during the hepatobiliary phase may reveal contrast enhancement in the fibrotic areas of the lesion (Figure 16.31). Analogously, no significant uptake is observed after SPIO administration due to the absence of Kupffer cells within ICC [100]. Main differential diagnoses for ICC are liver metastases, HCC, lymphoma, and carcinoid tumor. In particular, mixed CCC/HCC tumors can show intense, homogeneous enhancement during the arterial phase

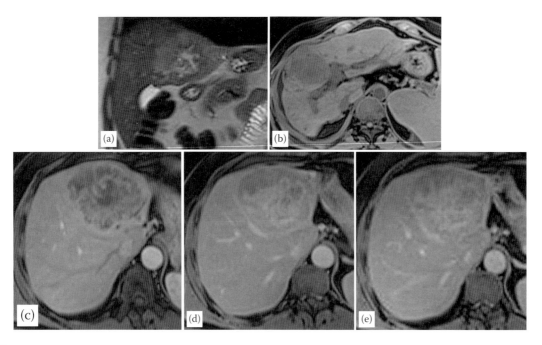

FIGURE 16.30

Intrahepatic cholangiocarcinoma: peripheral. (a and b) T_2- and T_1-weighted images demonstrate a heterogeneous liver mass in liver segment IV. (c through e) After contrast-medium administration, the lesion shows heterogeneous peripheral enhancement (c), which increases during venous (d) and delayed phases (e).

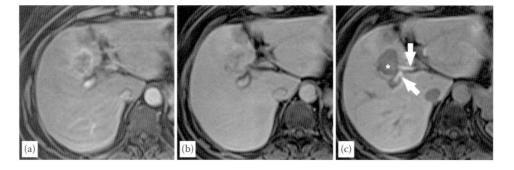

FIGURE 16.31

Cholangiocarcinoma: hilar. (a and b) Post-contrast images show a typical pattern of cholangiocarcinoma. (c) During the hepatobiliary phase the main lesion (asterisk) and the primary bile ducts (arrow) are clearly visible.

with prolonged enhancement in delayed phases due to marked hypervascularity.

Hilar cholangiocarcinoma (which generally manifests with jaundice) is usually classified into three macroscopic types: infiltrating, exophytic, and papillary. On MR-cholangiography (MRC) images, hilar cholangiocarcinoma appears as a moderately irregular thickening of the wall, with asymmetric upstream dilation of the intrahepatic ducts. Infiltrating form may be seen as a stretch of narrowed lumen; the exophytic type tends to produce complete obstruction, meanwhile papillary type is visualized as a protuberant-shaped morphology. On MR images, the lesion appears hypointense to the surrounding liver on T_1 and moderately hyperintense on T_2. After contrast-medium injection, it is generally hypovascular showing a heterogeneous enhancement that gradually increases during the delayed phases (Figure 16.32). Satellite nodules as well as the central scar are rarely seen in hilar cholangiocarcinomas [101,102]. The use of MRC in conjunction with MR is highly effective in predicting tumor extent and vascular involvement. In particular, the radiologist may be careful in the evaluation of the second-order biliary radicles, lobar artery, and the main portal branch. Generally, a tumor is not resectable if there is bilateral involvement of one or more of the above structures [103]. Main differential diagnoses for hilar cholangiocarcinoma are primary sclerosing cholangitis, inflammatory, pseudotumor, Mirizzi syndrome, recurrent pyogenic cholangitis, and xanthogranulomatous cholangitis [104].

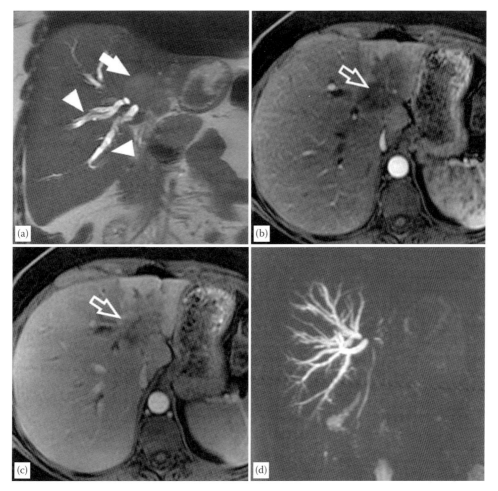

FIGURE 16.32
Intrahepatic cholangiocarcinoma: hilar. (a) Coronal T_2-weighted image shows a live mass with slightly higher signal intensity than the surrounding liver (arrow) and bile ducts dilatation (arrowheads). (b) Axial T_1-weighted image during the arterial phase reveals a hypointense lesion (open arrow) at hilum hepatis with delayed enhancement (c). (d) MRCP shows bile ducts dilatation with obstruction of the right bile duct intrahepatic.

16.5.3 Metastases

Metastases are the most common malignant tumors of the liver and most common indication for liver imaging. The liver is the primary organ of neoplastic diffusion after loco-regional lymph nodes [105]. Detection of liver metastases is important to properly stage the disease and determine the best treatment. Several morphologic patterns have been described for liver metastases: on T_1-weighted images, metastases are very often hypointense and they also have a "doughnut" appearance (hypointense with hyperintense rim) or rarely hyperintense due to the presence of melanin or proteinaceous content. On T_2-weighted images, they are slightly hyperintense with different appearance such as halo, target, light bulb or amorphous (Figure 16.33) [106]. After contrast-medium injection, metastases are classified as hypovascular or hypervascular compared with the surrounding liver parenchyma. Hypovascular metastases generally result from primary neoplasm of

the gastrointestinal tract (pancreas, colon, stomach), lung, and some subtypes of breast cancer; they are well visualized in the venous phase due to the highest contrast-to-noise ratio between the lesion and the liver parenchyma (Figure 16.34). Most common differential diagnoses for hypovascular metastases are peripheral cholangiocarcinoma, abscesses, and fibrous tumors. Hypervascular metastases derive from highly vascular tumors such as clear cell renal carcinoma, melanoma, carcinoid, islet cull tumor, thyroid carcinoma, pheochromocytoma, and breast carcinoma, and they are better detected in the arterial phase (Figures 16.35 and 16.36) [107]. In 10% of hypervascular metastases the wash-out sign has also been described [108]. Most common differential diagnoses for hypervascular metastases are FNH, "flash filling" hemangioma, and focal fatty area. On delayed hepatobiliary phase images after administering a paramagnetic liver-specific contrast agent, all metastases demonstrate marked hypointensity compared with the surrounding enhanced normal liver parenchyma

Liver metastases morphologic pattern

T_2
  – Amorphous   45%
  – Target   19%
  – Halo   26%
  – Light bulb   10%

T_1
  – Hypointense with halo *doughnut*   80%
  – Hyperintense (protein, melanin, and metahemoglobin)   10%

FIGURE 16.33
Scheme morphologic pattern of liver metastases on pre-contrast sequences.

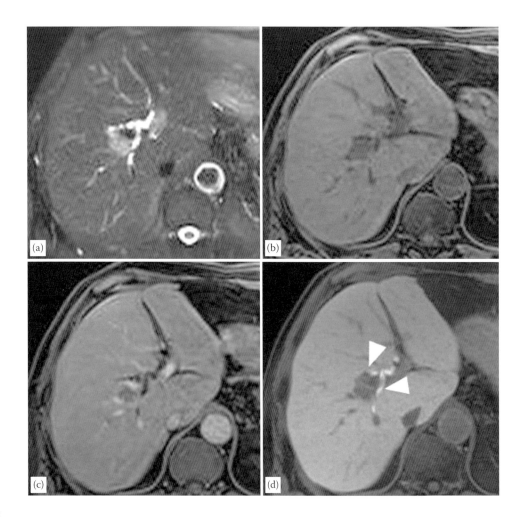

FIGURE 16.34
Liver metastases: hypovascular from colorectal cancer. (a) Axial T_2-weighted image shows a hyperintense lesion at hilum hepatis, which is hypointense on T_1 (b). The lesion is hypointense to the surrounding liver parenchyma (c) and it does not uptake the contrast medium during the hepatobiliary phase (d). Note in the hepatobiliary phase the adjacency with the main bile ducts (arrowheads).

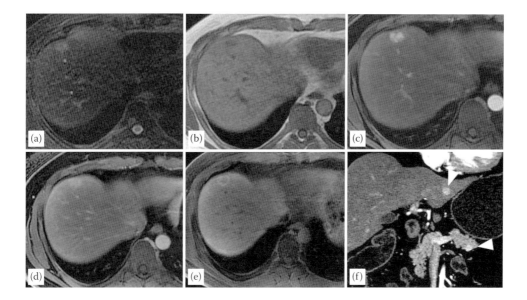

FIGURE 16.35
Liver metastases: hypervascular from neuroendocrine tumor. (a) Axial T_2-weighted image shows a faintly hyperintense lesion in liver segment eight, which is hypointense in T_1 image (b). On T_1-weighted post-contrast image during the arterial phase, the lesion is hypervascular and is isointense in the delayed phase (c and d). The hepatobiliary phase demonstrates loss of signal intensity of the lesion with a thin peripheral rim. (e) Coronal CT scan acquired during the arterial phase reveals a tiny hypervascular lesion in the tail of the pancreas with one more hypervascular lesion in liver segment two (arrowheads).

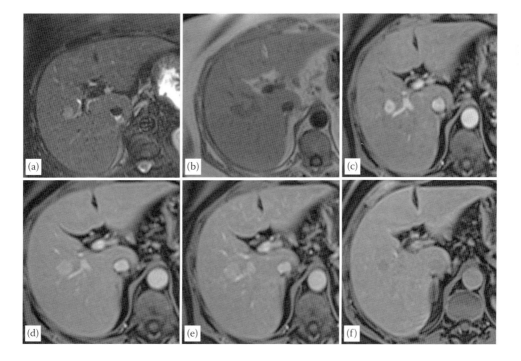

FIGURE 16.36
Liver metastases: hypervascular from melanoma. (a through f) As for metastases from neuroendocrine tumor, metastases from melanoma show the same MR pattern.

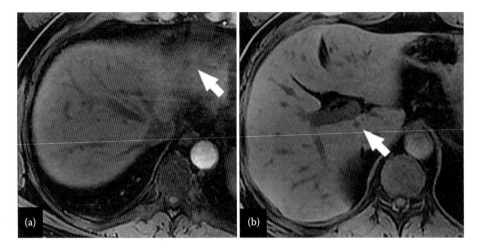

FIGURE 16.37
Liver metastases: role of the hepatobiliary phase: (a and b) example of two tiny hypvascular metastasis from colo-rectal cancer detectable only during the hepatobiliary phase.

(Figure 16.37) [109]. However, some retention of the contrast agent may occasionally be observed [110,111].

16.6 Other Malignant Lesions

16.6.1 Hepatic Epithelioid Hemangioendothelioma

Hepatic epithelioid hemangioendothelioma (HEHE) is a rare borderline tumor composed of epithelioid, endothelial, or dendritic cells [112]. HEHE is not specific to soft tissue and has also occurred in other organs, such as the lung, bone, brain, heart, salivary gland, veins, and pleura [113,114]. Patients affected by HEHE are generally women (60%), with peak incidence occurring between 30 and 50 years of age. Although the majority of patients present with multifocal growth, and metastases are found in one-third of patients at the time of initial diagnosis, the prognosis is considered much more favorable than that of other hepatic malignancies [115,116].

MRI radiological findings are well described in the literature. There are two predominant imaging features: (a) the target sign, which is characterized by a central area low intensity and peripheral hyperintensity or isointensity and (b) the "halo" sign, which consists of a hypointense center and periphery with an intermingled hyperintense layer in between. A "capsular retraction" sign can occur near the lesions [117,118].

In the MR hepatobiliary phase, the lesion does not uptake the contrast agent, showing no functional hepatocytes within itself (Figure 16.38). Noninvasive diagnosis of HEHE is difficult, and it could be misdiagnosed as hepatic metastatic disease, cholangiocarcinoma (peripheral), focal confluent fibrosis, hemangioma; hence, histopathologic confirmation is necessary.

16.6.2 Primary Hepatic Lymphoma

Primary hepatic lymphoma (PHL) is a very rare malignancy usually presenting with no specific symptoms, leading to late diagnosis. Of all primary extranodal NHL, only 0.4% arise in the liver. PHL can present in three ways: (1) as a solitary liver mass, (2) as multiple liver lesions, and (3) as a diffusely infiltrated liver. The most common manifestation of PHL is a single well-defined liver lesion [119]. On MR imaging, hepatic lymphoma is classically described as hypointense on T_1 images and hyperintense on T_2 compared with the surrounding liver parenchyma. After contrast-medium injection, PHL appears as a hypointense lesion with a thin peripheral rim, which is probably related to vasculitis induced in the adjacent liver parenchyma (Figure 16.39) [120,121].

16.7 Posttreatment Evaluation

Several therapeutic strategies are available for treatment of primary and secondary tumors of the liver. Knowledge of normal appearance and pathologic findings is crucial for the proper management and follow-up of the patient.

16.7.1 Radio-frequency Ablation

Radio-frequency (RF) ablation is the most promising nonsurgical technique for the treatment of primary and secondary tumors of the liver [122,123]. Current clinical experience suggests that it is effective, safe, and relatively simple and presents a lower complication rate, treatment time, and hospital costs compared with surgery [124,125]. The major limits of RF are related to the

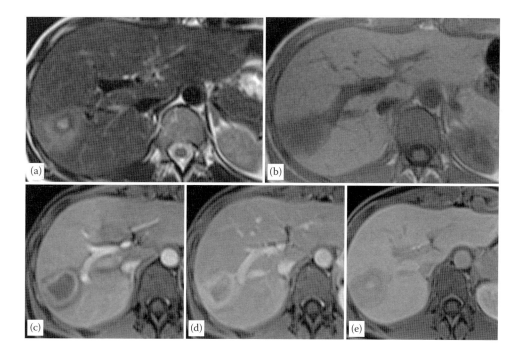

FIGURE 16.38

Hepatic hemangioendothelioma. (a) T_2-weighted image shows a heterogeneous focal liver mass in liver segment eight, which is hypointense on T_1 (b). After contrast-medium administration during the arterial and delayed phases (c and d) the lesion shows a target pattern, which is characteristic of hemangioendothelioma. (e) The hepatobiliary phase does not reveal uptake of the liver-specific contrast agent by the lesion.

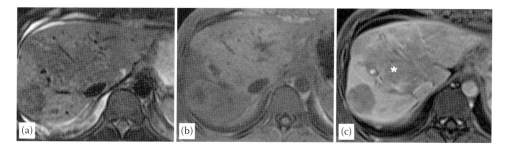

FIGURE 16.39

Hepatic lymphoma. (a and b) T_1- and T_2-weighted pre-contrast sequences show a rounded, well-defined, lesion in liver segment seven. (c) Post-contrast MR image shows a large hypointense mass. Note a diffuse and heterogeneous area in the central segment because of the presence of hepatitis (asterisk).

ultrasound visualization (lesion not visible), the lesion should be less than 4 cm to significantly reduce the risk of recurrence, and the proximity of the tumors to large vessels, which may prevent adequate "heating" (heat dispersion due to the blood flow), as well as proximity to central bile ducts, which predisposes the patient to a risk of biliary complications. The ablation procedure creates an area of necrosis within the liver parenchyma. The shape of the RF defect is variable, often round or ovoid, sometimes with a more complex geometric shape determined from needle placements or the proximity of a large vessel. On T_2-weighted images, the treated tumor is characterized by low signal intensity, whereas viable tumor tissue produces high signal intensity [126]. At an early follow-up, necrotic debris within the defect often produces heterogeneous high signal intensity

and a hypointense rim on T_2-weighted MR images. After administration of the contrast medium, completely ablated lesions show no enhancement within the necrotic area, but enhancement at the periphery of the defect may be a normal finding that can be a source of diagnostic difficulties (Figure 16.40) [127]. Another important criterion for complete ablation is the evolution over time of the size of the defect. While the evolution is unpredictable, if the ablation is complete, the size can only remain stable or decrease. Any enlargement represents tumor growth [128]. Incomplete ablation and recurrence disease are defined as new lesions discovered at follow-up imaging within 2 cm from the treated lesion: it occurs in a range between 2% and 10% of treated lesions. The main causes of recurrence disease are high histological grade, microvascular

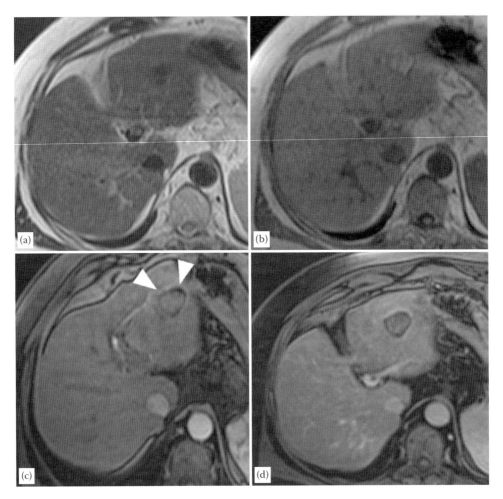

FIGURE 16.40

Radio-frequency treatment of HCC. (a and b) Pre-contrast sequences reveal a hypointense area on T_2 and hyperintense on T_1 due to hemorrhagic necrosis. (c and d) Post-contrast images do not demonstrate either hypervascular area or wash-out sign within or nearby the lesion despite a thin rim enhancement is present due to posttreatment hyperemia. (arrowheads).

invasion, and nearness of vascular structure (Figure 16.41). Complications of RF have an incidence of 9% and a mortality rate of 0.5: they should be either immediate or delayed. Some of the important immediate complications are vascular thrombosis (hepatic and portal veins), hematoma, infarction, arterio-venous fistula, biliary obstruction, pleural effusion, and pneumothorax (Figure 16.42). Delayed complications include infection and tumor seeding [129,130].

16.7.2 Transarterial Chemoembolization

Transarterial chemoembolization (TACE) is a palliative therapy for hypervascular primary and secondary tumors of the liver based on the infusion of chemotherapeutic drugs into the hepatic artery [131]. It is the treatment of choice in patients with the intermediate stage of HCC and who are not suitable for surgery, liver transplantation, or RF ablation. It should also be used to downstage disease in patients out of Milan or USCF criteria, or as bridging therapy in patients with long time interval for liver transplant. TACE is a catheter-based technique that combines both regional chemotherapy and embolization to increase the dwell time of cytotoxic agents and induce ischemia in the tumor. The use of drug-eluting microspheres in a new variation of the TACE method is designed to improve the precision of drug delivery. Another recent advance, a form of brachytherapy, involves the administration of yttrium-90 (90Y) microspheres via the hepatic artery, which preferentially are deposited within hypervascular tumors and emit beta radiation (radioembolization). Limits of the TACE technique are related to the liver function (not suitable for patients with Child–Pugh score >8 or with high value of bilirubin) and portal vein thrombosis. The oncologic standard for determining tumor response is the tumor size as categorized according to the modified Response Evaluation Criteria in Solid Tumors

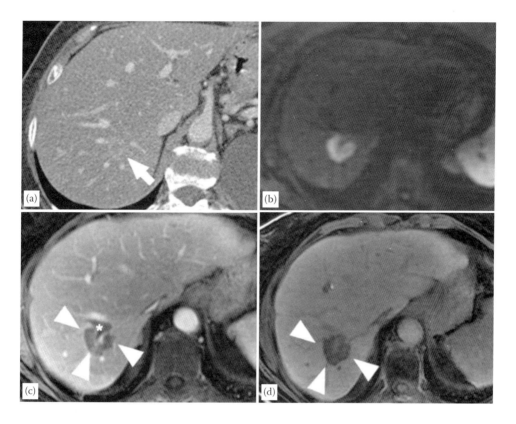

FIGURE 16.41

Radio-frequency treatment. Recurrence of disease: (a) CT scan reveals slightly hypovascular metastases on liver segment VII. (b) At follow-up, DW-MR imaging shows an area of restriction of molecular diffusion around the treated lesion, which is a sign of malignancy. (c and d) Post-contrast images during the venous and hepatobiliary confirm the presence of pathologic tissue around the necrosis (asterisk) of the treated lesion (arrowhead).

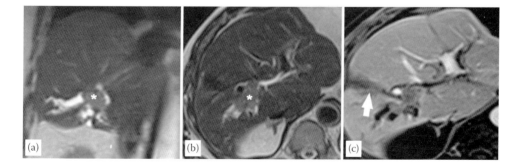

FIGURE 16.42

Radio-frequency complication. Bile ducts dilatation. (a and b) Axial and coronal T_2-weighted images show an area of liver necrosis at "hilum hepatis" (asterisk) with dilation of the bile ducts. (c) After contrast-medium injection, there is no enhancement of the treated lesion. The track released by the needle is also visible (arrow).

(mRECIST) parameters [132]. Recurrence disease after TACE has a frequency of 37% after 6 months and 67% after 1 year: the principal factors are the size of the lesion (>3 cm), the multinodular tumor, and the supply from extra-liver vessels (internal mammary, pericardiacophrenic, and adrenal phrenic arteries) (Figure 16.43) [133]. Main complications of TACE are liver infarction, rupture of filler artery, abscesses, bleeding, and arterioportal shunt; minor complication may be fever and pain [134,135].

16.7.3 Surgery

Among the different therapeutic strategies available in patients with primary and secondary tumor of the liver, in a subset of patients, surgical resection can offer the best chance of long-term survival and potentially even cure [136].

Liver resection is the most available, efficient treatment for patients with single nodule HCC with healthy or early stage of liver disease (Child–Pugh A). Better

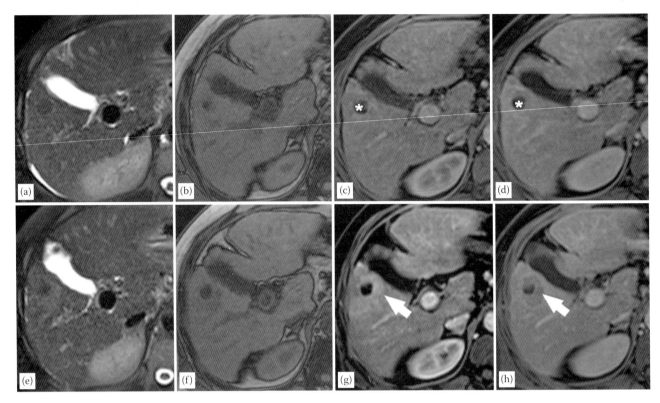

FIGURE 16.43
Transarterial chemoembolization. Recurrence of disease. (a through d) Three-month follow-up images after TACE do not reveal any enhancement or pathologic tissue within or nearby the area of necrosis (asterisk). (e through h) Six-month follow-up images demonstrate a hypervascular tissue near the gallbladder fossa that shows wash-out during the delayed phase, which is typical of HCC.

liver function assessment, increased understanding of segmental liver anatomy using more accurate imaging studies, and surgical technical progress are the most important factors that have led to reduced mortality, with an expected 5 year survival of 70%. Surgical resection is also useful in the treatment of hepatic metastases of primary colorectal carcinomas and may contribute to the 5 year survival of 24%–58% of affected patients [137]. However, the approach to the surgical treatment of hepatic metastases from primary tumors other than colorectal cancer is less obvious and is highly dependent upon the type of primary tumor. The main surgical technique are anatomic (segmental) and nonanatomic (wedge) resection: the first one refers to removal of liver segment according to Couinaud classification; the second is performed with the intent to remove macroscopic tumors independent of segmental anatomy (i.e., resection in liver segment VIII). Major liver resection refers to resection of three or more segments. The role of MR in the preoperative workup is to stage liver disease, to localize the neoplasm, to determine its relationship with major hepatic vessels (hepatic arteries, portal vein, and hepatic veins branches) and biliary ducts, and finally to evaluate if the postoperative remnant liver is enough for liver function [138]. Postoperative MR is helpful for the

detection of surgical complications such as hematoma/hemorrhage, abscesses, bile leak, or biloma (Figures 16.44 and 16.45) [139]. Biloma or abscess formations are characterized by a hypo- to hyperintense signal on T_1-weighted images and by typically high signal intensity on T_2-weighted images. A lack of central enhancement is seen both with biloma and abscess formations; however if a pronounced peripheral enhancement, indicating an abscess wall, can be identified, the patient may require interventional drainage or surgical revision [140]. In the case of biliary leaks, the precise location of the leak can frequently be ascertained with the use of hepatobiliary MR contrast agents. In this setting, it is possible to reveal the location of the leak through contrast extravasation at the site of bile duct damage (Figure 16.45). Recurrence of disease is unfortunately common, and it can develop along the resection margin, into the remnant parenchyma or with extrahepatic metastases [141].

16.7.4 Liver Transplant

Liver transplantation (LT) is the treatment of choice in patients with liver dysfunction due to chronic liver

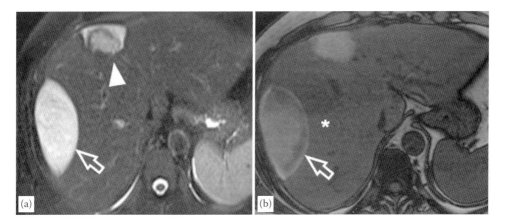

FIGURE 16.44
Surgery complication: hematoma. (a) T_2-weighted sequence shows the effect wedge resection in liver segment IV and a fluid collection under the liver surface, which is hyperintense on T_1-weighted image and (b) typical of hemorrhage. Note the low signal intensity of liver parenchyma on T_1 out-of-phase image due to liver steatosis (asterisk).

disease and acute liver failure [142]. Over the past several decades, advances in surgical techniques, organ preservation, immunosuppressive therapy, and early detection of postoperative complications have increased survival rates after LT, since it has as a 1 year survival rate of about 85% [143]. Early detection of postoperative complications is essential for graft and patient survival. Graft loss is a serious problem because of the complexity of the surgical procedures and the shortage of livers available for transplantation. Rejection is the most common cause of graft failure. In this setting, imaging plays a secondary role because the diagnosis is suspected at clinical and laboratory tests and can be achieved only with histologic analysis of a liver biopsy specimen. Some nonspecific imaging findings may be present such as edema of peri-porto-biliary space, irregular shape of hepatic margins

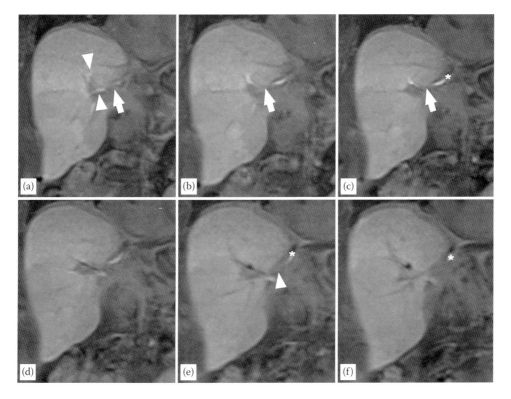

FIGURE 16.45
Surgery complication: bile leak. (a through f) Post-contrast serial coronal images using a hepatobiliary contrast agent demonstrate the track (arrow) between the biliary tree (arrowheads) and the fluid collection (asterisks).

and nonhomogeneous enhancement after contrast-medium administration [144]. The main complications of LT involve the surgical anastomotic sites: hepatic artery, portal vein, inferior vena cava and hepatic veins, and bile ducts. Hepatic artery complications include thrombosis, stenosis, and pseudoaneurysm (Figures 16.46 and 16.47). Hepatic artery thrombosis and stenosis occur in 2%–11% of liver transplant recipients. They can lead to biliary ischemia, since the hepatic artery is the only source of vascular supply to the bile ducts. Portal vein complications are relatively rare and include thrombosis and stenosis. Portal vein thrombosis occurs in about 1%–2% of cases [145,146]. Complications of the IVC and the hepatic vein have a low combined incidence (<1%). Technical factors, such as size discrepancy between donor and recipient vessels or suprahepatic caval kinking from organ rotation, may cause acute IVC stenosis [147,148]. After rejection, biliary disorders represent the most common

cause of graft disfunction. They occur in an estimated 25% of liver transplant recipients, usually within the first 3 months [149]. These complications include bile duct obstruction, anastomotic and nonanastomotic bile duct stricture, stone formation, bile leak, and biloma (Figures 16.48 through 16.51). Bile dust stricture is the most common biliary complication: at anastomotic site it is generally secondary to scar formation, while nonanastomotic biliary stricture occurs secondary to ischemia (often as a result of hepatic artery thrombosis or stenosis), infectious cholangitis, or pretransplantation sclerosing cholangitis [150]. In this setting, MR cholangiography plays a leading role in the noninvasive diagnosis of biliary strictures. Neoplastic complication after LT may include Kaposi sarcoma, melanoma, or non-Hodgkin lymphoma. Recurrence of hepatocellular is also visible after transplantation, and the liver is the second most common site after lung [151].

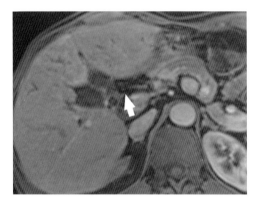

FIGURE 16.46
Liver transplant complication. T_1-weighted image during the arterial phase shows a thrombus within the stent placed into the hepatic artery.

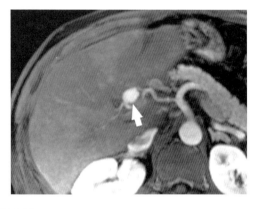

FIGURE 16.47
Liver transplant complication. T_1-weighted image during the arterial phase demonstrates a pseudoaneurism of the hepatic artery.

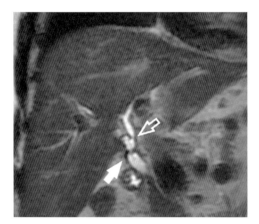

FIGURE 16.48
Liver transplant complication. Coronal T_2-weighted image shows an anastomotic stricture (arrow). Also appreciable is a nonanastomotic stricture (open arrow) at the confluence of the left hepatic ducts.

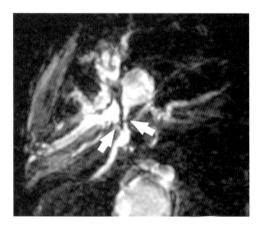

FIGURE 16.49
Liver transplant complication. Thick-slab MR-cholangiography demonstrates a nonanastomotic stricture (arrows) at the confluence of the right and left hepatic ducts.

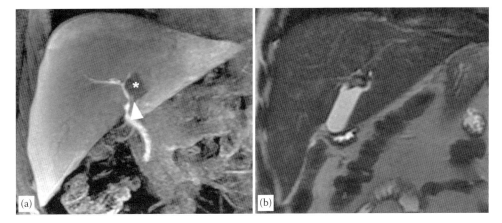

FIGURE 16.50

Liver transplant complication. (a) Coronal T_1-weighted image during the hepatobiliary phase shows a fluid collection (asterisk) that does not communicate with the biliary tree (arrowhead). (b) Corresponding T_2-weighted image reveals a "biloma at hilum hepatis."

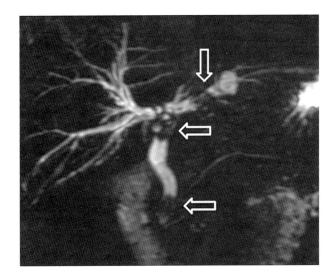

FIGURE 16.51

Liver transplant complication. Thick-slab MR-cholangiogram shows multiple biliary stones in the left intrahepatic ducts, at the anastomosis and in the periampullary region.

References

1. Brancatelli G1, Federle MP, Vilgrain V et al. Fibropolycystic liver disease: CT and MR Imaging Findings. *RadioGraphics*. 2005;25: 659–670.

2. Desmet VJ. What is congenital hepatic fibrosis? *Histopathology*. 1992;20:465–477.

3. Mortelé KJ, Ros PR. Cystic focal liver lesions in the adult: Differential CT and MR Imaging Features. *RadioGraphics*. 2001;21:895–910.

4. Lev-Toaff AS, Bach AM, Wechsler RJ, Hilpert PL, Gatalica Z, Rubin R. The radiologic and pathologic spectrum of biliary hamartomas. *AJR Am J Roentgenol*. 1995165:309–313.

5. Mortele B et al: Hepatic bile duct hamartomas (von Meyenburg Complexes): MR and MR cholangiography findings. *J Comput Assist Tomogr*. 2002;126:438–443.

6. Semelka RC, Hussain SM, Marcos HB, Woosley JT. Biliary hamartomas: Solitary and multiple lesions shown on current MR techniques including gadolinium enhancement. *J Magn Reson Imaging*. 1999;10:196–201.

7. Wohlgemuth WA, Böttger J, Bohndorf K. MRI, CT, US and ERCP in the evaluation of bile duct hamartomas (von Meyenburg complex): A case report. *Eur Radiol*. 1998;8:1623–1626.

8. Song JS, Noh SJ, Cho BH, Moon WS. Multicystic biliary hamartoma of the liver. *Korean J Pathol*. 2013;47:275–278.

9. Iha H, Nakashima Y, Fukukura Y et al. Biliary hamartomas simulating multiple hepatic metastasis on imaging findings. *Kurume Med J*. 1996;43:231–235.

10. van Sonnenberg E, Wroblicka JT, D'Agostino HB et al. Symptomatic hepatic cysts: Percutaneous drainage and sclerosis. *Radiology*. 1994;190:387–392.

11. Zangger P, Grossholz M, Mentha G, Lemoine R, Graf JD, Terrier F. MRI findings in Caroli's disease and intrahepatic pigmented calculi. *Abdom Imaging*. 1995;20:361–364.

12. Pavone P, Laghi A, Catalano C, Materia A, Basso N, Passariello R. Caroli's disease: Evaluation with MR cholangiopancreatography (MRCP). *Abdom Imaging*. 1996;21:117–119.

13. Baghbanian M, Salmanroghani H, Baghbanian A. Cholangiocarcinoma or Caroli disease: A case presentation. *Gastroenterol Hepatol Bed Bench*. 2013;6:214–216.

14. Rahimian J, Wilson T, Oram V, Holzman RS. Pyogenic liver abscess: Recent trends in etiology and mortality. *Clin Infect Dis*. 2004;39:1654–1659.

15. Qian LJ, Zhu J, Zhuang ZG, Xia Q, Liu Q, Xu JR. Spectrum of multilocular cystic hepatic lesions: CT and MR imaging findings with pathologic correlation. *RadioGraphics*. 2013;33:1419–1433.

16. Van Sonnenberg E, Mueller PR, Schiffman HR et al. Intrahepatic amebic abscesses: Indications for and results of percutaneous catheter drainage. *Radiology*. 1985;156:631–635.

17. Marani SA, Canossi GC, Nicoli FA et al. Hydatid disease: MR imaging study. *Radiology.* 1990;175:701–706.

18. Buetow PC, Midkiff RB. MR imaging of the liver. Primary malignant neoplasms in the adult. *Magn Reson Imaging Clin N Am.* 1997;5:289–318.

19. Kim HG. Biliary cystic neoplasm: Biliary cystadenoma and biliary cystadenocarcinoma. *Korean J Gastroenterol.* 2006;47:5–14.

20. Kele PG, van der Jagt EJ. Diffusion weighted imaging in the liver. *World J Gastroenterol.* 2010;16:1567–1576.

21. Fuks D, Voitot H, Paradis V, Belghiti J, Vilgrain V, Farges O. Intracystic concentrations of tumour markers for the diagnosis of cystic liver lesions. *Br J Surg.* 2014;101:408–416.

22. Lewis KH, Chezmar JL. Hepatic metastases. *Magn Reson Imaging Clin N Am.* 1997;5:241–253.

23. Semelka RC, Sofka CM. Hepatic hemangiomas. *Magn Reson Imaging Clin N Am.* 1997;5:241–253.

24. Vilgrain V, Boulos L, Vullierme MP, Denys A, Terris B, Menu Y. Imaging of atypical hemangiomas of the liver with pathologic correlation. *RadioGraphics.* 2000;20:379–397.

25. Brancatelli G, Federle MP, Blachar A et al. Hemangioma in the cirrhotic liver: Diagnosis and natural history. *Radiology.* 2001;219:69–74.

26. Soyer P, Gueye C, Somveille E et al. MR diagnosisf hepatic metastases from neuroendocrine tumors versus hemangiomas: Relative merits of dynamic gadolinium chelate–enhanced gradient-recalled echo and unenhanced spin-echo images. *AJR Am J Roentgenol.* 1995;165:1407–1413.

27. Darlak JJ, Moshowitz M, Kattan KR. Calcifications in the liver. *Radiol Clin North Am.* 1990;18:209–219.

28. Scatarige JC, Fishman EK, Saksouk FA, Siegelman S. Computed tomography of calcified liver masses. *J Comput Assist Tomogr.* 1983;7:83–89.

29. Nguyen BN, Flejou JF, Terris B et al. Focal nodular hyperplasia of the liver: A comprehensive pathologic study of 305 lesions and recognition of new histologic forms. *Am J Surg Pathol.* 1999;23:1441–1454.

30. International Working Party. Terminology of nodular hepatocellular lesions. *Hepatology.* 1995;22:983–993.

31. Wanless IR, Mawdsley C, Adams R. On the pathogenesis of focal nodular hyperplasia of the liver. *Hepatology.* 1985;5:1194–1200.

32. Marin D, Brancatelli G, Federle MP et al. Focal nodular hyperplasia: Typical and atypical MRI findings with emphasis on the use of contrast media. *Clin Radiol.* 2008;63:577–585.

33. Mortele´ KJ, Praet M, Van Vlierberghe H, Kunnen M, Ros PR. CT and MR imaging findings in focal nodular hyperplasia of the liver: Radiologic-pathologic correlation. *AJR Am J Roentgenol.* 2000;175:687–692.

34. Hussain SM, Terkivatan T, Zondervan PE et al. Focal nodular hyperplasia: Findings at state-of-the-art MR imaging, US, CT, and pathologic analysis. *RadioGraphics.* 2004;24:3–17.

35. Vilgrain V. Focal nodular hyperplasia. *Eur J Radiol.* 200;58:236–245.

36. Grazioli L, Bondioni MP, Haradome H et al. Hepatocellular adenoma and focal nodular hyperplasia: Value of gadoxetic acid-enhanced MR imaging in differential diagnosis. *Radiology.* 2012; 262:520–529.

37. Grazioli L, Morana G, Kirchin MA, Schneider G. Accurate differentiation of focal nodular hyperplasia from hepatic adenoma at gadobenate dimeglumine-enhanced MR imaging: Prospective study. *Radiology.* 2005;236:166–177.

38. Ba-Ssalamah A, Schima W, Schmook MT et al. Atypical focal nodular hyperplasia of the liver: Imaging features of nonspecific and liver-specific MR contrast agents. *AJR Am J Roentgenol.* 2002;179:1447–1456.

39. Terkivatan T, van den Bos IC, Hussain SM et al. Focal nodular hyperplasia: Lesion characteristics on state-of-the-art MRI including dynamic gadolinium-enhanced and superparamagnetic iron-oxide-uptake sequences in a prospective study. *J Magn Reson Imaging.* 2006;24:464–472.

40. Grazioli L, Federle MP, Brancatelli G et al. Hepatic adenomas: Imaging and pathologic findings. *RadioGraphics.* 2001;21:877–892.

41. Katabathina VS, Menias CO, Shanbhogue AKP et al. Genetics and imaging of hepatocellular adenomas: 2011 update. *RadioGraphics.* 2011;31:1529–1543.

42. Bioulac-Sage P, Rebouissou S, Thomas C et al. Hepatocellular adenoma subtype classification using molecular markers and immunohistochemistry. *Hepatology.* 2007;46:740–748.

43. Bioulac-Sage P, Laumonier H, Laurent C, Zucman-Rossi J, Balabaud C. Hepatocellular adenoma: What is new in 2008. *Hepatol Int.* 2008;2:316–321.

44. Lewin M, Handra-Luca A, Arrivé L et al. Liver adenomatosis: Classification of MR imaging features and comparison with pathologic findings. *Radiology.* 2006;241:433–440.

45. Laumonier H, Bioulac-Sage P, Laurent C, Zucman- Rossi J, Balabaud C, Trillaud H. Hepatocellular adenomas: Magnetic resonance imaging features as a function of molecular pathological classification. *Hepatology.* 2008;48:808–818.

46. Wachsberg RH, Cho KC, Adekosan A. Two leiomyomas of the liver in an adult with AIDS: CT and MR appearance. *J Comput Assist Tomogr.* 1994;18:156–157.

47. Santos I, Valls C, Leiva D, Serrano T, Martinez L, Ruiz S. Primary hepatic leiomyoma: Case report. *Abdom Imaging.* 2011;36:315–317.

48. Marin D, Catalano C, Rossi M et al. Gadobenate dimeglumine-enhanced magnetic resonance imaging of primary leiomyoma of the liver. *J Magn Reson Imaging.* 2008;28:755–758.

49. Ahmadi T, Itai Y, Takahashi M et al. Angiomyolipoma of the liver: Significance of CT and MR dynamic study. *Abdom Imaging.* 1998;23:520–526.

50. Balci NC, Akinci A, Akun E et al. Hepatic angiomyolipoma: Demonstration by out of phase MRI. *Clin Imaging.* 2002;26:418–420.

51. Hussain S. *Liver MR.* Springer, Berlin, Germany, 2007.

52. International Agency for Research on Cancer, World Health Organization (WHO). GLOBOCAN 2008: Cancer incidence, mortality and prevalence worldwide in 2008. Available at: http://globo can.iarc.fr. Accessed April 19, 2012.

53. Stroffolini T, Andreone P, Andriulli A et al. Characteristics of hepatocellular carcinoma in Italy. *J Hepatol.* 1998;29:944–952. 54. El-Serag HB, Mason AC. Rising incidence of hepatocellular carcinoma in the United States. *N Engl J Med.* 1999;340:745–750.

55. Deuffic S, Poynard T, Buffat L, Valleron AJ. Trends in primary liver cancer. *Lancet.* 1998;351:214–215.

56. Taylor-Robinson SD, Foster GR, Arora S, Hargreaves S, Thomas HC. Increase in primary liver cancer in the UK, 1979–1994. *Lancet.* 1997;350:1142–1143.

57. International Agency for Cancer Reseach. GLOBOCAN 2002. Available at: http://www-dep.iarc.fr. Accessed January 20, 2010.

58. Coleman WB. Mechanisms of human hepatocarcinogenesis. *Curr Mol Med.* 2003;3: 573–588.

59. Efremidis SC, Hytiroglou P. The multistep process of hepatocarcinogenesis in cirrhosis with imaging correlation. *Eur Radiol.* 2002;12:753–764.

60. Okuda K, Noguchi T, Kubo Y, Shimokawa Y, Kojiro M, Nakashima T. A clinical and pathological study of diffuse type hepatocellular carcinoma. *Liver.* 1981;1:280–289.

61. Bruix J, Scherman M. Management of hepatocellular carcinoma: An update. *Hepatology.* 2011;53:1020–1022.

62. Peterson MS, Baron RL, Marsh JW, Oliver III JH, Confer SR, Hunt LE. Pretransplantation surveillance for possible hepatocellular carcinoma in patients with cirrhosis: Epidemiology and CT-based tumor detection rate in 430 cases with surgical pathologic correlation. *Radiology.* 2000;217:743–749.

63. Trojan J, Raedle J, Zeuzem S. Serum tests for diagnosis and follow-up of hepatocellular carcinoma after treatment. *Digestion.* 1998;59(Suppl. 2):72–74.

64. Lok AS, Sterling RK, Everhart JE et al. Des-gamma-Carboxy Prothrombin and alpha-fetoprotein as biomarkers for the early detection of hepatocellular carcinoma. *Gastroenterology.* 2010;138:493–502.

65. Reinhold C, Hammers L, Taylor CR et al. Characterization of focal hepatic lesions with Duplex sonogrphy: Findings in 198 patients. *AJR Am J Roentgenol.* 1995;164:1131–1135.

66. Di Martino M, De Filippis G, De Santis A et al. Hepatocellular carcinoma in cirrhotic patients: Prospective comparison of US, CT and MR imaging. *Eur Radiol.* 2013;23:887–896.

67. Kadoya M, Matsui O, Takashima T, Nonomura A. Hepatocellular carcinoma: Correlation of MR imaging and histopathologic findings. *Radiology.* 1992;183:819–825.

68. Ebara M, Fukuda H, Kojima Y et al. Small hepatocellular carcinoma: Relationship of signal intensity to histopathologic findings and metal content of the tumor and surrounding hepatic parenchyma. *Radiology.* 1999;210:81–88.

69. Takayama Y1, Nishie A, Nakayama T et al. Hypovascular hepatic nodule showing hypointensity in the hepatobiliary phase of gadoxetic acid-enhanced MRI in patients with chronic liver disease: Prediction of malignant transformation. *Eur J Radiol.* 2012;81:3072–3078.

70. Lencioni R, Cioni D, Crocetti L et al. Magnetic resonance imaging of liver tumors. *J Hepatol.* 2004;40:162–171.

71. Shinmura R, Matsui O, Kobayashi S et al: Cirrhotic nodules: Association between MR imaging signal intensity and intranodular blood supply. *Radiology.* 2005;237:512–519.

72. Kim CK, Lim JH, Park CK et al. Neoangiogenesis and sinusoidal capillarization in hepatocellular carcinoma: Correlation between dynamic CT and density of tumor microvessels. *Radiology.* 2005;237:529–534.

73. Brancatelli G, Baron RL, Peterson MS, Marsh W. Helical CT screening for hepatocellular carcinoma in patients with cirrhosis: Frequency and causes of false-positive interpretation. *AJR Am J Roentgenol.* 2003;180:1007–1014.

74. Marrero JA, Hussain HK, Nghiem HV et al. Improving the prediction of hepatocellular carcinoma in cirrhotic patients with an arterially-enhancing liver mass. *Liver Transpl.* 2005;11:281–289.

75. Lim JH, Chooi D Kim SH et al. Detection of hepatocellular carcinoma:Value of adding delayed phase imaging to dual-phase helical CT. *AJR Am J Roentgenol.* 2002;179:67–73.

76. Iannaccone R, Laghi A, Catalano C et al. Hepatocellular carcinoma of unenhanced and delayed phase multidetector row helical CT in patients with cirrhosis. *Radiology.* 2005;234:460–467.

77. Di Martino M, Marin D, Guerrisi G et al. Intraindividual comparison of gadoxetate disodium-enhanced MR imaging and 64-section multidetector CT in the detection of hepatocellular carcinoma in patients with cirrhosis. *Radiology.* 2010;256:806–816.

78. Vogl TJ, Stupavsky A, Pegios W et al. Hepatocellular carcinoma: Evaluation of dynamic and static gadobenate dimeglumine-enhanced MR imaging and histopatologic correlation. *Radiology.* 1997;205:721–728.

79. Grazioli L, Morana G, Caudana R et al. Hepatocellular carcinoma: Correlation between gadobenate dimeglumine-enhanced MRI and pathologic findings. *Invest Radiol.* 2000;35:25–34.

80. Imai Y, Murakami T, Yoshida S et al. Superparamagnetic iron oxide-enhanced magnetic resonance images of hepatocellular carcinoma: Correlation with histological grading. *Hepatology.* 2000;32:205–212.

81. Mannelli L, Bhargava P, Osman SF et al. Diffusion-weighted imaging of the liver: A comprehensive review. *Curr Probl Diagn Radiol.* 2013;42:77–83.

82. Kim DJ, Ju JS, Kim JH, Chung JJ, Kim KW. Small hypervascular hepatocellular carcinomas: Value of diffusion weighted imaging compared with "washout" appearance on dynamic MRI. *Br J Med.* 2012;9:1–8.

83. Mannelli L, Kim S, Hajdu CH et al. Assessment of tumor necrosis of hepatocellular carcinoma after chemoembolization: Diffusion-weighted and contrast-enhanced MRI with histopathologic correlation of the explanted liver. *AJR Am J Roentgenol.* 2009;193:1044–1052.

84. LeMoigne F, DurieuxM, Baincel B et al. Impact of diffusion weighted MRimaging on the characterization of small hepatocellular carcinoma in the cirrhotic liver. *Magn Reson Imaging.* 2012;30:656–665.

85. Di Martino M, Di Miscio R, De Filippis G, Lombardo CV, Saba L, Geiger D, Catalano C. Detection of small (≤2 cm) HCC in cirrhotic patients: Added value of diffusion MR-imaging. *Abdom Imaging.* 2013;38:1254–1262.

86. Park MS, Kim S, Patel J et al. Hepatocellular carcinoma: Detection with diffusion-weighted versus contrast-enhanced magnetic resonance imaging in pretransplant patients. *Hepatology.* 2012;56:140–148.

87. Di Martino M, Saba L, Bosco S et al. Hepatocellular carcinoma (HCC) in non-cirrhotic liver: Clinical, radiological and pathological findings. *Eur Radiol.* 2014;24:1446–1454.

88. Iannaccone R, Piacentini F, Murakanmi T et al. Hepatocellular carcinoma in patients with nonalcoholic fatty liver disease: Helical CT and MR Imaging with clinical-pathologic correlation findings. *Radiology.* 2007;243:422–430.

89. Brancatelli G, Federle MP, Grazioli L, Carr BI. Hepatocellular carcinoma in non-cirrhotic liver: CT, clinical and pathological findings in 39 U.S. residents. *Radiology.* 2002;222:89–94.

90. Berman MA, Burnham JA, Sheahan DG. Fibrolamellar carcinoma of the liver: An immunohistochemical study of 19 cases and a review of the literature. *Hum Pathol.* 1988;19:784–794.

91. Ichikawa T, Federle MP, Grazioli L, Madariaga J, Nalesnik M, Marsh W. Fibrolamellar hepatocellular carcinoma: Imaging and pathologic findings in 31 recent cases. *Radiology.* 1999;213:352–361.

92. McLarney JK, Rucker PT, Bender GN, Goodman ZD, Kashitani N, Ros PR. Fibrolamellar carcinoma of the liver: Radiologic-pathologic correlation. *RadioGraphics.* 1999;19:453–471.

93. Won JL, Lim HK, Jang KM et al. Radiologic spectrum of cholangiocarcinoma: Emphasis on unusual manifestations and differential diagnoses. *RadioGraphics.* 2001;21:S97–S116.

94. Soyer P, Bluemke DA, Reichle R et al Imaging of intrahepatic cholangiocarcinoma:1. Peripheral cholangiocarcinomo. *AJR Am J Roentgenol.* 1995;165:1427–1431.

95. Liver Cancer Study Group of Japan. *Classification of Primary Liver Cancer.* Kanehara, Tokyo, Japan, 1997; pp. 6–8.

96. Worawattanakul S, Semelka RC, Noone TC et al. Cholangiocarcinoma: Spectrum of appearances on MR images using current techniques. *Magn Reson Imaging.* 1998;16:993–1003.

97. Maetani Y, Itoh K, Watanabe C et al. MR imaging of intrahepatic cholangiocarcinoma with pathologic correlation. *AJR Am J Roentgenol.* 2001;176:1499–1507.

98. Vilgrain V, Van Beers BE, Flejou JF et al. Intrahepatic cholangiocarcinoma: MRI and pathologic correlation in 14 patients. *J Comput Assist Tomogr.* 1997;21:59–65.

99. Park HJ, Kim YK, Park MJ, Lee WJ. Small intrahepatic mass-forming cholangiocarcinoma: Target sign on diffusion-weighted imaging for differentiation from hepatocellular carcinoma. *Abdom Imaging.* 2013;38:793–801.

100. Schneider G, Grazioli L, Saini S. *MR Imaging of the Live,* 2nd ed. Springer, Berlin, Germany, 2006.

101. Manfredi R, Masselli G, Maresca G, Brizi MG, Vecchioli A, Marano P. MR imaging and MRCP of hilar cholangiocarcinoma. *Abdom Imaging.* 2003;28:319–325.

102. Guthrie JA, Ward J, Robinson PJ. Hilar cholangiocarcinomas: T2-weighted spin-echo and gadolinium-enhanced FLASH MR imaging. *Radiology.* 1996;201:347–351.

103. Chryssou E, Guthrie JA, Ward J, Robinson PJ. Hilar cholangiocarcinoma: MR correlation with surgical and histological findings. *Clin Radiol.* 2010;65:781–788.

104. Menias CO, Surabhi VR, Prasad SR, Wang HL, Narra VR, Chintapalli KN. Mimics of cholangiocarcinoma: Spectrum of disease. *RadioGraphics.* 2008;28:1115–1129.

105. Pedro MS, Semelka RC, Braga L. MR imaging of hepatic metastases. *Magn Reson Imaging Clin N Am.* 2002;10:15–29.

106. Semelka RC, Braga L, Armao D et al. Liver. In: Semelka RC, ed. *Abdominal-Pelvic MRI,* 1st ed. Wiley-Liss, New York, 2002; pp. 101–134.

107. Danet IM, Semelka RC, Leonardou P, Spectrum of MRI appearances of untreated metastases of the liver. *AJR Am J Roentgenol.* 2003;181:809–817.

108. Nino-Murcia M, Olcott EW, Jeffrey RB Jr, Lamm RL, Beaulieu CF, Jain KA. Focal liver lesions: Pattern-based classification scheme for enhancement at arterial phase CT. *Radiology.* 2000;215:746–751.

109. Caudana R, Morana G, Pirovano GP et al. Focal malignant hepatic lesions: MR imaging enhanced with gadolinium benzyloxypropionictetra-acetate (BOPTA)-preliminary results of phase II clinical application. *Radiology.* 1996;199:513–520.

110. Ha S, Lee CH, Kim BH et al. Paradoxical uptake of Gd-EOB-DTPA on the hepatobiliary phase in the evaluation of hepatic metastasis from breast cancer: Is the "target sign" a common finding? *Magn Reson Imaging.* 2012;30:1083–1090.

111. Kim A, Lee CH, Kim BH et al. Gadoxetic acid-enhanced 3.0T MRI for the evaluation of hepatic metastasis from colorectal cancer: Metastasis is not always seen as a "defect" on the hepatobiliary phase. *Eur J Radiol.* 2012;81:3998–4004.

112. Ishak KG, Sesterhenn IA, Goodman ZD, Rabin L, Stromeyer FW: Epithelioid hemangioendothelioma of the liver: A clinicopathologic and follow-up study of 32 cases. *Hum Pathol.* 1984;15:839–852.

113. Kopniczky Z, Tsimpas A, Lawson DD et al. Epithelioid hemangioendothelioma of the spine: Report of two cases and review of the literature. *Br J Neurosurg.* 2008;22:793–797.

114. Lee YJ, Chung MJ, Jeong KC et al. Pleuralepithelioid hemangioendothelioma. *Yonsei Med J.* 2008;49:1036–1040.

115. Mehrabi A, Kashfi A, Fonouni H et al. Primary malignant hepatic epithelioid hemangioendothelioma: A comprehensive review of the literature with emphasis on the surgical therapy. *Cancer.* 2006;107:2108–2121.

116. Weitz J, Klimstra DS, Cymes K et al. Management of primary liver sarcomas. *Cancer.* 2007;109:1391–1396.

117. Van Beers B, Roche A, Mathieu D et al. Epithelioid hemangioendothelioma of the liver: MR and CT findings. *J Comput Assist Tomogr.* 1992;16:420–424.

118. Lin J, Ji Y. CT and MRI diagnosis of hepatic epithelioid hemangioendothelioma. *Hepatobiliary Pancreat Dis Int.* 2010;9:154–158.

119. Maher MM, McDermott SR, Fenlon HM et al. Imaging of primary non-Hodgkin's lymphoma of the liver. *Clin Radiol.* 2001;56:295–301.

120. Gazelle GS, Lee MJ, Hahn PF, Goldberg MA, Rafaat N, Mueller PR. US, CT and MRI of primary and secondary liver lymphoma. *J Comput Assist Tomogr.* 1994;18:412–415.

121. Weissleder R, Stark DD, Elizondo G. MRI of hepatic lymphoma. *Magn Reson Imaging.* 1988;6:675–681.

122. Garra BS, Shawker TH, Chang R, Kaplan K, White RD. The ultrasound appearance of radiation-induced hepatic injury. Correlation with computed tomography and magnetic resonance imaging. *J Ultrasound Med.* 1988 Nov;7(11):605–609.

123. McGahan JP, Dodd GD 3rd. Radiofrequency ablation of the liver. *AJR Am J Roentgenol.* 2001;176:3–16.

124. Gazelle GS, Goldberg SN, Solbiati L, Livraghi T. Tumor ablation with radio-frequency energy. *Radiology.* 2000;217:633–646.

125. Wood TF, Rose DM, Chung M, Allegra DP, Foshag LJ, Bilchik AJ. Radiofrequency ablation of 231unresectable hepatic tumors: Indications, limitations, and complications. *Ann Surg Oncol.* 2000;7:593–600.

126. Livraghi T, Meloni F, Di Stasi M et al. Sustained complete response and complications rates after radiofrequency ablation of very early hepatocellular carcinoma in cirrhosis: Is resection still the treatment of choice? *Hepatology.* 2008;47:82–89.

127. Sironi S, Livraghi T, Meloni F, De Cobelli F, Ferrero CG, Del Maschio A. Small hepatocellular carcinoma treated with percutaneous RF ablation: MR imaging follow-up. *AJR Am J Roentgenol.* 1999;173:1225–1229.

128. Choi H, Loyer EM, DuBrow RA et al. Radio-frequency ablation of liver tumors: Assessment of therapeutic response and complications. *RadioGraphics.* 2001;21:S41–S54.

129. Kuszyk BS, Boitnott JK, Choti MA et al. Local tumor recurrence following hepatic cryoablation: Radiologic-histopathologic correlation in a rabbit model. *Radiology.* 2000;217:477–486.

130. Rhim H, Dodd GD 3rd, Chintapalli KN et al. Radiofrequency thermal ablation of abdominal tumors: Lessons learned from complications. *RadioGraphics.* 2004;24:41–52.

131. Kalva SP, Thabet A, Wicky S. Recent advances in trans-arterial therapy of primary and secondary liver malignancies. *RadioGraphics.* 2008;28:101–117.

132. Lencioni R, Llovet JM. Modified RECIST (mRECIST) assessment for hepatocellular carcinoma. *Semin Liver Dis.* 2010;30:52–60.

133. Kim HC, Chung JW, Lee W, Jae HJ, Park JH. Recognizing extrahepatic collateral vessels that supply hepatocellular carcinoma to avoid complications of transcatheter arterial chemoembolization. *RadioGraphics.* 2005;25:S25–S39.

134. Gates J, Hartnell GG, Stuart KE, Clouse ME. Chemoembolization of hepatic neoplasms: Safety, complications, and when to worry. *RadioGraphics.* 1999;19:399–414.

135. Ramsey DE, Kernagis LY, Soulen MC, Geschwind JF. Chemoembolization of hepatocellular carcinoma. *J Vasc Interv Radiol.* 2002;13:S211–S221.

136. Shin DS, Ingraham CR, Dighe MK et al. Surgical resection of a malignant liver lesion: What the surgeon wants the radiologist to know. *AJR Am J Roentgenol.* 2014;203:W21–W33.

137. Donadon M, Ribero D, Morris-Stiff G, Abdalla EK, Vauthey JN. New paradigm in the management of liver-only metastases from colorectal cancer. *Gastrointest Cancer Res.* 2007;1:20–27.

138. Morris-Stiff G, Gomez D, Prasad R. Quantitative assessment of hepatic function and its relevance to the liver surgeon. *J Gastrointest Surg.* 2009;13:374–385.

139. Huynh-Charlier I, Taboury J, Charlier P, Vaillant J, Grenier P, Lucidarme O. Imaging of the postsurgical liver. *J Radiol.* 200;90:888–904.

140. Sadamori H, Yagi T, Shinoura S et al. Risk factors for major morbidity after liver resection for hepatocellular carcinoma. *Br J Surg.* 2013;100:122–129.

141. Arii S, Teramoto K, Kawamura T et al. Characteristics of recurrent hepatocellular carcinoma in Japan and our surgical experience. *J Hepatobiliary Pancreat Surg.* 2001;8:397–403.

142. Mazariegos GV, Molmenti EP, Kramer DJ. Early complications after orthotopic liver transplantation. *Surg Clin North Am.* 1999;791:109–129.

143. Caiado AH, Blasbalg R, Marcelino AS et al. Complications of liver transplantation: Multimodality imaging approach. *RadioGraphics.* 2007;27:1401–1417.

144. Nghiem HV. Imaging of hepatic transplantation. *Radiol Clin North Am.* 1998;36:429–443.

145. Singh AK, Nachiappan AC, Verma HA et al. Postoperative imaging in liver transplantation: What radiologists should know. *RadioGraphics.* 2010;30:339–351.

146. Wozney P, Zajko AB, Bron KM, Point S, Starzl TE. Vascular complications after liver transplantation: A 5-year experience. *AJR Am J Roentgenol.* 1986;147:657–663.

147. Glockner JF, Forauer AR, Solomon H et al. Vascular complications after orthotopic liver transplantation. *Am J Surg.* 1991;161:76–82.

148. Varma CR, Perman WH. Three-dimensional gadolinium-enhanced MR angiography of vascular complications after liver transplantation. *AJR Am J Roentgenol.* 2000;174:1447–1453.

149. Haberal M. Liver transplantation: Experience at our center. *Transplant Proc.* 2006;38:2111–2116.

150. Fulcher AS, Turner MA. Orthotopic liver transplantation: Evaluation with MR cholangiography. *Radiology.* 1999;211:715–722.

151. Aseni P, Vertemati M, De Carlis L et al. De novo cancers and post-transplant lymphoproliferative disorder in adult liver transplantation. *Pathol Int.* 2006;56:712–715.

17

Gallbladder

Ganeshan Dhakshinamoorthy, Nicolaus Wagner-Bartak, Rafael Andres Vicens, Shelby Kent, Neeraj Lalwani, and Priya Bhosale

CONTENTS

17.1 Introduction

The gallbladder (GB) is a pyriform sac shaped hollow viscus that helps store the bile produced by the liver. Although gallstones are one of the commonest conditions affecting the GB, numerous other pathologies may involve the GB. Right upper quadrant pain and/or jaundice may be seen in many GB pathologies, but often patients present with nonspecific clinical symptoms and imaging plays a critical role in diagnosis and guiding subsequent management. Ultrasonography (US) is the most common primary imaging modality used in evaluating GB pathology, but computed tomography (CT) and magnetic resonance imaging (MRI) are being increasingly used for improving diagnostic accuracy.

In this chapter, we discuss the embryology and normal anatomy of the GB, and review the various congenital, inflammatory, infectious, neoplastic, iatrogenic, and other miscellaneous conditions affecting the GB, with emphasis on MRI for diagnosis and management.

17.2 Normal Anatomy

17.2.1 Embryology

The GB, biliary tract, ventral pancreas, and the liver are all derived from the hepatic diverticulum, which appears in the ventral wall of the primitive caudal foregut during the fourth week of embryogenesis. The liver, common hepatic duct (CHD), and intrahepatic bile ducts arise from the cranial part of the diverticulum (pars hepatica), whereas the GB and the cystic duct arise from the caudal portions (pars cystica) [1]. By the beginning of the fifth week of embryogenesis, all components of the hepatobiliary tract including the GB, cystic duct, hepatic ducts, common bile duct (CBD), and ventral pancreas are recognizable. During the fifth week, there is rapid elongation of the extrahepatic bile ducts with associated epithelial proliferation resulting in plugging of the lumen with solid core of cells. By the sixth week, the lumen of CBD initially starts recanalizing distally and then extends proximally, with the cystic duct being recanalized by the eighth week.

17.2.1.1 Anatomy

The GB is a pear shaped sac seen along the inferior surface of the right hepatic lobe (Figure 17.1). It has a variable size but can measure up to 10×4 cm^2 and hold around 50 mL of bile. Parts of the GB include fundus, body, infundibulum, and neck. Hartmann's pouch refers to the bulge on the inferior surface of the infundibulum, wherein gallstones may become impacted. The cystic duct, which is about 4 cm long, runs from the neck of the GB to join the CHD and form the CBD. The CBD courses into the head of the pancreas and joins the main pancreatic duct to form the ampulla of Vater at the major duodenal papilla. The sphincter of Oddi, a circular band of smooth muscle, surrounds the distal end of the CBD and main pancreatic duct at the level of the ampulla of Vater. A normal anatomical variant is the duct of Luschka, which is a small bile duct, running in the bed of the GB, seen in 50% of the population. This has surgical significance, as it may be injured during cholecystectomy and may result in iatrogenic bile fistula unless ligated.

Vascular supply of the GB is via the cystic artery (a branch of the right hepatic or common hepatic artery), and cystic vein (a tributary of portal vein) supplies the GB. The GB is innervated by the vagus nerve and celiac plexus. The lymphatic drainage of the GB is of special importance as this explains why radical surgery is often difficult due to its extensive lymphatic drainage routes. Numerous studies have reported three main routes of drainage [2].

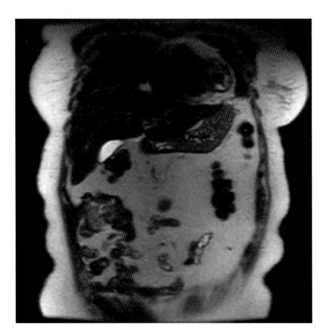

FIGURE 17.1
Coronal MRI T_2-weighted image shows pear shaped, normal gallbladder.

1. The cholecysto-retropancreatic pathway, the principal drainage route, runs both along the anterior and posterior surfaces of the CBD and drains into the retroportal node, which is at the posterior surface of the head of the pancreas.

2. The cholecysto-celiac pathway runs through the hepatoduodenal ligament to reach the celiac nodes.

3. The cholecysto-mesenteric pathway runs anterior to the portal vein and drains into the nodes at the superior mesenteric root.

These pathways further drain into the retroperitoneal nodes adjacent to the left renal vein and into aortico-caval nodes.

17.2.2 Anatomic Variants and Congenital Anomalies

Congenital malformations of the GB are rare, with a reported incidence around 0.15% [3]. Despite being uncommon, it is important to be aware of these as they are frequently associated with variants and anomalies in the bile ducts and vascular system, knowledge of which is crucial to avoid surgical complications during cholecystectomy. Further, these may also be associated with congenital malformations in other organs and predispose to complications.

Agenesis of the GB is rare, with an incidence of 0.02%–0.15% and occurs due to failure of the development of the caudal division of the primitive hepatic diverticulum [4]. Around 65% of the patients with GB agenesis also have other anomalies including congenital heart lesions, polysplenia, imperforate anus, and rectovaginal fistula [5]. Occasionally, completely intrahepatic GB may mimic agenesis of the GB, but crosssectional imaging, especially CT or MRI, can clearly identify the location of the GB and help to make the correct diagnosis. In fact, imaging can play an important role to avoid unnecessary surgeries in symptomatic GB agenesis by making the preoperative diagnosis [6,7]. GB hypoplasia is felt to be more frequent than agenesis. A small hypoplastic GB may be directly attached to the CHD or via a very short, cystic duct, and can be mistaken for the common duct.

GB duplication may be seen in about 1 in 4000 people (Figure 17.2). This occurs due to incomplete revacuolization of the primitive GB. Conditions that can mimic duplicated GB include bilobed or folded GB and GB diverticulum (Figure 17.3), but imaging, especially crosssectional imaging with CT or MRI, can help to identify the presence of two separate GB cavities, each having its own cystic ducts and each helps to make the right diagnosis. GB duplication may result in complications such as gallstones, secondary biliary cirrhosis, and GB carcinoma.

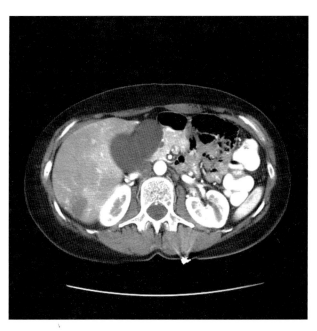

FIGURE 17.2
Axial CT scan shows gallbladder duplication in a 52-year-old male patient.

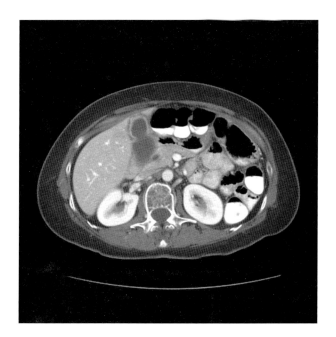

FIGURE 17.3
Axial CT scan shows gallbladder diverticulum in a 61-year-old male patient.

Abnormalities in the GB shape include Phrygian cap (named after its resemblance to the cap worn by slaves after liberation in ancient Greeks), septate GB, and GB diverticula. Although Phrygian cap (Figure 17.4) is a common variant seen in 1%–6% population and is of no clinical significance, septation and diverticulosis in the resultant cholelithiasis.

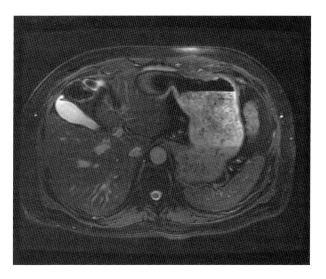

FIGURE 17.4
Axial T_2-weighted MRI shows Phrygian cap in the gallbladder fundus, a normal variant.

GB are uncommon, and can result in biliary stasis, with Congenital variants and anomalies resulting in abnormal location of the GB include "Wandering GB" and "Ectopic GB." Wandering GB refers to the presence of the GB with an unusually long mesentery, which allows the GB to "wander" to unusual locations including the pelvis, prevertebral space, and lesser sac. This condition is clinically significant as it may predispose to GB torsion. The GB may also occur in ectopic positions (Figure 17.5) and may be completely intrahepatic, causing diagnostic dilemmas or may be suprahepatic, retrohepatic, or even lie in the retroperitoneum. Left-sided GB (Figure 17.6) is a rare congenital anomaly, and may occur as part of situs inversus or be an isolated anomaly [8,9]. This may either occur as a result of GB migration toward the left side of the liver or of development of a second GB with atrophy of the original. They may be associated with anomalies in the portal venous system and in the pancreatobiliary tree.

17.2.3 Cystic Duct

The course and length of the cystic duct, as well as its site of insertion, may vary. For example, the cystic duct may run parallel to the CHD and insert medially. It is important to identify and report this variant anatomy on imaging, as there is an increased risk of injuring the CHD at cholecystectomy through clamping or traction, if not recognized during surgery.

Cystic duct may also have a very low insertion in duodenal papilla (8%–14%) or may insert high at the level of porta hepatis. Also, cysic duct may insert into the right or left hepatic duct or into the biliary confluence.

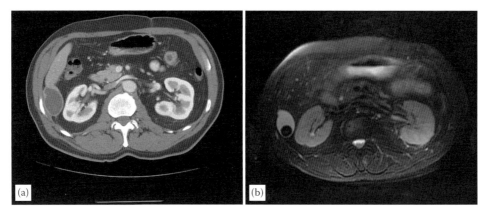

FIGURE 17.5
Axial CT scan (a) and Axial T_2-weighted MRI (b) shows the gallbladder lying in a retrohepatic location. Note that the MRI shows a gallstone, which is not clearly visualized on the CT scan.

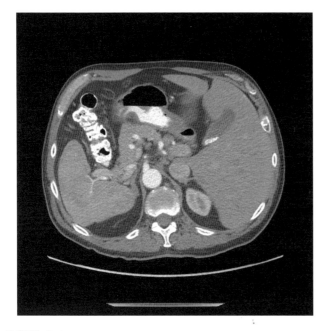

FIGURE 17.6
Left-sided gallbladder in a 47-year-old male patient with situs inversus

presence of gallstones, GB wall thickening, GB polyps and masses, pericholecystic fluid, and biliary ductal dilatation (Figures 17.7 through 17.9). US also helps evaluate for the presence of sonographic Murphy's sign, which is a very useful sign for acute cholecystitis.

Several studies have shown that cholescintigraphy is more sensitive and specific for acute cholecystitis (Figure 17.10). Recently, a meta-analysis reported 96% sensitivity and 90% specificity for cholescintigraphy compared to 81% sensitivity and 83% specificity [10]. However, despite its superior diagnostic performance in acute cholecystitis, ultrasound is preferred as initial test for several reasons—it can be performed rapidly, is more widely available, can give information on the status of biliary ducts, and is not associated with radiation exposure. Cholescintigraphy using Tc-99m-IDA (iminodiacetic acid) is also useful to diagnose acalculous cholecystitis, differentiate acute from chronic cholecystitis, and to evaluate for bile leaks.

Contrast-enhanced CT is very useful in evaluation of GB pathology including acute inflammatory conditions as well as GB malignancies. With regard to cholecystitis,

17.3 Imaging Techniques

Abdominal x-rays are not of much value in evaluating GB pathologies or in patients presenting with right upper quadrant pain. They have low sensitivity (15%–20%) for identifying gallstones. Further, even if X-rays demonstrate gallstones, they do not give any information about the presence or absence of cholecystitis. According to the current American College of Radiology (ACR) guidelines, ultrasound is the primary imaging modality of choice in evaluating patients with right upper quadrant pain [11]. It is widely available, cost-effective, can be performed rapidly and at bedside, and is excellent at assessing the

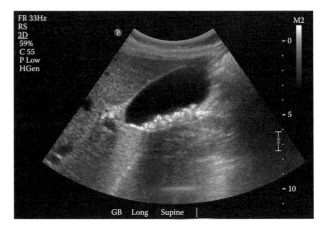

FIGURE 17.7
Longitudinal ultrasound view shows multiple gallstones.

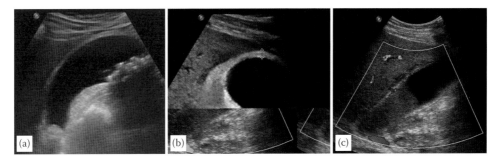

FIGURE 17.8
Longitudinal and transverse ultrasound views showing gallstones (a, b) and cholecystitis (c).

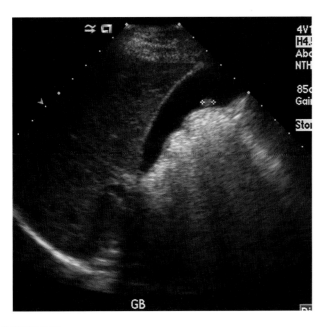

FIGURE 17.9
Longitudinal ultrasound view showing GB polyp.

CT is particularly useful for evaluating complications such as gangrenous cholecystitis, emphysematous cholecystitis, hemorrhagic cholecystitis, and GB perforation [12–16] (Figures 17.11 through 17.20). Studies indicate that CT can also help in preoperative planning, with the absence of GB wall enhancement and/or presence of a stone within the infundibulum increasing the necessity for converting from laparoscopic to open cholecystectomy [17]. Prior knowledge of these imaging findings may therefore help guide appropriate surgical approach. CT is also very useful for excluding alternate causes of right upper quadrant pathology.

MRI is currently not advocated as a primary imaging examination to evaluate acute right upper quadrant pain or other GB pathologies, but several studies have suggested that MRI is a reliable alternative in patients who are difficult to be evaluated with ultrasound or have equivocal results. In a meta-analysis, MRI is reported to have 85% sensitivity and 81% specificity for acute cholecystitis [10]. Also, MRI is particularly helpful in pregnant patients in whom there is a heightened concern regarding radiation exposure. MRI is also useful for identifying the presence of complications and in the evaluation of intrahepatic and extra hepatic biliary system. Further, MRI can help in differentiating benign and malignant pathologies and in the diagnosis and staging of GB malignancies.

MRI and MR cholangiopancreatography (MRCP) are noninvasive imaging techniques for evaluating GB and biliary pathologies (Figure 17.21). Heavily T_2-weighted sequences of the MRCP produce very high bile-to-background contrast, thereby improving the conspicuity of bile ducts and GB lumen. Fat saturation sequences may also help in better delineation of the GB and biliary system. Intravenous gadolinium enhanced images are useful for assessment of GB wall thickening, GB masses, and staging of GB malignancy (Figure 17.22). Preliminary studies have reported that diffusion-weighted MRI may be useful in differentiating benign from malignant GB polypoid lesions [18,19]. MRI protocol for evaluation of GB pathology may vary between institutions, but a standard protocol [20,21] would include a coronal T_2-weighted single-shot fast spin-echo sequence, an axial T_2-weighted fast spin-echo with respiratory gating (useful for evaluating soft tissue abnormalities involving the GB wall), additional heavily T_2-weighted fluid sensitive acquisition techniques such as half-Fourier acquisition single-shot turbo spin echo (HASTE), axial breath-hold gradient-recalled echo T_1-weighted in-phase and opposed-phase images (breath-hold spoiled gradient-echo techniques are superior to spin echo sequences as they decrease respiratory artifacts), MRCP sequences performed for biliary tract such as coronal oblique breath-hold thick-slab heavily T_2-weighted 2D fast spin-echo or heavily T_2-weighted fast-recovery fast spin-echo 3D sequence, and axial breath-hold 3D gradient echo sequence with fat saturation performed pre-contrast and dynamic acquisition

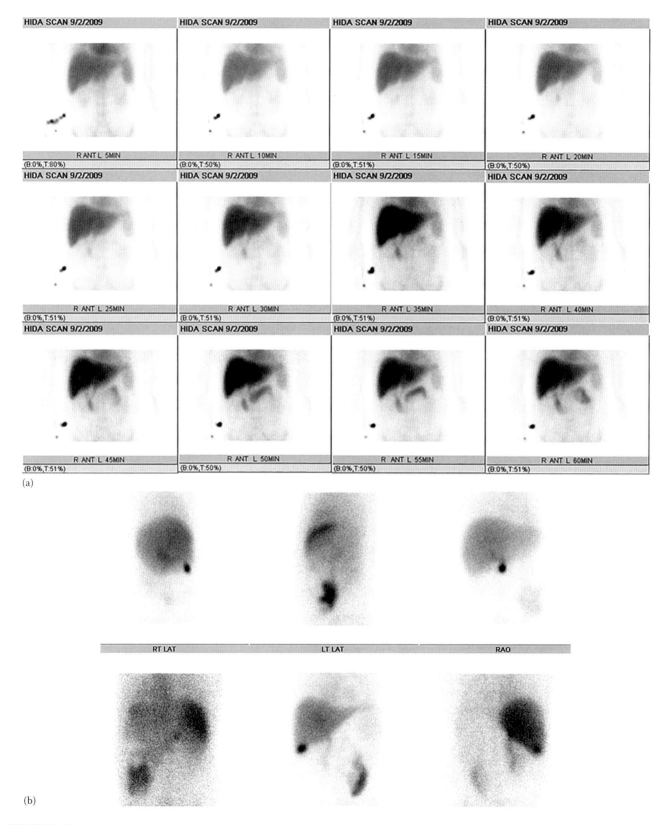

FIGURE 17.10
HIDA scan (a) shows no visualization of gallbladder for initial 60 min. Serial static images of the upper abdomen at 30 min after injection of 4 mg of morphine sulfate (b) demonstrated visualization of gallbladder. Prior ultrasound showed thickened gallbladder wall. Findings are consistent with chronic cholecystitis.

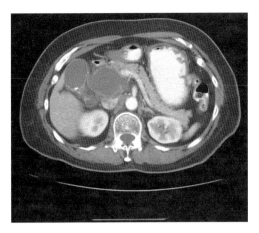

FIGURE 17.11
Axial CT scan shows perforated gallbladder with abscess formation.

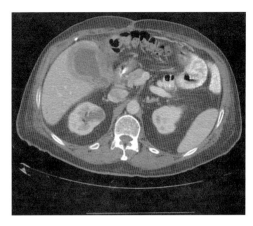

FIGURE 17.14
Axial CT scan shows gangrenous cholecystitis.

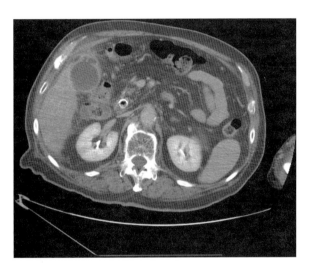

FIGURE 17.12
Axial CT scan shows complicated cholecystitis with microabscess.

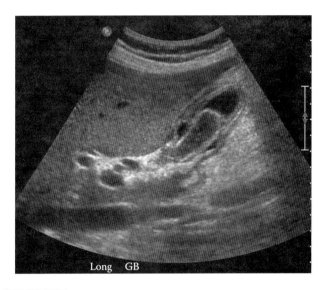

FIGURE 17.15
Longitudinal ultrasound view showing sloughed membranes in the gallbladder, consistent with gangrenous cholecystitis.

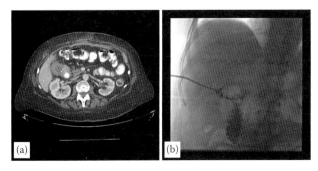

FIGURE 17.13
Axial CT (a) shows severe cholecystitis with markedly thickened duodenum. Follow-up cholecystostomy (b) showed cholecysto-duodenal fistula, with gallbladder directly communicating with the second portion of duodenum.

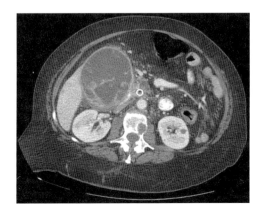

FIGURE 17.16
Axial CT scan shows CT gang showing the sloughed membranes in the gallbladder, consistent with gangrenous cholecystitis.

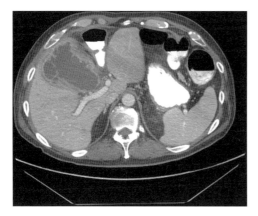

FIGURE 17.17
Axial CT scan shows gangrenous cholecystitis with associated perforation.

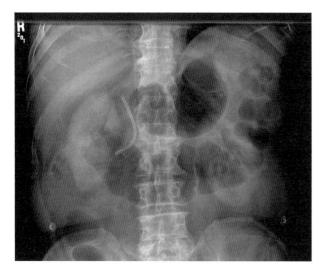

FIGURE 17.18
Plain film shows emphysematous cholecystitis.

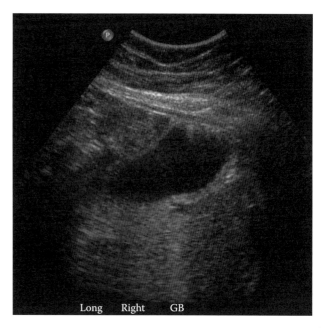

FIGURE 17.19
Longitudinal ultrasound view showing emphysematous cholecystitis.

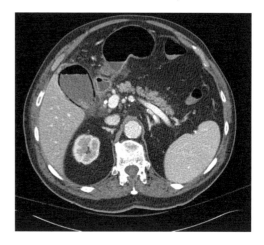

FIGURE 17.20
Axial CT scan shows emphysematous cholecystitis.

post-contrast [20,21]. Axial diffusion-weighted imaging using respiratory-triggered fat-suppressed single-shot echo planar imaging in the axial plane using multiple b values such as 0, 500, 1000 s/mm², and apparent diffusion coefficient (ADC) maps are also increasingly being used in evaluation of GB pathologies.

17.3.1 Normal MR Appearances of GB

The GB wall has intermediate signal intensity on T_1-weighted images, low signal intensity on T_2-weighted images and shows smooth and uniform enhancement on postgadolinium images (Figure 17.23). Normal thickness of the GB wall is up to 3 mm. The lumen of the GB appears hyperintense on T_2-weighted images and MRCP as it consists of static liquid bile (Figure 17.24). On T_1-weighted images, the bile may vary in signal intensity from hypointense to hyperintense depending

upon its concentration and composition (Figure 17.25). Prolonged fasting causes resorption of water from the bile thereby increasing the concentration of cholesterol, bile salts, and phospholipids. This concentrated bile appears intermediate to high intensity on T_1-weighted images [22,23].

17.3.2 Cholelithiasis

Cholelithiasis has an overall worldwide prevalence of 10%–20%. There is an increased risk of cholelithiasis in individuals with impaired GB motility and supersaturated bile. Risk factors for gallstones include pregnancy, obesity, rapid weight loss, diabetes mellitus, and alcohol

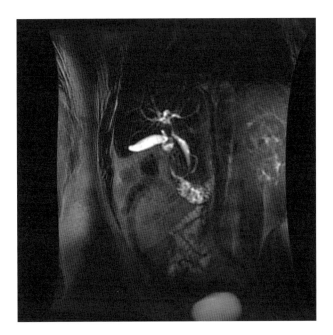

FIGURE 17.21
Coronal heavily T_2-weighted MRCP sequence demonstrating the biliary ducts and gallbladder.

use. Most patients with gallstones are asymptomatic. When clinically apparent, the commonest symptom is biliary colic, which refers to transient abdominal pain due to intermittent obstruction of the cystic duct due to a gallstone. Ultrasound is the modality of choice for evaluation of gallstone disease, but MRI may be helpful in diagnosing cholelithiasis and its associated complications and in differentiating gallstone adherent to the GB wall from the GB polyp.

On MRI, gallstones of all compositions appear hypointense relative to bile on T_2-weighted images and MRCP (Figure 17.26). They may have a central T_2 hyperintensity due to fluid filled clefts. As previously mentioned, gallstones may have a variable signal intensity on T_1-weighted images depending on their composition. Cholesterol stones are hypointense, whereas pigment stones are hyperintense relative to bile [24] (Figures 17.27 and 17.28). They are usually present in a dependent location in the GB unless impacted. A gallstone can be differentiated from a GB polyp by the lack of enhancement on T_1-weighted post-contrast images.

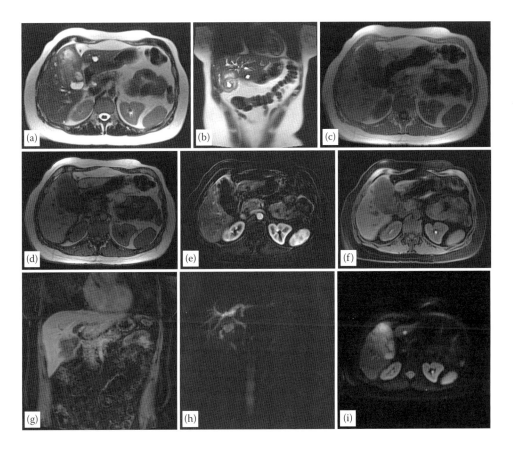

FIGURE 17.22
A 56-year-old patient with gallbladder cancer. (a) Axial T_2 HASTE, (b) Coronal HASTE, axial in phase (c) and out of phase (d) T_1-weighted images, early (e) and delayed (f) post-contrast VIBE images, coronal VIBE post-contrast (g), MRCP sequence (h) and axial diffusion-weighted image with b value 500 (i) shows an advanced gallbladder cancer involving the gallbladder wall diffusely.

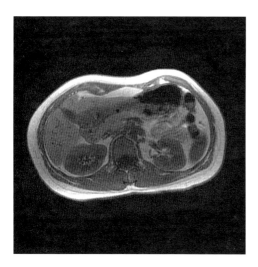

FIGURE 17.23
Axial T_1-weighted MRI shows intermediate signal intensity of the normal gallbladder wall.

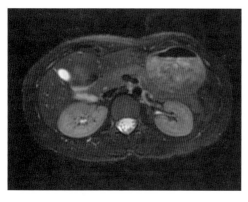

FIGURE 17.24
Axial T_2-weighted MRI shows hyperintensity in the lumen of the gallbladder.

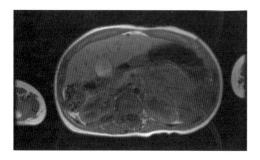

FIGURE 17.25
Axial T_1-weighted MRI shows high signal intensity of the bile. The signal intensity of the bile may vary on T_1-weighted images.

17.3.3 Cholecystitis

17.3.3.1 Acute Calculous Cholecystitis

Acute calculous cholecystitis refers to acute inflammation of the GB initiated by obstruction of the GB neck or cystic duct by a biliary calculus. This causes a build-up

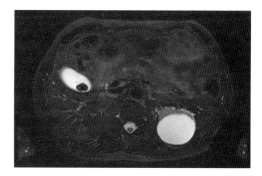

FIGURE 17.26
Axial T_2-weighted MRI shows gallstone with low signal intensity.

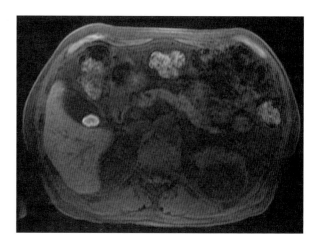

FIGURE 17.27
Axial T_1-weighted MRI in same patient shows gallstone with high to intermediate signal intensity.

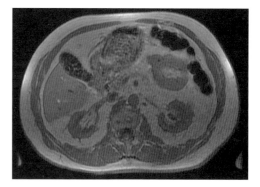

FIGURE 17.28
Axial T_1-weighted MRI shows multiple gallstones with low signal intensity. The signal intensity of the gallstone may vary on T_1-weighted images, according to its composition.

of pressure in the GB that leads to obstruction of the venous and lymphatic drainage of the GB. This may further progress to ischemia and perforation if not treated in time. Patients usually present with nausea, vomiting, and right upper quadrant abdominal pain. Most of the patients have had previous episodes of biliary colic. Ultrasound is usually the first imaging modality to be employed in the diagnosis of acute cholecystitis. MRI may be performed in doubtful cases [25].

MRI may demonstrate wall thickening in acute cholecystitis (Figure 17.29). The GB wall is considered to be thickened when its thickness exceeds 3 mm. The acutely inflamed wall appears hyperintense on T_2-weighted images (Figure 17.30). Pericholecystic fluid may be present. The impacted gallstone may be visualized on T_2-weighted or MRCP images. Postgadolinium images demonstrate hyper enhancement of the GB wall and transient enhancement of the adjacent liver parenchyma. These findings are considered to be more specific for the diagnosis of acute cholecystitis as these reflect the ongoing inflammatory processes resulting in an increased blood flow. Early post-contrast images may show initial enhancement of the inner layer of the GB, but the delayed images usually demonstrate more diffuse enhancement. The percentage of contrast enhancement of the GB wall may be useful in distinguishing acute from chronic cholecystitis [26]. In a recent study, Altun et al. reported that MRI had 95% sensitivity and 69% specificity for detection of acute cholecystitis [27].

On post-contrast T_1-weighted images, increased wall enhancement, increased GB wall thickness, and abnormal transient enhancement of the liver in the pericholecystic region were indicative of acute cholecystitis. On T_2-weighted images, apart from demonstrating the gallstones, wall thickening, and peri-cholecystic edema, T_2-weighted images may also depict the presence of intramural abscesses appearing as hyperintense foci in the GB wall. Increased GB wall enhancement and increased transient pericholecystic hepatic enhancement are reported to be the two most statistically significant factors that can help differentiate between acute and chronic cholecystitis. Hepatocyte specific contrast agents can help in the

diagnosis of acute calculous cholecystitis by demonstrating abrupt termination of the cystic duct and nonfilling of the GB with the contrast material on delayed images [28,29].

It is important to realize that GB wall thickening is not a specific finding for cholecystitis. Indeed several conditions unrelated to the GB can result in this finding including chronic liver disease, low albumin, and chronic renal failure. However, unlike in acute cholecystitis, post-contrast-enhanced T_1-weighted images in these conditions do not demonstrate the increased GB wall enhancement and the transient hepatic parenchymal enhancement in the pericholecystic region.

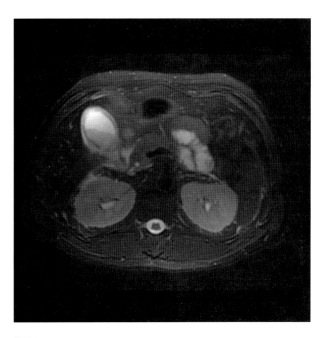

FIGURE 17.30
Axial T_2-weighted MRI shows thickened gallbladder wall in a patient with cholecystitis.

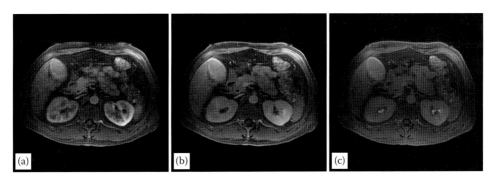

FIGURE 17.29
Postcontrast-enhanced axial T_1-weighted MRI (a–c) shows mildly thickened gallbladder wall, in keeping with cholecystitis.

Contrast-enhanced images are also helpful to detect complications of cholecystitis like perforation, abscess formation, and gangrenous cholecystitis. GB perforation may be indicated by detecting discontinuity in the wall enhancement of the GB [30]. It may be associated with rim-enhancing localized collections in the GB fossa. Gangrenous cholecystitis is diagnosed when there is diffuse or patchy nonenhancement of the GB wall. The nonenhancing portions correspond to necrotic areas. Presence of gas in the GB is seen in emphysematous cholecystitis and cholecysto-enteric fistula. It is identified as a nondependent hypointensity in the GB lumen with a fluid level. Presence of gas in the GB wall and in the pericholecystic tissues is highly suspicious for emphysematous cholecystitis. Presence of air is indicated by areas of hypointensity that arise due to signal loss caused by air-induced magnetic field inhomogeneity [31]. Hemorrhagic cholecystitis can be more easily diagnosed on MRI by detecting abnormal signal intensity of the blood products in the GB wall and lumen on both T_1- and T_2-weighted images. Depending upon the age of the haemorrhage, the blood breakdown products tend to show varying signal intensities on T_1- and T_2-weighted sequences.

17.3.3.2 Other Complications of Gallstones

Besides acute cholecystitis, gallstones cause several other acute complications including pancreatitis, biliary fistula, choledocholithiasis, gallstone ileus, and Mirizzi syndrome (Figure 17.31). In the Mirizzi syndrome, gallstone impacted in the cystic duct or the GB neck causes narrowing and subsequent biliary obstruction of the CHD. There are two types—type 1 results in simple obstruction of the CHD, whereas, in type 2, there is associated erosion of the wall of the CHD resulting in cholecystocholedochal fistula. Preoperative identification of this condition is important to prevent iatrogenic bile duct injury. MRI is particularly useful in this condition as unlike ultrasound and CT, it can demonstrate the cause and level of obstruction and help distinguish between type 1 and type 2 varieties. MRI can also help in preoperative identification of any anatomical variants in the course of the cystic duct, again a critical piece of information prior to surgery.

17.3.4 Acalculous Cholecystitis

Acalculous cholecystitis is uncommon, accounting only for 5%–15% of acute cholecystitis, but it is associated with a poorer prognosis. This usually occurs in critically ill patients in the ICU setting, and may occur due to decreased motility or reduced blood flow in the cystic artery or by bacterial infection. Patients present with unexplained sepsis. Diagnosis is not usually straight forward due to the presence of multiple comorbidities

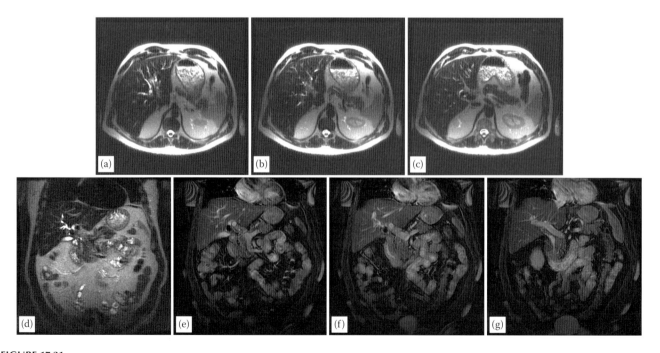

FIGURE 17.31
Axial (a–c) and coronal (d) SSFSE images and coronal FIESTA (e–h) demonstrate marked intrahepatic dilatation. The CHD is also dilated but the CBD is not dilated. Biliary obstruction arose from gallstone from the gallbladder, which eroded into the CHD, consistent with Mirrizzi syndrome. These findings were confirmed on surgery.

and nonspecific signs and symptoms. If not recognized in time, acalculous cholecystitis may be complicated by perforation and gangrenous cholecystitis. MRI demonstrates a dilated GB with thickened and T_2-hyperintense wall in the absence of GB calculi. Pericholecystic fluid may be present [15]. These critically ill patients are usually treated by percutaneous cholecystostomy.

17.3.4.1 Ischemic/Chemical Cholecystitis

Ischemic/chemical cholecystitis is a potential complication of hepatic chemoembolization due to reflux of the chemotherapeutic agent into the cystic artery. As the cystic artery originates from the right hepatic artery, this form of cholecystitis in most patients is more likely to be seen following chemoembolization of a right hepatic lesion. These patients may have no symptoms or nonspecific abdominal pain up to a few weeks posttherapy. The disease course is generally self-limiting, and thus treatment, if rarely, is necessary. MRI demonstrates GB distension. On contrast-enhanced images, there is marked wall edema with increased enhancement and transient pericholecystic enhancement. As this process is related to reflux of the chemotherapeutic agent, gallstones are not implicated.

17.3.5 Chronic Cholecystitis

Chronic cholecystitis refers to a chronically inflamed GB and is usually associated with cholelithiasis. Patients may be asymptomatic or have recurrent attacks of biliary colic. MRI demonstrates GB wall thickening and gallstones. The GB may be contracted in some patients due to fibrosis and may demonstrate delayed enhancement. Transient enhancement of the adjacent hepatic parenchyma is not seen in chronic cholecystitis, and this feature is useful in differentiating this condition from acute cholecystitis [15]. Further, calcifications

may develop in the GB wall (porcelain GB). Although calcification is more readily identified on CT and ultrasound, MRI may show signal void in regions of calcification. Further, patients with porcelain GB are at a higher risk for GB carcinoma and MRI, particularly the post-contrast images, can help identifying the tumor, by demonstrating irregular, nodular enhancement.

Xanthogranulomatous cholecystitis is an uncommon type of inflammatory disease of the GB, representing between 1% and 13% of GB disease and occurs predominantly in elderly women aged between 60 and 70 years [32–35]. The pathogenesis is unclear, but occlusion of Rokitansky–Aschoff sinuses is believed to be a precipitating factor, with resultant extravasation of bile into the GB wall, and subsequent inflammation and formation of intramural xanthogranulomatous nodules. It is important to diagnose this condition correctly, as often this mimics GB carcinoma, both clinically and radiologically. MRI shows focal or diffuse GB wall thickening on T_2-weighted images and increased wall enhancement on post-contrast images (Figure 17.32). Areas of iso- to slightly high signal intensity on T_2-weighted images, showing slight enhancement at early phase and strong enhancement at last phase on dynamic study, may reflect areas of abundant xanthogranulomas [36]. MRI may also depict necrosis as areas with very high signal intensity on T_2-weighted images, without any enhancement on post-contrast images. Differentiating from wall-thickening type of GB carcinoma may be difficult, but recent studies indicate diffusion-weighted imaging may be useful [37].

17.3.5.1 Adenomyomatosis

Adenomyomatosis is characterized by benign noninflammatory GB wall mucosal hyperplasia that invaginates into a hypertrophied muscular layer. This process leads to the formation of Rokitansky–Aschoff sinuses, which are mucosal herniations within the muscular

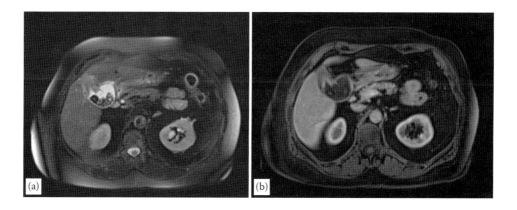

FIGURE 17.32
Axial fat suppressed T_2-weighted image (a) and contrast-enhanced T_1-weighted images (b) show thickening in the funal portion of the gallbladder and associated gallstones. Patient underwent cholecystectomy and surgery confirmed component of xanthogranulomatous cholecystitis.

layer. Although the focal form is frequently encountered, adenomyomatosis may also cause diffuse wall thickening or segmental wall thickening. The focal form is most commonly identified in the GB fundus with resultant crescentic thickening of the fundus. The segmental form may result in an "hourglass" appearance to the GB due to thickening of the mid gland. There is an association between adenomyomatosis and gallstones in the majority (>90%) of cases. As such, adenomyomatosis is not considered to be a premalignant condition [38]. However, there may be coexistent adenocarcinoma of the GB, which may be due to concurrent presence of chronic inflammation and gallstones, particularly in segmental forms of adenomyomatosis.

The Rokitansky–Aschoff sinuses frequently contain cholesterol crystal deposits and, in such cases, correlative ultrasound examinations will demonstrate comettail artifacts extending from a thickened GB wall. On MRI, the GB wall demonstrates focal or diffuse thickening. The Rokitansky–Aschoff sinuses are identified as T_2-hyperintense intramural foci in the GB wall. These are hypointense on T_1-weighted images and do not demonstrate enhancement on postgadolinium images. The ringlike arrangement of these sinuses on T_2-weighted images (Figure 17.33) and MRCP has been termed as "pearl necklace sign" or string of beads sign and is reported to be highly specific for adenomyomatosis (>90%) [38]. T_1-weighted post-contrast images may show early mucosal enhancement (Figure 17.34) and late homogeneous enhancement, but, in most patients, the contrast enhancement pattern of adenomyomatosis can mimic GB cancer.

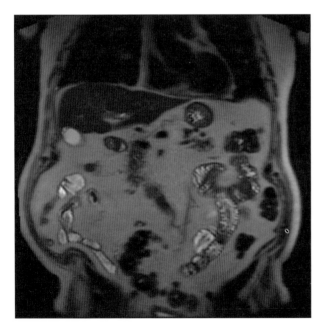

FIGURE 17.33
Coronal T_2-weighted MRI shows focal adenomyomatosis in the gallbladder fundus.

17.3.6 GB Polyps

GB polyps refer to lesions arising from the GB wall and projecting into the lumen. Polypoid lesions of the GB encompass a wide variety of pathology ranging from benign cholesterol polyps to premalignant adenomas and clearly malignant carcinomas. They are usually diagnosed incidentally at imaging or at pathology after cholecystectomy. Most of these tend to be benign; however, a small but significant proportion tend to be premalignant or malignant and need to be identified and treated early. This differentiation may not always be possible [39]. On T_2-weighted and MRCP images GB polyps appear as hypointense filling defects projecting into the lumen of the GB (Figure 17.35). Unlike calculi, they may be located in a nondependent location adjacent to the GB wall. On postgadolinium images, polyps demonstrate enhancement. Imaging features that should raise concern for malignancy are size greater than 1 cm, sessile morphology, solitary polyps, progressive growth, and coexistent gallstones [39]. Patients are referred for cholecystectomy if a polyp is greater than 10 mm. Polyps measuring 6–10 mm are often followed by ultrasound examination in 6 months. However, a patient with a 6–10 mm polyp may be referred for cholecystectomy if there has been interval growth, the patient is symptomatic, the patient is older than 50 years, or the patient has a history of primary sclerosing cholangitis.

17.3.6.1 GB Carcinoma

GB cancer is the fifth most common gastrointestinal malignancy in the United States and is usually associated with chronic inflammation and gallstones. It is the most common carcinoma of the biliary tree. Most GB cancers are adenocarcinoma (up to 90%), but squamous cell carcinoma and small cell carcinoma may also occur. There is an incidence of approximately 2.5 new cases per 100,000 people per year. Median survival time is 3 months as patients generally present with advanced disease. Overall survival at 5 years is 5%. GB cancer is more prevalent in women than men, likely due to the increased incidence of gallstones in women.

Patients may be completely asymptomatic, especially during early stages or present with nonspecific symptoms. When advanced, patients often present with vague abdominal pain, right upper quadrant pain, anorexia, weight loss, and/or fever. Jaundice may also be a presenting symptom in advanced disease due to biliary ductal obstruction. Patients are usually elderly with a mean age of 65 years at the time of presentation. Although less than 1% of patients with gallstones develop GB cancer, GB cancer is usually associated with gallstones in 90% of the patients. Risk factors include

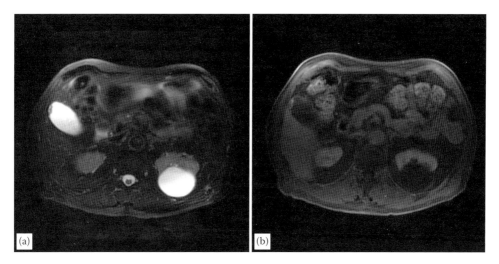

FIGURE 17.34
Axial T_2-weighted images (a) and post-contrast T_1-weighted images (b) show adenomyomatosis with early enhancement.

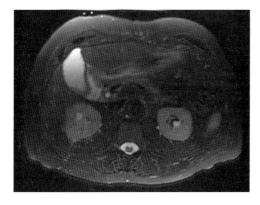

FIGURE 17.35
Axial T_2-weighted image shows numerous tiny low signal intensity foci in the gallballder wall, consistent with tiny polyps. Gallbladder polyps less than 6 mm are of no clinical significance.

chronic inflammation (often related to gallstones), porcelain GB (Figure 17.36), long common pancreatic–biliary channel, chronic biliary infections, obesity, and chronic typhoid carrier state.

GB cancer may take the form of diffuse or focal GB wall thickening (Figures 17.22 and 17.37), or a focal polypoid intraluminal mass. Advanced cases, which are more common than localized cases, often extend outside the GB wall and infiltrate into the adjacent liver parenchyma. The wall-thickening form represents 15%–30% of GB cancer. The polypoid form is the least common.

Staging is based on the American Joint Committee on Cancer tumour, node, metastasis (TNM) classification [40]. T1 tumors invade the lamina propria (1a) or muscular layer (1b). T2 tumors invade the perimuscular connective tissue but without extension beyond the serosa or into the liver. T3 tumors extend beyond the serosa and directly invade the liver and/or adjacent structure. T4 tumors invade the main portal vein or hepatic artery or invade at least two extrahepatic organs.

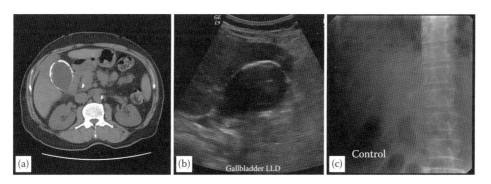

FIGURE 17.36
Axial CT (a), longitudinal ultrasound (b), and plain film (c) showing porcelain gallbladder.

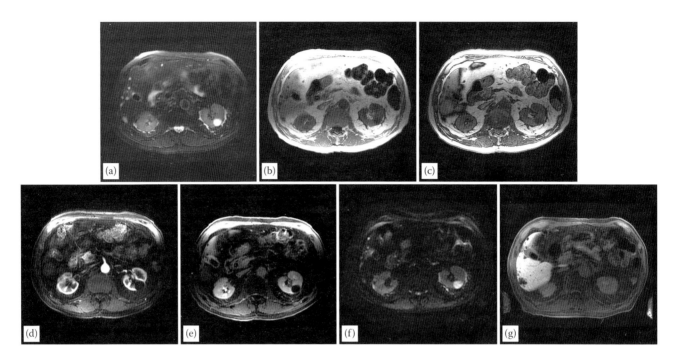

FIGURE 17.37

Axial T_2 (a), in-phase (b), out of phase (c), early (d) and delayed (e) post-contrast images, DWI with a b value 500 (f) shows focal irregular gallbladder wall thickening. Follow-up study with EOVIST demonstrated the extent of tumor nicely on the 20 min delayed sequence (g). Surgery confirmed gallbladder carcinoma.

N1 involves nodal metastases along the cystic duct, CBD, hepatic artery, and/or portal vein. N2 involves nodal metastases to periaortic, pericaval, superior mesenteric artery, and/or celiac artery lymph nodes. M1 involves distant metastases. (*AJCC Cancer Staging Manual*, 2010—available at www.cancer.gov)

Stage I (T1N0M0) is considered localized disease. The malignancy is confined to the GB wall and can be completely resected with regional lymphatics, and lymph nodes are also resected. Five-year survival is nearly 100% if the disease is confined to the mucosa, but drops to less than 15% if there is muscular invasion or beyond. Stage II–IV may involve T2–T4, N1–N2, and/or M1 disease. These patients represent the majority of cases and are generally unresectable. Treatment for stage II–IV disease is mainly palliative.

17.3.6.1.1 Imaging

MRI appearance is variable and may demonstrate a focal polypoid mass, GB wall thickening, or replacement of the entire GB by a mass.

> *Mass-like form.* The differential diagnosis for a mass-like lesion in the GB fossa includes primary malignancy, metastases, and pericholecystic abscess. The most common form of GB malignancy is the mass-like form. MRI demonstrates a heterogeneous mass that is T2 hyperintense and has T1 intermediate signal intensity.

There is early and prolonged enhancement of the mass (Figure 17.38). Gallstones may be seen within the mass. There may be associated lymphadenopathy, invasion of adjacent fat, and liver parenchyma. Post-contrast images may be particularly helpful in identifying the extent of invasion into adjacent organs and involvement of biliary ducts, and this information is critical for preoperative planning.

Wall-thickening form. The differential diagnosis for GB wall thickening is long and includes cholecystitis, malignancy, and adenomyomatosis and non-GB conditions such as hypoproteinemia and renal failure. Focal or diffuse thickening of the GB wall greater than 1 cm is a concerning feature for malignancy. Compared to the liver, the tumor may be heterogeneously hyperintense on T2 and hypointense to isointense on T1. The tumor may demonstrate irregular heterogeneous enhancement, which may help to distinguish this condition from chronic cholecystitis.

Polypoid form. The differential diagnosis for an intraluminal polypoid mass includes malignancy, polyps, focal adenomyomatosis, blood clot, and tumefactive sludge. Malignant polypoid masses are almost always greater than 1 cm. The lack of mobility can distinguish polypoid malignancy, adenomyomatosis, and polyps from tumefactive sludge. The polypoid

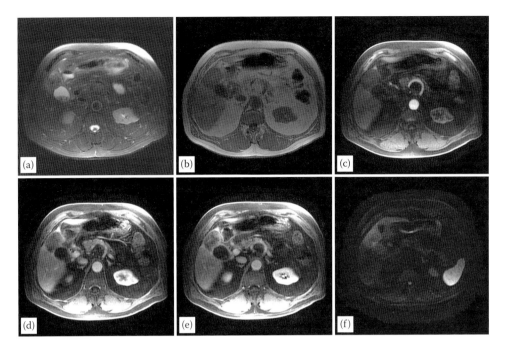

FIGURE 17.38
Axial T_2 (a), T_1-weighted image (b), post-contrast-enhanced arterial (c), portal venous (d) and delayed (e) phase images, DWI with a b value 500 (f) shows irregular mass in the fundal region of the gallbladder. Surgery confirmed gallbladder carcinoma.

mass is T1 intermediate and T2 hyperintense with moderate early and prolonged enhancement. On contrast-enhanced images, GB carcinomas demonstrate early and irregular enhancement. This enhancement persists in the delayed phase. Benign lesions demonstrate early enhancement that does not persist in the delayed phase. Recently, there has been a significant interest in the application of diffusion-weighted imaging in differentiating GB malignancies from benign conditions, and early studies are encouraging [41,42].

17.3.6.2 Other GB Malignancies

Other GB malignancies include metastatic lesions and lymphoma. Metastases to the GB from melanoma, breast cancer, and renal cell cancers have been reported. Melanoma metastases may appear hyperintense on T_1-weighted images. Similarly, metastases from renal cell carcinoma and breast cancer may present as an enhancing polypoidal lesion. Lymphomatous involvement of the GB is rare. Usually GB lymphoma tends to arise from secondary involvement of the GB in a patient with known widespread non-Hodgkin lymphoma but extremely rarely primary GB lymphoma may occur. It appears as thickening of the GB wall or solid mass in the region of the GB, and is usually associated with adenopathy in porta hepatis. However, imaging features cannot differentiate GB lymphoma from GB carcinoma.

References

1. Ando H. Embryology of the biliary tract. *Dig Surg* 2010; 27:87–89.
2. Sato T, Ito M, Sakamoto H. Pictorial dissection review of the lymphatic pathways from the gallbladder to the abdominal para-aortic lymph nodes and their relationships to the surrounding structures. *Surg Radiol Anat* 2013; 35:615–621.
3. Bronshtein M, Weiner Z, Abramovici H, Filmar S, Erlik Y, Blumenfeld Z. Prenatal diagnosis of gall bladder anomalies—Report of 17 cases. *Prenat Diagn* 1993; 13:851–861.
4. Bani-Hani KE. Agenesis of the gallbladder: Difficulties in management. *J Gastroenterol Hepatol* 2005; 20:671–675.
5. Senecail B, Nonent M, Kergastel I, Patin-Philippe L, Larroche P, Le Borgne A. Ultrasonic features of congenital anomalies of the gallbladder. *J Radiol* 2000; 81:1591–1594.
6. Balakrishnan S, Singhal T, Grandy-Smith S, El-Hasani S. Agenesis of the gallbladder: Lessons to learn. *JSLS* 2006; 10:517–519.
7. Piccolo G, Di Vita M, Zanghi A, Cavallaro A, Cardi F, Capellani A. Symptomatic gallbladder agenesis: Never again unnecessary cholecystectomy. *Am Surg* 2014; 80:E12–E13.
8. Idu M, Jakimowicz J, Iuppa A, Cuschieri A. Hepatobiliary anatomy in patients with transposition of the gallbladder: Implications for safe laparoscopic cholecystectomy. *Br J Surg* 1996; 83:1442–1443.

9. Carbajo MA, Martin del Omo JC, Blanco JI et al. Congenital malformations of the gallbladder and cystic duct diagnosed by laparoscopy: High surgical risk. *JSLS* 1999; 3:319–321.

10. Kiewiet JJ, Leeuwenburgh MM, Bipat S, Bossuyt PM, Stoker J, Boermeester MA. A systematic review and meta-analysis of diagnostic performance of imaging in acute cholecystitis. *Radiology* 2012; 264:708–720.

11. Yarmish GM, Smith MP, Rosen MP et al. ACR appropriateness criteria right upper quadrant pain. *J Am Coll Radiol* 2014; 11(3):316–322.

12. Bennett GL, Balthazar EJ. Ultrasound and CT evaluation of emergent gallbladder pathology. *Radiol Clin North Am* 2003; 41:1203–1216.

13. Bennett GL, Rusinek H, Lisi V et al. CT findings in acute gangrenous cholecystitis. *AJR Am J Roentgenol* 2002; 178:275–281.

14. Shakespear JS, Shaaban AM, Rezvani M. CT findings of acute cholecystitis and its complications. *AJR Am J Roentgenol* 2010; 194:1523–1529.

15. Smith EA, Dillman JR, Elsayes KM, Menias CO, Bude RO. Cross-sectional imaging of acute and chronic gallbladder inflammatory disease. *AJR Am J Roentgenol* 2009; 192:188–196.

16. Tsai MJ, Chen JD, Tiu CM, Chou YH, Hu SC, Chang CY. Can acute cholecystitis with gallbladder perforation be detected preoperatively by computed tomography in ED? Correlation with clinical data and computed tomography features. *Am J Emerg Med* 2009; 27:574–581.

17. Fuks D, Mouly C, Robert B, Hajji H, Yzet T, Regimbeau JM. Acute cholecystitis: Preoperative CT can help the surgeon consider conversion from laparoscopic to open cholecystectomy. *Radiology* 2012; 263:128–138.

18. Irie H, Kamochi N, Nojiri J, Egashira Y, Sasaguri K, Kudo S. High *b*-value diffusion-weighted MRI in differentiation between benign and malignant polypoid gallbladder lesions. *Acta Radiol* 2011; 52:236–240.

19. Ogawa T, Horaguchi J, Fujita N et al. High b-value diffusion-weighted magnetic resonance imaging for gallbladder lesions: Differentiation between benignity and malignancy. *J Gastroenterol* 2012; 47:1352–1360.

20. Elsayes KM, Oliveira EP, Narra VR, El-Merhi FM, Brown JJ. Magnetic resonance imaging of the gallbladder: Spectrum of abnormalities. *Acta Radiol* 2007; 48:476–482.

21. Tan CH, Lim KS. MRI of gallbladder cancer. *Diagn Interv Radiol* 2013; 19:312–319.

22. Demas BE, Hricak H, Moseley M et al. Gallbladder bile: An experimental study in dogs using MR imaging and proton MR spectroscopy. *Radiology* 1985; 157:453–455.

23. Bilgin M, Shaikh F, Semelka RC, Bilgin SS, Balci NC, Erdogan A. Magnetic resonance imaging of gallbladder and biliary system. *Top Magn Reson Imaging* 2009; 20:31–42.

24. Tsai HM, Lin XZ, Chen CY, Lin PW, Lin JC. MRI of gallstones with different compositions. *AJR Am J Roentgenol* 2004; 182:1513–1519.

25. Yusoff IF, Barkun JS, Barkun AN. Diagnosis and management of cholecystitis and cholangitis. *Gastroenterol Clin North Am* 2003; 32:1145–1168.

26. Loud PA, Semelka RC, Kettritz U, Brown JJ, Reinhold C. MRI of acute cholecystitis: Comparison with the normal gallbladder and other entities. *Magn Reson Imaging* 1996; 14:349–355.

27. Altun E, Semelka RC, Elias J, Jr. et al. Acute cholecystitis: MR findings and differentiation from chronic cholecystitis. *Radiology* 2007; 244:174–183.

28. Catalano OA, Sahani DV, Kalva SP et al. MR imaging of the gallbladder: A pictorial essay. *RadioGraphics* 2008; 28:135–155; quiz 324.

29. Choi IY, Cha SH, Yeom SK et al. Diagnosis of acute cholecystitis: Value of contrast agent in the gallbladder and cystic duct on Gd-EOB-DTPA enhanced MR cholangiography. *Clin Imaging* 2014; 38(2):174–178.

30. Pedrosa I, Guarise A, Goldsmith J, Procacci C, Rofsky NM. The interrupted rim sign in acute cholecystitis: A method to identify the gangrenous form with MRI. *J Magn Reson Imaging* 2003; 18:360–363.

31. Koenig T, Tamm EP, Kawashima A. Magnetic resonance imaging findings in emphysematous cholecystitis. *Clin Radiol* 2004; 59:455–458.

32. Casas D, Perez-Andres R, Jimenez JA et al. Xanthogranulomatous cholecystitis: A radiological study of 12 cases and a review of the literature. *Abdom Imaging* 1996; 21:456–460.

33. Duber C, Storkel S, Wagner PK, Muller J. Xanthogranulomatous cholecystitis mimicking carcinoma of the gallbladder: CT findings. *J Comput Assist Tomogr* 1984; 8:1195–1198.

34. Ros PR, Goodman ZD. Xanthogranulomatous cholecystitis versus gallbladder carcinoma. *Radiology* 1997; 203:10–12.

35. Shetty GS, Abbey P, Prabhu SM, Narula MK, Anand R. Xanthogranulomatous cholecystitis: Sonographic and CT features and differentiation from gallbladder carcinoma: A pictorial essay. *Jpn J Radiol* 2012; 30:480–485.

36. Shuto R, Kiyosue H, Komatsu E et al. CT and MR imaging findings of xanthogranulomatous cholecystitis: Correlation with pathologic findings. *Eur Radiol* 2004; 14:440–446.

37. Kang TW, Kim SH, Park HJ et al. Differentiating xanthogranulomatous cholecystitis from wall-thickening type of gallbladder cancer: Added value of diffusion-weighted MRI. *Clin Radiol* 2013; 68:992–1001.

38. Boscak AR, Al-Hawary M, Ramsburgh SR. Best cases from the AFIP: Adenomyomatosis of the gallbladder. *RadioGraphics* 2006; 26:941–946.

39. Andren-Sandberg A. Diagnosis and management of gallbladder polyps. *N Am J Med Sci* 2012; 4:203–211.

40. Edge SB, Byrd DR, Compton CC, Fritz AG, Greene FL, Trotti A eds. *AJCC Cancer Staging Manual*, 7th edn., 2010. New York, NY: Springer.

41. Kim SJ, Lee JM, Kim H, Yoon JH, Han JK, Choi BI. Role of diffusion-weighted magnetic resonance imaging in the diagnosis of gallbladder cancer. *J Magn Reson Imaging* 2013; 38:127–137.

42. Lee NK, Kim S, Kim TU, Kim DU, Seo HI, Jeon TY. Diffusion-weighted MRI for differentiation of benign from malignant lesions in the gallbladder. *Clin Radiol* 2014; 69:e78–e85.

18

Magnetic Resonance Imaging of Biliary Tract

Sachin Kumbhar, Manjiri K. Dighe, Ganeshan Dhakshinamoorthy, and Neeraj Lalwani

CONTENTS

18.1 Introduction

The biliary tract consists of the gallbladder (GB) and the (intra and extrahepatic) bile ducts. It is concerned with storage and transport of bile from the liver to its site of action in the duodenum. It may be affected by a wide range of disorders, from benign to malignant. Imaging is important to identify not only the pathology but also the exact site of disease.

Magnetic resonance imaging (MRI) has gained popularity in the diagnosis of biliary diseases, as it provides a complete assessment of the entire biliary tree irrespective of the site and severity of biliary duct obstruction. Standard MRI techniques are useful in the evaluation of the biliary duct wall and extraductal pathology. Intraductal pathologies are better evaluated by MR cholangiopancreatography (MRCP), which demonstrates the fluid-filled lumen of the GB and the bile ducts. Hepatocyte-specific MRI contrast agents have the potential of being used for dynamic evaluation of the bile ducts.

18.2 MRI of Bile Ducts: Sequences and Techniques

MRCP is used along with the T_1-weighted, T_2-weighted, and T_1-enhanced sequences in the evaluation of the bile ducts. There are multiple different MRCP techniques and all of them have one thing is common: They rely upon heavy T_2 weighting. On these heavily T_2-weighted sequences, background signal from the bile duct wall and extraductal structures is suppressed, and the high signal of the stationary bile within the biliary tract stands out. Fluid within the bowel and stomach also appears high signal on MRCP. Therefore, negative oral contrast can be used to suppress the signal arising from the gut [1]. Patients should be fasting for at least 4 h prior to the examination, so that the GB is well distended. MRCP may be performed using 2D or 3D acquisitions with non-breath-hold sequences that utilize respiratory triggering. Alternatively, there are breath-hold sequences like the single-shot fast spin-echo (SSFSE) and the half Fourier acquisition single-shot turbo spin-echo (HASTE). These may be performed with a thin slice collimation or thick slab acquisition [2].

18.2.1 Breath-Hold Sequences

The breath-hold acquisitions of MRCP have lesser scan times. They are performed in a single breath hold or during free breathing and hence can be utilized even in patients who cannot cooperate with breathing instructions. Artifacts due to respiratory motion and misregistration are avoided in these acquisitions. The SSFSE sequence utilizes an infinite repetition time (TR) and a very long echo train to decrease scan time and acquisition, which is completed usually in 30 s. A high echo time (TE) of more than 600 ms causes suppression of signal from extraductal solid tissues including fat. Hence, fat suppression is not required in this acquisition. Modified fast spin-echo sequences like rapid acquisition with rapid enhancement (RARE) and HASTE have become popular for breath-hold MRCP. All of these sequences have a high signal-to-noise ratio (SNR) due to lack of motion artifacts [3].

These breath-hold sequences can be acquired in the form of a single thick slab or multiple thin sections. In the thick slab acquisition, a single image is acquired that represents an average of the entire imaging volume. It takes less than 2 s to acquire this image. Usually, this acquisition is repeated at different orientations and in different planes to provide multiple *projection-like* images of the biliary tree. However, averaging can lead to obscuring of small or subtle pathologies. Hence, thick slab acquisitions have less sensitivity and have to be complemented by acquiring multiple thin sections. These are obtained using a slice thickness of 3–4 mm and usually in the oblique coronal plane. They are particularly helpful to depict small intraluminal filling defects in the bile ducts [3].

18.2.2 Non-Breath-Hold Sequences

Non-breath-hold sequences are acquired using a navigator sequence that accesses the position of the diaphragm. The navigator is positioned at the interface of the right diaphragm and the lung on coronal images. The MRI scanner monitors the position of this interface and triggers the MRCP acquisition when its position is within a preset window. The non-breath-hold sequence may be a 3D or 2D acquisition. From these images, maximum intensity projection (MIP) images can be generated. These consist of pixels with highest signal intensity throughout the imaging volume projected along a direction perpendicular to the plane of the projection. They are similar in appearance to the conventional cholangiogram images [1].

18.2.3 Secretin-Stimulated MRCP

Secretin is a hormone released by the duodenum that acts upon the pancreatic exocrine cells and stimulates them to release bicarbonate rich fluid. When the synthetic form of secretin is injected intravenously, it enhances the visualization of the pancreatic duct by increasing its caliber [4], and can be utilized in the diagnosis of early chronic pancreatitis. This effect of secretin peaks at about 2–5 min after injection and subsides by about 10 min after injection.

18.2.4 Contrast-Enhanced MRCP

Some of the gadolinium-based contrast media are partly excreted by the liver into the bile and partly by the kidneys. These are called hepatocyte-specific contrast media and include gadoxenate and gadobenate. After intravenous injection, these agents appear into the bile after a certain amount of time that depends upon the relative contribution of the liver to the excretion of the gadolinium agent. Also, their appearance in the bile is delayed if the liver function is decreased. Delayed imaging is performed usually after 20 min using a T_1-weighted fat-suppressed sequence. Contrast-enhanced MRCP is particularly helpful to demonstrate communication of cystic structures with the bile ducts and to demonstrate biliary leaks [5].

18.3 Normal Anatomy of the Bile Ducts

According to the Couinaud classification, the liver is divided into eight segments based on their relation to the portal vein and the hepatic veins. Each of these segments has their independent portal venous supply, hepatic venous drainage, and biliary drainage. The right hepatic duct drains hepatic segments 5–8. It is formed by the fusion of the right anterior bile duct, which drains segments 5 and 8, and the right posterior duct, which drains segments 6 and 7. The left hepatic duct drains segments 2–4. The right and left hepatic ducts join together to form the common hepatic duct (CHD) (Figure 18.1). The bile duct draining segment 1 may join either the right or the left hepatic duct. The CHD joins the cystic duct arising from the GB to form the common bile duct (CBD) (Figure 18.1). The pancreatic duct joins the CBD near the ampulla. However, variations in the anatomy of the bile ducts are not uncommon.

The peripheral intrahepatic biliary radicals are not commonly visualized by MRCP unless dilated. The normal CBD has a diameter of up to 6 mm. Although normal CBD may have a diameter of more than 6 mm, such cases should correlate with clinical features and liver function tests. Also, the CBD diameter goes on increasing with age. In patients aged above 50, it may reach up to 8 mm. In patients who have their GB removed, up to 10 mm CBD diameter should be considered normal [6].

18.3.1 Anatomic Variants of the Biliary Tract

Anatomic variations of the biliary drainage pathways are not uncommon. These are sometimes important to identify as they may predispose to disease. Also, it is vital to recognize any forms of anatomical variants prior to hepato-biliary surgery, so that potentially disastrous complications are avoided.

The most common anatomic variations involve the formation of the right hepatic duct, left hepatic duct, and the CHD. Normally, the right posterior and the right anterior biliary ducts join together to form the right hepatic duct which in turn joins the left hepatic duct to form the CHD (Figure 18.2). However, in some individuals, the right posterior duct drains into the left hepatic duct, which then joins the right anterior duct

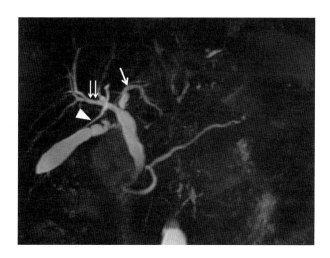

FIGURE 18.1
Normal biliary anatomy on MRCP: MIP image processed from the MRCP sequence demonstrates ERCP-like *projection* images of the bile ducts. The right posterior duct (arrowhead) joins the right anterior duct (double arrows) to form the right hepatic duct, which joins the left hepatic duct (arrow) to form the CHD.

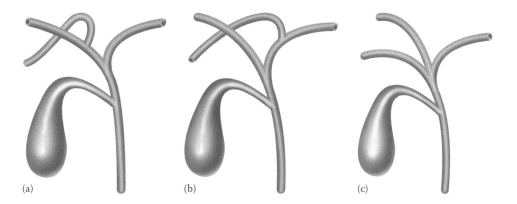

FIGURE 18.2
Common anatomic variants of the bile ducts: (a) The right posterior bile duct joins the right anterior bile duct to form the right hepatic duct. The right hepatic duct then joins the left hepatic duct to form the CHD. This is the most common biliary anatomy (b) The right posterior bile duct joins the left hepatic duct, which then joins the right anterior bile duct to form the CHD. The right hepatic duct is absent. (c) The right anterior bile duct, right posterior bile duct, and the left hepatic duct join simultaneously to form the CHD.

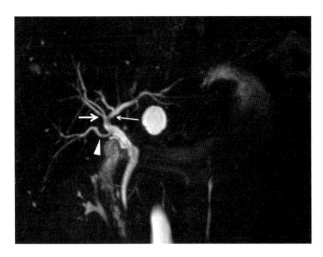

FIGURE 18.3
Anatomic variant of the bile ducts: The right anterior duct (thick arrow) joins the left hepatic duct (thin arrow) to form the CHD. The right posterior duct (arrowhead) drains directly into the CHD.

to form the CHD. This is the most commonly reported biliary variant [7]. Another common variant is trifurcation of the CHD. In this variation, the right anterior bile duct, the right posterior bile duct, and the left hepatic duct join together simultaneously to form the CHD. The right hepatic duct does not exist. Infrequently, the right posterior duct drains in CHD (Figure 18.3). It is vital to detect these anatomic variations in potential living liver donors, as more than one biliary anastomosis would be required, thus increasing the chance of complications [8]. Also, precise determination of biliary anatomy is helpful to the transplant surgeons to plan the surgical approach for resection [9].

Anatomic variants of the cystic duct are usually of no clinical consequence. However, as cholecystectomy is frequently performed laparoscopically, it is important to identify these variations to avoid potential damage to the CBD and the right hepatic duct during laparoscopy. The cystic duct may have a low insertion into the distal third of the CBD. It may also have a close parallel course to the CBD. Rarely, the cystic duct may drain into the right hepatic duct. If these variations are not identified prior to surgery, there is a risk of inadvertent ligation and transection of the right hepatic duct or the CBD [10,11]. This may result in a morbid postoperative course and potentially life-threatening complications. MRCP can easily identify these anatomic variations and avoid such an eventuality.

The CBD and the pancreatic duct normally join together with an approximately 4–5 mm of common channel before opening into the duodenum. The junction is surrounded by the sphincter of Oddi. Sometimes, the pancreatic duct and the CBD may have separate openings into the duodenum. Anomalous pancreaticobiliary junction (APBJ) is a rare condition in which the CBD and the pancreatic duct join together outside the duodenum and proximal to the spincter of Oddi. The common channel is usually more than 15 mm. As the junction is proximal to the spincter of Oddi, bile may reflux into the pancreatic duct and pancreatic secretions may reflux into the CBD. This has been postulated to be a reason for the high association of APBJ with choledochal cysts, cholangiocarcinoma, and recurrent pancreatitis [12,13].

18.3.2 Congenital Anomalies of the Biliary Tract

18.3.2.1 Choledochal Cysts

Choledochal cysts are congenital disorders of the bile ducts characterized by cystic or saccular dilatation of the extrahepatic and/or intrahepatic bile ducts. There is a female predilection and most of the cases are diagnosed in children. The classic triad of symptoms consists of right upper quadrant pain, palpable abdominal mass, and jaundice. However, it is not seen in all the patients. Choledochal cysts usually lead to complications such as cholangitis, pancreatitis, choledocholithiasis, biliary cirrhosis, and cholangiocarcinoma. Hence, they are treated by complete excision of the cysts with Roux-en-Y hepaticojejunostomy. Partial hepatic resection or liver transplant may be required if the intrahepatic bile ducts are involved [14,15].

Choledochal cysts are categorized according to the classification proposed by Todani as illustrated in Figure 18.4 [16]. Type I is the commonest type and consists of fusiform dilatation of the CBD (Figure 18.5), while type II consists of focal saccular dilatation of the CBD. These can be demonstrated on MRCP as diffuse dilatation of the CBD or a cystic structure communicating with the CBD. Type III choledochal cyst represents focal dilatation of the intraduodenal portion of the CBD (Figure 18.5). This is similar to an ureterocele. Differentials include duodenal duplications cysts and pancreatic cystic lesions. Type IV choledochal cysts involve multiple dilatations of the extrahepatic ducts either alone or in combination with the intrahepatic ducts. Type V choledochal cysts, also called *Caroli's disease*, has exclusive dilatations of the intrahepatic ducts. Imaging plays an important role in the diagnosis of the disease, delineation of its extent, and detection of complications [17].

18.4 Biliary Obstruction

Endoscopic retrograde cholangiopancreatography (ERCP) is considered the gold standard for diagnosis in suspected biliary obstruction. MRI is very useful in the evaluation of biliary obstruction. It demonstrates the entire biliary tree from the intrahepatic radicles to the ampulla and both intraductal and extraductal pathologies in a single examination. Unlike ERCP, MRCP demonstrates both

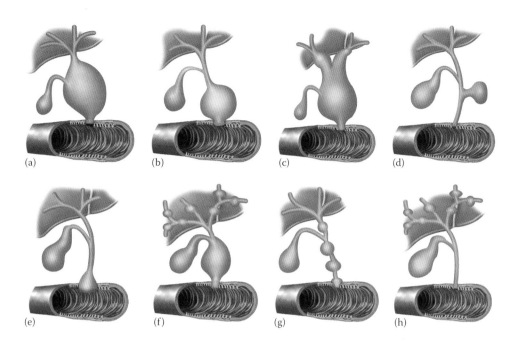

FIGURE 18.4
Todani's classification of choledochal cysts: (a) Type Ia choledochal cyst: Cystic dilatation of the CBD. (b) Type 1b choledochal cyst: Focal dilatation of the CBD. (c) Type 1c: Fusiform dilatation of the entire extrahepatic biliary tree. (d) Type II choledochal cyst: Diverticulum of the CBD. (e) Type III choledochalcyst (choledochocele): dilatation of the intraduodenal part of the CBD. (f) Type IVa choledochal cyst: multiple intrahepatic and extrahepatic duct dilatations. (g) Type IVb choledochal cysts: multiple extrahepatic dilatations of the bile ducts. (h) Type V choledochalcysts (Caroli's disease): multiple dilatations of the intrahepatic bile ducts.

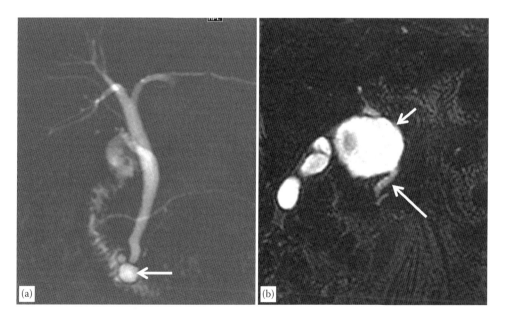

FIGURE 18.5
Choledochal cysts: (a) Type III choledochal cyst: choledochocele and (b) type I choledochal cyst.

proximal and distal ducts to the site of obstruction and produces images comparable to ERCP. At the same time, it also avoids radiation exposure and the procedural- and anaesthesia-related complications of ERCP. MRCP has sensitivity comparable to ERCP in the diagnosis of biliary obstruction and a specificity close to 100% [18]. MRI is able to confirm biliary obstruction and also demonstrates the exact site of obstruction. Intraductal disease such as calculi appears as filling defects in the hyperintense bile on T2-weighted images. Extraductal disease such as enlarged lymph nodes and pancreatic head masses compressing upon the bile ducts are also detected.

18.4.1 Choledocholithiasis

The presence of calculi in the bile ducts is termed as *choledocholithiasis*. It is the commonest cause of biliary obstruction. The accuracy of MRI to detect bile duct stones depends on the size of the calculi. The sensitivity, specificity, and accuracy of MRCP for detection of choledocholithiasis are 90%, 88%, and 89%, respectively. For calculi of 6 mm or more in size, the sensitivity, specificity, and accuracy increase to 100%, 99%, and 99%, respectively [19]. MRCP is comparable to ERCP for the diagnosis of CBD stones and better than ERCP for detection of stones in the intrahepatic bile ducts.

On T_2-weighted images and MRCP images bile duct stones appear as a hypointense intraluminal filling defect when compared to the adjacent hyperintensebile (Figure 18.6). Pigment stones may appear hyperintense to bile on T_1-weighted images. Small stones are difficult to detect, especially if they are present in the intrahepatic bile ducts or impacted at the ampulla. Thin section MRCP images and axial images are very useful for detecting these small calculi. Other differentials for filling defects in the bile duct are gas bubbles, blood clots, and tumors [20]. Stones can be differentiated from gas bubbles on axial images, as stones will be present in the dependent position, whereas the gas bubbles will rise to the non-dependent position in the bile duct [21]. Blood clots usually have an irregular shape unlike the round or oval shape of biliary calculi. Blood clots may also demonstrate hyperintense signal on T_1 images depending on the age of the bleed. Tumors demonstrate enhancement on postgadolinium T_1-weighted images.

Some MRI artifacts may appear as filling defects in the bile ducts and may be erroneously diagnosed as stones. Susceptibility artifacts arising from surgical clips and adjacent bowel gas and flow artifacts may simulate

biliary calculi. In case of flow artifacts, MR angiography might demonstrate flow in the bile duct. Arterial pulsations of adjacent vessels may cause an extrinsic narrowing of the bile duct, which can be misinterpreted as pathologic narrowing. Again MR angiography will demonstrate the causative artery [21].

Mirizzi syndrome deserves a special mention as it causes CHD obstruction due to the impact of the stone in the adjacent cystic duct. Preoperative diagnosis is helpful to avoid complications during surgery, as inflammatory adhesions in this region may obscure anatomy, and the CBD may be mistaken for cystic duct [22].

18.4.2 Cholangitis

18.4.2.1 Primary Sclerosing Cholangitis

Primary sclerosing cholangitis (PSC) is a chronic inflammatory disease of the bile ducts characterized by idiopathic progressive inflammation and fibrosis of the intrahepatic and extrahepatic bile ducts (Figure 18.7a). Although the exact etiology is unknown, it is believed to be an autoimmune disease as it has a high degree of association with inflammatory bowel disease. PSC eventually progresses to cirrhosis and liver failure. It may also be complicated by the development of portal hypertension orcholangiocarcinoma (Figure 18.7b). Liver transplantation is the only curative treatment [23].

ERCP has been the gold standard for the diagnosis of PSC. However, in recent studies, MRCP has been demonstrated for better visualization of extrahepatic and intrahepatic ducts when compared to ERCP. MRI also provides information about the condition of the liver and detects disease complications [24]. The commonest abnormality detected on MRI is strictures and dilatation of the intrahepatic bile ducts. Early in the disease process, the strictures are short and alternate with dilated ducts giving rise to a beaded appearance of the bile ducts. The ductal dilatation is usually less than expected according to the severity of the strictures. As the disease progresses, there is obliteration of the peripheral bile ducts giving rise to a *pruned tree* appearance. Thickening and enhancement of the bile duct wall can be demonstrated on other MRI sequences [25].

Morphological alterations in the liver depend on the stage of the disease. Early in the disease the liver parenchyma appears normal. As the disease progresses, high T_2 signal intensity of the peri-portal hepatic parenchyma may be demonstrated and represents inflammatory edema. Peripheral areas of increased signal intensity and arterial hyper-enhancement may also be present and are because of alterations in perfusion pattern due to inflammation and development of fibrosis. In advanced disease, there is atrophy of the right and left lobes of the liver with caudate lobe hypertrophy. Enlarged portal lymph nodes are not uncommon in these patients [25].

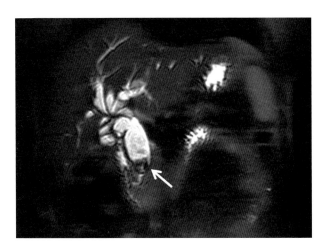

FIGURE 18.6
Choledocholithiasis: T_2-weighted coronal HASTE image demonstrates an impacted biliary calculus (arrow) in the lower CBD with dilatation of the proximal CBD and the intrahepatic bile ducts.

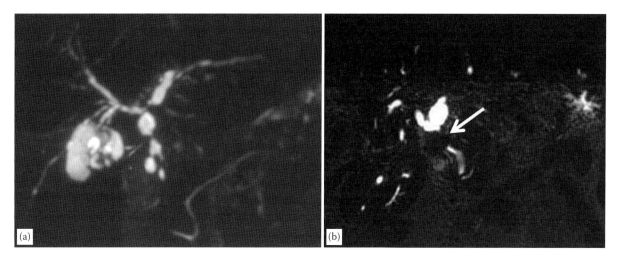

FIGURE 18.7
Primary sclerosing cholangitis: (a) MRCP image demonstrating beading of the bile ducts due to multiple strictures and dilatations of the intervening bile ducts. (b) MRCP image demonstrating disproportionately more dilatation of the right and left hepatic ducts due to the development of cholangiocarcinoma (arrow).

18.4.2.2 Infectious Cholangitis

Infectious cholangitis or ascending cholangitis is due to infection of the bile ducts by gram-negative enteric organisms in the setting of biliary obstruction. It is a potentially life-threatening condition that usually presents with fever, jaundice, and abdominal pain and may be complicated by sepsis. Treatment comprises supportive measures, antibiotics, and relief of obstruction. MRCP reveals biliary dilatation without beading or peripheral pruning. Choledocholithiasis may be present [26]. There is diffuse thickening of the CBD and enhancement of the walls of the intrahepatic bile ducts. Extension of inflammation into the hepatic parenchyma appears as peripheral wedge-shaped or peri-portal patchy areas of hyperintensity on T_2-weighted images that demon strate hyper-enhancement on postgadolinium images. Infectious cholangitis may progress to formation of hepatic abscesses that appear as peripherally enhancing focal lesions in the liver [27].

18.4.2.3 Recurrent Pyogenic Cholangitis

Recurrent pyogenic cholangitis or Oriental cholangitis is characterized by recurrent episodes of cholangitis, intrahepatic pigment stones, and biliary dilatation. It is associated with parasitic infestations such as *Ascaris lumbricoides* and *Clonorchis sinensis* (Figure 18.8). It is believed that chronic infestation of the bile duct with these parasites leads to inflammation, stricture formation, and calculi formation. Patients usually present with recurrent episodes of cholangitis [28]. MRCP reveals strictures of the intrahepatic bile ducts with abrupt tapering and dilatation of the central intrahepatic and

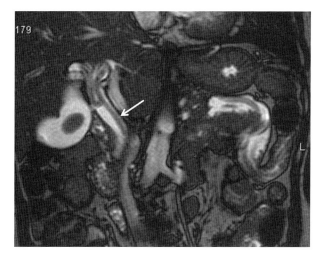

FIGURE 18.8
Biliary ascariasis: T_2-weighted coronal MRI section demonstrates *Ascaris lumbricoides* (round worm) in the bile duct as a long tubular filling defect (arrow) coiled at its upper end. Parasitic infestation of the bile ducts is associated with recurrent cholangitis.

extrahepatic bile ducts. Intrahepatic stones are visualized in 80% of patients, and these may appear hyperintense on T_1-weighted images because of their bilirubin content (Figure 18.9). Pneumobilia and hepatic abscesses may also be present. Liver atrophy develops in chronic cases [27].

18.4.2.4 AIDS Cholangiopathy

AIDS cholangiopathy comprises a group of diseases affecting the biliary tract in HIV-infected individuals. These diseases include opportunistic infections of the bile ducts by pathogens such as *Cryptosporidium parvum* and *Cytomegalovirus*, causing acalculous cholecystitis and lymphoma. MRI findings in AIDS

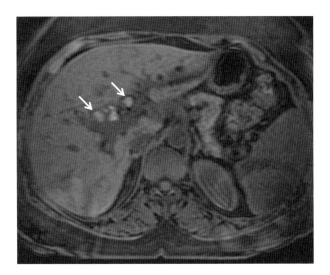

FIGURE 18.9
Recurrent pyogenic cholangitis: T_1-weighted axial MRI section demonstrates multiple T_1 hyperintense intrahepatic pigment stones (arrows) that are commonly visualized in recurrent pyogenic cholangitis.

cholangiopathy are biliary dilatation, biliary strictures, focal bile duct wall thickenings, and ampullary stenosis [29].

18.5 Neoplasms of the Bile Ducts

18.5.1 Cholangiocarcinoma

Cholangiocarcinomas are malignant neoplasms arising from the epithelium of the bile ducts. It was most frequent in the sixth and seventh decades. Some chronic diseases of the bile ducts are associated with an increased risk of developing cholangiocarcinomas. These include PSC, choledochal cysts, chronic choledocholithiasis, bile duct papillomas, and liver fluke infestation. However, in a majority of cases no risk factor can be identified. Patients usually present with anorexia, weight loss, abdominal pain, jaundice, and pruritus [30].

Based on their location, Cholangiocarcinomas are classified into three different forms. The intrahepatic cholangiocarcinomas arise from the peripheral intrahepatic bile ducts. These represent the second commonest primary hepatic malignancy after hepatocellular carcinomas. Hilar cholangiocarcinomas arise from the right or left hepatic ducts or their confluence. These are the commonest type of cholangiocarcinomas and are also called *Klatskin's tumor*. The third anatomical subtype of cholangiocarcinoma is the extrahepatic cholangiocarcinoma, which arises from the CHD or the CBD. Histologically, a majority of cholangiocarcinomas are adenocarcinomas of the sclerosing type [30].

Intrahepatic cholangiocarcinomas appear as a nonencapsulated solid focal mass on MRI. They are usually hypointense on T_1-weighted images and hyperintense on T_2-weighted images. In some patients, there may be a central hypointensity on T_2-weighted images that corresponds to fibrosis. When present, this central hypointensity can be used to distinguish cholangiocarcinomas from liver metastases. On dynamic postgadolinium images, there is perilesional marginal enhancement in the arterial and portal-venous phase with subsequent progressive centripetal enhancement (Figure 18.10). Delayed imaging is important as some tumors may only be visualized on the delayed scans. Other features that may be identified on MRI are focal retraction of the liver capsule and mild focal dilatation of the intrahepatic biliary radicles in the region of the tumor [31].

Hilar and extrahepatic cholangiocarcinomas are usually of the periductal infiltrating type. These are usually difficult to visualize on routine T_1- and T_2-weighted images. MRCP sequences indirectly indicate the presence and location of the tumor by demonstrating proximal biliary dilatation and abrupt bile duct narrowing due to a focal stricture. Focal thickening of the bile duct wall may or may not be visualized. Less commonly, they present as a small mass. In these cases, they follow the signal intensity characteristics and enhancement patterns similar to that of the intrahepatic cholangiocarcinomas. Rarely, the hilar and extrahepatic cholangiocarcinomas may have a predominantly intraluminal polypoidal growth and appear as a filling defect on MRCP. This type of growth has a relatively better prognosis [31].

MRI also helps in the staging of cholangiocarcinomas and determining the resectability and surgical approach to these lesions. Enlargement of the portal lymph nodes in itself is not specific for metastasis as it may be due to PSC, infection, or stenting. It is important to determine the presence of peritoneal and hepatic metastatic lesions. The resectability of the primary lesion is determined by assessing the degree and extent of involvement of the bile ducts, vessels, and the hepatic segments. In the case of intrahepatic cholangiocarcinomas, volumetric assessment of the involved segments relative to the rest of the liver is useful in planning the extent of surgical resection [31]. For hilar and extrahepatic cholangiocarcinomas, resectability and the type of operative procedure depends upon the proximity of the tumor from the biliary confluence and the status of the involvement of the hilum, the right and left hepatic ducts, and their secondary confluences. Based on these factors, the hilar and extrahepatic cholangiocarcinomas are classified according to the Bismuth classification (Figure 18.11) [32].

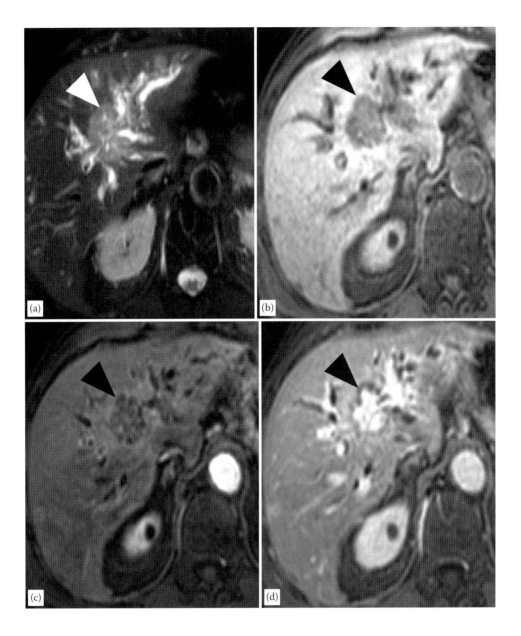

FIGURE 18.10
Hilar cholangiocarcinoma (arrowheads): (a) T_2-weighted axial MRI section demonstrates the T_2 hyperintense hilar cholangiocarcinoma. (b) The tumor appears hypointense on T_1-weighted images. (c) Arterial phase postgadolinium T_1-weighted images demonstrate peripheral enhancement of the cholangiocarcinoma. (d) Delayed phase postgadolinium T_1-weighted images demonstrate delayed enhancement of the tumor.

18.5.2 Ampullary Lesions

The ampulla of Vater or the major papilla is a protrusion in the medial part of the third portion of the duodenum that contains the combined opening of the CBD and the pancreatic duct. The normal ampulla may not be visualized by MRI. When visualized, it is less than 10 mm in size and has an enhancement pattern similar to the adjacent duodenal mucosa [33]. In cases of obstruction at the ampulla, MRCP demonstrates dilatation of the CBD down to the level of the ampulla. The ampullary lesion causing obstruction might itself not be visualized. Ampullary obstruction could be due to ampullary spasm, inflammatory stenosis, calculus, and benign or malignant neoplasms [34]. Imaging features frequently overlap.

In the case of ampullary spasm and ampullary stenosis, no other abnormality is identified apart from biliary obstruction at the ampulla. Acute inflammation of the papilla may appear as mild smooth wall thickening of the distal CBD with hyperenhancement of the papilla. Ampullary carcinomas are adenocarcinomas arising from the duodenal epithelium overlying the ampulla. Small ampullary carcinomas may not be visualized by MRI, in which case differentiation from other causes of ampullary obstruction may not be possible. However, when large enough ampullary carcinomas appear as hypoenhancing ampullary masses with irregularity of the papilla [33].

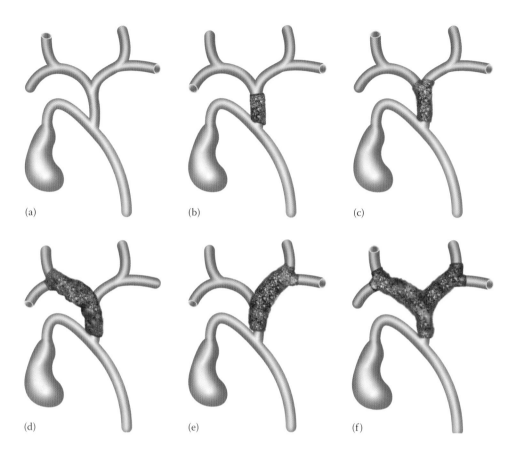

(a)   (b)   (c)

(d)   (e)   (f)

FIGURE 18.11

Bismuth–Corlette classification of perihilar cholangiocarcinomas: (a) Normal anatomy of the bile ducts. (b) Type I: cholangiocarcinoma limited to the CHD distal to the biliary confluence. (c) Type II: cholangiocarcinoma involving the biliary confluence. (d) Type III a: cholangiocarcinoma extending to the bifurcation of the right hepatic duct. (e) Type III b: cholangiocarcinoma extending to the bifurcation of the left hepatic duct. (f) Type IV: cholangiocarcinoma extending to the bifurcations of both the right and the left hepatic ducts.

18.6 MRI of the GB

18.6.1 Technique

It is recommended that patients fast for at least 4 h prior to MRI examination of the GB so that it is well distended. Routine sequences used for MRI of the abdomen like T_1- and T_2-weighted sequences are utilized to image the GB. MRCP sequence is also performed and is especially useful for visualization of the cystic duct and small calculi. Postgadolinium T_1-weighted images are acquired if required.

18.6.2 Normal Appearance

The lumen of the GB appears hyperintense on T_2-weighted images and MRCP as it consists of static liquid bile. On T_1-weighted images, bile may vary in signal intensity from hypointense to hyperintense depending upon its concentration and composition. Prolonged fasting causes resorption of water from the bile, thereby increasing the concentration of cholesterol, bile salts, and phospholipids. This concentrated bile appears intermediate to high intensity on T_1-weighted images [2]. The GB wall appears hypointense on T_2-weighted and intermediate intensity on T_1-weighted images and demonstrates smooth and uniform enhancement on postgadolinium images. Normal thickness of the GB wall is up to 3 mm [35]. The cystic duct connects the GB with the CHD to form the CBD. The length of the normal cystic duct is variable, and it measures up to 5 mm in diameter.

18.6.3 Cholelithiasis

Cholelithiasis has an overall worldwide prevalence of 10%–20%. There is an increased risk of cholelithiasis in individuals with impaired GB motility and supersaturated bile. Risk factors for gallstones include pregnancy, obesity, rapid weight loss, diabetes mellitus, and alcohol use. Most patients with gallstones are asymptomatic. When clinically apparent, the commonest symptom is biliary colic, which refers to transient abdominal pain

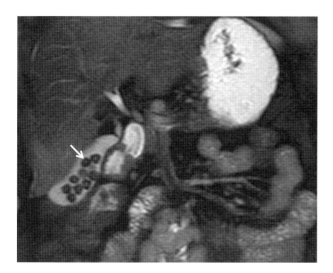

FIGURE 18.12
Gallstones: multiple hypointense calculi (arrow) within the gallbladder. Note the hyperintense cleft in these gallstones.

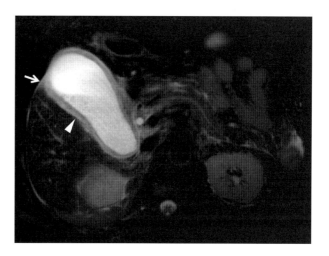

FIGURE 18.13
Acute cholecystitis: T_2-weighted axial fat-saturated image demonstrates markedly distended gallbladder with diffuse T_2 intermediate to hyperintense wall thickening (arrowhead) and pericholecystic fluid (arrow). These features are indicative of acute cholecystitis in the right clinical setting.

because of intermittent obstruction of the cystic duct due to a gallstone [36]. Ultrasound is the modality of choice for evaluation of gallstone disease.

On MRI gallstones of all compositions appear hypointense relative to bile on T_2-weighted images and MRCP (Figure 18.12). They may have a central T_2 hyperintensity due to fluid-filled clefts. As previously mentioned, gallstones may have a variable signal intensity on T_1-weighted images depending on their composition. Cholesterol stones are hypointense, whereas pigment stones are hyperintense relative to bile [37]. They are usually present in a dependent location in the GB unless impacted.

18.6.4 Cholecystitis

18.6.4.1 Acute Calculous Cholecystitis

Acute calculous cholecystitis refers to acute inflammation of the GB initiated by obstruction of the GB neck or cystic duct by a biliary calculus. This causes a build-up of pressure in the GB that leads to obstruction of the venous and lymphatic drainage of the GB. This may further progress to ischaemia and perforation if not treated in time. Patients usually present with nausea, vomiting, and right upper quadrant abdominal pain. Most of the patients have had previous episodes of biliary colic. Ultrasound is usually the first imaging modality to be employed in the diagnosis of acute cholecystitis. MRI may be performed in doubtful cases [38].

MRI may demonstrate wall thickening in acute cholecystitis (Figure 18.13). The GB wall is considered to be thickened when its thickness exceeds 3 mm. The acutely inflamed wall appears hyperintense on T_2-weighted images. Pericholecystic fluid may be present. All of

these features are nonspecific and may be seen in other conditions such as chronic liver disease. The impacted gallstone may be visualized on T_2-weighted or MRCP images. Postgadolinium images demonstrate hyper enhancement of the GB wall and transient enhancement of the adjacent liver parenchyma. These are considered to be more specific for the diagnosis of acute cholecystitis. Hepatocyte-specific contrast agents can help in the diagnosis of acute calculous cholecystitis by demonstrating abrupt termination of the cystic duct and nonfilling of the GB with contrast material on delayed images [35,39].

Contrast-enhanced images are also helpful to detect complications of cholecystitis such as perforation, abscess formation, and gangrenous cholecystitis. GB perforation may be indicated by detecting discontinuity in the wall enhancement of the GB. It may be associated with rim-enhancing localized collections in the GB fossa. Gangrenous cholecystitis is diagnosed when there is diffuse or patchy non-enhancement of the GB wall. The non-enhancing portions correspond to necrotic areas [40]. Presence of gas in the GB is seen in emphysematous cholecystitis and cholecysto-enteric fistula. It is identified as a nondependent hypointensity in the GB lumen with a fluid level. Presence of gas in the GB wall and in the pericholecystic tissues is highly suspicious for emphysematous cholecystitis. Presence of air is indicated by areas of hypointensity, which arise due to signal loss caused by air-induced magnetic field inhomogenities [41]. Hemorrhagic cholecystitis is diagnosed by detecting high-intensity blood products in the GB wall and lumen on both T_1- and T_2-weighted images.

18.6.4.2 Chronic Cholecystitis

Chronic cholecystitis refers to a chronically inflamed GB and is usually associated with cholelithiasis. Patients may be asymptomatic or have recurrent attacks of biliary colic. MRI demonstrates GB wall thickening and gallstones. The GB may be contracted in some patients due to fibrosis and may demonstrate delayed enhancement. Transient enhancement of the adjacent hepatic parenchyma is not seen in chronic cholecystitis, and this feature is useful in differentiating this condition from acute cholecystitis [42].

18.6.4.3 Acalculous Cholecystitis

Acalculous cholecystitis usually occurs in critically ill patients in the ICU setting. Patients present with unexplained sepsis. Diagnosis is not usually straightforward due to presence of multiple co-morbidities and nonspecific signs and symptoms. If not recognized in time, acalculous cholecystitis may be complicated by perforation and gangrenous cholecystitis. MRI demonstrates a dilated GB with thickened and T_2 hyperintense wall in the absence of GB calculi. Pericholecystic fluid may be present [42]. These critically ill patients are usually treated by percutaneous cholecystostomy.

18.6.4.4 Adenomyomatosis

Adenomyomatosis of the GB is characterized by proliferation of the mucosal layer, hypertrophy of the muscular layer and extension of mucosal diverticula in to the muscular layer. These intramural diverticula are called Rokitansky–Aschoff sinuses. It is associated with cholelithiasis and is not considered to be a premalignant condition [43]. However, there may be co-existent adenocarcinoma of the GB.

Adenomyomatosis is not uncommonly identified on MRI performed for unrelated reasons. The GB wall demonstrates focal or diffuse thickening. The Rokitansky–Aschoff sinuses are identified as T_2 hyperintense cystic structures in the GB wall. These are hypointense on T_1-weighted images and do not demonstrate enhancement on postgadolinium images. The ring-like arrangement of these sinuses on T_2-weighted images and MRCP has been termed as *pearl necklace sign* (Figure 18.14) [43].

18.6.5 Polyps

GB polyps refer to lesions arising from the GB wall and projecting into the lumen. Differentials include benign conditions such as cholesterol polyps and benign tumors as well as malignant tumors such as GB adenocarcinoma and metastases. They are usually diagnosed incidentally at imaging or at pathology after cholecystectomy.

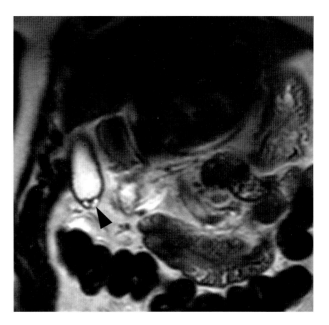

FIGURE 18.14
Adenomyomatosis: T_2-weighted coronal MRI section demonstrates focal thickening of the gallbladder wall at the fundus. Note the T_2 hyperintense intramural cysts (arrow) in this thickening. These are the intramural mucosal diverticulae of adenomyomatosis called *Rokitansky–Aschoff sinuses*.

When identified on imaging, it is important to distinguish benign polyps from malignant ones. This differentiation may not always be possible [44].

On T_2-weighted and MRCP images GB polyps appear as hypointense filling defects projecting into the lumen of the GB. Unlike calculi, they may be located in a nondependent location adjacent to the GB wall. On postgadolinium images, polyps demonstrate enhancement. Imaging features that should raise concern for malignancy are size greater than 1 cm, sessile morphology, solitary polyps, progressive growth, and coexistent gallstones [44].

18.6.6 GB Carcinoma

GB carcinoma is the most common biliary malignancy worldwide. It is more common in women and incidence increases with age. Unfortunately, most of the patients are diagnosed in the advanced stage as initial symptoms are nonspecific. Symptoms include vague abdominal pain, weight loss, and jaundice. Some carcinomas may be diagnosed at pathological examination of the GB post cholecystectomy. Risk factors for GB carcinoma include cholelithiasis, chronic biliary infections, obesity, and porcelain GB [45].

Early cancers appear as focal or asymmetrical thickening of the GB wall. Wall thickening of more than 10 mm should raise concern for malignancy. GB cancers appear hyperintense on T_2-weighted images and iso- to hypointense on T_1-weighted images relative to the adjacent hepatic parenchyma. On contrast-enhanced images GB carcinomas demonstrate early and irregular enhancement. This enhancement persists in the delayed phase. Benign lesions demonstrate early enhancement that does not persist in the delayed phase. However, imaging features show significant overlap between benign and malignant GB lesions [45].

Advanced cancers appear as a solid mass in the region of the GB. The GB itself may or may not be identifiable. The mass is usually associated with gallstones, which are engulfed by the mass. The mass may extend in to the segments 5 and 4a of the liver by direct extension. It may also extend along the cystic duct to the CBD. It appears T_1 hypointense and heterogenously T_2 hyperintense. MRI is also helpful to detect invasion of adjacent organs, vessels and the bile ducts and lymph nodal, peritoneal, and distant metastases [35].

18.6.7 Other GB Malignancies

Other GB malignancies include metastatic lesions and lymphoma. Metastases to the GB from melanoma, breast cancer, and renal cell cancers have been reported. Melanoma metastases may appear hyperintense on T_1-weighted images. Lymphomatous involvement of the GB is rare. It appears as thickening of the GB wall or solid mass in the region of the GB. However, imaging features cannot differentiate GB lymphoma from GB carcinoma.

References

1. Griffin N, Charles-Edwards G, Grant LA. Magnetic resonance cholangiopancreatography: The ABC of MRCP. *Insights Imaging*. Feb 2012;3(1):11–21.
2. Bilgin M, Shaikh F, Semelka RC, Bilgin SS, Balci NC, Erdogan A. Magnetic resonance imaging of gallbladder and biliary system. *Top Magn Reson Imaging*. Feb 2009;20(1):31–42.
3. Vitelas KM, Keogan MT, Spritzer CE, Nelson RC. MR cholangio-pancreatography of bile and pancreatic duct abnormalities with emphasis on the single-shot fast spin echo technique. *RadioGraphics*. 2000;20:939–957.
4. Sanyal R, Stevens T, Novak E, Veniero JC. Secretin-enhanced MRCP: Review of technique and application with proposal for quantification of exocrine function. *AJR Am J Roentgenol*. Jan 2012;198(1):124–132.
5. Gupta RT, Brady CM, Lotz J, Boll DT, Merkle EM. Dynamic MR imaging of the biliary system using hepatocyte-specific contrast agents. *AJR Am J Roentgenol*. Aug 2010;195(2):405–413.
6. Senturk S, Miroglu TC, Bilici A, Gumus H, Tekin RC, Ekici F et al. Diameters of the common bile duct in adults and postcholecystectomy patients: A study with 64-slice CT. *Eur J Radiol*. Jan 2012;81(1):39–42.
7. Mortelé KJ, Ros PR. Anatomic variants of the biliary tree: MR cholangiographic findings and clinical applications. *AJR Am J Roentgenol*. Aug 2001;177(2):389–394.
8. Lim JS, Kim M-J, Kim JH, Kim S IL, Choi J-S, Park M-S et al. Preoperative MRI of potential living-donor-related liver transplantation using a single dose of gadobenate dimeglumine. *AJR Am J Roentgenol*. Aug 2005;185(2):424–431.
9. Bassignani MJ, Fulcher AS, Szucs RA, Chong WK, Prasad UR, Marcos A. Use of imaging for living donor liver transplantation. *RadioGraphics*. 2001;21(1):39–52.
10. Yu J, Turner MA, Fulcher AS, Halvorsen RA. Congenital anomalies and normal variants of the pancreaticobiliary tract and the pancreas in adults: Part 1, Biliary tract. *AJR Am J Roentgenol*. Dec 2006;187(6):1536–1543.
11. Wu Y-H, Liu Z-S, Mrikhi R, Ai Z-L, Sun Q, Bangoura G et al. Anatomical variations of the cystic duct: Two case reports. *World J Gastroenterol*. Jan 7, 2008;14(1):155–157.
12. Itoh S, Fukushima H, Takada A, Suzuki K, Satake H, Ishigaki T. Assessment of anomalous pancreaticobiliary ductal junction with high-resolution multiplanar reformatted images in MDCT. *AJR Am J Roentgenol*. Sep 2006;187(3):668–675.
13. Rizzo RJ, Szucs RA, Turner MA. Congenital abnormalities of the pancreas and biliary tree in adults. *RadioGraphics*. Jan 1995;15(1):49–68.
14. Bhavsar MS, Vora HB, Giriyappa VH. Choledochal cysts: A review of literature. *Saudi J Gastroenterol*. 2005;18(4):230–236.
15. Baek S-J, Park J-Y, Kim D-Y, Kim J-H, Kim Y-M, Kim Y-T et al. Stage IIIC epithelial ovarian cancer classified solely by lymph node metastasis has a more favorable prognosis than other types of stage IIIC epithelial ovarian cancer. *J Gynecol Oncol*. Dec 2008;19(4):223–228.
16. Todani T, Watanabe Y, Narusue M, Tabuchi K, Okajima K. Congenital bile duct cysts: Classification, operative procedures, and review of thirty-seven cases including cancer arising from choledochal cyst. *Am J Surg*. Aug 1977;134(2):263–269.
17. Singham J, Yoshida EM, Scudamore CH. Choledochal cysts: Part 1 of 3: classification and pathogenesis. *Can J Surg*. Oct 2009;52(5):434–440.
18. Adamek HE, Albert J, Weitz M, Breer H, Schilling D, Riemann JF. A prospective evaluation of magnetic resonance cholangiopancreatography in patients with suspected bile duct obstruction. *Gut*. Nov 1998;43(5):680–683.
19. Guarise A, Baltieri S, Mainardi P, Faccioli N. Diagnostic accuracy of MRCP in choledocholithiasis. *Radiol Med*. Mar 2005;109(3):239–251.
20. Eason JB, Taylor AJ, Yu J. MRI in the workup of biliary tract filling defects. *J Magn Reson Imaging*. May 2013;37(5):1020–1034.
21. Irie H, Honda H, Kuroiwa T, Yoshimitsu K, Aibe H, Shinozaki K, Masuda K. Pitfalls in MR cholangiopancreatographic interpretation. *RadioGraphics*. Jan-Feb 2001; 21(1):23–37.
22. Becker CD, Hassler H, Terrier F. Preoperative diagnosis of the Mirizzi syndrome: Limitations of sonography and computed tomography. *AJR Am J Roentgenol*. Sep 1984; 143(3):591–596.

23. Enns RA, Spritzer CE, Baillie JM, Nelson RC. Radiologic manifestations of sclerosing cholangitis with emphasis on MR cholangiopancreatography. *RadioGraphics.* 2000;20(4):959–975.

24. Variability CI, Vaswani KK, Bennett WF, Tzalonikou M, Mabee C, Kirkpatrick R et al. MR cholangiopancreatography in patients with primary sclerosing. Aug 2002;179:399–407.

25. Ito K, Mitchell D, Outwater EK, Blasbalg R. Primary sclerosing MR imaging features. Jun 1999;172:1527–1533.

26. Bader TR, Braga L, Beavers KL, Semelka RC. MR imaging findings of infectious cholangitis. *Magn Reson Imaging.* Jul 2001;19(6):781–788.

27. Catalano OA, Sahani DV, Forcione DG, Czermak B, Liu C-H, Soricelli A et al. Biliary infections: Spectrum of imaging findings and management. *RadioGraphics.* Nov 2009;29(7):2059–2080.

28. Heffernan EJ, Geoghegan T, Munk PL, Ho SG, Harris AC. Recurrent pyogenic cholangitis: From imaging to intervention. *AJR Am J Roentgenol.* Jan 2009;192(1):W28–W35.

29. Bilgin M, Balci NC, Erdogan A, Momtahen AJ, Alkaade S, Rau WS. Hepatobiliary and pancreatic MRI and MRCP findings in patients with HIV infection. *AJR Am J Roentgenol.* Jul 2008;191(1):228–232.

30. Vanderveen KA, Hussain HK. Magnetic resonance imaging of cholangiocarcinoma. *Cancer Imaging.* Jan 2004;4(2):104–115.

31. Choi BI, Lee JM, Han JK. Imaging of intrahepatic and hilar cholangiocarcinoma. *Abdom Imaging.* 2004;29(5):548–557.

32. Sainani NI, Catalano OA, Holalkere NS, Zhu AX, Hahn PF, Sahani DV. Cholangiocarcinoma: Current and novel imaging techniques. *RadioGraphics* 2008;28:1263–1287.

33. Kim TU, Kim S, Lee JW, Woo SK, Lee TH, Choo KS et al. Ampulla of Vater: Comprehensive anatomy, MR imaging of pathologic conditions, and correlation with endoscopy. *Eur J Radiol.* Apr 2008;66(1):48–64.

34. Chung YE, Kim M-J, Kim HM, Park M-S, Choi J-Y, Hong H-S et al. Differentiation of benign and malignant ampullary obstructions on MR imaging. *Eur J Radiol.* Nov 2011;80(2):198–203.

35. Catalano OA, Sahani DV, Kalva SP, Cushing MS, Hahn PF, Brown JJ, Edelman RR. MR imaging of the gallbladder: A pictorial essay. *RadioGraphics* 2008;28:135–155.

36. Reshetnyak VI. Concept of the pathogenesis and treatment of cholelithiasis. *World J Hepatol.* Mar 27, 2012;4(2):18–34.

37. Tsai H, Lin X, Chen C, Lin P, Lin J. MRI of gallstones with different compositions. *AJR Am J Roentgenol* 2004;182:1513–1519.

38. Yusoff IF, Barkun JS, Barkun AN. Diagnosis and management of cholecystitis and cholangitis. *Gastroenterol Clin North Am.* Dec 2003;32(4):1145–1168.

39. Choi IY, Cha SH, Yeom SK, Lee SW, Chung HH, Je BK et al. Diagnosis of acute cholecystitis: Value of contrast agent in the gallbladder and cystic duct on Gd-EOB-DTPA enhanced MR cholangiography. *Clin Imaging.* Oct 30, 2013;38:1–5.

40. Pedrosa I, Guarise A, Goldsmith J, Procacci C, Rofsky NM. The interrupted rim sign in acute cholecystitis: A method to identify the gangrenous form with MRI. *J Magn Reson Imaging.* Oct 2003;18(3):360–363.

41. Koenig T, Tamm EP, Kawashima A. Magnetic resonance imaging findings in emphysematous cholecystitis. *Clin Radiol.* May 2004;59(5):455–458.

42. Smith EA, Dillman JR, Elsayes KM, Menias CO, Bude RO. Cross-sectional imaging of acute and chronic gallbladder inflammatory disease. *AJR Am J Roentgenol.* Jan 2009;192(1):188–196.

43. Boscak AR, Al-Hawary M, Ramsburgh SR. Best cases from the AFIP: Adenomyomatosis of the gallbladder. *RadioGraphics.* 2006;26(3):941–946.

44. Andrén-Sandberg A. Diagnosis and management of gallbladder polyps. *N Am J Med Sci.* May 2012;4(5):203–211.

45. Tan CH, Lim KS. MRI of gallbladder cancer. *Diagn Interv Radiol.* 2013;19(4):312–319.

19

Update in Magnetic Resonance Imaging of the Spleen

Luis Luna Alcalá, Antonio Luna, and Christine Menias

CONTENTS

19.1 Introduction

The spleen is probably the least studied abdominal organ in the radiological literature. Although it is an uncommon site for primary disease, the involvement of the spleen in trauma, vascular, inflammatory, hematologic, oncologic, or storage disorders is not rare. However, the frequency of splenic abnormalities is low, with the majority presenting as incidental findings without clinical significance. With this perspective, both the search and characterization of diseases involving the spleen are challenging for radiologists.

Ultrasound (US) and computed tomography (CT) have been the primary noninvasive modalities for the evaluation of the spleen rendering different results in the detection and characterization of focal and diffuse splenic lesions [1]. Recent advances in magnetic resonance imaging (MRI) have allowed the development of more accurate sequences for the evaluation of the spleen. The use of postcontrast dynamic imaging with immediate and delayed acquisitions allows for better detection of focal splenic lesions. In cases of incidental focal splenic lesions, contrast-enhanced MRI studies allow the distinction between cystic and solid lesions, and in a vast majority of cases, their classification in either benign or malignant [2].

In this chapter, we describe our MRI protocol and review the different types of diseases that involve the spleen (Table 19.1), including their usual and unusual MR features.

TABLE 19.1

MRI Characteristics of Splenic Lesions

	Presentation	T_1-WI	T_2-WI	Dynamic Postcontrast Series
Secondary hemochromatosis and sickle cell anemia	Diffuse	Hypo	Hypo	Diffuse hypovascularization on immediate postcontrast series
Abscess	Focal	Hypo to iso	Hyper	Mild peripheral enhancement
Candidiasis	Focal (multiple microabcesses)	Hypo to iso	Hyper	No enhancement on acute phase. Mild enhancement on chronic phase
Hydatid cyst	Focal	Variable	Hyper	Absence of enhancement
Cyst	Focal	Hypo	Hyper	Absence of enhancement
Pseudocyst	Focal	Hypo to hyper depending on proteinaceous or hemorrhagic content	Hyper/mixed	Absence of enhancement
Hemangioma	Focal	Hypo to iso	Typical: hyper Sclerosed: hypo Large: hyper with central scar	Progressive enhancement with variable patterns (1) peripheral and progressive to uniform in delayed phase (2) immediate homogeneous and persistent (3) centripetal filling
Littoral cell angioma	Focal/diffuse	Hypo	Variable	Mild heterogeneous enhancement on arterial phase and homogeneous on delayed images
Hemangioendothelioma	Focal/diffuse	Hypo	Hypo	Hypovascular Heterogeneous enhancement of solid portions
Sclrerosing angiomatoid nodular transformation (SANT)	Focal	Hypo	Hypo areas of susceptibility artifact (siderosis)	Peripheral with *spoke wheel* pattern on immediate phase and persistent on delayed phase
Angiosarcoma	Focal/diffuse	Variable	Variable	Intense and heterogeneous
Hemangiopericytoma	Focal/diffuse	Hypo	Hyper	Intense enhancement of solid areas
Infarcts	Focal/diffuse	Hypo (hyper in hemorrhagic infarcts)	Hyper heterogeneous (hypo in chronic infarcts)	Absence of enhancement
Hematoma	Focal	Acute: hyper Subacute: hyper Chronic: hypo	Acute: hypo Subacute: hyper Chronic: hypo	None or mild peripheral enhancement
Arteriovenous malformation	Focal	Multiple flow signal void	Multiple flow signal void	Early serpentine enhancement

(Continued)

TABLE 19.1 (*Continued*)

MRI Characteristics of Splenic Lesions

	Presentation	T_1-WI	T_2-WI	Dynamic Postcontrast Series
Hemangiomatosis	Diffuse	Variable	Variable	Intense and heterogeneous
Lymphangioma	Focal/diffuse	Hypo; hyper if proteinaceous or hemorrhagic content	Hyper	Absence of enhancement
Peliosis	Focal/diffuse	Variable	Hyper	Peripheral and septal enhancement
Hamartoma	Focal	Iso	Hyper heterogeneous	Heterogeneous enhancement on arterial phase and homogenous in delayed images
Lipoma	Focal	Hyper (loss of signal in fat-sat sequences)	Hyper (loss of signal in fat-sat sequences)	Absence of enhancement
Lymphoma	Focal/diffuse	Focal: iso	Focal: iso to hypo	Diffuse form: irregular enhanced areas of mixed low and high signal Focal: hypovacular on immediate phase, isointense within first minute with progressive enhancement on delayed phase
Leukemia (chloromas)	Diffuse/focal	Iso to hypo	Hyper	Hypovascular on immediate images with progressive enhancement on delayed phase
Metastases	Focal	Iso (hyper if acute hemorrhage or melanin content)	Iso to hyper	Hypovascular on immediate phase and isointense on delayed phase
Splenic pleomorphic undifferentiated sarcoma	Focal	Variable	Variable	Progressive and heterogeneous enhancement
Splenic cystoadenocarcinoma	Focal	Hypo	Hyper	Peripheral and septal and solid portions enhancement
Gaucher disease	Focal	Iso	White nodules: hypo; red lesions: hyper	Not published
Nieman-Pick disease	Focal		Variable	Delayed enhancement
Amyloidosis	Diffuse	Hypo	Hypo	Hypoperfusion
Sarcoidosis	Diffuse	Hypo	Hypo	Minimal and delayed enhancement
Gamna–Gandy bodies	Focal	Signal void (blooming artifact on in-phase GRE images)	Signal void	Absence of enhancement
Inflammatory pseudotumor	Focal	Iso to hypo	Hypo	Heterogeneous enhancement on delayed phase
Extramedullary hematopoiesis	Diffuse/focal	Active lesions: intermediate Chronic: hypo	Active lesions: hyper Chronic: hypo	Active lesions: minimal Chronic: absence of enhancement

19.2 Anatomy

The spleen is an encapsulated abdominal organ of vascular and lymphoid tissue, located between the gastric fundus and the diaphragm, posteriorly in the left upper quadrant. The normal size of the adult spleen is 12 × 7 × 4 cm, being the maximum craniocaudal diameter 10–12 cm [3]. This organ has a diaphragmatic and visceral surface and is crescent shaped, with a convex lateral border in contact with the abdominal wall and the left

hemidiaphragm and a concave medial border in contact with the stomach and left kidney. The splenic hilum is anteromedial, where vessels and nerves enter and exit.

The spleen is held in position by the splenorenal and the gastrosplenic ligaments [4]. The splenorenal ligament is derived from the peritoneum; it extends between the spleen and left kidney and connects the splenic hilum to the left kidney, containing the tail of the pancreas and the splenic artery and vein. The gastrosplenic ligament is formed by the union of the peritoneum of the lesser and greater sacs and connects

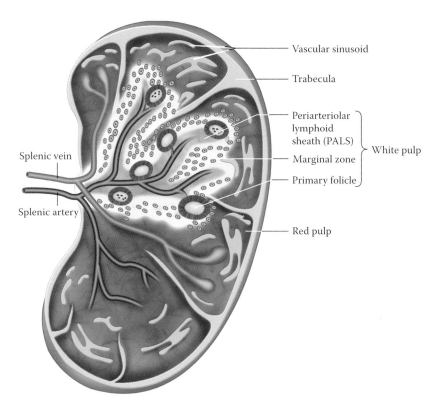

FIGURE 19.1
Schematic representation of the spleen structure.

the splenic hilum to the stomach, through which the short gastric and left gastroepiploic branches of the splenic artery. Finally, the phrenicocolic ligament connects the lower pole of the spleen to the splenic flexure of colon and the diaphragm.

The spleen is a blood-filtering organ and can be functionally divided into three compartments: red pulp, marginal zone, and white pulp (Figure 19.1). The red pulp, microscopically, is formed by numerous thin-walled fenestrated vascular sinusoids, allowing an easy passage of cells, separated by the splenic cords that consist of a labyrinth of macrophages, reticular cells, and reticular fibers. This framework provides a physical and functional filter to the systemic circulatory blood. In the splenic cords, there are lymphocytes and hematopoietic cells, and in the space between the splenic cords, there are different types of blood cells [5]. The white pulp consists of lymphoid follicles, mainly B cells, containing a central arteriole surrounded by the periarteriolar lymphoid sheath (PALS), predominantly T cells. The white pulp, which includes the marginal zone, is associated with the arterial tree and the red pulp, with the venous system draining the spleen.

In neonates, the spleen is predominantly composed of red pulp, and with age and progressive antigenic stimulation, the white pulp increases up to 20% of the splenic parenchyma in the adult [6].

19.3 MRI Technique

In many occasions, splenic pathology is found incidentally on MRI. Most of the time, a general abdominal MRI protocol is sufficient to evaluate the spleen but when possible, we prefer to use a specific protocol (Table 19.2), which includes (a) coronal T_2-weighted half-Fourier single-shot turbo spin-echo (HASTE) acquisition, (b) axial breath-hold gradient-echo (GE) T_1-weighted chemical shift in-phase and out-of-phase sequence, (c) axial breath-hold turbo spin-echo (TSE) T_2-weighted sequence (with or without fat saturation), and (d) an axial 3D fat-suppressed GRE dynamic series with precontrast and immediate (10–12 s after contrast administration), 1 and 5 min delayed postcontrast acquisitions. More commonly, we perform an enhanced T_1-weighted high-resolution isotropic volume excitation (e-THRIVE) as the dynamic postcontrast sequence.

A different approach to evaluate the splenic parenchyma is to use superparamagnetic iron oxide (SPIO) particles. This intravenous contrast is taken up by the reticuloendothelial system (RES) in the spleen and the liver. On T_2-weighted sequences, the normal parenchyma appears hypointense, revealing focal lesions with a higher signal intensity, which renders them more conspicuous to the background spleen [7]. The use of ultrasmall SPIO

TABLE 19.2
MRI Protocol

	HASTE	T₂ TSE	Dual/FFE	Diffusion	DCE-MRI
(a) In 1.5 T					
Orientation	Coronal	Transverse	Transverse	Transverse	Transverse
Voxel size (mm)	1.5/1.5/7	1.2/1.2/5	1.5/1.9/6	3/3/7	2/2.2/2.2
SENSE	1.5	1.5	1.5	2	1.7
TE (ms)	80	120	2.3/4.6	70	2.2
TR (ms)	540	2500	136	1636	4.4
Respiratory compensation	Breath-hold	Breath-hold	Breath-hold	No	Breath-hold
b values (s/mm²)				4 (b0; b250; b500; b1000)	
Total scan duration (seconds)	0:14.9	2:20	0:29.9	3:13	1:12.5
(b) In 3 T					
Voxel size (mm)	1/1.5/7	1/1.5/7	2/1/5	3/3/7	1.6/1.6/1.6
SENSE	2	1.3	2	2	2
TE (ms)	103	80	1.15/2.3	54	1.4
TR (ms)	924	778	150	1150	3
Respiratory compensation	Trigger	Breath-hold	Breath-hold	Trigger	Breath-hold
b values (s/mm²)				4 (b0; b250; b500; b1000)	
Total scan duration (seconds)	1:21	47	41	2:50	1:30

DCE-MRI, dynamic contrast enhanced MRI; FFE, fast field echo; HASTE, half Fourier acquisition single-shot turbo-spin echo; SENSE, sensitivity encoding; TE, echo time; TR, time to repetition; TSE, turbo spin echo.

(USPIO) particles also allows an increase in the detection of splenic lesions with the advantages of further characterization of hemangiomas [8] and widening of the time to perform imaging due to longer blood pool half-life [9]. The results of dynamic gadolinium-enhanced series and the use of SPIO or USPIO contrast are similar [8]. Although this approach has been promising, it has never been extensively used in the clinical arena, probably because of its complicated management.

19.4 Normal MRI Appearance

The normal splenic parenchyma appears low in signal intensity on T_1-weighted images and high in signal intensity on T_2-weighted images compared to liver parenchyma (Figure 19.2).

Immediate postcontrast sequences allow a better differentiation between normal and abnormal splenic areas [10], because the difference in the blood supply between them reaches its maximum. The immediate postcontrast acquisition in normal spleen usually shows an arciform pattern with areas of high and low signal intensity. On slightly delayed acquisitions (1 min or later), the splenic parenchyma becomes homogeneous. Other patterns of enhancement have been reported on immediate postcontrast acquisitions (Figure 19.3) [11].

In cases of inflammatory or neoplastic diseases or hepatic steatosis, a homogeneous enhancement of the spleen is commonly demonstrated as uniform high signal. A decrease of the signal intensity in the entire spleen during the immediate postcontrast phase has been related to transfusional iron overload. The recognition of these patterns of splenic enhancement on immediate postcontrast images is important to improve the detection of splenic abnormalities.

On diffusion-weighted images (DWI), the normal spleen shows restricted diffusion with high *b* values, because of the hypercellularity of this organ. The use of diffusion-weighted imaging for the spleen has not been validated in the literature. In our own experience, the normal spleen shows restriction of diffusion. Preliminary data suggests that apparent diffusion coefficient (ADC) decreases with age and increases in the presence of cirrhosis and portal hypertension [12,13]. At this point, DWI has still to show if it has any role in the assessment of the spleen.

19.5 Normal Variants

The spleen may show a nodular contour or may present a fractured appearance due to the presence of notches or clefts [1]. These remnants of fetal lobulations can be misinterpreted as splenic nodules on unenhanced CT or MRI.

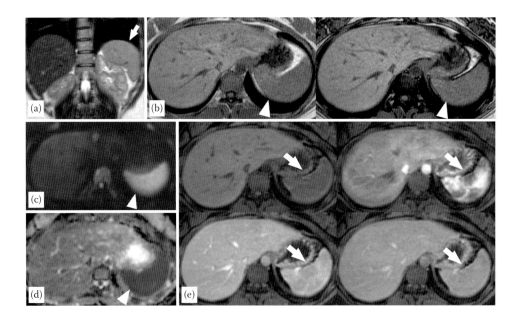

FIGURE 19.2
Appearance of normal spleen on MRI. Normal spleen (arrows and arrowheads) is slightly hyperintense on coronal T_2-weighted image (a) and hypointense on axial in-phase T_1-weighted image (b), with no signal loss on the out-of-phase sequence. On DWI with high *b* values, the spleen is hyperintense (c) and hypointense on the corresponding ADC map (d), demonstrating the physiological restriction in normal conditions of this organ. On dynamic postcontrast series, the spleen most commonly shows a serpiginous enhancement pattern on the immediate postcontrast image, with homogeneous delayed enhancement (e).

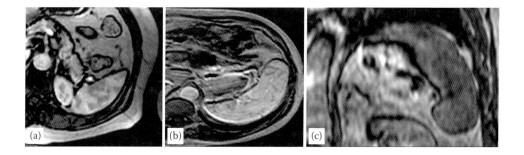

FIGURE 19.3

Variants in the splenic enhancement on immediate postcontrast phase. (a) Normal spleen usually shows an arciform pattern of early enhancement after contrast administration. (b) However, uniform early enhancement is more commonly found associated with liver diseases, as in this case shown of hepatic steatosis. (c) Diffuse low signal of the spleen during immediate postcontrast phase is secondary to transfusional iron overload involving the spleen.

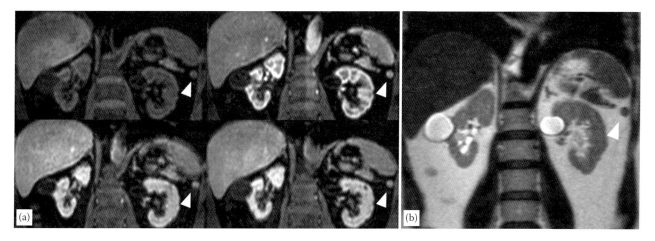

FIGURE 19.4

Splenule. A small nodular pseudolesion (arrowheads) adjacent to the inferior pole of the spleen is identified, showing identical signal intensity in all sequences to the adjacent spleen, being hyperintense on axial T_2-weighted image (b) and presenting the same enhancement pattern than the spleen on coronal THRIVE dynamic postcontrast series (a), consistent with a splenule.

Multiplanar capabilities of MRI and the parallel signal intensity of these pseudolesions in all sequences allow their differentiation.

Spleniculi or accessory spleens are a common embryologic variant, seen in up to 40% of the population (Figure 19.4) [14]. They are small spherical lesions with the same signal intensity and postcontrast pattern as the spleen. Most commonly, they are less than 4 cm in diameter and may be solitary or multiple [15]. Although they can be located anywhere in the abdomen, they are usually near either the splenic hilum or pancreatic tail [16].

19.6 Polysplenia Syndrome

Polysplenia syndrome is a congenital syndrome associated with complete situs inversus, multiple small splenic masses, and thoracoabdominal abnormalities (Figure 19.5).

In this syndrome, the splenic nodules usually appear along the greater curvature of the stomach [16]. A similar presentation with multiple splenules may be found in cases of previous splenectomy and traumatic splenic rupture with resultant splenosis (Figure 19.6). The MRI appearance of all these nodules is similar to that of spleniculi.

19.7 Splenomegaly

Enlargement of the spleen is most commonly found in cases of portal hypertension, although it can be also found associated with several disorders such as malignancy, infection, vascular, and metabolic diseases.

Portal hypertension is usually a complication of cirrhosis, as progressive hepatic fibrosis leads to increased vascular resistance at the level of the hepatic sinusoids [17].

This increased pressure gradient causes the development of collateral vessels and splenomegaly. MRI allows a comprehensive assessment of the liver and spleen in this setting (Figure 19.7). Therefore, MRI is also useful for other findings of portal hypertension, such as enlargement of the splenic vein, portosystemic shunts, and the formation of siderotic splenic nodules (Gamna–Gandy bodies). Although for the diagnosis of splenomegaly, US is the most common imaging technique performed, MRI has the advantage to provide exact volume and length measurements, which may be of interest in cases where strict follow-up is necessary.

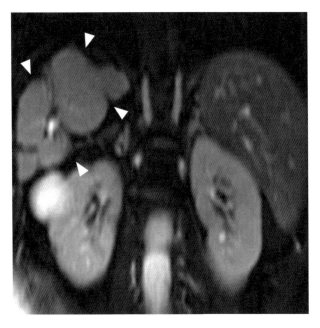

FIGURE 19.5
Polysplenia syndrome. Coronal fat-suppressed TSE T_2-weighted image shows mirror-image location of upper abdominal viscera and vessels. Right hypochondrium is occupied by multiple splenic nodules of different sizes (arrowheads). Findings correspond to polysplenia syndrome associated with situs ambiguous in a three-month-old girl.

19.8 Hemochromatosis (Hemosiderosis and Sickle Cell Disease)

Primary hemochromatosis is an autosomal recessive disorder resulting in iron deposition in multiple organs but spares the spleen. In secondary hemochromatosis (transfusional hemosiderosis), iron is deposited into the RES (Kupffer cells of the spleen, the bone marrow, and the liver), usually secondary to blood transfusions or rhabdomyolysis [18]. On MRI, a significant decrease of signal intensity, on both T_1-weighted images and T_2-weighted images, is observed, because of iron accumulation (Figure 19.8). MRI can accurately differentiate primary and secondary hemochromatosis due to the pattern of involved organs in each case [18].

Sickle cell disease is a common condition in the African-American population, resulting in low signal intensity of the spleen on all pulse sequences, because of iron deposition from blood transfusion. In the homozygous form, autosplenectomy can result in the spleen, presenting as a signal void [19]. Scarring and infarcts are also a common finding [6].

19.9 Trauma

The spleen is the most commonly injured abdominal organ in cases of blunt and penetrating trauma. Another possible mechanism is iatrogenic injury, which can occur intraoperatively [20]. The splenic parenchyma can be affected by contusion, laceration, subcapsular or intraparenchymal hematoma, or vascular injury [6]. In the acute setting, US and CT are performed to evaluate the spleen status, although most of the posttraumatic splenic lesions can be studied with MRI. MRI may also play a useful role in detecting small subcapsular fluid

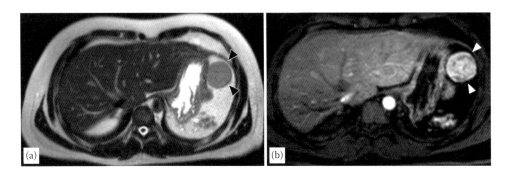

FIGURE 19.6
Splenosis. A nodular pseudolesion adjacent to the anterior superior pole of the spleen (arrowheads) is identified in a postsplenectomy patient, showing identical behavior than the one expected for the spleen in all sequences. This nodule is hyperintense on axial T_2-weighted images (a) and shows characteristic serpiginous enhancement pattern of the spleen on the immediate postgadolinium image (b).

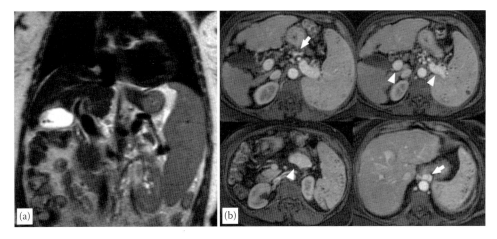

FIGURE 19.7
Portal hypertension. Severe splenomegaly is demonstrated on coronal T_2-weighted image (a), secondary to cirrhosis. Notice, cirrhotic changes of liver and huge dilatation of portal and splenic veins (arrowheads), secondary to portal hypertension, as shown on dynamic postgadolinium images at different levels (b). Periesophageal and gastrohepatic collateral veins are also seen (arrows). Multiple siderotic nodules scattered within the splenic parenchyma are also identified.

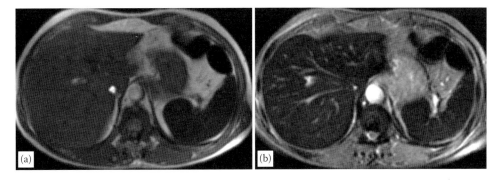

FIGURE 19.8
Secondary hemochromatosis involving the spleen. Diffuse hypointensity of the splenic and liver parenchyma compared to paraspinal muscles on both GE T_1-weighted (a) and TSE T_2-weighted (b) sequences, confirming involvement by secondary hemochromatosis.

collections or in detecting small foci of subacute hemorrhage, or even more, in the follow-up of young patients where radiation may be a concern or in patients with contraindications for iodinated-contrast material.

19.10 Infectious Lesions (Infection)

Viral infection may result in splenomegaly, but neither abscesses nor focal splenic involvement is a common feature as in infections of other etiologies. *Candida albicans*, *Aspergillus fumigatus*, and *Cryptococcus neoformans* compose the majority of fungal infections in immunocompromised patients [21]. MRI can depict hepatosplenic microabscesses, differentiating acute, subacute, and chronic lesions (Figure 19.9). Acute lesions are more apparent in the spleen, and subacute and chronic lesions

are predominant in the liver. Acute lesions present as very small hyperintense lesions on T_2-weighted images [22]. Fat-suppressed TSE T_2-weighted TSE sequences are useful in order to render these tiny lesions more conspicuous. DWI is also able to detect them, appearing as multiple bright foci with high *b* values and with low signal intensity on the ADC map. Enhancement of microabcesses is usually poor in all phases, due to the immunocompromised state of these patients in the acute phase or due to fibrotic changes in the chronic stage [23]. Histoplasmosis can involve the spleen, most commonly seen in immunocompromised patients. Acute and subacute lesions appear as small nodules with low signal intensity in all sequences (Figure 19.10). As in tuberculosis, in the chronic phase of histoplasmosis, calcified healed granulomas are common. They show typical blooming artifacts, better depicted on GRE sequences, reflecting the calcified nature of these lesions, which can be easily confirmed by CT [15].

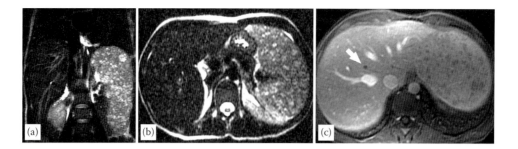

FIGURE 19.9

Acute candidiasis. Coronal and axial fat-suppressed TSE T_2-weighted images show multiple tiny hyperintense lesions on both the spleen and the liver, representing acute *C. albicans* microabscesses in an immunosuppressed patient. Delayed postcontrast GE T_1-weighted image demonstrates homogeneous enhancement of the liver, depicting only one of the lesions (arrow in c), and the absence of enhancement of the splenic lesions. Fat-suppressed TSE T_2-weighted images make lesions appear more conspicuous.

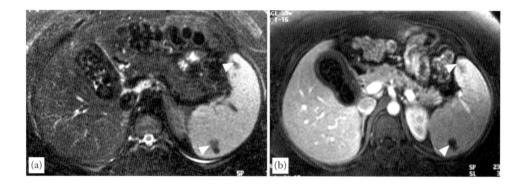

FIGURE 19.10

Chronic histoplasmosis. Axial fat-suppressed TSE T_2-weighted (a) and delayed postcontrast GE T_1-weighted images (b) show two hypointense nodules (arrowheads) with blooming artifact consistent with calcified granulomas in a patient with prior exposure to histoplasmosis.

Splenic abscesses are uncommon, but the increase of immunocompromised patients has raised the incidence. They are associated with high mortality rates due to delayed detection and treatment [24]. They reflect systemic infection, usually associated with endocarditis, although it may also result from direct extension of infection from adjacent organs or develop in patients with bacteremia and intrinsic splenic disease [1]. The primary etiology is aerobic organisms, followed by fungi, predominantly *Candida albicans*, and anaerobic organisms. Abscesses, which may be single or multiple, appear on MRI as slightly hypo- to isointense on T_1-weighted images and heterogeneous and mildly to moderately hyperintense on T_2-weighted acquisitions compared to splenic parenchyma. All focal infections tend to be slightly hypervascular in their periphery on immediate postcontrast images, reflecting the development of a capsule, and may show alteration of the perilesional perfusion in delayed acquisitions, [2]. On DWI, acute abscesses show restriction as in other organs, being hyperintense with high *b* values and markedly hypointense on ADC maps (Figure 19.11). This appearance may be more heterogeneous if there is gas or cystic components [25].

Splenic involvement in echinococcal infection varies between 0.9% and 8% [26]. Although the description of MRI of splenic hydatid disease is limited, it is considered to be an adequate imaging technique for its evaluation, with a role in the monitoring to treatment response [27]. On MRI, *Echinococcus granulosus* disease appears as single or multiple cysts with or without wall calcifications. Depending on the cystic content, the signal on T_1-weighted images can vary (Figure 19.12) [22].

19.11 Cysts

Cysts are the most common benign focal lesion of the spleen. They can be divided into epithelial or true cysts, lined by epithelium seen in 25% of cystic lesions [28], and false cysts or pseudocysts, not lined by epithelium and usually posttraumatic or sequelae of pancreatitis in origin [21]. The MRI features of cysts include sharp margins, with low signal intensity on T_1-weighted images and very high signal intensity on T_2-weighted images (Figure 19.13). Complicated pseudocysts with

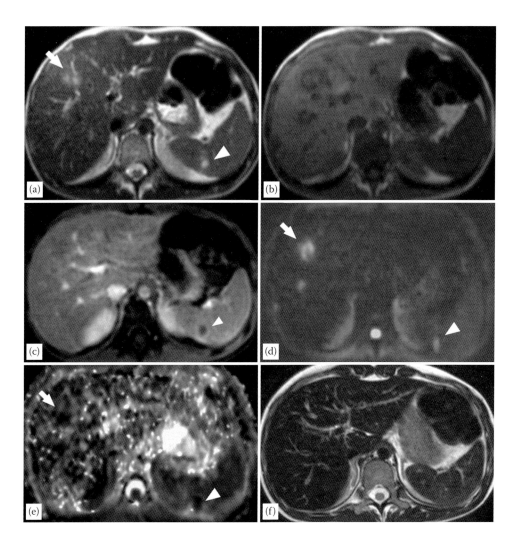

FIGURE 19.11
Aspergillus fumigatus hepatosplenic abscesses in a 14-year-old girl, under chemotherapy for osteosarcoma of the humerus. (a) Multiple cystic-appearing lesions are depicted on axial T_2-weighted image in the liver (white arrow) and in the spleen (arrowhead). (b) These lesions are barely seen on the precontrast GE T_1-weighted sequence. (c) The splenic lesion shows peripheral enhancement after gadolinium administration. (d, e) On DWI with high b value, liver and splenic lesions demonstrate high signal intensity, representing true restriction of diffusion, as both are hypointense on the corresponding ADC map. (f) After 2 months of antifungal treatment, the hepatic and splenic lesions have completely resolved, as shown in axial TSE T_2-weighted image.

proteinaceous or hemorrhagic content may present with areas of hyperintensity on T_1-weighted images, mixed signal intensity on T_2-weighted acquisitions, or both (Figure 19.14) [22,28]. In addition, true cysts may present with septations in their walls with occasional peripheral calcifications. Cysts do not show any enhancement on postcontrast series. These lesions do not show restricted diffusion, being hypointense with high b values and hyperintense on ADC maps. False cysts usually have a more heterogeneous appearance than true cysts [2]. MRI can be considered a valid alternative in the characterization of splenic cysts when US and CT findings are inconclusive, and in the differential diagnosis between primary splenic cysts, pseudocysts, and parasitic cysts [15,29].

19.12 Vascular Lesions

19.12.1 Focal Vascular Lesions

19.12.1.1 Benign Tumors

19.12.1.1.1 Hemangiomas

Hemangiomas are the second most common focal lesion and the most common benign tumor of the spleen [30]. They show a slight male predominance and commonly occur in persons aged between 30 and 50. Their origin is conceived to be congenital, arising from sinusoidal epithelium [24]. They may be single (Figure 19.15) or multiple (Figure 19.16) and may occur as part of a generalized angiomatosis (Klippel–Trenaunay–Weber and

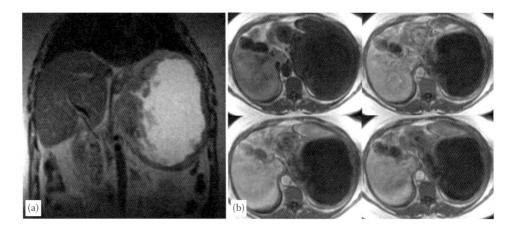

FIGURE 19.12
Splenic hydatid disease. Large complex cystic mass occupying most of the splenic parenchyma, which is hyperintense on (a) coronal HASTE, with an irregular internal margin, containing membranes and peripheral calcifications. (b) On the dynamic postcontrast series, the lesion does not show any internal enhancement.

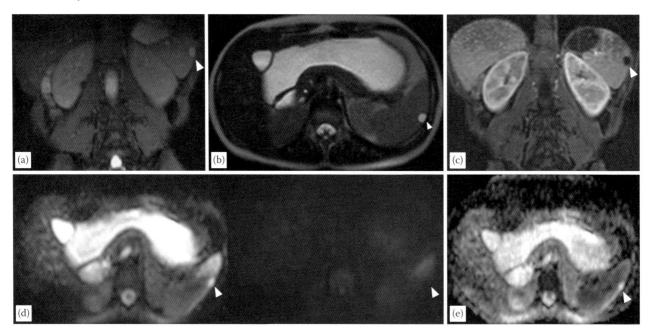

FIGURE 19.13
Epithelial splenic cyst. A well-defined hyperintense lesion on (a) coronal fat-suppressed SSFP image and (b) HASTE images is seen in the posterior aspect of the mid portion of the spleen. (c) This cyst shows absence of enhancement after contrast administration. (d) As expected in cystic lesions, it has no restriction on DWI, appearing hyperintense on b0 s/mm^2 (right image), hypointense on b1000 s/mm^2 (left image) and (e) hyperintense on the ADC map.

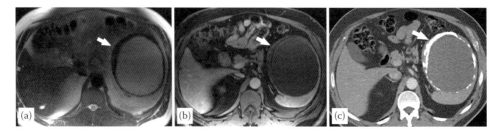

FIGURE 19.14
Posttraumatic pseudocyst. (a, b) Axial HASTE and delayed postcontrast fat-suppressed GE T_1-weighted images demonstrate a large subcapsular cystic lesion. Notice the presence of a thick hypointense rind around the lesion on HASTE (arrows), corresponding to peripheral calcification as shown on (c) enhanced CT.

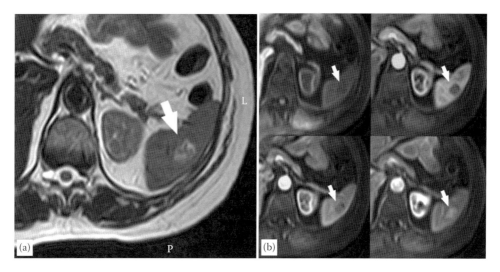

FIGURE 19.15
Classic small hemangioma. On axial T_2-weighted image (a) is seen as a hyperintense nodular lesion (arrow), which shows progressive centripetal enhancement on dynamic postcontrast series (b, arrows), with complete fill-in on the delayed postcontrast phase, typical features of a hemangioma.

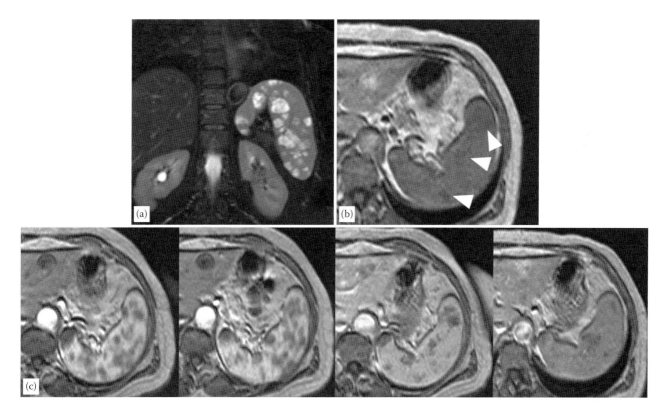

FIGURE 19.16
Multiple hemangiomas. An eight-year-old male with splenomegaly and several focal splenic lesions on routine ultrasonography. MRI allows a better characterization of focal splenic lesions, showing multiple hyperintense nodules scattered in the splenic parenchyma on (a) coronal fat-suppressed T_2-weighted image. (b) GE in-phase T_1-weighted image show most of these lesions isointense to the adjacent parenchyma, although some of them are slightly hypointense (arrowheads). (c) On immediate, 1, 5, and 10 min delayed postcontrast images all these lesions shows a centripetal and progressive enhancement, consistent with multiple hemangiomas.

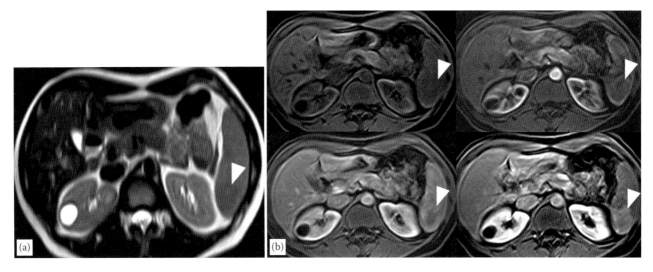

FIGURE 19.17
Sclerosing hemangioma. (a) Nodular hypointense lesion on axial T_2-weighted image (arrowhead) in the posterior aspect of the spleen. (b) On dynamic postcontrast series, this lesion shows a progressive centripetal enhancement (arrowheads) typical for hemangioma.

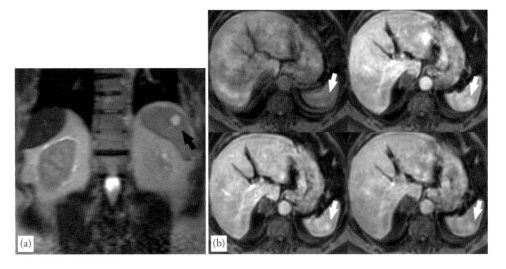

FIGURE 19.18
Hemangioma with homogeneous enhancement on immediate postcontrast image. A hyperintense nodular lesion on coronal T_2-weighted image (black arrow, a) is shown, which demonstrates homogeneous enhancement on immediate, 1 min and delayed postcontrast phases (white arrows, b).

hemangiomatous syndromes). They are divided into cavernous and capillary types depending on the size of their vascular channels [31].

Most hemangiomas are less than 2 cm and are usually asymptomatic. Calcifications may or may not be present. Large hemangiomas may be associated with anemia, thrombocytopenia, and coagulopathy (Kasabach–Merritt syndrome). The most frequent complication is rupture. Other associated complications include infarction, thrombosis, fibrosis, hypersplenism, and even malignant degeneration [32]. Their appearance on MRI varies depending on their size.

They are well-defined homogeneous intrasplenic lesions, hypo- to isointense on T_1-weighted images and mildly hyperintense on T_2-weighted images to the background spleen, although they can also be isointense or hypointense on T_2-weighted images [30]. The hypointensity on T_2-weighted images is normally related to sclerosed hemangiomas (Figure 19.17) [24]. After intravenous contrast administration, most commonly, they enhance peripherally with delayed progression to uniform enhancement, although occasionally, they show immediate homogeneous enhancement with persistent delayed enhancement (Figure 19.18). The typical early nodular peripheral enhancement with centripetal progression of hepatic hemangiomas, although can be seen, is an uncommon imaging feature [32]. Therefore, a small splenic lesion, hyperintense on T_2-weighted

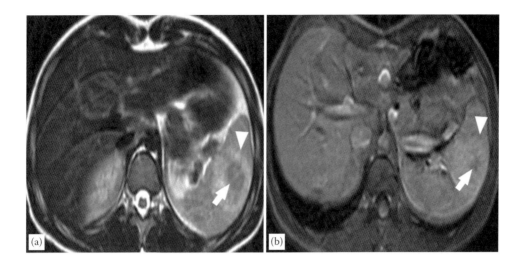

FIGURE 19.19
Large hemangioma with a central scar. A nodular hyperintense heterogeneous lesion (arrowhead) with a hypointense central area (arrow) is shown on (a) axial TSE T_2-weighted image, which demonstrates (b) peripheral homogeneous enhancement (arrowhead) and lack of enhancement of the central scar (arrow) on the delayed postcontrast image.

images, showing either a homogeneous or a peripheral enhancement on immediate images is highly suggestive of being a hemangioma.

Larger hemangiomas are more heterogeneous lesions and may show areas of hemorrhage and thrombosis. They enhance in a centripetal fashion either with persistent non-enhancing areas, representing a central scar, or with homogeneous enhancement on delayed images (Figure 19.19). Larger hemangiomas need follow-up or histological analysis to assure their origin, because similar appearances have been described in angiosarcomas [33].

In our experience, splenic hemangiomas show similar characteristics to liver hemangioma because of their long T_2. It is usually hypointense on high b values, although it may occasionally show high signal intensity with high b values due to T_2 shine-through effect, but invariably, they show high signal intensity on the ADC maps. Thus, there is no true restriction of diffusion.

19.12.1.1.2 Littoral Cell Angioma

Littoral cell angioma (LCA) is a rare vascular lesion, usually benign, although may have malignant components. It arises from littoral cells, which line the splenic sinus of red pulp, and it can be considered a relatively new clinicopathological entity. The lesion has also been associated with other neoplastic processes, including colorectal, renal, pancreatic adenocarcinoma, and meningioma [32]. Patients with LCA commonly present with signs of hypersplenism (anemia and thrombocytopenia).

LCA usually presents with multiple small nodules and splenomegaly. They can vary on imaging, depending on their histological features, of anatomizing vascular channels with irregular lumina featuring cyst-like spaces, lined by tall endothelial cells [24]. Most commonly, coalescent nodules filled with hemosiderin, because of

the cellular hematopahgocytic capacity of this tumor, are demonstrated [34]. On MRI, these nodules appear hypointense on T_1-weighted images and of variable signal intensity on T_2-weighted acquisitions, although they may be hypointense in relation to siderosis presence (Figure 19.20). On dynamic postcontrast images, LCA can have mild heterogeneous enhancement on the arterial phase and homogeneous enhancement on delayed phase, representing contrast pooling [35,36]. Splenectomy is often necessary for definitive diagnosis and treatment. Malignant variants show an infiltrative or solid growth pattern on pathology reflective of angiosarcoma.

19.12.1.1.3 Hemangioendothelioma

Hemangioendothelioma is a very rare primary vascular tumor composed of a proliferation of small vascular channels lined by endothelial cells. It is possibly an intermediate entity between hemangioma and angiosarcoma. Frequently, splenic hemangioendothelioma occurs in young adults and occasionally in children, presenting as a painful palpable mass [32].

On MRI examination, hemangioendothelioma appears as a heterogeneous solid mass that may exhibit hypointensity on T_1- and T_2-weighted images, suggesting the presence of hemosiderin [32]. It can show an infiltrative pattern, areas of necrosis or hemorrhage, and rarely capsular retraction, commonly seen in the liver [37].

19.12.1.1.4 Sclerosing Angiomatoid Nodular Transformation

Sclerosing angiomatoid nodular transformation (SANT) is a benign vascular splenic lesion consisting of multiple angiomatoid nodules surrounded by dense fibrous tissue that may be more prominent in the center forming a scar. The angiomatoid nodules can present different

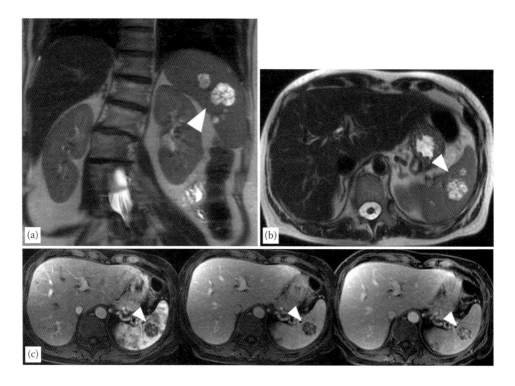

FIGURE 19.20
Littoral cell angioma. (a, b) Coronal and axial TSE T_2-weighted images demonstrate multiple hyperintense nodules with internal septations in a normal-sized spleen (arrowheads). (c) On immediate, 1 min and delayed postcontrast dynamic fat-suppressed GE T_1-weighted images, these nodules show different degree of internal enhancement demonstrating the vascular nature of this tumor (arrowheads).

types of blood vessels normally found in the red pulp. It is a recent entity described by Martel et al. in 2004 [38]. Most of them are an incidental finding during imaging, associated with malignancies elsewhere. SANT appears in middle-aged adults, with a slight female predominance [39]. Rarely, SANT can cause symptoms such as abdominal pain, anemia, or splenomegaly. Splenectomy has been curative in all reported cases. Recently, three small series have better defined the imaging characteristics of SANT [39–41], before limited to several case reports. SANT is a single nodular lesion without necrosis or cystic changes, with old foci of hemorrhage, which appear as areas of low signal intensity on T_2-weighted images and responsible for the loss of signal in chemical shift imaging [2,5,39,42]. Most commonly, it has a *spoke wheel* appearance in dynamic postcontrast series, defined as peripheral-enhancing radiating lines or rim enhancement during the arterial or portal phases, with progressive or persistent enhancement on delayed phases. The enhancement of the radiating lines corresponds to branches of fibrous scar, the peripheral enhancement to angiomatous nodules in the periphery of the lesion and the delayed enhancement represents the fibrous scar and angiomatous content of this lesion. This lesion is predominantly hypointense on T_2-weighted images. Calcifications may be occasionally present. Therefore, SANT has characteristic MRI findings that help in differentiating them from other benign vascular splenic lesions.

19.12.1.2 Malignant Neoplasms

19.12.1.2.1 Angiosarcoma

Primary angiosarcoma of the spleen is a very rare vascular neoplasm, but is the most common primary non-lymphoid malignant tumor of the spleen [22]. Their association with other malignancies is not uncommon and no previous exposure to thorotrast is required as in liver angiosarcomas [43]. Most commonly, it occurs in elderly persons, without any gender predilection. Angiosarcoma has been reported in patients treated with chemotherapy for lymphoma or radiation therapy for breast cancer. Splenomegaly is almost always associated with general symptoms related to hypersplenism. This tumor has an increased risk of spontaneous rupture in up to 30% of patients. Metastasis to the liver, lungs, bone, bone marrow, and lymphatic system is common [32].

On MRI, angiosarcoma typically appears as multiple nodular masses, with diverse signal intensity on T_1- and T_2-weighted images, consistent with the presence of blood products and necrosis (Figure 19.21). Hypointense signal may also be found secondary to the presence of siderotic nodules [44]. After contrast administration, this lesion shows intense and heterogeneous enhancement due to its vascular nature [22]. Angiosarcoma shows a heterogeneous appearance due to its inherent blood products in different stages, which may be indistinguishable to

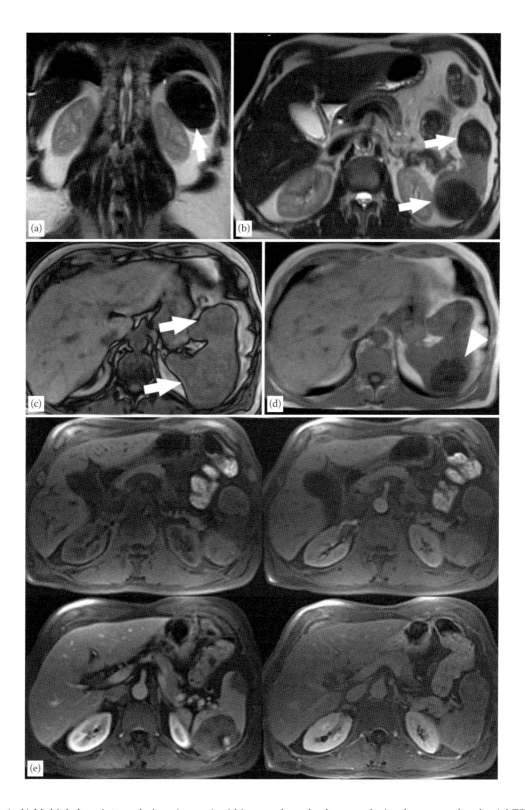

FIGURE 19.21

Angiosarcoma. (a, b) Multiple hypointense lesions (arrows) within an enlarged spleen are depicted on coronal and axial TSE T_2-weighted images. (c, d) On axial chemical-shift sequence, these lesions appear iso- to slightly hypointense to spleen (arrows). One of them presents important drop of signal on in-phase sequence (arrowhead), consistent with hemosiderin. On dynamic postgadolinium series (e), these masses present a heterogeneous and progressive pattern of enhancement.

that of benign lesions such as hemangioendothelioma, hemangioma, or littoral cell angioma. Except in cases where angiosarcoma demonstrates aggressive features as ill-defined borders or extrasplenic involvement, ultimately, splenectomy is necessary to establish a definitive histological diagnosis. MRI is the best imaging method for the assessment of all these entities, especially in the case of angiosarcoma [15].

19.12.1.2.2 Hemangiopericytoma

Hemangiopericytoma is a rare vascular benign tumor with relatively high malignant potential [45]. Furthermore, histologically, it is difficult to define the degree of biological aggressiveness of this tumor [46]. It rarely originates in the spleen as a primary tumor, and when present in the spleen, hemangiopericytoma is a highly vascular neoplasm that arises from capillary pericytes. It is typically asymptomatic and may be associated with splenomegaly. Just a few cases of primary hemangioperycytomas of the spleen have been reported in the literature [32,45,47–49]. Hemangiopericytoma appears with multiple and confluent nodules replacing the spleen with intense enhancement of their solid components and lack of enhancement of their cystic portions. Intratumoral hemorrhage has also been

described. On MRI, they demonstrate low signal intensity on T_1-weighted images and high signal intensity on T_2-weighted images, with intense enhancement of the solid areas [32,49]. Imaging surveillance is necessary after splenectomy due to high incidence of local recurrence [45].

19.12.1.3 Non-Neoplastic Lesions

19.12.1.3.1 Infarcts

Splenic infarcts are common and most frequently arise from emboli from a cardiac source, or result from the obstruction of the splenic artery or one of its branches. They may also arise from a hypercoagulable state or from an underlying hematologic disorder [1]. Other causes of splenic infarcts are thrombosis, vasculitis, and splenic torsion. Infarcts may present with acute left upper quadrant pain. Fever and laboratory findings showing an acute inflammatory event suggests an embolic origin. Infarcts are typically focal, but can be diffuse.

On MRI examination, infarcts appear usually as a peripheral wedge-shaped defect, but they can also be round or linear. Typically, they are most clearly depicted on late phase dynamic postcontrast series, as low signal intensity areas without enhancement (Figure 19.22).

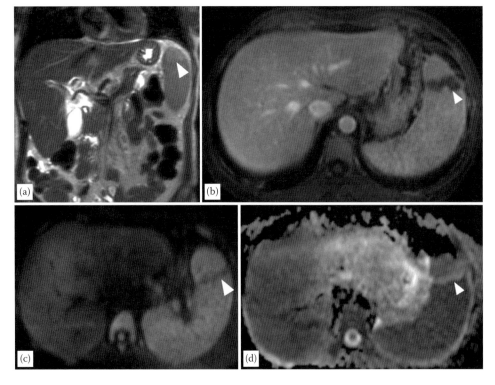

FIGURE 19.22
Old infarct. Peripheral lesion with band-like shape in the superior pole of the spleen (arrowheads) which is slightly hyperintense on coronal HASTE (a) and shows absence of enhancement on fat-suppressed GE T_1-weighted delayed postcontrast image (b). Note the absence of diffusion restriction of this lesion as shown on DWI with high b value (c) and corresponding ADC map (d).

The splenic capsule appears as a thin peripheral enhancing linear structure, called the *rim sign* [50]. The signal of infarcts can vary depending on their composition and age, but usually are hypointense on T_1-weighted images and heterogeneously hyperintense on T_2-weighted sequences. Hemorrhagic infarcts show high signal intensity on T_1- and T_2-weighted images. Caution is necessary, as the classical imaging appearance of acute splenic infarcts is absent in 50% of patients, presenting as multinodular, poorly marginated, and heterogeneous [22,51]. In these cases, they may be difficult to differentiate from other lesions such as tumors, hematomas, or abscesses. Global infarcts may show a complete lack of enhancement of the spleen, with only the *cortical rim sign*, consisting of a thin layer of peripheral residual capsular contrast enhancement, [52]. Chronic infarcts are of low signal intensity on all pulse sequences. In this stage, infarcts can disappear, or more frequently, they are peripheral avascular areas associated with capsular retraction and sometimes have calcifications [52].

19.12.1.3.2 Hematoma

On MRI, subcapsular or intraparenchymal hematomas show different signal intensity depending on the time of the injury due to the paramagnetic properties of degradation products of hemoglobin. In the acute phase, hematomas show high signal intensity on T_1-weighted images and low signal intensity on T_2-weighted images [50]. Subacute hemorrhage is particularly conspicuous because of its high signal intensity on both T_1- and T_2-weighted images. Chronic stages (of greater than 3 weeks) usually evolve with liquefaction and become cystic, sometimes with a hemosiderin ring (Figure 19.23) [21]. Devascularization is well shown on immediate postcontrast images, being very hypointense compared to the hyperintense normal vascularized splenic tissue. Old-healed hematomas are commonly hypointense on T_2-weighted images.

19.12.2 Diffuse

19.12.2.1 Arteriovenous Malformation

Arteriovenous malformations and fistulas are rare in the spleen. Arteriovenous fistulas may be congenital or acquired, with an increased risk in pregnant and multiparous women [52]. They usually present with a *machinery-type* bruit during auscultation, and the predominant clinical manifestation is portal hypertension. MRI can suggest the diagnosis, as multiple flow-voids are present on all unenhanced sequences, with an early serpentine enhancement being characteristic on dynamic postcontrast acquisitions (Figure 19.24) [15]. In addition, a tortuous and dilated splenic vein with early enhancement is very suggestive [53].

19.12.2.2 Hemangiomatosis

Hemangiomatosis is a very rare vascular tumor-like entity [54]. Histopathologically, hemangiomatosis should be distinguished from other tumor-like vascular lesions of the spleen that exhibit endothelial features typical of splenic sinuses, such as LCA or peliosis (Figure 19.25) [55]. Most commonly, this entity is a manifestation of systemic angiomatosis, associated with Klippel–Trenaunay–Weber, Turner, Kasabach–Merritt-like, Proteus and Beckwith–Wiedemann syndromes [15]. Its MRI appearance overlaps with that of angiosarcoma [2].

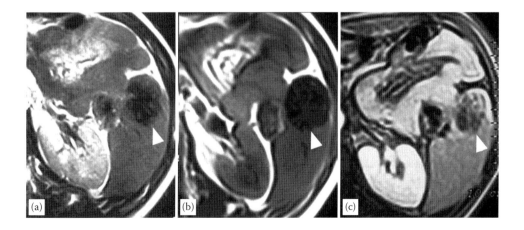

FIGURE 19.23
Splenic hematoma. (a) Axial TSE T_2-weighted and (b) GE T_1-weighted sequences demonstrate a round hypointense mass in the anterior aspect of the spleen, representing an old healed hematoma (arrowheads). (c) Axial delayed postcontrast GE T_1-weighted images demonstrates peripheral and heterogeneous internal enhancement of this mass (arrowhead).

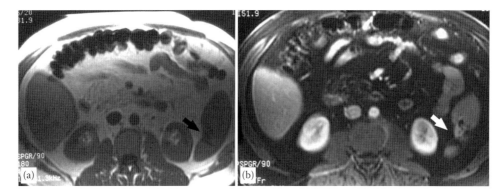

FIGURE 19.24
Arteriovenous malformation. Small area of signal void in the inferior aspect of the spleen with serpentine enhancement is demonstrated on (a) axial GE T_1-weighted and (b) postcontrast fat-suppressed VIBE image.

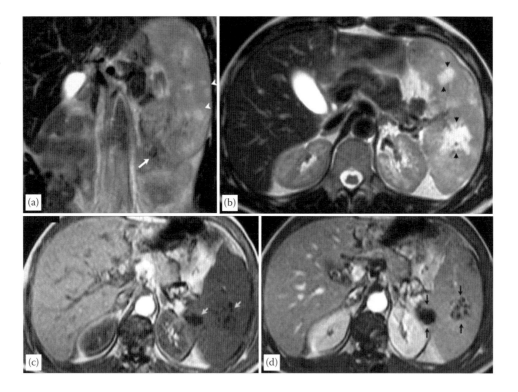

FIGURE 19.25
A 41-year-old woman with diffuse splenic hemangiomatosis. (a, b) Coronal and axial TSE T_2-weighted image shows multiple hyperintense focal splenic lesions (arrowheads) and hypointense nodules representing siderotic foci (arrow) in an enlarged spleen. (c) Axial unenhanced gradient-echo T_1-weighted image shows hypointense nodules with areas of magnetic susceptibility artifact (arrows) representing siderotic nodules. (d) Postcontrast (1 min) gradient-echo T_1-weighted image reveals subtle peripheral enhancement of both nodules (arrows). The superior nodule also shows heterogeneous internal enhancement. Heterogeneous splenic appearance is also possible in cases of angiosarcoma or littoral cell angioma. (From Luna, A. et al., *AJR Am. J. Roentgenol.*, 186, 1533, 2006. With permission.)

19.12.2.3 Lymphangioma

Lymphangiomas are composed of collections of small and cystic dilated lymphatic channels. Although splenic lymphangiomas are rare, they are the third most common benign tumor of the spleen. Lymphangiomas may be solitary or more often multiple, involving multiple organs (lymphangiomatosis). Most commonly, they are asymptomatic, although signs derived from mass effect, bleeding or hypersplenism, portal hypertension, and rupture have been reported. Symptomatic tumors may require partial or complete splenectomy.

On MRI, they appear as multiloculated cystic lesions separated by fibrous septa, with high signal intensity on T_2-weighted images and without enhancement on postcontrast imaging (Figure 19.26) [50]. These cysts

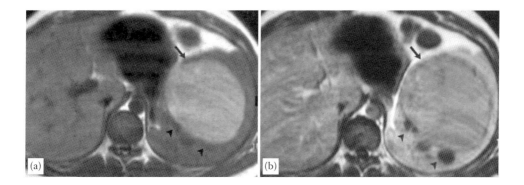

FIGURE 19.26
Lymphangioma. A 27-year-old woman with complex cystic mass detected on routine ultrasound study and confirmed later on CT. Axial unenhanced (a) and postcontrast (b) GE T_1-weighted images show a multilocular mass with hypointense (arrowheads) and hyperintense (arrows) unenhanced areas, confirming the cystic nature of the lesion. Hyperintense areas are secondary to proteinaceous content. Diagnosis was confirmed after splenectomy. (From Luna, A. et al., *AJR Am. J. Roentgenol.*, 186, 1533, 2006. With permission.)

show classical features on T_1-weighted images, although some of them may show high signal intensity, because of the proteinaceous nature of the fluid or, less commonly, because of their hemorrhagic content [21]. Although curvilinear calcifications are an unusual feature of lymphangioma, they are difficult to detect on MRI [45]. Malignant degeneration of lymphangioma is rare, but MRI is able to show intracystic solid components [32].

19.12.2.4 Peliosis

Peliosis is a rare disorder that consists in multiple blood-filled spaces involving the spleen and the liver. Most cases are associated with use of anabolic steroids, but can also be associated with anaplastic anemia, tuberculosis, acquired immunodeficiency syndrome (AIDS), and cancer. Peliosis is often discovered incidentally, unless rupture of peliotic lesions on the surface of the spleen occurs, leading to intraperitoneal hemorrhage. Rupture of these lesions may be spontaneous or secondary to minimal trauma. Oval to round markedly dilated blood-filled cystic spaces predominantly in the

parafollicular area are the characteristic pathologic features of peliosis, and it allows its differentiation from congested sinuses in splenic involvement [56].

On MRI examination, multiple foci may be seen with high signal intensity on T_2-weighted images and variable signal intensity on T_1-weighted sequences (Figure 19.27). Thrombosed cavities show mixed signal intensity on T_2-weighted images, due to the presence of deoxyhemoglobin and methemoglobin. Fluid–fluid levels and absence of calcifications are also findings in this lesion [32]. Recently, restriction on DWI of peliosis hepatis has been described [57]. The radiological differential diagnosis of peliosis can include hemangiomatosis, lymphangioma, and angiosarcoma.

19.12.3 Other Vascular Diseases

19.12.3.1 Splenic Artery Aneurysms

Splenic artery aneurysm is the most common type of visceral artery aneurysm. Most splenic artery aneurysms are small (less than 2 cm), saccular, and located in

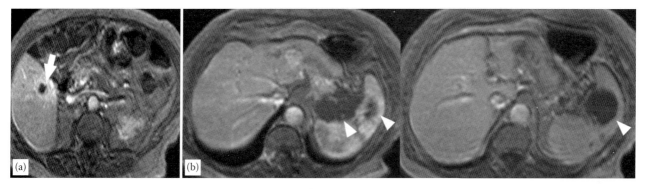

FIGURE 19.27
Hepatosplenic peliosis. A 72-year-old woman with disseminated tuberculosis and hepatosplenic peliosis. (a, b) postcontrast T_1 GE images on the arterial phase at three different levels show a subcapsular multicystic mass extending to the splenic hilum with septal and peripheral enhancement (arrowheads) and several cystic liver lesions also with peripheral enhancement (arrow).

the middle-to-distal splenic artery. In 20% of cases, they can be multiple [58]. Splenic artery aneurysms have a strong association with female gender, pregnancy (due to the hormonal effects of estrogen and progesterone on the arterial wall), and portal hypertension. Other etiologies include atherosclerosis, congenital, and mycotic causes and fibromuscular dysplasia. Splenic artery pseudoaneurysms are also a complication of pancreatitis, due to digestion of the arterial wall by pancreatic enzymes, or after a traumatic event [15]. The most serious complication is rupture, which is associated with a high mortality rate.

Splenic artery aneurysms are usually an incidental finding. MRI allows an effective diagnosis and characterization, using 3D-GRE sequences after gadolinium administration or even better depicted, with 3D MR angiographic sequences (Figure 19.28) [59]. Both of them will show early enhancement of the patent lumen, which will show low signal intensity on unenhanced sequences. Conversely, if a blood clot is present, it will show high signal intensity [52].

19.12.3.2 Splenic Vein Thrombosis

Splenic vein thrombosis has multiple causes, most commonly pancreatitis. It can appear in at least 20% of patients with chronic pancreatitis [15]. On MRI examinations, it appears as a filling defect in the splenic vein on angiographic sequences. Acute thrombosis will show high signal intensity on both T_1- and T_2-weighted

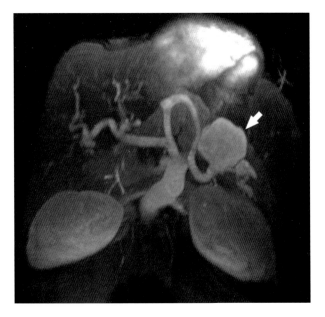

FIGURE 19.28
Splenic artery pseudoaneurysm. A large focal eccentric dilatation of the splenic artery is well depicted on axial-oblique MIP of 3D MR angiography on the arterial phase (arrowheads).

images, and low signal intensity on true fast imaging in steady-stage precession [52]. In cases of chronic splenic vein thrombosis, collateral venous circulation develops, appearing more commonly gastric varices. MR angiographic techniques with and without contrast allow an accurate map of the distribution of these portosystemic collaterals [60].

Splenic vein thrombosis can be bland or malignant, and they can be differentiated according to their MRI features. Bland thrombus is low in signal intensity on T_2-weighted images and has no enhancement after contrast administration. Tumor thrombus, usually seen in patients with pancreatic adenocarcinoma, shows high signal intensity on T_2-weighted images and enhances in postcontrast sequences (Figure 19.29).

19.13 Nonvascular Tumors

19.13.1 Benign Tumors

19.13.1.1 Hamartomas

Hamartoma is a rare non-neoplastic tumor composed of a mixture of different splenic tissues [30]. It can occur at any age, equally in men and women. Usually, it is an incidental finding that usually appears as a well-defined, rounded, single solid lesion [61]. This lesion can be also cystic and may show punctate calcifications. They are most likely to occur in the mid portion of the spleen. They can be classified as white pulp or red pulp hamartomas, but the majority is a mixture of the two cell types. It is rarely symptomatic, causing splenomegaly, pancytopenia, fatigue, and anorexia. Splenic hamartoma is commonly associated with tuberous sclerosis, and it may also be associated with Wiskott–Aldrich-like syndrome.

On MRI, typically, splenic hamartoma shows heterogeneous hyperintensity on T_2-weighted images and is isointense on T_1-weighted acquisitions [30]. Hamartoma can show hypointense areas on T_2-weighted images, representing fibrous tissue, as an occasional finding. This appearance on T_2-weighted images and its typical faint heterogeneous enhancement on immediate postcontrast acquisitions are the key features in the differentiation between hamartomas and hemangiomas. On delayed postcontrast images, hamartomas enhance in a relatively uniform and intense fashion [62] (Figure 19.30) with central hypovascular areas consistent with scarring or cystic components [30]. MRI is a more definitive imaging technique to suggest the diagnosis of hamartoma, although they may show overlapping features with hemangiomas and metastasis, being a challenging diagnosis for radiologists.

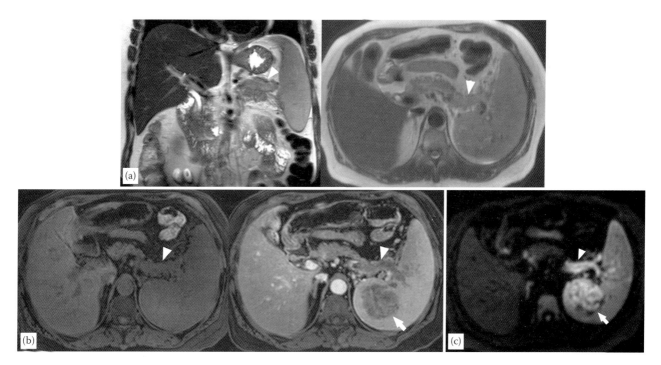

FIGURE 19.29

Malignant splenic vein thrombosis. A 57-year-old female with history of ovarian carcinoma. (a) On coronal and axial HASTE, the splenic vein appears thickened with hyperintense content. (b) Axial pre- and postcontrast venous phase fat-suppressed GE T_1-weighted images show enhancement of the thrombus after contrast administration (b), indicating its tumoral nature. In addition, a splenic hematogenous metastasis is also noted (white arrow). (c) On high b value DWI, both the thrombus and metastasis show high signal intensity, consistent with diffusion restriction and indicating their malignant origin.

19.13.1.2 Fatty Tumors: Lipoma and Angiomyolipoma

Primary fatty tumors of the spleen are extremely rare, although pathological descriptions of lipoma [15] and angiomyolipoma [63] have been described, without imaging descriptions reported. Lipoma is a rare benign tumor of the spleen, consisting of a soft tissue mass composed entirely of fat [28]. As in other body regions, MRI allows its accurate characterization. Lipoma shows parallel signal intensity to fat on all pulse sequences, with high signal on T_1- and T_2-weighted images and marked loss of signal intensity on fat saturated sequences. Typical uncomplicated lipoma has no enhancement after contrast administration [64].

19.13.2 Malignant Nonvascular Tumors

19.13.2.1 Lymphoma

Splenic lymphoma is the most common splenic malignancy [1]. Both Hodgkin and non-Hodgkin lymphoma may be present in the spleen either as a primary tumor or more commonly as part of a generalized (secondary) disease [65]. Primary splenic lymphoma is rare, representing less than 1% of lymphomas [66]; most primary lymphomas are of non-Hodgkin type (Figure 19.31). Secondary splenic involvement is present in up to 40% of patients with lymphoma [22]. These patients present with nonspecific symptoms, most commonly demonstrating left upper quadrant pain or constitutional symptoms. Associated splenomegaly and retroperitoneal adenopathy are frequent. Lymphomas may invade through the capsule into the surrounding organs.

There are four patterns of splenic involvement in primary and secondary lymphoma: (1) homogeneous splenomegaly without a focal lesion, (2) diffuse infiltration with miliary lesions of less than 5 mm, (3) multiple focal nodular lesions, and (4) a large solitary mass [21,65,67,68]. Diffuse, uniform infiltration is the most common form of secondary splenic lymphoma, and it may be present on imaging as a normal appearing spleen or as an isolated splenomegaly. Splenomegaly by itself is not specific, because the spleen can be normal in size despite tumor infiltration or may be enlarged without neoplastic involvement. Conversely, primary lymphoma presents itself as a solitary bulky mass [22]. The pattern of involvement has also been related to cell type; large and solitary masses are most typical of large cell lymphoma, and multiple masses or miliary infiltration is most commonly associated with other non-Hodgkin lymphomas [67,69].

On MRI, diffuse, infiltrative splenic involvement is difficult to detect with unenhanced images, appearing on immediate postcontrast images as irregular enhancing areas of mixed high and low signal intensity [22].

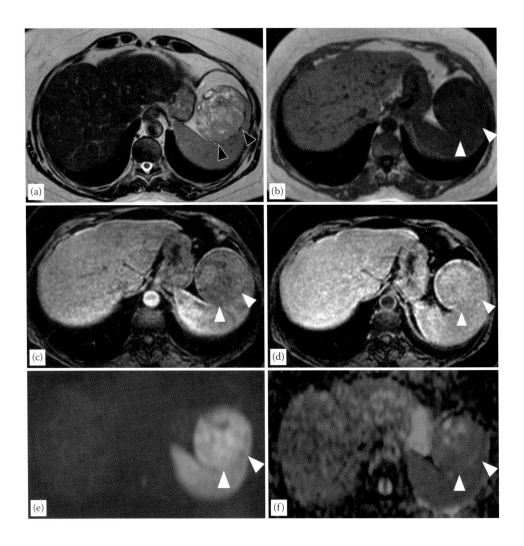

FIGURE 19.30
Splenic hamartoma (arrowheads). (a) Axial TSE T_2-weighted image shows a heterogeneous hyperintense nodular lesion in the anterior aspect of the spleen. (b) This lesion is slightly hypointense compared to spleen on GE T_1-weighted sequence. (c) This mass shows heterogeneous enhancement on immediate postgadolinium fat-suppressed GE T_1-weighted image and a relative uniform and intense enhancement (d) on delayed postcontrast fat-suppressed GE T_1-weighted image. (e) On DWI with high b value, this lesion appears hyperintense and hypointense on (f) the corresponding ADC map, indicating its hypercellularity and diffusion restriction.

Focal involvement is also difficult to detect, as most of the nodules are isointense to splenic parenchyma on T_1- and T_2-weighted images, although may be hypointense on T_2-weighted images, allowing its differentiation from metastasis, which are rarely hypointense on T_2-weighted images [2]. On immediate postcontrast images, the nodules are hypovascular, becoming isointense to the spleen within the first minute after contrast, with progressive and variable delayed enhancement (Figure 19.32) [50]. Lymphomatous nodules can show small irregularities suggestive of necrosis, fibrosis, edema, and hemorrhage [67]. Immediate postcontrast MRI images surpass CT in the evaluation of lymphoma; nevertheless, the role of MRI has not been established yet [24]. The use of SPIO makes splenic lymphoma more conspicuous, appearing as hyperintense nodules [7,70] compared to the low signal splenic parenchyma

on T_2-weighted images. On DWI, these lesions are hyperintense with high b values and hypointense on the corresponding ADC map, representing high diffusion restriction as in other body regions. Lymphomas show important 18 Fluoro-deoxy-glucose (FDG) uptake in PET, which makes any FDG-avid splenic area suspicious for lymphoma [71].

After chemotherapy, the imaging characteristic of the lymphomatous nodules may change, showing decreased signal intensity on T_1- and T_2-weighted images in relation with an induced fibrotic reaction and indicating a favorable response.

19.13.2.2 Leukemia

Acute myelogenous leukemia subtypes are the most common leukemic conditions affecting the spleen [64]. Splenic involvement of leukemia usually presents without

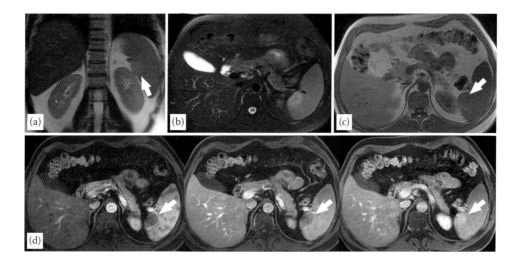

FIGURE 19.31
Primary mass-like splenic lymphoma. A 44-year-old patient in study for progressive splenomegaly and focal splenic lesion. (a) On coronal TSE T_2-weighted images a slightly hyperintense nodular lesion (arrow) is detected, which is more conspicuous (b) on axial fat-suppressed FSE T_2-weighted image. (c) On GE T_1-weighted image, this lesion is isointense to splenic parenchyma. (d) On immediate, 1 and 5 min delayed post-contrast THRIVE images, the lesion shows progressive enhancement, which is homogeneous on the delayed phase. Splenectomy confirmed primary splenic lymphoma.

any imaging abnormalities or with homogeneous spleno-megaly with prolonged T_2-values secondary to leukemic infiltration, and is associated with abdominal adenopathy. In cases of massive splenomegaly, spontaneous rupture may result. In leukemia patients, there is an increase incidence of splenic infarcts and abscesses [22].

Chloromas, most commonly associated with chronic lymphocytic leukemia, are rare and appear on MRI as multiple ill-defined masses without enhancement on immediate postcontrast images (Figure 19.33) [2].

19.13.2.3 Metastases

Splenic metastases of hematogenous origin are rare and usually occur in the late stages of disease. Most common primary tumors that metastasize to the spleen are breast, lung, ovary, stomach, melanoma, and prostate [21,22].

On MRI, splenic metastases appear isointense to splenic parenchyma on T_1-weighted and iso- or hyperintense on T_2-weighted sequences [72]. Metastases may be hyperintense on T_1-weighted sequences due to acute hemorrhage or melanin content in cases of melanoma metastases [21]. Other specific features according to the primary tumor are necrosis in metastases of colon adenocarcinoma or calcification if the primary tumor is cystadenocarcinoma [22]. The degree and characteristics of enhancement after contrast administration depend on the nature and type of the underlying primary tumor. Usually, splenic metastases are hypovascular on immediate postcontrast images becoming isointense on delayed images [10,73]. The lack of hyperintensity on T_1- or T_2-weighted sequences of lymphoma can help in

the differentiation between these two entities. The use of SPIO particles improves their detection [7]. On DWI splenic metastases shows diffusion restriction, appearing with high signal intensity with high b values and low signal intensity on the ADC map (Figure 19.34).

In cases of ovarian carcinoma, gastrointestinal adenocarcinoma and pancreatic cancer, peritoneal implants can result in splenic capsular scalloping with cystic (pseudomyxoma peritonei) or solid implants. Direct tumor invasion of the spleen is rare, as the spleen is usually displaced more than infiltrated by adjacent malignancies. Infiltration is most commonly found with pancreatic cancer, although it is also possible in cases of malignant mesothelioma, retroperitoneal sarcomas, and gastric, colonic, renal, and adrenal malignancies [22]. Lymphoma has also propensity to involve the spleen in continuity with other organs.

19.13.2.4 Other Primary Sarcomas of the Spleen

Splenic pleomorphic undifferentiated sarcoma, previously named malignant fibrous histiocytoma, is exceedingly rare, with only 15 cases reported in the literature [74]. This very aggressive tumor presents with splenomegaly and without specific imaging characteristics (Figure 19.35) [75]. It can present as a cystic, solid, or a complex mass (76). Surgical resection is the treatment of choice and necessary to confirm the diagnosis. Even more rare, primary fibrosarcoma and leiomyosarcoma of the spleen have been reported with overlapping clinical and imaging features with pleomorphic undifferentiated sarcoma [22].

Kaposi sarcoma of the liver is associated with AIDS [77]. This spindle cell neoplasm can also involve the spleen. Nodular splenomegaly has been described in

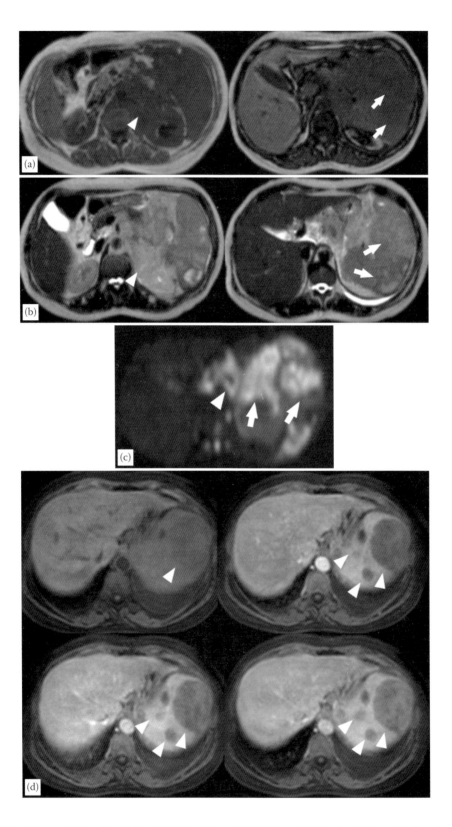

FIGURE 19.32
Systemic non-Hodgkin lymphoma with multinodular splenic involvement. A 57-year-old female with constitutional symptoms. (a) Axial GE T_1-weighted and (b) TSE T_2-weighted images at two different levels show a large aggressive retroperitoneal mass causing infiltration of left kidney body/tail of pancreas and stomach. In addition, diffuse and multinodular involvement of the spleen is demonstrated. Splenic focal lesions appear isointense to spleen on T_1-weighted and hyperintense on T_2-weighted images. (c) On DWI with high b value, the retroperitoneal mass and splenic focal lesions (arrowheads and arrows, respectively) demonstrate important diffusion restriction. (d) On dynamic postcontrast series, the splenic lesions are hypovascular in all phases (arrowheads), although with progressive faint internal enhancement.

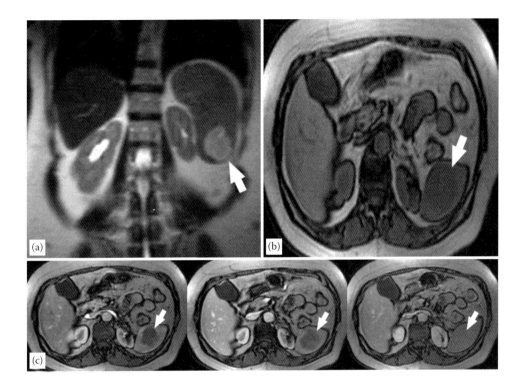

FIGURE 19.33
Chloroma. A 5-year-old boy with acute myeloid leukemia. (a) On coronal TSE T_2-weighted image, a hyperintense nodule in the lower pole of the spleen is shown (arrow), being slightly hypointense (arrow) on (b) the out-of-phase GE T_1-weighted image. (c) On the dynamic postcontrast images (arrows), it appears as a hypovascular nodule on immediate postcontrast image (left image), which becomes isointense to splenic parenchyma on delayed postcontrast phase (right image).

this neoplasm. On CT, the splenic nodules are iso- or hypoattenuating on delayed imaging, and mimic hemangiomas [77].

19.13.2.5 Splenic Cystoadenocarcinoma

Primary splenic mucinous cystoadenocarcinoma is an extremely rare tumor. The main complaint of patients with this condition is upper abdominal pain and the palpation of a left upper quadrant mass. On cross-sectional imaging, primary splenic cystoadenocarcinoma appear as cystic lesions containing large uni- or multilocular cysts, similarly to its counterpart in the pancreas [64]. It can be associated with pseudomyxoma peritonei, with multiple lesions of cystic appearance inside abdominal cavity [78]. Diagnosis is only confirmed after surgical resection of the lesion or splenectomy.

19.14 Miscellaneous

19.14.1 Storage Diseases: Gaucher and Niemann–Pick Diseases

Gaucher disease is an autosomal recessive lysosomal metabolic disorder, caused by a deficiency of the enzyme glucocerebrosidase, promoting abnormal accumulation of a glycolipid (glucocerebroside) in the RES. Patients have hepatosplenomegaly, thrombocytopenia, and anemia. In the series by Hill et al., including 46 patients, 30% of patients presented with splenic nodules [79]. These nodules were of red and white types, formed in the first case, by Gaucher cells and dilated sinusoids filled with blood, and predominantly by Gaucher cells alone, on the second type. On MRI, the white splenic nodules, which are more frequent, appear isointense on T_1-weighted images and hypointense on T_2-weighted sequences compared to splenic parenchyma, and the red lesions show high signal intensity on T_2-weighted sequences. This disorder is also associated with the presence of splenic infarcts and fibrosis, which may give a multifocal pattern to the spleen [80].

Similar to Gaucher disease, Niemann–Pick diseases causes deposit of lipids in cells of the RES, causing hepatosplenomegaly and splenic nodules, which can exhibit variable signal intensity on T_2-weighted images with delayed enhancement on postcontrast series (Figure 19.36) [2,81].

19.14.2 Amyloidosis

Primary amyloidosis may also involve the spleen. A diffuse form showing generalized decrease of signal intensity

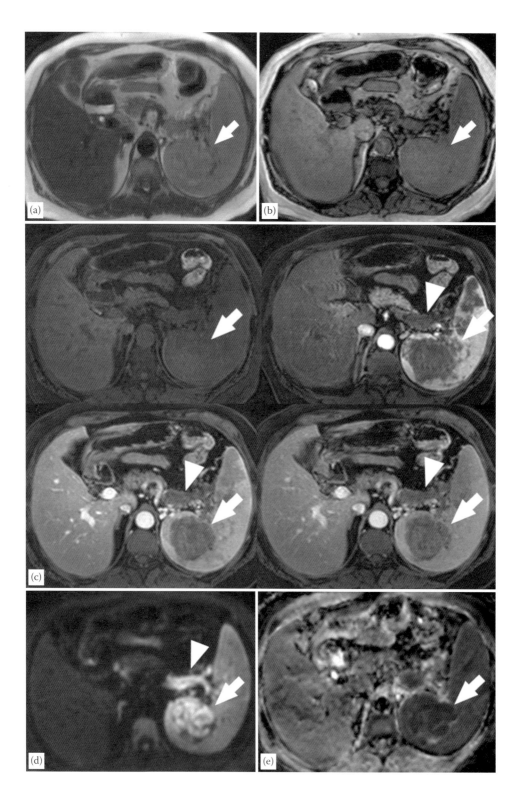

FIGURE 19.34

Metastases. A 57-year-old woman with ovarian carcinoma. (a, b) On axial TSE T_2-weighted image, a slightly hyperintense mass (arrows) in the posterior region of the spleen is identified, which is isointense to spleen (b) on GE T_1-weighted image. (c) On dynamic postcontrast series this mass is hypovascular with progressive and heterogeneous delayed enhancement. (d, e) DWI and ADC map prove restriction diffusion appearing hyperintnese with high *b* values and hypointense on the ADC map. Notice malignant splenic vein thrombosis (arrowheads) with some degree of enhancement and diffusion restriction.

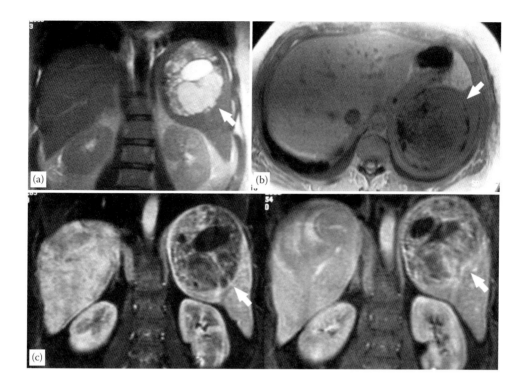

FIGURE 19.35

Pleomorphic sarcoma. (a) Coronal HASTE and (b) axial GE T_1-weighted images show a complex and heterogeneous mass located in the upper pole of the spleen. Coronal fat-suppressed dynamic VIBE postcontrast images on the arterial and delayed phases (c) demonstrate heterogeneous enhancement on immediate phase with progressive enhancement on delayed phase. There are persistent areas with lack of enhancement representing cystic change and necrosis.

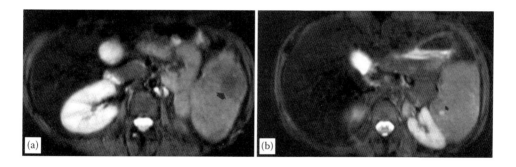

FIGURE 19.36

Niemann–Pick disease. A 22-year-old woman with splenomegaly in patient with Niemann–Pick type B disease. (a) Axial fat-suppressed TSE T_2-weighted image shows a well-defined hypointense nodule (arrow). (b) Also, hyperintense nodules on T_2-weighted images are noticed (arrow). Diagnosis was established according to imaging and clinical criteria, and the results of hepatic and bone marrow biopsies and levels of glucocerebrosidase and sphingomyelinase.

on T_2-weighted images and splenic hypoperfusion has been described [82,83]. These imaging features are similar to those of iron deposition in the RES secondary-to-multiple blood transfusions [84].

19.14.3 Sarcoidosis

Sarcoidosis is a granulomatous systemic disease that can involve numerous organs, including the spleen. Microscopic evidence of splenic sarcoidosis is demonstrated in

approximately 24%–59% of patients with this systemic disease, but imaging findings are much less frequent [1]. Sarcoidosis is usually associated with systemic symptoms, as fever, malaise, and weight loss. Nonspecific hepatosplenomegaly and abdominal lymphadenopathy are commonly observed [85].

Splenic sarcoidosis usually presents itself as diffuse, multiple small nodules up to 3 cm, sometimes presenting with punctate calcifications [86]. On MRI, these lesions are low in signal intensity on T_1- and T_2-weighted sequences, with minimal and delayed enhancement on postcontrast images, being less conspicuous in delayed postcontrast images (Figure 19.37) [87]. These imaging features distinguish them from those of acute splenic candidiasis. If caseating granulomas are present, they appear as hyperintense lesions on T_2-weighted sequences with peripheral hypointense signal.

19.14.4 Gamna–Gandy Bodies

Gamna–Gandy bodies or siderotic nodules are tiny foci of iron deposits that result from splenic microhemorrhages in the splenic parenchyma. These lesions are composed of fibrous tissue associated with hemosiderin and calcium. They usually occur in patients with cirrhosis and portal hypertension or multiple blood transfusions. These nodules are smaller than 1 cm and have a characteristic MRI appearance of signal void in all pulse sequences (Figure 19.38) [88]. T_2^* sequences and the lack of enhancement on postgadolinium T_1-weighted sequences increase their conspicuity, making MRI superior to US and CT in their detection [89]. On in-phase GRE images, these lesions show a susceptibility artifact, known as *blooming artifact*, pathognomonic for this entity. Gamna–Gandy bodies appear smaller on shorter TE sequences because of a diminution of this susceptibility artifact, which allow its differentiation from fibrotic nodules that remain unchanged on shorter TE sequences. This finding helps in their distinction from

granulomas of military tuberculosis, histoplasmosis, and disseminated Pneumocystis carinii infection [88].

19.14.5 Inflammatory Pseudotumor

Inflammatory pseudotumor is an extremely rare benign lesion of the spleen, which occurs in middle-aged or older persons of both genders. It presents as a well-circumscribed solitary mass composed of localized areas of inflammatory cells. This lesion is usually discovered as an incidental solitary lesion, but left flank pain, fever, or splenomegaly can also be present. On MRI, inflammatory pseudotumor has been described as hypointense on T_2-weighted, iso- to hypointense on T_1-weighted sequences and showing heterogeneous enhancement on delayed postcontrast images, reflecting its fibrous nature [90,91]. High signal intensity on T_2-weighted image has also been reported [92]. MRI and other imaging techniques show nonspecific features, making splenectomy necessary to make an accurate diagnosis.

19.14.6 Extramedullary Hematopoiesis

Extramedullary hematopoiesis is a physiological compensatory phenomenon secondary to insufficient bone marrow function. Extramedullary hematopoiesis predominantly affects the spleen and liver, presenting as hepatosplenomegaly, with diffuse microscopical infiltration. Rarely, mass-like foci of hematopoiesis can be formed. On MRI, the signal intensity of these masses depends on

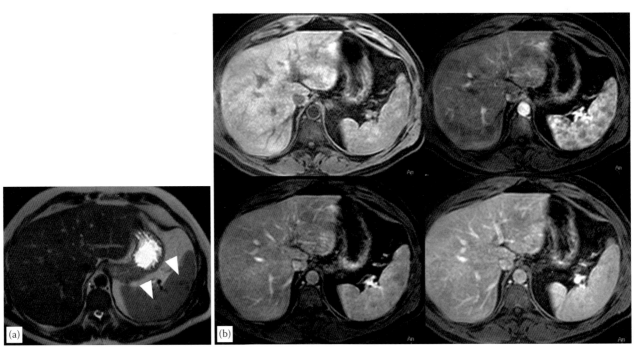

FIGURE 19.37
Sarcoidosis. A 37-year-old male with sarcoidosis, presenting on MRI multiple small hypointense nodules (arrowheads) on TSE T_2-weighted image (a), scattered throughout the splenic parenchyma. On dynamic postgadolinium series (b), these lesions are hypovascular on immediate phase, with progressive enhancement being isointense to spleen on delayed phase.

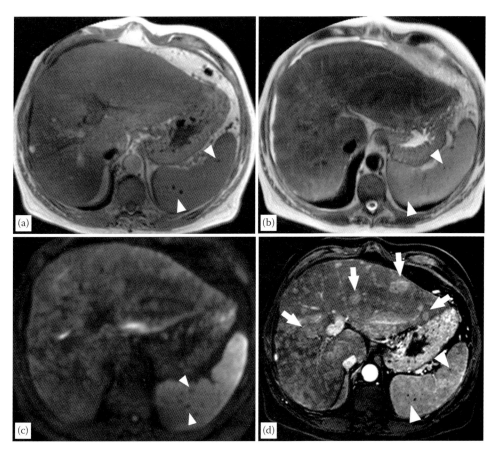

FIGURE 19.38
Gamna–Gandy bodies (arrowheads). Multiple tiny hypointense lesions on (a) GE T_1-weighted and (b) TSE T_2-weighted images are shown in the spleen. These lesions have no enhancement after gadolinium administration (d) and have no restriction on DWI with high b value (c), being hypointense with high *b* values. Several hypervascular nodular lesions (arrows) that show diffusion restriction are seen on the hepatic parenchyma, consistent with multicentric hepatocellular carcinoma.

the evolution of the hematopoiesis. Active lesions show intermediate signal intensity on T_1-weighted and high signal intensity on T_2-weighted images with some degree of enhancement after contrast administration. Chronic lesions usually show low signal intensity on T_1- and T_2-weighted images, without enhancement in postcontrast sequences (Figure 19.39) [15]. These lesions also show a significant drop in signal on in-phase T_1-weighted GRE images compared with that on opposed-phase images, due to the presence of iron [93]. Therefore, in cases of an incidental splenic mass in patients with a hematological disorder, extramedullary hematopoiesis should be included in the differential diagnosis [94].

19.15 Conclusions

Although splenic focal and diffuse lesions are uncommon and usually discovered incidentally, their study is challenging for the radiologist. MRI is a useful tool for the detection of focal and diffuse splenic lesions, surpassing CT when using dynamic postcontrast images. In the case of splenomegaly, MRI allows a comprehensive assessment of chronic hepatopathy, portal hypertension and its complications. Most common focal lesions such as infarcts, small hemangiomas or cystic lesions can be characterized by MRI. Although MRI allows the differentiation of hamartomas and large hemangiomas in many cases, follow-up imaging or histological analysis are required. A variety of unusual vascular lesions such as angiosarcoma, hemangiomatosis, or littoral cell angioma show a similar MRI appearance, making biopsy necessary. Lymphoma and metastases are better depicted on immediate postcontrast images as hypovascular nodules, but their different T_2-values and clinical histories help in their differentiation. On infectious, inflammatory, and storage diseases, MRI also improves the detection and characterization of lesions. Correlation of the clinical setting with the MRI features of the lesions allows a presurgical diagnosis in the vast majority of cases, although histological findings remain necessary in cases with atypical features

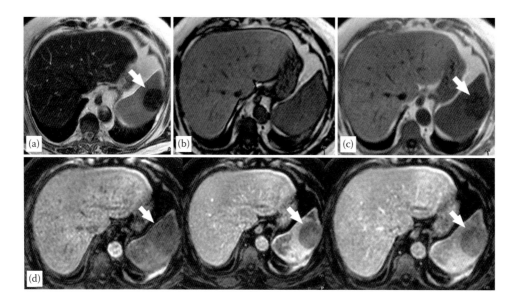

FIGURE 19.39

Extramedullary hematopoiesis. A 58-year-old male with myelodysplastic syndrome. (a) Axial TSE T_2-weighted image shows a well-defined hypointense mass (arrows), which has reduced signal intensity on (c) the in-phase image compared with that on (b) the out-of-phase image. On dynamic postcontrast series, (d) this lesion demonstrates discrete internal enhancement. Surgical resection proved to be extramedullary hematopoiesis.

or where malignancy is suspected. Therefore, because of the increased use of MRI in upper abdomen assessment, it is necessary for radiologists to know the appearance of the most common splenic diseases.

References

1. Robertson F, Leander P, Ekberg O. (2001) Radiology of the spleen. *Eur Radiol*; 11:80–95.
2. Luna A, Ribes R, Caro P et al. (2006) *AJR Am J Roentgenol*; 186:1533–1547.
3. Benter T, Klühs L, Teichgräber U. (2011) Sonography of the spleen. *JUM*; 30(9):1281–1293.
4. Gray H. (2000) The spleen. In: Lewis WH (ed.). *Anatomy of the Human Body*. 20th edn. Philadelphia, PA: Bartleby.com: 1282–1286.
5. Cesta MF (2006) Normal structure, function, and histology of the spleen. *Toxicol Pathol*; 34:455–465.
6. Altun E, Elias J Jr, Kim YH, Semelka RC. (2010) Spleen. In: Semelka RC (ed.). *Abdominal-Pelvic MRI*. 3rd edn. NJ:Wiley-Blackwell, pp. 677–724.
7. Weissleder R, Hahn PF, Stark DD et al. (1988) Superparamagnetic iron oxide: Enhanced detection of focal splenic tumors with MR imaging. *Radiology*; 169:399–403.
8. Harisinghami MG, Saini S, Weissleder R et al. (2001) Splenic imaging with USPIO Ferumoxtran-10 (AMI-7227): Preliminary observations. *J Comput Assist Tomogr*; 25(5):770–776.
9. Bremer C, Allkemper T, Baergmig J, Reimer P. (1999) Res-specific imaging of the liver and spleen with iron oxide particles designed for blood pool MR angiography. *J Magn Reson Imaging*; 10(3):461–467.
10. Semelka RC, Shoenut JP, Lawrence PH. (1992) Spleen: Dynamic enhancement patterns on gradient echo MR imaging enhanced with gadopentetate dimeglumide. *Radiology*; 185:479–482.
11. Hamed MM, Hamn B, Ibrahim ME, Taupitz M, Mahfouz AE. (1992) Dynamic MR imaging of the abdomen with gadopentate dimeglumide: Normal enhancement patterns of liver, spleen, stomach and pancreas. *AJR Am J Roentgenol*; 158:479–482.
12. Li G, Xu P, Pan X et al. (2014)The effect of age on apparent diffusion coefficient values in normal spleen: A preliminary study. *Clin Radiol*; 69(4):e165–e167.
13. Klasen J, Lanzman RS, Wittsack HJ et al. (2013) Diffusion-weighted imaging (DWI) of the spleen in patients with liver cirrhosis and portal hypertension. *Magn Reson Imaging*; 31(7):1092–1096.
14. Storm BL, Abbitt PL, Allen DA, Ros PR. (1992) Splenosis: Superparamagnetic iron oxide enhanced MR imaging. *AJR Am J Roentgenol*; 159:333–335.
15. Elsayes KM, Narra VR, Mukundan G et al. (2005) MR imaging of the spleen: Spectrum of abnormalities. *RadioGraphics*; 25(4):967–982.
16. Gayer G, Apter S, Jonas T et al. (1999) Polysplenia syndrome detected in adulthood: Report of eight cases and review of the literature. *Abdom Imaging*; 24(2):178–184.
17. Brancatelli G, Federle MP, Ambrosini R et al. (2007) Cirrhosis: CT and MR imaging evaluation. *Eur J Radiol*; 61(1):57–69.

18. Nakamoto DA, Onders RP. (2003) The spleen. In: Haaga JR, Lanzieri CF, Gilkeson RC (eds.). *CT and MRI of the whole body*. Mosby (ed.) *CT and MR Imaging of the Whole Body*, 4th edn. St. Louis, MO: Wiley-Blackwell, pp. 1487–1508.

19. Adler DD, Glazer GM, Aisen AM. (1986) MRI of the spleen: Normal appearance and findings in sickle cell anemia. *AJR Am J Roentgenol*; 147:843–845.

20. Taylor FC, Frankl HD, Riemer KD. (1989) Late presentation of splenic trauma after routine colonoscopy. *Am J Gastroenterol*; 84:442–443.

21. Urrutia M, Mergo PJ, Ros LH, Torres GM, Ros PR. (1996) Cystic masses of the spleen: Radiologic-pathologic correlation. *RadioGraphics*; 16:107–129.

22. Rabushka LS, Kawashima A, Fishman EK. (1994) Imaging of the spleen: CT with supplemental MR examination. *RadioGraphics*; 14(2):307–332.

23. Semelka RC, Kelekis NL, Sallah S, Worawattanakul S, Ascher SM. (1997) Hepatosplenic fungal disease: Diagnostic accuracy and spectrum of appearances on MR imaging. *AJR Am J Roentgenol*; 169:1311–1316.

24. Palas J, Matos AP, Ramalho M. (2013) The spleen revisited: An overview on magnetic resonance imaging. *Radiol Res Pract*. doi: 10.1155/2013/219297.

25. Ng KK, Lee TY, Wan YL et al. (2002) Splenic abscess: Diagnosis and management. *Hepatogastroenterology*; 49: 567–571.

26. Akhan O, Koroglu M. (2007) Hydatid disease of the spleen. *Semin Ultrasound CT MR*; 28(1):28–34.

27. von Sinner WN, Stnidbeck H. (1992) Hydatid disease of the spleen: Ultrasonography, CT and MR imaging. *Acta Radiol*; 33:459–446.

28. Giovagnoni A, Giorgi C, Goteri G. (2005) Tumours of the spleen. *Cancer Imaging*; 5(1):73–77.

29. Labruzzo C, Haritopoulos KN, El Tayar AR et al. (2002) Posttraumatic cyst of the spleen: A case report and review of the literature. *Int Surg*; 87:152–156.

30. Ramani M, Reinhold C, Semelka RC. (1997) Splenic hemangiomas and hamartomas: MR imaging characteristics of 28 lesions. *Radiology*; 202:166–172.

31. Ros PR, Moser RP Jr, Dachman AH, Murari PJ, Olmsted WW. (1987) Hemangioma of the spleen: Radiologic-pathologic correlation in ten cases. *Radiology*; 162:73–77.

32. Abbott RM, Levy AD, Aguilera NS et al. (2004) From the archives of the AFIP: Primary vascular neoplasms of the spleen: Radiologic–pathologic correlation. *RadioGraphics*; 24:1137–1163.

33. Ha HK, Kim HH, Kim BK, Han JK, Choi BI. (1994) Primary angiosarcoma of the spleen. CT and MR imaging. *Acta Radiol*; 35(5):455–458.

34. Levy AD, Abbott RM, Abbondanzo SL. (2004) Littoral cell angioma of the spleen: CT features with clinicopathologic comparison. *Radiology*; 230:485–490.

35. Schneider G, Uder M, Altmeyer K, et al. (2000) Littoral cell angioma of the spleen: CT and MR imaging appearance. *Eur Radiol*; 10:1395–1400.

36. Tatli S, Cizginer S, Wieczorek TJ et al. (2008) Solitary littoral cell angioma of the spleen: Computed tomography and magnetic resonance imaging features. *J Comput Assist Tomogr*; 32:772–775.

37. Miller WJ, Dodd GD, 3rd, Federle MP, Baron RL. (1992) Epithelioid hemangioendothelioma of the liver: Imaging findings with pathologic correlation. *AJR Am J Roentgenol*; 159:53–57.

38. Martel M, Cheuk W, Lombardi L, Lifschitz-Mercer B, Chan JK, Rosai J. (2004) Sclerosing angiomatoid nodular transformation (SANT): Report of 25 cases of a distinctive benign splenic lesion. *Am J Surg Pathol*; 28:1268–1279.

39. Lewis RB, Lattin GE Jr, Nandedkar M, Aguilera NS. (2013) Sclerosing angiomatoid nodular transformation of the spleen: CT and MRI features with pathologic correlation. *AJR Am J Roentgenol*; 200(4):353–360.

40. Raman SP, Singhi A, Horton KM, Hruban RH, Fishman EK. (2013) Sclerosing angiomatoid nodular transformation of the spleen (SANT): Multimodality imaging appearance of five cases with radiology-pathology correlation. *Abdom Imaging*; 38(4):827–834.

41. Kim HJ, Kim KW, Yu ES et al. (2012) Sclerosing angiomatoid nodular transformation of the spleen: Clinical and radiologic characteristics. *Acta Radiol*; 53(7):701–716.

42. Thacker C, Korn R, Millstine J, Harvin H, Van Lier Ribbink JA, Gotway MB. (2010) Sclerosing angiomatoid nodular transformation of the spleen: CT, MR, PET, and 99mTc-sulfur colloid SPECT CT findings with gross and histopathological correlation. *Abdom Imaging*; 35:683–689.

43. Falk S, Krishnan J, Meis JM. (1993) Primary angiosarcoma of the spleen: A clinic-pathologic study of 40 cases. *Am J Surg Pathol*; 17:959–970.

44. Vrachliotis TG, Bennett WF, Vaswani KK et al. (2000) Primary angiosarcoma of the spleen—CT, MR and sonographic characteristics: Report of 2 cases. *Abdom Imaging*; 25(3): 283–285.

45. Ferrozzi F, Bova D, Draghi F, Garlaschi G. (1996) CT findings in primary vascular tumors of the spleen. *AJR Am J Roentgenol*; 166(5):1097–1101.

46. Hatva E, Bohling T, Jaaskelainen J, Persico MG, Haltia M, Alitalo K. (1996) Vascular growth factors and receptors in capillary hemangioblastomas and hemangiopericytomas. *Am J Pathol*; 148:763–775.

47. Guadalajara Jurado J, Turegano Fuentes F, Garcia Menendez C, Larrad Jimenez A, Lopez de la Riva M. (1989) Hemangiopericytoma of the spleen. *Surgery*; 106(3):575–577.

48. Hosotani R, Momoi H, Uchida H, et al. (1992) Multiple hemangiopericytomas of the spleen. *Am J Gastroenterol*; 87(12):1863–1865.

49. Yilmazlar T, Kirdak T, Yerci O et al. (2005) Splenic hemangiopericytoma and serosal cavernous hemangiomatosis of the adjacent colon. *World J Gastroenterol*. 14;11(26):4111–4113.

50. Ito K, Mitchell DG, Honjo K et al. (1997) MR Imaging of acquired abnormalities of the spleen. *AJR Am J Roentgenol*; 168:697–702.

51. Goerg C, Schwerk WB. (1990) Splenic infarction: Sonographic patterns, diagnosis, follow-up, and complications. *Radiology*; 174:803.

52. Vanhoenacker FM, Op de Beeck B, De Schepper AM, Salgado R, Snoeckx A, Parizel PM. (2007) Vascular disease of the spleen. *Semin Ultrasound CT MRI*; 28:35–51.

53. De Schepper AM, Vanhoenacker F, Op de Beeck B et al. (2005) Vascular pathology of the spleen, part I. *Abdom Imaging*; 30:96–104.

54. Dufau JP, le Tourneau A, Audouin J, Delmer A, Diebold J. (1999) Isolated diffuse hemangiomatosis of the spleen with Kasabach-Merritt-like syndrome. *Histopathology*; 35(4):337–344.

55. Ruck P, Horny HP, Xiao JC, Bajinski R, Kaiserling E. (1994) Diffuse sinusoidal hemangiomatosis of the spleen. A case report with enzyme-histochemical, immunohistochemical, and electron-microscopic findings. *Pathol Res Pract*; 190(7):708–714.

56. Rege JD, Kavishwar VS, Mopkar PS. (1998) Peliosis of spleen presenting as splenic rupture with haemoperitoneum—A case report. *Indian J Pathol Microbiol*; 41(4):465–467.

57. Battal B, Kocaoglu M, Atay AA, Bulakbasi N. (2010) Multifocal peliosis hepatis: MR and diffusion-weighted MR-imaging findings of an atypical case. *Ups J Med Sci.* 115(2):153–156.

58. Madoff DC, Denys A, Wallace MJ et al. (2005) Splenic arterial interventions: Anatomy, indications, technical considerations and potential complications. *RadioGraphics*; 25:S191–S211.

59. Kehagias DT, Tzalonikos MT, Moulopoulos LA et al. (1998) MRI of a giant splenic artery aneurysm. *Br J Radiol*; 71:444–446.

60. Kreft B, Strunk H, Flacke S et al. (2000) Detection of thrombosis in the portal venous system: Comparison of contrast-enhanced MR angiography with intraarterial digital subtraction angiography. *Radiology*; 216:86–92.

61. Pinto PO, Advigo P, Garcia H et al. (1995) Splenic hamartomas: A case report. *Eur J Radiol*; 5:93–95.

62. Ohotmo K, Fukoda H, Mori K et al. (1992) CT and MR appearances of splenic hamartoma. *J Comput Assist Tomogr*; 16:425–428.

63. Tang P, Alhindawi R, Farmer P. (2001) Case report: Primary isolated angiomyolipoma of the spleen. *Ann Clin Lab Sci*; 31(4):405–410. (Abstract).

64. Gupta S, Deshmukh SP, Ditzler MG, Elsayes KM. (2014) Splenic lesions. In: Luna A et al. (eds.) *Functional Imaging in Oncology*, vol. 2. Berlin, Germany: Springer-Verlag, 2014.

65. Freeman JL, Jafri SZ, Roberts JL, Mezwa DG, Shirkhoda A. (1993) CT of congenital and acquired abnormalities of the spleen. *RadioGraphics*; 13(3):597–610.

66. Kamaya A, Weinstein S, Desser TS. (2006) Multiple lesions of the spleen: Differential diagnosis of cystic and solid lesions. *Semin Ultrasound CT MR*; 27:389–403.

67. Warshauer DM, Hall HL. (2006) Solitary splenic lesions. *Semin Ultrasound CT MR*; 27:370–388.

68. Peddu P, Shah M, Sidhu PS. (2004) Splenic abnormalities: A comparative review of ultrasound, microbubble-enhanced ultrasound and computed tomography. *Clin Radiol*; 59:777–792.

69. Lindfors KK, Meyer JE, Palmer EL 3rd, Harris NL. (1984) Scintigraphic findings in large-cell lymphoma of the spleen: Concise communication. *J Nucl Med*; 25:969–971.

70. Kreft BP, Tanimoto A, Leffler S et al. (1994) Contrast enhanced MR imaging of diffuse and focal splenic disease with use of magnetic starch microspheres. *AJR Am J Roentgenol*; 4:373–379.

71. Mainenti PP, Iodice D, Cozzolino I et al. (2012) Tomographic imaging of the spleen: The role of morphological and metabolic features in differentiating benign from malignant diseases. *Clin Imaging*; 36:559–567.

72. Hahn PF, Weissleder R, Stark DD et al. (1988) MR imaging of focal splenic tumors. *AJR Am J Roentgenol*; 150:823–827.

73. Mirowitz SA, Brown JJ, Lee JKT et al. (1991) Dynamic gadolinium-enhanced MR imaging of the spleen: Normal enhancement patterns and evaluation of splenic lesions. *Radiology*; 179:681–686.

74. Dawson L, Gupta O, Garg K. (2012) Malignant fibrous histiocytoma of the spleen: An extremely rare entity. *J Cancer Res Ther.* 8:117–119.

75. Fotiadis C, Georgopoulos I, Stoidis C, Patapis P. (2009) Primary tumors of the spleen. *Int J Biomed Sci*; 5:85–91.

76. Amatya BM, Sawabe M, Arai T et al. (2011) Splenic undifferentiated high grade pleomorphic sarcoma of a small size with fatal tumor rupture. *J Pathol Nepal.* 1:151–153.

77. Restrepo CS, Martinez S, Lemos JA et al. (2006) Imaging manifestations of Kaposi sarcoma. *RadioGraphics*; 26:1169–1185.

78. Ohe C, Sakaida N, Yanagimoto Y et al. (2010) A case of splenic low-grade mucinous cystadenocarcinoma resulting in pseudomyxoma peritonei. *Med Mol Morphol*; 43:235–240.

79. Hill SC, Damaska BM, Ling A et al. (1992) Gaucher disease: Abdominal MR imaging findings in 46 patients. *Radiology*; 184:561–566.

80. Poll LW, Koch JA, vom Dahl S et al. (2000) Gaucher disease of the spleen: CT and MR findings. *Abdom Imaging*; 25:286–289.

81. Omarini LP, Frank-Burkhardt SE, Seemayer TA, Mentha G, Terrier, F. (1995) Niemann-Pick disease type C: Nodular splenomegaly. *Abdom Imaging*; 20(2):157–160.

82. Monzawa S, Tsukamoto T, Omata K, Hosoda K, Araki T, Sugimura K. (2002) A case with primary amyloidosis of the liver and spleen: Radiologic findings. *Eur J Radiol*; 41(3):237–241.

83. Mainenti PP, Camera L, Nicotra S, et al. (2005) Splenic hypoperfusion as a sign of systemic amyloidosis. *Abdom Imaging*; 30(6):768–772.

84. Paterson A, Frush DP, Donnelly LF, Foss JN, O'Hara SM, Bisset GS 3rd. (1999) A pattern-oriented approach to splenic imaging in infants and children. *RadioGraphics*; 19(6):1465–1485.

85. Warshauer DM, Molina PL, Hamman SM et al. (1995) Nodular sarcoidosis of the liver and spleen: Analysis of 32 cases. *Radiology*; 195(3):757–762.

86. Scott GC, Berman JM, Higgins JL. (1997) CT patterns of nodular hepatic and splenic sarcoidosis: A review of the literature. *J Comput Assist Tomogr*; 21:369–372.

87. Warshauer DM, Semelka RC, Ascher SM. (1994) Nodular sarcoidosis of the liver and spleen: Appearance on MR images. *J Magn Reson Imaging*; 4(4):553–557.

88. Sagoh T, Itoh K, Togashi K et al. (1989) Gamna-Gandy bodies of the spleen: Evaluation with MR imaging. *Radiology*; 172:685.

89. Chan YL, Yang WT, Sung JJ, Lee YT, Chung SS. (2000) Diagnostic accuracy of abdominal ultrasonography compared to magnetic resonance imaging in siderosis of the spleen. *J Ultrasound Med*; 19(8):543–547.

90. Irie H, Honda H, Kaneko K, et al. (1996) Inflammatory pseudotumors of the spleen: CT and MRI findings. *J Comput Assist Tomogr*; 20(2):244–248.

91. Ma PC, Hsieh SC, Chien JC, Lao WT, Chan WP. (2007) Inflammatory pseudotumor of the spleen: CT and MRI findings. *Int Surg*; 92(2):119–122.

92. Noguchi H, Kondo H, Kondo M, Shiraiwa M, Monobe Y. (2000) Inflammatory pseudotumor of the spleen: A case report. *Jpn J Clin Oncol*; 4:196–203.

93. Granjo E, Bauerle R, Sampaio R et al. (2002) Extramedullary hematopoiesis in hereditary spherocytosis deficient in ankyrin: A case report. *Int J Hematol*; 76:153–156.

94. Gabata T, Kadoya M, Mori A, Kobayashi S, Sanada J, Matsui O. (2000) MR imaging of focal splenic extramedullary hematopoiesis in polycythemia vera: Case report. *Abdom Imaging*; 25(5):514–516.

20

Pancreas

Fernanda Garozzo Velloni, Ersan Altun, Miguel Ramalho, and Richard Semelka

CONTENTS

20.1 Introduction

Magnetic resonance imaging (MRI) is a useful tool for detecting and characterizing pancreatic disorders. Recent technical innovations have also improved the evaluation of the pancreatic parenchyma and ducts, peripancreatic soft tissue, adjacent organs, and vascular structures.

20.2 Normal Anatomy

The pancreas is a mildly lobulated mixed exocrine and endocrine gland that is located retroperitoneally adjacent to the major vascular structures. It is located adjacent to the stomach anteriorly, second portion of the duodenum medially and major vascular structures/vertebral column posteriorly. The pancreatic tail is adjacent to the spleen. The pancreas is located anterior to the anterior pararenal fascia bilaterally. Its anterior surface is covered by peritoneum. The anatomic divisions of the pancreas include the head, uncinate process, neck, body, and tail. The broad head is embraced by the curve of the duodenum. An extension of the head, the uncinate process hooks behind the superior mesenteric artery and vein. The border between the head and the body is a slightly narrowed region, called neck and this portion is also located at the level of portal vein confluence.

The posterior aspect of the pancreas does not have a serosal covering, which accounts for the extensive dissemination of fluid in pancreatitis and the early spread of pancreatic ductal cancer into retroperitoneal fat. Although considerable variation in the size of head occurs, the normal pancreatic head is 2–2.5 cm in diameter, with the remainder of the gland approximately 1–2 cm thick.

The main pancreatic duct is termed as the duct of Wirsung and measures 1–2 mm in diameter in normal subjects. It extends from the tail of the pancreas through the head and empties into the second part of the duodenum, at the major papilla, via the sphincter of Oddi. A smaller accessory duct, the duct of Santorini, is frequently present and extends from the body of the pancreas through the neck, and enters separately into the duodenum, in a more proximal location at the minor papilla.[1]

20.3 MRI Technique

New MRI techniques including faster and breathing-independent sequences and parallel imaging that limits artifacts in the abdomen, have increased the role of MRI in detection and characterization of pancreatic disease. A standard protocol includes the use of coronal and transverse single-shot echo train spin-echo (SS-ETSE) T_2-weighted sequences, transverse fat-suppressed SS-ETSE T_2-weighted sequences, transverse in-phase and out-of-phase 2D or 3D-GE T_1-weighted sequences, transverse fat-suppressed 3D-GE T_1-weighted sequences, and dynamic postgadolinium fat-suppressed 3D-GE sequences including transverse (three consecutive phases), coronal, and sagittal (optional) planes (the last two phases). The combined evaluation of these sequences, which result in high image quality, enables us to detect and characterize focal pancreatic mass lesions even smaller than 1 cm in diameter, and evaluate diffuse pancreatic disease.[2–5]

The use of high spatial resolution MRI at 3.0 T may also improve the detection of small focal lesions particularly. MR cholangiopancreatography (MRCP) permits good demonstration of the biliary and pancreatic ducts to assess ductal obstruction, dilatation, and abnormal duct pathways.[6–8] Therefore, MRCP examination should always be the part of MRI evaluation/protocol of the pancreas as it provides comprehensive information to evaluate the full range of pancreatic disease. MRCP sequences are heavily T_2-weighted sequences and should be acquired in two different ways including thick-slab and thin-slab methods. Thick-slab MRCP is acquired as a breath-hold sequence in a very short time (in a few seconds) with a thick slice (e.g., 40 mm) covering a large area including the pancreatic ductal system and biliary system. This technique allows us to acquire images without motion artifacts despite low spatial resolution. Thin-slab technique is acquired as a breathing-independent sequence with navigator echo (preferably) or respiratory triggered techniques. This sequence is a 3D technique enabling us to acquire very thin slices (such as 1 mm) of the pancreatic and biliary ductal system. Since this sequence is a 3D sequence, 3D maximum intensity projection reconstruction could also be done that is also valuable for the evaluation and helpful for the clinicians to understand the 3D anatomy of the pancreatic and biliary ductal anatomy.

T_2-weighted SS-ETSE sequences of the standard protocol provide a sharp anatomic display of the common bile duct (CBD) on coronal plane images and of the pancreatic duct on transverse plane images. T_2-weighted images also provide information on the fluid content of the solid lesions, which is especially important in the presence of accompanying chronic pancreatitis; internal structure of the cystic lesions including septations/solid components; and complexity of the fluid in pancreaticpseudocysts/collections which may reflect the presence of complications such as necrotic debris or infection.[1]

The normal pancreas shows high T_1 signalon T_1-weighted GE sequences because of the presence of

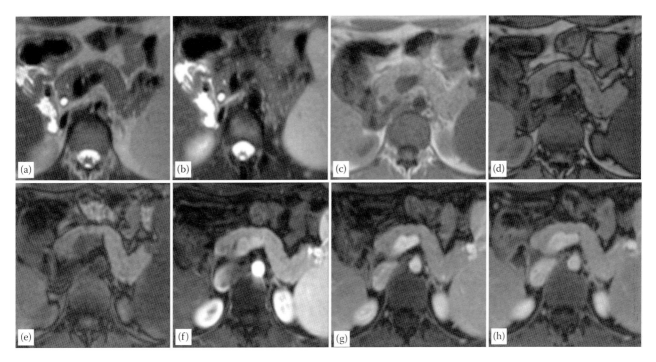

FIGURE 20.1

Normal pancreas. T_2-weighted SS-ETSE (a), fat-suppressed T_2-weighted SS-ETSE (b), T_1-weighted in-phase (c) and out-of-phase (d) SGE, T_1-weighted fat-suppressed spoiled gradient echo (SGE) (e), T_1-weighted postgadolinium hepatic arterial dominant phase (f), hepatic venous phase (g) and interstitial phase (h). The normal pancreas shows high signal on T_1-weighted GE sequences, particularly on T_1-weighted fat-suppressed images, and prominent enhancement on hepatic arterial dominant phase, which decreases during the hepatic venous and interstitial phases.

aqueous protein in the acini of the pancreas and this is particularly prominent on T_1-weighted fat-suppressed images[2,9,10] (Figure 20.1). T_1-weighted in-phase and out-of-phase GE sequences, which are particularly important for the identification of solid lesions in the absence of chronic pancreatitis and identification of fat and iron in the pancreas. Fat-suppressed 3D-GE sequence is particularly important for the identification of solid lesions in the pancreas especially if there is no accompanying chronic pancreatitis. In elderly patients, the signal intensity of the pancreas may diminish and be lower than that of liver.[3] This may reflect decrease in the functional status of the pancreas secondary to the aging process with or without associated changes of fibrosis.[1]

Postgadolinium imaging is performed on three different phases including the capillary phase (hepatic arterial dominant phase), hepatic venous phase (portal venous phase), and interstitial phase with 3D-GE T_1-weighted fat-suppressed sequences. The hepatic arterial dominant phase is characterized by the presence of intravenous (IV) contrast in hepatic arteries and portal veins but not in hepatic veins. This phase is usually acquired after 35–38 s following the administration of IV contrast. This phase is not only important for the evaluation of the liver lesions, particularly the hypervascular ones, but also for the evaluation of pancreas parenchyma and pancreatic lesions. The normal pancreas shows prominent enhancement in this phase and

the enhancement of pancreatic parenchyma is comparable to the renal cortex and much more than the liver parenchyma in this phase (Figure 20.1). Additionally, hypovascular lesions can be delineated more easily in this phase due to their relative low signal in the background of prominently enhancing pancreas if there is no associated chronic pancreatitis. Hypervascular lesions also enhance prominently in this phase. The arterial vasculature may also be evaluated. However, the pancreas does not show prominent enhancement in this phase when there is chronic pancreatitis. On the hepatic venous phase, the enhancement of normal pancreatic parenchyma decreases and becomes equal to the liver (Figure 20.1). This phase is acquired between 60 and 90 s after the administration of IV contrast. This phase is especially important for the further evaluation of enhancement characteristics and venous structures. Progressive central enhancement of hypovascular lesions, washout of hypervascular lesions, enhancement of septa or solid components of cystic lesions, and enhancement of walls of fluid collections can be evaluated in this phase. Interstitial phase is again important for the evaluation of enhancement features described for the hepatic venous phase since some lesions may show these features not only in the hepatic venous phase but also in the interstitial phase. Interstitial phase is acquired in 90–120 s after the IV contrast administration. The pancreas affected by chronic pancreatitis could show increased enhancement in interstitial phase due to the presence of associated fibrotic changes. These three phases are acquired on the transverse planes. Coronal acquisition is also performed after the acquisition of these three phases. Sagittal acquisition is optional but could be helpful especially for the evaluation of vascular invasion, and surgical resectability and planning.[5]

In noncooperative patients, who are unable to hold their breath, breathing-independent sequences including T_2-weighted SS-ETSE sequences, thick-/thin-slab MRCP, pre-contrast T_1-weighted magnetization prepared rapid gradient echo (MPRAGE) sequences, and pre-contrast/post-contrast T_1-weighted fat-suppressed 3D-GE sequence with radial sampling are used.

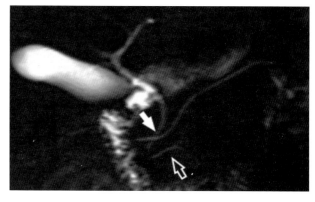

FIGURE 20.2
Pancreas divisum. MRCP image demonstrates separate entry of the ducts of Santorini (white arrow) and Wirsung (white open arrow) into the duodenum with no communication between the ductal systems.

20.4 Developmental Anomalies

20.4.1 Pancreas Divisum

Pancreas divisum is the most common anatomic variant of the pancreas and its ductal system. Conventional pancreatic ductal anatomy is characterized by the presence of main pancreatic duct and accessory pancreatic duct. The main pancreatic duct arises from the dorsally derived pancreas and fuses with the ventrally derived pancreatic duct during the embryologic development. The duct of Wirsung is a portion of ventrally derived pancreatic duct fusing with the dorsally derived main pancreatic duct in the head of the pancreas and drains to the ampulla with the CBD. The accessory pancreatic duct (the duct of Santorini) derives from the dorsally derived pancreas and is the continuation of dorsal pancreatic duct draining into the minor papilla. This conventional anatomy is seen in 70% of the cases. A rudimentary duct of Santorini is seen in 30% of the cases.

Pancreas divisum is seen approximately in 10% of the population. This results from the failure of ventral and dorsal pancreatic ducts during the embryologic development.[11] This anomaly leads to the drainage of ventral pancreatic anlage by the duct of Wirsung into the major papilla and the drainage of the dorsal pancreatic anlage (the majority of the gland) through the dorsal pancreatic duct into the minor papilla via the duct of Santorini. The result of this congenital abnormality is that portions of the pancreas have separate ductal systems without any connection: a very short ventral duct of Wirsung drains only the lower portion of the head, whereas the dorsal duct of Santorini drains the tail, body, neck, and upper aspect of the head. This anomaly is the most common type of pancreas divisum and has been reported to increase the risk of pancreatitis. On MRCP images, separate entries of the ducts of Santorini

and Wirsung into the duodenum are consistently demonstrated because of the good conspicuity of the high-signal-intensity ductal structures (Figure 20.2).

The absent ventral duct with a dominant duct of Santorini/inadequate functional connections between the dorsal and ventral ducts, and ansa pancreatica/ loop type of pancreatic ducts are also other rare types of pancreas divisum and pancreatic ductal variations, respectively.

20.4.2 Annular Pancreas

Annular pancreas is an uncommon congenital anomaly in which glandular pancreatic tissue, in continuity with the head of the pancreas, encircles the duodenum. In most cases, the annular portion surrounds the second part of the duodenum. Patients may present with duodenal obstruction/stenosis. The identification of the continuity of duodenum inside the pancreatic parenchyma and its encasement by the pancreatic tissue is sufficient for the diagnosis and this is easily seen particularly in non-fat-suppressed T_2-weighted sequences and T_1-weighted pre- and post-contrast sequences. Fat-suppressed T_1-weighted sequences may be particularly helpful since the background pancreas appears as hyperintense compared to the hypointense duodenum (Figure 20.3).[12]

20.4.3 Congenital Absence of the Dorsal Pancreas

Congenital absence of the dorsal pancreas is a very rare anomaly. The head of the pancreas terminates with a rounded contour, unlike surgical or post-traumatic absence of the distal pancreas, which has more squared-off or irregular terminations.[1]

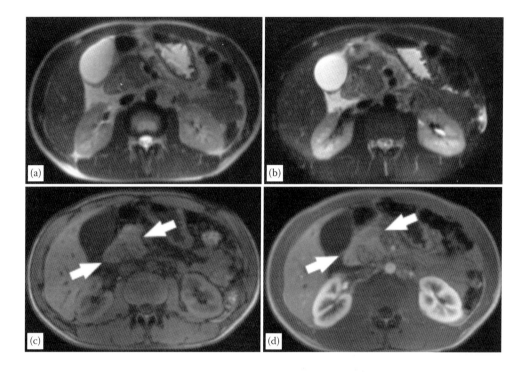

FIGURE 20.3
Annular pancreas. T_2-weighted SS-ETSE (a), fat-suppressed T_2-weighted SS-ETSE (b), T_1-weighted fat-suppressed SGE (c), T_1-weighted postgadolinium hepatic arterial dominant phase (d). Normal pancreatic parenchyma (white arrows) surrounds the second portion of the duodenum, diagnostic for annular pancreas. This is best shown on noncontrast T_1-weighted fat-suppressed (c) and immediate postgadolinium (d) images.

20.4.4 Short Pancreas in the Polysplenia Syndrome

Polysplenia syndrome is a congenital syndrome characterized by multiple, misplaced small spleens without a parent spleen. This syndrome is characterized by features of bilateral left-sidedness and these splenules may be bilateral.[13] This syndrome may be associated with semiannular or congenitally short pancreas.[1]

20.5 Genetic Disease

20.5.1 Cystic Fibrosis

It is an autosomal recessive multisystem disease characterized by a dysfunction of the secretory process of all exocrine glands and reduced mucociliar transport that results in mucous plugging of the exocrine glands.

Patients usually have clinical manifestations of recurrent bronchopulmonary infections leading to chronic lung disease, malabsorption secondary to pancreatic insufficiency, and an increased sweat sodium concentration. Three basic imaging patterns of pancreatic abnormalities have been described: pancreatic enlargement with complete fatty replacement (most common) with or without loss of the lobulated contour (Figure 20.4), atrophic pancreas with partial fatty replacement, and diffuse atrophy of the pancreas without fatty replacement.[14–16]

Fatty replacement is high in signal intensity on T_1- and T_2-weighted images and demonstrates loss of signal intensity in corresponding fat-suppressed images. Another manifestation of cystic fibrosis is pancreatic cysts secondary to duct obstruction by secretion. Pancreatic cystosis is a rare manifestation characterized by large cysts.[1]

20.5.2 Primary Hemochromatosis

Primary hemochromatosis is an autosomal recessive hereditary disease in which there is excessive accumulation of body iron, most of which is deposited in the parenchyma of various organs. The liver, pancreas, and heart are primarily affected. Iron deposition results in a loss of signal intensity on T_2-weighted sequences and T_1-weighted in-phase GE sequences. Iron deposition causes signal decrease in the liver, pancreas, and myocardium on T_1-weighed in-phase GE sequences compared to T_1-weighted out-of-phase GE sequence due to longer TE times. In contrast to iron deposition seen in the spleen secondary to blood transfusion, no iron deposition is seen in the spleen in primary hemochromatosis.

20.5.3 von Hippel–Lindau Syndrome

von Hippel–Lindau syndrome is an autosomal dominant condition with variable penetration. This condition is characterized by hemangioblastomas in the

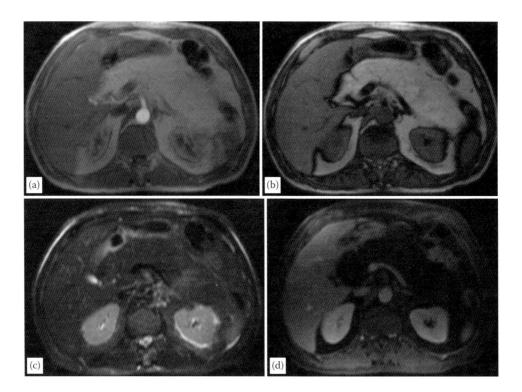

FIGURE 20.4
Cystic fibrosis. T_1-weighted in-phase (a) and out-of-phase (b) SGE, fat-suppressed T_2-weighted SS-ETSE (c) and T_1-weighted postgadolinium hepatic arterial dominant phase (d). The pancreas is markedly enlarged and hyperintense on T_1-weighted images (a) and (b) and hypointense on fat-suppressed images (c) and (d), consistent with replacement of the pancreatic parenchyma by adipose tissue. The complete fatty replacement of the pancreas is the most common manifestation of cystic fibrosis.

cerebellum and retina. Patients may have cysts of the liver and kidney, with a strong propensity to develop renal cell carcinoma. Patients with von Hippel–Lindau syndrome may develop pancreatic cysts (most common), islet cell tumors, or microcystic cystadenoma.

20.6 Inflammatory Disease

20.6.1 Pancreatitis

Pancreatitis is defined as the inflammation of the pancreas and considered the most common pancreatic disease in children and adults. It can be acute, representing an acute inflammatory process of the pancreas, or chronic, progressing slowly with continued, permanent inflammatory injury to the pancreas.[17]

Over half of the cases of acute pancreatitis in adults are related to cholelithiasis or alcohol consumption, whereas trauma, viral infections, and systemic diseases account for the majority of cases in children. Alcohol consumption accounts for the majority (80%) of cases of chronic pancreatitis in adults in developed countries, whereas malnutrition is the most common cause worldwide.[18]

20.6.1.1 Acute Pancreatitis

Acute pancreatitis results from the exudation of fluid containing activated proteolytic enzymes into the interstitium of the pancreas and leakage of this fluid into the surrounding tissues.

Two distinct phases of acute pancreatitis were introduced: first, or early, phase that occurs within the first week of the onset of disease; and second, or late, phase that takes place after the first week of the onset.[19]

During the early or first phase (less than 1 week), pancreatic or peripancreatic ischemia or edema may completely resolve, develop fluid collections, or progress to permanent necrosis and liquefaction. The late or second phase (after 1 week from onset) occurs mostly in patients with moderate to severe acute pancreatitis and may extend for weeks to months. It is characterized by the presence of local complications, systemic manifestations (due to ongoing inflammation) and/or transient or persistent organ failure.[17]

MRI plays a significant role in the diagnosis of acute pancreatitis in clinically suspected cases or suggesting alternative diagnoses. It helps determine the causes of pancreatitis: gallstones, biliary duct obstruction, or structural abnormalities. It also helps in grading the severity of the disease and identifying pancreatic or peripancreatic complications.

MRI is particularly sensitive for the detection of subtle changes of acute pancreatitis, particularly minor peripancreatic inflammatory changes, even in the setting of a morphologically normal pancreas on CT imaging, which may appear normal in up to 15%–30% of patients with clinical features of acute pancreatitis.[20]

20.6.1.2 Morphologic Forms of Acute Pancreatitis

20.6.1.2.1 Interstitial Edematous Pancreatitis

Interstitial edematous pancreatitis (IEP) is a milder form of acute pancreatitis that usually resolves over the first week. IEP is characterized by diffused or localized enlargement of the pancreas secondary to interstitial or inflammatory edema without necrosis.

On MRI, enlargement of the pancreas, parenchymal/peripancreatic edema, and fat stranding are well demonstrated on T_1- and T_2-weighted images. Fat suppressed T_2-weighted sequences are very sensitive for detecting edema or minimal fluid and therefore have a role in detecting even milder forms of pancreatitis (Figure 20.5).[21]

20.6.1.2.2 Necrotizing Pancreatitis

Necrotizing pancreatitis is the inflammation of the pancreas with obvious pancreatic and peripancreatic tissue necrosis.

On MRI, necrosis appears as hypointense areas on T_1-weighted images corresponding to areas of increased signal on fat-suppressed T_2-weightedimages, associated with areas of nonenhancing pancreatic parenchyma on postgadolinium sequences.[22–24]

Pancreatic duct disruption is an important prognostic factor. It is seen in 30% of the patients of necrotizing pancreatitis[25] when necrosis involves the central

gland.[26,27] MRCP, as a noninvasive imaging method, is highly accurate in detecting pancreatic duct disruption, thus helping in identifying patients who might benefit from early treatment.

20.6.1.3 Complications of Acute Pancreatitis

An important distinction is made between collections that are composed of fluid alone and those that arise from necrosis and appear in complex nature (Table 20.1).

20.6.1.3.1 Acute Peripancreatic Fluid Collections

Acute peripancreatic fluid collections (APFCs) occur in less than 4 weeks after IEP, in the peripancreatic region, lack a discrete wall and show homogeneous simple appearance. They can be single or multiple.

T_2-weighted sequences are very sensitive in detecting APFCs that demonstrate high T_2 signal intensity (Figure 20.5). On T_1-weighted GE images, APFCs demonstrate low signal intensity in a background of high

TABLE 20.1

Types of Collections After Acute Pancreatitis

Type of Pancreatitis	Fluid Collection	Appearance
<4 weeks		
IEP	APFC	Homogeneous, fluid attenuation, no debris, not encapsulated
Necrotizing pancreatitis	ANC	Heterogeneous, debris, loculated, not encapsulated
>4 weeks		
IEP	Pseudocyst	Homogeneous, fluid attenuation, no debris, encapsulated
Necrotizing pancreatitis	WON	Heterogeneous, debris, loculated, encapsulated

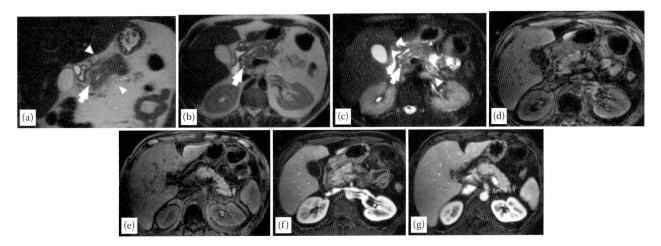

FIGURE 20.5
Interstitial edematous pancreatitis. Coronal (a) and transverse (b) T_2-weighted SS-ETSE images, fat-suppressed T_2-weighted SS-ETSE (c), T_1-weighted fat-suppressed SGE (d) and (e), T_1-weighted fat-suppressed postgadolinium hepatic arterial dominant phase (f) and (g). There is mild diffuse increased T_2 signal involving the pancreatic parenchyma (white arrows), associated with a small amount of peripancreatic fluid (white arrow heads), with fairly normal enhancement of the pancreas on the post-contrast images (f) and (g).

signal intensity fat. Perceptible enhancement is not usually depicted on postgadolinium fat-suppressed T_1-weighted images.

The majority of fluid collections are typically confined to the lesser sac and anterior pararenal space or may track down to the pelvis and superiorly into mediastium.[22] Most acute fluid collections remain sterile and usually resolve spontaneously without intervention.[23]

20.6.1.3.2 Pancreatic Pseudocysts

Fluid collections that persist for more than 4 weeks after an episode of IEP, located in the peripancreatic region, with a well-defined wall and no internal solid components. Pseudocysts demonstrate low signal intensity on T_1-weighted gradient-echo images and relatively homogeneous high signal intensity on T_2-weighted images. Pseudocysts walls enhance minimally on early postgadolinium images and may show progressively prominent enhancement on the interstitial phase postgadolinium images, due to the presence of fibrous tissue (Figure 20.6).

Pseudocysts may sometime have communication with pancreatic duct and detecting this communication is helpful in the further patient's management. The majority of pseudocysts resolve spontaneously.

Infection and hemorrhage may also complicate simple pseudocysts. Infected pseudocyst may contain gas bubbles that demonstrate susceptibility artifacts, particularly on T_1-weighted images.

20.6.1.3.3 Acute Necrotic Collections

During the first 4 weeks, a collection containing variable amounts of fluid, and necrotic and hemorrhagic tissue is termed as an acute necrotic collection (ANC). Unlike APFCs, ANCs are present within the pancreas and peripancreatic regions. ANC's may also be in communication with the main pancreatic duct or one of its sidebranches.

On MRI, the necrotic debris may appear as irregularly shaped regions of low signal intensity within the necrotic collections on T_1- and T_2-weighted images. Hemorrhagic changes may also show increased T_1 and T_2 signal in these fluid collections. These necrotic/hemorrhagic changes do not show enhancement. These fluid collections may also show peripheral enhancement without identifiable wall (Figure 20.7).

20.6.1.3.4 Walled-Off Necrosis

ANC's mature and develop thick non-epithelialized wall, acquiring the term walled-off necrosis (WON), particularly after 4 weeks. However, these acute necrotic

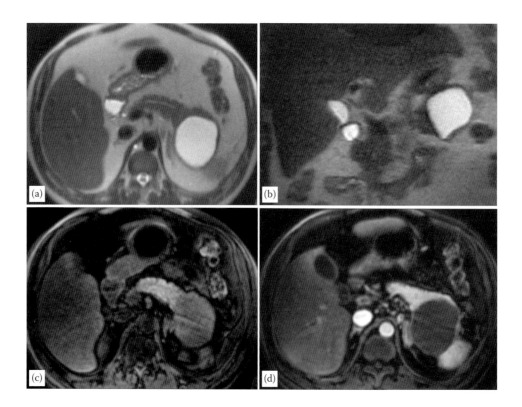

FIGURE 20.6
Pseudocyst. Transverse (a) and coronal (b) T_2-weighted SS-ETSE images, T_1-weighted fat-suppressed SGE (c), and T_1-weighted postgadolinium hepatic arterial dominant phase (d). There is a large thin-walled cyst located posterior to the pancreatic tail thatdemonstrates mild uniform wall enhancement.

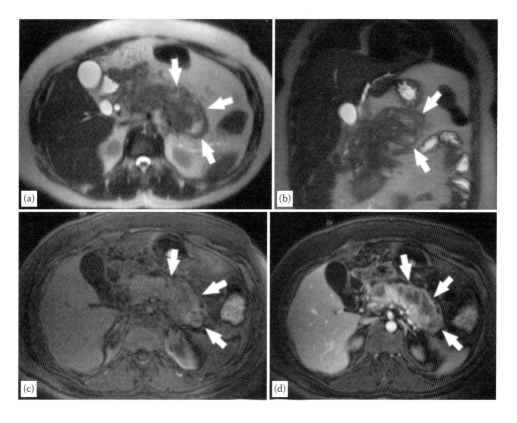

FIGURE 20.7

Acute pancreatitis with necrotic collections. Transverse (a) and coronal (b) T_2-weighted SS-ETSE images, T_1-weighted fat-suppressed SGE (c), and T_1-weighted postgadolinium hepatic arterial dominant phase (d). There are necrotic debris (white arrows) involving the pancreatic parenchyma and the peripancreatic region, with low signal intensity on T_1- and T_2-weighted images, showing peripheral enhancement on postgadolinium images.

fluid collections may also transform into WON before 4 weeks. Management for WON is different from pseudocyst as it contains non-liquefied debris, which may need to be drained or surgically removed. These fluid collections are usually loculated and contain necrotic/hemorrhagic material showing variable T_1 and T_2 signal depending on the content. These fluid collections show progressive enhancement on postgadolinium images.

20.6.1.3.5 Infected Pancreatic Necrosis

Pancreatic and peripancreatic necrosis/collections can remain sterile or become infected. The development of secondary infection in pancreatic necrosis is associated with increased morbidity and mortality.[28]

The diagnosis of infected ANC or WON can be suspected in the presence of extraluminal gas. This extraluminal gas is present in areas of necrosis and may or may not form a gas/fluid level depending on the amount of gas/fluid content present at that stage of the disease.

20.6.1.4 Fistula Formation

Disruption of pancreatic ducts and fistulous connections to the fluid collections and gastrointestinal tract could be visualized with T_2-weighted sequences, MRCP images, and postgadolinium fat-suppressed 3D-GE sequences. Fistulous tracts can be directly visualized with T_2-weighted sequences including MRCP images and indirectly with increased peripheral enhancement.

20.6.1.4.1 Chronic Pancreatitis

Chronic pancreatitis is defined pathologically as continuous or relapsing inflammation of the organ leading to irreversible morphologic injury and typically leading to permanent impairment of both exocrine and endocrine functions. Chronic pancreatitis can occur as a consequence of multiple factors, including biliary stone disease, alcohol consumption, malignancy, metabolic disorders, and various genetic and environmental insults, including trauma.[29] It can cause abdominal pain, weight loss, steatorrhea, and diabetes mellitus.

The histopathological changes in chronic pancreatitis evolve from unevenly distributed fibrosis in early chronic pancreatitis to diffuse fibrosis involving the entire gland in late stages. In advanced disease, large areas of acinar parenchyma are replaced with sclerotic tissue causing atrophy. Ductal irregularities such as strictures, dilatation, and side-branches ectasia occur due to surrounding fibrosis. Other characteristic findings of severe chronic pancreatitis are calcifications and presence of complications such as pseudocyst, vascular aneurysms, and venous thrombosis.

20.6.1.4.2 Role of Imaging

Imaging plays a significant role in detecting parenchymal and ductal abnormalities in chronic pancreatitis and helps in differentiating early from advanced phases to a certain extent, which further guides the management of these patients (Table 20.2).

20.6.1.4.3 Early Chronic Pancreatitis

Ultrasound and CT are insensitive in diagnosis of early chronic pancreatitis, as they often show no abnormalities. A recent study showed that parenchymal changes might precede ductal changes in chronic pancreatitis, thus depicting the importance of MRI compared to MRCP in the early diagnosis of disease.[30]

MRI detects not only morphologic characteristics, but also early fibrotic changes. Fibrosis is shown by diminished signal intensity on T_1-weighted fat-suppressed images and diminished enhancement on immediate postgadolinium gradient-echo images (Figure 20.8).[31] Some investigators reported that patients with abnormal MRI findings but normal MRCP might benefit from

dynamic secretin-MRCP (S-MRCP), which may reveal ductal abnormalities due to improved visualization otherwise not detected on MRCP.[30]

20.6.1.4.4 Late Chronic Pancreatitis

The findings of late or advanced chronic pancreatitis include pancreatic atrophy with diminished signal intensity of the pancreas on T_1-weighted fat-suppressed images, and abnormally low percentage of contrast enhancement on immediate post-contrast images, and progressive parenchymal enhancement on the interstitial phase post-contrast images, reflecting the pattern of enhancement of fibrous tissue. MRCP demonstrates dilatation of the main pancreatic duct with ectasia of the side branches, giving chain of lakes appearance manifested as pancreatic ductal strictures, irregularities, and intraductal calculi, appearing as hypointense filling defects (Figure 20.9).

The imaging findings of early and late chronicpancreatitis are summarized in Table 20.2.

20.6.1.4.5 Complications of Chronic Pancreatitis

The most common non-neoplastic complication of chronic pancreatitis include pseudocysts, pseudoaneurysms (due to erosion of the arterial wall), splenic vein thrombosis with subsequent development of collaterals, biliary obstruction (due to pseudocysts), and gastrointestinal complications such as gastric outlet obstruction or bowel ischemia.[32,33] Chronic pancreatitis uncommonly may show inflammatory mass lesions and also increase the risk of pancreatic adenocarcinoma.

20.6.2 Special Types of Pancreatitis

20.6.2.1 Autoimmune Pancreatitis

Autoimmune pancreatitis (AIP) is a distinct form of pancreatitis characterized clinically by obstructive jaundice (with or without pancreatic mass), histologically by a lymphoplasmacytic infiltrate and fibrosis, and therapeutically by a dramatic response to steroids.[34] It has been reported to be a part of the spectrum of IgG4-related

TABLE 20.2

Imaging Findings of Chronic Pancreatitis

Early Findings	Diminished signal intensity on T_1-weighted fat-suppressed
	Diminished enhancement on immediate post-gadolinium gradient-echo images
	Ductal abnormalities revealed by MRCP or dynamic secretin-MRCP
Late Findings	Parenchymal atrophy
	Dilatation and beading of the main pancreatic duct with ectasia of the side branches
	Complications: pseudocysts, pseudoaneurysms, splenic vein thrombosis with collaterals, biliary obstruction

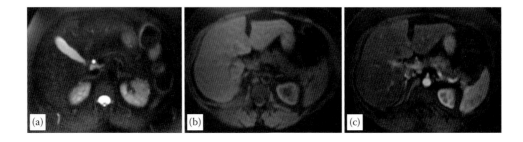

FIGURE 20.8
Early chronic pancreatitis. Fat-suppressed T_2-weighted SS-ETSE (a), T_1-weighted fat-suppressed SGE (b), and T_1-weighted postgadolinium hepatic arterial dominant phase (c). Fibrotic pancreatic parenchyma is shown by diminished signal intensity on T_1-weighted fat-suppressed images (b) and diminished enhancement on immediate postgadolinium gradient-echo images (c). The mains pancreatic duct is normal (a).

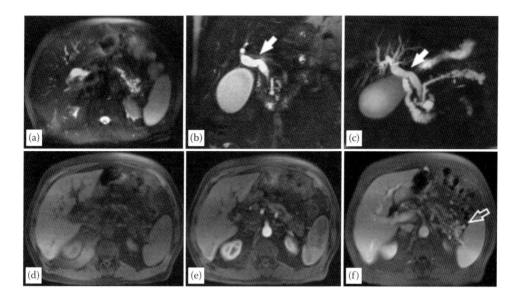

FIGURE 20.9
Chronic pancreatitis with ductal changes. Transverse (a) and coronal (b) fat-suppressed T_2-weighted SS-ETSE images, MRCP (c), T_1-weighted fat-suppressed SGE (d), T_1-weighted postgadolinium hepatic arterial dominant phase (e), and hepatic venous phase (f). The pancreatic parenchyma is atrophic with diminished signal intensity on T_1-weighted fat-suppressed images (d), and diminished enhancement on immediate post-contrast images (e). The main pancreatic duct is dilated, in association with ectasia of the side branches (a), (b), and (c). Complications as biliarydilatation (white arrows) and splenic vein thrombosis with subsequent development of collaterals (white open arrow) can also be seen.

disease that is characterized by increased IgG4 levels. IgG4-related disease is a systemic disease that involves multiple organs including but not limited to the pancreas, bile ducts, kidneys, salivary glands, and lymph nodes. Manifestations of IgG4-related disease include but not limited to AIP, sclerosing cholangitis, and enlarged lymph nodes.

The MR appearance of AIPis characterized by enlarged pancreas with moderately decreased signal intensity on T_1-weighted images, mildly high signal intensity on T_2-weighted images, decreased early enhancement and delayed postgadolinium enhancement of the pancreatic parenchyma. Additional findings that may be observed in AIP include: (1) capsule-like rim surrounding the diseased parenchyma that is hypointense/mildly hyperintense on T_2-weighted images and may show delayed postgadolinium enhancement;[35] (2) absence of parenchymal atrophy, (3) ductal beading and ductal dilatation proximal to the site of stenosis, (4) absence of peripancreatic fluid, and (5) clear demarcation of the abnormality[36] (Figure 20.10).

MRCP depicts diffused or segmental narrowing, and irregularity of the main pancreatic duct as characteristic findings. Sclerosing cholangitis is frequently seen in association with AIP. Intrahepatic and extrahepatic bile ducts frequently show beading, strictures, and wall thickening with increased enhancement. AIP has three types based on morphologic patterns: diffuse, focal, and multifocal. Diffuse disease is the most common type. A swollen, sausage-like pancreas with poorly demonstrated borders and a capsule-like rim is characteristic for this type.[37]

The diffuse form of AIP may mimic lymphoma.

20.6.2.2 Groove/Paraduodenal Pancreatitis

Groove pancreatitis is a rare form of the focal form of pancreatitis involving the anatomic groove between the pancreatic head, duodenum, and CBD. Groove pancreatitis may progress to chronic pancreatitis.

Pathogenesis remains controversial but may result from obstruction of the accessory pancreatic duct as it drains into the second portion of the duodenum through the minor ampulla. It is commonly seen in patients with history of alcohol abuse.[38]

Groove pancreatitis is categorized into two forms: pure, involving exclusively the groove, and segmental, involving the groove and extending into the pancreatic head.[39]

Presence of cystic changes, frequently located in the expected region of the pancreatic accessory duct, is considered a prominent feature of this process, likely related to accessory duct obstruction.[40]

The features of groove pancreatitis also depend on the age of inflammation: acute versus chronic. Classic features in the pure form can range from fluid, ill-defined fat stranding to frank soft tissue within the pancreaticoduodenal groove with increased delayed enhancement due to fibrosis in the chronic form. Thickening of medial duodenal wall on coronal images and presence of cysts can be appreciated sometimes.[39,41] Groove pancreatitis may also form mass lesions.

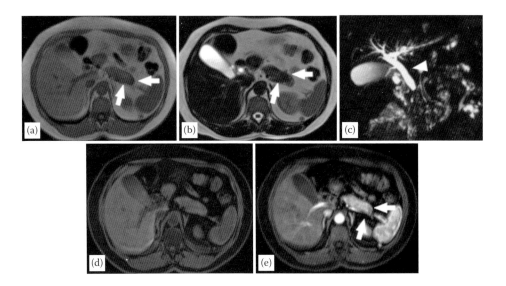

FIGURE 20.10

IgG4 autoimmune pancreatitis. T_1-weighted in-phase SGE (a), T_2-weighted SS-ETSE (b), MRCP (c), T_1-weighted fat-suppressed SGE (d) and T_1-weighted postgadolinium hepatic arterial dominant phase (e). Enlarged pancreas with moderately decreased signal intensity on T_1-weighted images and mildly high signal intensity on T_2-weighted images. Additional findings as hypointense capsule-like rim surrounding the parenchyma (white arrows) and biliarydilatation are also seen (white arrow head).

20.6.2.3 Hereditary Pancreatitis

Hereditary pancreatitis is an autosomal dominant disease presenting as multiple episodes of pancreatitis in the absence of any predisposing factors. Imaging findings include parenchymal and intraductal calcifications and parenchymal atrophy. However, in hereditary pancreatitis, imaging plays an important role to rule out structural causes of pancreatitis and closely monitor the development of pancreatic cancer, the risk of which is increased by many folds in these patients.

20.7 Neoplasms

According to the World Health Organization classification, pancreatic tumors are classified depending on the cell lineage they arise from. The tumors may have an epithelial or nonepithelial origin.

Tumors with epithelial origin include the exocrine pancreas (1) *ductal cells*, including ductal adenocarcinoma with its different histopathological variants, and mucinous and serous cystic tumors; (2) *acinar cells*, including acinar cell carcinoma (ACC) and mixed acinar-endocrine carcinoma; or (3) *uncertain origin*, including solid pseudopapillary tumor (SPT) and pancreatoblastoma or the endocrine pancreas (functioning and nonfunctioning tumors).

The nonepithelial tumors include tumors such as primary lymphoma and tumors of mesenchymal cell origin (hemangioma, lymphangioma, sarcoma, lipoma, etc).

There are also nonpancreatic tumor lesions in origin that might involve the pancreas, including malignant lesions such as metastasis or secondary lymphoma, and benign lesions such as intrapancreatic splenule.

20.8 Solid Neoplasms

20.8.1 Pancreatic Ductal Adenocarcinoma

Ductal adenocarcinoma is the most common malignant pancreatic neoplasm accounting for about 90% of all pancreatic tumors. Males are affected twice as often as women and the peak age of occurrence is in the seventh to eighth decades of life.[42] Pancreatic adenocarcinoma has a poor prognosis, with a five-year survival rate of only 5%, and surgery remains the sole curative treatment.[43]

The appearance of the typical ductal adenocarcinoma is an irregular, small focal solid mass (2–3 cm) without necrosis or hemorrhage. It is a heterogeneous, poorly enhancing, focal solid mass with a tendency for local invasiveness, including vascular encasement.

Approximately 60%–70% of pancreatic adenocarcinomas involve the pancreatic head, 10%–20% are located in the body and 5%–10% in the tail. Diffuse glandular involvement occurs in 5% of the cases.[44]

On T_2-weighted images, tumors are usually mildly hyperintense relative to pancreas and therefore difficult to visualize. Pancreatic adenocarcinoma sappear as low-signal-intensity masses on noncontrast fat-suppressed

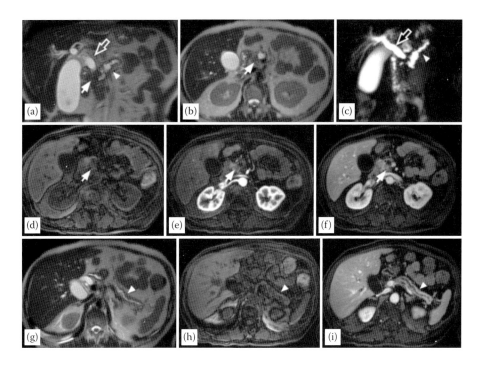

FIGURE 20.11

Pancreatic adenocarcinoma arising in the head. Coronal (a) and transverse (b) and (g) T_2-weighted SS-ETSE images, MRCP (c), T_1-weighted fat-suppressed SGE (d) and (h), T_1-weighted postgadolinium hepatic arterial dominant phase (e) and hepatic venous phase (f) and (i). There is a tumor arising in the pancreatic head (white arrow), which appears hypointense on T_1- (d) and T_2-weighted images (a) and (b). On immediate postgadolinium images (e), the tumor exhibits diminished enhancement compared to normal adjacent pancreatic parenchyma. On the hepatic venous phase gadolinium-enhanced fat-suppressed image (f), the tumor has decreased in conspicuity because of progressive tumor enhancement and pancreatic parenchymal washout. The MRCP image (c) demonstrates obstruction of the CBD (white open arrow) and pancreatic duct (white arrow head) creating the *double duct* sign. Obstruction of the main pancreatic duct caused by the neoplasm results in atrophy of the parenchyma, which shows low signal intensity on T_1-weighted fat-suppressed images (h) (tumor-associated chronic pancreatitis).

T_1-weighted images and usually well delineated from normal pancreatic tissue, which is high in signal intensity if there is no associated chronic pancreatitis that could result in decreased parenchymal signal (Figures 20.11 and 20.12).[5,45–48]

Detection of adenocarcinoma is best performed on immediate postgadolinium T_1-weighted gradient-echo images, where the lesion will enhance to a lesser extent than the surrounding normal pancreatic tissue (Figures 20.11 and 20.12), due to the abundant fibrous stroma and sparse tumor vascularity of the lesion.[47] The appearance of adenocarcinoma on the interstitial phase (>1 min postgadolinium) reflects the increased volume of the extracellular space and the venous drainage of cancers compared to normal pancreatic tissue such that the signal difference is often minimal.[47]

Obstruction of the main pancreatic duct caused by the neoplasm results in tumor-associated chronic pancreatitis, especially with large tumors. Pancreatic tissue distal to pancreatic cancer is characteristically atrophic and low in signal intensity compared to normal pancreatic parenchyma, due to chronic inflammation associated with progressive fibrosis and diminished proteinaceous fluid of the gland (Figure 20.11).[47] In these cases, depiction of cancer is poor on noncontrast T_1-weighted fat-suppressed images; however, arterial phase images can define the size and extent of adenocarcinomas that obstruct the pancreatic duct, as tumors almost always enhance less than adjacent chronically inflamed pancreas.[47,48]

Although these were the typical features of hypovascular adenocarcinomas, which are the most common form, pancreatic adenocarcinomas may be rarely isovascular. Isovascular tumors could not be differentiated from the background pancreas, especially when there is associated chronic pancreatitis resulting in decreased enhancement of the background pancreas. When there is associated chronic pancreatitis, the pancreas parenchyma shows decreased signal on T_1-weighted GE images and this also makes the detection of small pancreatic adenocarcinomas from the background pancreas. Under these circumstances, the combination of findings and findings on T_2-weighted images are particularly important for the identification.

A characteristic imaging appearance of pancreatic carcinoma consists of enlargement of the head of the pancreas with dilatation of the pancreatic and CBD and atrophy of the body and tail of the pancreas (Figure 20.11). However, enlargement of the head of the pancreas with obstruction of both ducts is not a feature unique

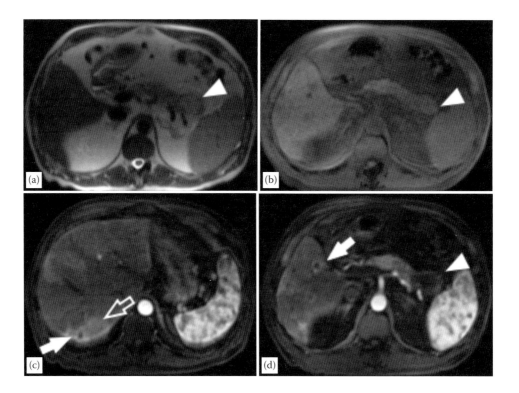

FIGURE 20.12

Pancreatic adenocarcinoma arising in the tail with liver metastases. T_2-weighted SS-ETSE (a), T_1-weighted fat-suppressed SGE (b), and T_1-weighted postgadolinium hepatic arterial dominant phase (c) and (d). There is a tumor arising in the pancreatic tail (white arrow head), which appears hypointense on T_1-weighted images (b) and mildly hyperintense on T_2-weighted images (a). On immediate postgadolinium images (d), the tumor exhibits diminished enhancement compared to normal adjacent pancreatic parenchyma. Liver metastases are present (white arrow, c and d) and are most clearly defined on immediate postgadolinium images as focal low-signal intensity masses with irregular rim enhancement. Transient, ill-defined, increased perilesional enhancement in the hepatic parenchyma is observed on immediate postgadolinium images (white open arrow, c).

to pancreatic cancer, as this same appearance may be appreciated, although less commonly, in patients with focal pancreatitis/AIP.

There are several MRI findings that radiologists should evaluate regarding respectability and describe in the MRI report such as: (1) distant metastases: liver, peritoneum, lung, and para-aorticlymph nodes; (2) infiltration of adjacent organs: stomach, colon, spleen; (3) invasion into the peripancreatic arteries: celiac trunk, hepatic artery, superior mesenteric artery; and (4) invasion into peripancreatic veins: portal and superior mesenteric vein.[5,46,47]

State of the art MRI is suitable to detect and characterize focal pancreatic ductal adenocarcinoma smaller than 1 cm,[45,46,49] which tend to appear as small non-contour-deforming pancreatic lesions. Detection of this early manifestation of disease is difficult or impossible to identify even with multiphasic current-generation CT.[50,51]

Liver metastases from pancreatic adenocarcinoma are generally irregular in shape, are low in signal intensity on conventional or fat-suppressed T_1-weighted images and minimally hyperintense on T_2-weighted images, and demonstrate irregular rim enhancement on immediate post-contrast gradient-echo images. The low fluid content and hypovascular nature of these metastases permit the distinction between these lesions from cysts and hemangiomas, respectively, even when lesions are 1 cm in diameter. Transient, ill-defined, increased perilesional enhancement in the hepatic parenchyma may be observed on immediate postgadolinium images (Figure 20.12).[1]

Lymph nodes are well shown on T_2-weighted fat-suppressed images and interstitial-phase gadolinium-enhanced fat-suppressed T_1-weighted images. Lymph nodes are moderately high in signal intensity in a background of low-signal-intensity suppressed fat with both of these techniques. T_2-weighted fat-suppressed imaging is particularly useful for the demonstration of lymph nodes in close approximation to the liver because of the signal intensity difference between moderately high-signal-intensity nodes and moderately low-signal-intensity liver.[1]

Vascular encasement by tumor is best shown with thin-section 3D gradient-echo images. Reformatted images can also be acquired. The combined evaluation of all postgadolinium images on different planes with

TABLE 20.3

Main Features of Pancreatic Ductal Adenocarcinoma

Gender	Male >female
Peak age	Seventh to eighth decades of life.
T_2-weighted images	Minimally hyperintense relative to pancreas
T_1-weighted images	Hypointense and usually well delineated from normal pancreatic tissue
Postgadolinium T_1-weighted images	Enhances less than the surrounding normal pancreatic tissue
Other features	Dilatation of the pancreatic and CBD and atrophy of the distal pancreas
Spread	Distant metastases to liver, peritoneum, lung, and para-aortic lymph nodes; infiltration of adjacent organs; and invasion into the peripancreatic arteries and veins

pre-contrast sequences is very helpful for local staging. Immediate postgadolinium gradient-echoimages are useful for evaluating arterial patency, and immediate and hepatic venous phase postgadolinium gradient-echo images for evaluating venous patency.[1] Main features of pancreatic ductal adenocarcinoma are summarized in Table 20.3.

20.8.2 Pancreatic Neuroendocrine Tumors

Pancreatic neuroendocrine tumors (NET) were the second most common solid tumors, previously called islet cell tumor, because they were thought to have originated from the islets of Langerhans; however recent evidence suggest that these tumors originate from pluripotential stem cells in the ductal epithelium.[52] Most cases are sporadic, but association with syndromes such as multiple endocrine neoplasia type 1,von Hippel–Lindau syndrome, neurofibromatosis type 1, and tuberous sclerosis have been observed. NETs are classified into functioning and nonfunctioning tumors. The prevalence of functioning and non-functioning tumors is not well established although it has been reported that functioning tumors are more common compared to nonfunctioning tumors. Nonfunctional tumors account for at least 15%–20% of pancreatic endocrine tumors and tend to present with symptoms owing to large tumor mass or metastatic disease. The functioning tumors may present with an endocrine abnormality resulting from the secretion of hormones.[53] The diagnosis of functioning NETs is almost always established biochemically, and the role of imaging is to depict the precise location of the tumor. The most common functioning pancreatic endocrine tumors are insulinomas and gastrinomas, Insulinomas are usually benign and gastrinomas are usually malignant. In general, functioning tumors manifest early in the course of disease when they are small, due to the clinical manifestations of excessive hormone production.Tumor morphologic features are variable. Small tumors are generally solid and homogeneous, whereas larger tumors are heterogeneous with presence of cystic degeneration. Calcifications are seen in 20% of these tumors.

On MRI, endocrine tumors are moderately low in signal intensity on T_1-weighted fat-suppressed and moderate to markedly high signal intensity on T_2-weighted images.[54] These tumors are typically highly hypervascular and therefore they enhance intensely on the hepatic arterial dominant phase after contrast administration (Figure 20.13). This distinctive feature must be interpreted cautiously, as although they may enhance more rapidly and intensely than the normal pancreas, they may also appear iso-intense during the arterial phase, as the normal pancreatic parenchyma is also highly vascularized. On the hepatic venous or interstitial phases, they may show washout or fade, if their sizes are particularly smaller than 2 cm.

Insulinomas are usually seen as small tumors (<2 cm), with intense and homogeneous enhancement on immediate postgadolinium images, whereas gastrinomas most commonly are larger lesions (3–4 cm approximately), with peripheral ring-like enhancement on immediate postgadolinium images.[1,55] Gastrinomas generally occur in a distinctive location, termed as the gastrinoma triangle, bordered superiorly by the confluence of the cystic and CBDs; inferiorly, by the second and third portions of the duodenum; and medially, by the neck and body of the pancreas.

The likelihood of malignancy rises in parallel with tumor size, and tumors larger than 5 cm are frequently malignant. Even when malignant, these tumors are slow-growing and the prognosis is better than for ductal adenocarcinoma.[42,56] Metastases to lymph nodes and solid organs such as the liver may have an enhancement pattern similar to that of the primary tumor (Figure 20.13).

It is important to differentiate NETs from other neoplasms of the pancreas, particularly ductal adenocarcinoma, since the prognoses and treatment options are different for both entities. Features that distinguish most pancreatic endocrine tumors from pancreatic adenocarcinoma include the high signal intensity on T_2-weighted sequences, increased homogeneous enhancement on postgadolinium hepatic arterial dominant phase images, hypervascular liver metastases, and lack of pancreatic duct obstruction or vascular encasement.[55] Venous thrombosis, peritoneal, and regional node enlargement, which are characteristic features of pancreatic ductal adenocarcinoma, are generally not present in endocrine tumors.

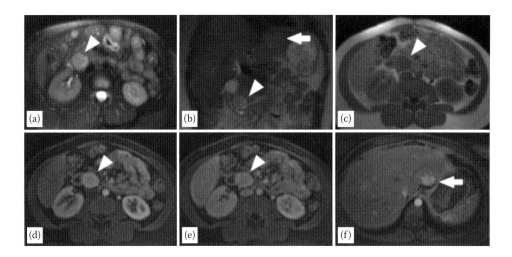

FIGURE 20.13
Pancreatic neuroendocrine tumor arising in the head with liver metastases. Transversefat-suppressed T_2-weighted SS-ETSE (a), coronal T_2-weighted SS-ETSE (b), T_1-weighted in-phase SGE (c), T_1-weighted postgadolinium hepatic arterial dominant phase (d) and (f), and hepatic venous phase (e). There is a tumor arising in the pancreatic head/uncinate process (white arrow head), which appears hypointense on T_1 (c) and hyperintense on T_2, with prominent enhancement on hepatic arterial dominant phase (d) that persists on hepatic venous phase (e). Liver metastases are present (white arrow), showing mildly hyperintense signal on T_2-weighted images (b) and hypervascular enhancement pattern, similar to that of the primary tumor (f).

20.8.3 Acinar Cell Carcinoma

ACC is a rare primary tumor of the exocrine gland of the pancreas, and although acinar cells constitute most of the pancreatic parenchyma, ACC represents only 1% of all exocrine pancreatic cancers. Tumors generally occur between the fifth and seventh decades.[1]

These cancers are generally exophytic, oval or round, well marginated, and hypovascular. Small tumors are generally solid, whereas larger tumors almost invariably contain cystic areas representing regions of necrosis, hemorrhage, and occasionally amorphous intratumoral calcifications, seen as signal voids.[57,58] The signal intensity on MRI is predominantly T_1 hypointense and T_2 iso to hyperintense.

ACC should always be considered when a largehypovascular pancreatic solid mass with variably sized central cystic areas or cystic masses is seen.[1,57,59] The tumor marker CA 19.9 that is generally increased in pancreatic adenocarcinomais rarely elevated in ACC.

20.8.4 Solid Pseudopapillary Tumor

SPT of the pancreas is a rare, low-grade epithelial malignancy of the exocrine pancreas that most often occurs in adolescent and young adult females,[1,44,60] and accounts for about 1%–2% of all pancreatic tumors.[61] This age presentation is rarely seen in ductal adenocarcinoma. It appears to have a predilection for Asian and African-American women, although rare cases have been reported in children and men. These tumors show low malignant potential.[62,63] Complete surgical removal

is the treatment of choice. Metastasis is rare but local recurrence has been described. The prognosis is excellent after resection.

The mass occurs most frequently in the head or tail. SPT is often discovered incidentally and presents as a large well-demarcated and encapsulated pancreatic mass, surrounded by a peripheral thick capsule and with variable relative proportions of intralesional solid, cystic, and hemorrhagic components. The peripheral capsule is an almost uniformly present finding in SPT on MRI, with an incidence ranging from 95%to 100%.[62] Vascular encasement by SPT may be seen in large tumors.

MRI typically demonstrates a well-defined lesion with heterogeneous signal intensity on T_1- and T_2-weighted images, which reflects the complex nature of the mass. Areas of high signal intensity on T_1-weighted images, and low or inhomogeneous signal intensity on T_2-weighted images can help identify blood products and may also help differentiate SPTs from endocrine tumors, whose cystic components generally are not hemorrhagic and therefore not typically possessing moderately increased signal intensity on T_1-weighted images.

20.8.5 Mesenchymal Tumors

Mesenchymal tumors of the pancreas are rare, accounting for 1%–2% of all pancreatic tumors.[64,65] They derive from various connective tissue components and are classified according to their histologic origin.

Primary pancreatic lymphoma, although unusual, is the most common malignant mesenchymal tumor appearing in the pancreas. Benign mesenchymal adipose tissue tumors, such as lipomas or teratomas are extremely rare and show diagnostic features on MRI, with homogeneous encapsulated mature fat or with fat-fluid levels, respectively.[64,65] Other mesenchymal tumors, such as lymphangiomas, leiomyoma, leiomyosarcoma, schwannoma, hemangioma, or hemangioendothelioma, have also been reported; however, they are exceedingly rare, existing in the literature mainly in the form of isolated case reports.

20.8.5.1 Pancreatic Lymphoma

Pancreatic lymphoma may be primary and secondary.

Non-Hodgkin lymphoma may involve peripancreatic lymph nodes or may secondarily invade the pancreas. Intermediate-signal-intensity peripancreatic lymph nodes are distinguished from high-signal-intensity normal pancreas on T_1-weighted fat-suppressed images.[66]

Two morphologic patterns of primary pancreatic lymphoma are recognized: focal and diffuse form. The focal form occurs in the pancreatic head in 80% of the cases and may mimic adenocarcinoma.

At MRI, lymphoma has a low signal intensity on T_1-weighted images and intermediate signal intensity on T_2-weighted images.

Pancreatic lymphoma carries a better prognosis than pancreatic ductal adenocarcinoma because first line treatment with chemotherapy is generally effective, producing long-term disease regression or remission. Surgery is not required in most cases.

Several features that may help distinguish pancreatic lymphoma from ductal adenocarcinoma are the presence of a bulky localized tumor in the pancreatic head with absent or minimal main pancreatic duct dilatation, enlarged lymph nodes below the level of the renal vein, and invasive tumor growth with infiltration of retroperitoneal and upper abdominal organs. Vascular invasion is less common in lymphoma than in pancreatic adenocarcinoma.[1] Additionally, CA 19.9 levels are usually not elevated in primary or secondary pancreatic lymphoma.

20.8.5.2 Nonpancreatic Tumor Lesion

20.8.5.2.1 Metastases of the Pancreas

Involvement of the pancreas by metastatic tumor may be the result of spread by direct extension or hematogenous metastases. Direct invasion from stomach and transverse colon carcinoma and GIST tumors are rare, but the most common forms of direct extension.

Metastases are most frequently from renal cell carcinoma and lung cancer followed by breast, colon, prostate, and malignant melanoma. Three morphological patterns of metastatic involvement of the pancreas have been described: solitary lesion (50%–70% of cases), multifocal (5%–10%), and diffuse (15%–44%).[67]

Metastases generally have low signal intensity on T_1-weighted, and mildly high signal intensity on T_2-weighted images. The enhancement of the majority of metastases follows a ring pattern, with a variable degree of enhancement depending on the angiogenic properties of the primary neoplasm. Ductal obstruction is uncommon, even with larger tumors, which is an important feature distinguishing from pancreatic ductal adenocarcinoma.[1,67]

20.8.5.2.2 Intrapancreatic Splenule

The presence of accessory splenules may arise within the substance of solid organs, notably the pancreas. Intrapancreatic splenule is a relatively uncommon location for splenules. These lesions typically are <2cm in size and occur within 3 cm of the tip of the tail of the pancreas.[68] The presence of a well-marginated rounded mass located in the distal tail of the pancreas with signal intensity features of the spleen on all MR sequences suggests the diagnosis of intrapancreatic accessory spleen. A distinctive feature of these masses is that when greater than 2 cm they may exhibit serpiginous enhancement on arterial phase images, as typically seen in the spleen.[68]

20.9 Cystic Neoplasms

Pancreatic cysts have become a common incidental finding due to the increasing use of cross-sectional imaging. Although secondary cystic change can be seen in most types of pancreatic neoplasms, cystic pancreatic lesions are characterized by their consistent, invariably present, cystic configuration.

Since cystic lesions may represent neoplastic and non-neoplastic processes, the accurate imaging-based characterization is necessary to distinguish cystic neoplasms of the pancreas from pseudocysts.

The evaluation of pancreatic cystic lesions involves a detailed analysis of cyst morphology and fluid content, as well as delineation of any communication with the pancreatic ductal system. Assessment of the entire pancreatic parenchyma provides additional important diagnostic information.[63] The soft-tissue imaging contrast capabilities of MRI are best suited for evaluating these features.[69]

Although different histologic types of cystic pancreatic neoplasms have been reported in the literature, serous cystadenomas, mucinous cystic neoplasms, and intraductal papillary mucinous neoplasms (IPMNs) account for 90% of all primary cystic pancreatic neoplasms.[70]

20.9.1 Serous Cystadenoma

This tumor frequently occurs in female patients (4:1), usually in the middle age to elderly (after 60 years). There is also an increased association with von Hippel–Lindau disease.[71] Serous cystadenomas most commonly affect the pancreatic head and usually have microcystic and multilocular appearance, demonstrating multiple small cysts less than 1–2 cm in diameter. Uncommonly, serous cystadenomas may be macrocystic (cysts measuring from 2 to 8 cm) including multilocular, oligolocular, or unilocular subtypes. These tumors do not show communication with the pancreatic ducts.

Microcystic serous cystadenoma is well demarcated and occasionally contains a central fibrotic scar. Tumors range in size from 1 to 12 cm, with an average diameter at presentation of 5 cm. The lesion may exhibit either a smooth or a nodular contour. On cut surface, small, closely packed cysts are filled with clear, watery (serous) fluid and separated by fine, fibrous septa, creating a honeycomb appearance. Calcifications may sometimes be present in the central scar. On MR images, the tumors are well-defined and do not demonstrate invasion of fat or adjacent organs.[72] On T_2-weighted images, the small cysts and intervening septations may be well shown as a cluster of small grape-like high-signal-intensity cysts. Tumor septations usually enhance minimally with gadolinium on early and late post-contrast images, although moderate enhancement on early post-contrast images may occur (Figure 20.14).[1] Delayed enhancement of the central scar may occasionally be observed,[2] and is more typical of large tumors. The presence of these features are generally sufficient for the diagnosis of this benign tumor.

Macrocystic serous cystadenomas exhibit distinctly different macroscopic features from microcystic lesions and may pose diagnostic difficulties for both radiologist and pathologist. They have high signal intensity on T_2-weighted images, and low signal intensity on T_1-weighted images. They may be multilocular, oligolocular, or unilocular. The cyst wall and septations demonstrate progressive mild to moderate enhancement on postgadolinium T_1-weighted images. The central scar is usually absent.[1] The diagnosis is usually made by follow-up and/or cyst aspiration if the lesion is smaller than 3 cm. If the lesion is larger than 3 cm or shows interval growth, surgical resection is usually preferred.

The main features of pancreatic serous cystoadenoma are summarized in Table 20.4.

20.9.2 Serous Cystadenocarcinoma

This malignant pancreatic tumor is extremely rare. The presence of thick septations and solid components are suggestive signs for serous cystadenocarcinoma.[1] They mimic macrocystic serous cystadenomas.

TABLE 20.4

Main Features of Pancreatic Serous Cystadenoma

Gender	Female >male
Peak age	After 60 years
Location	Pancreatic head
Imaging features	Usually microcystic and multilocular
	Honeycomb appearance
	Central scar with or without calcifications

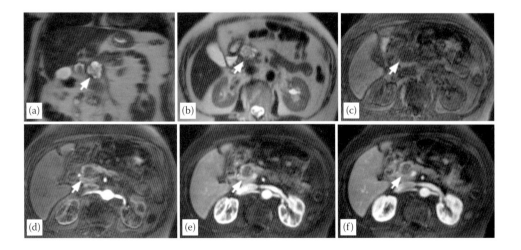

FIGURE 20.14

Pancreatic serous cystoadenoma arising in the head. Coronal (a) and transverse (b) T_2-weighted SS-ETSE images, T_1-weighted fat-suppressed SGE (c), T_1-weighted postgadolinium hepatic arterial dominant phase (d), hepatic venous phase (e) and interstitial phase (f). There is a cystic mass in the pancreatic head (white arrow). The lesion is well defined and shows a low T_1-signal intensity in a background of high-signal-intensity pancreas (c). T_2-weighted images (a) and (b) show fine septations creating a cluster of small grape-like high-signal-intensity cysts. The septations enhance minimally on immediate postgadolinium images (d) with progressive enhancement on late images (e) and (f).

20.9.2.1 Mucinous Cystadenoma/Cystadenocarcinoma

Mucinous cystic neoplasms occur more frequently in females (6:1), and approximately 50% occur in patients between the ages of 40 and 60 years.[73]

These tumors are characterized by the formation of large unilocular or multilocular cysts filled with abundant, thick gelatinous mucin, located in the body and tail of the pancreas.[1] They may be large (mean diameter of 10 cm), often multiloculated, and encapsulated.[74,75] Of these tumors, 10% may have scattered calcifications. These tumors do not show communication with the pancreatic ducts.

Mucinous cystadenomas have malignant potential, and adequate sampling of the cyst lining must be performed for pathologic analysis to determine whether foci of dysplasia or carcinoma *in situ* are present.[63]

Mucin produced by these tumors may result in high signal intensity on T_1- and T_2-weighted images of the primary tumor and liver metastases. These cystic lesions are usuallycomplex lesions with thick walls, septations, and solid components, which are usually low signal on T_2-weighted images. On gadolinium-enhanced T_1-weighted fat-suppressed images, the thick walls, septations, and solid components demonstrate increased enhancement.[2] Cyst walls and septations are often thicker in mucinous cystadenocarcinomas than those of mucinous cystadenomas. Mucinous cystadenomas are well circumscribed and show no evidence of metastases or invasion of adjacent tissues. Mucinous cystadenocarcinoma may be very locally aggressive malignancies with extensive invasion of adjacent tissues and organs. Absence of demonstration of tumor invasion into surrounding tissues does not, however, exclude malignancy.

Liver metastases are generally hypervascular showing intense ring enhancement on immediate gadolinium images, and may contain mucin, which results in mixed low and high signal intensity on T_1- and T_2-weighted images.[1]

The main features of pancreatic mucinous cystoadenoma/cystadenocarcinoma are summarized in Table 20.5.

TABLE 20.5

Main Features of Pancreatic Mucinous Cystadenoma/Cystadenocarcinoma

Gender	Female >male
Peak age	Fourth to fifth decades
Location	Pancreatic body and tail
Imaging features	Large unilocular or multilocular cyst
	High signal intensity on T_1- and T_2-weighted (mucin contents)
	Signs of malignancy: thicker walls or septations, papillary solid projections, invasion into surrounding tissue, liver metastases

20.9.2.2 Intraductal Papillary Mucinous Neoplasms

IPMNs of the pancreas were first described relatively recently and occur most frequently in men (mean age, 65 years).[63,76] IPMNs are mucinous cystic tumors of the pancreas that are clinically and histopathologically distinct from mucinous cystadenomas. The tumors are characterized by a mucinous transformation of the pancreatic ductal epithelium that usually demonstrates papillary projections at histologic analysis.[77,78] Duct obstruction is secondary to tenacious plugs of mucin, elaborated by the epithelium, or ductal compression by cystic masses.[79]

The lesions can represent a spectrum of abnormalities from simple hyperplasia to dysplasia, papillary adenoma, and carcinoma. This spectrum of abnormalities may coexist. In general hyperplasia, dysplasia and adenoma may undergo malignant transformation and transform into carcinoma.[1]

IPMNs may be classified according to whether the disease process involves the main pancreatic duct or isolated side branches. They also may be characterized according to whether they produce a diffuse pattern of ductal dilatation or a segmental cystic appearance.[77] The location of the tumor is an important factor for the prognosis.[80] Main duct IPMNs are more likely to be malignant, with approximately 60%–70% of cases demonstrating invasive carcinoma,[81,82] whereas only 22% of branch duct IPMNs demonstrate foci of carcinoma.[83] IPMNs are frequently multifocal, and 5%–10% involve the entire pancreas.[84]

20.9.2.3 IPMN—Main Duct Type

Main pancreatic duct involvement presents diffuse ductal dilatation, copious mucin production, and papillary growth. On MR images, a greatly expanded main pancreatic duct is demonstrated on T_2-weighted or MRCP images. Irregular-enhancing tissue along the ductal epithelium is appreciated on postgadolinium images, confirming that underlying tumor is the cause of the ductal dilatation.[1]

20.9.2.4 IPMN—Side-Branch Type

IPMN involving predominantly side-branch ducts appear as oval-shaped cystic masses in proximity to the main pancreatic duct, most commonly located in the head of the pancreas. Septations are generally present, creating a cluster of grapes appearance.[1] MRCP images are able to show communication of the cystic tumor with the main pancreatic duct in the majority of cases (Figure 20.15).[85–89]

A combination of the clinical (presence or absence of symptoms) and imaging characteristics (size of the cyst,

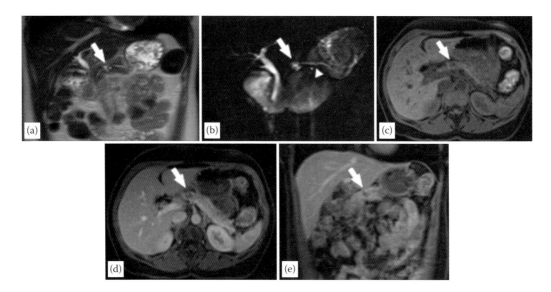

FIGURE 20.15

Side-branch IPMN arising in the pancreatic neck. Coronal T_2-weighted SS-ETSE (a), MRCP (b), T_1-weighted fat-suppressed SGE (c), transverse (d), and coronal (e) T_1-weighted postgadolinium hepatic venous phase (f). There is a cluster of small cysts (white arrow) in the pancreatic head that exhibit communication with the main pancreatic duct, well shown on the MRCP image (white arrow head, b).

TABLE 20.6

Main Features of Pancreatic IPMN

Gender	Male>female
Peak age	Sixth to seventh decades
Location	Main and/or side-branch ducts
Image features	*Main Duct:* dilatation of the main pancreatic ducton T_2-weighted or MRCP images, papillary projections, irregular-enhancing tissue along the ductal epithelium on postgadolinium images
	Side Branches: oval-shaped cystic masses (cluster of grapes), most commonly located in the head of the pancreas, showing communication with the main pancreatic ducton MRCP images

presence or absence of solid component, and caliber of the main pancreatic duct) is important to take a decision between resection and conservative management.[90] If they do not contain any solid components, thick septa or internal growth, they can be followed under the size of 3 cm. The main features of intraductal papillary intraductal mucinous neoplasms are summarized in Table 20.6.

20.10 Trauma

The pancreas is vulnerable to crushing injury in blunt trauma due to impact against the adjacent vertebral column. In adults, over 75% of blunt injuries to the pancreas are due to motor vehicle collisions. In children, bicycle injuries are common, and child abuse may result in pancreatic injuries in infants.[91] Two-thirds of pancreatic injuries occur in the pancreatic body, and the remainder occur equally in the head, neck, and tail.[92] Isolated pancreatic injuries are rare, and associated injuries, especially to the liver, stomach, duodenum, and spleen, occur in over 90% of cases.[92,93]

The main source of delayed morbidity and mortality from pancreatic trauma is disruption of the pancreatic duct. Injuries that spare the pancreatic duct rarely result in morbidity or death.[93] Disruption of the pancreatic duct is treated surgically or by therapeutic endoscopy with stent placement, while injuries without duct involvement are usually treated nonsurgically. As such, it is critical that imaging focus on the integrity of the duct or findings that suggest damage to the pancreatic duct.[94]

In the past, endoscopic retrograde cholangiopancreatography (ERCP) was the only method available for evaluating pancreatic duct integrity. MR pancreatography has emerged as an attractive alternative for direct imaging of the pancreatic duct.[95] MR pancreatography has the advantage of being noninvasive, faster, and more readily available than ERCP, may demonstrate other abnormalities, such as fluid collections, and is helpful in assessing parenchymal injury.[96]

As a sequel of trauma, stenosis of the pancreatic duct with distal ductal dilatation and distal changes of chronic pancreatitis may be observed. This condition is not rare, and should be entertained when a sharp transition is observed in the midbody of the pancreas, overlying the vertebral column, with normal pancreatic head and midbody, and distal atrophy and ductal dilatation. History of trauma, typically motor vehicle accident, even if remote, should be sought.[1]

20.11 Pancreatic Transplants

Pancreas transplantation is an established therapy for severe type-1 diabetes mellitus complicated by end-stage renal failure or, less often, for poorly controlled severe diabetes.[97] Transplantation offers freedom from insulin therapy and reduces the frequency of complications such as nephropathy, retinopathy, and vasculopathy.[97]

Postoperative complications can be categorized as early or late, with most early complications being surgical or technical. Surgical complications include anastomotic breakdown with bowel leak, hemorrhage, infection, and vascular thrombosis. Nonsurgical complications are usually immunologic, with rejection being the single most common cause of graft loss.[97]

High-resolution 3D contrast-enhanced MR angiography is an accurate diagnostic technique for evaluation of the arterial and venous anatomy of the pancreas transplant.[98-101] On T_1-weighted images, the pancreatic parenchyma is homogeneous and should be hyperintense relative to the liver. On T_2-weighted images, the normal pancreas allograft shows signal intensity between that of fluid and muscle. T_2-weighted images are most sensitive to abnormalities of the pancreas transplant because the majority of pathologic processes increase glandular water content.[97]

20.12 Conclusion

The combination of all MRI sequences available enables a full evaluation of most types of pancreatic diseases and their consequences, offering potential advantages over CT images.

References

1. Semelka RC. *Abdominal-Pelvic MRI*, 3rd edn. Hoboken, New Jersey: John Wiley & Sons; 2010.
2. Semelka RC, Ascher SM. MR imaging of the pancreas. *Radiology*. Sep 1993;188(3):593–602.
3. Winston CB, Mitchell DG, Outwater EK, Ehrlich SM. Pancreatic signal intensity on T1-weighted fat saturation MR images: Clinical correlation. *Journal of Magnetic Resonance Imaging: JMRI*. May–Jun 1995;5(3):267–271.
4. Mitchell DG, Vinitski S, Saponaro S, Tasciyan T, Burk DL, Jr., Rifkin MD. Liver and pancreas: Improved spin-echo T1 contrast by shorter echo time and fat suppression at 1.5 T. *Radiology*. Jan 1991;178(1):67–71.
5. Semelka RC, Kroeker MA, Shoenut JP, Kroeker R, Yaffe CS, Micflikier AB. Pancreatic disease: Prospective comparison of CT, ERCP, and 1.5-T MR imaging with dynamic gadolinium enhancement and fat suppression. *Radiology*. Dec 1991;181(3):785–791.
6. Takehara Y, Ichijo K, Tooyama N et al. Breath-hold MR cholangiopancreatography with a long-echo-train fast spin-echo sequence and a surface coil in chronic pancreatitis. *Radiology*. Jul 1994;192(1):73–78.
7. Bret PM, Reinhold C, Taourel P, Guibaud L, Atri M, Barkun AN. Pancreas divisum: Evaluation with MR cholangiopancreatography. *Radiology*. Apr 1996;199(1):99–103.
8. Soto JA, Barish MA, Yucel EK et al. Pancreatic duct: MR cholangiopancreatography with a three-dimensional fast spin-echo technique. *Radiology*. Aug 1995;196(2):459–464.
9. Semelka RC, Simm FC, Recht MP, Deimling M, Lenz G, Laub GA. MR imaging of the pancreas at high field strength: Comparison of six sequences. *Journal of Computer Assisted Tomography*. Nov–Dec 1991;15(6):966–971.
10. Mitchell DG, Winston CB, Outwater EK, Ehrlich SM. Delineation of pancreas with MR imaging: Multiobserver comparison of five pulse sequences. *Journal of Magnetic Resonance Imaging: JMRI*. Mar–Apr 1995;5(2):193–199.
11. Cruikshank AH, Benbow EW. *Pathology of the Pancreas*. 2nd edn. Berlin, Germany: Springer; 1995.
12. Desai MB, Mitchell DG, Munoz SJ. Asymptomatic annular pancreas: Detection by magnetic resonance imaging. *Magnetic Resonance Imaging*. 1994;12(4):683–685.
13. Applegate KE, Goske MJ, Pierce G, Murphy D. Situs revisited: Imaging of the heterotaxy syndrome. *RadioGraphics: A Review Publication of the Radiological Society of North America, Inc.* Jul–Aug 1999;19(4):837–852; discussion 853–834.
14. Tham RT, Heyerman HG, Falke TH et al. Cystic fibrosis: MR imaging of the pancreas. *Radiology*. Apr 1991;179(1):183–186.
15. Ferrozzi F, Bova D, Campodonico F et al. Cystic fibrosis: MR assessment of pancreatic damage. *Radiology*. Mar 1996;198(3):875–879.
16. King LJ, Scurr ED, Murugan N, Williams SG, Westaby D, Healy JC. Hepatobiliary and pancreatic manifestations of cystic fibrosis: MR imaging appearances. *RadioGraphics: A Review Publication of the Radiological Society of North America, Inc.* May–Jun 2000;20(3):767–777.
17. Busireddy KK, AlObaidy M, Ramalho M et al. Pancreatitis-imaging approach. World journal of gastrointestinal pathophysiology. Aug 15 2014;5(3):252–270.
18. Shanbhogue AK, Fasih N, Surabhi VR, Doherty GP, Shanbhogue DK, Sethi SK. A clinical and radiologic review of uncommon types and causes of pancreatitis. *Radiographics: A Review Publication of the Radiological Society of North America, Inc.* Jul–Aug 2009;29(4):1003–1026.
19. Banks PA, Bollen TL, Dervenis C et al. Classification of acute pancreatitis–2012: Revision of the Atlanta classification and definitions by international consensus. *Gut*. Jan 2013;62(1):102–111.
20. Balthazar EJ. CT diagnosis and staging of acute pancreatitis. *Radiologic Clinics of North America*. Jan 1989;27(1):19–37.

21. Kim YK, Ko SW, Kim CS, Hwang SB. Effectiveness of MR imaging for diagnosing the mild forms of acute pancreatitis: Comparison with MDCT. *Journal of Magnetic Resonance Imaging: JMRI.* Dec 2006;24(6):1342–1349.

22. Balthazar EJ, Freeny PC, vanSonnenberg E. Imaging and intervention in acute pancreatitis. *Radiology.* Nov 1994;193(2):297–306.

23. Lenhart DK, Balthazar EJ. MDCT of acute mild (non-necrotizing) pancreatitis: Abdominal complications and fate of fluid collections. *AJR American Journal of Roentgenology.* Mar 2008;190(3):643–649.

24. Matos C, Cappeliez O, Winant C, Coppens E, Deviere J, Metens T. MR imaging of the pancreas: A pictorial tour. *Radiographics: A Review Publication of the Radiological Society of North America, Inc.* Jan–Feb 2002;22(1):e2.

25. Lau ST, Simchuk EJ, Kozarek RA, Traverso LW. A pancreatic ductal leak should be sought to direct treatment in patients with acute pancreatitis. *American Journal of Surgery.* May 2001;181(5):411–415.

26. Sandrasegaran K, Tann M, Jennings SG et al. Disconnection of the pancreatic duct: An important but overlooked complication of severe acute pancreatitis. *RadioGraphics: A Review Publication of the Radiological Society of North America, Inc.* Sep–Oct 2007;27(5):1389–1400.

27. Tann M, Maglinte D, Howard TJ et al. Disconnected pancreatic duct syndrome: Imaging findings and therapeutic implications in 26 surgically corrected patients. *Journal of Computer Assisted Tomography.* Jul–Aug 2003;27(4):577–582.

28. Petrov MS, Shanbhag S, Chakraborty M, Phillips AR, Windsor JA. Organ failure and infection of pancreatic necrosis as determinants of mortality in patients with acute pancreatitis. *Gastroenterology.* Sep 2010;139(3):813–820.

29. Peery AF, Dellon ES, Lund J et al. Burden of gastrointestinal disease in the United States: 2012 update. *Gastroenterology.* Nov 2012;143(5):1179–1187 e1171–e1173.

30. Balci NC, Alkaade S, Magas L, Momtahen AJ, Burton FR. Suspected chronic pancreatitis with normal MRCP: Findings on MRI in correlation with secretin MRCP. *Journal of Magnetic Resonance Imaging: JMRI.* Jan 2008;27(1):125–131.

31. Semelka RC, Shoenut JP, Kroeker MA, Micflikier AB. Chronic pancreatitis: MR imaging features before and after administration of gadopentetatedimeglumine. *Journal of Magnetic Resonance Imaging: JMRI.* Jan–Feb 1993;3(1):79–82.

32. Bollen TL. Imaging of acute pancreatitis: Update of the revised Atlanta classification. *Radiologic Clinics of North America.* May 2012;50(3):429–445.

33. Remer EM, Baker ME. Imaging of chronic pancreatitis. *Radiologic Clinics of North America.* Dec 2002;40(6):1229–1242, v.

34. Shimosegawa T, Chari ST, Frulloni L et al. International consensus diagnostic criteria for autoimmune pancreatitis: Guidelines of the International Association of Pancreatology. *Pancreas.* Apr 2011;40(3):352–358.

35. Irie H, Honda H, Baba S et al. Autoimmune pancreatitis: CT and MR characteristics. *AJR American Journal of Roentgenology.* May 1998;170(5):1323–1327.

36. Van Hoe L, Gryspeerdt S, Ectors N et al. Nonalcoholic duct-destructive chronic pancreatitis: Imaging findings. *AJR American Journal of Roentgenology.* Mar 1998;170(3):643–647.

37. Takahashi N, Fletcher JG, Fidler JL, Hough DM, Kawashima A, Chari ST. Dual-phase CT of autoimmune pancreatitis: Amultireader study. *AJR American Journal of Roentgenology.* Feb 2008;190(2):280–286.

38. Chatelain D, Vibert E, Yzet T et al. Groove pancreatitis and pancreatic heterotopia in the minor duodenal papilla. *Pancreas.* May 2005;30(4):e92–e95.

39. Blasbalg R, Baroni RH, Costa DN, Machado MC. MRI features of groove pancreatitis. *AJR American Journal of Roentgenology.* Jul 2007;189(1):73–80.

40. Triantopoulou C, Dervenis C, Giannakou N, Papailiou J, Prassopoulos P. Groove pancreatitis: A diagnostic challenge. *European Radiology.* Jul 2009;19(7):1736–1743.

41. Raman SP, Salaria SN, Hruban RH, Fishman EK. Groove pancreatitis: Spectrum of imaging findings and radiology-pathology correlation. *AJR American Journal of Roentgenology.* Jul 2013;201(1):W29–W39.

42. Mergo PJ, Helmberger TK, Buetow PC, Helmberger RC, Ros PR. Pancreatic neoplasms: MR imaging and pathologic correlation. *RadioGraphics: A Review Publication of the Radiological Society of North America, Inc.* Mar–Apr 1997;17(2):281–301.

43. Ros PR, Mortele KJ. Imaging features of pancreatic neoplasms. *JBR-BTR: organe de la Societeroyalebelge de radiologie.* 2001;84(6):239–249.

44. Low G, Panu A, Millo N, Leen E. Multimodality imaging of neoplastic and nonneoplastic solid lesions of the pancreas. *RadioGraphics: A Review Publication of the Radiological Society of North America, Inc.* Jul–Aug 2011;31(4):993–1015.

45. Lee ES, Lee JM. Imaging diagnosis of pancreatic cancer: A state-of-the-art review. *World Journal of Gastroenterology: WJG.* Jun 28, 2014;20(24):7864–7877.

46. Benassai G, Mastrorilli M, Quarto G et al. Factors influencing survival after resection for ductal adenocarcinoma of the head of the pancreas. *Journal of Surgical Oncology.* Apr 2000;73(4):212–218.

47. Gabata T, Matsui O, Kadoya M et al. Small pancreatic adenocarcinomas: Efficacy of MR imaging with fat suppression and gadolinium enhancement. *Radiology.* Dec 1994;193(3):683–688.

48. Semelka RC, Kelekis NL, Molina PL, Sharp TJ, Calvo B. Pancreatic masses with inconclusive findings on spiral CT: Is there a role for MRI? *Journal of Magnetic Resonance Imaging: JMRI.* Jul–Aug 1996;6(4):585–588.

49. Morgan KA, Adams DB. Solid tumors of the body and tail of the pancreas. *The Surgical Clinics of North America.* Apr 2010;90(2):287–307.

50. Saisho H, Yamaguchi T. Diagnostic imaging for pancreatic cancer: Computed tomography, magnetic resonance imaging, and positron emission tomography. *Pancreas.* Apr 2004;28(3):273–278.

51. Vellet AD, Romano W, Bach DB, Passi RB, Taves DH, Munk PL. Adenocarcinoma of the pancreatic ducts: Comparative evaluation with CT and MR imaging at 1.5 T. *Radiology.* Apr 1992;183(1):87–95.

52. Oberg K, Eriksson B. Endocrine tumours of the pancreas. Best practice & research. *Clinical Gastroenterology*. Oct 2005;19(5):753–781.

53. Mozell E, Stenzel P, Woltering EA, Rosch J, O'Dorisio TM. Functional endocrine tumors of the pancreas: Clinical presentation, diagnosis, and treatment. *Current Problems in Surgery*. Jun 1990;27(6):301–386.

54. Semelka RC, Custodio CM, CemBalci N, Woosley JT. Neuroendocrine tumors of the pancreas: Spectrum of appearances on MRI. *Journal of Magnetic Resonance Imaging: JMRI*. Feb 2000;11(2):141–148.

55. Semelka RC, Cumming MJ, Shoenut JP et al. Islet cell tumors: Comparison of dynamic contrast-enhanced CT and MR imaging with dynamic gadolinium enhancement and fat suppression. *Radiology*. Mar 1993;186(3):799–802.

56. Sheth S, Hruban RK, Fishman EK. Helical CT of islet cell tumors of the pancreas: Typical and atypical manifestations. *AJR American Journal of Roentgenology*. Sep 2002;179(3):725–730.

57. Tatli S, Mortele KJ, Levy AD et al. CT and MRI features of pure acinar cell carcinoma of the pancreas in adults. *AJR American Journal of Roentgenology*. Feb 2005;184(2):511–519.

58. Hsu MY, Pan KT, Chu SY, Hung CF, Wu RC, Tseng JH. CT and MRI features of acinar cell carcinoma of the pancreas with pathological correlations. *Clinical Radiology*. Mar 2010;65(3):223–229.

59. Hu S, Hu S, Wang M, Wu Z, Miao F. Clinical and CT imaging features of pancreatic acinar cell carcinoma. *La Radiologiamedica*. Aug 2013;118(5):723–731.

60. Guerrache Y, Soyer P, Dohan A et al. Solid-pseudopapillary tumor of the pancreas: MR imaging findings in 21 patients. *Clinical Imaging*. Jul–Aug 2014;38(4):475–482.

61. Ng KH, Tan PH, Thng CH, Ooi LL. Solid pseudopapillarytumour of the pancreas. *ANZ Journal of Surgery*. Jun 2003;73(6):410–415.

62. Cooper JA. Solid pseudopapillary tumor of the pancreas. *RadioGraphics: A Review Publication of the Radiological Society of North America, Inc*. Jul–Aug 2006;26(4):1210.

63. Kalb B, Sarmiento JM, Kooby DA, Adsay NV, Martin DR. MR imaging of cystic lesions of the pancreas. *RadioGraphics: A Review Publication of the Radiological Society of North America, Inc*. Oct 2009;29(6):1749–1765.

64. Ferrozzi F, Zuccoli G, Bova D, Calculli L. Mesenchymal tumors of the pancreas: CT findings. *Journal of Computer Assisted Tomography*. Jul–Aug 2000;24(4):622–627.

65. Carlo Procacci AJM. *Imaging of the Pancreas: Cystic and Rare Tumors*. Berlin, Germany: Springer 2003.

66. Zeman RK, Schiebler M, Clark LR et al. The clinical and imaging spectrum of pancreaticoduodenal lymph node enlargement. *AJR American Journal of Roentgenology*. Jun 1985;144(6):1223–1227.

67. Tsitouridis I, Diamantopoulou A, Michaelides M, Arvanity M, Papaioannou S. Pancreatic metastases: CT and MRI findings. *Diagnostic and Interventional Radiology*. Mar 2010;16(1):45–51.

68. Heredia V, Altun E, Bilaj F, Ramalho M, Hyslop BW, Semelka RC. Gadolinium- and superparamagnetic-iron-oxide-enhanced MR findings of intrapancreatic accessory spleen in five patients. *Magnetic Resonance Imaging*. Nov 2008;26(9):1273–1278.

69. Martin DR, Semelka RC. MR imaging of pancreatic masses. *Magnetic Resonance Imaging Clinics of North America*. Nov 2000;8(4):787–812.

70. Fernandez-del Castillo C, Warshaw AL. Cystic tumors of the pancreas. *The Surgical Clinics of North America*. Oct 1995;75(5):1001–1016.

71. Ros PR, Hamrick-Turner JE, Chiechi MV, Ros LH, Gallego P, Burton SS. Cystic masses of the pancreas. *RadioGraphics: A Review Publication of the Radiological Society of North America, Inc*. Jul 1992;12(4):673–686.

72. Lewandrowski K, Warshaw A, Compton C. Macrocystic serous cystadenoma of the pancreas: A morphologic variant differing from microcystic adenoma. *Human Pathology*. Aug 1992;23(8):871–875.

73. Compagno J, Oertel JE. Mucinous cystic neoplasms of the pancreas with overt and latent malignancy (cystadenocarcinoma and cystadenoma). A clinicopathologic study of 41 cases. *American Journal of Clinical Pathology*. Jun 1978;69(6):573–580.

74. Minami M, Itai Y, Ohtomo K, Yoshida H, Yoshikawa K, Iio M. Cystic neoplasms of the pancreas: Comparison of MR imaging with CT. *Radiology*. Apr 1989;171(1):53–56.

75. Friedman AC, Lichtenstein JE, Dachman AH. Cystic neoplasms of the pancreas. Radiological-pathological correlation. *Radiology*. Oct 1983;149(1):45–50.

76. Campbell F, Azadeh B. Cystic neoplasms of the exocrine pancreas. *Histopathology*. Apr 2008;52(5):539–551.

77. Lack EE. *Pathology of the Pancreas, Gallbladder, Extrahepatic Biliary Tract, and AmpullaryRegion*. New York: Oxford University Press; 2003.

78. Adsay NV. Cystic neoplasia of the pancreas: Pathology and biology. *Journal of Gastrointestinal Surgery: Official Journal of the Society for Surgery of the Alimentary Tract*. Mar 2008;12(3):401–404.

79. Silas AM, Morrin MM, Raptopoulos V, Keogan MT. Intraductal papillary mucinous tumors of the pancreas. *AJR American Journal of Roentgenology*. Jan 2001;176(1):179–185.

80. Terris B, Ponsot P, Paye F et al. Intraductal papillary mucinous tumors of the pancreas confined to secondary ducts show less aggressive pathologic features as compared with those involving the main pancreatic duct. *The American Journal of Surgical Pathology*. Oct 2000;24(10):1372–1377.

81. Balzano G, Zerbi A, Di Carlo V. Intraductal papillary mucinous tumors of the pancreas: Incidence, clinical findings and natural history. *JOP: Journal of the Pancreas*. Jan 2005;6(1 Suppl.):108–111.

82. Salvia R, Fernandez-del Castillo C, Bassi C et al. Main-duct intraductal papillary mucinous neoplasms of the pancreas: Clinical predictors of malignancy and long-term survival following resection. *Annals of Surgery*. May 2004;239(5):678–685; discussion 685–687.

83. Rodriguez JR, Salvia R, Crippa S et al. Branch-duct intraductal papillary mucinous neoplasms: Observations in 145 patients who underwent resection. *Gastroenterology*. Jul 2007;133(1):72–79; quiz 309–310.

84. Salvia R, Crippa S, Falconi M et al. Branch-duct intraductal papillary mucinous neoplasms of the pancreas: To operate or not to operate? *Gut*. Aug 2007;56(8):1086–1090.

85. Buetow PC, Rao P, Thompson LD. From the Archives of the AFIP. Mucinous cystic neoplasms of the pancreas: Radiologic-pathologic correlation. *RadioGraphics: A Review Publication of the Radiological Society of North America, Inc.* Mar–Apr 1998;18(2):433–449.

86. Procacci C, Megibow AJ, Carbognin G et al. Intraductal papillary mucinous tumor of the pancreas: A pictorial essay. *RadioGraphics: A Review Publication of the Radiological Society of North America, Inc.* Nov–Dec 1999;19(6):1447–1463.

87. Koito K, Namieno T, Ichimura T et al. Mucin-producing pancreatic tumors: Comparison of MR cholangiopancreatography with endoscopic retrograde cholangiopancreatography. *Radiology.* Jul 1998;208(1):231–237.

88. Onaya H, Itai Y, Niitsu M, Chiba T, Michishita N, Saida Y. Ductectatic mucinous cystic neoplasms of the pancreas: Evaluation with MR cholangiopancreatography. *AJR American Journal of Roentgenology.* Jul 1998;171(1):171–177.

89. Irie H, Honda H, Aibe H et al. MR cholangiopancreatographic differentiation of benign and malignant intraductal mucin-producing tumors of the pancreas. *AJR American Journal of Roentgenology.* May 2000;174(5):1403–1408.

90. Tanaka M, Fernandez-del Castillo C, Adsay V et al. International consensus guidelines 2012 for the management of IPMN and MCN of the pancreas. *Pancreatology: Official Journal of the International Association of Pancreatology.* May–Jun 2012;12(3):183–197.

91. Ilahi O, Bochicchio GV, Scalea TM. Efficacy of computed tomography in the diagnosis of pancreatic injury in adult blunt trauma patients: A single-institutional study. *The American Surgeon.* Aug 2002;68(8):704–707; discussion 707–708.

92. Madiba TE, Mokoena TR. Favourable prognosis after surgical drainage of gunshot, stab or blunt trauma of the pancreas. *The British Journal of Surgery.* Sep 1995;82(9):1236–1239.

93. Bradley EL, 3rd, Young PR, Jr., Chang MC et al. Diagnosis and initial management of blunt pancreatic trauma: Guidelines from a multiinstitutional review. *Annals of Surgery.* Jun 1998;227(6):861–869.

94. Gupta A, Stuhlfaut JW, Fleming KW, Lucey BC, Soto JA. Blunt trauma of the pancreas and biliary tract: A multimodality imaging approach to diagnosis. *RadioGraphics: A Review Publication of the Radiological Society of North America, Inc.* Sep–Oct 2004;24(5):1381–1395.

95. Fulcher AS, Turner MA, Yelon JA et al. Magnetic resonance cholangiopancreatography (MRCP) in the assessment of pancreatic duct trauma and its sequelae: Preliminary findings. *The Journal of Trauma.* Jun 2000;48(6):1001–1007.

96. Soto JA, Alvarez O, Munera F, Yepes NL, Sepulveda ME, Perez JM. Traumatic disruption of the pancreatic duct: Diagnosis with MR pancreatography. *AJR American Journal of Roentgenology.* Jan 2001;176(1):175–178.

97. Vandermeer FQ, Manning MA, Frazier AA, Wong-You-Cheong JJ. Imaging of whole-organ pancreas transplants. *RadioGraphics: A Review Publication of the Radiological Society of North America, Inc.* Mar–Apr 2012;32(2):411–435.

98. Dillman JR, Elsayes KM, Bude RO, Platt JF, Francis IR. Imaging of pancreas transplants: Postoperative findings with clinical correlation. *Journal of Computer Assisted Tomography.* Jul–Aug 2009;33(4):609–617.

99. Hagspiel KD, Nandalur K, Burkholder B et al. Contrast-enhanced MR angiography after pancreas transplantation: Normal appearance and vascular complications. *AJR American Journal of Roentgenology.* Feb 2005;184(2):465–473.

100. Dobos N, Roberts DA, Insko EK, Siegelman ES, Naji A, Markmann JF. Contrast-enhanced MR angiography for evaluation of vascular complications of the pancreatic transplant. *RadioGraphics: A Review Publication of the Radiological Society of North America, Inc.* May–Jun 2005;25(3):687–695.

101. Hagspiel KD, Nandalur K, Pruett TL et al. Evaluation of vascular complications of pancreas transplantation with high-spatial-resolution contrast-enhanced MR angiography. *Radiology.* Feb 2007;242(2):590–599.

21

The Adrenal Gland

Shaunagh McDermott, Colin J. McCarthy, and Michael A. Blake

CONTENTS

With the increasing use of medical imaging, more unsuspected masses are being identified, with adrenal incidentalomas being seen in approximately 5% of all abdominal CT examinations [1,2]. Furthermore 9%–13% of patients being scanned for a known malignancy will have an adrenal nodule [1], although less than 36% of such lesions are metastatic [3]. It is the role of the radiologist to identify the lesions in which a definite diagnosis of benignity can be made based on imaging characteristics. The objective of this chapter is to review current and emerging magnetic resonance imaging (MRI) techniques in the imaging of the adrenal gland and to illustrate the MRI features of commonly encountered adrenal lesions.

21.1 MRI Sequences

21.1.1 In-Phase and Opposed-Phase Chemical Shift Imaging

Chemical shift imaging (CSI) takes advantage of the different resonance frequencies of fat and water. The hydrogen protons in fat precess at a lower frequency than those in water due to the presence of longer chemical side chains. For in-phase imaging, the echo time (TE) is selected so that the fat and water proton signals are additive. In contradistinction, for opposed-phase imaging, the TE is selected such that the fat and water signals are opposed by $180°$, which causes signal cancellation, manifesting as a drop in signal if both water and fat protons are present in the same voxel as are present in many adenomas. Lesions that contain either fat or water components in a single voxel, but not both, will not display signal cancellation. Almost complete signal intensity loss is seen on the opposed-phase image when equal concentrations of fat and water are present in most voxels. Conversely, if either lipid or water protons predominate, the signal intensity on the opposed-phase image is essentially unchanged, and the lesion remains indeterminate.

At 1.5 T, the frequency shift between water signals and fat signals is approximately 225 Hz, which results in in-phase signals at TEs of 4.4, 8.8, and 13.2 ms, and opposed-phase signals at TEs of 2.2, 6.6, and 11.0 ms. On 3 T MR scanners, the frequency difference is twice that of standard 1.5 T scanners, and therefore the TE pairs require adjustment. On 3 T scanners, the fat and water are in opposed-phase at 1.1, 3.3, and 5.5 ms, and in-phase at 2.2, 4.4, and 6.6 ms, respectively [4].

Ideally, the opposed-phase image should be obtained with the first echo of the dual echo sequence. In doing so, any loss of signal intensity between the in-phase and opposed-phase images is due to the presence of lipid and water protons within the same voxel. If the opposed-phase image is obtained using the second, longer TE, loss of signal may be due to either the presence of intracellular lipid or from T_2^* susceptibility effects [5,6] or a combination of both. The lowest TE pair of opposed- and in-phase values should be used to maximize the signal-to-noise ratio and minimize T_2^* susceptibility and T_2 effects.

There are different ways in which one can establish the presence of lipid within an adrenal lesion. One can qualitatively compare the in-phase and corresponding opposed-phase images on a workstation using identical window settings. If the adrenal lesion demonstrates a drop in signal on the opposed-phase image, it contains intravoxel lipid and water protons [7] (Figure 21.1). Another method is to subtract the opposed-phase image from the in-phase image, and thus generate a subtraction image. Any signal present on the subtracted image indicates the presence of lipid and water protons in the same voxel [8].

Quantitative analysis of adrenal lesion lipid can also be performed, with multiple methods using internal references available.

Adrenal signal intensity (SI) index is calculated as

$$\left[\frac{\left(SI_{\text{in phase}} - SI_{\text{opposed phase}} \right)}{SI_{\text{in phase}}} \right] \times 100\%$$

Adrenal-to-spleen chemical shift ratio is calculated as

$$\left[\frac{\left(\text{adrenal SI}_{\text{opposed phase}} / \text{spleen SI}_{\text{opposed phase}} \right)}{\left(\text{adrenal SI}_{\text{in phase}} / \text{spleen SI}_{\text{in phase}} \right)} - 1 \right] \times 100\%$$

The liver should not be used as an internal reference due to the frequent incidence of hepatic steatosis, which also results in signal loss on the opposed-phase image due to the presence of intracytoplasmic fat. Muscle can also be subject to fatty infiltration and therefore is typically not used either.

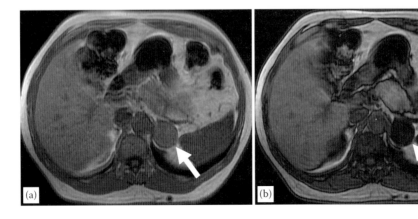

FIGURE 21.1
A 57-year-old man with a left adrenal nodule (arrow), which demonstrates a dramatic drop in signal between the in-phase (a) and opposed-phase (b) images compatible with a lipid-rich adenoma.

No optimal threshold for the adrenal SI index has been established with various thresholds reported for the differentiation of adenomas from nonadenomas. These reported adrenal SI indices range from 1% to 23% at 1.5 T [9–13], and from 1.7% to 20.2% at 3.0 T [6,14,15].

Dual-echo in- and opposed-phase imaging can also be used to identify macroscopic fat by causing an *India-ink* artifact between intralesional fat and adjacent water containing tissue [16]. Two- and three-point Dixon T_1 fat-only images can show both intravoxel lipid protons and macroscopic fat in adrenal lesions.

21.1.2 Fat-Suppression T_1-Weighted Imaging

By comparing an in-phase T_1-weighted image with a corresponding fat-suppressed T_1-weighted image, macroscopic fat within a lesion can be identified (Figure 21.2).

21.1.3 T_2-Weighted Imaging

Early studies found that adenomas tended to have signal intensity on T_2-weighted sequences comparable to liver or muscle, whereas adrenal metastases

have T_2 signal intensity more similar to spleen [17–20]. Pheochromocytomas can have relatively high signal intensity on T_2-weighted sequences (Figure 21.3); however, at least 30% of them demonstrate moderate or low T_2 signal intensity similar to other adrenal lesions [21].

21.1.4 Dynamic Contrast-Enhanced Imaging

The enhancement pattern of neoplasms on dynamic contrast-enhanced studies depends on their vasculature, vessel permeability, and the size of the extracellular spaces. The normal vascular supply of the adrenal cortex has a rich network of anastomosing capillary sinusoids. Adrenal carcinoma demonstrates reduced and slower enhancement because of reduced vascularization compared to adenomas and normal adrenal cortex and the invasion of the normal arterial system by malignant cells.

Older studies suggested that delayed contrast-enhanced MRI may be helpful in differentiating adenomas from nonadenomas, with most adenomas showing more rapid wash-out of contrast material than metastases [19,20,22]. However, the use of contrast wash-out

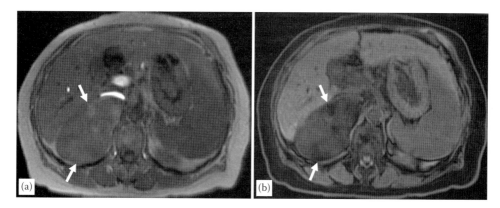

FIGURE 21.2
An 80-year-old man with a large heterogeneous right adrenal mass. Areas that are high signal intensity on the T_1-weighted sequences (a) demonstrate a drop in signal on the fat-suppressed sequence (b) (arrow), consistent with fat in a myelolipoma.

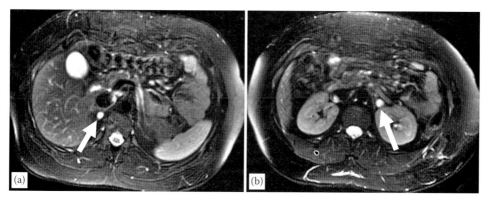

FIGURE 21.3
A 24-year-old man with VHL with right (a) and left (b) T_2 markedly hyperintense adrenal nodules (arrows) consistent with pheochromocytomas.

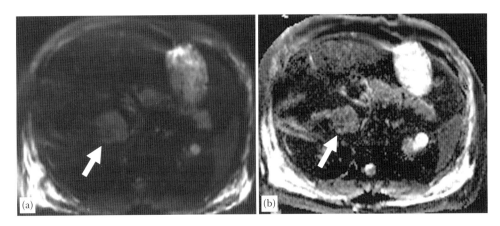

FIGURE 21.4
A 53-year-old man with a right adrenal mass (arrow), which is high in signal on DWI (a) and low in signal on ADC (b). Biopsy confirmed adrenocortical carcinoma.

patterns of lesions on delayed images is not generally assessed on MR as dedicated delayed scans are not routinely included in MR adrenal protocols.

More recently, studies have looked at the early enhancement patterns of adrenal lesions to overcome this problem. One study found that the majority of adenomas exhibited a capillary blush at the immediate postcontrast sequence (18th second) and a wash-out at the 45th second. In comparison, malignant lesions showed either negligible or weak enhancement at the 18th second and still showed irregular or peripheral enhancement at the 45th second [23]. Another study found that early phase contrast enhancement was more commonly homogenous or punctate in adenomas, compared with patchy or peripheral enhancement in malignant lesions. On the late phase images, all malignant lesions demonstrated heterogeneous enhancement, whereas over half of the adenomas demonstrated peripheral enhancement. The wash-in rates of the adenomas were significantly higher than those of the malignant masses and consequently, the time-to-peak enhancement of the malignant lesions was significantly longer than that of the adenomas [24].

21.1.5 Diffusion-Weighted Imaging

Diffusion-weighted imaging (DWI) generates image contrast based on the movement of water at a microscopic level. The most commonly used sequence is based on the use of two equal-sized gradients around a 180° refocusing pulse of a spin-echo sequence. The effect of these two gradients on water molecules depends on their movement. In stationary water molecules that have not moved substantially between the two gradients, the effect of these gradient pulses is complete rephasing with preservation of signal. In contrast, in nonstationary water, molecules that have moved between the first and second gradient result in dephasing and consequent loss of signal. The degree of signal loss is proportional to the degree of water motion.

By performing DWI using at least two different b values, quantitative analysis is possible by calculating apparent diffusion coefficient (ADC). ADC is independent of the magnetic field and can overcome the effects of T_2 shine-through. Areas of restricted diffusion will appear bright on the DWI and dark on the ADC map (Figure 21.4), as opposed to areas of simple T_2 prolongation (T_2 shine-through), which will demonstrate high signal intensity on both the DWI and ADC map.

Although DWI has been shown to be helpful in the characterization of tumors elsewhere in the body, the available literature suggests that neither DWI nor evaluation of the ADC of adrenal lesions is useful in differentiating benign from malignant adrenal lesions [13,25–28].

21.1.6 MR Spectroscopy

In MR spectroscopy (MRS), the frequency of a signal from a tissue after the application of a radio-frequency pulse is used to separate and characterize the actual metabolites or chemicals within a voxel. The signal intensity and the line width can be used to determine the relative quantity of the chemical.

To date there is limited experience with the use of MRS in the characterization of adrenal lesions. One possible role of MRS would be to differentiate adenomas from nonadenomas, as studies have found that adenomas have a larger lipid peak [29] and a greater lipid/creatinine peak [30] compared to nonadenomas. Another possible application of MRS is in the identification of pheochromocytomas. A small study on pathologically proven pheochromocytomas found they had a unique spectral peak at 6.8 ppm that corresponded with the presence of intralesional catecholamines [31].

However, MRS of the adrenal gland is challenging because of the small size of the lesions and respiratory artifacts from motion of the diaphragm.

21.1.7 PET/MR

As more PET/MR systems come into everyday clinical practice, its role or additional benefits over currently available imaging techniques are as yet unknown.

21.2 Specific Adrenal Lesions

21.2.1 Fat Containing Lesions

21.2.1.1 Adrenal Adenomas

The prevalence of adrenal adenomas is age related, with one study citing a frequency of 0.14% in patients aged 20–29 years and 7% in those older than 70 years of age [32]. Approximately 70% of adrenal adenomas are lipid-rich and contain sufficient amounts of intracytoplasmic fat to cause loss of signal on CSI (Figures 21.1 and 21.5). A study has shown that in patients with an indeterminate adrenal lesion on noncontrast CT (i.e., those with an HU value between 10 and 30), chemical shift MRI had a 89% sensitivity and 100% specificity for adenoma characterization [33]. Because up to 30% of adrenal adenomas are lipid-poor, chemical shift MRI is limited in the evaluation of these lesions. However, it should be noted that lesion other than adenomas, such as adrenal cortical carcinoma, pheochromocytoma, and clear cell renal cell carcinoma and hepatocellular carcinoma metastasis, can sometimes show signal loss on opposed-phase images.

21.2.1.2 Myelolipomas

Myelolipomas are benign tumors that are composed of mature adipose tissue with scattered hematopoietic elements. Myelolipomas account for up to 6% of incidentally discovered adrenal lesions [2]. The imaging diagnosis of myelolipomas is based on the detection of macroscopic fat. The fat component is hyperintense on nonfat-suppressed T_1-weighted images and demonstrates loss of signal intensity on fat suppressed images (Figure 21.2). The presence of an India ink artifact at the lesion–adrenal interface or within an adrenal lesion on opposed-phase images should indicate a myelolipoma.

The appearance on MRI is based on the amount of fat within the lesion. If predominantly composed of fat, the lesion is homogenous, hyperintense on T_1-weighted images with intermediate signal intensity on T_2-weighted images. If composed of mixed fatty and myeloid elements, the lesion is heterogeneous and contains foci with the same signal intensity of fat intermixed with focal high signal intensity on T_2-weighted and contrast-enhanced T_1-weighted images. However, if the lesion is composed primarily of myeloid cells, the lesion will be relatively hypointense to liver on T_1-weighted images, hyperintense to liver on T_2-weighted sequences and enhance after administration of contrast (Figure 21.6).

Although the presence of macroscopic fat is characteristic of myelolipoma it is not diagnostic of it, as case reports have described macroscopic fat in other tumors. Other imaging features, such as margins, heterogeneity, and invasion, should be considered to exclude a rare fat containing malignancy.

21.2.2 Cystic Masses

21.2.2.1 Simple Cysts

Simple cysts are hyperintense on T_2-weighted images, hypointense on T_1-weighted images, without soft-tissue components or internal enhancement (Figure 21.7).

21.2.2.2 Pseudocysts

Pseudocysts are usually the sequelae of a prior episode of hemorrhage or infarct. Adrenal pseudocysts can have a more complex appearance, including the presence of septations, blood products, or even a soft-tissue component secondary to hemorrhage or hyalinized thrombus (Figure 21.8).

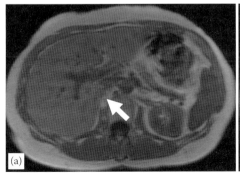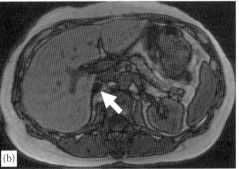

FIGURE 21.5
A 63-year-old woman with a small right adrenal nodule (arrow), which demonstrates a marked drop in signal between the in-phase (a) and opposed-phase (b) images compatible with a lipid-rich adenoma.

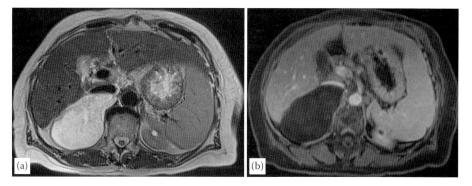

FIGURE 21.6
Same patient as in Figure 21.2. The right adrenal myelolipoma is of high signal intensity on T_2-weighted sequences (a) and demonstrates heterogeneous enhancement (b).

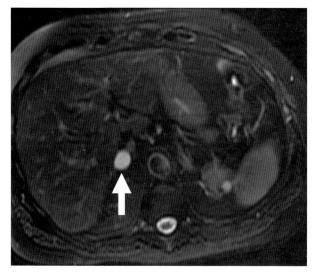

FIGURE 21.7
A 61-year-old man with a T_2 hyperintense right adrenal nodule (arrow). This was T_1 hypointense and did not demonstrate enhancement (not shown). Findings are consistent with an adrenal cyst.

21.2.2.3 Lymphangiomas

Cystic lymphangiomas of the adrenal gland are rare. On MRI, these lesions are thin-walled cystic structures that have low signal intensity on T_1-weighted sequences, high signal intensity on T_2-weighted sequences, and do not demonstrate significant internal enhancement.

21.2.3 Hypervascular Lesions

21.2.3.1 Pheochromocytomas

Pheochromocytomas are tumors composed of chromaffin cells and secrete catecholamines. On MRI, pheochromocytomas have low signal intensity on T_1-weighted sequences; however, if there is fat or hemorrhage present within the lesion, these will demonstrate high signal intensity. Classically, pheochromocytomas had been described as demonstrating marked high signal

intensity (*lightbulb bright*) on T_2-weighted sequences, in contrast to other lesions (Figures 21.3 and 21.9). However, more recently, it has been found that up to 30% of pheochromocytomas can show low signal intensity on T_2-weighted sequences. Atypical imaging features are particularly prevalent in genetically associated pheochromocytomas. There are, unfortunately, no specific features to predict malignancy in the absence of local invasion or metastases (Figures 21.9 and 21.10).

Until recently, only 10% of pheochromocytomas were considered genetically associated; however, more recent advances in molecular genetics indicate that at least one-third of patients with a pheochromocytoma harbor a genetic mutation [34,35]. Pheochromocytomas can be seen in various syndromes, including multiple endocrine neoplasia (MEN) syndrome types IIa and IIb, neurofibromatosis type 1, von Hippel–Lindau syndrome, tuberous sclerosis, Sturge–Weber syndrome, succinate dehydrogenase (SDH) mutations B, C, and D, and Carney triad (Figure 21.3).

21.2.4 Malignant Lesions

21.2.4.1 Adrenocortical Carcinoma

Adrenocortical carcinoma (ACC) is an aggressive tumor arising from the adrenal cortex. ACCs are typically heterogeneous on MRI due to the presence of fat and/or necrosis (Figure 21.11). On T_1-weighted sequences, these tumors are usually iso- to hypointense to liver, with high T_1 signal in areas or hemorrhage. These tumors may also contain intracytoplasmic fat, which can result in a drop in signal on opposed-phase sequences. ACCs have a propensity to invade veins.

21.2.4.2 Lymphoma

Lymphoma can, occasionally, involve the adrenal glands either primary or secondary. Primary adrenal lymphoma is extremely rare. Primary adrenal lymphoma

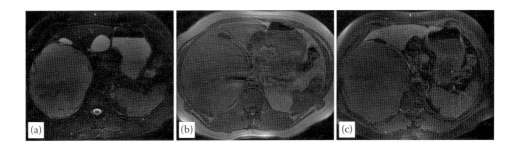

FIGURE 21.8
A 46-year-old man with a large right adrenal mass that is heterogeneous in signal on T_2-weighted sequence (a), and does not demonstrate enhancement between the pre- (b) and post-contrast sequences (c). On resection, this was confirmed to be a pseudocyst.

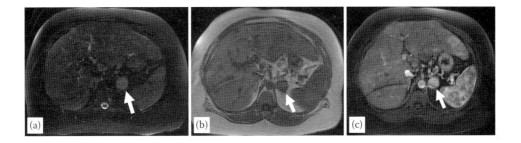

FIGURE 21.9
A 61-year-old woman with a left adrenal mass (arrow), which is mildly T_2 hyperintense (a), T_1 hypointense (b), and demonstrates marked enhancement (c). This proved to be a pheochromocytoma and there are also multiple hepatic metastases.

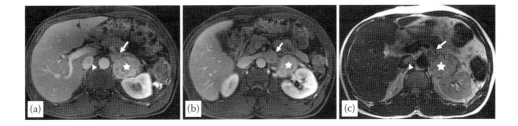

FIGURE 21.10
A 52-year-old man with a large left adrenal mass (star), which demonstrates heterogeneous enhancement post-contrast and extension into the left renal vein (arrow) and inferior vena cava (arrow head) (a and b). Similar findings are seen on the T_2-weighted sequences (c). This mass proved to be a pheochromocytoma.

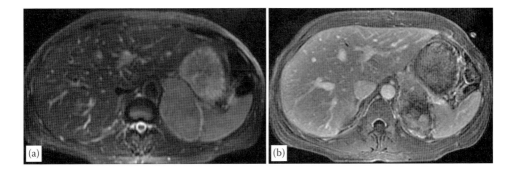

FIGURE 21.11
A 42-year-old woman with a left adrenal mass that is heterogeneous on T_2-weighted (a) and demonstrates heterogeneous enhancement (b). This mass proved to be an ACC.

usually maintains an adreniform appearance; however, the presence of necrotic or cystic components and the heterogeneous enhancement makes them difficult to distinguish from other tumors such as ACC, pheochromocytoma, or metastases [36]. Secondary adrenal involvement is typically seen in non-Hodgkins lymphoma. Early secondary involvement may only result in diffuse adrenal enlargement with no change in gland configuration. More nodular enlargement is seen in progressive disease and MR findings are nonspecific and similar to metastases. Typically, the lesion is hypointense to liver on T_1-weighted sequences and heterogeneously hyperintense to liver on T_2-weighted sequences.

21.2.4.3 Metastases

The adrenal gland is the fourth most common site of metastatic disease after lung, liver, and bone. Lung, breast, colon, melanoma, and thyroid cancers are most commonly associated with adrenal metastases [37]. Adrenal metastases can present as either soft tissue masses or as a diffusely enlarged gland. Metastases tend to be large, heterogeneous and ill-defined (Figures 21.12 and 21.13). Rarely, metastases from primary malignancies that contain intracytoplasmic fat, such as clear cell renal cell carcinoma or hepatocellular carcinoma, can cause signal loss on opposed-phase images and, therefore, potentially result in an incorrect diagnosis of an adenoma (Figure 21.14).

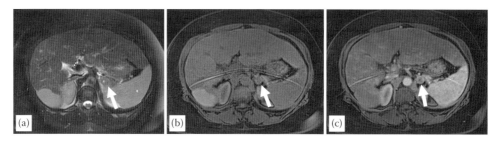

FIGURE 21.12
A 63-year-old woman with hepatocellular carcinoma (HCC) with a left adrenal nodule (arrow). This is slightly T_2 hyperintense (a) and demonstrates enhancement between the pre- (b) and post-contrast sequences (c). This was confirmed as a HCC metastasis on biopsy.

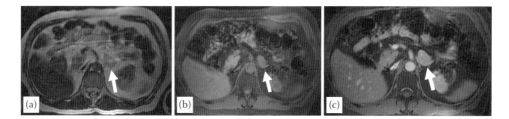

FIGURE 21.13
A 74-year-old man with a history of renal cell carcinoma (RCC), with an enlarging left adrenal nodule (arrow). This is slightly T_2 hyperintense (a) and demonstrates enhancement between the pre- (b) and post-contrast sequences (c). This was confirmed as a RCC metastasis on biopsy.

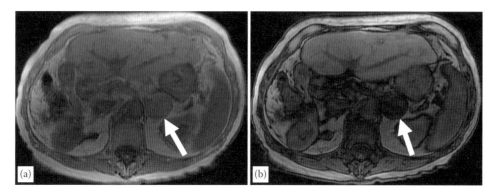

FIGURE 21.14
A 78-year-old man with HCC with a rapidly enlarging left adrenal nodule (arrow), which demonstrates a drop in signal between the in-phase (a) and opposed-phase (b) sequences. However, due to its rapid growth, this was considered a metastasis.

21.2.4.4 Collision Tumor

Collision tumors are defined as the coexistence of two adjacent but histologically distinct tumors, both of which can be benign or one may be malignant [38].

21.2.5 Pediatric Neoplasms

21.2.5.1 Neuroblastoma

Neuroblastoma is the third most common malignant tumor seen in children; however, it is rare in adults. Adrenal involvement presents as soft tissue masses that are often calcified. On MRI, neuroblastomas are typically heterogeneous, variably enhancing, and of relative low signal intensity on T_1-weighted images and high-signal intensity on T_2-weighted images [39,40] (Figure 21.15). Neuroblastomas may encase and/or compress adjacent vessels, with vascular invasion being rare. An advantage of MRI over other cross-sectional techniques is the ability to detect intraspinal extension and/or bone marrow involvement.

21.2.5.2 Ganglioneuroma

Ganglioneuromas are rare, benign neurogenic tumors that are composed of Schwann cells and ganglion cells that arise from the sympathetic ganglia. Ganglioneuromas are homogenously hypointense on T_1-weighted sequences, with varying signal intensity on T_2-weighted sequences, depending on the myxoid, cellular, and collagen components (Figure 21.16). One of the MRI characteristics is curvilinear bands of low signal intensity on T_2-weighted sequences, which gives the tumor a whorled appearance [41]. On contrast-enhanced dynamic MRI, ganglioneuromas usually demonstrate gradually increasing enhancement rather than early enhancement [42].

21.2.5.3 Ganglioneuroblastoma

Ganglioneuroblastoma are intermediate-grade tumors that have elements of benign ganglioneuromas and malignant neuroblastomas. Imaging appearances vary, ranging from a predominantly solid mass to a predominantly cystic mass with a few thin strands of solid tissue.

21.2.6 Other

21.2.6.1 Adrenal Hemorrhage

Adrenal hemorrhage can be secondary to traumatic and nontraumatic etiologies. It occurs more commonly in the right adrenal gland, with bilateral involvement in 20% of cases. Nontraumatic causes include coagulopathy, stress, venous hypertension, and hemorrhagic tumors.

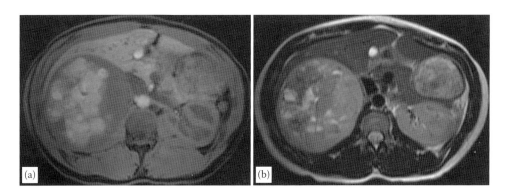

FIGURE 21.15
A 24-year-old woman with a large right adrenal mass that is heterogeneous on both T_1- (a) and T_2-weighted sequences (b). This mass was confirmed as a neuroblastoma on resection.

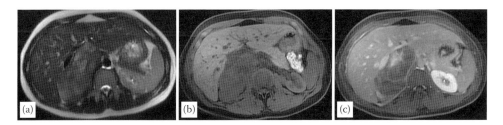

FIGURE 21.16
A 20-year-old woman with a large right adrenal mass that is heterogeneous on T_2-weighted sequences (a), hypointense on T_1-weighted sequences (b), and demonstrates heterogeneous enhancement on postgadolinium sequences (c). This mass was confirmed as a ganglioneuroma on resection.

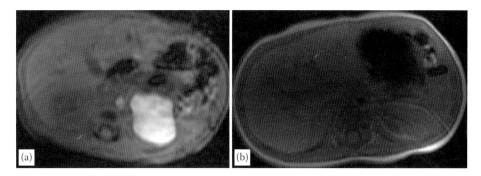

FIGURE 21.17
A 10-day-old girl with sepsis with a large T_1 hyperintense lesion in the region of the left adrenal gland (a). Follow-up MRI 5 months later demonstrates complete resolution of the hemorrhage (b).

MRI findings depend on the phase of hemorrhage. In the acute phase, less than 7 days, the hematoma appears iso- to slightly hypointense on T_1-weighted sequences and markedly hypointense on T_2-weighted sequences. In the subacute phase, 1–7 weeks, the hematoma appears hyperintense on T_1- and T_2-weighted sequences (Figure 21.17). The high T_1 signal appears at the periphery of the hematoma and fills in centrally over several weeks. In the chronic phase, after 7 weeks, a hypointense rim is present on both T_1- and T_2-weighted sequences, secondary to hemosiderin deposition, which demonstrates *blooming* on gradient echo sequences.

In patients without risk factors for traumatic or nontraumatic hemorrhage, contrast-enhanced and possibly follow-up imaging is required to exclude an underlying mass.

21.2.6.2 Infection

Tuberculosis, histoplasmosis, and other granulomatous disease can affect the adrenal glands, often with bilateral involvement. Imaging features are nonspecific, and may include soft tissue masses, cystic changes, and calcification [43].

References

1. Bovio S, Cataldi A, Reimondo G et al. (2006) Prevalence of adrenal incidentaloma in a contemporary computerized tomography series. *J Endocrinol Invest* 29:298–302.
2. Song JH, Chaudhry FS, Mayo-Smith WW (2008) The incidental adrenal mass on CT: Prevalence of adrenal disease in 1,049 consecutive adrenal masses in patients with no known malignancy. *AJR Am J Roentgenol* 190:1163–1168.
3. Oliver TW, Jr., Bernardino ME, Miller JI et al. (1984) Isolated adrenal masses in nonsmall-cell bronchogenic carcinoma. *Radiology* 153:217–218.
4. Chang KJ, Kamel IR, Macura KJ et al. (2008) 3.0-T MR imaging of the abdomen: Comparison with 1.5 T. *RadioGraphics* 28:1983–1998.
5. Tsushima Y, Dean PB (1995) Characterization of adrenal masses with chemical shift MR imaging: How to select echo times. *Radiology* 195:285–286.
6. Schindera ST, Soher BJ, Delong DM et al. (2008) Effect of echo time pair selection on quantitative analysis for adrenal tumor characterization with in-phase and opposed-phase MR imaging: Initial experience. *Radiology* 248:140–147.
7. Mayo-Smith WW, Lee MJ, McNicholas MM et al. (1995) Characterization of adrenal masses (< 5 cm) by use of chemical shift MR imaging: Observer performance versus quantitative measures. *AJR Am J Roentgenol* 165:91–95.
8. Savci G, Yazici Z, Sahin N et al. (2006) Value of chemical shift subtraction MRI in characterization of adrenal masses. *AJR Am J Roentgenol* 186:130–135.
9. Tsushima Y, Ishizaka H, Kato T et al. (1992) Differential diagnosis of adrenal masses using out-of-phase FLASH imaging. A preliminary report. *Acta Radiol* 33:262–265.
10. Namimoto T, Yamashita Y, Mitsuzaki K et al. (2001) Adrenal masses: Quantification of fat content with double-echo chemical shift in-phase and opposed-phase FLASH MR images for differentiation of adrenal adenomas. *Radiology* 218:642–646.
11. Fujiyoshi F, Nakajo M, Fukukura Y et al. (2003) Characterization of adrenal tumors by chemical shift fast low-angle shot MR imaging: Comparison of four methods of quantitative evaluation. *AJR Am J Roentgenol* 180:1649–1657.
12. Israel GM, Korobkin M, Wang C et al. (2004) Comparison of unenhanced CT and chemical shift MRI in evaluating lipid-rich adrenal adenomas. *AJR Am J Roentgenol* 183:215–219.
13. Sandrasegaran K, Patel AA, Ramaswamy R et al. (2011) Characterization of adrenal masses with diffusion-weighted imaging. *AJR Am J Roentgenol* 197:132–138.
14. Marin D, Soher BJ, Dale BM et al. (2010) Characterization of adrenal lesions: Comparison of 2D and 3D dual gradient-echo MR imaging at 3 T—Preliminary results. *Radiology* 254:179–187.

15. Nakamura S, Namimoto T, Morita K et al. (2012) Characterization of adrenal lesions using chemical shift MRI: Comparison between 1.5 Tesla and two echo time pair selection at 3.0 Tesla MRI. *J Magn Reson Imaging* 35:95–102.

16. Hood MN, Ho VB, Smirniotopoulos JG et al. (1999) Chemical shift: The artifact and clinical tool revisited. *RadioGraphics* 19:357–371.

17. Baker ME, Blinder R, Spritzer C et al. (1989) MR evaluation of adrenal masses at 1.5 T. *AJR Am J Roentgenol* 153:307–312.

18. Kier R, McCarthy S (1989) MR characterization of adrenal masses: Field strength and pulse sequence considerations. *Radiology* 171:671–674.

19. Krestin GP, Freidmann G, Fishbach R et al. (1991) Evaluation of adrenal masses in oncologic patients: Dynamic contrast-enhanced MR vs CT. *J Comput Assist Tomogr* 15:104–110.

20. Slapa RZ, Jakubowski W, Januszewicz A et al. (2000) Discriminatory power of MRI for differentiation of adrenal non-adenomas vs adenomas evaluated by means of ROC analysis: Can biopsy be obviated? *Eur Radiol* 10:95–104.

21. Blake MA, Kalra MK, Maher MM et al. (2004) Pheochromocytoma: An imaging chameleon. *RadioGraphics* 24(Suppl. 1):S87–S99.

22. Krestin GP, Steinbrich W, Friedmann G (1989) Adrenal masses: Evaluation with fast gradient-echo MR imaging and Gd-DTPA-enhanced dynamic studies. *Radiology* 171:675–680.

23. Chung JJ, Semelka RC, Martin DR (2001) Adrenal adenomas: Characteristic postgadolinium capillary blush on dynamic MR imaging. *J Magn Reson Imaging* 13:242–248.

24. Inan N, Arslan A, Akansel G et al. (2008) Dynamic contrast enhanced MRI in the differential diagnosis of adrenal adenomas and malignant adrenal masses. *Eur J Radiol* 65:154–162.

25. Tsushima Y, Takahashi-Taketomi A, Endo K. (2009) Diagnostic utility of diffusion-weighted MR imaging and apparent diffusion coefficient value for the diagnosis of adrenal tumors. *J Magn Reson Imaging* 29:112–117.

26. Miller FH, Wang Y, McCarthy RJ et al. (2010) Utility of diffusion-weighted MRI in characterization of adrenal lesions. *AJR Am J Roentgenol* 194:W179–W185.

27. Cicekci M, Onur MR, Aydin AM et al. (2013) The role of apparent diffusion coefficient values in differentiation between adrenal masses. *Clin Imaging* 38:148–153.

28. Halefoglu AM, Altun I, Disli C et al. (2012) A prospective study on the utility of diffusion-weighted and quantitative chemical-shift magnetic resonance imaging in the distinction of adrenal adenomas and metastases. *J Comput Assist Tomogr* 36:367–374.

29. Leroy-Willig A, Roucayrol JC, Luton JP et al. (1987) In vitro adrenal cortex lesions characterization by NMR spectroscopy. *Magn Reson Imaging* 5:339–344.

30. Faria JF, Goldman SM, Szejnfeld J et al. (2007) Adrenal masses: Characterization with in vivo proton MR spectroscopy—Initial experience. *Radiology* 245:788–797.

31. Kim S, Salibi N, Hardie AD et al. (2009) Characterization of adrenal pheochromocytoma using respiratory-triggered proton MR spectroscopy: Initial experience. *AJR Am J Roentgenol* 192:450–454.

32. Kloos RT, Gross MD, Francis IR et al. (1995) Incidentally discovered adrenal masses. *Endocr Rev* 16:460–484.

33. Haider MA, Ghai S, Jhaveri K et al. (2004) Chemical shift MR imaging of hyperattenuating (>10 HU) adrenal masses: Does it still have a role? *Radiology* 231:711–716.

34. Neumann HP, Bausch B, McWhinney SR et al. (2002) Germ-line mutations in nonsyndromic pheochromocytoma. *N Engl J Med* 346:1459–1466.

35. Benn DE, Gimenez-Roqueplo AP, Reilly JR et al. (2006) Clinical presentation and penetrance of pheochromocytoma/paraganglioma syndromes. *J Clin Endocrinol Metab* 91:827–836.

36. Kato H, Itami J, Shiina T et al. (1996) MR imaging of primary adrenal lymphoma. *Clin Imaging* 20:126–128.

37. Taffel M, Haji-Momenian S, Nikolaidis P et al. (2012) Adrenal imaging: A comprehensive review. *Radiol Clin North Am* 50:219–243, v.

38. Schwartz LH, Macari M, Huvos AG et al. (1996) Collision tumors of the adrenal gland: Demonstration and characterization at MR imaging. *Radiology* 201:757–760.

39. Lonergan GJ, Schwab CM, Suarez ES et al. (2002) Neuroblastoma, ganglioneuroblastoma, and ganglioneuroma: Radiologic-pathologic correlation. *RadioGraphics* 22:911–934.

40. Hiorns MP, Owens CM. (2001) Radiology of neuroblastoma in children. *Eur Radiol* 11:2071–2081.

41. Rha SE, Byun JY, Jung SE et al. (2003) Neurogenic tumors in the abdomen: Tumor types and imaging characteristics. *RadioGraphics* 23:29–43.

42. Zhang Y, Nishimura H, Kato S et al. (2001) MRI of ganglioneuroma: Histologic correlation study. *J Comput Assist Tomogr* 25:617–623.

43. Wilson DA, Muchmore HG, Tisdal RG et al. (1984) Histoplasmosis of the adrenal glands studied by CT. *Radiology* 150:779–783.

22

Diseases of the Upper Gastrointestinal Tract and Small Bowel

Davide Bellini, Carlo Nicola De Cecco, Domenico De Santis, Marco Rengo, Justin Morris, and Andrea Laghi

CONTENTS

22.1 Introduction

The use of cross-sectional imaging techniques for the noninvasive evaluation of upper gastrointestinal tract and small-bowel disorders is increasing. Computed tomography (CT) is still the modality of first choice for the study of esophageal and gastric malignancy, while barium swallow tests are used for the assessment of esophageal motility dysfunction. In any case, magnetic resonance imaging (MRI) applications are expanding the capability of obtaining both anatomical and functional information with a single imaging technique. Additionally, MR enterography already plays a significant role not only in Crohn's disease (CD), but also for the evaluation of various benign and malignant neoplasms, celiac disease, infectious diseases, and small-bowel obstructions. Advantages of MRI over CT include high contrast resolution, lack of radiation exposure, and use of intravenous contrast media with better safety profiles. MR enterography also allows dynamic assessment of small-bowel peristalsis and distensibility of lumen narrowing, providing functional information. On the other hand, MRI has limitations related to higher cost and a test quality that is dependent upon a patient's cooperation and their ability to hold their breath.

The scope of this chapter is to provide an overview of the actual role of MRI in upper gastrointestinal and small-bowel disorders, illustrating in particular the actual clinical indications, the acquisition protocols, and the classical imaging findings of the most common diseases.

22.2 Anatomy

22.2.1 Esophagus

The esophagus is a section of the alimentary canal that follows the pharynx and continues in the stomach.

The upper part of the esophagus is located in the neck, at the level of the sixth cervical vertebra. It then descends into the chest and, through the diaphragm, passes into the abdomen, with a total length of about 25–26 cm.

In its cervical portion, the esophagus is related to the trachea, the thyroid gland, and right and left recurrent laryngeal nerves.

Further down the esophagus lies in the posterior mediastinum and is crossed by the left bronchus.

In the mediastinum, the esophagus lies posterior to the trachea and the mediastinal lymph nodes. The posterior surface has a close relationship to the thoracic vertebrae up to C4, then gently curves forward and is related to the azygos vein, the hemiazygos vein, the thoracic duct, and the descending aorta. The esophagus is laterally related to the mediastinal pleura and the fibers of the vagus nerve.

Immediately below, the esophagus passes through the esophageal hiatus of the diaphragm into the abdominal cavity and continues in the stomach. The abdominal part of the esophagus is covered by the peritoneum on its anterior face.

22.2.2 Stomach

The stomach is a sac-like dilated portion of the alimentary canal, located in the abdominal cavity immediately below the diaphragm between the esophagus and the duodenum. Its average capacity in the adult is about 1200 mL. The position of the stomach varies with the posture, with the amount of the stomach contents and with the condition of the intestines on which it rests.

The stomach is formed by

- *Two surfaces:* anterosuperior and posteroinferior.
- *Two borders:* the lesser curvature which forms the right or posterior border of the stomach, and the greater curvature which forms an arch backward, upward, and to the left. Near the pyloric end of the lesser curvature there is a well-marked notch, the incisura angularis.
- *Two openings:* the cardia orifice superiorly and the pyloric orifice inferiorly.

A plane passing through the incisura angularis on the lesser curvature and the left limit of the opposed dilatation on the greater curvature divides the stomach into an upper portion, or body, and a lower, or pyloric portion. The upper portion of the body is known as the fundus, and is marked off from the remainder of the body by a plane passing horizontally through the cardiac orifice. The pyloric portion is formed sequentially by the pyloric antrum and the pyloric canal.

22.2.3 Small Bowel

The small bowel lies between the stomach and the large bowel and includes the duodenum, jejunum (proximal), and ileum (distal). It varies in length from 3 to 10 m, with an average length of 6 m. The duodenum continues into the jejunum at the duodenojejunal flexure, fixed to the retroperitoneum by the ligament of Treitz. The demarcation between the jejunum and the ileum is not very clear. The jejunum constitutes about one third of the small bowel, has a thicker wall and a wider lumen than the ileum, and mainly occupies the left upper abdomen. Circular mucosal folds, known as valvulae conniventes or plicae semilunaris, are the typical morphological feature of the jejunum. They are more prominent in the jejunum than in the ileum and are easily identifiable during an MR enterography. The ileum constitutes about two third of the small bowel and occupies the central and right lower abdomen and pelvis.

The normal caliber of the jejunal and ileal lumens in an MR enterography study without the administration of spasmolytic agents must be <25 mm, and the wall thickness of each is typically <2 mm.

The Ileum continues into the large intestine at the ileocecal junction.

One of the main differences between the small and the large bowel is the presence of a mesentery. Its root extends from the left of L2 to the right sacroiliac joint and is only 15 cm long.

The digestive tract anatomy is depicted in Figure 22.1.

22.3 MRI Technique

22.3.1 Esophagus

The esophagus is an organ that is affected by image artifacts due to breathing cardiac pulsation, resulting in poor image quality and motion artifacts on spoiled gradient-echo (SGE) sequences.

Nowadays, MRI of the mediastinum and the esophagus has been improved with Gd-enhanced 3D gradient-echo (GRE) sequences, acquired during a very short breath hold that minimizes motion artifacts and yields excellent images of the mediastinum and the esophageal wall [1].

Compared to a traditional fast spin-echo sequence, the 3D GRE T_1 is characterized by shorter echo time (TE) and repetition time (TR), which allow for the reduction of paramagnetic artifacts typical of the gas–tissue interface.

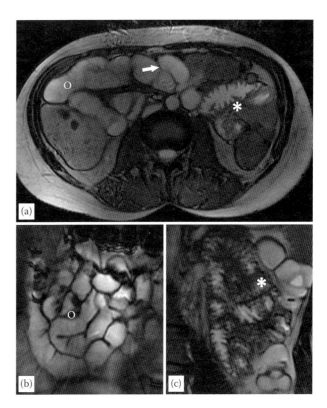

FIGURE 22.1
A 27-year-old man with normal MRI findings at enterography. T_2-weighted axial (a) and coronal (b)–(c) images of normal small bowel distended using biphasic oral contrast agent. Note the wall of the small bowel is thin, measuring 1–2 mm (arrow). Moreover, the normal fold pattern of jejunum with many folds (asterisk) is distinguished from that of the ileum, which is characterized by few folds (circle).

In addition to T_1-weighted imaging, T_2-weighted sequences with single-shot turbo spin-echo are useful in detecting the presence of free fluid or cystic lesions.

Balanced gradient-echo sequences can also be used; with these sequences that have both T_1- and T_2-weighted properties, ultrafast (roughly one second per slice) and robust images with excellent liquid/tissue contrast and signal-to-noise ratio can be obtained. Cine-magnetic resonance imaging with a steady-state free precession sequence can be implemented to assess motility function by calculating the amplitude and frequency of contractions.

Nevertheless, the role of these sequences for a routine evaluation of the esophagus is still not established.

22.3.2 Stomach

Wall distension and hypotonia provide better results in the MRI of the stomach. The wall distension is achieved by oral administration of contrast agents. They can be divided into positive (producing high signal intensity on T_1- and T_2-weighted images), negative (producing low signal intensity on T_1- and T_2-weighted images), or biphasic contrast agents (producing opposite signal intensities on T_1- and T_2-weighted images) [2]. Water is biphasic and the simplest contrast agent available. It must be administered within 5 min before MRI (800–1000 mL). Water is superior to air or other negative contrast agents in avoiding magnetic sensitivity artifacts, especially for diffusion-weighted imaging (DWI).

Gastric hypotonia is obtained by injection of a spasmolytic agent that is commonly administered intravenously shortly before performing the study; the two main spasmolytic agents used in clinical routine are butyl scopolamine (Buscopan®) and glucagon (GlucaGen®).

Patients must be trained to breath normally and hold their breath before MR examination [3].

The MRI protocol should include breath-hold T_1-weighted 2D, T_2-weighted turbo spin echo (TSE) (fat-suppressed and non-fat-suppressed), DWI sequences, and contrast-enhanced T_1-weighted sequences with fat suppression.

22.3.3 Small Bowel

A good evaluation of small bowel requires a combination of two important conditions: fast imaging technique and good luminal distension throughout administration of enteric contrast agents. Furthermore, the use of intravenous injection of contrast medium is mandatory for the assessment of bowel walls, lesions enhancement, and mesenteric vessels [4].

22.3.3.1 Contrast Media

Several enteric contrast agents have been described, including plain water, methylcellulose, solutions containing locust bean gum, mannitol, barium sulfate, and polyethylene glycol [5]. These agents work by retarding the resorption of water in the intestine and can cause mild diarrhea.

Enteric contrast agents can be classified according to the action on the signal intensity of bowel lumen into positive (*bright* lumen), negative (*dark* lumen), or biphasic (*bright* lumen on T_2 and *dark* on T_1). The choice of a single agent presents advantages and disadvantages. The radiologist should choose the appropriate contrast medium according to the clinical setting, MRI experience, availability of the agent, and patient tolerance.

Positive contrast agents cause a reduction in T_1 relaxation time; consequently, these agents act on T_1-weighted images by increasing the signal intensity of the bowel lumen. The use of these contrast agents has been largely abandoned.

Negative contrast agents are based on superparamagnetic particles and act by inducing local field inhomogeneities, which results in shortening of both T_1 and T_2 relaxation times. T_2-weighted effects are predominant and are useful for the evaluation of wall edema.

Biphasic contrast agents are substances that have different signal intensities on different sequences, depending on the concentration at which they are administered. They include several nonabsorbable solutions that produce a *water-like effect*; they are hyperintense on T_2-weighted images and hypointense on T_1-weighted images. Following the administration of intravenous contrast media, the differences between the lumen and the wall are enhanced, a perfect condition for evaluating different patterns of enhancement.

The ideal protocol for small-bowel filling before MR enterography is still under debate; enteric contrast agents can be administered orally (MR enterography) or injected through a naso-jejunal tube (MR enteroclysis). Generally, the use of MR enteroclysis is reserved for the evaluation of low-grade small-bowel obstruction and for patients who are unable to ingest oral contrast material for MR enterography. Some studies have shown that MR enteroclysis provides much better depiction of mucosal abnormalities than enterography because of improved distention [6]. However this degree of distention may not be required to adequately assess the small bowel for certain disorders. In fact, other studies have shown that enterography shows a sensitivity similar to that of enteroclysis for the detection of active inflammation in CD [7,8].

22.3.3.2 Enterography

Several different dosage algorithms have been proposed and are determined by the agent used and the amount administered [7,9]. The most accepted filling strategy involves fasting for 6 h prior to the examination and a large volume of contrast material (1500–2000 mL) given during the 40 min before the examination. Spasmolytics are useful for reducing bowel peristalsis and motion artifacts.

22.3.3.3 Enteroclysis

A nasoenteric tube should be placed during fluoroscopy beyond the duodenojejunal junction. Baseline images (true fast imaging with steady state precession [TrueFISP]) are obtained to assess bowel caliper. About 1500 mL of enteric contrast material are administered with a manual injection, at a rate of 80 mL/min, monitoring the filling of the small bowel every 2 min. The entire procedure must be monitored by a radiologist and the flow rate should be adjusted accordingly. When the contrast agent reaches the colon, spasmolytics can be administered intravenously (20 mg Buscopan® or 0.3 mg GlucaGon® intravenously) [5].

22.3.3.4 MRI Protocols

Patients can be imaged in either the supine or the prone position.

Usually, with the use of biphasic enteric contrast agents, T_2-weighted and contrast-enhanced gradient-echo T_1 sequences are the most helpful.

First of all, a coronal and axial T_2-weighted sequences based on steady-state precession (TrueFISP) images are obtained to allow an overview of the entire abdomen for the assessment of bowel distention. If distention is adequate, T_2-weighted sequences based on single-shot technique (half-Fourier acquisition single-shot turbo spin echo [HASTE]) are obtained in the coronal and axial planes [5].

Chemical shift and susceptibility artifacts may appear on images obtained with steady-state precession sequences and may complicate the assessment of bowel wall thickness; however, they are relatively insensitive to motion artifacts and allow a better depiction of mesenteric vessels and lymph nodes. On the other hand, HASTE sequences provide excellent depiction of wall thickening and changes in the fold pattern by producing high contrast between the bowel lumen and the wall. Fat suppression on T_2-weighted images is important for improving the conspicuity of high-signal-intensity inflammation in the bowel wall and perienteric fat.

Cine-magnetic resonance imaging, with a steady-state free precession sequence can be implemented to monitor and assess small-bowel motility function calculating the amplitude and frequency of contractions. Cine-MRI can provide added information, especially for equivocal bowel segments, and also aids in the confirmation of true strictures.

To reduce small-bowel peristalsis, a dose of 0.5 mg glucagon should be administered intravenously. One minute later, coronal and axial two-dimensional or 3D T_1-weighted SGE sequences should be acquired before and 70 s after the administration of gadoterate dimeglumine (Dotarem; Guerbet, Roissy, France; injection rate of 2 mL/s, 0.2 mg/kg).

DWI sequences can be applied. However, there is no consensus on the ideal b-values to use in clinical practice. Active disease or neoplastic conditions can demonstrate restricted diffusion and DWI sequences can help to identify pathological segments of bowel.

Obviously, MRI of the small bowel can be performed also with higher field strength (3 T system) after an adequate modification in the pulse sequences that are used at 1.5 T.

22.4 Upper Gastrointestinal Tract Pathology

22.4.1 Achalasia

22.4.1.1 Clinical Presentation

Achalasia is a chronic esophageal motility disorder characterized by failure of the lower esophageal sphincter (LES). The result is impaired flow of ingested food

onto the stomach at the level of the gastroesophageal junction, stagnation of food, and consequent esophageal dilatation. The main symptom is dysphagia [10].

22.4.1.2 Pathophysiology

The pathogenesis of this disease is related to death of neurons in the myenteric (Auerbach's) plexus at the lower third of the esophagus, which causes hypotonia of the LES and lack of motor coordination, with disappearance of the primary (swallow induced) and secondary (distension-induced) peristalsis and the onset of uncoordinated and nonpropulsive segmental contractions (tertiary peristalsis). Primary achalasia may develop from causes including infection, such as Chagas disease, but may also occur secondary to neoplasm.

22.4.1.3 MRI Findings

MRI allows visualization of the soft tissues in and adjacent to the wall of the distal esophagus and cardia.

The main findings are the stenotic appearance of the distal esophagus with and/or without wall thickening and, in advanced stages, a dilatation of the segments above. Typical finding is a *bird-beak* appearance of the junction with a dilated esophageal body. Cine-MRI can show both dilated esophagus and the inability of the LES. Usually there is an increased bolus transit time (>20 s) [11].

22.4.1.4 Differential Diagnosis

Malignancies, normal peristalsis, asymmetric stenosis, strictures from gastroesophageal reflux disease.

22.4.2 Diffuse Esophageal Spasm

22.4.2.1 Clinical Presentation

Patients with diffuse esophageal spasm (DES) may complain of dysphagia, chest pain, regurgitation, and heartburn while swallowing a bolus, especially those rich in collagen fibers (such as meat), cold or carbonated drinks. Often the spasm occurs independently of swallowing, causing a feeling of constriction, sometimes violent, or retrosternal discomfort.

22.4.2.2 Pathophysiology

The main cause is an impairment of inhibitory innervation leading to both premature and rapidly propagated contractions in the distal esophagus.

In the distal esophagus, contractions are mediated by cholinergic (excitatory) and nitric oxide (inhibitory) ganglionic neurons. Furthermore, a neural gradient exists such that there is an increasing proportion of inhibitory ganglionic neurons progressing distally. Hence, the deglutitive response initiates a period of quiescence (deglutitive inhibition) in the distal esophagus that is progressively prolonged approaching the LES [12]. Distal contractile latency, measured from the onset of the pharyngeal swallow to the onset of the contraction in the distal esophagus, is shorter in patients with simultaneous contractions than in those with normal peristaltic propagation. In patients with DES, there is also an increased esophageal smooth muscle thickness [13]. Finally, some cases of DES are characterized (such as achalasia) by impaired LES relaxation.

22.4.2.3 MRI Findings

In DES, functional alteration affects the lower third of the esophagus. The bolus transit triggers a spastic and uncoordinated contraction of the muscles of the esophagus, thus altering the morphology of the esophageal lumen such that it resembles a *corkscrew* or *rosary bead*. The bolus becomes trapped in the spastic segment because the lower segment contracts prematurely with inadequate time for the bolus to migrate.

Obviously spasm does not occur during every ingestion of food and it is mandatory to repeat this procedure several times.

22.4.3 Gastroesophageal Reflux Disease

22.4.3.1 Clinical Presentation

Gastroesophageal reflux disease (GERD) is one of the most prevalent gastrointestinal disorders in Western Countries [14] and is defined as backflow of gastric and sometimes duodenal contents into the esophagus.

Heartburn and acid regurgitation are the principal symptoms and are considered reasonably specific symptoms for the diagnosis of GERD.

GERD can also be associated with asthma, a history of pneumonia, atypical chest pain, chronic coughing, hoarseness, globus sensation, dyspepsia, and dysphagia (atypical symptoms).

Heartburn is produced by the contact of refluxed material with the sensitized or ulcerated esophageal mucosa. Persistent dysphagia suggests the development of a peptic stricture. Many patients with peptic stricture have a clinical history of several years of heartburn.

GERD may be associated with complications such as bleeding, strictures, Barrett's esophagus, and esophageal adenocarcinoma.

22.4.3.2 Pathophysiology

The normal anti-reflux mechanisms consist of the LES and the anatomical location of the gastroesophageal junction; reflux occurs when the pressure gradient between the LES and the stomach is lost (weakness of the LES due

to scleroderma-like diseases, other myopathies, hiatal hernia, obesity, drugs, surgical damages, gastric stasis).

The reflux of gastric content leads to injury of the esophagus and esophageal inflammation.

Symptoms are triggered by the activation of intramucosal chemosensitive nociceptors and, at the same time, the inflammatory cascade generated by luminal acid diffusing into the tissue [15]. It has only recently been suggested that gastric juice reflux does not directly damage the esophageal mucosa, but stimulates the esophageal epithelial cells to secrete chemokines that attract and activate immune cells, causing damage to the esophageal squamous epithelial cells [16].

With the decreasing prevalence of *Helicobacter pylori* infection, a concomitant decrease in peptic ulcer prevalence has been observed and this correlates with an increased prevalence of GERD. There are suggestions that infection with *H. pylori* may protect the esophagus from reflux esophagitis, probably because the infection causes a chronic gastritis that can decrease gastric acid secretion [17].

However, the connection between *H. pylori* eradication and the development of GERD is still questionable [18].

22.4.3.3 MRI Findings

Common MRI finding is thickness of the esophageal wall. After intravenous injection of gadolinium, the inflamed and fibrosed wall can show enhancement on delayed images.

22.4.4 Hiatal Hernia

22.4.4.1 Clinical Presentation

A hiatal hernia is an upward dislocation of part of the stomach into the thoracic cavity through the esophageal hiatus of the diaphragm. Depending on the relationship between the cardia, the diaphragm, and the herniated portion of the stomach, hiatal hernias are classified in four different types [19]:

Type I—Sliding hiatal hernia, is by far the most common type. It is characterized by the sliding of the gastro-esophageal junction and part of the fundus of the stomach into the mediastinum.

*Types II, III and IV—*are all varieties of *Paraesophageal hiatal hernia*. Type II occurs when the esophago-gastric junction stays normally positioned and the fundus herinates through the hiatus; Type III is a combination of Type I and Type II hernia as well as erniation of the fundus/body of stomach; Tape IV is characterised by displacement of the stomach along with other organs into the thorax.

22.4.4.2 Pathophysiology

A sliding hernia may result from weakening of the anchors of the gastro-esophageal junction to the diaphragm, from longitudinal contraction of the esophagus, or from increased intra-abdominal pressure.

22.4.4.3 MRI Findings

Enlarged esophageal hiatus and supra-diaphragmatic dislocation of the gastro-esophageal junction.

22.4.5 Diverticula

Diverticula are outpouchings of the esophageal wall. They are not frequent and occur in the esophagus quite rarely compared to other areas of the gastrointestinal tract.

They can be classified according to localization in three groups: diverticula that occur in the upper esophagus, typically Zenker diverticula; diverticula that occur in the middle of the esophagus; and epiphrenic diverticula, which occur just above the LES.

A *Zenker diverticulum* appears in the natural zone of weakness in the posterior hypo-pharyngeal wall (Killian dehiscence), just proximal to the upper esophageal sphincter, and causes halitosis and regurgitation of saliva and food particles consumed several days previously. When it becomes large and filled with food, it can compress the esophagus and cause dysphagia or complete obstruction.

A *mid-esophageal diverticulum* and *epifrenic diverticulum* may be caused by traction from old adhesions or by propulsion associated with esophageal motor abnormalities. Small- and medium-size diverticula are usually asymptomatic.

MRI shows the outpunching of esophageal wall and allows to define dimensions, localizations, and relationships with other mediastinal structures.

22.4.6 Tumors

22.4.6.1 Esophageal Cancer

22.4.6.1.1 Pathophysiology

Esophageal cancer is a serious malignancy with regards to mortality and prognosis. It is the eighth most common malignancy in the world and the incidence is rapidly increasing [20]. Despite many advances in diagnosis and treatment, the five-year survival rate for all patients diagnosed with esophageal cancer ranges from 15% to 20% [21].

The two most common histological types of esophageal carcinoma include squamous cell carcinoma (SCC) and adenocarcinoma.

SCC is the most common type of esophageal cancer worldwide and manifests with equal frequency in the middle and lower esophagus. Tobacco and alcohol use are the major risk factors for SCC and appear to have a compounding effect in putting a patient at risk.

Adenocarcinoma is the second most common malignant tumor of the esophagus in most countries. However, the incidence of esophageal adenocarcinoma has increased dramatically in some western countries over the past several decades. It occurs in the distal esophagus approximately three-fourth of the time and is related to Barrett's esophagus, a condition defined as gastric metaplasia of the esophageal epithelium, secondary to chronic gastric reflux [22]. Esophageal adenocarcinomas have a marked tendency to invade the gastric cardia and fundus by direct extension across the gastro-esophageal junction [23].

22.4.6.1.2 Clinical Presentation

The most common symptoms of SCC are progressive dysphagia, odynophagia, and weight loss. Dysphagia initially occurs with solid foods and gradually progresses to include semisolids and liquids. Patients with mediastinal tumor invasion may have chest pain. Unfortunately, symptomatic patients usually have advanced disease at the time of diagnosis [24].

Most patients affected by esophageal adenocarcinoma have advanced disease at the time of presentation and their symptoms are similar to those in patients with SCC. Regardless of staging, esophageal adenocarcinoma has a better prognosis than SCC [25].

22.4.6.1.3 MRI Findings

Diagnostic work-up usually includes a combination of endoscopic ultrasonography (EUS) and CT. EUS is effective for discrimination of stages T_1 and T_2 from stages T_3 and T_4. CT plays an important role in assessing the extent of invasion of surrounding and in detecting distant metastases [26]. Endoluminal MRI is also being developed as a potential method of local staging [27], although the difficulties with stenotic tumors persist with this technique.

High-resolution MRI can demonstrate the three main layers of the esophageal wall: the mucosa shows intermediate signal intensity and is surrounded by high-signal-intensity submucosa and intermediate- to low-signal-intensity muscularis propria [28].

T_1 tumors are difficult to differentiate from the surrounding normal wall.

T_2 tumors show intermediate signal intensity involving the submucosal layer and extending into but not beyond the muscularis propria layer.

The loss of the smooth outer margin of the muscularis propria or nodular extension of intermediate signal intensity indicates that the tumor has extended into the adventitia (T_3). Nodular extension of intermediate signal intensity into the perioesophageal tissues also indicates T3 extension. Tumor spread to different organ indicates a T_4 disease [29].

MRI is more accurate than CT (which has limited soft tissue contrast) in determining the resectability of a tumor. The criteria used to determine resectability with MRI is that there must be a margin of ≥1 mm between the tumor and a surrounding structure not resected at the time of surgery [30].

Accurate preoperative assessment of the extent of lymph node metastases is crucial. The overall five-year survival rate after surgical resection in patients without nodal involvement is 70%–92%, compared with 18%–47% for patients with lymph node metastasis [31].

The incidence of lymphatic metastasis in esophageal cancer is higher than in any other gastrointestinal cancer. Distribution of these lymph node metastases is very wide, extending from the neck to abdominal regions, and the sizes of lymph node metastases are frequently very small [32]. Characterization of lymph nodes based on T_1 or T_2 relaxation times has limited value; the size is not always a reliable predictor of metastasis because large nodes can be benign and small nodes can be malignant [33]. The staging system for esophageal cancer is reported in Table 22.1 [34].

TABLE 22.1

Staging System for *Esophageal Carcinoma* According to the Seventh Edition of the *AJCC Cancer Staging Manual*

Primary Tumor (T)

TX	Primary tumor cannot be assessed.
T0	No evidence of primary tumor.
Tis	High-grade dysplasia.
T1	Tumor invades lamina propria, muscularis mucosae, or submucosa.
T1a	Tumor invades lamina propria or muscularis mucosae.
T1b	Tumor invades submucosa.
T2	Tumor invades muscularispropria.
T3	Tumor invades adventitia.
T4	Tumor invades adjacent structures.
T4a	Resectable tumor invading pleura, pericardium, or diaphragm.
T4b	Unresectable tumor invading other adjacent structures, such as aorta, vertebral body, trachea, etc.

Regional Lymph Nodes (N)

NX	Regional lymph nodes cannot be assessed.
N0	No regional lymph node metastasis.
N1	Metastases in 1–2 regional lymph nodes.
N2	Metastases in 3–6 regional lymph nodes.
N3	Metastases in ≥7 regional lymph nodes.

Distant Metastasis (M)

M0	No distant metastasis.
M1	Distant metastasis.

22.4.6.2 Gastric Adenocarcinoma

22.4.6.2.1 Pathophysiology

Gastric cancer is the fourth most common malignancy in the world after lung, breast, and colorectal cancers. More than 70% of gastric cancer cases occur in developing countries with half the world's total cases occurring in Eastern Asia. Overall, gastric cancer incidence and mortality have fallen dramatically over the past decades. Despite its recent decline, it is still the second leading cause of cancer death worldwide [35].

Helicobacter pylori infection is a risk factor related to gastric cancer. This bacterium causes chronic gastritis and atrophic gastritis, which is characterized by chronic inflammation of gastric mucosa that leads to a condition of gastric intestinal metaplasia and dysplasia [36].

Gastric carcinomas spread by direct extension through the gastric wall to the perigastric tissues, sometimes adhering to adjacent organs such as the pancreas, colon, or liver. The disease also spreads via lymphatics or by seeding of peritoneal surfaces. The liver is the most common site for the hematogenous spread of tumor.

22.4.6.2.2 Clinical Presentation

Gastric cancer in the early stage is usually asymptomatic. As the tumor becomes more extensive, patients may complain of an upper abdominal discomfort varying from a vague, postprandial fullness to a severe steady pain. Anorexia and nausea are very common. Weight loss may eventually be observed. Nausea and vomiting are very common with tumors of the pylorus. If the lesion involves the cardia, dysphagia may be the major symptom.

The Borrmann classification is based on shape and infiltration margin and it divides advanced gastric cancer into the following four types:

- Type 1: polyploid fungating.
- Type 2: excavated.
- Type 3: ulcerated infiltrating.
- Type 4: diffusely infiltrating.

It is generally accepted that the prognosis of Borrmann type-4 gastric cancer is worse than that of the other three types [37].

22.4.6.2.3 MRI Findings

MRI can show focal thickening and enhancement of the inner layer of the gastric wall or an irregular mass with heterogeneous enhancement.

The pathological tissue is mainly isointense on T_1-weighted imaging and hyperintense on T_2-weighted imaging, but these features are not present in all cases, resulting in poor signal consistency. However, on DWI the tumor demonstrates high signal intensity, so the addition of diffusion-weighted MRI to T_2-weighted and dynamic contrast-enhanced MRI significantly improves the overall T-staging accuracy of gastric cancer. Lymph node metastasis can be easily identified on DW compared with T_2-weighted and CE MRI.

There are two difficulties in staging advanced gastric cancers. First, pT2 lesions can be overestimated because of the rough boundary of the tumor. Second, pT3 lesions are easily overestimated as stage T4 in some parts of the stomach close to the liver, pancreas, and diaphragm, due to a thin layer of perivisceral fat that surrounded these structures [3].

A characteristic appearance named *sandwich sign* can be widely and definitely observed on DW images, suggesting a stage ≥pT3. This sign is characterized by high signal intensity of the mucosa/submucosa, an intermediate low signal intensity of the muscularispropria and high signal intensity of the subserosa/serosa [38].

Compared to those using MDCT, there are only a small number of MRI studies of gastric cancer patients. Despite the excellent soft-tissue contrast of MRI, cumulative value on diagnostic accuracy indicate that MDCT and MRI show similar performance in T- and N-staging [39]. The staging system for gastric cancer is reported in Table 22.2 [34].

22.4.6.3 Gastric Gastrointestinal Stromal Tumors

22.4.6.3.1 Pathophysiology

Gastrointestinal stromal tumors (GISTs) are the most common gastrointestinal mesenchymal tumor, they are characterized by the expression of a specific immunohistochemical marker: c-kit [40]. GISTs can occur anywhere in the gastrointestinal tract, but the most frequent location of growth is the stomach (60%–70%) [41]. Around 20% of GISTs are malignant and the most common sites of metastasis are the liver and the peritoneum; metastases to the lymph nodes are rare [42]. The treatment of choice is surgical resection. Combination chemotherapy should be reserved for patients with metastatic disease.

22.4.6.3.2 Clinical Presentation

GISTs are often clinically silent. If symptomatic, the most common symptoms of gastric GISTs are bleeding, dyspepsia, anorexia, and abdominal pain.

22.4.6.3.3 MRI Findings

Radiologic features of GISTs vary depending on tumor size. They most commonly have an exophytic

TABLE 22.2

Staging System for *Gastric Adenocarcinoma* According to the Seventh Edition of the *AJCC Cancer Staging Manual*

Primary Tumor (T)

TX	Primary tumor cannot be assessed
T0	No evidence of primary tumor
Tis	Carcinoma in situ: intraepithelial tumor without invasion of thelamina propria
T1	Tumor invades lamina propria, muscularis mucosae, or submucosa
T1a	Tumor invades lamina propria or muscularis mucosae
T1b	Tumor invades submucosa
T2	Tumor invades muscularis propria
T3	Tumor penetrates subserosal connective tissue without invasion of visceral peritoneum or adjacent structures. T3 tumors also include those extending into the gastrocolic or gastrohepatic ligaments, or into the greater or lesser omentum, without perforation of the visceral peritoneum covering these structures
T4	Tumor invades serosa (visceral peritoneum) or adjacent structures
T4a	Tumor invades serosa (visceral peritoneum)
T4b	Tumor invades adjacent structures such as spleen, transverse colon, liver, diaphragm, pancreas, abdominal wall, adrenal gland, kidney, small intestine, and retroperitoneum

Regional Lymph Nodes (N)

NX	Regional lymph nodes cannot be assessed.
N0	No regional lymph node metastasis.
N1	Metastases in 1–2 regional lymph nodes.
N2	Metastases in 3–6 regional lymph nodes.
N3	Metastases in ≥7 regional lymph nodes.

Distant Metastasis (M)

M0	No distant metastasis.
M1	Distant metastasis.

growth pattern and manifest as dominant masses outside the stomach. Intramural and intraluminal masses are less common.

Small tumors are often well-defined and have homogeneous signal whereas large tumors tend to have lobulated margins, mucosal ulceration, central necrosis, hemorrhage, cavitation, and heterogeneous enhancement [43]; obstruction is rare despite the large size [44]. Patterns of growth are characterized by displacement of adjacent organs and vessels, direct invasion of the adjacent structures is uncommon and sometimes seen with advanced disease.

On MRI, the tumor usually shows low signal intensity on T_1-weighted images, intermediate to high signal intensity on T_2-weighted images, and enhancement after administration of gadolinium. A formal staging system for GIST appeared for the first time in the seventh edition of the *AJCC Cancer Staging Manual* (Table 22.3) [34].

TABLE 22.3

Staging System for Gastrointestinal Stromal Tumors (GIST) According to the Seventh Edition of the *AJCC Cancer Staging Manual*

Primary Tumor (T)	
TX	Primary tumor cannot be assessed.
T0	No evidence for primary tumor.
T1	Tumor ≤2 cm.
T2	Tumor >2 cm but not >5 cm.
T3	Tumor >5 cm but not >10 cm.
T4	Tumor >10 cm in greatest dimension.

Regional Lymph Nodes (N)	
NX	Regional lymph nodes cannot be assessed.
N0	No regional lymph node metastasis.
N1	Regional lymph node metastasis.

Distant Metastasis (M)	
M0	No distant metastasis.
M1	Distant metastasis.

22.5 Small-Bowel Pathology

22.5.1 Congenital Anomalies

Embryologically, the small intestine and the proximal large intestine develop from the midgut.

The midgut is supplied by the mesenteric artery and extends from the apex of the duodenal loop, which is fixed to the large liver anlage via the bile duct, to the last third of the transverse colon.

At an early stage of development, the midgut communicates with the yolk sac via a vitellointestinal (omphalomesenteric) duct, which disappears later. Differentiation occurs in a cranio-caudal sequence within one month (from 31 to 59 days).

Two common conditions that can be detected during an MRI examination are intestinal malrotation and Merckel diverticulum.

22.5.1.1 Intestinal Malrotation

Intestinal malrotation is a congenital anomaly that results in the abnormal positioning of the small and/ or the large bowel. It occurs when the midgut fails to complete the required 270° counter-clockwise rotation during embryologic development.

Usually, it is isolated but can also be associated with congenital heart disease or situs viscerum inversus.

In adults this is usually an asymptomatic condition and an incidental finding; rarely it can lead to an acute colonic obstruction due to volvulus.

The diagnosis of this condition is relatively simple. Typical findings include jejunum localized in the right lower abdomen, left side cecum, aplasia of the uncinate process of pancreatic gland, and abnormal relationship between superior mesenteric artery and vein.

22.5.1.2 Meckel Diverticulum

Meckel diverticulum is the most common congenital anomaly of the gastrointestinal tract, occurring in 2%–3% of the population and accounting for 90% of all omphalomesenteric duct anomalies [45,46]. It is a true diverticulum composed of all layers of the intestinal wall, resulting from an incomplete atrophy of the omphalo-mesenteric duct [47–49].

Its point of origin is commonly found within 100 cm of the ileo-cecal valve, at the anti-mesenteric border of the terminal ileum. It has no gender predilection, it is asymptomatic, and it is generally discovered incidentally during radiological or surgical procedures. About 25% of Meckel diverticula become symptomatic. Symptoms are secondary to complications, with the most common ones being hemorrhage, diverticulitis, and intestinal obstruction from intussusception, torsion or volvulus.

22.5.2 Tumors

Tumors of the small bowel are relatively uncommon. Differentiation between benign and malignant small-bowel lesions at MRI may prove difficult, particularly when lesions are smaller [50,51].

22.5.2.1 Benign Tumors

GIST is the most common benign tumor of the small intestine. It is now thought to arise from a precursor of the interstitial cells of Cajal, normally present in the myenteric plexus. GISTs can occur anywhere along the gastrointestinal tract and affect the small bowel in 33% of cases.

Primary GISTs are typically large rounded masses, showing intense enhancement after administration of contrast medium. The signal is often heterogeneous because of necrosis, hemorrhage, or cystic degeneration.

Smaller tumors (<2 cm) are generally considered benign with a very low risk of recurrence and may have a more homogeneous signal. They can be exophytic, intramural, or intraluminal.

Lipoma is usually an incidental finding during MRI-studies. In most cases it is asymptomatic, but can sometimes can cause an intussusception. Most often affecting the ileum and jejunum, it is well recognized by MRI due to its fat content.

Other benign cancers of the small intestine are adenomas, hamartomas, hyperplastic polyps, and

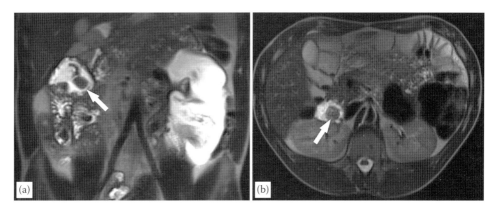

FIGURE 22.2
Surveillance of polyps in a 35-year-old man with Peutz–Jeghers syndrome. (a) Coronal HASTE and (b) axial HASTE images from MR enterography show one large low signal polyp (arrow) in the small bowel.

neurofibromas. All these types of cancer have no typical features on MRI. Multiple adenomas can occur in patients affected by familial adenomatous polyposis syndrome (FAP) and multiple hamartomas characterize patients with Peutz–Jeghers syndrome (Figure 22.2).

22.5.2.2 Malignant Tumors

Malignant tumors of the small bowel account for 1%–2% of all gastrontestinal tract neoplasms and are usually misdiagnosed [52]. Adenocarcinoma is the most common primary tumor of the small bowel. Up to 70% of adenocarcinomas of the small intestine are found in the duodenum and jejunum. Patients typically arise when the cancer is already at an advanced stage. Commonly it appears as a polypoid mass or as a luminal stenosis with annular wall thickening, causing bowel obstruction.

Carcinoid tumors are usually malignant and are the most common primary tumors of the small bowel. They arise from neuroendocrine cells in the submucosa and are typically small. The diagnosis is obtained taking into account the typical appearance of metastatic lymph nodes at the root of the mesentery that appear as spiculated masses, often calcified in the center. The vasoactive amines, serotonin and tryptophan are produced by the tumor and cause a desmoplastic reaction with local retraction of the mesentery, kinking of small-bowel loops, and adjacent wall thickening. These tumors spread locally and via the lymphatic system. Patients may present a carcinoid syndrome characterized by flushing, headache, diarrhea, nausea, and vomiting [53].

Lymphoma of the small intestine is a B-cell lymphoma in two third of the cases, arising in the distal ileum, and a T cell lymphoma in one third of cases, typically occurring in the duodenum and jejunum. Lymphomas can appear as rounded, cavitated, or polypoid masses of the wall [54]. Usually coexist mesenteric and retroperitoneal adenopathy. Vessels are typically not involved.

Serosa of the small bowel and mesentery can be a common localization of peritoneal metastasis, most commonly from ovarian cancer, adenocarcinoma of colon and gastric cancer. MRI has good sensitivity in identifying small peritoneal nodules.

22.5.3 Inflammatory Condition

22.5.3.1 Crohn's Disease

22.5.3.1.1 Clinical Presentation

CD is an inflammatory bowel disease that may affect any part of the gastrointestinal tract from mouth to anus. It is observed predominantly in the developed countries of the world.

The usual onset is between 15 and 30 years of age, but can occur at any age.

The course of the disease is characterized by remitting and relapsing episodes. Symptoms vary depending on the location, severity of disease, and behavior. The most common presenting symptom is chronic diarrhea (decrease in faecal consistency for more than 6 weeks). Other symptoms involved abdominal pain, weight loss, malaise, anorexia, or fever. Blood and/or mucus in the stool may be seen in up to 40%–50% of patients with Crohn's colitis.

When the terminal ileum is involved, an acute presentation may occur and may be mistaken for appendicitis. Patients may also present extra-intestinal manifestations involving the musculoskeletal system, eye, gallbladder, skin, and endocrine system.

A single gold standard for the diagnosis of CD is not available; a combination of clinical evaluation, endoscopy, histological results, biochemical investigations, and radiological examinations is important to confirm the diagnosis.

22.5.3.1.2 Pathophisiology

The precise aetiology of CD is unknown. Accumulating evidence suggests that inflammatory bowel diseases (IBD) results from an inappropriate inflammatory response to intestinal microbes in a genetically susceptible host. An altered gut mucosal immunity leads to cytokine overproduction and increased bowel wall leukocyte infiltration. Neutrophils and mononuclear cells infiltrate the crypts, leading to inflammation (crypitits) or abscess (crypt abscess).

Pathological and Histological samples show a transmural pattern of inflammation (the inflammation involves the entire depth of the intestinal wall) with an abrupt transition between unaffected and affected tissue (skip lesions); ulcerations and granulomas are common findings in CD.

The advanced stage of pathology shows chronic mucosal damage, characterized by metaplasia and fibrosis.

22.5.3.1.3 MRI Findings

MR enterography examinations are used to evaluate the entire spectrum of CD manifestation [4]. It provides a comprehensive look at intraluminal and extraluminal pathology and it is able to assess the activity of disease, complications, providing a valid support for clinical decision-making. The disease usually starts in the mucosal lining and extends across the bowel wall, involving the perienteric structures. The most common accepted imaging classification system proposed divides CD into three subtypes: active inflammatory, fibrostenotic, and fistulising/perforating. Each one of these conditions includes typical radiological findings.

- *Active inflammatory subtype:*
  - Wall thickening: wall thickness of 3 mm or greater correlates with biologically active disease. The thickening is due to edema and infiltration of inflammatory cells.
  - Wall post contrast hyper-enhancement: The enhancement in active inflammation is transmural and stratified. This layered appearance is relative to an inner layer of enhancing mucosa, an intermediate layer represented by edematous submucosa, and an outer layer of enhancing serosa and bowel wall musculature.
  - Increased T_2-weighted signal: this is due to the presence of wall enema.
  - Increased mesenteric vascularity: this is called a *comb sign* and it is strongly correlated with active inflammation. Vascular arcades appear prominent secondary to an increased flow, outlined against a background of inflammation and fibrofatty proliferation.

- Enlarged lymph nodes: adjacent to the active inflamed segment; axial dimension of 5 mm or greater.
- Mucosal ulcerations

These typical findings are shown in Figure 22.3

- *Fibrostenotic subtype*

Imaging findings in fibrostenosing disease rely more on lack of findings typically associated with active inflammation rather than direct visualization of fibrotic tissues. Differentiation between inflammation and fibrosis is difficult by imaging. However, there are some findings suggesting a fibrostenotic subtype to be considered (Figure 22.4):

- Wall thickening (due to deposition of collagen): >3 mm.
- Homogeneous mural enhancement.
- Luminal narrowing with associated upstream dilatation: even if strictures may be secondary to inflammation, fibrosis, or a combination of both.

- *Fistulizing/Perforating Subtype*

In the perforating subtype, transmural inflammation extends beyond the bowel wall to involve adjacent mesentery, organs, or bowel loops. Typical findings (Figure 22.5) are

- Fistula: inflammatory erosion into adjacent bowel loops. It can show enhancement after contrast administration.
- Sinus tract: a blind ending connection between bowel and muscle or mesentery.
- Abnormal kinking of adjacent loops: highly indicative of fistulae.

Frequently, patients with longstanding CD may have a mixed type of disease characterized by fibrostenosing subtype with superimposed active inflammation (Figure 22.6). However, an accurate assessment of tissue fibrosis is difficult; in this condition, the best way to describe imaging findings is to report two main patterns: the presence of stenosis with radiographic signs of inflammation or the presence of stenosis without radiographic signs of inflammation.

22.5.3.2 Celiac Disease

22.5.3.2.1 Clinical Presentation

Celiac disease is a gluten-sensitive enteropathy that affects the small intestine in genetically susceptible individuals. Generally considered a pediatric condition,

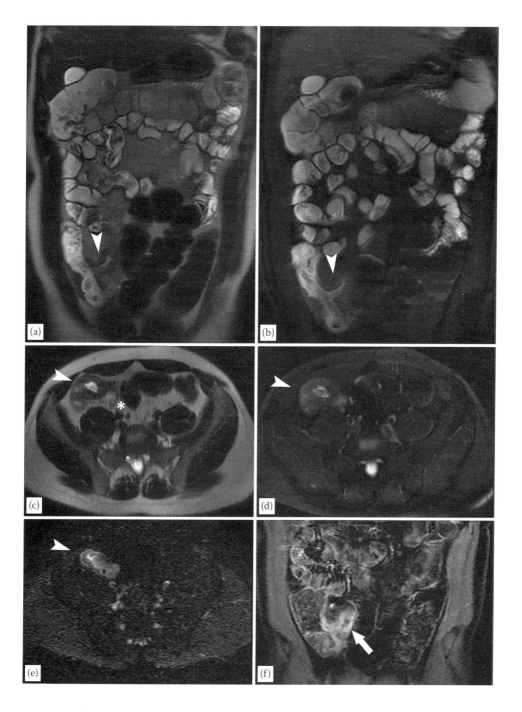

FIGURE 22.3

A 27-year-old female patient with active inflammatory CD. (a) Coronal and (c) axial T_2-weighted single shot fast spin-echo imaging demonstrates circumferential bowel wall thickening involving the terminal ileum. (b) Coronal and (d) axial T_2-weighted with fat saturation show an increase in signal intensity in the bowel wall, indicating bowel wall edema (arrowhead). (e) Axial DWI image demonstrates increased signal intensity due to restricted diffusion (arrowhead). (f) Coronal T_1-weighted contrast-enhanced GRE imaging demonstrates the stratified bowel wall enhancement (arrow) with engorged mesenteric vessels consistent with active inflammatory disease.

it has also two peaks of incidences in the fourth and sixth decade of life.

Celiac disease is now recognized as a common disease, occurring in about 1 in every 200 individuals. Common presentations include abdominal pain, iron deficiency anemia, and guaiac-positive stools. However, the symptoms of celiac disease are non specific and include diarrhea (20%), constipation (15%) steatorrhea, abdominal pain, abdominal distention, and vomiting. Weight loss is rare, affecting only 5% of patients [55,56].

Long-term complications include malnutrition, which can lead to anemia, osteoporosis, and miscarriage,

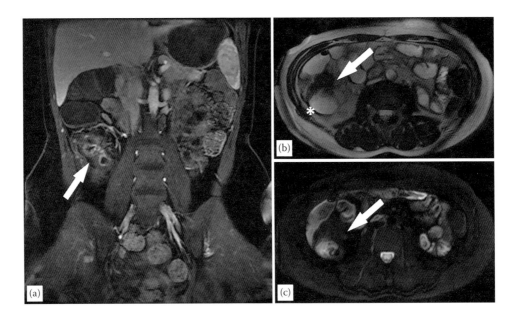

FIGURE 22.4
Chronic inflammation. A 20-year-old patient with CD. (a) Coronal delayed postcontrast 3D T_{1W} image demonstrating thickening of the terminal ileum with abnormal enhancement limited to the mucosa (arrow), suggesting the presence of mural fibrosis. (b) Axial T_{2W} image of corresponding segment; pre-stenotic dilatation is clearly visible (asterisk). (c) Axial T_{2W} image with fat saturation shows low signal intensity in the bowel wall, indicating the absence of bowel wall edema.

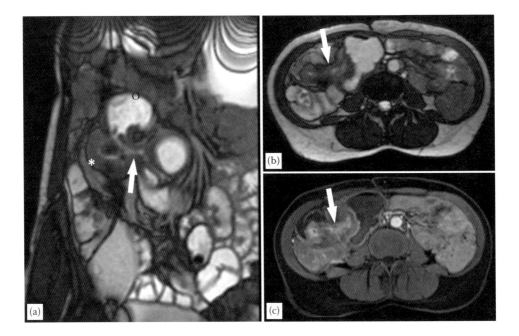

FIGURE 22.5
A 30-year-old male patient with penetrating CD. (a) Coronal T_2-weighted SSFSE imaging demonstrates circumferential bowel wall thickening (asterisk) involving the terminal ileum with associated luminal narrowing and fistula tract arising proximal to the level of the stricture (arrows in a, b). A pre-stenotic dilatation is also clearly visible (circle in a). (c) Axial contrast-enhanced T1-weighted GRE images demonstrate the enhancing fistulous tract (arrows).

among other problems such us liver disease, intestinal lymphoma, and intestinal carcinoma. The latter should be suspected when previously asymptomatic patients with celiac disease develop rapid changes in bowel habits.

Diagnosis of celiac disease is verified with a biopsy of the small intestine. The use of MR enterography is important for the evaluation of non specific intestinal disorders and for the detection of complications such as intestinal intussusception and neoplastic conditions.

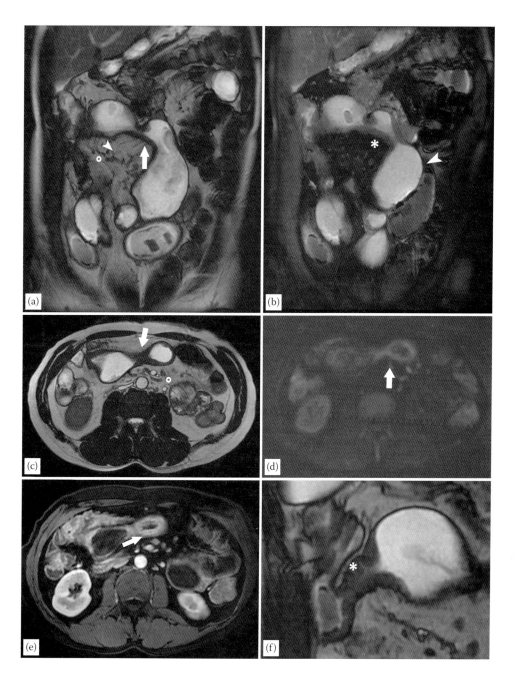

FIGURE 22.6

A 26-year-old male patient with active inflammation in a stenotic subtype of CD. Coronal (a) and axial (c) T_2-weighted (FIESTA) imaging demonstrates circumferential bowel wall thickening involving the terminal ileum (arrow), prominent mesenteric vessels supplying the abnormal bowel segment (arrowhead), and enlarged lymph nodes (circle). (b) Coronal T_2-weighted images with fat saturation shows increased signal intensity in the bowel wall indicating bowel wall edema (asterisk) with pre-stenotic bowel dilatation (arrowhead). (d) DW imaging shows increasing bowel wall restricted diffusion (arrow). (e) Axial T_1-weighted contrast-enhanced (GRE) images demonstrate the stratified bowel wall enhancement (arrow) with engorged mesenteric vessels consistent with active inflammatory disease. (f) Mucosal ulcer is clearly visible in an adjacent segment (asterisk).

22.5.3.2.2 Pathophisiology

Celiac disease predominantly involves the duodenum and proximal jejunum.

Histological findings are graded by the Marsh grading system and involve autoimmune inflammatory infiltrate and villous atrophy. In stage 0 (quiescent phase), there are no significant changes and biopsy results are normal. Stages 1 and 2 (active stage) are characterized by progressive lymphocytic infiltration associated with thickened duodenal and jejunal folds.

Stage 3 (destructive stage) is the typical pattern; there is infiltration of the surface epithelium with lymphocytes, partial villous atrophy, and crypt hyperplasia. Stage 4 (atrophic stage) is quite rare and often occurs in the context of very complicated celiac disease. There is a complete villous atrophy and wall thinning.

With the progression of the disease, ileal folds become inflamed and thickened, producing the jejunoileal fold reversal (increased number of ileal folds and decreased number of jejunal folds).

Lymph nodes progressively enlarge, producing large, low-attenuation or even cavitating, fat-filled lymph nodes.

22.5.3.2.3 *MRI Findings*

Alterations indicative of celiac disease, detectable by MRI, are present in about 75% of patients. Typical findings (Figure 22.7) involve:

- Dilatation of small-bowel loops (>3 cm)
- Jejunoileal fold reversal: reversal of the normal jejunoileal fold pattern, with a decreased number of jejunal folds ($n<3$ folds in 2.5 cm) and an increased number of ileal folds ($n>4$ folds in 2.5 cm).
- Inflammatory thickening of the bowel wall: the lymphocytic infiltration thickens both the jejunal folds and wall. As the inflammatory process advances (phases 3 and 4), thickening of the wall and folds progresses into the ileum.
- Intramural Fat: can be seen in duodenal or jejunal wall; it is believed to be the result of chronic inflammatory processes stimulating fat deposition.

- Lymphadenopathy: nodal enlargement of upper mesenteric lymph nodes, due to follicular hyperplasia caused by proliferation of reactive B and T lymphocytes. Lymph nodes can be considered prominent when their cumulative axial area is greater than the axial area of adjacent blood vessels.
- Mesenteric vascular engorgement: the small-bowel mesentery may appear hypervascular, enlarged; fat mesentery may appear edematous, hyperdence (misty mesentery).
- Nonobstructive intussusception: the mesenteric fat and vessels of one bowel loop are seen within the lumen of an adjacent bowel loop.
- Splenic atrophy.
- Cavitary mesenteric lymph node syndrome: characterized by fat-fluid levels inside lymph node, it is infrequent and found in patients with advanced symptomatic disease.

22.6 Summary

MRI is a clinically established technique for the study of small bowel. In particular, it provides a comprehensive assessment of intraluminal and extraluminal

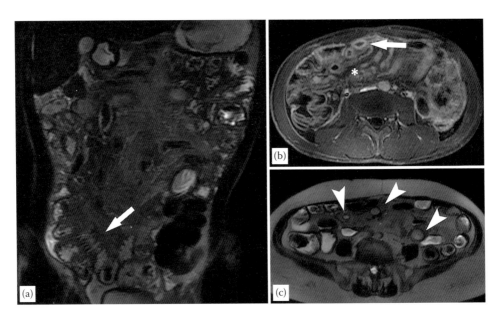

FIGURE 22.7

(a) Celiac disease in a 23-year-old man. Coronal T_2 image from MR enterography shows an increased number of folds in ileal loops (arrow). (b) Axial GRE T_1 image after intravenous injection of contrast media shows increased wall thickness and enhancement (arrow). Enlarged lymph nodes are also clearly visible (asterisk). (c) Cavitary mesenteric lymph node syndrome with enlarged and necrotic lymph nodes (arrowheads) in 36-year old woman.

pathologies. For this reason, it is considered the imaging technique of choice for the evaluation of CD, capable of assessing disease activity and related complications. MRI can play a role for the upper gastrointestinal tract diseases evaluations, in particular for the dynamic assessment of esophageal motility.

References

1. Kulinna-Cosentini C, Schima W, Lenglinger J et al. Is there a role for dynamic swallowing MRI in the assessment of gastroesophageal reflux disease and oesophageal motility disorders? *EurRadiol.*2011;22(2):364–370.

2. Yang DM, Kim HC, Jin W et al. 64 multidetector-row computed tomography for preoperative evaluation of gastric cancer: Histological correlation. *J Comput Assist Tomogr.* 2007;31(1):98–103.

3. Liu S, He J, Guan W et al. Added value of diffusion-weighted MR imaging to T2-weighted and dynamic contrast-enhanced MR imaging in T staging of gastric cancer.*JClin Imaging*: Elsevier Inc., 2014;38(2):122–128.

4. Anupindi SA, Terreblanche O, Courtier J. Magnetic resonance enterography: Inflammatory bowel disease and beyond. *MagnReson Imaging Clin N Am.* 2013;21(4):731–750.

5. Fidler JL, Guimaraes L, Einstein DM. MR imaging of the small bowel.*RadioGraphics.* 2009;29(6):1811–1825.

6. Masselli G, Casciani E, Polettini E, Gualdi G. Comparison of MR enteroclysis with MR enterography and conventional enteroclysis in patients with Crohn's disease. *EurRadiol.* 2008;18(3):438–447.

7. Negaard A, Paulsen V, Sandvik L et al. A prospective randomized comparison between two MRI studies of the small bowel in Crohn's disease, the oral contrast method and MR enteroclysis. *EurRadiol.* 2007;17(9):2294–2301.

8. Schreyer AG, Geissler A, Albrich H et al. Abdominal MRI after enteroclysis or with oral contrast in patients with suspected or proven Crohn's disease. *ClinGastroenterolHepatol.* 2004;2(6):491–497.

9. Frokjaer JB, Larsen E, Steffensen E, Nielsen AH, Drewes AM. Magnetic resonance imaging of the small bowel in Crohn's disease.*Scand J Gastroenterol.* 2005;40(7):832–842.

10. O'Neill OM, Johnston BT, Coleman HG. Achalasia: A review of clinical diagnosis, epidemiology, treatment and outcomes. *World J Gastroenterol.* 2013;19(35):5806–5812.

11. Richter JE. Oesophageal motility disorders.*Lancet.* 2001;358(9284):823–828.

12. Roman S, Kahrilas PJ. Management of spastic disorders of the esophagus. *GastroenterolClin North Am.* 2013;42(1):27–43.

13. Pehlivanov N,Liu J,Kassab GS et al. Relationship between esophageal muscle thickness and intraluminal pressure in patients with esophageal spasm. *Am J PhysiolGastrointest Liver Physiol.* 2002;282:1016.

14. Dent J, El-Serag HB, Wallander MA, Johansson S. Epidemiology of gastro-oesophageal reflux disease: A systematic review. *Gut.* 2005;54(5):710–717.

15. Altomare A. Gastroesophageal reflux disease: Update on inflammation and symptom perception. *WJG.*2013;19(39):6523.

16. Tutuian R, Castell DO. Review article: Complete gastro-oesophageal reflux monitoring—Combined pH and impedance. *Aliment PharmacolTher.* 2006;24(Suppl. 2):27–37.

17. Koike T, Ohara S, Sekine H et al. *Helicobacter pylori* infection prevents erosive refluxoesophagitis by decreasing gastric acid secretion. *Gut.* 2001;49(3):330–334.

18. Malfertheiner P, Sipponen P, Naumann M et al. *Helicobacter pylori* eradication has the potential to prevent gastric cancer: A state-of-the-art critique. *Am J Gastroenterol.* 2005;100(9):2100–2115.

19. Kahrilas PJ, Kim HC, Pandolfino JE. Approaches to the diagnosis and grading of hiatal hernia. *Best Pract Res ClinGastroenterol.* 2008;22(4):601–616.

20. van Hagen P, Hulshof MC, van Lanschot JJ et al. Preoperative chemoradiotherapy for esophageal or junctional cancer. *N Engl J Med.* 2012;366(22):2074–2084.

21. Pennathur A, Gibson MK, Jobe BA, Luketich JD. Oesophageal carcinoma. *Lancet.* 2013;381(9864):400–412.

22. Zhang Y. Epidemiology of esophageal cancer. *World J Gastroenterol.* 2013;19(34):5598–5606.

23. Levine MS, Caroline D, Thompson JJ, Kressel HY, Laufer I, Herlinger H. Adenocarcinoma of the esophagus: Relationship to Barrett mucosa. *Radiology.* 1984;150(2):305–309.

24. Lewis RB, Mehrotra AK, Rodriguez P, Levine MS. From the radiologic pathology archives: Esophageal neoplasms: Radiologic-pathologic correlation. *RadioGraphics.* 2013;33(4):1083–1108.

25. Siewert JR, Stein HJ, Feith M, Bruecher BL, Bartels H, Fink U. Histologic tumor type is an independent prognostic parameter in esophageal cancer: Lessons from more than 1,000 consecutive resections at a single center in the Western world. *Ann Surg.* 2001;234(3):360–367; discussion 368–369.

26. Sakurada A, Takahara T, Kwee TC et al. Diagnostic performance of diffusion-weighted magnetic resonance imaging in esophageal cancer. *EurRadiol.* 2009;19(6):1461–1469.

27. Dave UR, Williams AD, Wilson JA et al. Esophageal cancer staging with endoscopic MR imaging: Pilot study. *Radiology.* 2004;230(1):281–286.

28. Riddell AM, Hillier J, Brown G et al. Potential of surface-coil MRI for staging of esophageal cancer. *AJR Am J Roentgenol.* 2006;187(5):1280–1287.

29. Riddell AM, Allum WH, Thompson JN, Wotherspoon AC, Richardson C, Brown G. The appearances of oesophageal carcinoma demonstrated on high-resolution, T2-weighted MRI, with histopathological correlation. *EurRadiol.* 2007;17(2):391–399.

30. Lehr L, Rupp N, Siewert JR. Assessment of resectability of esophageal cancer by computed tomography and magnetic resonance imaging. *Surgery.* 1988;103(3):344–350.

31. Kayani B, Zacharakis E, Ahmed K, Hanna GB. Lymph node metastases and prognosis in oesophageal carcinoma—A systematic review. *Eur J SurgOncol.* 2011;37(9):747–753.

32. Nishimura H, Tanigawa N, Hiramatsu M, Tatsumi Y, Matsuki M, Narabayashi I. Preoperative esophageal cancer staging: Magnetic resonance imaging of lymph node with ferumoxtran-10, an ultrasmall superparamagnetic iron oxide. *J Am Coll Surg*. 2006;202(4):604–611.

33. Lahaye MJ, Engelen SME, Kessels AGH et al. USPIO-enhanced MR imaging for nodal staging in patients with primary rectal cancer: Predictive criteria 1. *Radiology*. 2008;246(3):804–811.

34. Washington K. 7th edition of the AJCC cancer staging manual: Stomach. *Ann SurgOncol*. 2010;17(12):3077–3079.

35. Rahman R, Asombang AW, Ibdah JA. Characteristics of gastric cancer in Asia. *World J Gastroenterol*. 2014;20(16):4483–4490.

36. Lechago J, Correa P. Prolonged achlorhydria and gastric neoplasia: Is there a causal relationship? *Gastroenterology*. 1993;104(5):1554–1557.

37. An JY, Kang TH, Choi MG, Noh JH, Sohn TS, Kim S. Borrmann type IV: An independent prognostic factor for survival in gastric cancer. *J Gastrointest Surg*. 2008;12(8):1364–1369.

38. Zhang XP, Tang L, Sun YS et al. Sandwich sign of Borrmann type 4 gastric cancer on diffusion-weighted magnetic resonance imaging. *Eur J Radiol*. 2012;81(10):2481–2486.

39. Tokuhara T, Tanigawa N, Matsuki M et al. Evaluation of lymph node metastases in gastric cancer using magnetic resonance imaging with ultrasmallsuperparamagnetic iron oxide (USPIO): Diagnostic performance in postcontrast images using new diagnostic criteria. *Gastric Cancer*. 2008;11(4):194–200.

40. Chourmouzi D, Sinakos E, Papalavrentios L, Akriviadis E, Drevelegas A. Gastrointestinal stromal tumors: A pictorial review. *J Gastrointestin Liver Dis*. 2009;18(3):379–383.

41. De Vogelaere K, Aerts M, Haentjens P, De Grève J, Delvaux G. Gastrointestinal stromal tumor of the stomach: Progresses in diagnosis and treatment. *ActaGastroenterol Belg*. 2013;76(4):403–406.

42. Gong J, Kang W, Zhu J, Xu J. CT and MR imaging of gastrointestinal stromal tumor of stomach: A pictorial review. *Quant Imaging Med Surg*. 2012;2(4):274–279.

43. Lee NK, Kim S, Kim GH et al. Hypervascularsubepithelial gastrointestinal masses: CT-pathologic correlation. *RadioGraphics*. 2010;30(7):1915–1934.

44. Sandrasegaran K, Rajesh A, Rushing DA, Rydberg J, Akisik FM, Henley JD. Gastrointestinal stromal tumors: CT and MRI findings. *EurRadiol*. 2005;15(7):1407–1414.

45. Moore TC. Omphalomesenteric duct malformations. *SeminPediatr Surg*. 1996;5(2):116–123.

46. Bauer SB, Retik AB. Urachal anomalies and related umbilical disorders. *UrolClin North Am*. 1978;5(1):195–211.

47. Yahchouchy EK, Marano AF, Etienne JC, Fingerhut AL. Meckel's diverticulum. *J Am Coll Surg*. 2001;192(5):658–662.

48. Mackey WC, Dineen P. A fifty year experience with Meckel's diverticulum. *SurgGynecol Obstet*. 1983;156(1):56–64.

49. Ymaguchi M, Takeuchi S, Awazu S. Meckel's diverticulum. Investigation of 600 patients in Japanese literature. *Am J Surg*. 1978;136(2):247–249.

50. Kamaoui I, De-Luca V, Ficarelli S, Mennesson N, Lombard-Bohas C, Pilleul F. Value of CT enteroclysis in suspected small-bowel carcinoid tumors. *AJR Am J Roentgenol*. 2010;194(3):629–633.

51. Van Weyenberg SJ, Meijerink MR, Jacobs MA et al. MR enteroclysis in the diagnosis of small-bowel neoplasms. *Radiology*. 2010;254(3):765–773.

52. Masselli G, Polettini E, Casciani E, Bertini L, Vecchioli A, Gualdi G. Small-bowel neoplasms: Prospective evaluation of MR enteroclysis. *Radiology*. 2009;251(3):743–750.

53. Horton KM, Kamel I, Hofmann L, Fishman EK. Carcinoid tumors of the small bowel: Amultitechnique imaging approach. *AJR Am J Roentgenol*. 2004;182(3):559–567.

54. Lohan DG, Alhajeri AN, Cronin CG, Roche CJ, Murphy JM. MR enterography of small-bowel lymphoma: Potential for suggestion of histologic subtype and the presence of underlying celiac disease. *AJR Am J Roentgenol*. 2008;190(2):287–293.

55. Schuppan D, Dennis MD, Kelly CP. Celiac disease: Epidemiology, pathogenesis, diagnosis, and nutritional management. *NutrClin Care*. 2005;8(2):54–69.

56. Dickey W, Kearney N. Overweight in celiac disease: Prevalence, clinical characteristics, and effect of a gluten-free diet. *Am J Gastroenterol*. 2006;101(10):2356–2359.

23

MRI of the Retroperitoneum

Abed Ghandour, Verghese George, Rakesh Sinha, and Prabhakar Rajiah

CONTENTS

Magnetic resonance imaging (MRI) has emerged as an important imaging modality in disorders of the abdomen. Retroperitoneal pathologies are uncommon, and their diagnosis and characterization can be challenging. MRI has several advantages in the evaluation of these disorders including high inherent contrast resolution that allows tissue characterization, good spatial resolution, wide field-of-view, and multiplanar imaging-capabilities. However, the modality is expensive, not widely available, and has several contraindications and limitations, the latter including patient claustrophobia. In addition, caution should be exercised in administering gadolinium-based contrast agents to patients with severe renal dysfunction due to the potential risk of nephrogenic systemic fibrosis (NSF).

In this chapter, we discuss the various MRI sequences that are useful in the evaluation of retroperitoneal disorders and the MRI features of these retroperitoneal pathological processes.

and the presence and nature of vascular thrombosis or encasement. If vascular mapping is required (for example, in surgical planning), magnetic resonance angiography (MRA) is performed typically with a T_1-weighted 3D spoiled gradient-echo sequence. In patients with severe renal dysfunction, who are at a risk of NSF, there are several noncontrast MRA techniques, including SSFP, VIPR, and NATIVE SPACE. A recent technique with potential application in the retroperitoneum is PET/MRI that enables simultaneous acquisition of both modalities. This hybrid imaging technique combines the excellent morphological and tissue characteristic information provided by MRI with the metabolic information provided by positron emission tomography (PET). FDG (18-F Fluorodeoxyglucose), the most commonly used PET radiopharmaceutical, is an analog of glucose and accumulates in inflamed and neoplastic cells, which overexpress glucose transporter cellular receptors.

23.1 MRI Sequences in the Evaluation of Retroperitoneum

There are several MRI sequences available for evaluation of the retroperitoneum. T_1-weighted sequences are used to characterize fat or hemorrhage (high signal intensity), and to evaluate lymphadenopathy and vascular invasion. In-phase and out-of-phase dual-echo T_1-weighted images can be performed to detect the presence of microscopic fat and fat at interfaces, which are present with signal drop in the out-of-phase images. T_2 images with and without fat saturation are useful in evaluation of lymphadenopathy, contiguous invasion by a disease process, cystic or necrotic changes, fluid collections, and bone marrow edema. Single-shot turbo/fast spin-echo (HASTE, SSTSE) and steady-state free precession (SSFP) are fast sequences, where fluid appears bright. Diffusion-weighted images with apparent diffusion coefficient (ADC) maps are useful in evaluating the restriction of diffusion caused by the retroperitoneal abnormalities; increased restriction is an indicator of cellularity and is typically seen in neoplasms and inflammatory processes. T_1- and T_2-weighted three-point dixon sequences enable simultaneous generation of four sets of contrasts, namely in-phase, out-of-phase, water only, and fat-only images, all in a single acquisition, thus enabling rapid lesion characterization. Contrast enhancement of retroperitoneal lesions is assessed using volumetric, fat-saturated T_1-weighted gradient-echo sequences (VIBE, THRIVE, or LAVA). Contrast enhancement is useful in distinguishing solid from nonenhancing cystic or necrotic lesions, and to evaluate the extent of disease,

23.2 Pathological Conditions

Table 23.1 lists a summary of retroperitoneal neoplastic and non-neoplastic pathologies, while a summary of their common MRI findings and the associated differential diagnoses can be found in Table 23.2.

23.2.1 Neoplastic Processes

The four major categories of retroperitoneal neoplasms are: (1) mesodermal, (2) germ cell and sex cord stromal, (3) neurogenic, and (4) lymphoid tumors. Primary retroperitoneal neoplasms arise within the retroperitoneum but separate from the major retroperitoneal organs. Features suggestive of primary retroperitoneal origin are—anterior displacement of retroperitoneal organs; negative *beak* sign (rounded edges of an adjoining retroperitoneal organ); negative embedded organ sign (crescenteric deformation by tumor); and absence of a definite organ of origin (Phantom organ sign) [1,2]. Further characterization of a primary retroperitoneal mass may be possible in some neoplasms based on imaging features and demographics, but often histological sampling is required.

23.2.1.1 Mesodermal Neoplasms

One third of malignant retroperitoneal lesions are sarcomas, usually seen in the sixth and seventh decades of life, often presenting with large tumors due to late onset of symptoms [3,4]. There is a wide variety of tumors,

TABLE 23.1

Common Retroperitoneal Neoplasms by Origin and Non-Neoplastic Entities

Lesions	Type of Neoplasm
Neoplastic Lesions	
Mesodermal Origin	
Adipose tissue	Lipoma, liposarcoma
Smooth muscle	Leiomyoma, leiomyosarcoma
Connective tissue	Fibroma, malignant fibrous histiocytoma, chondrosarcoma, synovial cell sarcoma
Striated muscle	Rhabdomyoma, rhabdomyosarcoma
Blood vessels	Hemangioma, angiosarcoma
Perivascular epithelioid cells	Perivascular epithelioid cell tumor (PEComa) Group (e.g., angiomyolipoma, lymphangioleiomyomatosis), sarcoma of perivascular cells
Miscellaneous	Fibromatosis
Uncertain	Xanthogranuloma
Neurogenic Origin	
Nerve sheath	Schwannoma, neurofibroma, malignant schwannoma, neurogenicsarcoma, neurofibrosarcoma
Sympathetic nerves	Ganglioneuroma, ganglioneuroblastoma, neuroblastoma
Chromaffin tissue	Paraganglioma, pheochromocytoma, malignant paraganglioma, or pheochromocytoma
Germ Cell, Sex Cord, and Stromal Cell Origin	Mature teratoma, immature teratoma, malignant teratoma, mixed germ cell tumor
Lymphoid or Hematologic Origin	Lymphoma, extramedullary plasmacytoma
Non-Neoplastic Lesions	
Pseudotumoral lipomatosis, retroperitoneal fibrosis, Erdheim–Chester disease, and extramedullary hematopoiesis	

TABLE 23.2

Common MRI Findings and Their Associated Differential Diagnoses

Pure fat-containing mass	Lipoma, well-differentiated liposarcoma
Heterogeneous mass with fat	Dedifferentiated liposarcoma, angiomyolipoma
Fat-fluid level	Teratoma, well-differentiated liposarcoma
Myxoid stroma	Myxoid liposarcoma, neurogenic tumor, myxoid malignant fibroushistiocytoma
Large mass, extensive necrosis, invasion of IVC	Leiomyosarcoma
Fluid–fluid level caused by hemorrhage	Paraganglioma
Hypovascular	Lymphoma, low-grade liposarcoma, benign tumor
T_2 hypointensity	Lymphoma, desmoid tumor, retroperitoneal fibrosis, Erdheim–Chester disease
Mantle-like mass around aorta or IVC	Lymphoma, retroperitoneal fibrosis, Erdheim–Chester disease
Floating aorta or CT angiogram sign	Lymphoma

depending on the cell of origin. These are generally treated by surgical resection [5].

23.2.1.1.1 Liposarcoma

Liposarcoma is a malignant tumor originating from primitive mesenchymal cells [6], and accounts for 40% of retroperitoneal sarcomas [4,5,7,8]. Thirty-five percent of tumors originate in the perirenal fat [8,9]. There are five subtypes:well-differentiated, myxoid, dedifferentiated, round cell, and pleomorphic [10]. They often present

late with large-sized masses. Metastatic rate is less than 10% and local recurrence is the most common cause of morbidity and mortality [4,5,8,11,12].

Well-differentiated liposarcoma, the most common type of liposarcoma (Figure 23.1) contains components that are similar in signal intensity to macroscopic fat, that is, they show high signal intensity on T_1-weighted images; intermediate to high intensity on T_2 and loss of signal on fat-suppressed images. It typically has smooth margins and a lobular

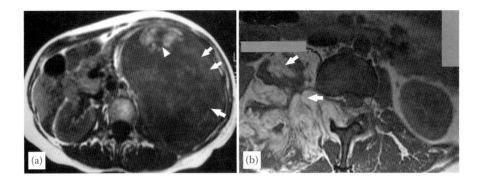

FIGURE 23.1
Well-differentiated liposarcoma—(a) T_1-weighted axial MRI image shows a large mass in the left flank of the abdomen, which has predominantly intermediate signal (arrows) with areas of high signal intensity (arrow tip) in the anterior portion suggestive of fat. The mass is seen displacing the kidney anteriorly. (b) Axial T_1-weighted MRI image in another patient shows a well-differentiated liposarcoma containing large areas of fat (arrows), displacing the right kidney anteriorly. The mass is seen extending into the right neural foramen into the spinal canal.

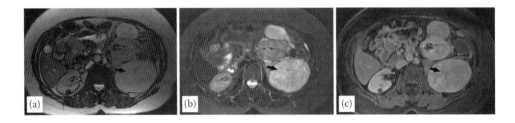

FIGURE 23.2
Dedifferentiated liposarcoma. Axial T_2 (a), STIR (b) and post-contrast VIBE (c) images show a well-defined soft tissue mass that has high signal in T_2 and STIR and showing intense contrast enhancement (arrow). The mass was proven to be a dedifferentiated liposarcoma, containing large areas of solid component.

contour [13] with little or no contrast enhancement of the fatty component. Differential diagnosis includes lipoma or angiomyolipoma. Imaging differentiation of well-differentiated liposarcoma from lipoma can be challenging. Large size (>10 cm), nodular or globular components, thick septa (>2 mm), soft tissue components, and a low proportion of fat in the mass (<75%) favor liposarcoma over a lipoma. It is a slow growing tumor, with local recurrence and minimal metastatic potential. Dedifferentiation occurs in 25% of tumors to a higher grade tumor [14].

Myxoid liposarcoma has a pseudocystic appearance due to the mucopolysaccharide contents in the myxoid matrix, resulting in low signal on T_1-weighted and high signal on T_2-weighted images. Lacy, linear, or amorphous areas of high signal intensity on T_1-weighted images and intermediate signal intensity on T_2-weighted images may be seen because of the intratumoral fat content [1,15]. It has gradual, heterogenous, and often incomplete enhancement due to slow, progressive accumulation of contrast in the extracellular space [13]. It has an aggressive clinical course and a marked propensity for metastasis, often to unusual sites such as bone and skin [8].

Dedifferentiated liposarcoma (Figure 23.2) may be primary or develop from a pre-existing well-differentiated liposarcoma, thus containing a well-differentiated lipogenic component and a dedifferentiated non-lipogenic component. On MRI, it appears as a non-lipomatous soft tissue mass with intermediate signal intensity and contrast enhancement within, adjacent to, or encompassing a fatty mass [16,17]. It is associated with rapid growth and metastasis [8].

Pleomorphic liposarcoma typically occurs in the elderly and is very aggressive with high metastatic potential [8]. They tend to present as heterogeneous tumors with little or no detectable fat, and it is often indistinguishable from other malignant soft tissue masses [13].

Round cell liposarcoma presents as heterogeneous soft tissue mass with areas of necrosis.

23.2.1.1.2 Leiomyosarcoma

Leiomyosarcoma is the second most common retroperitoneal sarcoma (30%), with two thirds occurring in women [5,11], usually diagnosed in the sixth decade [18] (Figures 23.3 and 23.4). Clinical presentation depends on the location and extent of involvement of the inferior

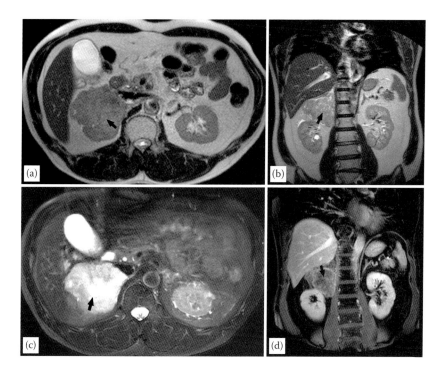

FIGURE 23.3

Leiomyosarcoma. (a) Axial T_1-weighted MR image shows a large intermediate signal intensity mass in the retroperitoneum that is infiltrating the right kidney (arrow). (b) Coronal T_2 single-shot (HASTE) image shows the mass having heterogeneous signal with areas of intermediate and high signal intensity (arrow). (3) Axial T_2-weighted image with fat saturation shows high signal intensity of the mass, which is indenting, but not invading the IVC (arrow). (d) Coronal 3D T_1-weighted MRI image shows heterogeneous contrast enhancement of the mass, which was biopsy proven to be a leiomyosarcoma (arrow).

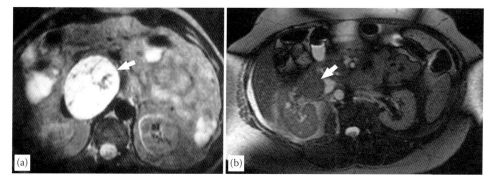

FIGURE 23.4

Intracaval leiomyosarcoma. (a) T_2 STIR image in another patient shows a leiomyosarcoma that is partially extending into the IVC (arrow). (b) T_2-weighted image in another patient shows a leiomyosarcoma extending into the IVC (arrow).

vena cava (IVC), and tumors may be completely external to the IVC lumen, have both intraluminal and extra-luminal components or be purely intraluminal. The tumor is usually well circumscribed, and can have extensive areas of necrosis (low to intermediate signal on T_1-weighted images, heterogeneous intermediate to high signal on T_2-weighted images) and variable amount of enhancement depending on the amount of muscular and fibrous components [17,19–21]. When occurring within IVC, differential diagnosis includes a bland thrombus, extension of renal cell carcinoma, lymphoma, liposarcoma, and leiomyomatosis [22].

23.2.1.1.3 Malignant Fibrous Histiocytoma

Malignant fibrous histiocytoma (MFH) is the third most common retroperitoneal sarcoma (15%) [11], more commonly seen in men [23]. Pathological types are pleomorphic (the most common), myxoid, giant cell, and inflammatory [10,24]. On MRI, MFH has low to intermediate T_1-signal intensity and heterogeneously increased T_2-signal intensity relative to muscle (Figure 23.5) [11,17]. A mosaic of mixed low, intermediate, and high T_2-signal intensity that correlates with the presence of intratumoral solid components, cystic degeneration, hemorrhage, myxoid stroma, and fibrous tissue leads to an

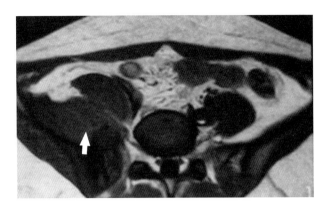

FIGURE 23.5
Malignant fibrous histiocytoma. Axial T_1-weighted MRI image shows an intermediate signal intensity mass in the right iliac fossa (arrow), which was shown to be a malignant fibrous histiocytoma.

appearance of a *bowl of fruit* sign [25]. Contrast enhancement is heterogeneous.

23.2.1.1.4 Less Common Sarcomas

There are several other types of sarcomas, most of which do not have specific distinguishing MRI appearances. *Rhadomyosarcoma* is more common in the pediatric population, and involves the retroperitoneum in 7% of cases [26]. There are several histological types, and usually show, intermediate signal intensity on MRI with variable contrast enhancement [27]. *Angiosarcoma* is an aggressive tumor that presents as a mass expanding the involved vessel (Figure 23.6), with heterogeneous enhancement

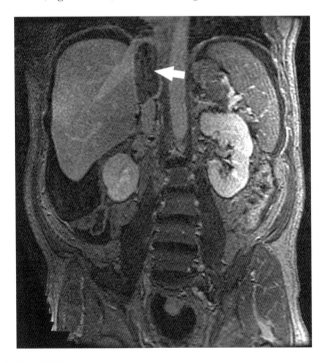

FIGURE 23.6
Angiosarcoma. Coronal post-contrast T_1-weighted MRI image shows a hypointense mass (arrow) that is extending and expanding the IVC.

due to necrosis [28]. *Chondrosarcoma* occurs in extraskeletal locations in 2%, with the mean age of presentation of 50 years [29]. It shows typical MRI features of a chondroid neoplasm, with very high signal intensity on T_2-weighted images, and only mild peripheral to septal enhancement after contrast. Mesenchymal chondrosarcoma has lower water content and hence intermediate signal on T_2-weighted images and diffuse, heterogeneous enhancement [30]. *Synovial cell sarcoma* has nonspecific MRI appearances, but may present as multicystic masses with characteristic triple signal intensity (interspersed high, intermediate, and low signal areas) on T_2-weighted MRI [31]. Also, synovial sarcoma should be considered when MRI shows a well-defined but inhomogeneous hemorrhagic lesion. Fluid–fluid levels on T_2-weighted sequences also support the diagnosis [32].

23.2.1.1.5 Perivascular Epithelioid Cell Tumor

Perivascular epithelioid cell tumors (PEComas) are perivascular mesenchymal tumors, that express melanocytic (HMB 45, melan-A) and muscle markers (SMA,desmin) [33]; these include angiomyolipoma, clear cell *sugar* tumor, lymphangioleiomyomatosis, and clear cell myomelanocytic tumor of the ligamentum teres/falciform ligament. Usually, the tumor is large and invasive, and it may recur and metastasize [33]. On MRI, tumors are hypointense to isointense on T_1-weighted images, and heterogeneously hyperintense on T_2-weighted images, with heterogeneous contrast enhancement (Figure 23.7). Hemorrhage is more common in larger tumors [34]. Enlarged vessels coursing through the lesion, aneurysms, and associated hemorrhage are features that enable differentiation from liposarcoma [35].

23.2.1.1.6 Desmoid Tumor

Desmoid tumors (deep fibromatosis, aggressive fibromatosis) are composed of spindle cells and develop from muscle connective tissue, fascia, and aponeuroses. Desmoid tumors are estrogen-dependent and are therefore more common in young women, with a peak occurrence in the third decade [36,37]. They are associated with familial polyposis coli and Gardner syndrome and may be single or multiple. On MRI, they have well-defined borders, but may occasionally be infiltrative. Signal characteristics depend on the tissue composition (spindle cells, collagen, and myxoid matrix) and vascularity, and may change over time. Early stage lesions are cellular, with high T_2 signal, while late-stage lesions have loss of cellularity and collagen deposition resulting in hypointense T_2 signal (Figure 23.8) [38]. Hypointense tracks may also be seen that are due to dense collagen bands [39]. Enhancement is moderate to marked [40]. They are locally aggressive with high recurrence rate (50%) even after wide surgical excision [41].

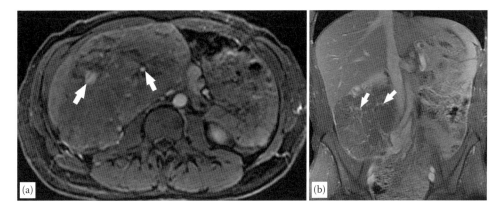

FIGURE 23.7

PEComa. Axial (a) and Coronal (b) contrast enhanced MRI scans show a large retroperitoneal mass, containing areas of enhancing vessels within it, which was biopsy proven to be a PEComa (arrow).

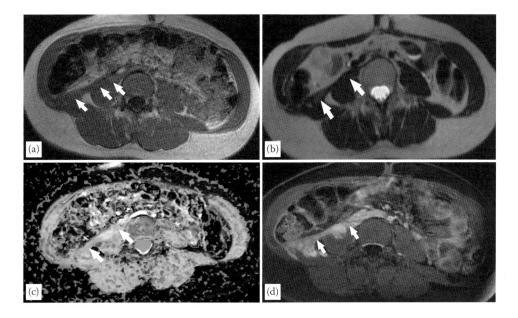

FIGURE 23.8

Desmoid. (a) Axial T_1-weighted image shows an ill-defined intermediate signal soft tissue mass in the right retroperitoneum anterior to the right psoas muscle (arrow). (b) Axial T_2-weighted MRI image at the same level shows the lesion having intermediate to low signal (arrow); (c) Color coded ADC map overlaid on axial T_2-weighted image shows mild to moderate diffusion restriction in the lesion (arrow). (d) Contrast-enhanced T_1 image, 5 min post-contrast injection, shows intense delayed enhancement (arrow), consistent with retroperitoneal desmoids (fibromatoses).

23.2.1.1.7 *Lipoma*

Lipoma is a benign tumor composed of well-encapsulated benign fat cells. Lipoma is rare in the retroperitoneum and such a diagnosis should be made with caution since most retroperitoneal fat-containing lesions are not lipomas but lipoma-like areas of an undersampled, well-differentiated liposarcoma. However, if proven to be a lipoma, it is worth reassuring the patient as malignant transformation of lipoma to liposarcoma is virtually unknown [35]. Symptoms are usually due to mass effect and compression. On MRI, lipoma has fat signal intensity and contains few if any septations. They do not enhance and have no soft tissue components [35].

23.2.1.2 *Germ Cell and Sex Cord-Stromal Tumors*

Germ cell and sex cord-stromal tumors are less commonly seen in the retroperitoneum.

23.2.1.2.1 *Primary Retroperitoneal Extragonadal Germ Cell Tumor*

Primary retroperitoneal extragonadal germ cell tumors (EGCTs) arise from primordial midline germ cell remnants of the genital ridge that had failed to migrate during embryological development, without any apparent gonadal primary lesion [11,42,43]. The diagnosis should be arrived at with caution since metastatic germ cell tumors are more common. Retroperitoneum is the second most common site of extragonadal germ cell tumor after the mediastinum. Their prevalence is greater in males [42]. Retroperitoneal EGCTs may be seminomatous or non-seminomatous (such as embryonal carcinoma, yolk sac tumor, choriocarcinoma, teratoma, and mixed germ cell tumors). On MRI, primary EGCTs are large, midline, enhancing masses that have low to intermediate T_1-signal intensity and intermediate to high T_2-signal intensity relative to skeletal muscle. Seminomas appear as uniformly solid, lobulated masses with fibrovascular septa that enhance intensely. Non-seminomatous GCTs appear as heterogeneous masses with areas of necrosis, hemorrhage, or cystic degeneration [43].

23.2.1.2.2 Primary Retroperitoneal Teratoma

Teratomas are comprised of mixed elements derived from the three germ cell layers [44]. Primary retroperitoneal teratomas represent 1%–11% of retroperitoneal neoplasms [45]. Teratomas can be histologically classified as mature (well-differentiated), immature (poorly differentiated), and malignant (non-germ cell malignancy arising from one of the three embryologic components). Macroscopically, there are two variants: cystic teratomas that are composed of fully mature elements; they contain sebaceous material and hair, and are usually benign; and solid teratomas that are more likely to be malignant and formed of immature elements [46]. On MRI, the presence of fat or fluid level within a tumor is highly suggestive. Calcification has low signal on MRI. Contrast enhancement is variable and heterogeneous (Figure 23.9).

23.2.1.2.3 Primary Sex Cord Stromal tumors

Primary sex cord stromal tumors are rare in the retroperitoneum, and more common in women. Granulosa cell tumor is the most common, while Sertoli–Leydig cell tumors, thecomas, and other types are less common. Elevated estrogen is seen in granulosa cell tumor and thecoma. MRI findings are nonspecific and include a

heterogeneous solid tumor and heterogeneous enhancement [34].

23.2.1.3 Neurogenic Tumors

23.2.1.3.1 Schwannoma

Schwannomas are neural sheath tumors that originate from Schwann cells of peripheral nerve fibers, with 0.7% of benign and 1.7% of malignant schwannomas reported in the retroperitoneum [47,48]. The *target sign*—central—low to intermediate T_2-signal intensity fibrous tissue surrounded by peripheral high-signal-intensity myxoid tissue [49], and the *fascicular sign* indicating the presence of fascicular bundles [50] are specific imaging features of schwannoma, but only rarely seen in the retroperitoneum. Another important feature seen on MRI is destruction of adjacent bony structures [51]. Contrast enhancement is heterogeneous, with nonenhancing cystic components and enhancing solid components.

23.2.1.3.2 Neurofibroma

Neurofibromas are benign nerve sheath tumors, more common in men, especially in the second to fourth decades of life [49], and have a strong association with neurofibromatosis-1 (NF-1). Neurofibromas associated with NF-1 present at a younger age, are typically multifocal, large (>5 cm), and more likely to be symptomatic. They often undergo malignant degeneration, particularly in cases of neurofibromatosis [49]. Retroperitoneal neurofibromas are usually bilateral and symmetric in parapsoas or presacral locations, and follow the distribution of the lumbosacral plexus [52]. On T_1-weighted MRI images, the central portion has higher signal intensity than the periphery, whereas on T_2-weighted images the periphery has higher signal intensity, due to central nerve tissue and peripheral myxoid degeneration [53]. Contrast enhancement is variable. With spinal nerve

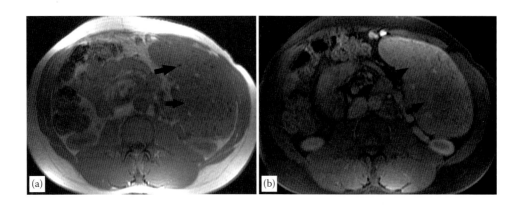

FIGURE 23.9
Teratoma. (a) Axial T_1 weighted image shows a large retroperitoneal mass that has intermediate T_1 signal (arrow). (b) Post-contrast image shows heterogeneous enhancement of the mass (arrow).

root neurofibromas that extend through and enlarge a neural foramen, a dumbbell shape is seen [17,25,54]. Plexiform neurofibromas show a *bag of worms* appearance with large number of infiltrative masses that diffusely thicken the parent nerve [10].

23.2.1.3.3 Malignant Peripheral Nerve Sheath Tumor

Malignant peripheral nerve sheath tumor (MPNST) refers to malignant schwannoma, neurogenic sarcoma, and neurofibrosarcoma [55]. MPNST can arise in patients with or without associated neurofibromatosis [49]. Men and women are equally affected, and this tumor is a disease of adulthood [49]. MRI features are nonspecific, but irregular infiltrative tumor borders and internal inhomogeneity are suggestive [49] (Figure 23.10). Rapid enlargement of tumor and pain are suspicious clinical features [49,56,57].

23.2.1.3.4 Ganglioneuroma

Ganglioneuroma is a benign neoplasm composed of nerve fibers and mature ganglion cells. It is often seen before the age of 20, more common in males, with 32%–52% occurring in the retroperitoneum [58]. Some ganglioneuromas are functionally active and secrete catecholamines, vasoactive intestinal polypeptides, or androgenic hormones, explaining symptoms such as hypertension, diarrhea, and virilization [49]. On MRI, they appear as well-circumscribed oval, crescentic, or lobulated masses surrounding major blood vessels but causing little or no compromise of the lumen [49]. Ganglioneuromas have homogeneous T_1-hypointensity and variable T_2 signal. Tumors with intermediate to high signal intensity on T_2-weighted MRI images contain abundant cellular and fibrous components and little myxoid stroma. Those with markedly high intensity on T_2-weighted images contain significant myxoid stroma and little cellular and fibrous components [49]. Like MPNST, curvilinear bands of low signal intensity on T_2-weighted images give this tumor a whorled appearance, making differentiation from MPNST difficult [49].

Ganglioneuroblastoma is an intermediate-grade tumor, with elements of benign ganglioneuroma and malignant neuroblastoma. It occurs in the 2–4-year age group, and equally between boys and girls. Its prognosis and response to therapy are significantly more favorable than neuroblastoma [59]. On MRI, it could be solid or cystic with solid components [34,49].

23.2.1.3.5 Neuroblastoma

Neuroblastoma is a malignant tumor that consists of primitive neuroblasts, typically seen in the first 10 years of life, more common in girls [60], and located anywhere along the sympathetic plexus or in the adrenal medulla. On MRI, neuroblastoma is irregular, lobulated, and heterogeneous and may show invasion of adjacent organs and encasement of vessels with luminal compression [34,49].

23.2.1.3.6 Paraganglioma

Paraganglioma is a chromaffin cell tumor originating in the parasympathetic ganglia, and is typically more aggressive than a pheochromocytoma originating from the adrenal medulla. Most paragangliomas are active (up to 60%) and secrete epinephrine or norepinephrine, with hormonal activity correlating poorly with the likelihood of malignancy. On MRI, they demonstrate low to intermediate T_1-signal intensity and moderately high T_2-signal intensity. They are hypervascular and show intense enhancement [2] (Figures 23.11 through 23.13). An intense high T_2 signal, referred to as *light bulb* sign is seen in 80% [61].

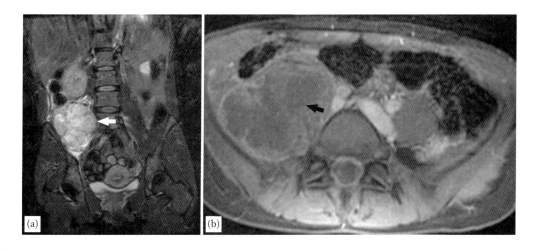

FIGURE 23.10
MPNST. (a) Coronal STIR image shows a mass with heterogeneous high signal in the retroperitoneum (arrow), adjacent to the vertebral bodies. (b) Axial post-contrast T_1-weighted image shows heterogeneous enhancement of the mass (arrow).

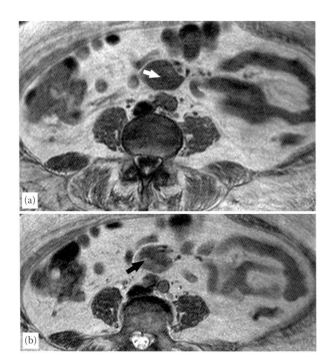

FIGURE 23.11
Paraganglioma. Axial T_1- (a) and T_2- (b) weighted MRI images show a heterogeneous mass in the retroperitoneum (arrow) adjacent to the organ of Zuckerkandl, which was shown to be a paraganglioma.

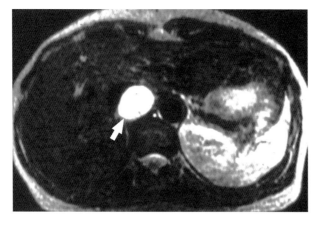

FIGURE 23.12
Extra adrenal pheochromocytoma. Axial T_2-weighted STIR image shows an intensely hyperintense mass (arrow) posterior to the IVC and displacing it anteriorly.

23.2.1.4 Lymphoid Neoplasms

23.2.1.4.1 Lymphoma

Lymphoma is the most common malignant retroperitoneal tumor, accounting for one third of these lesions. Abdominal Hodgkin's lymphoma tends to be confined to the spleen and retroperitoneum with spread of disease to contiguous lymph nodes, while non-Hodgkin's lymphoma involves discontiguous nodal groups and extranodal sites [62,63]. On MRI, lymphomas are relatively

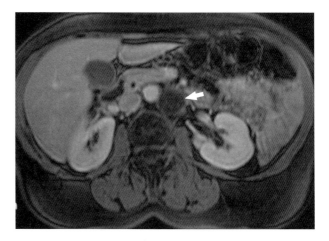

FIGURE 23.13
Cystic pheochromocytoma. Axial T_1-weighted image of a cystic pheochromocytoma shows hypointense signal intensity mass in the retroperitoneum (arrow).

homogeneous, hypointense to fat and slightly hyperintense to muscle on T_1-weighted images, but isointense to fat and hyperintense to muscle in T_2-weighted images (Figure 23.14).

23.2.1.4.2 Metastatic Lymphadenopathy

Metastatic lymph nodes are common in the retroperitoneum. A size criterion is often used in making the diagnosis, using 1.0 cm as the cutoff. However, this is a nonspecific finding and occasionally a neoplasm may be found in a nonenlarged lymph node. Metastatic adenopathy typically shows spherical nodes with lobulated contours, in association with heterogeneous contrast enhancement and T_2 signal intensity due to necrosis. Benign nodes on the other hand appear ovoid with smooth contours and show homogeneous T_2 signal and contrast enhancement. With ultrasmall superparamagnetic iron oxide particles, malignant adenopathy shows lower signal than benign nodes [64]. Metastasis in the retroperitoneum can be due to other causes, including primary solid tumors of other organs. Metastatic deposits can be seen in musculature (Figure 23.15).

23.2.1.4.3 Extramedullary Plasmacytoma

Extramedullary plasmacytoma is characterized by monoclonal proliferation of plasma cells in an extramedullary site and can be either primary or secondary to diffuse myeloma [65]. It is more common in males, in sixth and seventh decades [66]. Primary plasmacytoma can be diagnosed only after excluding multiple myeloma in bone marrow with absence of bone marrow plasmacytosis, and a serum or urinary paraprotein level of less than 2 g/dL [67]. On MRI, it is a solid tumor, isointense on T_1-weighted images and isointense to hyperintense on T_2-weighted images, with mild to marked

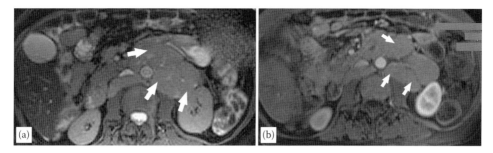

FIGURE 23.14
Lymphoma. Axial T_2-weighted images (a, b) show a large conglomerate heterogeneous mass encasing the abdominal aorta and displacing it anteriorly, which was consistent with a lymphoma (arrow).

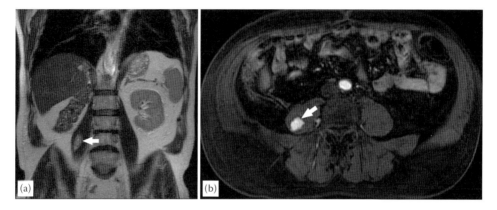

FIGURE 23.15
Metastasis. (a) Coronal single-shot fast spin-echo image in a patient with renal carcinoma shows a high-signal mass in the right psoas muscle (arrow). (b) Axial STIR image in the same patient shows the lesion having high signal in the psoas muscle (arrow). This was consistent with a metastatic lesion in the psoas from renal cell carcinoma.

heterogeneous enhancement. Large tumors may show areas of necrosis [68].

23.2.2 Non-Neoplastic Processes

23.2.2.1 Retroperitoneal Fibrosis

Idiopathic retroperitoneal fibrosis (RPF) is characterized by chronic inflammatory tissue with fibrosis in the retroperitoneum that can expand to entrap various abdominal organs, most commonly the ureters. RPF is now considered an important abdominal component of the group of IgG4-related sclerosing disorders [69]. Other causes include aneurysms, drugs (methysergide, LSD, bromocriptine, beta blockers, methyldopa, and hydralazine), infectious and inflammatory processes, retroperitoneal hemorrhage [70], smoking, and asbestos exposure [71]. Malignancy manifesting as retroperitoneal desmoplastic response, may be associated with up to 8% of cases; primary tumors include breast, lung, thyroid, gastrointestinal, and genitourinary cancers, as well as lymphomas and some sarcomas [70,72]. Most patients present in their 50s to 70s; the condition is two to three times more common in men [72,73].

On MRI, RPF manifests as hypointensity on T_1-weighted images. T_2 signal, however, varies with the degree of disease activity, reflected by inflammation and edema. Chronic, inactive fibrosis will contain little edema and thus displays low-intensity signal on both T_1 and T_2, while active disease displays high signal in T_2 and shows intense contrast enhancement that correlates with disease activity [69,74] (Figures 23.16 through 23.19). RPF usually begins as a confluent fibrotic plaque, usually below the aortic bifurcation at the L4–5 level. It then extends contiguously along the midline retroperitoneum, usually in a superior direction, encasing the aorta, IVC, and the ureters. Typically, the process does not show lateral extension beyond the lateral margins of the psoas muscles [70].

23.2.2.2 Retroperitoneal Fluid Collections

23.2.2.2.1 Hemorrhage/Hematoma

Although trauma is the most common cause, retroperitoneal hemorrhage may be associated with bleeding diathesis, anticoagulant therapy, leaking aneurysm, adrenal and renal disorders (especially neoplastic), and

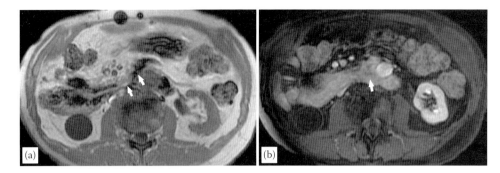

FIGURE 23.16
Retroperitoneal fibrosis. Active—(a) Axial T_1 MRI image shows a retroperitoneal soft tissue mass with intermediate signal intensity (arrow). (b) There is intense contrast enhancement of the retroperitoneal mass in delayed phase images, indicating that the fibrosis is active (arrow).

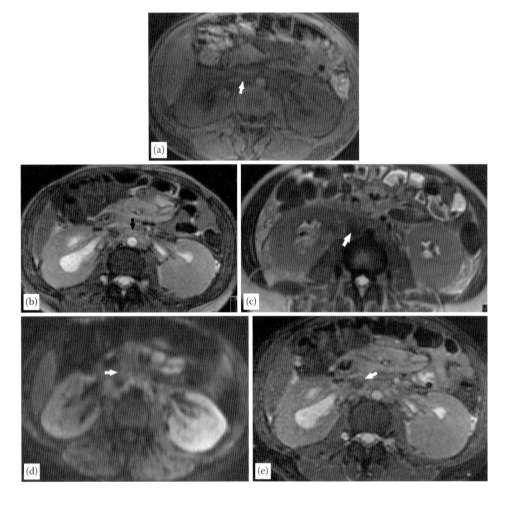

FIGURE 23.17
Ig4-related fibrosing disease and small vessel leucocytoclastic vasculitis. (a) T_1-weighted image shows low signal soft tissue mass encasing the aorta and IVC (arrow). (b) SSFP image shows high signal of the mass (arrow). (c) T_2-weighted image (T_2 FSE) shows intermediate signal of the mass (arrow). (d) Diffusion-weighted image shows mild restricted diffusion (arrow). (e) Heterogeneous contrast enhancement is seen (arrow).

trauma [75]. A hematoma may occur in any retroperitoneal space, and its position may be a clue to the etiology. Spontaneous hemorrhage classically begins in the posterior pararenal space, and extends into the properitoneal fat, the pelvis, psoas muscle, or abdominal wall musculature [76]. Blood from a leaking aortic aneurysm or graft first surrounds the aorta and then extends into the anterior pararenal space and frequently into the psoas muscle [75]. Adrenal and renal hemorrhages are usually confined within the perirenal space [77]. MRI signal varies with time. High signal intensity on T_1-weighted images is attributable to the presence of methemoglobin

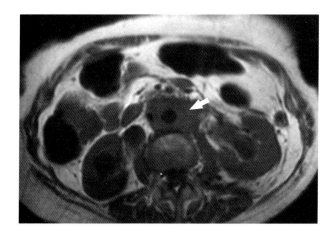

FIGURE 23.18

Retroperitoneal fibrosis. Inactive—Axial T_1-weighted image shows a hypointense mass circumferentially encasing the abdominal aorta (arrow). STIR (not shown here) showed low signal and there was no contrast enhancement.

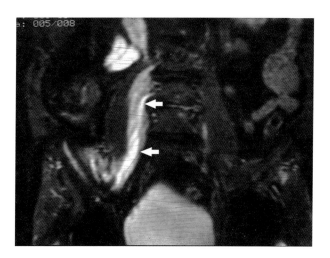

FIGURE 23.20

Psoas hematoma. Fat saturated T_1-weighted image shows high signal collection in the right psoas muscle (arrow), consistent with a hematoma.

within the hematoma [78]. Active contrast extravasation indicates ongoing arterial hemorrhage and necessitates immediate supportive, angiographic, or surgical intervention (Figure 23.20) [79].

23.2.2.2.2 Lymphocele

Lymphoceles are fluid-filled cysts without an epithelial lining that usually occur after pelvic or retroperitoneal lymphadenectomy or renal transplant surgery. Retroperitoneal lymphoceles may cause venous obstruction, with subsequent edema and thromboembolic complications [55]. They may also cause compression of surrounding structures such as bladder and ureter, and may lead to further complications such as renal dysfunction or hydroureter [80]. Usually they are well-circumscribed water-signal intensity collections adjacent to surgical clips.

23.2.2.2.3 Urinoma

Urinomas are encapsulated collections of urine that lie outside the renal collecting system; they most commonly result from obstructive uropathy resulting in rupture of the collecting system. Other causes are abdominal trauma, surgery, or diagnostic instrumentation [81].

Most urinomas reside in the perinephric space [82]. On MRI, an urinoma is a water-signal intensity collection, with obvious leak of contrast into the collection on the delayed post-contrast sequences. Percutaneous aspiration and drainage may allow confirmation of the diagnosis and treatment [55].

23.2.2.2.4 Inflammatory Collections

Most inflammatory fluid collections begin in the anterior pararenal space, arising from the extraperitoneal portions of the alimentary tract (the pancreas, ascending and descending colon, duodenum, and the retroperitoneal appendix) [76,83], with the pancreas being the most common cause of these collections. Inflammatory perirenal collections are much less frequent than in the anterior pararenal space. They are mostly secondary to direct extension from the kidney. A predisposing factor, such as diabetes mellitus, renal trauma, urinary tract obstruction, or renal calculus is usually present, although at other times, no source can be found [77]. Inflammatory effusions in the posterior pararenal space are rare because this space has no organs, and so most effusions are secondary to severe infection in another space. The properitoneal fat and the psoas muscle may be secondarily

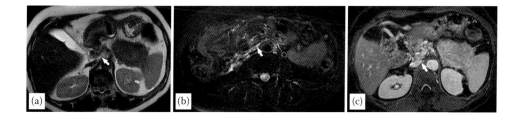

FIGURE 23.19

Variant retroperitoneal fibrosis. (a) HASTE image shows soft tissue mass in the retroperitoenum (arrow). (b) STIR image shows intermediate high signal of the mass (arrow). (c) Heterogeneous intense contrast enhancement is seen (arrow).

involved [77]. A collection may be well defined or irregular, tending to assume the boundaries of the retroperitoneal compartment it occupies which is the more common case. Adjacent structures may be obliterated or displaced by the inflammatory fluid collection, and there is frequently thickening of the surrounding soft tissue planes. A fulminant infection may spread transfascially [77]. On MRI, inflammatory collections show variable T_1-signal intensity, intermediate to high T_2-signal intensity, and thick, peripheral rim of enhancement [84].

23.2.2.2.5 Psoas Abscess

Psoas abscess is an uncommon condition; the etiology is usually tuberculous in developing countries and non-tuberculous in developed countries, most commonly related to *Staphylococcus aureus* [85]. It can be associated with several sources of infection such as gastrointestinal (most common), renal, or extension from lumbar osteomyelitis [10]. Clinical diagnosis of a psoas abscess can be difficult and the classical symptoms of flank pain, fever, and limping may not always be present [86]. On imaging, it appears similar to other inflammatory collections. Treatment is catheter drainage and surgery is reserved for failure of percutaneous drainage [86].

23.2.3 Miscellaneous Retroperitoneal Conditions

23.2.3.1 Xanthogranulomatosis/ Erdheim–Chester Disease

Xanthogranulomatosis is the mass-like accumulation of non-Langerhans lipid-laden histiocytes; Erdheim–Chester disease refers to multisystem xanthogranulomatosis. The condition is associated with osteosclerosis, periostitis, partial epiphyseal involvement, and medullary infarction of the long bones [87]. Histologically, there is xanthogranulomatous infiltration with foamy histiocytes surrounded by fibrosis, without Birbeck granules or S-100 immunostaining. On MRI (Figure 23.21), xanthogranulomatosis presents as enhancing infiltrative soft tissue with intermediate T_1- and T_2-weighted signal intensity relative to skeletal muscle [88,89]. Retroperitoneal involvement characteristically produces a soft-tissue rind of fibrous perinephritis surrounding the kidneys and ureters, which can result in renal failure [90]. Circumferential periaortic involvement with associated bilateral symmetrical perirenal space involvement but with sparing of the IVC and pelvic ureters help distinguishing Erdheim–Chester disease from RPF [88].

23.2.3.2 Extramedullary Hematopoiesis

Extramedullary hematopoiesis (EMH) is the body's response to deficient erythropoiesis by the bone marrow,

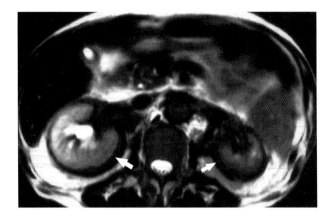

FIGURE 23.21
Erdheim–Chester disease. Axial T_2-weighted image shows bilateral circumferential encasement of the kidneys by soft tissue component that has low signal (arrow).

and it refers to deposits of erythroid precursors in sites outside the bone marrow. EMH in the retroperitoneal space is uncommon [91]. On MRI, low signal intensity on T_2-weighted images and slight contrast enhancement in a patient with a hematological condition is suggestive [91] (Figure 23.22). Following blood transfusion therapy, the lesions tend to shrink with resultant loss of enhancement [92]. Liver and splenic masses as well as skeletal changes due to the chronic conditions that cause the marrow failure may be seen on imaging [91].

23.2.3.3 Lipomatosis

Lipomatosis is a benign overgrowth of mature fat, more commonly seen in black males. Patients present with urinary tract and gastrointesinal symptoms, low back ache, and flank pain [35]. On MRI, the pelvis appears

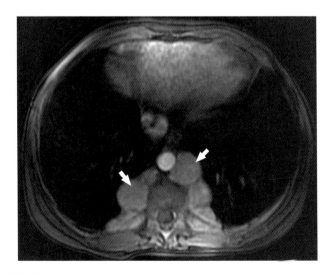

FIGURE 23.22
Extramedullary hematopoiesis. Axial post-contrast image shows intensely enhancing lobulated masses (arrow), adjacent to expanded ribs, which is consistent with extramedullary hematopoiesis.

crowded because of overabundant amounts of symmetrically distributed fat, with occasional strands of fibrous tissue [35]. The fat does not enhance and the soft-tissue planes of the pelvis are preserved [93].

23.2.3.4 Fat Necrosis

The most common cause of retroperitoneal fat necrosis is acute pancreatitis [94–97]. On MRI, it appears as fat signal intensity lesion, with surrounding soft tissue. Due to the association with pancreatitis, its distribution is mostly peripancreatic, but it may also extend elsewhere [98]. Fat necrosis tends to remain stable in size with time or may even shrink.

23.2.4 Retroperitoneal Nonparenchymal Cysts and Cystic Lesions

23.2.4.1 Lymphangioma

Lymphangioma is a developmental malformation caused by the failure of communication of retroperitoneal lymphatic tissue with the main lymphatic vessels and the remainder of the lymphatic system [99].It is a unilocular or multilocular cyst containing clear or milky fluid and are lined with a single layer of flattened endothelium [99]. It is more common in men [99,100]. Lymphangioma has variable (related to protein content) signal on T_1 and intermediate to high signal on T_2.

23.2.4.2 Lymphangiomatosis

Diffuse lymphangiomatosis is characterized by proliferation of irregular lymphatic channels, involving soft tissue, viscera, retroperitoneum, eyes, and skeletal system [101], and is most commonly seen in children and adolescents. On MRI, the lesions appear hypointense on T_1-weighted sequences and hyperintense or isointense on T_2-weighted images (Figure 23.23). Some lymphangiomas also demonstrate high signal on T_1-weighted images, probably because of protein, fat, or blood content. Diffusion-weighted MRI is described to help in diagnosis of atypical abdominal lymphangiomatosis [102]; magnetic resonance lymphangiography may also be obtained with a heavily T_2-weighted 3D-TSE sequence and a T_1-weighted 3D-VIBE sequence after intracutaneous contrast media application. Complications of the condition might include chylothorax, chylopericardium, hepatosplenomegaly, chylous ascites, and protein loosing enteropathy, development of lymphedema, lymphphorrhea, and infection [101,103].

23.2.4.3 Nonpancreatic Pseudocyst

Nonpancreatic pseudocysts arise from the mesentery and omentum [104]. They have a thick, fibrous wall and they

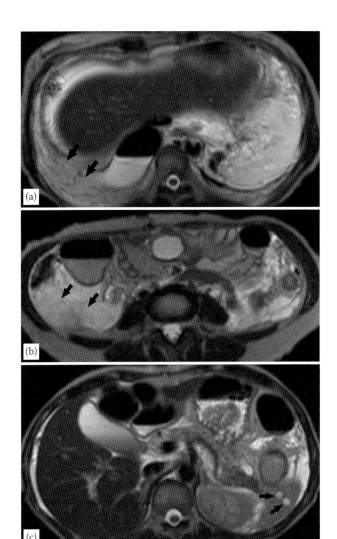

FIGURE 23.23
Lymphangiomatosis. (a) Axial T_2-weighted image shows diffuse, heterogeneous high signal intensity infiltrative lesion in the retroperitoneum (arrow). (b) Image at lower level confirms the finding (arrow). (c) Splenic lymphangiomas are seen in an image at a higher level (arrow).

usually contain hemorrhage, pus, or serous fluid. They are not associated with high levels of amylase or lipase in the cystic fluid, which makes them different from pancreatic pseudocysts. On MRI, they appear as unilocular or multilocular fluid-filled masses with thick walls [55].

23.3 Abdominal Great Vessels

23.3.1 Aorta

23.3.1.1 Atherosclerosis

Atherosclerosis is a diffuse inflammatory process, characterized by deposition of lipid and fibrous products in the arterial wall. The main components of the atherosclerotic plaques are (1) connective tissue extracellular matrix, including collagen, proteoglycans, and fibronectin elastic fibers; (2) crystalline cholesterol, cholesteryl esters, and phospholipids; and (3) cells such as monocyte-derived macrophages, T lymphocytes, and smooth muscle cells [105]. A vulnerable plaque has a thin fibrous cap, large lipid core, and significant inflammatory cell infiltration [106]. Plaques in the abdominal aorta are associated with increasing age and blood pressure, and tend to be severe in patients with coronary artery disease and cardiac events [107]. On regular MRI, the plaque appears as an irregular lesion along the aortic wall. Lipid components are identified as hyperintense on T_1-weighted and hypointense on T_2-weighted images. Fibrocellular components are identified as hyperintense regions on T_1 and T_2. Calcium deposits are identified as hypointense regions on both T_1 and T_2 contrasts [107]. Targeted MRI contrast agents (superparamagnetic particles of iron oxides) that are phagocytosed by macrophages in atherosclerotic plaques are promising tools to image the vulnerable plaques [106] (Figures 23.24 and 23.25).

23.3.1.2 Aortoiliac Occlusive Disease

Aortoiliac occlusive disease is seen in chronic atherosclerotic disease, occurring at arterial bifurcation points. Symptoms depend on the extent of collateral flow and include intermittent thigh, hip, or buttock claudication and impotence (in 30%–50%) [108,109]. MRA helps in the evaluation of the length and multiplicity of the involved segments, the degree of stenosis, the development of collaterals, and the evaluation of coexistent femoropopliteal disease (Figures 23.26 and 23.27). There are several treatment options including endovascular interventions such as balloon angioplasty and stenting, or surgeries such as aortobifemoral or axillofemoral with femo-femoral bypass [108].

23.3.1.3 Aneurysm

An aneurysm is a focal dilation of a segment of the abdominal aorta greater than 50% of its normal diameter (or >3 cm). Prevalence is 4%–8% in men and 1% in women [108]. Risk factors include increased age, smoking, male gender, and family history [110]. Theories of aneurysm formation include—proinflammatory mediators triggering enzymes that digest the extracellular matrix and impair its subsequent repair or turbulent flow resulting in shear forces, leading to aneurysmal dilatation and eventual rupture [108,111]. Most aneurysms are asymptomatic until rupture, but some may present with back pain, abdominal pain, or a palpable abdominal mass. On MRI, the aneurysm is seen as a saccular or fusiform dilation of the abdominal aorta. Surgery is recommended when the aneurysm is >5.5 cm. Features that are assessed in presurgical evaluation include—maximal anteroposterior diameter of the aneurysmal sac; the axial length of the aneurysmal neck (the distance between the lowermost renal artery to the start of the aneurysm); the shape and angulation of the neck; the diameter of the iliac arteries (for access through the groin), and the potential length and condition of the distal iliac arteries (Figure 23.28). Aneurysms can be managed by surgery or endovascular repair (Table 23.3).

Pseudoaneurysms (false aneurysms) do not involve all three layers of the vessel, and are often associated with intimal and medial disruption, with containment of blood by the adventitia or surrounding tissue. They are caused by trauma, infection (Figure 23.29), surgery, or penetrating atherosclerotic ulcers. They are more common in the suprarenal abdominal aorta due to the fact

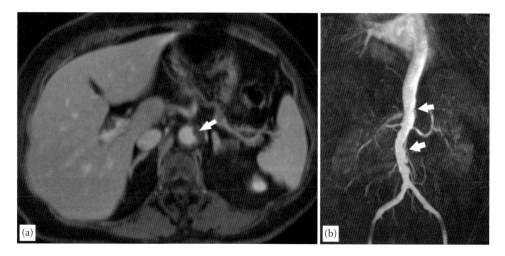

FIGURE 23.24
Severe atherosclerotic plaque of aorta. (a) Axial post-contrast VIBE image shows a large, irregular hypointense plaque in the periphery of the abdominal aorta (arrow). (b) MRA shows irregular contour of the abdominal aorta, consistent with severe atherosclerotic disease (arrow).

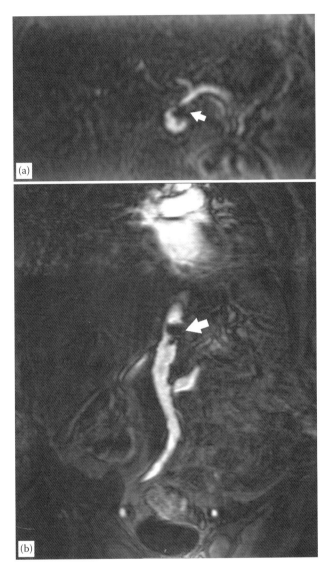

FIGURE 23.25
Calcified plaque. (a) Axial MRA image shows a hypointense nonenhancing lesion in the aortic wall close to the celiac artery origin, which is consistent with a densely calcified atheromatous plaque (arrow). (b) Coronal MRA image shows extensive atherosclerotic changes of thoracic aorta, with a large hypointense plaque, likely calcified (arrow).

that there is a higher likelihood of tamponade in this location due to density and rigid fixation of the aorta to the surrounding tissues [112]. The treatment of pseudoaneurysms is surgery due to risk of rupture and the high mortality rates [113].

23.3.1.4 Ruptured Aneurysm

Rupture is one of the complications of abdominal aortic aneurysm and can be either contained or free, and acute or chronic. Rupture is seen in aneurysms >5.5 cm or those that expand >0.5 cm over 6 month follow up [114]. The 5-year overall cumulative rupture rate is 25%–40%

for aneurysms larger than 5.0 cm, compared with 1%–7% for aneurysms 4.0–5.0 cm in diameter [115,116]. Rupture presents with shooting abdominal or back pain and hypotension in a patient with a pulsatile abdominal mass. Syncope, constipation, urinary retention, and urge to defecate are other presentations. Only approximately 50% of patients with ruptured abdominal aortic aneurysms (AAA) reach the hospital alive; of which 50% do not survive repair [117]. Rupture of AAA has a dismal prognosis, with a community mortality of about 79% and a perioperative mortality of about 40% [118].

AAA rupture most commonly occurs through the posterolateral wall into the retroperitoneum [119], with blood extending to the perirenal space, pararenal space, or psoas muscle. Intraperitoneal rupture is less common [108]. MRI is not typically performed for rupture, but can show the hemorrhage extending from the ruptured site into the retroperitoneum. A subtle rupture is seen as a discontinuity in the circumferential calcification [120]. Signs of impending rupture include the *crescent sign*, which is an area of hemorrhage within the mural thrombus [120] and *drape sign* where the posterior wall of aorta is irregular and drapes over the vertebral body without a distinct fat plane [121]. A chronic contained rupture is diagnosed in a patient with history of abdominal aortic aneurysm, resolved pain, stable hemodynamics, a normal hematocrit, and retroperitoneal hemorrhage [122].

23.3.1.5 Aortoenteric Fistula

Aortoenteric fistulas is a communication between abdominal aorta and bowel, which can be primary, due to a penetrating ulcer, diverticulitis, foreign bodies, aortitis, appendicitis, and gastrointestinal malignancies [123–126] or more commonly secondary due to prior surgery or intervention. Chronic perigraft infection or prolonged pressure upon the bowel by a graft is believed to be the common etiology [127] of a secondary fistula. The classic location (60%) of the fistula is the transverse portion of the duodenum. Less common locations include the remainder of the duodenum, jejunum and ileum, stomach, sigmoid colon, and ascending/descending colon [123]. Presenting symptoms include a herald gastrointestinal hemorrhage, which is followed by a massive hemorrhage. Other symptoms might include pain, pulsatile mass, and a groin mass. MRI findings of aortoenteric fistula include—effacement of the periaortic fat plane, focal thickening, and tethering of a bowel loop immediately adjacent to the aorta, periaortic free fluid and soft tissue thickening, and disruption of a graft or significant graft migration. Ectopic gas and extravasation of contrast from the aorta into the bowel lumen are less common findings [128]. Differential diagnosis includes perigraft infection, aortitis, mycotic

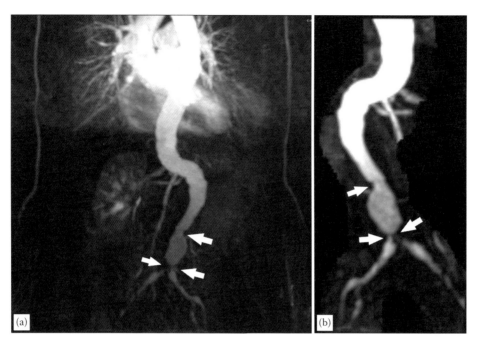

FIGURE 23.26
Severe aortoiliac stenotic disease. (a) Coronal MRA image shows severe stenosis of the infrarenal portion of abdominal aorta (arrow) and severe stensois of the origins of the common iliac arteries bilaterally (arrow). (b) Sagittal oblique MRA image of the same patient demonstrates stenotic lesions in the distal abdominal aorta and bilateral proximal common iliac arteries (arrow).

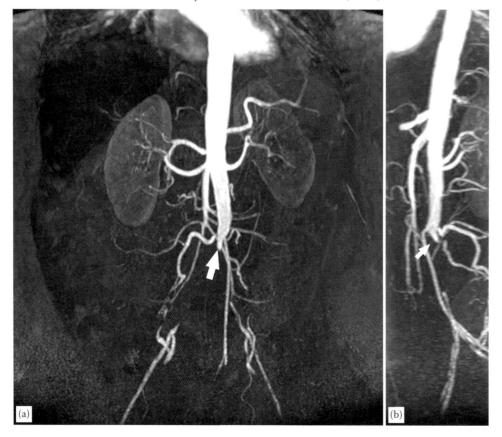

FIGURE 23.27
Aortoiliac occlusive disease. Coronal (a) and sagittal (b) reconstructed Maximal intensity projection (MIP) image shows extensive athero-sclerotic changes with complete occlusion of the distal abdominal aorta (arrow) and the entire length of the common iliac arteries. There is reconstruction of the external iliac arteries and the proximal internal iliac arteries through collateral vascular supply.

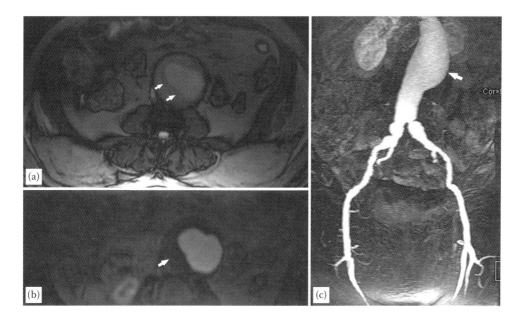

FIGURE 23.28
Aneurysm. Axial SSFP (a), axial post-contrast MRA (b), and coronal post-contrast MRA (c) images show a large aneurysm of the infrarenal portion of abdominal aorta, which had maximal dimensions of 7.2 × 4.6 cm (arrow).

TABLE 23.3

Types of Endoleak

Type	Description

I	Leak of the stent-graft at the site of its attachment with the native aortic wall
	  IA—Proximal attachment
	  IB—Distal attachment

II	Leak into the aneurysmal sac due to reversal of flow in branch vessels that connect to it
III	Leak due to mechanical failure of the graft (junction leak, mid graft hole, disconnection, and fractions)
IV	Leak due to graft porosity
V	Endotension (expansion of the native aneurysm without a visible leak → diagnosis of exclusion)

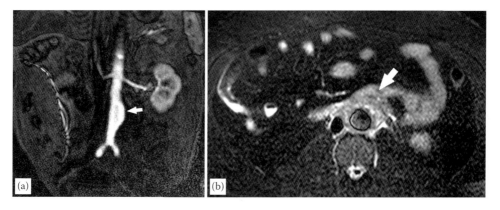

FIGURE 23.29
Inflammatory aneurysm. (a) Coronal MIP image from MRA shows a focal aneurysm (arrow) in the infrarenal abdominal aorta. (b) Axial T_2-weighted STIR image in the same patient shows high signal of the aortic wall thickening (arrow) consistent with an inflammatory aortic aneurysm.

aneurysms, and perianeurysmal fibrosis [129]. The treatment for aortoenteric fistulas is removal of the infected graft, bowel resection, and creation of an extra-anatomic bypass graft [124]. Recently, the trend is toward the use of endovascular techniques [128].

23.3.1.6 Aortocaval Fistula

Primary aortocaval fistula is a rare complication of abdominal aortic aneurysm and involves less than 1% of all AAAs, but 2%–6.7% of ruptured aneurysms [130]. Eighty percent of these fistulas are a result of a spontaneous rupture of an atherosclerotic aneurysm directly into the adjacent vena cava; the remainders are either traumatic (15%) or iatrogenic (5%) [131]. It is more common in men, with an average age at presentation of 64 years [132]. Acute fistula can present with a pulsatile abdominal mass, continuous abdominal bruit or thrill, and high output congestive cardiac failure. Chronic presentation of aortocaval fistula is high-output cardiac failure, abdominal bruit and thrill, palpable abdominal aneurysm, oliguria, and consequences of regional venous hypertension [133]. On MRA, there is early and intense opacification of the IVC and common iliac veins at the same time and with the same intensity as that of abdominal aorta [134]. The fistulous communication may be demonstrated. Unlike an aortic rupture, the IVC is distended. Absence of an obstructing lesion distinguishes a fistula from other causes of IVC dilation. Management is through surgical repair or endovascular exclusion of the AAA [133].

23.3.1.7 Aortic Dissection

Dissection is a disruption of the aortic wall, with resulting separation of the intima and inner media from the outer media and adventitia, resulting in an intimomedial flap with a true lumen, and a false lumen that can have either flow or thrombosis [135]. Abdominal aortic dissection is usually an extension of thoracic aortic dissection, but focal abdominal aortic dissection may also be seen, extending proximally between the renal arteries and the inferior mesenteric artery (IMA) (33%) or the celiac trunk and the renal arteries (23%) terminating

distally at the common iliac arteries. It is usually caused by hypertension, but other causes include connective tissue disorders such as Marfan and Ehler Danlos syndrome, fibrodysplasia, smoking, diabetes, hypercholesterolemia, and previous aneurysm surgery [136]. In a focal dissection, the proximal extent is between either renal arteries and the inferior mesenteric artery (IMA) (33%), or celiac trunk and renal arteries (23%) and distal extent is proximal to the common iliac artery [136]. Clinical features are acute onset of abdominal, chest or back pain. Other findings include mesenteric ischemia, renovascular hypertension, or extremity ischemia [108,136].

On MRI, a flap is seen within the aortic lumen that may extend to the branch vessels (Figure 23.30). The false lumen is often larger, has a *beak* sign due to wedge of hematoma at the leading edge of propagation of false lumen, and the *cobweb* sign that represent low-density linear areas due to incompletely sheared media. Occlusion of branch vessels may be due to either extension of the flap within the wall of the branch vessel or dynamic prolapse of the flap across branch-vessel origin [137]. Differential diagnosis of a dissection includes a thrombus within an aneurysm: thrombus has an irregular border, a constant circumferential relationship with aorta and calcification deep to the thrombus, while a dissection has a spiroidal configuration, smooth inner border, and calcification superficial to the false lumen [138,139]. Although open repair is used for the treatment of aortic dissection and trauma, endovascular techniques are gaining popularity, including aortic fenestration or stent graft repair [140].

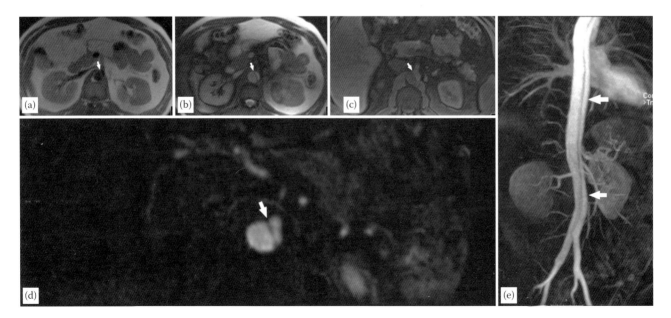

FIGURE 23.30
Dissection. Axial HASTE (a), SSFP (b), post-contrast VIBE (c), axial MRA (d), and coronal MRA (e) images in a patient who presented with acute chest pain shows a dissection flap that extends throughout the length of abdominal aorta.

23.3.1.8 Aortic Intramural Hematoma

Intramural hematoma (IMH) forms due to hemorrhage within the media of the aortic wall, which results in weakening of the wall, without rupture of the intima or formation of the intimal flap of aortic dissection. Hypertension and atherosclerosis play a major role in the formation of aortic IMH [141]. Theories of formation include spontaneous rupture of the aortic vasa vasorum, pathological neovascularization, microscopic tears in the intima, or extension from a penetrating aortic ulcer [142,143]. IMH is a disease of the seventh, eighth, and ninth decades of life, presenting with acute chest/abdominal pain. On MRI, the presence of hematoma in the wall of the aorta can be detected using black blood imaging, with T_1- and T_2-weighted sequences. High signal in T_1-weighted imaging is seen due to methemoglobin [144] and there is no contrast enhancement or flap (Figure 23.31). The appearance and continuity of the intima and the presence of an intimal flap helps distinguish acute aortic dissection from IMH and penetrating atherosclerotic ulcer. However, a thrombosed false lumen of acute aortic dissection, with intimal calcification or atherosclerotic plaque in the aortic wall, can be difficult to differentiate from IMH [143].

23.3.1.9 Penetrating Atherosclerotic Ulcer

A penetrating aortic ulcer is an atherosclerotic ulcer that penetrates the internal elastic lamina and media [145]. It is usually associated with hematoma formation within the media layer of the aortic wall [146]. This ulcer might progress into a saccular aneurysm, aortic dissection, or aortic rupture [147]. It often presents

with acute abdominal pain, similar to the presentation of aortic dissection. MRI is helpful in the diagnosis because the signal intensity of the aortic wall increases in the presence of the ulcer on both T_1- and T_2-weighted images [148]. After contrast administration, a focal area of ulceration is seen extending through the intima into the aortic wall, and a thick enhancing aortic wall may also be observed due to the associated IMH (Figure 23.32). A penetrating ulcer should be distinguished from an intimal atherosclerotic ulcer, which is confined to the intima, asymptomatic, and without an IMH [138]. This is treated surgically [147].

23.3.1.10 Traumatic Aortic Injury

Blunt abdominal aortic injuries are rare occurring in 0.05% of blunt abdominal trauma [149]. Significant forces are required to injure the abdominal aorta, most commonly seen in motor vehicle accidents [150]. Most of the injuries (33%) occur at the level of the inferior mesenteric

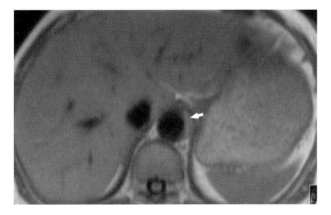

FIGURE 23.31
Well-differentiated liposarcoma—(a) T_1-weighted axial MRI shows a large mass in the left flank of the abdomen, which has predominantly intermediate signal (arrows) with areas of high signal intensity (arrow tip) in the anterior portion suggestive of fat. The mass is seen displacing the kidney anteriorly. (b) Axial T_1-weighted MRI image in another patient shows a well-differentiated liposarcoma containing large areas of fat (arrows), displacing the right kidney anteriorly. The mass is seen extending into the right neural foramen into the spinal canal.

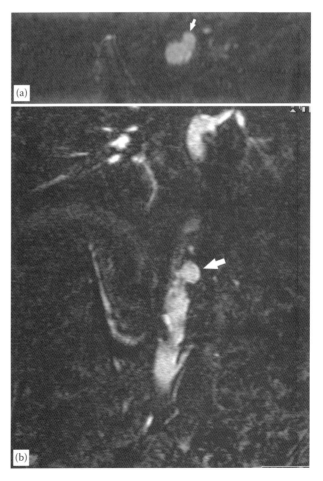

FIGURE 23.32
Penetrating ulcer. (a) Axial MRA image shows a focal penetrating ulcer from the anterior aspect of the upper abdominal aorta (arrow). (b) Coronal MRA image shows the penetrating ulcer from the upper abdominal aorta (arrow).

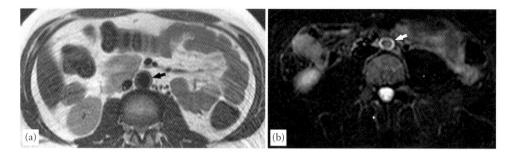

FIGURE 23.33
Aortitis. (a) Axial T_2-weighted image shows circumferential thickening of the wall of the aorta (arrow). (b) STIR image in the same patient at the same level shows high signal in the thickened wall (arrow), which is indicative of edema seen in active arteritis.

artery, 24% near the renal arteries, 19% between the inferior mesenteric artery and aortic bifurcation, and the rest at other locations [151]. Although most patients are asymptomatic, those with symptoms have an acute abdomen, neurological deficits, and end organ and limb ischemia [152]. On MRI, the diagnosis is suggested by the presence of an intimal flap, thrombosis, or pseudoaneurysm [151]. Treatment includes surgical or endovascular repair [153].

23.3.1.11 Aortitis

Aortitis is inflammation of the aortic wall that can be infectious or noninfectious. Infectious causes include bacterial or viral causes. Noninfectious causes include large, medium, and small vessel vasculitis, idiopathic inflammatory aneurysm, or RPF.

23.3.1.11.1 Noninfectious Aortitis

Takayasu arteritis (also known as pulseless disease and aortic-arch syndrome), the most common type of thoracoabdominal aortic vasculitis is a large-vessel vasculitis usually found in young Asian women. Type-1 disease is confined to the aortic arch and branches; type-2 involves the descending thoracic and abdominal aorta and visceral branches of the abdominal aorta; type-3 involves type 1 and 2 components; and type-4 combines types 1, 2, or 3 with associated pulmonary artery vasculitis. Takayasu arteritis exhibits an inflammatory cellular infiltrate of the aortic media, adventitia, and vasa vasorum that contains a predominance of lymphocytes, macrophages, and multinucleated giant cells. Over time, there is scarring of the aortic media and destruction of the elastic lamina [154,155]. The pathogenesis is unknown, but it is most probably an antigen-driven cell mediated autoimmune process [156]. The average age at diagnosis is 25–30 years, and 75%–97% of patients are female. Common presenting symptoms are due to arterial occlusive disease of the aorta, aortic

arch, and large vessels (hypertension due to aortic or renal artery occlusive disease, pulse deficits and/or vascular bruits, and upper and/or lower extremity claudication) [157,158]. The abdominal aorta is the most common site of involvement, and on imaging stenotic lesions in the aorta are frequently detected, although aortic aneurysms are also common [157–160]. On MRI, there is wall thickening of the aorta. Active arteritis shows high signal in short tau inversion recovery (STIR) (Figure 23.33) and contrast enhancement [161]. Chronic arteritis presents with wall thickening, no edema or contrast enhancement, arterial stenosis, occlusion, or aneurysm formation [108].

Giant cell arteritis is a large and medium vessel systemic granulomatous arteritis characterized by a similar histopathological appearance to Takayasu arteritis and a similar autoimmune etiology [156], but associated with advanced age and Caucasian women [162]. Usually it presents with headache, temporal artery abnormalities on physical examination, elevated markers of inflammation, polymyalgia rheumatica, scalp tenderness, jaw claudication, and visual field changes [163]. MRI findings are similar to those of Takayasu arteritis. Some studies have showed that giant cell arteritis is associated with increased risk of abdominal aortic aneurysm [164].

Idiopathic inflammatory arteritis is characterized by wall thickening and fibrosis with adhesions to surrounding structures. Inflammation is limited to the aorta and peri-aortic tissues, rather than a manifestation of a widespread vasculitis [156]. Patchy necrosis of the aortic media is the primary histologic finding, along with an inflammatory cellular infiltrate, which may include multinucleated giant cells [165]. Usually the presentation is subclinical in nature and diagnosis is incidental at the time of histopathology review following aortic aneurysm surgery [166]. MRI shows aneurysm with periaortic inflammatory wall thickening, adventitial fibrosis, and turbulent flow.

23.3.1.11.2 Infectious Aortitis

Infectious aortitis is most commonly caused by *Salmonella*, *Staphylococcus*, or *Streptococcus* pneumonia, typically by seeding of a prediseased aortic wall by bacteria via vasa vasorum [154,167]. Chronic inflammatory infiltrate of the medial and adventitial vasa vasorum is seen, which ultimately leads to medial necrosis [154]. MRI features are similar to noninfectious aortitis, with clinical signs of bacteremia and soft tissue extension on imaging. A combination strategy of intensive antibiotic therapy and surgical debridement, with aneurysm repair if necessary, is generally recommended [167,168].

23.3.1.11.3 Mycotic (Infected) Aneurysm

Mycotic aneurysm is an infected aneurysm, either a sequel of infectious aortitis, superinfection of atherosclerotic surface or extension from adjacent neurovascular structures. *Staphylococcus* and *Salmonella* are the most common organisms [165]. MRI features of mycotic aneurysm are a saccular aneurysm, soft tissue mass surrounding the aneurysm, and leakage of contrast or rapid growth [165]. Complications may be seen such as vertebral body destruction, psoas abscess, renal abscess, rupture, hemorrhage, sepsis, and death. Treatment includes prolonged systemic antibiotics or surgical resection with debridement and abscess drainage in case of severe disease [169].

23.3.1.12 Aortic Tumors

Primary tumors of the aorta are rare and mostly malignant [170]. They are classified as tumors involving primarily the intima, which present with thromboembolism and tumors arising from media and adventitia, which exhibit mass effect [171]. On MRI, the displacement of flowing blood provides inherent contrast between vessel lumen and vessel wall or surrounding soft tissue that allows for better diagnosis. Treatment is unclear but usually involves surgery [170].

23.3.1.13 Hypoplastic Abdominal Aorta

Hypoplastic abdominal aorta is a diffuse narrowing of the abdominal aorta, typically caused by a developmental defect, due to overfusion of primitive dorsal aorta, including origins of posterior aortic branches [172]. Other proposed etiologies include viral infections or inflammatory processes such as arteritis [173].

Radiation or trauma can also produce similar appearances. It is more common in females, typically seen in second or third decades of life [174]. Symptoms include hypertension, lower extremity claudication, and early development of atherosclerosis in the aortoiliac distribution [173]. When all symptoms are present, midaortic syndrome is the diagnosis. On MRA, there is a long segment of circumferential narrowing. MRI can also evaluate for the presence of vascular inflammation. Treatment is surgical revascularization, with resection of affected segment and end-to-end anastomastosis or placement of interposition graft [173].

23.3.2 Inferior Vena Cava

23.3.2.1 Congenital Anomalies

IVC develops by sequential union of posterior cardinal, subcardinal, and supracardinal veins, and abnormal persistence or regression of these veins results in several congenital anomalies. Most of these anomalies are asymptomatic and are incidental findings in imaging studies.

Left-sided IVC results from regression of the right supracardinal vein and persistence of the left supracardinal vein. The prevalence is 0.2%–0.5% [175]. The major clinical significance of this anomaly is the potential for misdiagnosis as left-sided para-aortic adenopathy [176]. This abnormality is also important when placing an IVC filter, especially with a transjugular approach [177] (Figure 23.34).

Double IVC is due to the persistence of both right and left supracardinal veins. Prevalence is 1%–3% [175]. The most common form of this abnormality is one in which two distinct IVCs arise from each iliac vein without a normal confluence [178]. Diagnosis is suspected in cases of recurrent pulmonary embolism following placement of an IVC filter. IVC filters should be placed in both the IVC or at their confluence.

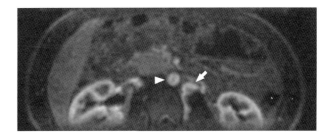

FIGURE 23.34
Left-sided IVC. Axial post-contrast MRI image shows a left-sided IVC (straight arrow) located to the left of the abdominal aorta (arrow tip).

Interruption of the IVC is due to failure of the right subcardinal-hepatic anastomosis and atrophy of the right subcardinal vein. Consequently, blood is shunted from the suprasubcardinal anastomosis through the retrocrural azygos vein, which is partially derived from the thoracic segment of the right supracardinal vein. The prevalence is 0.6% [179]. The hepatic portion of the IVC is interrupted, but the IVC continues as a dilated azygos that connects to the SVC on the right side of the chest. There are no pathological consequences *per se*, but this anomaly is associated with cardiac malformations such as heterotaxy syndromes, polysplenia, AV septal defects, partial anomalous pulmonary venous connection, and pulmonary atresia. Also, accidental ligation of the azygos vein during surgery may prove to be fatal [180] (Figure 23.35).

Retro-aortic left renal vein is due to persistence of the dorsal portion of the circumaortic collar and regression of the ventral portion [181]. It has prevalence of 0.8%–3.7% [182]. Retroaortic left renal vein can sometimes cause clinical symptoms such as hematuria and abdominal/flank pain [183]. Knowledge of this anomaly is necessary when a left renal transplant and/or splenorenal shunt are considered. Failure to recognize these anomalies may lead to severe hemorrhage [184] (Figure 23.36).

A circumaortic left renal vein, is due to the persistence of the dorsal and ventral portions of the circumaortic collar. The major clinical significance is in preoperative planning prior to nephrectomy and in renal vein catheterization for venous sampling. Care should be taken to avoid misdiagnosis as retroperitoneal adenopathy [177]. Similar to retroaortic left renal vein, compression of the renal vein by the aorta causes clinical symptoms such as hematuria and abdominal/flank pain [183].

Retrocaval ureter is due to development of the IVC from the posterior cardinal rather than the

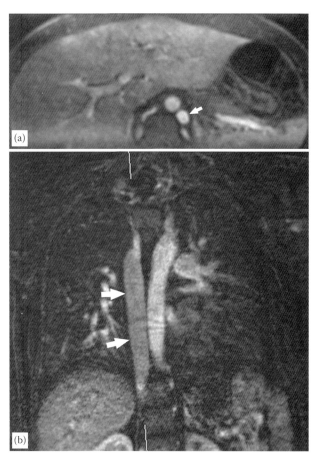

FIGURE 23.35
Interruption of IVC. (a) Axial post-contrast MRI image shows a left-sided IVC (arrow). In addition, there is absence of the intrahepatic segment of the IVC, which is consistent with an interrupted IVC. (b) Coronal MRA image through the chest shows a large and dilated azygos vein (arrow), which drains the veins into the SVC.

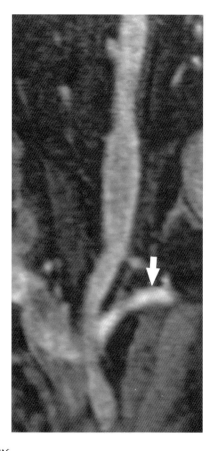

FIGURE 23.36
Retroaortic left renal vein. Coronal MRA image shows a retroaortic left renal vein (arrow), extending behind the abdominal aorta (AA).

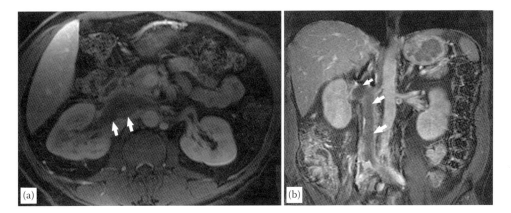

FIGURE 23.37
IVC thrombus. Axial (a) and coronal (b) post-contrast T_1-weighted images shows a large thrombus (arrows) that is extending into the IVC and iliac veins from right renal cancer.

normal supracardinal vein. This makes the right ureter posterior and medial to the IVC, resulting in ureteral compression with complications such as hydronephrosis or recurrent UTIs [177]. It is reported to be present in 0.06%–0.17% of autopsies [185]. On imaging, it presents as a fish hook or a reverse J appearance of the proximal ureter at the level of lumbar pedicles [186].

23.3.2.2 IVC Thrombus and Tumors

Bland thrombus is seen in the IVC usually after extension of the thrombus in the lower limb or pelvic veins. Acute thrombus expands the IVC, while chronic thrombus does not expand the IVC. Bland thrombus has heterogeneous intermediate to low signal in T_1- and T_2-weighted images and does not show any contrast enhancement, whereas tumor thrombus shows contrast enhancement and is typically contiguous with the primary tumor.

Tumors of the IVC are more often secondary than primary. Leiomyosarcoma is the most common primary malignant tumor of the IVC. It usually affects women in the fifth and sixth decades of life [187]. The tumor can be completely intraluminal or can have both extra- and intraluminal components; it has a poor prognosis [187]. Secondary malignant tumors that extend into the IVC from neighboring structures include renal cell carcinoma, hepatocellular carcinoma, and adrenocortical carcinoma [188]. Benign tumors extending into the IVC include pheochromocytoma, angiomyolipoma, and intravenous leiomyomatosis [189]. Imaging findings dictate surgical technique to be used: thrombus extension into the supradiaphragmatic IVC

requires cardiopulmonary bypass surgery and is associated with increased morbidity and mortality rates [188] (Figure 23.37).

23.4 Summary

MRI is a valuable imaging tool in the evaluation of retroperitoneal lesions, including vascular abnormalities of the aorta and IVC. The advantages of MRI include its ability to tissue characterization and evaluation of extent of lesions without the need of ionizing radiation. MRI sequences should be optimized for appropriate evaluation and characterization of retroperitoneal lesions.

References

1. Nishino M, Hayakawa K, Minami M et al. Primary retroperitoneal neoplasms: CT and MR imaging findings with anatomic and pathologic diagnostic clues. *RadioGraphics* 2003;23(1):45–57.
2. Sanyal R, Remer EM. Radiology of the retroperitoneum: Case-based review. *AJR Am J Roentgenol* 2009;192(6Suppl.):S112–S127; (Quiz S118–S121).
3. Jemal A, Siegel R, Ward E et al. Cancer statistics, 2006. *CA Cancer J Clin* 2006;56(2):106–130.
4. Shibata D, Lewis JJ, Leung DH et al. Is there a role for incomplete resection in the management of retroperitoneal liposarcomas? *J Am Coll Surg* 2001;193(4):373–379.
5. Lewis JJ, Leung D, Woodruff JM et al. Retroperitoneal soft-tissue sarcoma: Analysis of 500 patients treated and followed at a single institution. *Ann Surg* 1998;228(3):355–365.

6. Sung MS, Kang HS, Suh JS et al. Myxoid liposarcoma: Appearance at MR imaging with histologic correlation. *RadioGraphics* 2000;20(4):1007–1019.

7. Cormier JN, Pollock RE. Soft tissue sarcomas. *CA Cancer J Clin* 2004;54(2):94–109.

8. Vijay A, Ram L. Retroperitoneal liposarcoma: A comprehensive review. *Am J Clin Oncol* 2015;38(2):213–219.

9. DasGupta TK. Tumors and tumor-like conditions of adipose tissue. *Curr Probl Surg* 1970;7:1–60.

10. Torigian DA, Ramchandani P. The retroperitoneum. In: Haaga JR, Dogra VS, Forsting M et al. (eds.), 5th edn., *CT and MRI of the whole body.* Philadelphia, PA: Mosby Elsevier, 2009, pp. 1953–2040.

11. Engelken JD, Ros PR. Retroperitoneal MR imaging. *Magn Reson Imaging Clin N Am* 1997;5(1):165–178.

12. Lane RH, Stephens DH, Reiman HM. Primary retroperitoneal neoplasms: CT findings in 90 cases with clinical and pathologic correlation. *AJR Am J Roentgenol* 1989;152(1):83–89.

13. Kim T, Murakami T, Oi H et al. CT and MR imaging of abdominal liposarcoma. *AJR Am J Roentgenol* 1996;166:829–833.

14. Henricks WH, Chu YC, Goldblum JR et al. Dedifferentiated liposarcoma: A clinicopathological analysis of 155 cases with a proposal for an expanded definition of dedifferentiation. *Am J Surg Pathol* 1997;21:271–281.

15. Jelinek JS, Kransdorf MJ, Shmookler BM et al. Liposarcoma of the extremities: MR and CT findings in the histologic subtypes. *Radiology* 1993;186(2):455–459.

16. Tateishi U, Hasegawa T, Beppu Y et al. Primary dedifferentiated liposarcoma of the retroperitoneum. Prognostic significance of computed tomography and magnetic resonance imaging features. *J Comput Assist Tomogr* 2003;27:799–804.

17. Neville A, Herts BR. CT characteristics of primary retroperitoneal neoplasms. *Crit Rev Comput Tomogr* 2004;45(4):247–270.

18. Kieffer E, Alaoui M, Piette JC et al. Leiomyosarcoma of the inferior vena cava: Experience in 22 cases. *Ann Surg* 2006;244(2):289–295.

19. McLeod AJ, Zornoza J, Shirkhoda A. Leiomyosarcoma: Computed tomographic findings. *Radiology* 1984;152(1):133–136.

20. Hartman DS, Hayes WS, Choyke PL et al. From the archives of the AFIP. Leiomyosarcoma of the retroperitoneum and inferior vena cava: Radiologic-pathologic correlation. *RadioGraphics* 1992;12(6):1203–1220.

21. La Fianza A, Alberici E, Meloni G et al. Extraperitoneal pelvic leiomyosarcoma. MR findings in a case. *Clin Imaging* 2000;24(4):224–226.

22. Ganeshalingam S, Rajeswaran G, Jones RL, Thway K, Moskovic E. Leiomyosarcomas of the inferior vena cava: Diagnostic features on cross-sectional imaging. *Clin Radiol* Jan 2011;66(1):50–56.

23. Kransdorf MJ. Malignant soft-tissue tumors in a large referral population: Distribution of diagnoses by age, sex, and location. *AJR Am J Roentgenol* 1995;164(1):129–134.

24. Ko SF, Wan YL, Lee TY et al. CT features of calcifications in abdominal malignant fibrous histiocytoma. *Clin Imaging* 1998;22(6):408–413.

25. Nishimura H, Zhang Y, Ohkuma K et al. MR imaging of soft-tissue masses of the extraperitoneal spaces. *RadioGraphics* 2001;21(5):1141–1154.

26. Maurer HM, Beltangady M, Gehan EA et al. The Intergroup Rhabdomyosarcoma Study-I. A final report. *Cancer* Jan 15, 1988;61(2):209–220.

27. Kransdorf MJ, Murphey MD. Muscle tumors. In: Kransdorf M, Murphey M (eds.), *Imaging of Soft Tissue Tumors.* Philadelphia, PA: WB Saunders, 1997, pp. 3–36 and 57–102.

28. Best AK, Dobson RL, Ahmad AR. Cardiac angiosarcoma. *RadioGraphics* 2003;23(Spec Issue): S141–S145.

29. Enzinger FM, Shiraki M. Extraskeletal myxoid chondrosarcoma: An analysis of 34 cases. *Hum Pathol* 1972;3:421–435.

30. Murphey MD, Walker EA, Wilson AJ, Kransdorf MJ, Temple HT, Gannon FH. From the archives of the AFIP: Imaging of primary chondrosarcoma: Radiologic-pathologic correlation. *RadioGraphics* Sep–Oct 2003;23(5):1245–1278.

31. Nakanishi H, Araki N, Sawai Y, Kudawara I, Mano M, Ishiguro S, Ueda T, Yoshikawa H. Cystic synovial sarcomas: Imaging features with clinical and histopathologic correlation. *Skeletal Radiol* Dec 2003;32(12):701–707.

32. Jones BC, Sundaram M, Kransdorf MJ. Synovial sarcoma: MR imaging findings in 34 patients. *AJR Am J Roentgenol* Oct 1993;161(4):827–830.

33. Wu JH, Zhou JL, Cui Y, Jing QP, Shang L, Zhang JZ. Malignant perivascular epithelioid cell tumor of the retroperitoneum. *Int J Clin Exp Pathol* Sep 15, 2013;6(10):2251–2256. eCollection 2013.

34. Rajiah P, Sinha R, Cuevas C et al. Imaging of uncommon retroperitoneal masses. *RadioGraphics* 2011;31(4):949–976.

35. Craig WD, Fanburg-Smith JC, Henry LR, Guerrero R, Barton JH. Fat-containing lesions of the retroperitoneum: Radiologic-pathologic correlation. *RadioGraphics* 2009;29(1):261–290.

36. Castellazzi G, Vanel D, Le Cesne A et al. Can the MRI signal of aggressive fibromatosis be used to predict its behavior? *Eur J Radiol* 2009;69(2):222–229.

37. Kreuzberg B, Koudelova J, Ferda J et al. Diagnostic problems of abdominal desmoid tumors in various locations. *Eur J Radiol* 2007;62(2):180–185.

38. Vandevenne JE, De Schepper AM, De Beuckeleer L et al. New concepts in understanding evolution of desmoid tumors: MR imaging of 30 lesions. *Eur Radiol* 1997;7:1013–1019.

39. Kransdorf MJ, Jelinek JS, Moser Jr RP et al. Magnetic resonance appearance of fibromatosis: A report of 14 cases and review of the literature. *Skeletal Radiol* 1990;19:495–499.

40. Lee JC, Thomas JM, Phillips S, Fisher C, Moskovic E. Aggressive fibromatosis: MRI features with pathologic correlation. *AJR Am J Roentgenol* 2006;186:247–254.

41. Dinauer PA, Brixey CJ, Moncur JT et al. Pathologic and MR imaging features of benign fibrous soft-tissue tumors in adults. *RadioGraphics* 2007;27(1):173–187.

42. Choyke PL, Hayes WS, Sesterhenn IA. Primary extragonadal germ cell tumors of the retroperitoneum: Differentiation of primary and secondary tumors. *RadioGraphics* 1993;13(6):1365–1375; quiz 1377–1378.

43. Ueno T, Tanaka YO, Nagata M et al. Spectrum of germ cell tumors: From head to toe. *RadioGraphics* 2004;24(2):387–404.

44. Gatcombe HG, Assikis V, Kooby D et al. Primary retroperitoneal teratomas: A review of the literature. *J Surg Oncol* 2004;86(2):107–113.

45. Wang RM, Chen CA. Primary retroperitoneal teratoma. *Acta Obstet Gynecol Scand* 2000;79(8):707–708.

46. Bruneton JN, Diard F, Drouillard JP et al. Primary retroperitoneal teratoma in adults: Presentation of two cases and review of the literature. *Radiology* 1980;134:613–616.

47. Hayasaka K, Tanaka Y, Soeda S et al. MR findings in primary retroperitoneal schwannoma. *Acta Radiol* 1999;40(1):78–82.

48. Li Q, Gao C, Juzi JT et al. Analysis of 82 cases of retroperitoneal schwannoma. *ANZ J Surg* 2007;77(4): 237–240.

49. Rha SE, Byun JY, Jung SE et al. Neurogenic tumors in the abdomen: Tumor types and imaging characteristics. *RadioGraphics* 2003;23(1):29–43.

50. Kinoshita T, Naganuma H, Ishii K, Itoh H. CT features of retroperitoneal neurilemmmoma. *Eur J Radiol* 1998;27:67–71.

51. Wong CS, Chu TY, Tam KF. Retroperitoneal Schwannoma: A common tumor in an uncommon site. *Hong Kong Med J* 2010;16:66–68.

52. Bass JC, Korobkin M, Francis IR et al. Retroperitoneal plexiform neurofibromas: CT findings. *AJR Am J Roentgenol* 1994;163(3):617–620.

53. Sakai F, Sone S, Kiyono K et al. Intrathoracic neurogenic tumors: MR-pathologic correlation. *AJR Am J Roentgenol* 1992;159:279–283.

54. Hughes MJ, Thomas JM, Fisher C et al. Imaging features of retroperitoneal and pelvic schwannomas. *Clin Radiol* 2005;60(8):886–893.

55. Yang DM, Jung DH, Kim H et al. Retroperitoneal cystic masses: CT, clinical, and pathologic findings and literature review. *RadioGraphics* 2004;24(5):1353–1365.

56. Hrehorovich PA, Franke HR, Maximin S et al. Malignant peripheral nerve sheath tumor. *RadioGraphics* 2003;23(3):790–794.

57. Korf BR. Malignancy in neurofibromatosis type 1. *Oncologist* 2000;5(6):477–485.

58. Felix EL, Wood DK, Das Gupta TK. Tumors of the retroperitoneum. *Curr Probl Cancer* Jul 1981;6(1):1–47.

59. Kilton LJ, Aschenbrener C, Burns CP. Ganglioneuroblastoma in adults. *Cancer* 1976;37:974–983.

60. Bousvaros A, Kirks DR, Grossman H. Imaging of neuroblastoma: An overview. *Pediatr Radiol* 1986;16:89–106.

61. Francis IR, Korobkin M. Pheochromocytoma. *Radiol Clin North Am* 1996;34:1101–1112.

62. Blackledge G, Best JJ, Crowther D et al. Computed tomography (CT) in the staging of patients with Hodgkin's Disease: A report on 136 patients. *Clin Radiol* 1980; 31(2):143–147.

63. Neumann CH, Robert NJ, Canellos G et al. Computed tomography of the abdomen and pelvis in non-Hodgkin lymphoma. *J Comput Assist Tomogr* 1983;7(5):846–850.

64. Froehlich JM, Triantafyllou M, Fleischmann A, Vermathen P, Thalmann GN, Thoeny HC. Does quantification of USPIO uptake-related signal loss allow differentiation of benign and malignant normal-sized pelvic lymph nodes? *Contrast Media Mol Imaging* May–Jun 2012;7(3):346–355.

65. Alexiou C, Kau RJ, Dietzfelbinger H et al. Extramedullary plasmacytoma: Tumor occurrence and therapeutic concepts. *Cancer* 1999;85:2305–2314.

66. Monill J, Pernas J, Montserrat E et al. CT features of abdominal plasma cell neoplasms. *Eur Radiol* 2005;15:1705–1712.

67. Galieni P, Cavo M, Avvisati G et al. Solitary plasmacytoma of bone and extramedullary plasmacytoma: Two different entities? *Ann Oncol* 1995;6:687–691.

68. Oh D, Kim CK, Park BK, Ha H. Primary extramedullary plasmacytoma in retroperitoneum: CT and integrated PET/CT findings. *Eur J Radiol Extra* 2007;62:57–61.

69. George V, Tammisetti VS, Surabhi VR, Shanbhogue AK. Chronic fibrosing conditions in abdominal imaging. *RadioGraphics* Jul–Aug 2013;33(4):1053–1080.

70. Amis ES Jr. Retroperitoneal fibrosis. *AJR Am J Roentgenol* 1991;157(2):321–329.

71. Uibu T, Oksa P, Auvinen A et al. Asbestos exposure as a risk factor for retroperitoneal fibrosis. *Lancet* 2004;363(9419):1422–1426.

72. Lepor H, Walsh PC. Idiopathic retroperitoneal fibrosis. *J Urol* 1979;122(1):1–6.

73. Cronin CG, Lohan DG, Blake MA, Roche C, Mc-Carthy P, Murphy JM. Retroperitoneal fibrosis: A review of clinical features and imaging findings. *AJR Am J Roentgenol* 2008;191(2):423–431.

74. Mehta A1, Blodgett TM. Retroperitoneal fibrosis as a cause of positive FDG PET/CT. *J Radiol Case Rep* 2011;5(7):35–41.

75. Sagel SS, Siegel MJ, Stanley RJ, Jost RG. Detection of retroperitoneal hemorrhage by computed tomography. *AJR Am J Roentgenol* Sep 1977;129(3):403–407.

76. Meyers MA. *Dynamic Radiology of the Abdomen.* New York: Springer-Verlag, 1976, pp.I 13–95.

77. Alexander ES, Colley DP, Clark RA. Computed tomography of retroperitoneal fluid collections. *Semin Roentgenol* Oct 1981;16(4):268–276.

78. Syuto T, Hatori M, Masashi N, Sekine Y, Suzuki K. Chronic expanding hematoma in the retroperitoneal space: A case report. *BMC Urol* Nov 18, 2013;13:60.

79. Shanmuganathan K, Mirvis SE, Sover ER. Value of contrast-enhanced CT in detecting active hemorrhage in patients with blunt abdominal or pelvic trauma. *AJR Am J Roentgenol* 1993;161(1):65–69.

80. Thaler M, Achatz W, Liebensteiner M, Nehoda H, Bach CM. Retroperitoneal lymphatic cyst formation after anterior lumbar interbody fusion: A report of 3 cases. *J Spinal Disord Tech* Apr 2010;23(2):146–150.

81. Kawashima A, Sandler CM, Corriere JN Jr, Rodgers BM, Goldman SM. Ureteropelvic junction injuries secondary to blunt abdominal trauma. *Radiology* 1997;205:487–492.

82. Gore RM, Balfe DM, Aizenstein RI, Silverman PM. The great escape: Interfascial decompression planes of the retroperitoneum. *AJR Am J Roentgenol* Aug 2000; 175(2):363–370.

83. Altemeier WA, Alexander JW. Retroperitoneal abscess. *Arch Surg* 1961;83:512–524.

84. Callen PW. Computed tomographic evaluation of abdominal and pelvic abscesses. *Radiology* 1979; 131(1):171–175.

85. Wells RD, Bebarta VS. Primary iliopsoas abscess caused by community-acquired methicillin-resistant *Staphylococcus aureus*. *Am J Emerg Med* 2006;24: 897–898.

86. Charalampopoulos A, Macheras A, Charalabopoulos A, Fotiadis C, Charalabopoulos K. Iliopsoas abscesses: Diagnostic, aetiologic and therapeutic approach in five patients with a literature review. *Scand J Gastroenterol* 2009;44(5):594–599.

87. Dion E, Graef C, Miquel A et al. Bone involvement in Erdheim-Chester disease: Imaging findings including periostitis and partial epiphyseal involvement. *Radiology* 2006;238(2):632–639.

88. Dion E, Graef C, Haroche J et al. Imaging of thoracoabdominal involvement in Erdheim-Chester disease. *AJR Am J Roentgenol* 2004;183(5):1253–1260.

89. Fortman BJ, Beall DP. Erdheim-Chester disease of the retroperitoneum: A rare cause of ureteral obstruction. *AJR Am J Roentgenol* 2001;176(5):1330–1331.

90. Gottlieb R, Chen A. MR findings of Erdheim-Chester disease. *J Comput Assist Tomogr* 2002;26(2): 257–261.

91. Mesurolle B, Sayag E, Meingan P, Lasser P, Duvillard P, Vanel D. Retroperitoneal extramedullary hematopoiesis: Sonographic, CT, and MR imaging appearance. *AJR Am J Roentgenol* Nov 1996;167(5):1139–1140.

92. Tsitouridis J, Stamos S, Hassapopoulou E et al. Extramedullary paraspinal hematopoiesis in thalassemia: CT and MRI evaluation. *Eur J Radiol* 1999;30(1):33–38.

93. Waligore MP, Stephens DH, Soule EH, McLeod RA. Lipomatous tumors of the abdominal cavity: CT appearance and pathologic correlation. *AJR Am J Roentgenol* 1981;137:539–545.

94. Andac N, Baltacioglu F, Cimsit NC et al. Fat necrosis mimicking liposarcoma in a patient with pelvic lipomatosis. CT findings. *Clin Imaging* 2003;27(2):109–111.

95. Haynes JW, Brewer WH, Walsh JW. Focal fat necrosis presenting as a palpable abdominal mass: CT evaluation. *J Comput Assist Tomogr* 1985;9(3):568–569.

96. Ross JS, Prout GR, Jr. Retroperitoneal fat necrosis producing ureteral obstruction. *J Urol* 1976;115(5):524–529.

97. Takao H, Yamahira K, Watanabe T. Encapsulated fat necrosis mimicking abdominal liposarcoma: Computed tomography findings. *J Comput Assist Tomogr* 2004;28(2):193–194.

98. Jeffery GM, Theaker JM, Lee AH et al. The growing teratoma syndrome. *Br J Urol* 1991;67(2):195–202.

99. Davidson AJ, Hartman DS. Lymphangioma of the retroperitoneum: CT and sonographic characteristic. *Radiology* 1990;175(2):507–510.

100. Konen O, Rathaus V, Dlugy E et al. Childhood abdominal cystic lymphangioma. *Pediatr Radiol* 2002;32(2):88–94.

101. Foeldi M, Foeldi E, Kubik S. *Textbook of Lymphology*. 2nd edn. Munich, Germany: Elsevier, 2007.

102. Humphries PD, Wynne CS, Sebire NJ, Olsen ØE. Atypical abdominal paediatric lymphangiomatosis: Diagnosis aided by diffusion-weighted MRI. *Pediatr Radiol* 2006;36:857–859.

103. Lohrmann C, Foeldi E, Langer M. Assessment of the lymphatic system in patients with diffuse lymphangiomatosis by magnetic resonance imaging. *Eur J Radiol* Nov 2011;80(2):576–581.

104. Ros PR, Olmsted WW, Moser RP, Dachman AH, Hjermstad BH. Mesenteric and omental cysts: Histologic classification with imaging correlation. *Radiology* 1987;164:327–332.

105. Fayad ZA, Fuster V. Clinical imaging of the high-risk or vulnerable atherosclerotic plaque. *Circ Res* Aug 17, 2001;89(4):305–316.

106. Kramer CM. Magnetic resonance imaging to identify the high-risk plaque. *Am J Cardiol* Nov 21, 2002;90(10C):15L–17L.

107. Momiyama Y, Fayad ZA. Plaque imaging and monitoring atherosclerotic plaque interventions. *Top Magn Reson Imaging* 2007;18:349–355.

108. Budovec JJ, Pollema M, Grogan M. Update on multi-detector computed tomography angiography of the abdominal aorta. *Radiol Clin N Am* 2010;48:283–309.

109. Brewster DC. Clinical and anatomical considerations for surgery in aortoiliac disease and results of surgical treatment. *Circulation* 1991;83(2 Suppl.):I42–I52.

110. Pande RL, Beckman JA. Abdominal aortic aneurysm: Populations at risk and how to screen. *J Vasc Interv Radiol* 2008;19(Suppl. 6):S2–S8.

111. Khanafer KM, Bull JL, Upchurch GR Jr et al. Turbulence significantly increases pressure and fluid shear stress in an aortic aneurysm model under resting and exercise flow conditions. *Ann Vasc Surg* 2007;21(1):67–74.

112. Veith FJ, Gupta S, Daly V. Technique for occluding the supraceliac aorta through the abdomen. *Surg Gynecol Obstet* 1980;151:426–428.

113. Chase CW, Layman TS, Barker DE et al. Traumatic abdominal aortic pseudoaneurysm causing biliary obstruction: A case report and review of the literature. *J Vasc Surg* 1997;25(5):936–940.

114. Hirsch AT, Haskal ZJ, Hertzer NR et al. ACC/AHA 2005 Practice Guidelines for the management of patients with peripheral arterial disease (lower extremity, renal, mesenteric, and abdominal aortic): AU. *Circulation* 2006;113(11):e463.

115. Nevitt MP, Ballard DJ, Hallett JW, Jr. Prognosis of abdominal aortic aneurysms. A population-based study. *N Engl J Med* 1989;321:1009–1014.

116. Lederle FA, Johnson GR, Wilson SE et al. Rupture rate of large abdominal aortic aneurysms in patients refusing or unfit for elective repair. *JAMA* 2002;287:2968–2972.

117. Harris LM, Faggioli GL, Fiedler R, Curl GR, Ricotta JJ. Ruptured abdominal aortic aneurysms: Factors affecting mortality rates. *J Vasc Surg* 1991;14:812–818.

118. Yankelevitz DF, Gamsu G, Shah A et al. (2000) Optimization of combined CT pulmonary angiography with lower extremity CT venography. *AJR Am J Roentgenol* 174(1):67–69.

119. Schwartz SA, Taljanovic MS, Smyth S et al. CT findings of rupture, impending rupture, and contained rupture of abdominal aortic aneurysm. *AJR Am J Roentgenol* 2007;188:W57–W62.

120. Siegel CL, Cohan RH, Korobkin M et al. Abdominal aortic aneurysm morphology; CT features in patients with ruptured and non-ruptured aneurysms. *AJR Am J Roentgenol* 1994;163:1123–1129.

121. Rakita D, Newatia A, Hines JJ et al. Spectrum of CT findings in rupture and impending rupture of abdominal aortic aneurysms. *RadioGraphics* 2007;27:497–507.

122. Jones CS, Reilly MK, Dalsing MC et al. Chronic contained rupture of abdominal aortic aneurysms. *Arch Surg* 1986;121:542–546.

123. Sevastos N, Rafailidis P, Kolokotronis K et al. Primary aortojejunal fistula due to foreign body: A rare cause of gastrointestinal bleeding. *Gastroenterol Hepatol* 2002;14:797–800.

124. Tse DML, Thompson ARA, Perkins J et al. Endovascular repair of a secondary aorto-appendiceal fistula. *Cardiovasc Interv Radiol* 2011;34(5):1090–1093.

125. Kappadath SK, Clarke MJ, Stormer E, Steven L, Jaffray B Primary aortoenteric fistula due to a swallowed twig in a threeyear-old child. *Eur J Vasc Endovasc Surg* 2010;39:217–219.

126. Skourtis G, Papacharalambous G, Makris S et al. Primary aortoenteric fistula due to septic aortitis. *Ann Vasc Surg* 2010;24(6):825.e7–825.e11.

127. Hagspiel KD, Turba UC, Bozlar U et al. Diagnosis of aortoenteric fistulas with CT angiography. *J Vasc Interv Radiol* 2007;19:497–504.

128. Raman SP, Kamaya A, Federle M, Fishman EK. Aortoenteric fistulas: Spectrum of CT findings. *Abdom Imaging* Apr 2013;38(2):367–375.

129. Vu QDM, Menias CO, Bhalla S, Peterson C, Wang LL, Balfe DM. Aortoenteric fistulas: CT features and potential mimics. *RadioGraphics* 2009;20:197–209.

130. Schmidt R, Bruns C, Walter M, Erasmi H. Aorto-caval fistula—An uncommon complication of infrarenal aortic aneurysms. *Thorac Cardiovasc Surg* 1994;42:208–211.

131. Abbadi AC, Deldime P, Van Espen D, Simon M, Rosoux P. The spontaneous aortocaval fistula: A complication of the abdominal aortic aneurysm. Case report and review of the literature. *J Cardiovasc Surg (Torino)* 1998;39:433–436.

132. Miani S, Giorgetti PL, Arpesani A, Giuffrida GF, Biasi GM, Ruberti U. Spontaneous aorto-caval fistulas from ruptured abdominal aortic aneurysms. *Eur J Vasc Surg* 1994;8:36–40.

133. Brightwell RE, Pegna V, Boyne N. Aortocaval fistula: Current management strategies. *ANZ J Surg* Jan 2013;83(1–2):31–35.

134. Alexander JJ, Imbembo AL. Aorta-vena cava fistula. *Surgery* 1989;105:1–12.

135. Liu PS, Platt JF. CT angiography in the abdomen: A pictorial review and update. *Abdom Imaging* Feb 2014;39(1):196–214.

136. Jonker FH, Schlosser FJ, Moll FL et al. Dissection of the abdominal aorta. Current evidence and implications for treatment strategies; a review and meta analysis of 92 patients (comment). *J Endovasc Ther* 2009;16(1):71–80.

137. Sebastia C, Pallisa E, Quiroga S et al. Aortic dissection:Diagnosis and follow-up with helical CT. *RadioGraphics* 1999;19:45–60.

138. Cataner D, Andreu M, Gallardo X et al. CT in non-traumatic acute thoracic aortic disease: Typical and atypical features and complications. *RadioGraphics* 2003;23:S93–S110.

139. Cambria RP, Brewster DC, Gertler J et al. Vascular complications associated with spontaneous aortic dissection. *J Vasc Surg* 1988;7:199–209.

140. Barnes DM, Williams DM, Dasika NL et al. A single-center experience treating renal malperfusion after aortic dissection with central aortic fenestration and renal artery stenting. *J Vasc Surg* 2008;47(5):903–910.

141. Ganaha F, Miller DC, Sugimoto K, Do YS, Minamiguchi H, Saito H, Mitchell RS, Dake MD. Prognosis of aortic intramural hematoma with and without penetrating atherosclerotic ulcer: A clinical and radiological analysis. *Circulation* 2002;106:342–348.

142. Coady MA, Rizzo JA, Elefteriade JA. Pathologic variants of thoracic aortic dissections:Penetrating atherosclerotic ulcer and intramural hematoma. *Cardiol Clinic* 1999;17:637–657.

143. Alomari IB, Hamirani YS, Madera G, Tabe C, Akhtar N, Raizada V. Aortic intramural hematoma and its complications. *Circulation* Feb 11, 2014;129(6):711–716.

144. Chao CP, Walker TG, Kalva SP. Natural history and CT appearances of aortic intramural hematoma. *RadioGraphics* 2009;29:791–804.

145. Stanson AW, Kazmier FJ, Hollier LH. Ulceres atheromateux penetrants de l'aorte thoracique: Histoire naturelle et correlations anatomo-cliniques. *Ann Chir Vasc* 1986;1:15–23.

146. Georgiadis GS, Trellopoulos G, Antoniou GA, Georgakarakos EI, Nikolopoulos ES, Pelekas D, Pitta X, Lazarides MK. Endovascular therapy for penetrating ulcers of the infrarenal aorta. *ANZ J Surg* Oct 2013;83(10):758–763.

147. Sato M, Imai A, Sakamoto H, Sasaki A, Watanabe Y, Jikuya T. Abdominal aortic disease caused by penetrating atherosclerotic ulcers. *Ann Vasc Dis* 2012;5(1):8–14.

148. Tsuji Y, Tanaka Y, Kitagawa A. Endovascular stent-graft repair for penetrating atherosclerotic ulcer in the infrarenal abdominal aorta. *J Vasc Surg* 2003;38:383–388.

149. Teruya TH, Bianchi C, Abou-Zamzam AM, Ballard JL. Endovascular treatment of a blunt traumatic abdominal aortic injury with a commercially available stent graft. *Ann Vasc Surg* 2005;19(4):474–478.

150. Rosengart MR, Zierler RE. Fractured aorta—A case report. *Vasc Endovascular Surg* 2002;36(6):465–467.

151. Nucifora G, Hysko F, Vasciaveo A. Blunt traumatic abdominal aortic rupture: CT imaging. *Emerg Radiol* 2008;15(3):211–213.

152. Gunn M, Campbell M, Hoffer EK. Traumatic abdominal aortic injury treated by endovascular stent placement. *Emerg Radiol* 2007;13(6):329–331.

153. Lock JS, Huffman AD, Johnson RC. Blunt trauma to the abdominal aorta. *J Trauma* 1987;27(6):674–677.

154. Virmani R, Burke A. Nonatherosclerotic diseases of the aorta and miscellaneous disease of the main pulmonary arteries and large veins. In: Silver M, Gotlieb A, Schoen F (eds.), *Cardiovascular Pathology*, 3rd edn. Philadelphia, PA: Churchill Livingstone, 2001, pp. 107–137.

155. Gravanis MB. Giant cell arteritis and Takayasu aortitis: Morphologic, pathogenetic and etiologic factors. *Int J Cardiol* 2000;75(Suppl. 1):S21–33. discussion S35–S36.

156. Gornik HL, Creager MA. Aortitis. *Circulation* 2008;117(23):3039–3051.

157. Kerr GS, Hallahan CW, Giordano J, Leavitt RY, Fauci AS, Rottem M, Hoffman GS. Takayasu arteritis. *Ann Intern Med* 1994;120:919–929.

158. Mwipatayi BP, Jeffery PC, Beningfield SJ, Matley PJ, Naidoo NG, Kalla AA, Kahn D. Takayasu arteritis: Clinical features and management: Report of 272 cases. *ANZ J Surg* 2005;75:110–117.

159. Sueyoshi E, Sakamoto I, Hayashi K. Aortic aneurysms in patients with Takayasu's arteritis: CT evaluation. *AJR Am J Roentgenol* 2000;175:1727–1733.

160. Matsumura K, Hirano T, Takeda K, Matsuda A, Nakagawa T, Yamaguchi N, Yuasa H, Kusakawa M, Nakano T. Incidence of aneurysms in Takayasu's arteritis. *Angiology* 1991;42:308–315.

161. Flamm SD, White RD, Hoffman GS. The clinical application of 'edema-weighted' magnetic resonance imaging in the assessment of Takayasu's arteritis. *Int J Cardiol* 1998;66(Suppl. 1):S151–S159.

162. Salvarani C, Crowson CS, O'Fallon WM, Hunder GG, Gabriel SE. Reappraisal of the epidemiology of giant cell arteritis in Olmsted County, Minnesota, over a fifty-year period. *Arthritis Rheum* 2004;51:264–268.

163. Salvarani C, Cantini F, Boiardi L, Hunder GG. Polymyalgia rheumatica and giant-cell arteritis. *N Engl J Med* 2002;347:261–271.

164. Evans JM, O'Fallon WM, Hunder GG. Increased incidence of aortic aneurysm and dissection in giant cell (temporal) arteritis. A population-based study. *Ann Intern Med* 1995;122:502–507.

165. Miller DV, Isotalo PA, Weyand CM, Edwards WD, Aubry MC, Tazelaar HD. Surgical pathology of noninfectious ascending aortitis: A study of 45 cases with emphasis on an isolated variant. *Am J Surg Pathol* 2006;30:1150–1158.

166. Rojo-Leyva F, Ratliff NB, Cosgrove DM, 3rd, Hoffman GS. Study of 52 patients with idiopathic aortitis from a cohort of 1,204 surgical cases. *Arthritis Rheum* 2000;43:901–907.

167. Foote EA, Postier RG, Greenfield RA, Bronze MS. Infectious aortitis. *Curr Treat Options Cardiovasc Med* 2005;7:89–97.

168. Reddy DJ, Ernst CB. Infected aneurysms. In: Rutherford RB (ed.), *Vascular Surgery*, 4th edn. Philadelphia, PA: WB Saunders, 1995, pp. 1139–1153.

169. Restrepo CS, Ocazionez D, Suri R, Vargas D. Aortitis: Imaging spectrum of the infectious and inflammatory conditions of the aorta. *RadioGraphics* 2011;31:435–451.

170. Daas AK, Reddy KS, Suwanjindar P, Fulmer A, Siquiera A, Floten S, Starr A. Primary tumors of the aorta. *Ann Thorac Surg* 1996;62:1526–1528.

171. Mason MD, Wheeler JR, Gregory RT et al. Primary tumors ofthe aorta: Report of a case and review of the literature. *Oncology* 1982;39:167–172.

172. Graham LM, Zelenock GB, Erlandson EE, Coran AG, Lindenauer SM, Stanley JC. Abdominal aortic coarctation and segmental hypoplasia. *Surgery* 1979;86:519–529.

173. Terramani TT, Salim A, Hood DB, Rowe VL, Weaver FA. Hypoplasia of the descending thoracic and abdominal aorta: A report of two cases and review of the literature. *J Vasc Surg* Oct 2002;36(4):844–848.

174. Bashour T, Jokhadar M, Cheng TO, Nasri M, Kabbani S. Hypoplasia of descending aorta as a rare cause of hypertension. A report of 5 cases. *Angiology* 1982;33:790–799.

175. Phillips E. Embryology, normal anatomy, and anomalies. In: Ferris EJ, Hipona FA, Kahn PC, Phillips E, Shapiro JH (eds.), *Venography of the Inferior Vena Cava and Its Branches*. Baltimore, MD: Williams & Wilkins, 1969, pp. 1–32.

176. Siegfried MS, Rochester D, Bernstein JR, Milner JW. Diagnosis of inferior vena cava anomalies by computerized tomography. *Comput Radiol* 1983;7:119–123.

177. Bass JE, Redwine MD, Kramer LA, Huynh PT, Harris JH Jr. Spectrum of congenital anomalies of the inferior vena cava: Cross-sectional imaging findings. *RadioGraphics* May–Jun 2000;20(3):639–652.

178. Pineda D, Moudgill N, Eisenberg J, DiMuzio P, Rao A. An interesting anatomic variant of inferior vena cava duplication: Case report and review of the literature. *Vascular* Jun 2013;21(3):163–167.

179. Ginaldi S, Chuang VP, Wallace S. Absence of hepatic segment of the inferior vena cava with azygous continuation. *J Comput Assist Tomogr* 1980;4:112–114.

180. Mazzucco A, Bortolotti U, Stellin G, Gallucci V. Anomalies of the systemic venous return: A review. *J Card Surg* 1990;5(2):122–133.

181. Karaman B, Koplay M, Ozturk E, Basekim CC, Ogul H, Mutlu H, Kizilkaya E, Kantarci M. Retroaortic left renal vein: Multidetector computed tomography angiography findings and its clinical importance. *Acta Radiol* Apr 2007;48(3):355–360.

182. Karkos CD, Bruce IA, Thomson GJ, Lambert ME. Retroaortic left renal vein and its implications in abdominal aortic surgery. *Ann Vasc Surg* 2001;15:703–708.

183. Cuellar i Calabria H, Quiroga Gomez S, Sebastia Cerqueda C, Boye de la Presa R, Miranda A, Alvarez-Castells A. Nutcracker or left renal vein compression phenomenon: Multidetector computed tomography findings and clinical significance. *Eur Radiol* 2005;15:1745–1751.

184. Brancatelli G, Galia M, Finazzo M, Sparacia G, Pardo S, Lagalla R. Retroaortic left renal vein joining the left common iliac vein. *Eur Radiol* 2000;11:1724–1725.

185. Uthappa MC, Anthony D, Allen C. Retrocaval ureter: MR appearances. *Br J Radiol* 2002;75:177–179.

186. Talner LB, Reilly PHO, Wasserman NF. Specific causes of obstruction. In: Pollack HM, McClennan BL (eds.), *Clinical Urography*, 2nd edn. Philadelphia, PA: WB Saunders, 2000,pp. 1967–2136.

187. Jenkins S, Marshall GB, Gray R. Leiomyosarcoma of the inferior vena cava. *Can J Surg* 2005;48(3):252–253.

188. Kaufman LB, Yeh BM, Joe BN, Qayyum A, Coakley F. Inferior vena cava filling defects on CT and MRI. *AJR Am J Roentgenol* 2005;185:717–726.

189. Kutcher R, Rosenblass R, Mitsudo S, Goldman M, Kogan S. Renal angiomyolipoma with sonographic demonstration of extension into the inferior vena cava. *Radiology* 1982;143:755–756.

24

Abdominal Wall and Hernias

Sonja M. Kirchhoff

CONTENTS

24.1 Introduction

Intra-abdominal adhesions mostly involving the abdominal wall are very common as well as an unpreventable secondary effect of surgery. In approximately 93% of the patients, adhesions develop after major abdominal or pelvic surgery [1]. Primarily, these adhesions consist of fibrous bands interconnecting the abdominal wall and subperitoneal organs [2,3]. Despite the presence of adhesions many patients remain asymptomatic; however, a significant proportion of patients present with adhesion-related symptoms. The incidence of adhesion-related small bowel obstruction accounts for approximately 65%–75% [4]. In general, operations or interventions of the lower abdomen and pelvis or those procedures resulting in damage to a large peritoneal surface tend to put patients at a higher risk for developing adhesions [5]. The highest extent of adhesions is expected after re-operations.

Thus, it is mandatory to determine a correct diagnosis of adhesions before surgery or re-surgery to manage the treatment properly. In this context, the available imaging modalities should determine the site and extent of adhesions with a high accuracy.

Usually computed tomography (CT) and/or ultrasound (US) are performed in the acute situation, but magnetic resonance imaging (MRI) including functional cine imaging provides a valuable alternative.

24.2 MRI Protocols

MRI protocols should generally be adjusted to the patients' clinical conditions and acquisition time should be minimized.

In the current literature, there exist no standard MR protocols for imaging of the abdomen, especially for diagnosing adhesions and/or hernias. However, usually gadolinium-based contrast media is applied except during pregnancy and patients with known renal impairment. It is possible to choose from either free-breathing or breath-holding protocols. In summary, a typical MRI protocol for evaluating the abdomen may include the following:

- Axial and coronal T_2-weighted images using the so-called half Fourier single-shot spin-echo (HASTE) technique
- T_1-weighted breath-hold gradient-echo sequence including in-phase and out-of phase images
- Breath-hold unenhanced or contrast-enhanced axial and coronal 3D T_1-weighted fat-saturated images
- Coherently balanced steady-state free precession sequence in the coronal orientation

24.2.1 Functional Cine MRI

Functional cine MRI should not be mistaken for dynamic MRI using intravenous or arterial contrast media but explained as acquisition of consecutive MR images during respiration freezing motion. To overcome the impossibility of other even 3D MRI techniques to visualize the visceral slide a special MRI protocol was developed by the workgroup of Lienemann et al. [6].

Functional MRI examinations are performed on high-field systems of at least 1.5 T. The patients are lying in a supine position using body-array surface coils covering the abdomen. No administration of contrast media or any premedication is necessary. At first, a coronal localizer is acquired with a superimposed grid used as reference for screening the entire abdomen in the sagittal and axial orientations, covering the area from the diaphragm to the pelvis. Consecutively, one cycle consisting of 10 consecutive measurements at the same position is performed at every point of the grid (see Figure 24.1). At every examination position, the patient is asked to increase the intra-abdominal pressure by straining and subsequently relaxing during each cycle to induce the visceral slide to be able to diagnose or rule out adhesions (see Figure 24.2). The average distance between two consecutive positions of the grid is 3 cm, so that approximately 300–400 images are acquired with an approximate examination time of 30 min.

To facilitate the localization of adhesions in many institutions a nine-segment-map of the abdomen is used with two bilateral lines along the borders of the rectus abdominis muscle, one transverse line across the inferior costal margins and another transverse line across the iliac crest (see Figure 24.3).

24.3 Abdominal Wall

24.3.1 Adhesions

In general, adhesions may be congenital or acquired, even though most adhesions are acquired after injury to the peritoneum, infection, or abdomino-pelvic surgery. Fortunately, most patients with adhesions do not experience any severe clinical symptoms, whereas for others adhesions may cause significant increased morbidity and mortality [7].

Adhesions in the peritoneal cavity are predisposing factors to injury intra-abdominal organs during trocar insertion for laparoscopy for a re-operation. Thus, abdominal surgeons who plan a surgical approach in an already-operated patient have to consider the presence of adhesions, so that an accurate diagnosis of intra-abdominal adhesions is highly valuable [8]. Regarding imaging modalities real-time US has been proposed in the literature for the assessment of adhesions but also for the initial needle or trocar placement for laparoscopy to avoid visceral injury [9]. However, US also provides several severe disadvantages such as examiner-dependant results, difficult examination conditions due to patients' physique, and overlay of intestinal gas.

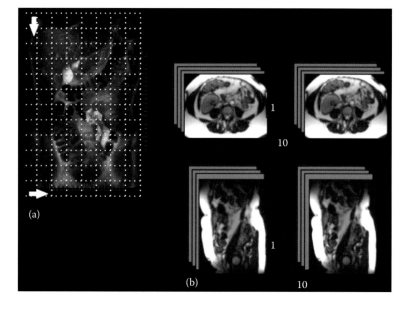

FIGURE 24.1
Coronal MR—localizer (a) showing the superimposed grid allowing for a precise positioning of the slices. During the examination the entire patient's abdomen is scanned from right to left and from cranial to caudal (see arrows) with an approximate gap in between the slices of 3 cm. At each examination point, consecutive image cycles consisting of 10 images are performed in the axial and sagittal orientations (b).

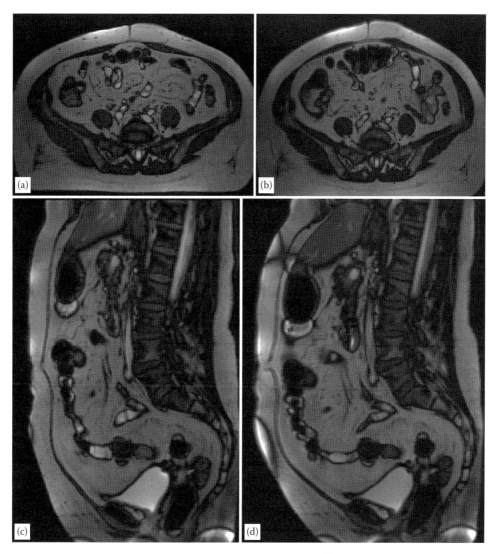

FIGURE 24.2
These figures present an example of the normal visceral slide in the axial orientation during rest (a) and during (b) straining. Figures (c) and (d) also show the normal visceral slide in the sagittal orientation with a single bowel loop moving back and forth in mainly cranio-caudal direction, also during rest (c) and during (d) straining.

Besides US, other imaging methods such as small bowel enteroclysis are rather insensitive regarding the detection of fibrous bands manifesting as luminal narrowing or failure to separate adjacent bowel loops by manually applied pressure to the abdomen [9]. In 2000, the workgroup of Lienemann et al. [6] introduced functional cine MRI performing the visceral slide for the reliable, accurate, and noninvasive detection of adhesions. The results of our workgroup [10] suggest that functional cine MRI is the best imaging modality for the detection of intra-abdominal adhesions. In this recent study, the results of functional cine MRI of 89 patients with postoperative adhesion-related complaints were compared to intra-operative findings resulting in a sensitivity of 93%, an overall accuracy of 90%, and a positive predictive value of 96% of MRI. Another study on detection of

intra-abdominal adhesions and correlation with surgical findings by our workgroup [11] revealed a total of 71 adhesions with the most common type between small bowel loops and the abdominal wall (see Figures 24.4 through 24.6) followed by adhesions between small bowel loops and pelvic organs (see Figure 24.7).

Signs of distortion of neighbored organs including bowel loops during the Valsalva maneuver are considered as another direct sign for the presence of adhesions.

24.3.2 Tumors

Desmoid tumors also known as aggressive type of fibromatosis are rather benign but locally aggressive soft tissue tumors that can occur in a variety of anatomic sites being classified according to their location as

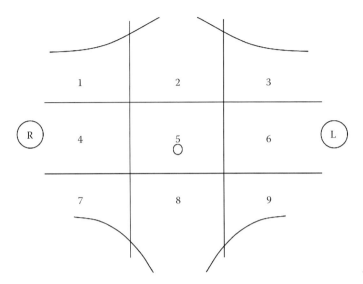

FIGURE 24.3
This scheme presents the nine-segment map of the abdomen for an adequate and exact localization of adhesions in the entire abdomen.

intra-abdominal, abdominal wall, or extra-abdominal tumors [12]. These tumors arise from connective tissue in the fascia, aponeurosis, and muscle and present with a high tendency of recurrence after primary resection but usually do not metastasize. Abdominal wall desmoid tumors usually occur during gestation or in the first year after pregnancy. The appearance of desmoid tumors on MRI varies according to their composition. According to their evolution desmoid tumors tend to become less cellular and more fibrous, so that in their early evolution they present with rather high signal on T_2-weighted images becoming lower in signal intensity on T_2-weighted images as they evolve and correspondingly the collagen extent increases. Typically, desmoid tumors show moderate to significant enhancement after gadolinium administration (see Figure 24.8).

A relatively common gynecological problem in women of reproductive age is endometriosis. The abdominal wall has been described as the most common extrapelvic location of endometriosis, although it may occur in nearly all body cavities and organs [13]. Usually cases of endometriosis of the abdominal wall are associated with previous operations opening the uterus and spread of the corresponding cells [14]. Most patients present with a palpable mass at the site of the maximum tenderness usually in the area of the surgical scar. Endometriosis of the abdominal wall presents with a wide morphologic spectrum, varying from purely chocolate cystic to solid deposits or fibrosis [15]. MRI findings in case of abdominal wall endometriosis are rather unspecific, and thus MRI is usually performed not for the initial diagnosis but for depicting the extent of the disease preoperatively. MRI of abdominal wall endometriosis may however present differently from ovarian endometriosis with isointense or mild hyperintensity to muscle on

T_1- as well as on T_2-weighted images. After intravenous gadolinium administration, they usually enhance quite homogeneously.

In the case of a mass of the abdominal wall, metastatic disease should also be considered as differential diagnosis because many malignancies may disseminate to the superficial soft tissue. Breast cancer is the most common primary malignancy spreading abdominal wall in women. For men, malignant melanoma presents the primary tumor most commonly spreading in this location even presenting as subcutaneous nodules. Metastatic lesions are usually low in signal on T_1-weighted images and low to moderate signal on T_2-weighted images, showing a rather rim enhancement after gadolinium administration (see Figure 24.9).

When soft tissue lesions occur in the subcutaneous tissue of the ventral abdominal wall, especially along prior incision, hematoma should be considered as differential diagnosis, especially with a typical history of trauma, coagulopathy, or even severe exertion. The MR appearance of hematomas varies with the degradation products of the red blood cells. On T_1-weighted images, acute hematoma tends to present a higher signal compared to muscle along with low signal intensity on T_2-weighted sequences. In the evolvement of hematoma, the signal on T_2-weighted images may vary depending on the concentration of intracellular deoxyhemoglobin and hemosiderin causing hypointense signal (see Figure 24.10). Hematomas do not show significant contrast enhancement.

Frequent injection granulomas are seen in the subcutaneous fat at the upper outer quadrant of the buttock but injection granuloma can appear at any injection site also in the abdominal wall, but rather in terms of incidental findings. Typically on T_1-weighted sequences, they appear hypointense, whereas the signal on T_2-weighted images

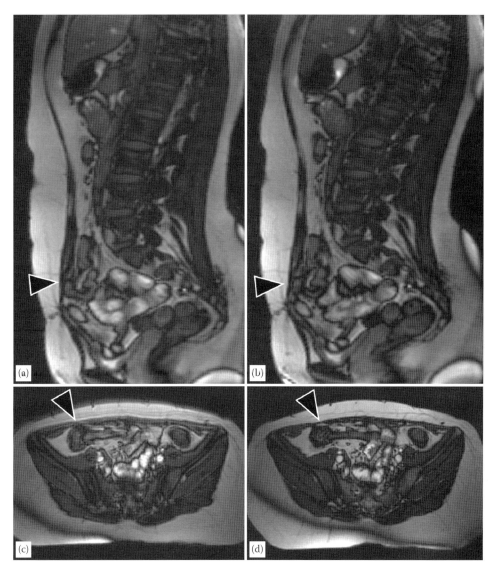

FIGURE 24.4
Functional cine MRI (TrueFISP) in the sagittal and axial planes, respectively, during rest (a, c) and Valsalva maneuver (b, d) of a 34-year-old woman suffering from unspecific abdominal pain following previous adenectomy. Several small bowel loops seem adherent to the ventral abdominal wall (see arrow) since during straining, a lack of separation and excursion of the bowel loops is noticeable.

depends on the evolution of the granulomas, so that the signal may be hyperintense if there is an inflammatory reaction, or hypointense in case the reaction is of fibrous character [16].

24.4 Hernia

Diagnosis of abdominal wall hernias is usually made during physical exams; however, the diagnosis may be difficult, especially in patients suffering from obesity, pain, or scarring of the abdominal wall. In these cases, abdominal imaging may provide the first clue to the correct diagnosis. In the past for the diagnosis of hernia, conventional radiography or studies using barium were predominantly performed. Nowadays, CT plays a predominant role in the acute mostly emergency situation, but MRI presents a valuable alternative.

Especially functional cine MRI is also useful in detecting hernias in untypical or unsuspected sites as well as in obese patients or after surgery. In 2009, our study group [17] was able to reliably detect and evaluate implanted meshes and typical complications such as mesh dislocation, in 43 patients after incidental hernia repair.

Based on the anatomic origin and its orifice, hernias can generally be divided as external (e.g., inguinal and femoral) and internal (e.g., paraduodenal).

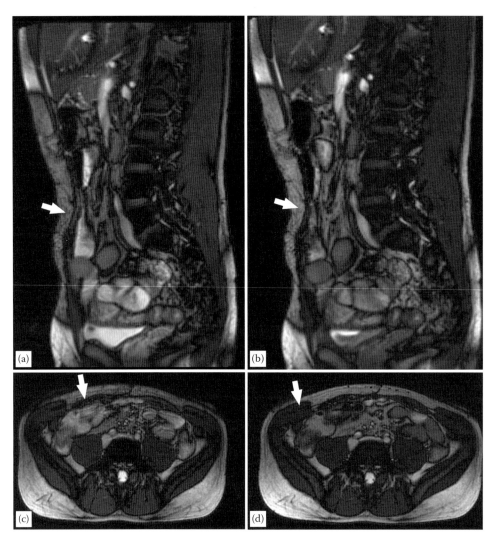

FIGURE 24.5
Functional cine MRI of a 41-year-old male patient suffering from chronic abdominal pain following appendectomy. In the sagittal (a, b) as well as in axial (c, d) orientations, several small bowel loops can be recognized sticking to the abdominal wall during rest (a, c) as well as during performance of the Valsalva maneuver (b, d) (see arrows). Also, adhesions between small bowel loops in the pelvic area were detected.

24.4.1 External Hernia

For external hernia, the orifice is mostly located in specific sites of either congenital weakness or previous surgery. Thus, defects of the abdominal wall account for the most common type of external hernia [18].

Groin hernia: Inguinal hernia can be differentiated as direct and indirect hernia; the indirect type is by far the most common abdominal wall hernia prevalent in the United States [19], resulting from a herniation through a patent processus vaginalis located lateral to the inferior epigastric vessels. Indirect inguinal hernia are usually acquired and caused by weakness along with a dilatation of the internal inguinal ring [20]. In contrast, direct hernias are caused by weakness

of the transverse fascia and thus located medial to the inferior epigastric vessels.

However, femoral hernias are significantly less frequent compared to inguinal hernia. This type of hernia occurs due to a defect in the attachment of the transverse fascia to the pubis, thus occurring medial to the femoral vein and posterior to the inguinal ligament. Femoral hernia presents with a high tendency to incarcerate and are often difficult to differentiate from inguinal hernia.

Ventral hernia (including all hernias through the anterior/lateral abdominal wall): Umbilical, epigastric, and hypogastric hernias are considered as midline defects. In children, this hernia type is usually congenital in contrast to adults when this

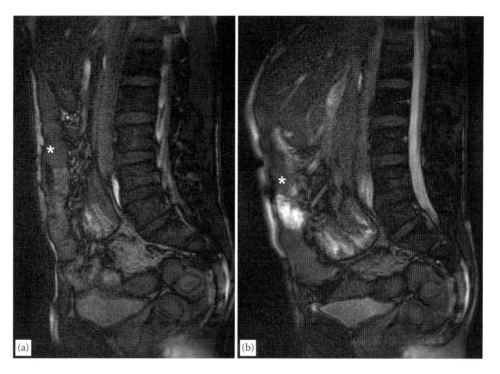

FIGURE 24.6
Functional cine MRI in the sagittal plane at rest (a) and during straining (b) next to a postoperative scar several small bowel loops can be noticed (see asterisk) without any excursion during straining.

hernia type is usually acquired, with multiple pregnancies, ascites, and obesity being considered as risk factors [20]. A high prevalence of incarceration and strangulation is associated with umbilical hernia, and they usually do not reduce spontaneously.

Lateral defects of the abdominal wall: The most common lateral defect of the abdominal wall in terms of hernia is the so-called Spigelian hernia occurring through a defect in the linea semilunaris. These hernias are often secondary to acquired weakness of the aponeurosis or surgical incision and typically parts of the omentum and short bowel segments protrude here with a correspondingly high tendency of incarceration [18].

Incisional hernias: In general, delayed complications of abdomino-pelvic surgery are considered as incisional hernias, mostly developing during the first months after surgery. These hernias develop due to vertical than due to transverse incisions; however, they may also develop at small laparoscopic puncture sites [20]. Age, obesity, postoperative wound infection, and ascites are considered as typical risk factors for incisional hernia development (see Figure 24.11).

In case a hernia develops adjacent to a stoma, it presents a so-called parastomal hernia and is considered a form of incisional hernia. Factors such as obesity and chronic cough aggravate the extension of this hernia type [20].

24.4.2 Internal Hernia

In general, the herniation of bowel loops through a developmental or surgically created defect of the peritoneum, omentum, or mesentery is considered as internal hernia. These types of hernias are less frequent than external hernias. Diagnosis is always based on radiological findings mostly based on CT images since internal hernias often present as emergency situation.

24.5 Conclusion

The detection of intra-abdominal adhesions is usually based on clinical examinations mostly presented by acute or even more frequent chronic abdominal pain. However, to be able to accurately plan and adjust therapy, imaging methods have to be considered. Of course, in the acute, mostly emergency, situations, US or CT are the most suitable imaging modalities, but for the subacute conditions, MRI including functional cine MRI presents a noninvasive alternative regarding the detection and

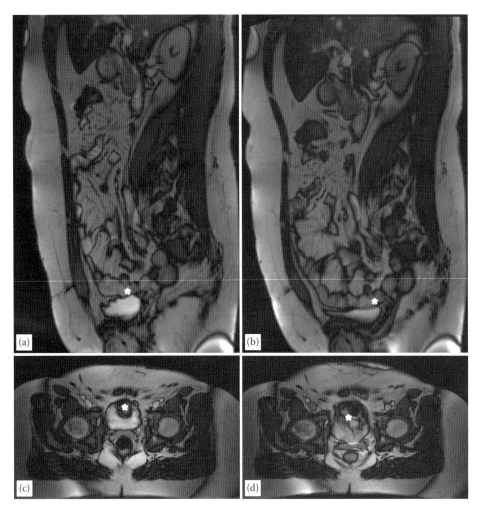

FIGURE 24.7

This functional cine MRI in the axial and sagittal planes shows a 46-year-old woman suffering from acute lower abdominal pain following hysterectomy. During rest (a, c) as well as during the Valsalva maneuver (b, d), there is no excursion or separation between several small bowel loops and the organs of the pelvis such as the urinary bladder and the uterus (see asterisk).

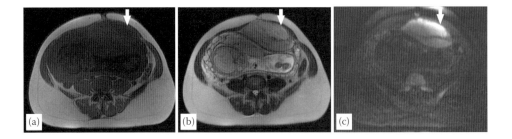

FIGURE 24.8

MRI of the abdomen of a 36-year-old pregnant woman demonstrated in a T_1-weighted sequence (a) to the surrounding muscle rather hypointense mass in the left ventral abdominal wall with corresponding rather hyperintense signal in the T_2-weighted HASTE sequence (b) and positive diffusion findings (c). These findings are compatible with the diagnosis of a desmoid tumor of the ventral abdominal wall.

evaluation of adhesions and hernias, which has been proven by several publications in the current literature. Functional cine MRI provides several significant advantages compared to real-time US, such as the possibility of evaluating the entire abdomen including the pelvic area, no dependency on the examiner's experience, and the possibility of examining obese patients or with severe gas overlay.

In general, radiologist should also evaluate functional cine MR exams for the presence of clinically

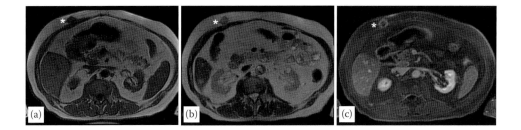

FIGURE 24.9

MRI of the abdomen of an 85-year-old male patient suffering from malignant melanoma as primary malignancy. The axial T_1-weighted image (a) shows to the surrounding tissue hypointense mass located in the right-sided ventral abdominal wall located in the subcutaneous fat (see asterisk). The correspondingly acquired T_2-weigthed (b) HASTE sequence shows a moderate hyperintense signal with rather rim enhancement (c) on T_1-weighted images after intravenous gadolinium administration. In accordance with the primary tumor, a metastasis of the ventral abdominal wall was diagnosed.

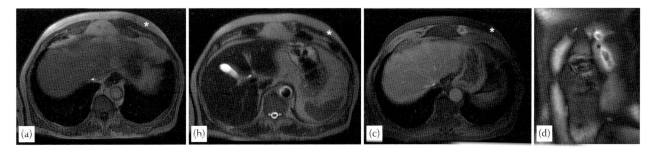

FIGURE 24.10

A 65-year-old man following abdominal trauma: On the axial T_1-weighted image (a), a moderate hyperintense lesion is recognizable in the left ventral abdominal wall (see asterisk) with corresponding rather hypointense signal on the axial T_2-weigthed image (b), showing rather diffuse contrast enhancement on the T_1-weigthed fat-saturated images in the axial (c) and coronal planes (d). In synopsis with the anamnesis, these MR findings are compatible with an abdominal wall hematoma.

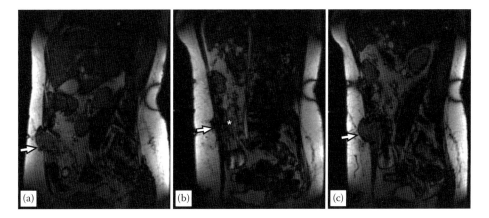

FIGURE 24.11

Functional cine MRI of a 45-year-old man suspected for incidental hernia. In the sagittal plane during Valsalva maneuver (a, c), a hernia of the ventral abdominal wall is recognizable (see arrow) with bowel entering the hernia sac, whereas during rest (b), the orifice is visible, showing missing excursion/separation of the abdominal wall and bowel loops (see asterisk).

occult hernia. If a hernia is detected, it is important to delineate its size, contents, shape, location, and related complications. Functional cine MRI provides a great perspective on abdominal anatomy, shows defects of the abdominal wall, and adds important information regarding interpretation and planning of the adequate treatment, especially when considering surgery.

References

1. Menzies D, Ellis H (1990) Intestinal obstruction from adhesions: How big is the problem? *Ann R Coll Surg Engl* 72:60–63.
2. Levrant SG, Bieber EJ, Barnes RB (1997) Anterior abdominal wall adhesions after laparotomy or laparoscopy. *J Am Assoc Gynecol Laparosc* 4:353–356.

3. Cox MR, Gunn IF, Eastman MC, Hunt RF, Heinz AW (1993) The operative aetiology and types of adhesions causing small bowel obstruction. *Aust N Z J Surg* 63:848–852.

4. Ellis H (1998) The magnitude of adhesion-related problems. *Ann Chir Gynaecol* 87:9–11.

5. Dijkstra FR, Nieuwenhuijzen M, Reijnen MM et al. (2000) Recent clinical developments in pathophysiology, epidemiology, diagnosis, and treatment of intraabdominal adhesions. *Scand J Gastroenterol Suppl* 232:52–59.

6. Lienemann A, Sprenger D, Steitz HO et al. (2000) Detection and mapping of intraabdominal adhesions by using functional cine MR imaging: Preliminary results. *Radiology* 217:421–425.

7. Ellis H, Moran BJ, Thompson JN et al. (1999) Adhesion-related hospital readmission after abdominal and pelvic surgery: A retrospective cohort study. *Lancet* 353:1476–1480.

8. Freys SM, Fuchs KH, Heimbucher J, Thiede A (1994) Laparoscopic adhesiolysis. *Surg Endosc* 8:1202–1207.

9. Bartram CI (1980) Radiologic demonstration of adhesions following surgery for inflammatory bowel disease. *Br J Radiol* 53:650–665.

10. Lang RA, Buhmann S, Hopman A et al. (2008) Cine-MRI detection of intraabdominal adhesions: Correlation with intraoperative findings in 89 consecutive cases. *Surg Endosc* 22:2455–2461.

11. Buhmann S, Lang RA, Kirchhoff C et al. (2008) Functional cine MR imaging for the detection and mapping of intraabdominal adhesions: Methods and surgical correlation. *Eur Radiol* 18:1215–1223.

12. Goldblum J, Fletcher JA (2002) Desmoid-type fibromatoses. In: Fletcher CD, Unni KK, Mertens F (eds) *World Health Organization Classification of Tumours: Pathology and Genetics of Tumours of Soft Tissue and Bone*. Lyon, France: IARC Press.

13. Ideyi SC, Schein M, Niazi M et al. (2003) Spontaneous endometriosis of the abdominal wall. *Dig Sur* 20:246–248.

14. Blanco RG, Parithivel VS, Shah AK et al. (2003) Abdominal wall endometriosis. *Am J Surg* 185:596–598.

15. Woodward PJ, Sohaey R, Mezzetti TP (2001) Endometriosis: Radiologicpathologic correlation. *RadioGraphics* 21:193–216.

16. Salgado R, Alexiou J, Engelhorn JL (2006) Pseudotumoral lesions. In: De Schepper AM, Vanhoenacker F, Gielen J, Parizel PM (eds) *Imaging of Soft Tissue Tumors*. Berlin, Germany: Springer Verlag.

17. Kirchhoff S, Ladurner R, Kirchhoff C et al. (2010) Detection of recurrent hernia and intraabdominal adhesions following incisional hernia repair: A functional cine MRI study. *Abdom Imaging* 35:224–231.

18. Miller PA, Mezwa DG, Feczko PJ et al. (1995) Imaging of abdominal hernias. *RadioGraphics* 15:333–347.

19. Rutkow IM (2003) Demographic and socioeconomic aspects of hernia repair in the United States in 2003. *Surg Clin North Am* 83:1045–1051.

20. Harrison LA, Keesling CA, Martin NL et al. (1995) Abdominal wall hernias: Review of herniography and correlation with cross sectional imaging. *RadioGraphics* 15:315–322.

25

Diseases of the Colon and Rectum

Maria Ciolina, Carlo Nicola De Cecco, Marco Rengo, Justin Morris, Franco Iafrate, and Andrea Laghi

CONTENTS

25.1 Introduction

During the past few decades, the use of cross-sectional imaging modalities for the noninvasive evaluation of lower gastrointestinal tract has significantly increased due to improvements in diagnostic technique and the dissemination of computed tomography (CT) and magnetic resonance (MR) scanners. CT is still the modality of choice for the assessment of acute colonic inflammation, for the evaluation of metastatic disease from colorectal malignancy, and for colorectal cancer screening; in each case, CT colonography (CTC) provides an excellent evaluation of the inner colonic surface. Because of superior acquisition speed, increased spatial resolution, and superior image quality robustness, multislice CT

appears to be more suitable for colorectal cancer screening than MR. However, MR colonography (MRC) can be a good alternative to noninvasive investigation of the colon, especially in young patients, because of the absence of ionizing radiation.

Although CT is a step ahead of MR imaging (MRI) for the evaluation of the colon, MRI is doubtless the gold standard for rectal and perineal diseases.

MRI allows for high spatial resolution and high contrast resolution images of the pelvis, representing the gold standard for rectal disease cancer assessment at the primary staging, for restaging after chemoradiotherapy (CRT), and for the oncological follow-up, especially adding other imaging MR biomarkers (as diffusion-weighted imaging [DWI] and perfusion MRI) to standard protocol. MRI also plays a crucial role in the assessment of fistulous track studies, giving detailed radiological information to choice among the different treatment options.

The scope of this chapter is to provide an overview of the actual role of MRI in the study of the lower gastrointestinal tract, illustrating in particular the actual clinical indications, the acquisition protocols, and the classical imaging findings of the most common diseases.

25.2 Disease of the Colon

25.2.1 Anatomy of the Colon

The large intestine extends from the ileocecal valve to the anus. It is approximately 1.5 m long in adults, although there is considerable variation in its length [1].

The external colonic aspect includes the three teniae, the haustra, and the appendices epiploicae.

The teniae are longitudinal bands, approximately 8 mm in width and run along the total length of the colon and represent the outer longitudinal muscular layer. We can distinguish three teniae with different topographical positions:

1. The tenia mesocolica lies dorsally where the mesocolon attaches to the transverse colon. It also lays dorsomedially on the ascending and descending colons, following the mesentery attachment to the colonic wall (mesenteric tenia).

2. The tenia omentalis (or epiploic tenia) lies on the ventrocranial surface of the transverse mesocolon, corresponding to the line of attachment of the greater omentum. It also runs along the dorsolateral surfaces of the ascending and descending colons.

3. The tenia libera, which means *free teniae*, runs along the inferior surface of the transverse colon and the anterior surfaces of the ascending and descending colons.

Where the appendix joins the cecum and the sigmoid colon passes into the rectum, the three teniae merge into one uniform coat of longitudinal muscle.

The haustra are sacculations formed in the spaces between the teniae, separated from each other by circular grooves of variable depth. The degree of their prominence depends on contraction of the teniae.

The appendices epiploicae consist of subserosal pockets filled with fat. They have a grape-like appearance and vary in size according to the individual's nutritional status. On the ascending and descending colons, they are generally distributed in two rows. On the transverse colon, they form only one row along the tenia libera. On an endoluminal view, the grooves correspond to folds called plicae semilunaris. The length of these folds corresponds to the distance between two teniae.

At the ileocecal junction, the terminal ileum invaginates the large intestine, creating a sphincter, the ileocecal valve (*valvula Bauhini*) (Figure 25.1d) [2].

The ileocecal valve can present a labial form with upper and lower lips. At the end of the lips, two mucosal ridges extend horizontally in the lumen of the large intestine, resembling the crescent-shaped folds of the colon. These ridges, known as the frenula of the valve, form the dividing plane between the cecum and the ascending colon. The ileocecal valve can also present as a large papilla protruding into the colon with a star-like orifice.

The location of the appendix (Figure 25.1d) on the skin surface corresponds to a point one-third of the way along an imaginary line drawn from the anterosuperior iliac spine to the umbilicus (*McBurney's point*).

In early fetal life, the cecum extends caudally forming a cone, the tip of which creates the vermiform appendix. Later, due to the growth of the cecal walls, the point where the appendix arises shifts on the dorsomedial wall to where the three teniae of the large intestine merge into one uniform coat of longitudinal muscle [2].

The ascending colon extends about 15 cm (Figure 25.1a), from the cecum to the surface of the right liver lobe, where it turns forward and to the left, giving rise to the hepatic flexure. It has a retroperitoneal location and is covered on three sides by peritoneum. It is narrower than the cecum and, in one-third of cases, has a narrow mesocolon.

The hepatic flexure connects the ascending and transverse colons (Figure 25.1c), but its position is variable. It has anatomical relationships with the right kidney posteriorly, the right liver lobe superiorly and laterally, the descending part of the duodenum medially, and the gallbladder anteromedially. Posteriorly, it is not covered by peritoneum but lies directly over the pararenal fascia.

The transverse colon connects the hepatic flexure to the splenic flexure (Figure 25.1b) and it is approximately 50 cm long, although its length and position are variable. It often resembles an inverted arch with posterior and superior concavities. The transverse colon is almost completely covered by peritoneum.

It is suspended in the peritoneal cavity by the transverse mesocolon. The gastrocolic ligament connects the transverse colon to the greater curvature of the stomach and continues with the greater omentum [1,2].

The splenic flexure (Figure 25.1e) connects the transverse and descending colons and is located in the left hypochondrium, and it has anatomical relationships posteriorly with the left kidney and the pancreatic tail

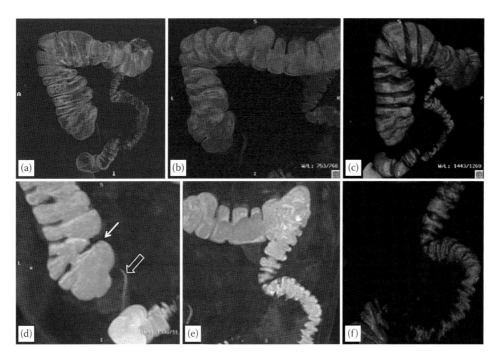

FIGURE 25.1
Coronal single-shot MR colonography images (a–c) and MR virtual colonography image (c) showing normal cecum, ascending colon, hepatic flexure (a–c), and transverse colon (b). Coronal single-shot MR colonography (d) showing ileocecal valve (arrow) and vermiform appendix (empty arrow). Coronal single-shot MR colonography (e) showing splenic flexure and descending colon. MR virtual colonography (f) showing descending and sigmoid colon.

and superiorly, posteriorly, and laterally with the spleen. Compared with the hepatic flexure, the splenic flexure is superior and posterior and is attached by the phrenicocolic ligament to the left hemidiaphragm.

The descending colon (Figure 25.1e and f) runs for about 25 cm downward from the left hypochondrium through the lumbar region up to the level of the iliac crest. At this point it turns inferomedially becoming the sigmoid colon.

The sigmoid colon (Figure 25.1f) is a mobile bowel loop beginning below the left iliac crest (inlet of the lesser pelvis) and ending at the rectum and, extended, it is about 40–50 cm long in adults and 18 cm in children.

Its position and shape vary among patients according to its length and the length and mobility of the mesocolon. Its position and shape also depend on whether it is empty or distended and on the status of nearby organs such as the bladder (full or empty) and the uterus. The sigmoid mesocolon is usually covered by peritoneum. It presents an inverted *V* attachment to the posterior abdominal wall.

The left arm of the root of the sigmoid mesocolon begins at the inner edge of the left psoas major, bypassing the left gonadic vessels and the left ureter. The right arm of the attachment line of the sigmoid mesocolon crosses over the left common iliac vessels, just above the division of the artery. At the rectosigmoid junction, the circular muscle layer forms a prominence called the

third sphincter of O'Beirne, located about 2–3 cm above the superior valve. At this level the lumen of the rectum decreases in caliber to match that of the sigmoid (rectosigmoid junction). It is not a true sphincter, but similar in action [1,2].

25.2.1.1 Vascular Supply

The arterial supply to the large bowel derives from both the superior and the inferior mesenteric arteries. The cecum, the ascending colon, and two-thirds of the transverse colon (segments derived from the midgut) are supplied by the ileocolic, right colic, and middle colic arteries, which are all branches of the superior mesenteric artery. The left part of the transverse colon, the descending colon and the sigmoid colon, and the upper rectum (segments derived from the hindgut) are supplied by the left colic artery and the sigmoid and superior rectal arteries, all branches of the inferior mesenteric artery.

The marginal artery (of Drummond) is the vessel that lies closest and parallel to the colonic wall and it is formed by arcades arising from the ileocolic, right colic, middle colic, and left colic arteries.

The arc of Riolan is an anastomosis that is sometimes present in the mesentery between the right side of the transverse colon and the upper descending colon [1]. It is formed from a large branch of the middle colic artery that runs parallel and posteriorly to the middle colic artery in the transverse mesocolon and anastomoses with an ascending branch of the left colic artery. It provides a direct connection between the superior mesenteric artery and the inferior mesenteric artery.

The terminal arterial branches divide into the vasa longa and vasa brevia, which either enter the colonic wall directly or run through the subserosa. These vessels then cross the circular smooth layer, giving rise to the epiploic arteries.

The venous drainage is as follows: the part of the colon derived from the midgut (cecum, appendix, ascending colon, and the right two-thirds of the transverse colon) drain into the superior mesenteric vein. The part of the colon derived from the hindgut (the left part of the transverse colon, the descending colon, rectum, and upper anal canal) drain into the inferior mesenteric vein [1].

25.2.2 Magnetic Resonance Colonography

MRC is a noninvasive diagnostic imaging modality, first introduced in the late 1990s, which has enabled the evaluation of the colon and colonic diseases, like CTC.

The acquisition is free from exposure to ionizing radiation, provides excellent contrast resolution to study the colonic wall and acceptable execution time (about 20–23 min) in comparison to other MR abdominal examinations [3]. The reported rate of perforation (about 0.0009%) is low in comparison to conventional colonoscopy (CC) that is about 0.3% [4,5] due to the lower pressure applied by fluid distension and to the lower numbers of MRC examinations performed in comparison to CC and CTC.

Main disadvantages of MRC include poor availability of the technique that remains limited to few specialist centers, high cost in comparison to CT, without as much therapeutic function as CC.

The usual contraindications of MRC are those applied for general MRI (claustrophobia, cardiac pacemaker or other implanted metallic device, renal failure, and allergy to contrast media).

In reality, there is no consensus about the optimal bowel preparation required, or the intraluminal agent or about the dual or single positioning or whether stool tagging should be employed.

25.2.2.1 Bowel Preparation, Types of Enemas, and Fecal Tagging

Bowel cleansing is required to eliminate residual stool, which can potentially be misinterpreted as an endoluminal filling defect, such as a polyp or a carcinoma.

Patients undergoing MRC require accurate bowel cleansing, which is usually achieved with polyethylene glycol (PEG) and sodium phosphate assumed at a dose of 1 L and 4 L of oral fluids prescribed the day before the examination.

After bowel cleansing, colonic distension can be performed using one of the three following methods.

1. Positive contrast agent (bright lumen MRC) involves the instillation of a gadolinium-spiked enema (approximately between 1.5 and 2 L) into the colon with a concentration of 5 10 mmol/L is obtained by mixing gadolinium with water. Dual positioning is usually required to displace air and stool, which will move according to gravity. Colonic lumen appears hyperintense both on T_2-weighted and T_1-weighted sequences. Polyps and masses appear as T_1-hypointense filling defects against the bright, contrast-filled lumen. Occasionally, filling defects, such as air or impacted residual stool, will not displace with the change of positioning and can be incorrectly identified as intraluminal polyps, resulting in false-positive examinations. Another important disadvantage of this technique is the difficulty in distinguishing polypoid contrast enhancement from bright intraluminal contrast [6]. Per-patient sensitivity for colon cancer screening with bright lumen MRC is only 75% with a better per-patient specificity of 95% [7].

2. Negative contrast agent (dark lumen MRC) is obtained through negative intraluminal contrast agent such as room air, carbon dioxide, or water. Air distends the colon better than water enema, and this is not usually associated with increased susceptibility artifacts [8]. However, susceptibility artifacts can occur at the air/tissue interface, decreasing image quality and the sensitivity of the examination, but short echo times may be used to limit this artifact. Dark lumen MRC is free from risk of rectal spillage. Results about accuracy of dark lumen MRC for colon cancer screening are controversial with a per-polyp/lesion sensitivity of 10.5% for lesions measuring less than 5 mm, 57.6% for polyps measuring between 5 mm and 10 mm, and 73.9% for lesions greater than 10 mm. However, many of the polyps that were missed were hyperplastic polyps, which are not the target of colorectal cancer screening.

3. Biphasic luminal contrast agent is obtained with an enema containing a PEG in a water solution. Using this method the colonic lumen appears hyperintense on T_2-weighted imaging and hypointense on T_1-weighted imaging. The disadvantages of using water is rectal spillage (or the need to defecate) during the examination

[9] and the difficulties in identification of intra-abdominal abscesses (especially in Crohn's disease or diverticulitis) that appear fluid within the colonic lumen in T_2-weighted sequences.

Stool tagging can be added to reduce or to even eliminate cathartic bowel cleansing [10]. Stool tagging can alter the signal intensity of the stool residues, making them invisible and with the same signal intensity of the enema used (bright stool for bright lumen MRC and dark stool for dark lumen MRC). For the bright lumen MRC, the tagging agent typically used is the gadolinium but it implies increasing costs and may cause constipation [11]. For dark lumen MRC an oral contrast agent containing small iron particles and named ferumoxsil (Lumirem, Guerbet Group, Paris, France) is usually administrated [12].

Another method to change the signal intensity of the fecal residues, used for dark lumen MRC, is fecal cracking. Based on a combination of lactulose and docusate sodium rectal enema (0.5%), it increases the water content of stool with consequent decreasing of signal intensity on dark lumen MRC [13].

25.2.2.2 MRC Technique

One or two surface coils can be used with the built-in phased-array coil for signal reception.

Once complete filling and adequate distension of the entire colon is obtained, a three-dimensional (3D) spoiled gradient-echo sequence is acquired with the patient in a prone and then supine position. Each imaging sequence is performed in the coronal plane within a single breath hold of less than 30 s. The imaging protocol also includes 2D single-shot fast spin-echo (SS-FSE or HASTE) pulse sequence and a contrast-enhanced 2D spoiled gradient-echo sequence to evaluate extra-colonic findings (i.e., liver metastases and lymphadenopathy).

These sequences are also preferably acquired in the coronal plane. A dedicated software finally enables to perform multiplanar reformations and endoluminal fly-through in both antegrade and retrograde fly-throughs. However, the endoluminal fly-through is more helpful for the detection of polyps than for the assessment of inflammatory bowel disease. Direct comparison between pre-contrast T_1-weighted imaging and contrast-enhanced SE T_1-weighted sequences is recommended in cases of a suspected polypoid lesions to differentiate them from stool. T_2-weighted imaging enables better evaluation of submucosal edema and pericolic inflammatory changes in cases of IBD, diverticulitis, and colitis.

25.2.2.3 MRC: Clinical Applications

- *Colon cancer screening*
  - MRC is accurate in detecting colon cancer (100% sensitive) and polyps (the cancer precursor) greater than 10 mm (per patient sensitivity of 88%) (Figures 25.2 and 25.3). Variable sensitivities and specificities have been found for the detection of polyps measuring 6–9 mm [14]. Since MRC provides a dataset covering the entire abdomen and pelvis, incidental extracolonic abnormalities and potentially serious lesions can be detected at an early stage. Incidental findings should be reported in a standardized manner, similar to the CTC Reporting And Data System (C-RADS), used in CTC [15] based on the severity of the findings.

- *Incomplete colonoscopy*
  - In case of incomplete colonoscopy due to a narrowed segment (occurring in up to 13% of

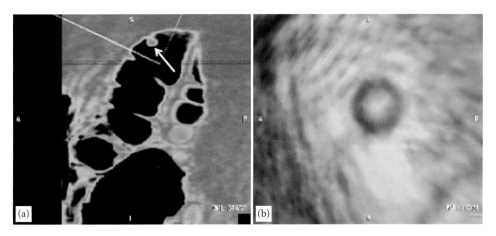

FIGURE 25.2
Coronal T_1-weighted MR colonography image (a) shows a 8-mm polypoid sessile lesion (*arrow*) into the splenic flexure. The sessile polyp is also confirmed through MR virtual image (b).

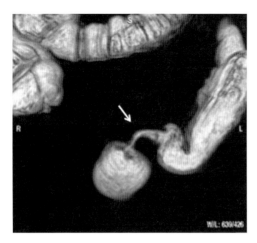

FIGURE 25.3
MR virtual colonography image an *apple coring* appearance (*arrow*) of proximal sigmoid colon (*arrow*) due to the presence of a circumferential cancer determining a severe luminal stenosis.

patients undergoing CC [16]), only air or fluid has to pass through the stenotic segment to distend the proximal colon. MRC thus enables the evaluation of synchronous tumors or extrinsic compression determined by extracolonic disease, thereby enabling the differentiation of benign strictures from malignant strictures [17] through a morphological evaluation, post-contrast sequences, and DWI.

- *Evaluation of colonic anastomosis*
  - MRC has a sensitivity of 84% and specificity of 100% for evaluating colonic anastomoses and [18] adds the potential to differentiate between inflammation, IBD recurrence, or tumor recurrence at the anastomoses.

- *Inflammatory bowel disease*
  - MRC is particularly appealing in evaluating chronic diseases because of the possibility to perform more frequent evaluations without ionizing radiation, especially in cases of young patients. MRC is useful in the assessment of bowel thickening, transmural and extraluminal disease, bowel enhancement (trasmural enhancement or *target* appearance) (Figure 25.4), submucosal edema, stricture, vascular engorgement (Comb's sign), enhancing mesenteric lymph nodes, fibrofatty proliferation fistulae, and abscess or the assessment of adequate responses to medical treatment particularly relevant in acute inflammatory bowel disease [19].

  - MRC allows also differentiation between active inflammatory disease and fibrostenotic disease in CD, which is clinically important for the different surgical and medical therapeutic approaches for fibrostenotic disease and inflammatory disease, respectively.

  - Ajaj et al. reported a sensitivity and a specificity for detection of inflammation of 87% and 100%, respectively [20]. Nevertheless, MRC seems to be less sensitive for detecting mild inflammation than severe inflammation in both CD and UC with a sensitivity of only 31.6% in CD and of 58.8% in UC [21].

- *Diverticulitis*
  - Although abdominal CT (not CTC) remains the most commonly used technique in cases

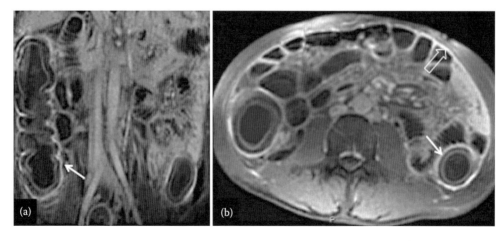

FIGURE 25.4
Coronal (a) and axial (b) T_1-weighted MR colonography images after gadolinium administration showing diffuse colonic wall thickening (*arrows*) with a *target* appearance due to hyperhemic, hypointense edematous submucosa and the inflamed hyper-enhanced outer serosa (*empty arrow*).

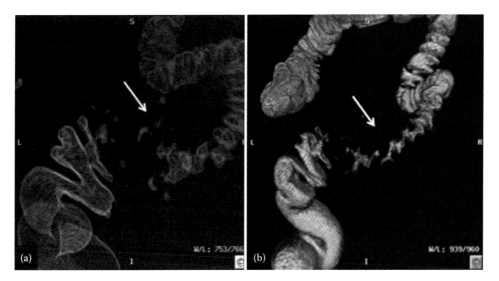

FIGURE 25.5
Coronal MR colonography (a and b) showing a narrowing of the sigmoid lumen with multiple parietal sacculations (*arrow*) due to colonic diverticulosis.

of suspected diverticulitis, MRC was found to be accurate in detecting diverticulosis (Figure 25.5) and acute diverticulitis [22] with an overall sensitivity of 86% and specificity of 92% for dark lumen MRC, enabling a more accurate differentiation between acute diverticulitis and colon cancer in three cases [23].

- *Endometriosis*
  - When MRC is compared to high-resolution MRI pelvis, the sensitivity and specificity for detection of colorectal endometriosis increases from 76% and 96% to 95% and 97%, respectively, for an experienced MR reader [24].

25.3 Disease of the Rectum

25.3.1 Anorectal Anatomy

25.3.1.1 Rectal Anatomy

The rectum extends from the level of the third sacral vertebral body (S3) to the anorectal line, located slightly below the tip of the coccyx and the apex of the prostate in males [1] (Figure 25.6).

The rectum is usually 15–20 cm in length above the anal verge and can be divided in three parts: the lower (from 0 cm to 6 cm above the anal verge), the middle (from 7 cm to 11 cm above the anal verge), and the upper (from 12 cm to 15 cm) [1,25].

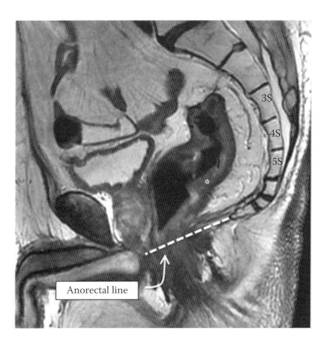

FIGURE 25.6
Sagittal T_2-weighted MR images showing the rectum usually extending from the level of the third sacral vertebral body (S3) to the anorectal line that passes from the tip of the coccyx to the apex of the prostate in males. This patient also presents a neoplastic circumferential wall thickening (*asterisks*) extending along the middle and upper third of the rectum and a mesorectal lymphadenopathy (*arrow*).

The circumference varies from 1.5 cm at the rectosigmoid junction to 3.5 cm or more at its widest ampullary portion [1,2].

The rectum deviates along its course in three lateral curves: the upper, convex to the right; the middle (the most prominent), convex to the left; the lower, convex to the right. Both ends of the rectum are in the median plane (Figure 25.7) [2].

The rectum lacks haustra, but it usually has three semilunar transverse folds named Houston valves: the superior valve, the middle valve, and the inferior valve (Figure 25.8). The superior fold, located either on one side or around the rectal lumen, marks endoluminally the transition point between the rectum and the sigmoid colon. At the rectosigmoid junction, considered to be located at the level of S3, the luminal caliber of the rectum decreases, matching that of the sigmoid

(rectosigmoid junction) [1,2]. Apart from the caliber, the sigmoid colon differs from the rectum by the nastriform aspect of the longitudinal muscular layer forming the taenia coli. The middle fold, the thickest, lies immediately above the rectal ampulla, originates from the anterior and right walls, and is also called *Kohlraush's valve*. The inferior fold is usually found on the left, around 2.5 cm below the middle fold [2,25].

The rectum is predominantly an extra-peritoneal organ, with the exception of the upper rectum that is anteriorly and laterally covered by a thin layer of

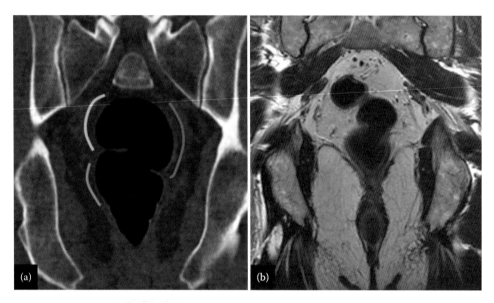

FIGURE 25.7
Coronal CT colonography image (a) perfectly showing the rectum deviating in three lateral curves: the upper, convex to the right (*green line*); the middle (the most prominent, *pink line*), convex to the left; the lower, convex to the right (*blue line*). (b) Coronal MR image showing that both ends of the rectum are in the median plane.

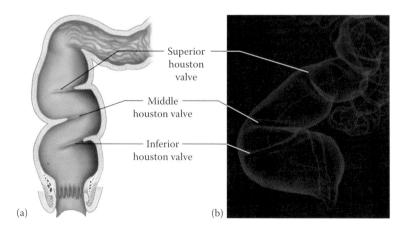

FIGURE 25.8
Drawing (a) and 3D volume rendered "similar Barium Enema" view CT colonography image (b) illustrating that the rectum usually has three semilunar transverse folds named Houston valves: the superior valve, the middle valve, and the inferior valve. The superior fold marks endoluminally the transition point between the rectum and the sigmoid colon. The middle fold lies immediately above the rectal ampulla, originates from the anterior and right walls and it is also called Kohlraush's valve. The inferior fold is usually found on the left, around 2.5 cm below the middle fold. (Courtesy of Dr. M Ciolina.)

visceral peritoneum. The peritoneal reflection is usually found between 7 cm and 9 cm from the anal verge, but in women it may be lower at 5–7.5 cm (Figure 25.9) [26].

25.3.1.2 Anatomy of Anal Canal

The anal canal has an average length of 4.2 cm in adults and begins at the anorectal junction (at the level of puborectalis sling) ending at the anal verge. The definition above refers to the surgical anal canal while the anatomical anal canal is considered from the dentate line to the anal verge and has an average length of about 2.1 cm (Figure 25.10) [25]. The anal canal appears to be

angulated in relation to the rectum because of the pull determined by the puborectalis sling, producing the anorectal angle. Posteriorly, the anal canal is attached to the coccyx by the anococcygeal ligament, which runs between the posterior aspect of the middle portion of the external sphincter and the coccyx. The anococcygeal ligament merges with the raphe of the levator ani (Figure 25.11) [1,3].

Anteriorly, the middle third of the anal canal is attached by dense connective tissue to the *perineal body*, which separates it from the membranous urethra and penile bulb in males or from the lower vagina in females (Figure 25.12).

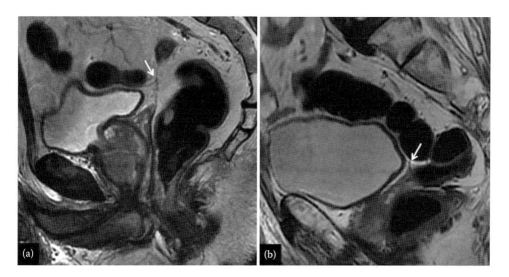

FIGURE 25.9
Sagittal T_2-weighted MR image (a) showing the peritoneal reflection of the vesicorectal pouch as a thin isointense line located anteriorly to the rectum at 7–9 cm from the anal verge. Sagittal MR image (b) showing that peritoneal reflection can be located lower, at about 5 cm from the anal verge.

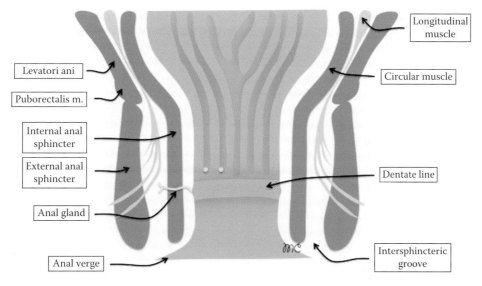

FIGURE 25.10
Drawing illustrates schematic anatomy of the anal canal. (Courtesy of Dr. M Ciolina.)

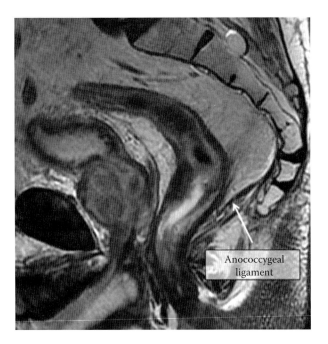

FIGURE 25.11
Sagittal T_2-weighted MR image showing that the anal canal is posteriorly attached to the coccyx by the anococcygeal ligament running between the posterior aspect of the middle portion of the external sphincter and the coccyx. The anococcygeal ligament merges with the raphe of the levetor ani, along the midline.

Laterally and posteriorly, the anal canal is surrounded by loose adipose tissue within the ischioanal fossae (Figure 25.12) [1].

The anal sphincter is composed of a smooth muscle internal sphincter and an external sphincter complex composed of skeletal muscle (Figure 25.12). The internal sphincter represents the muscularis propria and is a continuation of the circular muscle layer of the rectum,

whereas the longitudinal muscle of the rectum merges with the distal portion of the striated fibers of levator ani at the level of the puborectalis sling, then forming the external sphincter (Figure 25.13). The term *external sphincter complex* refers to the most inferior part of the levator ani muscle, the puborectalis sling, and the external sphincter muscles [26]. At the level of puborectalis muscle, the rectal adventitia fuses with a thin layer of connective tissue between the internal and external sphincters, the intersphincteric space (Figure 25.12) [27].

The mucosa of the anal canal is composed of the proximal 10 mm by columnar epithelium-like rectal mucosa. The following 15 mm constitute the transition zone marking the passage between the columnar epithelium and the stratified epithelium, and include the anal valves and the dentate line. The most distal 5–10 mm of anal lumen is lined by hairy skin stratified epithelium [26].

25.3.1.3 Blood Supply

The upper two-thirds of the rectum are the distal portion of the embryological hindgut that are supplied by the superior mesenteric artery, a vascular branch arising from the inferior mesenteric artery. At the level of S3 the superior mesenteric artery bifurcates into two vessels, the right and the left. The right is the larger one and supplies the posterior and lateral surface of the rectum. The left branch supplies the anterior rectal wall. The middle rectal arteries may be absent, but usually originate from the internal iliac arteries of each side through their anterior divisions or through the inferior vesicle vessels and anastomose with the superior and inferior arteries [26].

Inferior rectal arteries arise from the internal pudenda arteries supplying the external and internal sphincter

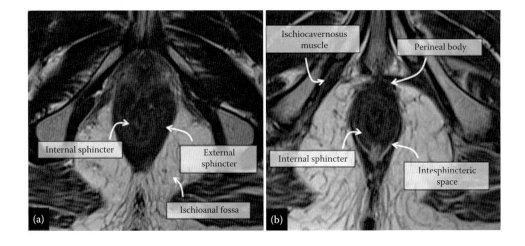

FIGURE 25.12
Axial T_2-weighted MR images (a and b) showing the inner annular muscular layer of muscularis propria constituting the internal sphincter and the outer muscular striated fibers of the external sphincter. The thin layer of connective tissue between the internal and external sphincters represents the intersphincteric space.

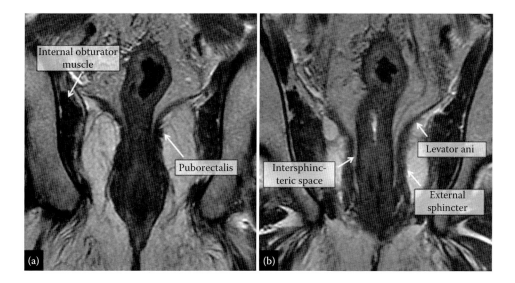

FIGURE 25.13

Coronal T_2-weighted MR images (a and b) showing the so-called external sphincter complex referring to the most inferior part of the levator ani muscle, the puborectalis sling, and the external sphincter muscles. The longitudinal muscle of the rectum merging with the distal portion of the striated fibers of levator ani at the level of puborectalis sling and forms then the external sphincter.

and finally reaching the mucosa and the submucosa of the anal canal [1].

The venous drainage of the hindgut is guaranteed by the superior rectal veins, draining into the portal system through the inferior mesenteric vein.

Veins from the lower third of the rectum are drained by the internal iliac veins through the middle and inferior rectal veins. This venous drainage explains why tumors of the lower rectum and anal canal can directly cause pulmonary metastases without hepatic metastases [28,29].

25.3.1.4 Mesorectum

The mesorectum is a subperitoneal compartment constituting the mesentery of the rectum and extending from the peritoneal reflection to the puborectalis muscle. Mesorectum contains the superior rectal artery and its branches, the superior rectal vein and its tributaries, the lymphatic vessels, the nodes, branches of the inferior mesenteric plexus, and loose adipose connective tissue and septa [1,2]. The connective tissue becomes thinner inferiorly and vanishing beyond the puborectalis muscle. At this level, beyond the puborectalis muscle, there are virtually no more lymphatic vessels or lymph nodes [27].

The mesorectal fascia, derived from the visceral peritoneum, covers the mesorectum. The mesorectal fascia is the visceral layer of the endopelvic fascia. The fascia posteriorly binds the mesorectum to the presacral fascia, that is the parietal layer of the endopelvic fascia covering the sacrum, the coccyx, and the middle sacral artery and presacral veins; superiorly, it blends with the connective tissue bounding the sigmoid mesentery;

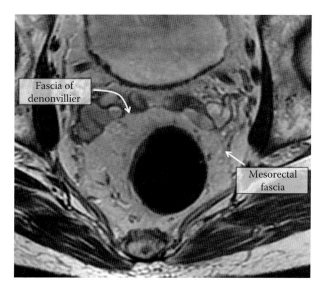

FIGURE 25.14

Axial T_2-weighted MR image showing the mesorectal fascia that covers the mesorectum. In males, the anterior fascia is known as the rectovesical fascia of Denonvillier separating the mesorectum from the seminal vesicles, and in females it forms the fascia of the rectovaginal septum.

laterally, it extends around the rectum and the mesorectum and becomes continuous with a denser condensation of fascia anteriorly. In males, the anterior fascia is known as the rectovesical fascia of Denonvillier separating the mesorectum from the seminal vesicles, and in females it forms the fascia of the rectovaginal septum (Figure 25.14). Below the fourth sacral vertebra (S4) the mesorectal fascia condenses with the rectosacral fascia, knows as Waldeyer's fascia, which is a fascial reflection running anteroinferiorly to the presacral fascia [1,26].

25.3.2 MR in Rectal Cancer

25.3.2.1 Clinical Aspects and Indications

In industrialized countries rectal adenocarcinoma is one of the most common tumors, accounting for approximately 40 cases in every 100,000 individuals with a slight male predilection, and a prevalence increasing steadily after the fifth decade [30]. Other rectal tumors are relatively rare and include carcinoid tumors (0.1% of cases), lymphoma (1.3%), and gastrointestinal stromal tumors (<1%) [31]. In the preoperative assessment of rectal tumors, MRI plays a crucial role, especially because endoscopy and ultrasonography (US) do not adequately depict the breadth of a tumor's coverage or lymph node involvement, which are important prognostic features for the preoperative management of rectal carcinoma [32–36].

Total mesorectal excision (TME) with or without neoadjuvant CRT represents the principal treatment for rectal cancer. In the past few decades, diffusion of TME has resulted in a dramatic decline in the prevalence of local recurrence from 38% to less than 10% [37]. This surgical technique entails en bloc resection of the primary tumor and the mesorectum by means of dissection along the mesorectal fascia plane representing the circumferential resection margin (CRM) [37]. The presence of a tumor or malignant node within 1 mm of the CRM remains an important predisposing factor for local recurrence [38].

The main roles of MRI in the preoperative staging for rectal cancer are

1. To stratify the patients who might benefit from a neoadjuvant CRT (advanced T3 and T4 tumors) [39,40].

2. To avoid overtreatment and toxicity due to extended course CRT, thus identifying patients who can benefit from local therapy (e.g., transanal excision and transanal endoscopic microsurgery), which are usually patients with stage T1 cancers [41–43] and those patients requiring TME (mainly those with stage T2 and early stage T3 tumors) [44].

3. To achieve a detailed surgical planning using a proforma-based reporting [38].

4. To assess tumor downstaging during and after CRT [39,40].

Stratification of patients who need short- or long-course neoadjuvant CRT is primarily based on evaluation of tumor stage (T), nodal (N) staging, depth of tumor invasion outside the muscularis propria (early vs. advanced stage T3 tumors), and the relationship of the tumor to the potential CRM. Recent studies have shown that high-resolution MRI enables the acquisition of reproducible data about the relationship of the tumor to the CRM, and the depth of tumor invasion outside the muscularis propria with a high specificity (92%) for predicting a negative CRM [45–47].

25.3.2.2 Technical Aspects

MR examinations usually require a high field magnet (performed using at least a 1.5 T scanner) with phased-array surface coils and patient in supine position [31]. Rectal cleansing is suggested to limit image misinterpretation, because residual stool may alter the distance between the tumor and the mesorectal fascia and may leave residues, while in some cases use of a small amount of endorectally injected, US gel (60–120 mL, amount depending on tumor location) allows better depiction of polypoid lesions, small rectal lesions (>2–3 cm), or residual tumor after neoadjuvant CRT [38,48]. However, the use of the endorectal array or an amount of endorectal gel, greater than the amount indicated above, is not recommended (especially in the evaluation of low rectal tumors) because rectal distension appears to alter the measurement of the real distance between the tumor and the CRM. Additionally, it can excessively compress the perirectal fat obscuring lymph nodes and infiltrated vessels [31,38]. The patient is positioned supine and coverage of the pelvis is recommended from the origin of the inferior mesenteric artery, located at the L3 vertebral level, to the cutaneous plane of perineum [31,38,48].

Imaging protocol includes high-resolution T_2-weighted fast spin-echo sequences acquired in the sagittal plane, in an oblique coronal plane oriented parallel to the major tumoral axis, and in an oblique axial plane oriented perpendicular to the major tumoral axis (Figure 25.15). This multiplanar approach improves accuracy in the assessment of tumor stage because the relationship of the tumor to the CRM and other structures can be confirmed in three planes. This appears to be especially helpful in case of tortuous rectal tumors [31].

High spatial resolution axial images acquired orthogonal to the major axis of the tumor allow a superior evaluation of mesorectal fascia and CRM. In particular, incorrect plane obliquity in the axial plane leads to blurring of the muscularis propria on the anterior aspect or a pseudo-spiculated appearance of the rectal wall that may lead to overstaging of the tumor (Figure 25.16) [48,49].

High-resolution sagittal images provide additional information about CRM and permit better evaluation of the positioning of tumor to the peritoneal reflection, which is a crucial prognostic factor because invasion of the peritoneal reflection upgrades the tumor to a stage T4 lesion (Figure 25.17) [38].

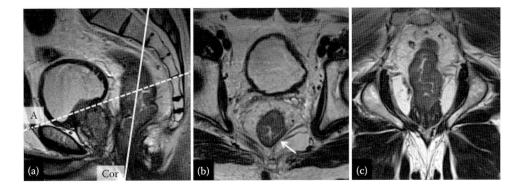

FIGURE 25.15
High-resolution T_2-weighted fastspin-echo sequences should be acquired primarily on sagittal plane (a). On the basis of the T_2-weighted sagittal image, an oblique axial plane (b) oriented perpendicular to the major tumor axis and an oblique coronal plane (c) oriented parallel to the major tumor axis can be obtained. T_2-weighted oblique axial plane (b) permits to obtain a good delineation of the outer profile of the muscularis propria (*arrow*).

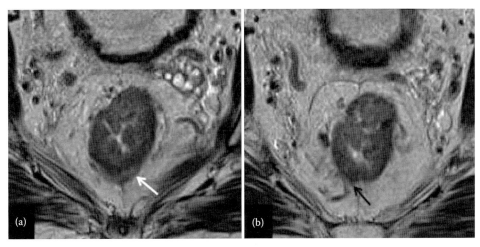

FIGURE 25.16
Axial T_2-weighted MR image (a), obtained on an incorrect plane, shows blurring (*white arrow*) of the muscularis propria that may lead to tumor overstaging. Oblique axial T_2-weighted MR image (b) well oriented perpendicular to the tumor axis showing a better delineation of the outer profile of the muscularis propria (*black arrow*) without blurring artifact.

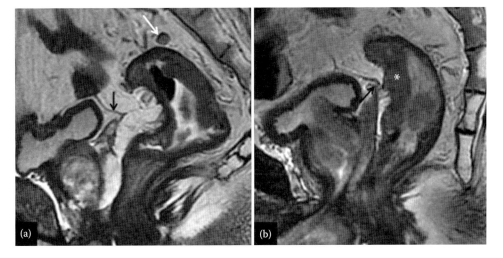

FIGURE 25.17
Sagittal T_2-weighted MR image (a) permits a perfect visualization of the peritoneal reflection (*black arrow*) in a patient with a T3b rectal tumor of the rectal upper third with a mesorectal lymphadenopathy. Sagittal T_2-weighted MR scan is fundamental to better depict peritoneal involvement (*black arrow*) in case of cancer of the rectal upper third (*asterisk*) as shown in image (b).

High-resolution coronal images can be acquired parallel to the major axis of rectal tumor (as reported in several previous articles), but the straight coronal plane seems to provide a superior definition of the anatomy of the anal sphincter and of the relationship of the tumor to the sphincter, helping also in the assessment of tumor relationship to the pelvic sidewall and peritoneal reflection, as well as the evaluation of pelvic sidewall nodal morphology [49].

High-resolution multiplanar imaging also allows for the morphological and dimensional evaluation of all mesorectal nodes and pelvic nodes, with the possibility of encountering them to define the N parameter [50].

DWI assists localization of tumor and lymph nodes increasing the contrast between these structures and the surrounding normal tissue. However, DWI has limited value for characterizing nodes due to the overlapping between apparent diffusion coefficient (ADC) values of malignant and that one of hyperplastic benign nodes [51]. In a recent study, low ADC values correlated with tumor biological aggressiveness, resulting higher in neoplasm limited to bowel wall and lower in staged T3N2 cancer, with involved mesorectal fascia and in tumors with moderate or poor differentiation [52]. Moreover, DWI can be potentially useful in predicting responses to chemotherapy on the basis of ADC values. Some studies reported also that a low ADC value reflects a good responsiveness to CRT because highly cellular tumors seem to respond better than necrotic and hypovascularized lesions [53–56]. Consequently, ADC values seem to be

an early MR biomarker of tumor response to chemotherapy–radiation therapy: increasing of ACD values, due to cell death, precedes volumetric reduction in tumor size [53–56]. Postgadolinium GRE T_1 fat-saturated sequences are not mandatory in preoperative MR staging of rectal cancer, but recent studies have demonstrated the role of these sequences in detecting residual tumor after chemotherapy or tumor recurrence. Additional dynamic perfusional acquisition obtained with MR provides K^{trans} values comparable to those obtained with perfusional CT [57] and will actually represent a biological imaging marker to predict tumor response to chemotherapy before starting the treatment (Tables 25.1 and 25.2) [58].

25.3.2.3 Pretreatment Planning

Assessment of the primary tumor with MRI consists of the evaluation of the following parameters [59–61]:

1. Morphology and distance from inferior margin of the tumor to the transitional skin of the anal verge
2. T staging
3. Anal sphincter complex, puborectalis, and levator ani muscles
4. Nodal staging
5. Extramural vascular invasion
6. CRM: safety of surgical plane (TME plane or more extensive resection)

TABLE 25.1

Scheme of the MRI Standardized Protocol for Primary Staging of Rectal Cancer According to Our Experience on a 3.0 Tesla System

	Sequence	W	Plane	TE (ms)	TR (ms)	FA	Matrix	NEX	Thick (mm)
Mandatory	2DFRFSE	T2	Sag	119.5	4172	—	512 × 512	2	4
	2DFRFSE	T2	Ax	122.3	3056		512 × 512	2–4	3
	2DFRFSE	T2	Cor	111.4	2086		512 × 512	2–4	4
Optional	3D GRE	T1	Ax	3.3	13.6	15	512 × 512	2	2
Optional	SSEPI	DW	Ax	81.4	4400	—	256 × 256	2	4
Optional	Dyn3D FSPGR	T1	Ax	3.3	13.6	15	512 × 512	2	2

TABLE 25.2

Scheme of the MRI Protocol for Restaging of Rectal Cancer According to Our Experience on a 1.5 and 3.0 Tesla System

	Sequence	W	Plane	TE (ms)	TR (ms)	FA	Matrix	NEX	Thick (mm)
Mandatory	2DFRFSE	T2	Sag	119.5	4172	—	512 × 512	2	4
	2DFRFSE	T2	Ax	122.3	3056		512 × 512	2–4	3
	2DFRFSE	T2	Cor	111.4	2086		512 × 512	2–4	4
Mandatory	3D GRE	T1	Ax	3.3	13.6	15	512 × 512	2	2
Mandatory	SSEPI	DW	Ax	81.4	4400	—	256 × 256	2	4
Optional	Dyn3D FSPGR	T1	Ax	3.3	13.6	15	512 × 512	2	2

25.3.2.3.1 Morphologic Features and Localization of Rectal Cancer

Rectal cancer can be morphologically described as (Figure 25.18)

- Annular or semiannular lesions with a wide and large margin of infiltration.
- Polypoid lesions protruding into the lumen through a stalk and presents an invasive front into perirectal fat that are usually shorter if compared to annular or semiannular lesions [61].

- Ulcerating lesions, typically presenting as deep ulcers from which malignant cells can infiltrate the mesorectum [61].

Localization of the rectal cancer is expressed by the distance measured from the anal verge to the most caudal aspect of the raised rolled edge of the tumor (Figure 25.19). Traditional division of the rectum into thirds is useful in defining the location of the tumor influencing surgical management (Figure 25.20) [59,61]:

Upper: The inferior margin of the tumor is more than 10 cm from the anal verge. At this level the anterior wall of the rectum is covered by the peritoneal reflection and careful assessment of the peritoneal invasion or perforation must be performed to alert the surgeon about the risk of tumor spillage. Nevertheless, the peritoneal reflection attachment occurs at a variable height, particularly in women.

Middle: The inferior margin of the tumor is located between 5 cm and 10 cm from the anal verge. This segment of the rectum lies below the peritoneal reflection and is completely covered by mesorectum and mesorectal fascia; this is the plane of dissection in TME surgery.

Lower: The inferior margin of the tumor is less than 5 cm from the anal verge. At this level, the mesorectum becomes thinner and, below the levator ani plane, the puborectalis muscle marks the anorectal junction corresponding to the surgical anal canal. Low rectal tumor, developing

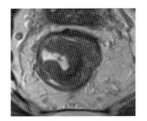

Semi-annular or annular lesion: Characterized by a wide front of invasion

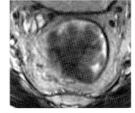

Polypoid lesion: Characterized by a small front of invasion through the stalk

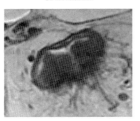

Ulcerated lesion: Infiltration occurs mainly through the deep ulceration

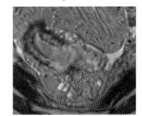

Mucinous lesion: Showing fluid collection containing mucin, appearing hyperintense in T_2-weighted images, histological type with high biological aggressiveness and poor prognosis

FIGURE 25.18
Morphologic features of rectal cancer.

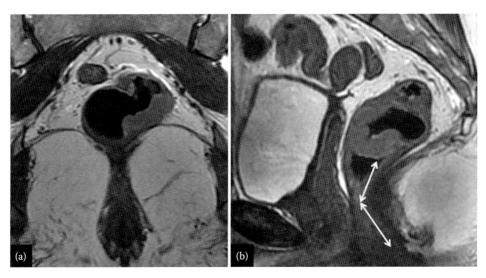

FIGURE 25.19
Tumor distance from the anal verge is measured from the most caudal aspect of the raised rolled edge of the tumor to the anal verge.

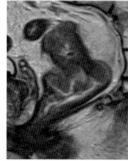

Upper third
The inferior margin of the tumor is
>10 cm from the anal verge

Higher risk of peritoneal
infiltration

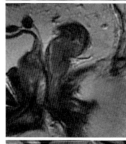

Middle third
The inferior margin of the tumor is
located 5–10 cm from the anal verge

Completly rounded by the
mesorectal fascia

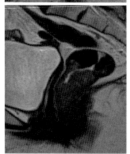

Lower third
The inferior margin of the tumor is
<5 cm from the anal verge

Higher risk of sphincters
infiltration

FIGURE 25.20
Localization of the rectal tumor can also be based on the traditional division of the rectum into thirds.

below the puborectalis sling, can invade the anal muscularis propria, corresponding to the internal sphincter and the external sphincter complex, composed of the most inferior part of the levator ani muscle, the puborectalis muscle, and the external sphincter muscle, separated from the internal sphincter by a thin groove of fat, forming the intersphincteric plane.

25.3.2.3.2 T Staging

As well demonstrated in MERCURY trial, MRI appears to be an accurate imaging modality for rectal cancer staging, giving reproducible results across institutions [45]. Following the American Joint Committee on Cancer (tumor-node-metastasis [TNM]) guidelines), MRI criteria for the staging of primary rectal tumors have been developed (Table 25.3) [38,62]. The real goal of these MRI criteria in the assessment of T staging is not to differentiate T2 tumors from early T3 tumors (which have the same good prognosis of T2), but to stratify T3 tumors, whose extramural spread measures on the order of

TABLE 25.3

TNM Guidelines for the Staging of Rectal Cancer

Tis	*In situ* carcinoma
T1	Invasion of submucosa
T2	Invasion of muscularis propriae
T3	Invasion through the muscular layer into sub-sierosa or into perirectal tissues
T3a	Tumor extends <5 mm beyond the muscularis propria*
T3b	Tumor extends 5–10 mm beyond the muscularis propria*
T3c	Tumor extends >10 mm beyond the muscularis propria*
T3d	Tumor extends >15 mm beyond the muscularis propria*
T4a	Tumor penetrates the visceral peritoneum
T4b	Tumor directly invades or is adherent to other organs or structures

Source: Staging system adapted from the American Joint Committee on Cancer (AJCC) Cancer Staging Manual 7th edition. New York, Springer, 2010.

Note: (*) The clinical staging classification according to Smith and Brown (*Acta. Oncol.*, 2008) differs slightly from the AJCC staging: T3a < 1 mm; T3b ≥ 1–5 mm; T3c > 5–15 mm, T3d > 15 mm.

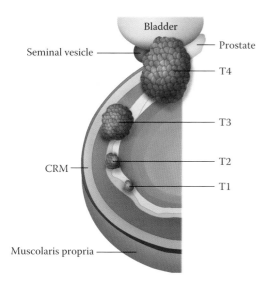

FIGURE 25.21
New MRI criteria in the assessment of T staging to stratify T3 tumors are based on measuring in millimeters the depth of extramural spread. (Courtesy of Dr. M Ciolina.)

TABLE 25.4

Stratification of Patients with Rectal Cancer in Three Risk Categories Considering the Probability of Recurrent Disease and Poor Outcome [62,63]

Risk Features	Low	Moderate	High
Extramural spread	≤5 mm	>5 mm	>5 mm
Nodal status	N0	N1–2	N2
CRM	Not at risk	Not at risk	At risk
Position of tumor	High	Low or high	Low
EMVI	Absent	Present	Present

CRM, circumferential resection margin; EMVI, extramural vascular invasion.

millimeters (Figure 25.21) and consequently subclassifying the population in three risk categories: low risk, moderate risk, and high risk of recurrent disease and poor outcome (Table 25.4) [38,62,63]. The 5-year survival rate is strongly dependent on the depth of tumor invasion outside the muscularis propria. According to the survival rate, early stage T3 tumors (<5 mm invasion) and stage T2 tumors may be grouped together into the low-risk category with a 5-year survival rate of approximately 85% and a total local recurrence rate of 3% [63,64] and may benefit only from surgery. However, tumors in which the depth of invasion outside the muscularis propria exceeds 5 mm, the 5-year survival rate, independently of nodal involvement, decreases to 54% [63–65]. These patients should be grouped in the moderate and high-risk category and are candidates for neoadjuvant CRT before surgery.

For the evaluation of the T staging, radiologists have to remember the following recommendations:

1. Planes of acquisition must be strictly perpendicular to the tumor to avoid blurring of the muscularis propria leading to overstaging.

2. The depth of extramural spread represents as an independent prognostic factor and must be measured in millimeters beyond the muscularis propria to differentiate patients with low risk of local recurrent disease (<5 mm) from those with higher risk of recurrence (>5 mm) (Figure 25.22).

3. Tumor infiltration into the mesorectum appears as a thicker tissue, with intermediate signal intensity and broad-based bulge or nodular appearance (Figure 25.23).

4. Although it is not always possible to differentiate it from real tumoral infiltration, tumoral desmoplastic reaction is often present and is a typical inflammatory reaction of the perirectal fat, seen as fine spicules along the tumor border with low signal intensity in T_2-weighted sequences.

5. In upper rectal tumor, peritoneal reflection must be evaluated using sagittal T_2-weighted images and appears as a low-signal-intensity linear structure from the posterior aspect of the dome of the bladder to the ventral aspect of the rectum (Figure 25.17).

6. Peritoneal involvement must be considered a T4a, but does not implicate a CRM involvement.

25.3.2.3.3 Relationship with the Anal Sphincter

Because of the more complex anatomic relationship, the rates of positive resection margins and rate of local recurrence are usually higher for low rectal tumor that consequently have a poor prognosis [66]. MR plays a crucial role in low rectal cancer in deciding if patients have a limited involvement of anal sphincter complex requiring only a partial sphincter resection with coloanal reconstruction, or present a more extensive involvement that must be treated with CRT or sphincter amputation [67–69]. For the evaluation of the low rectal cancer T staging, the radiologist needs to remember the following recommendations:

1. Use high spatial resolution T_2-weighted fast spin-echo coronal and axial images to better depict the tumor relationship with the levator and puborectal muscles, sphincter complex, and intersphincteric plane.

2. The puborectalis sling corresponds to the superior margin of the anal canal; if, on sagittal and coronal images, the inferior margin of the

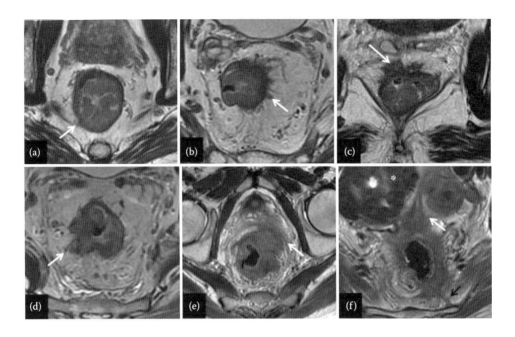

FIGURE 25.22
Axial T_2-weighted MR image (a) showing a diffuse circumferential thickening of rectal walls with minimal infiltration (<5 mm, T3a) of peri-rectal fat on the posterior right-side aspect (*arrow*); axial T_2-weighted MR image (b) showing a circumferential rectal cancer infiltrating the left rectal wall (*arrow*) with a depth of 5–10 mm (T3b). Axial T_2-weighted MR image (c) shows a low rectal cancer infiltrating anteriorly the mesorectum (*arrow*) for more than 10 mm (T3c). Axial T_2-weighted MR image (d) shows rectal cancer with pseudonodular infiltration (*arrow*) extending for >15 mm into mesorectal fat (T3d). Axial T_2-weighted MR image shows voluminous rectal cancer with infiltration of mesorectal fascia (*arrow*) and consequently the circumferential resection margin (CRM) (T4a). Axial T_2-weighted image shows a rectal cancer invading the peritoneum and the uterine posterior wall (T4b) containing also a large leiomyoma (*asterisk*).

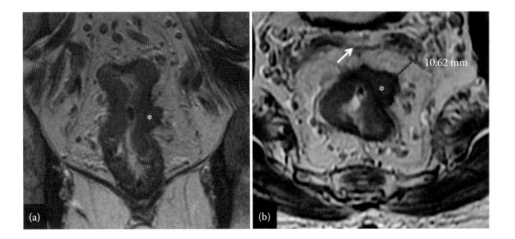

FIGURE 25.23
Coronal T_2-weighted MR image (a) shows a tumor infiltration into the mesorectum appearing as a broad-based bulge tissue (*asterisk*) or as a tissue with nodular appearance with intermediate signal intensity. A distance >1 mm between the tumor and the CRM correlates with a decrease rate of local recurrence; axial T_2-weighted MR image (b) shows the distance between the tumor and the mesorectal fascia (*arrow*).

tumor lies above puborectalis sling, sphincter involvement can be easily excluded.

3. If, on sagittal and coronal images, the inferior margin of the tumor lies above the puborectalis sling, low rectal cancer T staging by Shihab et al. [70,71] permits differentiation of tumors that (a) partially infiltrates the muscularis propria,

remaining within 1 mm to the outer border of the inner sphincter (Stage 1), (b) involves the full thickness of the inner sphincter (Stage 2), (c) the intersphincteric planes (Stage 3), or (d) invades the external sphincteric complex (external sphincter, levator ani, and puborectalis, Stage 4) (Figure 25.24).

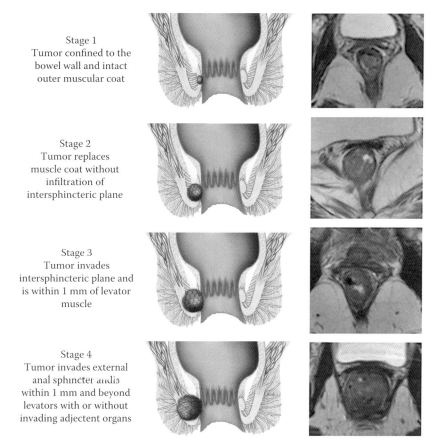

FIGURE 25.24
Scheme for low rectal cancer staging. (Courtesy of Dr. M Ciolina.)

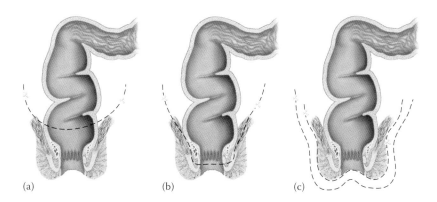

FIGURE 25.25
Types of surgical interventions for low rectal cancer. Low anterior resection (a). Low anterior resection with intersphincteric resection (b). Conventional abdominoperineal resection (APER) (c, *blue line*). Extralevator APER (c, *red line*). (Courtesy of Dr. M Ciolina.)

This MRI staging proposed by Shihab et al. for low rectal cancer enables one to determine if the plane of excision required is safe from tumor with a positive and negative predictive values of 57% and 96%, respectively [70]. Zhang et al. using a 3T MRI reach an accuracy of 96.9% to predict sphincter-sparing surgery [72]. However, accuracy of the staging strictly depends on the radiologist's expertize with a moderate possibility of overstaging or understaging [70]. Information given from this staging allows surgeons to choose one of the three major surgical strategies for low rectal cancer (Figure 25.25):

1. Low anterior resection (AR) for low rectal tumor without sphincter infiltration: en bloc resection of the rectum and of the mesorectum (i.e., TME) to the level of the pelvic floor.

2. Low AR with intersphincteric resection for low rectal tumor extending into the intersphincteric plane. To obtain tumor-free margins, the tumor should remain confined into the muscularis propria, laying within 1 mm to the outer border of the internal sphincter (Stage 1).

3. *Extralevator abdominoperineal resection (APER)*: it differs from conventional APER because the entire levator muscle is resected en bloc with the lower rectum and anal canal. This surgical intervention guarantees a low rate of positive resection margins and a better outcome in comparison with conventional APER [73]. This procedure is performed when the tumor has involved the full thickness of muscularis propria, the intersphincteric space, or extends into or beyond the levator muscles (Stages 2, 3, or 4).

25.3.2.3.4 N Staging

Preoperative assessment of lymph node status in patients with rectal cancer is important because it influences the prognosis of the patient. In particular, metastatic lymph nodes near the mesorectal fascia (the conventional surgical CRM) and malignant nodes outside the mesorectal fascia may remain unresected despite apparently clear surgical resection margins and increasing the risk of recurrence [50].

1. The following nodal groups must be assessed: mesorectal, superior rectal, and inferior mesenteric; internal, external, and common iliac; retroperitoneal; and superficial inguinal (Figure 25.26).

2. Lymph nodes' location and size will inform the radiotherapists to use an adequate radiation therapy field or the surgeons to perform an extended resection to extra-mesorectal lymph nodes. In particular pelvic sidewall lymphadenopathies are associated with a significant

worsening of the disease-free survival that can improve with the use of preoperative therapy. The MERCURY study group found patients with suspicious pelvic sidewall nodes on MRI had a 5-year disease-free survival of only 42% in comparison to patients without suspicious pelvic sidewall nodes (70%) [74].

3. Nodal size criterion (i.e., lymph nodes >5 mm) has a limited role to differentiate malignant from nonmalignant nodes; 30%–50% lymph nodal metastases from rectal cancer occur in nodes ranging between 2 mm and 5 mm [75,76].

4. Features that are suggestive of malignancy include irregular or spiculated nodal margins and heterogeneous signal intensity (mottled intensity pattern) (Figure 25.27). An interesting result of the study conducted by Brown et al. was the high specificity (97%) and moderate sensitivity (85%) in rectal lymph nodes assessment evaluating the border characteristics and the mixed signal intensity on MR images [77]. There are no known signal contrast enhancement properties that can reliably distinguish metastatic from reactive nodes [60,78].

5. DWI can facilitate lymph node detection, permitting the detection of 6% more nodes than T_2-weighted MRI but alone it is not reliable for differentiating between benign and malignant lymph nodes (Figure 25.28) [79]. ADC as a stand alone cannot identify metastatic nodes and is even worse than MRI [80].

6. If a malignant node is located less than 1 mm from the CRM, involvement of CRM must be suspected (Figure 25.28) [81].

25.3.2.3.5 Extramural Vascular Invasion

Vascular invasion does not influence clinical management but is an independent risk factor of local and distant recurrence with a bad prognostic value. It is present

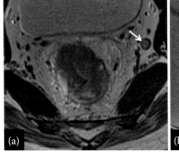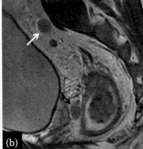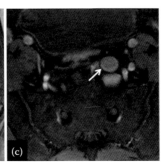

FIGURE 25.26
Axial T_2-weighted (a) and sagittal T_2-weighted images show a rectal cancer with infiltration of perirectal fat and lymphadenopathies along the left internal iliac vessels (a, *arrow*), along the superior rectal arteries (b, *white arrow*), and in the mesorectum (b, *black arrow*). Axial T_1-weighted fat-saturated postgadolinium image (c) shows also the presence of a metastatic lymphadenopathy along the left common iliac artery (*arrow*).

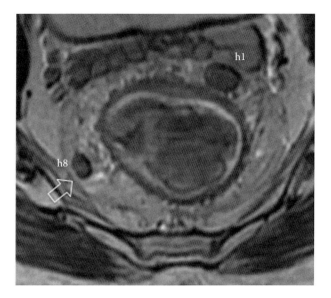

FIGURE 25.27
Axial T_2-weighted image shows two lymphadenopathies (8 o'clock and 1 o'clock) presenting features suggestive of malignancy including irregular or spiculated nodal margins and heterogeneous signal intensity with a mottled intensity pattern due to hypointense and hyperintense spots. When the metastatic lymph node is ≤1 mm from the mesorectal fascia (*empty arrow*), the CRM must be considered involved.

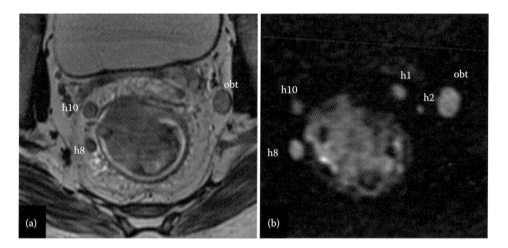

FIGURE 25.28
DWI due to its high contrast resolution can facilitate lymph node detection; axial T_2-weighted image (a) permits detection of an obturator lymphadenopathy (*obt*) and of mainly two mesorectal lymphadenopathies (one at 10 o'clock and one at 8 o'clock). DW imaging (b) improves number of lymph nodes detected adding visualization of a lymph node at 1 o'clock and at 2 o'clock.

at hystopathological examination in 10%–54% of patients with rectal cancer [82]. MR sensitivity and specificity in assessment of extramural vascular invasion (EMVI) are 62% and 88%, respectively [60].

EMVI should always be evaluated at imaging in case of T3 stage tumors and always suspected in case of vessels lying strictly close to the tumor. Generally, small-vessel involvement is difficult to assess, while involvement of larger vessels is suggested by tortuous and dilated vessels which may be recognized as [38]

1. Slightly hyperintense solid tissue within dilated and tortuous vessels

2. Abnormal signal intensity within vessels which remain normal in size

3. Vessels dilated with disrupting borders presenting pseudonodular morphology (Figure 25.29)

25.3.2.3.6 When CRM Must Be Considered Infiltrated

The relationship between the tumor and mesorectal fascia is critical for surgical planning and represents an important prognostic factor [83]. Taylor et al. reported a 5-year overall survival rate and a 5-year disease survival rate, respectively, of 62.2% and 67.2% in patients with MRI-clear CRM compared with 42.2% and 47.3% in patients with MRI-involved CRM [81]. A distance

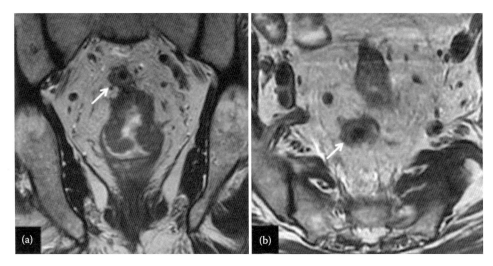

FIGURE 25.29
Coronal (a) and axial (b) T_2-weighted images of an annular rectal cancer, showing a nodular soft tissue with intermediate signal intensity (arrow) that surrounds an extramural mesorectal vessel appearing as a tubular hypointense structure due to flow void phenomenon.

>1 mm between the tumor and the CRM correlates with a decreased local recurrence. When the tumor is ≤1 mm from the mesorectal fascia, the CRM must be considered involved.

The distance to the mesorectal fascia is measured from all these structures: (a) the tumor margin, (b) a tumor deposit in the mesorectum, (c) tumor thrombus within a vessel, or (d) a malignant node.

The reported accuracy of MR images for prediction of the CRM in comparison with histologic findings ranges between 92% and 100% [47,84] with a high interobserver agreement [47].

Anteriorly, the mesorectal fat can be thin, and the rectum can be close to the CRM. In that case evaluation of CRM may be challenging for radiologists.

25.3.2.3.7 Relationship with Peritoneal Reflection and Pelvic Organs

The peritoneal reflection is best identified on sagittal or coronal high-resolution T_2-weighted images. If the tumor invades the peritoneal reflection it must be considered a stage T4a lesion.

Anteriorly, the pelvic structures most commonly involved by primary rectal cancer are the uterus, vagina, prostate gland, and seminal vesicles (Figure 25.22F).

Latero-posteriorly, other structures along the pelvic sidewall near the rectum include the common, external, and internal iliac vessels; the ureters; the pyriformis muscle; the internal obturator muscle; and, in the region of the sciatic foramen, exiting sacral nerve roots. These structures are covered by the parietal layer of the endopelvic fascia and are fused with the visceral layer of the endopelvic fascia (mesorectal fascia) at the level of the upper rectum. Posteriorly, the parietal pelvic fascia fuses with the presacral fascia that remains, separating

from the mesorectal fascia by a potential retrorectal space, which forms the plane of dissection in TME. At the level of the middle third of the rectum, the visceral and parietal layers of the endopelvic fascia can generally be identified as hypointense linear structures; they are separate and distinct on MR images. However, the same structures may be indistinguishable at the levels of the lower and upper rectum, and involvement of the mesorectal fascia in these regions may correspond to pelvic sidewall involvement.

The assessment of the relationship between tumor and the pelvic sidewall is best made on coronal or sagittal high-resolution images. The tumor may be unresectable, if it extends into the proximal sacrum or involves nerve root involvement above the S2 vertebral level [38].

25.3.2.4 Post-CRT Assessment

CRT allows one to obtain a downstaging of the rectal tumor, improving resectability and sphincter preservation and increasing the number of patients with a low risk of local and distant recurrence and with a best prognosis in term of survival. Complete tumor response rate after CRT was 10%–20% in several studies [39,85].

Main indications for CRT are

1. Locally advanced T3 rectal tumor with >5 mm of spread beyond the muscularis propria
2. Tumor with a distance ≤1 mm from the CRM
3. Threatened of involved anal sphincter
4. Nodal involvement
5. EMVI

25.3.2.4.1 MR Findings

- Areas of fibrosis have very low signal intensity that is similar to signal intensity of muscularis propria, while areas of viable residual tumor have intermediate signal intensity similar to that of baseline tumor (Figure 25.30) [86].

- Desmoplastic reaction is a sort of *reactive fibrosis* usually seen as low intensity radial spicules or strands departing from the tumor in the perirectal fat. Misinterpretation of desmoplastic spicules for residual tumor results in overstaging (Figure 25.31) [86].

- In some cases, necrosis of the tumor can lead to mucinous degeneration, appearing as pools of hyperintensity on T_2-weighted sequences within the residual tumor [86].

- In cases of primary mucinous-type tumors, the presence of hyperintense pools on T_2-weighted

scans can be a sign of nonresponsiveness to CRT with a poorer prognosis and increased risk of local recurrence [87].

- Before the CRT the non-neoplastic mucosa near the tumor appears redundant bulging into the lumen with a pseudotumor appearance due to reactive inflammation. After CRT, this pseudotumoral effect of the healthy mucosa adjacent to the tumor increases, acquiring intermediate signal intensity due to the submucosal edema (Figure 25.32). These pitfalls lead to potentially false misinterpretation and can be avoided by comparing the pre- and posttreatment scans, paying attention to the baseline positioning of the tumor and demonstrating that the portion of the rectal wall circumference showing a pseudotumoral appearance after CRT has not been involved by tumor before CRT [86].

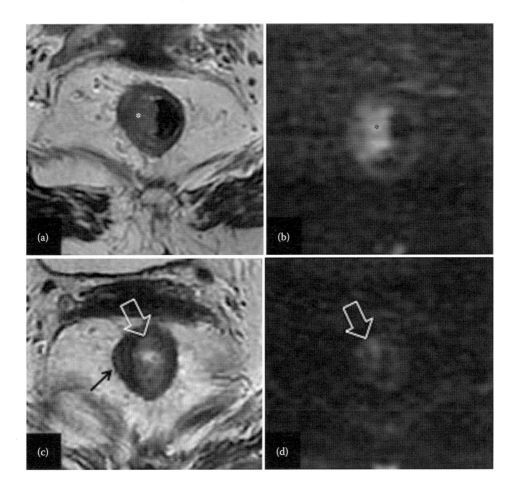

FIGURE 25.30
Axial T_2-weighted MR image (a) shows an eccentric wall thickening (*white asterisk*) corresponding to rectal cancer and presenting impeded diffusion of water molecules in DWI (b, *black asterisk*). After CRT, axial T_2-weighted image (c) shows diffuse hypointensity (*black arrow*) of the right rectal wall due to fibrotic change. DWI after-CRT (d) allows to detect the small tumor residue (*empty arrow*), a minimal hyperintensity caused by impeded diffusion, and an intermediate signal intensity on T_2-weighted image (c).

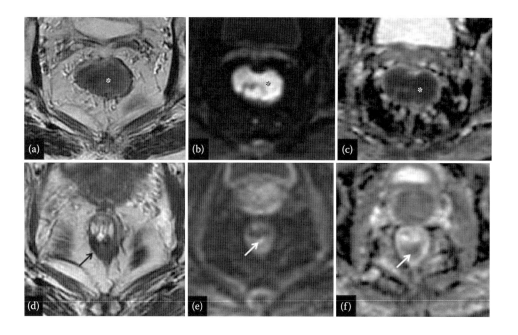

FIGURE 25.31
MR images of a polypoid rectal cancer before-CRT showing intermediate signal intensity on axial T_2-weighted (a, *asterisk*) and real impeded diffusion on DWI and ADC (b, c, *asterisks*). MR images of the same lesion after-CRT show the presence of low intensity radial spicules (d, *black arrow*) due to *reactive fibrosis* departing from the treated tumor in the perirectal fat. DWI (e) and ADC map (f) show only a focal residual area (*white arrows*) of impeded diffusion appearing hyperintense in DWI and hypointense in ADC map.

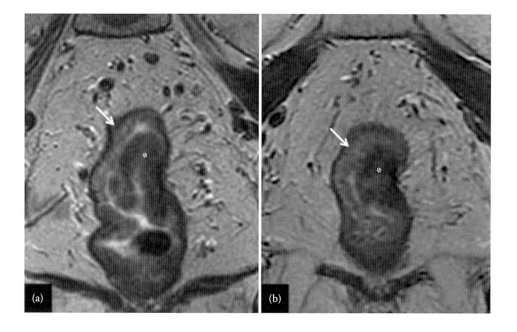

FIGURE 25.32
Before the CRT (coronal T_2-weighted image, a) the non-neoplastic mucosa (*arrow*) near the tumor (*asterisk*) appears redundant bulging into the lumen with a pseudotumor appearance due to reactive inflammation. After CRT (coronal T_2-weighted image, b), this pseudotumoral effect of the healthy mucosa (*arrow*) adjacent to the tumor (*asterisk*) increases acquiring intermediate signal intensity due to the submucosal edema.

25.3.2.4.2 Assessment of Tumor Size after CRT

Assessment of tumor length is always a means to quantify tumor response but lacks interobserver reproducibility. However, volumetry combined with MR morphologic changes seems to correlate well with pathologic tumor response in terms of downstaging and tumor regression grade (TRG) [88–90]. Nevertheless, volumetry needs dedicated software to sum every cross-sectional volumes manually traced on axial high spatial resolution T_2-weighted MR images. Many authors consider a tumor volume reduction of 70% or more after CRT to be associated with a good TRG at pathologic examination and higher disease-free survival [91,92]. However, on morphologic post-CRT MR images, measurement of cancer volumes is not simple. In particular, there is great difficulty in defining which of the fibrotic areas are still suspicious for tumor and therefore should be included in the volume measurements and which should not. Recently, Curvo-Semedo et al. have found that combining DWI with tumor volumetry evaluated on morphological images (T_2- weighted sequences) the quality of MRI response assessment increases, with an overall accuracy of 88% [93].

Height of treated tumor from the anal verge must be assessed and compared with that on baseline pretreatment scans.

25.3.2.4.3 T Staging

The reported overall accuracy of MRI in T staging assessment of non-irradiated rectal cancer is 71%–91% (mean − 85%). After CRT, T staging assessment is usually difficult due to the post-therapeutic changes of the irradiated tissues; overall accuracy of MRI decreases to 50% after treatment [94].

As for baseline scans, depth of maximum extramural spread after CRT must be recorded to stratify patients with a tumor extending less than 1 mm beyond the muscularis propria (prognostically identical to T2) and tumor with major infiltration. After CRT low-signal-intensity areas usually represent fibrotic scar but differentiating them from residual viable tumor remains difficult.

Added DW MRI has a consolidated role in differentiating viable tumor from fibrosis [95,96]. Fibrosis typically has a low cellular density, and consequently shows low signal intensity on high b-value DW images, while residual tumors have a relatively high cellular density resulting in high signal intensity on high b-values (Figure 25.31 and 25.33) [97,98]. T_2-weighted imaging combined with DWI in assessment of rectal cancer after CRT has a sensitivity of 93%–95% and a specificity of 95%–100% in comparison to T_2-weighted imaging alone (sensitivity, 82%–84%; specificity, 85%–90%) [99].

Moreover, considering the significant individual variation of response to CRT (9%–25% of patients show complete pathologic response, 54%–75% of patients show downstaging of tumor, and others would show no response) [100,101], DW MRI seems to be a promising noninvasive technique for predicting and monitoring early therapeutic responses in patients with rectal carcinoma as an imaging biomarker [102]. Early detection and assessment may be beneficial to identify nonresponders before treatment or during the first weeks of CRT, so that treatment may be intensified or modified [102]. Recent studies have demonstrated that low pretherapy mean ADC and early increases in mean tumor ADC during CRT in rectal carcinoma correlate with a

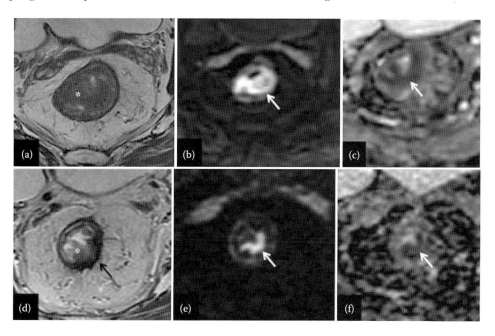

FIGURE 25.33

MR images of an annular rectal cancer before-CRT showing intermediate signal intensity on axial T_2-weighted (a, *asterisk*), hyperintensity on DWI (b, *arrow*) due to restriction of water molecules, confirmed by the hypointensity on ADC (c, *arrow*). MR image (axial T_2-weighted, d) of the same lesion after-CRT shows the presence of low intensity radial spicules (*black arrow*) due to *reactive fibrosis* departing from the treated tumor wall in the perirectal fat and an inner tissue with intermediate signal intensity (*asterisk*) suspected to be a residual tumor or the edematous mucosa. DWI allows to solve any doubt perfectly depicting the residual tumor as a semilunar hyperintensity (e, *arrow*) of the rectal wall confirmed as an hypointensity semilunar area (f, *arrow*) on ADC.

good final response. Patients with a final good response to CRT and a final downstaging have the tendency to show baseline ADC values lower than baseline ADC values of non-responding patients [96]. The consequent increase of ADC values reported during CRT both for downstaged group and non-downstaged group seems to be more consistent for downstage group. A reduction in mean tumor ADC was observed toward the end of CRT in all patients, mainly because CRT led to interstitial fibrosis, and radiation-induced inflammation regressed gradually [96].

Kim et al. reported an ADC cut-off value after CRT of 1.20×10^{-3} for discriminating the complete responders from the non-complete responders with a negative predictive value of 100%. However, the ADC cut-off value to predict patients' response before CRT is not already determined and needs a larger population to be assessed [103].

Moreover, the role of ADC measurements in the determination of complete response in patients with mucinous-type tumors is still a challenge due to the difficulty of differentiating residual tumors with hyperintense mucinous pools from inactive mucinous pools [95].

25.3.2.4.4 Tumor Regression Grade

MR evaluation of tumor regression grade (TRG) combined with the assessment of T staging after CRT is useful to stratify those patients with a small tumor remnant (ypT1-2N0) that are candidates for a local excision and patients with a complete response (ypT0N0) that could opt for a wait and see strategy [104,105].

The MERCURY study has demonstrated that the use of oblique high spatial resolution images allows for the development of an MRI-based tumor regression grading system according to the Dworak tumor regression grading system [106], which depends on the quantification of fibrotic tissue in comparison to tumoral tissue at the hystopathological analysis. The radiologic interpretation requires use of high spatial resolution oblique images after CRT compared with MR scans before CRT and aims to assess the proportion of fibrotic low signal intensity and the remaining residual intermediate signal intensity (Figure 25.34) [86]. The grade scale evaluated on T_2-weighted sequences includes

- *Grade 1*: complete pathological response, 100% hypointense fibrotic tissue, tumor absent.
- *Grade 2*: good pathological response, >75% of hypointense fibrotic tissue and <25% of intermediate signal intensity corresponding to tumor.
- *Grade 3*: intermediate pathological response, 50% of hypointense fibrotic tissue and 50% of intermediate signal intensity corresponding to tumor.
- *Grade 4*: poor pathological response, <75% of hypointense fibrotic tissue and >25% of intermediate signal intensity corresponding to tumor.
- *Grade 5*: no pathological response, fibrotic tissue absent, and 100% of intermediate signal intensity corresponding to tumor.

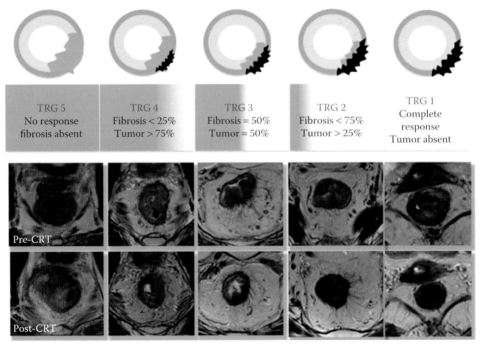

FIGURE 25.34
Scheme of tumor regression grade.

In the MERCURY Study, MRI assessment of TRG was shown to be a significant independent prognostic predictor. After CRT, patients with a good TRG (grades 1–3) were found to have a mean overall survival rate of 72%, a mean 5-year disease-free survival rate of 64%, and a mean local recurrence rate of 14%. On the other hand, patients with a final poor TRG (grades 4 and 5) were found to have a mean overall survival rate of 27%, a mean 5-year disease-free survival rate of 31%, and a mean local recurrence rate of 29% [86].

25.3.2.4.5 N Staging

Accurate nodal restaging after chemoradiation may have a significant impact on therapeutic decision-making. Minimally invasive surgical treatments (local excision) or a *wait-and-see* approach could be considered when patients show at MR images a node-negative status and a good or complete response of the primary tumor. MR accuracy of lymph node assessment after CRT ranges between 64% and 68% [94]. It is impossible to accurately distinguish malignant from reactive lymph nodes using only size criteria. A cut-off value of 10 mm permits a high specificity but yields a low sensitivity, while the reverse is true applying a cut-off value of 3 mm [50]. Differentiation of a metastatic lymph node from a lymph node with irradiation changes on post-CRT MR images is better achieved using morphologic criteria rather than size. These morphological criteria are border contour (sharply demarcated or irregular border) and inhomogeneity signal intensity of lymph nodes (Figure 25.35). Using these criteria, Koh et al. [107] showed MRI has a 90% negative predictive value, and 88% accuracy in detecting nodal disease after neoadjuvant treatment.

In the MERCURY study [108], patients with nodal disease detected on post-treatment MRI had a disease-free 5-year survival rate of 46% compared with 63% for those without malignant nodes. Two of the main pitfalls are that fibrotic may present spiculated borders appearing hypointense on T_2-weighted images, while slight hyperintensity of the signal intensity of the nodes can be due to mucinous degeneration induced by CRT.

Combining DWI and MRI with ADC and morphologic imaging criteria seems to be is useful in distinguishing malignant from benign small (sub-centimetric) lymph nodes as demonstrated in a few studies, mainly in head/neck and uterine/cervical cancer [109,110].

Recently, DW MRI has permitted a sensitivity of 80%, a specificity of 76.9%, and an accuracy of 78.3% in detecting regional metastatic lymph nodes [102].

Promising results were obtained with contrast agent containing ultra-small super-paramagnetic iron oxide contrast agent to assess lymph nodes involvement after CRT, but this agent is not available on commerce [111].

25.3.2.4.6 Assessment of CRM after CRT

MR accuracy of predicting CRM involvement of irradiated rectal cancer is 66% because of the difficulty in differentiating infiltration from a fibrotic scar attached to the mesorectal fascia [112]. However, MERCURY study group has obtained a specificity of 92% to predict after CRT a negative radial margin using MRI [86].

25.3.2.4.7 Role of Dynamic Contrast-Enhanced MRI

Dynamic contrast-enhanced MRI (DCE-MRI) has been recently applied in order to assess the perfusional features of tumor tissue reflecting the microvascular status of the cancer with values of trans-endothelial transfer constant K^{trans} significantly comparable to those derived from perfusional CT [57]. An increased microcirculation of the tumor tissue is related to the presence of abnormal vessels with increased endothelial permeability, increased angiogenic activity, and presence of arteriovenous shunts, findings suggestive of a more aggressive tumor.

DCE-MRI has been investigated as a potential imaging biomarker for predicting complete pathological

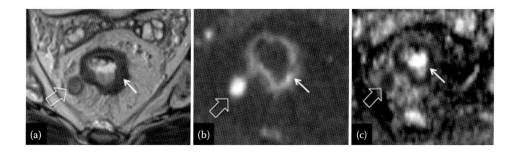

FIGURE 25.35
MR image (axial T_2-weighted, a) of a circumferential rectal cancer after-CRT shows diffuse low intensity of the outer layer of the rectal walls with an inner layer of intermediate signal intensity (*arrows*) due to residual tumor and confirmed by impeded diffusion on DWI (b) and on ADC map (c). The treated malignant mesorectal lymph node (*empty arrow*) appears to have irregular margins and heterogeneous signal intensity on T_2-weighted image. The same node presents restricted signal due to hypercellularity on DWI and ADC map (b and c, *empty arrows*), but this feature cannot be used to differentiate metastatic nodes from reactive nodes.

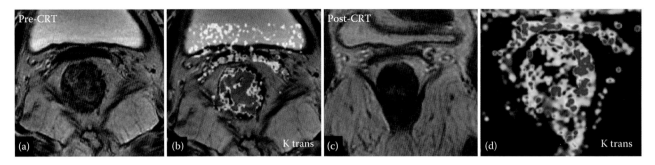

FIGURE 25.36

A case of rectal cancer (axial T_2-weighted MR image, a *asterisk*) that showed complete response to CRT resulted to be a pTx after surgical intervention. Before-CRT this cancer presented a higher initial flow/permeability as shown by the presence of several red spots in perfusional MR image (b). After CRT, on axial T_2-weighted image (c) there is no evidence of tumor and low perfusional K^{trans} value was found as documented by the decreased red spots (d).

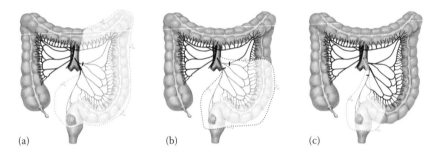

FIGURE 25.37

Types of surgical interventions for rectal cancer. High anterior resection (a). Extended low anterior resection with TME (b). Standard low anterior resection with TME (c). (Courtesy of Dr. M Ciolina.)

response before or during CRT, but heterogeneous results have been reported. Some investigators have found higher initial flow/permeability for good responders (Figure 25.36) [113] while others have found lower initial flow/permeability [114,115].

DCE-MRI could also be a useful tool after CRT, adding information that helps to assess if the patient required classical surgical intervention or a minimally invasive surgical intervention (local excision), or may follow a *wait-and-see* strategy avoiding surgery. In this regard, Gollub et al. reported that a significantly greater decrease in flow/permeability between pre- and post-treatment DCE-MRI correlates with good or complete response to CRT [58].

25.3.2.5 Posttreatment Assessment and Follow-Up

25.3.2.5.1 Types of Surgical Interventions for Rectal Cancer

AR: This type of surgical intervention is indicated when rectal carcinoma is more than 8 cm above the pectinate line [116] and includes three subtypes of techniques:

- *High AR*: Involving resection of distal descending colon, sigmoid colon, and upper rectum above peritoneal resection (Figure 25.37a).

- *Standard low AR*: Involving resection of rectosigmoid, upper and mid rectum below the peritoneal reflection with excision of rectal mesentery, the TEM (Figure 25.37c).

- *Extended low AR*: Including also resection of distal sigmoid colon and distal rectum (Figure 25.37b).

Partial mesorectal excision can be performed for high rectal cancer, whereas TME is necessary for mid and low rectal cancers [117,118]. This surgery has been associated with reduced local recurrence [119]. Anastomosis can be a colorectal or a coloanal accompanied by a proximal usually transient colostomy or ileostomy. At the time of surgical recanalization, types of anastomosis are

1. End-to-end
2. End-to-side
3. *Colonic J Pouch*: Reduced incidence of anastomotic complications increasing the volume of the neo-rectum, but it is more difficult CT evaluation for the complex anatomy. It is constructed from a double loop of colon formed into a J configuration by means of a linear stapling device used to maintain the two parallel colonic loops (Figure 25.38) [120].

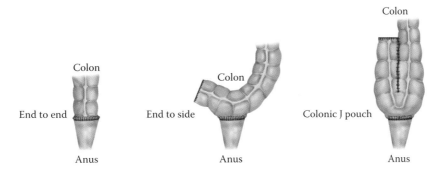

FIGURE 25.38
Types of colorectal or colo-anal anastomosis. (Courtesy of Dr. M Ciolina.)

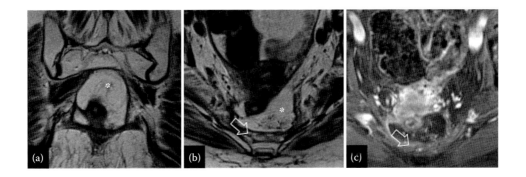

FIGURE 25.39
MR findings of anterior resection are the presence of small amount of soft tissue (*empty arrow*) in the presacral space that is widened (*asterisks*), usually larger 2 cm than normal as shown on coronal (a) and axial (b) T_2-weighted sequences. Axial post-gadolinium T_1-weighted image (c) demonstrates absence of significant enhancing of the presacral tissue (*empty arrow*).

In pT4 tumors (tumors directly invading other organs or structures and/or perforating visceral peritoneum), extended dissection with removal of adjacent organs must be performed if necessary.

Normal MRI findings after surgical procedures are (Figure 25.39) [121]

- Wall thickening at the site of anastomosis for postoperative edema
- Midline collection of fluid or a small amount of soft fibrotic tissue in the presacral space that is usually 2 cm larger than normal
- Small amounts of extra-peritoneal air or fluid around iliac vessels: a benign finding

An increased amount of soft tissue >5 cm should raise suspicion for anastomotic leak or recurrent tumor [122].
APER: Indications for this type of intervention are

- Rectal tumors less than 8 cm above the pectinate line, invading the sphincters or levator ani and bulky tumors within a narrow pelvis
- Anal malignancies

This type of surgical intervention involves excision of a portion of sigmoid colon, the entire rectum, and anus with an abdominal–perineal approach, the lack of sphincter preservation and creation of a permanent end colostomy [117,120]. Methods of managing the perineal wound vary between simple closure of the abdominoperineal wound is realized through gluteal [123] and rectus abdominis flaps (Taylor flaps) or using omentum [124].

MRI demonstrated the following anatomical changes (Figure 25.40) [125]:

- Absence of sphincters and colostomy, usually located in left iliac fossa.
- The bladder, the vesicles, the uterus, and the small bowel move posteriorly occupying a presacral or pre-coccygeal location.
- a presacral soft tissue often representing postsurgical fibrosis or ganulation tissue, with a maximum diameter of 3–5 cm at initial evaluation, comes more infiltrative and ill defined over time. Lack of growth over 1–2 years, clinical stability, and normal carcinoembryonic antigen levels are additional elements to confirm that the mass represents normal postoperative changes [126].

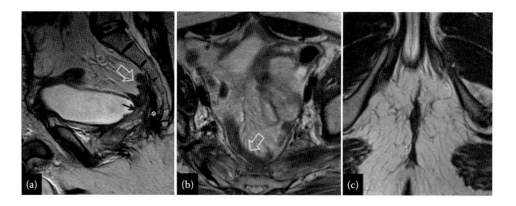

FIGURE 25.40
Sagittal (a) and axial (b) T_2-weighted MR images after APER showing posterior dislocation of the bladder (*black arrow*) and the small bowel loops (*empty arrow*) that move posteriorly to the sacrum and a presacral hypointense fibrotic tissue (*asterisk*). Axial T_2-weighted image (c) shows also absence of sphincters in perineal region.

If recurrency is suspected, positron emission tomography (PET)/CT allowed differentiation of benign from cancerous lesions with a sensitivity of 100% and a specificity of 96% [127].

25.3.2.5.1.1 Postsurgical Early Complications—Leakage, Fistulae, and Abscesses Anastomotic Leak: Anastomotic leaks usually manifest during the first two postoperative weeks. They are the most common postsurgical complications occurring in 5%–10% of cases with a mortality rate of about 50% if the anastomotic leak is not identified and treated [128]. They may result in complications including peritonitis and sepsis. Risk factors depending on the surgery: sepsis, perioperative active bleeding, and excessive anastomotic tension. The low anterior rectal resection has the highest risk of anastomotic leak. CT has been reported to confirm leakage in 48%–100%, while MR is usually not performed in the early postoperative period [129,130]. Anastomotic leaks can be clinically classified as symptomatic and asymptomatic [121].

- *Clinical leak*: Symptoms include fever, and leukocytosis, by the fourth postoperative day, intense abdominal pain, rebound tenderness, tenesmus, and constipation.

- *Subclinical leak*: This category includes occult leaks discovered at routine postoperative imaging 5.7%–10.7% and which can persist for 6 months or more following surgery.

Imaging findings of anastomotic leak are [121]

- Discontinuity of the staple line
- Gas within or adjacent to the bowel wall
- Pneumoperitoneum or free fluid usually located on the posterior aspect of the anastomosis

In rectal anastomotic leaks, the double rectum sign is usually seen and is attributable to the anterior rectum coupled with a posterior fluid collection presenting a inhomogeneous high signal intensity on T_2-weighted sequences or as a posterior air-filled cavity and which may or may not show peripheral enhance [122,130] (Figure 25.41).

Radiologists should describe if anastomotic leaks are contained or free leaks.

Contrast enema examination and CT with contrast endorectal administration are performed as a first-line investigation to evaluate the presence of a postoperative fluid collection. MRI is not routinely postoperatively performed but it may be an alternative to other imaging techniques, especially when a fistula is suspected and to delineate septic complication and inflammatory process [128]. Leaks from low anastomoses are generally contained within the pelvis, while leaks from a high AR or a more proximal colonic anastomosis: usually associated with free spillage into the peritoneal space and a consequently much higher morbidity [121].

Management on anastomotic leaks includes surgical drainage or imaging-guided drainage for low and proximal contained leaks and eventually with total parenteral nutrition. However, there is no urgency to operate again in the absence of sepsis and if a fistula arises at a later time.

Significant leaks with intraperitoneal spillage of bowel contents require immediate surgery and intra-abdominal washout. The anastomosis is repaired but is temporarily protected by ostomy [121]. Main differential diagnosis in case of suspected anastomotic leak is the normal MR aspect of end-to-side anastomosis. Multiplanar images are crucial to help avoid misinterpretation of anastomotic leak, while absence of symptoms, details of surgical intervention and absence of significant enhance of the fluid collection allow excluding anastomotic dehiscence (Figure 25.42).

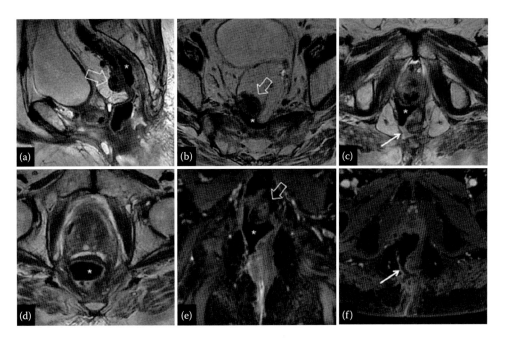

FIGURE 25.41
Sagittal and axial T_2-weighted MR images (a–d) showing a collection referring to an anastomotic leak, containing air (*asterisks*) and fluid material (*arrow*) appearing as a *double rectum*, and located posterior to the colonic lumen (*empty arrow*). Coronal and axial T_1-weighted images (e and f), acquired after IV contrast administration, showing enhancing wall collection due to inflammation and the presence of a fistulous track (f, *arrow*).

Fistulae: Fistulae result from anastomotic leaks and can be continuous with the skin or genitourinary tract, particularly the vagina, or can be continuous with the presacral space forming chronic collections. An active fistulous track appears as hyperintense linear structures in T_2-weighted sequences enhancing on gadolinium in fat-suppressed T_1-weighted images and which can communicate with an abscess or a fluid collection (Figure 25.43) [131]. Also abscesses present high signal intensity throughout on STIR and T_2-weighted images and an enhancing outer ring on T_1-weighted images [131]. Fistulae can be treated with the abscess drainage, diversion, and antibiotics. Management includes bowel rest and parenteral nutrition [121].

25.3.2.5.1.2 Postsurgical Late Complications: Perineal Hernia, Peritoneal Inclusion Cyst, Anastomotic Strictures, and Local Recurrence

Perineal hernia: Perineal hernia is a rare, usually late complication of APER due to the prolapse of the pelvic organ (small bowel and omentum) [132]. Symptoms include perineal pressure, fullness, pain, or a feeling as if sitting on a lump.

Peritoneal inclusion cyst: Peritoneal inclusion cysts are typically seen in premenopausal women who have undergone restorative proctocolectomy for ulcerative colitis, but other pelvic surgical intervention or inflammatory complications can contribute to the formation of peritoneal adhesions delimiting loculated fluid. MRI is the method of choice for the diagnosis of a peritoneal inclusion cysts, appearing as loculated fluid, which conforms to the peritoneal space and surrounds the ovary [133].

Anastomotic stricture: It has been defined by one's inability to pass a 12-mm-diameter sigmoidoscope through the narrowed area [118] and is an infrequent complication, resulting from fibrotic changes due to a postoperative leak or ischemia in the site of anastomosis and occurring within 2–12 months following surgery but is an infrequent complication [121]. Patients may present with increasing constipation, tenesmus, fecal soiling, urgency, diarrhea, and signs and symptoms of large bowel obstruction, but may be also asymptomatic, especially if they have been defunctioned with a proximal diverting stoma. MRI role in evaluation of anastomotic stricture is generally not useful unless there is a suspicion of recurrence as the cause of the stricture. The colon above the stricture will be dilated. Treatment on anastomotic strictures includes serial balloon dilatation, insertion of transrectal endoprotesis, or invasive surgical revision.

Tumor recurrence local recurrence: The local recurrence rate after potentially curative surgery for rectal cancer is widely variable (3%–30%). In 80% of cases, rectal cancer recurrence occurs within 2 years depending on: (a) T staging and the

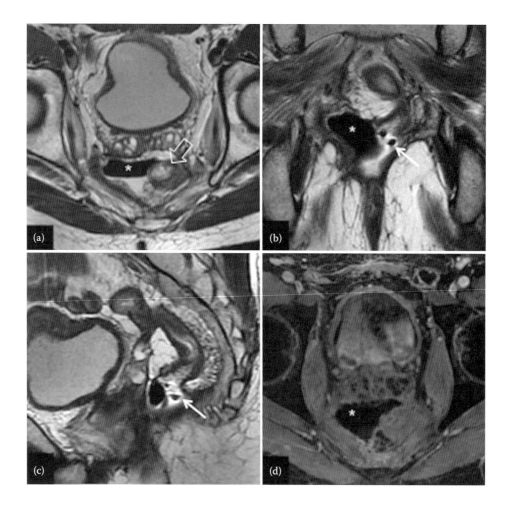

FIGURE 25.42
Axial, coronal, and sagittal T_2-weighted MR images (a–c) show a normal end-to-side anastomosis (*arrows* indicated suture ferromagnetic arti-facts) between the colon (*empty arrow*) and the rectum (*asterisk*). Multiplanar images are fundamental to avoid misinterpretation of anastomotic leak, while absence of symptoms, details of surgical intervention, and absence of significant enhance of the suspected leak (*asterisk*) in the axial post-gadolinium T_1-weighted image (d) permit to confirm that is a normal postsurgical finding.

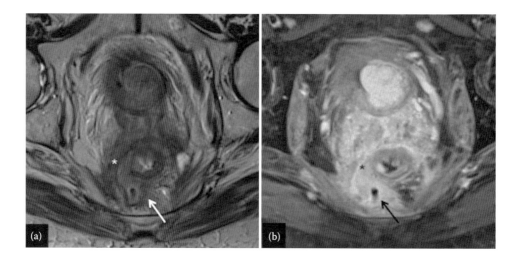

FIGURE 25.43
MRI after rectal anterior resection shows diffuse inhomogeneous thickening (*asterisk* on axial T_2-weighted image, a) and significant enhancement (*asterisk* on T_1-weighted image, b) of peri-anastomotic fat tissue and the presence of a fistula (*arrow*) due to an anastomotic dehiscence.

degree of differentiation, (b) lymphovascular invasion, (c) involvement of the CRM, (d) tumors in the lower third of the rectum, (e) tumors which have perforated or caused obstruction, (f) preoperative neoadjuvant treatment performed, (g) quality of the surgery [121].

Imaging of recurrent rectal cancer: Radiological evaluation of the pelvic recurrence from rectal cancer is complex due to postsurgical changing of pelvic anatomy and the difficulties in differentiating between residual fibrosis and recurrent tumor.

The majority of anastomotic recurrences are often not accessible by endoscopic means because usually originate outside the bowel lumen and required CT-guided biopsy.

MRI is essential in the assessment of recurrent rectal cancer because through T_2-weighted axial gradient-echo sequences provides anatomical detail of pelvis and usually allows distinguish pathological tissue appearing slight hyperintense from hypointense fibrotic tissue. The use of DWI and T_1-weighted fat-saturated sequences pre- and post-contrast performed using a dynamic acquisition in all three planes are fundamental to assess the relationship of tumor with the pelvic side wall (Figure 25.44). The type of enhancement help in distinguishing tumoral tissue from fibrosis because the tumoral enhancement is usually heterogeneous or with a rim pattern. On the other hand, fibrosis has generally a homogeneous delayed enhancement. Use of dynamic acquisition is based on the principle that the contrast peak occurs earlier in recurrent tumor due to the greater blood supply.

Description of recurrence might be reported within the following compartments [134]:

1. *Anterior*: Relationship to the bladder and vagina, cervix, and uterus in women or prostate and seminal vesicles in men.

2. *Lateral*: Relationship to the pelvic sidewall, the greater sciatic foramen, and evaluation of ureter involvement.

3. *Posterior*: Relationship to pelvic floor muscles, ischioanal fossa, presacral fascia, and the sacrum.

In clinical practice, MRI evaluation should be completed with PET/CT that is very sensitive to detect local recurrence. However, false-negative PET/CT can occur in cases of mucinous tumors, a small peritoneal disease, or small liver metastases due to low-level FDG uptake, and limited special resolution of PET scanner detectors [127].

25.3.2.5.2 Local Excision: Transanal Endoscopic Microsurgery

There is growing scientific evidence that local excision can be sufficient in patients with early rectal cancer of the mid and distal rectum with good histologic features and preoperative MR showing no lymph nodal involvement [135]. Transanal endoscopic microsurgery was introduced in the 1980s by Professor Gehard Buess from Tubingen, Germany. The operating equipment includes the use of an operating rectoscope (12–20 cm length), and a long-handled instruments for dissection, excision, and suturing. The endosurgical unit provides carbon dioxide (CO_2) insufflation, suction, irrigation, and continuous monitoring of intrarectal pressure [136].

25.3.2.5.2.1 Indications for TEM TEM is indicated for the removal of benign lesions such adenomas, GIST, or high anorectal fistulae, while other current indications for TEM include treatment of malignant lesions in the following cases [136]:

- Treatment of early stage rectal cancer (T1N0 and T1/T2N0 after CRT).
- Palliation in cases of advanced rectal cancer in patients who refuse radical excision or in those who are poor surgical candidates.

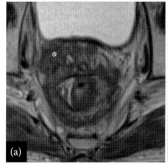 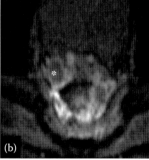 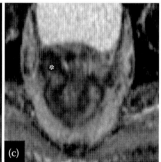

FIGURE 25.44
MR image (axial T_2-weighted image, a), 2-year-after anterior rectal resection, shows a diffuse soft tissue of intermediate signal intensity (*asterisk*) around the anastomosis, presenting restricted signal due to hypercellularity on DWI and on ADC map (b and c, *asterisks*). These findings are attributable to tumor recurrence.

- Patients with incidental carcinoma following polypectomy when there is concern about margin positivity.

For potentially curative resection of malignant lesions, both endorectal ultrasound and MRI can be used to determine the lesion depth of invasions, while lymph node assessment is mandatory with MRI. After TEM, MRI shows multiple fibrotic hypointense spicules on T_2-weighted images, departing from the rectal wall at the site of the surgical intervention (Figure 25.45).

25.3.2.5.2.2 Complications of TEM The overall complication after TEM has been reported to range from 6% to 31% [137–147]. Perioperative complications include

hemorrhage and intraperitoneal perforation, the rate of which varies from 0% to 9% [137,141,144]. Postoperative hemorrhage rate has been reported to be 1%–13% [137,140,145,148–152]. Most hemorrhages after TEM resolve spontaneously or are managed conservatively with blood transfusion [152–154]. In several randomized trials, the other early and late complications after TEM were similar or lower than for patients undergoing open resection (Figure 25.46). In particular, Lezoche and colleagues found no significant difference in complication rates between patients randomized to either TEM ($n = 35$) or laparoscopic TME ($n = 35$) for T2N0 rectal cancer following neoadjuvant treatment [104].

Wind and colleagues found that early morbidity after TEM was 21% in comparison to an early morbidity of

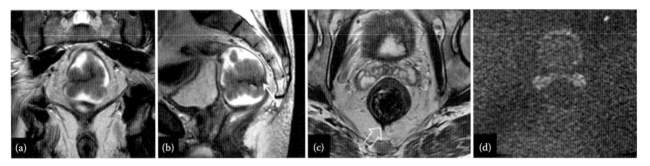

FIGURE 25.45
Coronal and sagittal T_2-weighted MR images (a and b) show a polypoid rectal cancer with a visible vascular stalk (*arrow*) attached to the posterior rectal wall. After TEM, on axial T_2-weighted MR image (c), some hypointense fibrotic spicules (*empty arrow*) are visible at the site of the surgical intervention, on the posterior aspect of rectal circumference. No signs of hyperintense areas, suspected for tumor residues, are present on DWI (d).

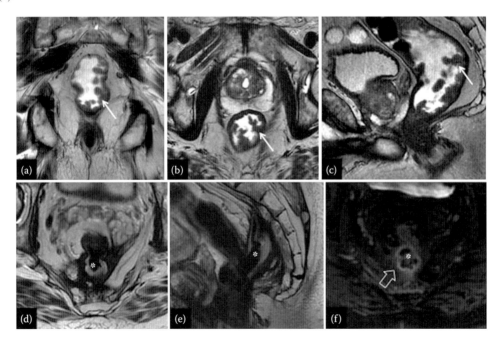

FIGURE 25.46
Coronal, axial, and sagittal T_2-weighted MR images (a–c) showing multiple nodular thickenings (*arrows*) of rectal mucosa resulting to be a lateral spreading tumor (pT2); after TEM, axial and sagittal MR images (d and e) show an air–fluid collection posterior to the rectal lumen representing a leak. Axial T_1-weighted MR image (f) shows a mild enhancement of the collection's walls (*empty arrow*).

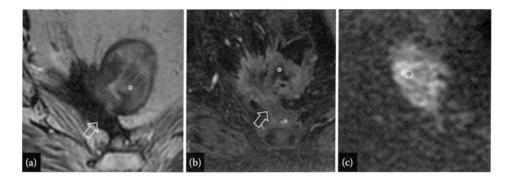

FIGURE 25.47
Axial T_2-weighted MR image (a), 1-year-after TEM, shows a nodular soft tissue of intermediate signal intensity (*asterisk*) above the fibrotic hypointense tissue (*empty arrow*) formed after the local excision. This nodular soft tissue is attributable to tumor recurrence and presents heterogeneous contrast enhancement (*asterisk*, b) and restricted signal due to hypercellularity on DWI (*asterisk*, c). Fibrotic postsurgical changes not usually appear to have significant enhancement (*empty arrow*, b).

35% in AR group. TEM procedure has demonstrated also shorter average operating times (103 min vs. 149 min, $p < 0.05$), lower blood loss ($p < 0.001$), shorter length of stay (5.7 days [standard deviation {SD}] 1.8 days vs. 15.4 days [SD 1.5 days], $p < 0.0001$) and a lower postoperative analgesia requirement [152].

25.3.2.5.2.3 Recurrence   Use of TEM as definitive and curative treatment should be limited to early stage T1N0 cancer. The recurrence rate following TEM in case of T1 lesions ranges from 0% to 11% [139,144–146,152] without demonstrated statistically significant difference in recurrence rate or survival rate comparing TEM to radical surgery [151,152]. If unfavorable histologic characteristics are found following TEM excision (pT2, lymphovascular invasion, or involved margins), immediate radical surgery should be performed (Figure 25.47).

However, the use of neoadjuvant or adjuvant therapy has been demonstrated to significantly reduce local recurrence rates, and for this reason, local excision with TEM can be considered for patients with a T2 lesions who haves received neoadjuvant therapy, as well. Duek and colleagues found a 0% local recurrence rate in patients who underwent radiotherapy following TEM compared to a 50% recurrence rate in patients who refused adjuvant radiotherapy following TEM [154]. Lezoche and al. had found no significant difference in local recurrence or disease-free survival during a minimum 5-year follow-up comparing TEM and laparoscopic TME in a population that received CRT [104].

Patients with T3 lesions cannot be candidates for TEM because of the high risk of local recurrence, lymph node metastases, and limited available data on the oncologic outcome.

25.3.2.5.3 Wait-and-See *Policy*
Tumor downstaging obtained with CRT may lead to complete pathological response (defined as absence of viable tumor cells after full pathologic examination of the resected specimen, pT0N0M0) in 10%–30% of patients [155–160]. In this condition, referred as stage 0 disease, patients can be carefully followed by clinical, endoscopic, and radiologic studies without immediate surgery avoiding unnecessary surgical complications. In a study conducted by Habr-Gama et al., 71 patients (26%) had complete clinical response following CRT and were treated by observation alone. Among this population, only two patients developed a late endorectal recurrence successfully treated by full-thickness transanal excision (pT1) or brachytherapy, while three patients developed systemic metastases treated by systemic chemotherapy [161]. Patients' selection for *wait-and-see* policy needs evaluation of primary tumor response and evaluation of lymph nodes after CRT and during follow-up. Assessment of primary tumor response should always be conducted using T_2-weighted sequences plus DWI. The reported positive predictive value (PPV) of T_2-weighted sequences for tumor residual after CRT or local recurrence ranging between 50% and 56% [95,97,162] while DWI seems to improve sensitivity of MRI by 17%–46%, with a reported PPV ranging between 66% and 92% to predict tumor residual after CRT or local recurrence during follow-up [95,97]. However, correlation between MRI and endoscopy remains fundamental, increasing the probability of detecting residual viable tumor or local recurrence by 36% and helping in DWI interpretation. Assessment of lymph node after CRT or during follow-up needs MRI imaging alone evaluating the signal intensity and the contours of the nodes in high spatial resolution T_2-weighted images [see paragraph 25.3.2.4.5]. Size criterion, as already explained, shows an accuracy of only 64%–68% [94], because metastases can occur also in 2–3 mm lymph nodes. DWI permits to easily detect lymph nodes due to the high contrast resolution but does not allow differentiate N+ from N0. However, absence of high signal intensity corresponding to lymph nodes at DWI confirms N0 (Figure 25.48).

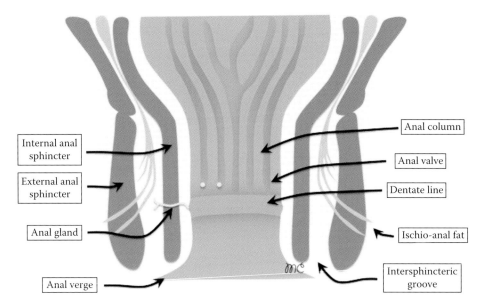

FIGURE 25.48
Schematic representation of the anal anatomy. (Courtesy of Dr. M Ciolina.)

25.3.3 MR in Rectal Fistula

25.3.3.1 Clinical Aspects and MR Indications

A fistula is defined as an abnormal tract connecting two epithelized organs or an organ and a surface of the body, which in case of perianal fistulae connects the anal canal with the skin of the perianal region involving the sphincter complex. Prevalence of perianal fistulae is 0.01%, occurring mostly in young males and showing male to female ratio of 2:1. Symptoms include discharge (65%) and perianal pain due to inflammation [163]. Perianal fistulae represent a cause of perianal sepsis that requires surgery and a preoperative detailed MR assessment to describe the anatomical relationship of the fistulous tract with the anal sphincter complex to help the surgeons to choose the best surgical treatment [164]. Success of surgical management of perianal fistulae is strictly related to the prevalence of recurrence [165].

The proximal half of the anal canal is covered by columnar epithelium raised in longitudinal mucosal folds, the anal columns of Morgagni. The distal portion of each columns of Morgagni is linked to the nearest one by a semilunar fold forming the anal valves end. The anal valves form all together an undulate mucosal rise which constitutes the dentate line (pectinate line) and marks the distal part of the transition zone (2 cm proximal to the anal verge), from which the columnar epithelium gives way to the squamous epithelium [164].

The superior profile of the anal valves forms in turn some pockets called the crypts of Morgagni. At the base of the crypts of Morgagni, there are the ducts opening of the anal glands, distributed around the circumference of the anal canal along the dentate line and lined by the columnar epithelium. These anal glands are 6–10 branched glandular structures, occupying an area of about 1 cm². Most of anal glands are subepithelial, while others have glandular branches with a deeper course, passing through the internal sphincter and ending into the intersphincteric space. These deep glandular branches represent channels through which the infection can spread from the anal lumen, reaching deeply into the sphincter planes, and consequently the skin or the pelvis. If an abscess involves a subepithelial glands the infection spontaneously discharges into the anal canal. Likewise, if the abscess reaches into a deeper gland, with glandular branches deeply placed beyond the internal sphincter, the same internal sphincter muscle acts as a barrier and the infection spread along the path of least resistance, travelling along the intersphincteric plane or passing through the external sphincter and entering the ischioanal fossa [166].

25.3.3.2 MR in Rectal Fistula: Technical Aspects

MRI protocol includes high spatial resolution multiplanar imaging performed with a phase array coil to cover the pelvis. After the localizer, the first scan must be the sagittal fast spin-echo (TSE) T_2-weighted image, which is useful in providing an overview of the pelvis and to show the extension and the orientation of the anal canal [164].

From the extension and orientation of the anal canal the correct position of the other sequences can be derived. Oblique axial T_2 TSE sequence is obtained orienting the plane of acquisition perpendicular to the long axis of the anal canal, while oblique coronal T_2 TSE sequence is obtained orienting the plane of acquisition parallel to the long axis of the anal canal. These planes should be repeated for the other sequences, as well: oblique axial T_2 TSE-weighted images and oblique coronal T_2 TSE-weighted images with fat suppression; oblique axial T_1 TSE-weighted images without fat suppression; oblique axial and coronal T_1 GRE with fat suppression before and after gadolinium IV administration [167,168]. Fat-suppressed T_2-weighted sequences are crucial to generate high contrast between the hyperintensity of the fistulous track due to edema and granulation tissue and surrounding fat tissue [169,170]. Fat suppression in T_2-weighted sequences is usually obtained with frequency selective fat saturation (FAT SAT) to obtain an image with high spatial resolution [171]. However, this method of fat suppression is more susceptible to ferromagnetic artifacts and in those cases does not guarantee the homogeneity of the fat suppression. To solve this problem, in presence of suture artifacts or seton in the fistula, an effective alternative is represented by STIR sequences which provide a good and homogeneous fat suppression and are more stable in presence of susceptibility artifacts [172–174].

New 3D T_2-weighted sequences offer the possibility to obtain data for the post-processing reformation of high-resolution images in sagittal, oblique axial, and oblique coronal planes by acquiring only a single 3D T_2 sequence with a significant shortening of acquisition time, thereby expanding the possibility of obtaining thinner section and reducing the errors dependent by technical operator [175–177].

Other advantages in MRI of perianal fistulae are the digital subtraction of T_1-weighted post-contrast images to obtain an MR fistulography of active fistulae or abscesses and the use of DWI to obtain restricted signal intensity of the purulent tracks and to better depict abscesses, especially in cases of previous allergic reaction or other contraindication to contrast agent [178].

25.3.3.3 MR in Rectal Fistula: Preoperative Imaging Findings

Active perianal fistulae appear as tracks of high signal intensity in T_2-weighted images with perifistulous edema that is better seen in fat-saturated T_2-weighted images where there is more contrast in comparison to the surrounding saturated fat tissue and to the hypointense muscular sphincteric structures. On gadolinium-enhanced fat-suppressed T_1-weighted images, fistulae and abscesses demonstrate intense peripheral enhancement due to active granulation tissue and inflammation, while the center of the track or of the abscess remains hypointense [164].

Reporting anal fistula is essential to describe the precise site of internal opening, the radial direction, and the fistulous path [165].

The internal opening and radial direction are described according to an *anal clock* scheme with the anterior perineum located at 12 o'clock and the natal cleft at 6 o'clock, the left lateral wall at 3 o'clock, and the right lateral wall at 9 o'clock [179].

Paths of fistulae in ano are described according to Park's classification (1976) [180]:

- *Intersphincteric*: Passes through the internal sphincter to the intersphincteric space and communicating with the perineum. These account for 70% of anal fistulae (Figure 25.49 and 25.50).
- *Transsphincteric*: Through both internal and external sphincters reaching the ischioanal fossa and communicating with the perineum. These account for 25% of anal fistulae (Figure 25.51).
- *Suprasphincteric*: Through the intersphincteric space extends superiorly over the top of the puborectalis muscle into ischiorectal fossa and then to the perineum. These account for about 5% of anal fistulae.

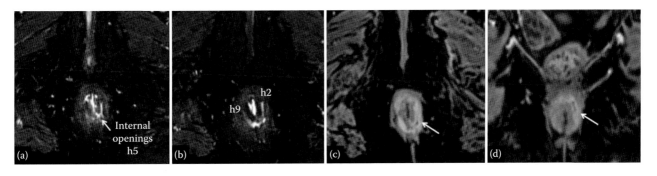

FIGURE 25.49
Axial fat-saturated T_2-weighted MR images (a and b) show an intersphincteric hyperintense linear structure due to a fistula extending from 2 o'clock to 9 o'clock with a horseshoe morphology and presenting an internal opening at 5 o'clock. Axial (c) and coronal (d) T_1-weighted image, acquired after IV contrast administration, shows enhancement of fistulous walls (*arrows*) due to active inflammation.

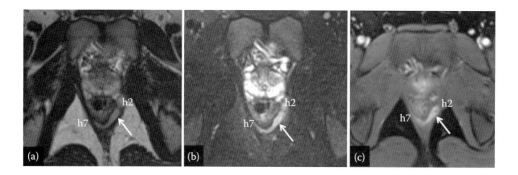

FIGURE 25.50
Axial T_2-weighted (a) and axial fat-saturated T_2-weighted (b) MR images show a hyperintense tubular structure, representing an intersphincteric fistula, and extending from 2 o'clock to 7 o'clock. Axial (c) T_1-weighted image, after IV contrast administration, shows enhancement of fistulous walls (*arrow*) due to active inflammation.

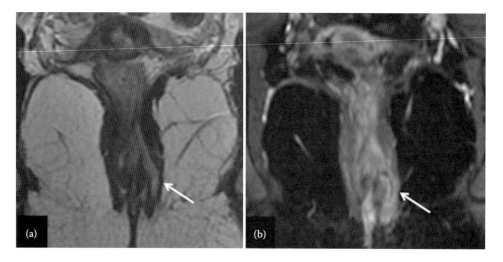

FIGURE 25.51
Coronal T_2-weighted MR images (a) show a hyperintense tubular structure representing an intersphincteric fistula that overpasses the left external sphincter at the level of low anal canal (*arrow*). Coronal (b) T_1-weighted image, acquired after IV contrast administration, shows enhancement of fistulous walls (*arrow*) due to active inflammation.

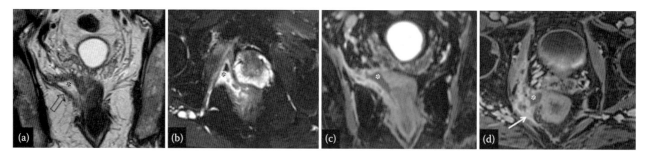

FIGURE 25.52
Coronal T_2-weighted MR image (a) and axial T_2-weighted fat-saturated MR image (b) show a collection (*asterisks*) with air–fluid level representing an abscess derived from a supralevator fistula and located over the right levator ani (*empty arrow*). Coronal (c) and axial (d) T_1-weighted image, acquired after IV contrast administration, shows enhancement of the abscess, of the fistulous walls, and of surrounding fat tissue (*arrow*) due to active inflammation.

- *Extrasphincteric*: From the rectal lumen to the perianal skin passing through the levator ani muscles and completely outside the external sphincter. These account for 1% of anal fistulae (Figure 25.52).

Another classification proposed by radiologist is the St James's University Hospital classification (Table 25.5) including some important MR findings to differentiate simple fistulae from complicated fistulae with abscesses or secondary tracks [179].

TABLE 25.5

Classification of Perianal Fistulae Proposed by St James's University Hospital

Grade	Type	Description
1	Simple linear intersphincteric fistulae	Confined into the intersphincteric space as Parks A without secondary track or abscesses
2	Intersphincteric fistula with an abscess or secondary track	Confined into the intersphincteric space as Parks A with abscess, secondary branches in the ipsilateral intersphincteric space or extending in the contralateral intersphincteric space with a horseshoe morphology
3	Trans-sphincteric fistula	It passes beyond the intersphincteric space crossing the external sphincter as Parks B, without abscesses or secondary track in ischiorectal or ischioanal fossae
4	Trans-sphincteric fistula with abscess or secondary track in the ishiorectal or ischioanal fossae	It passes beyond the intersphincteric space crossing the external sphincter as Parks B, but it is complicated by abscesses or secondary track in ischiorectal or ischioanal fossae
5	Supralevator and translevator disease	It extends upward through the intersphincteric plane, passing over the top of the insertion point of levator ani muscle and puborectalis muscles and descends in ischiorectal and ischioanal fossae to reach the perianal skin

This classification is mainly based on the differentiation, for each type of Park's fistula, between uncomplicated simple fistulae and complicated fistulae with abscesses or secondary tracks [179].

25.3.3.4 Therapy and Postoperative MRI Findings

The management of patients with perianal fistulae needs an accurate preoperative assessment with MRI to describe the relationship of the fistulous track to the anal sphincters preserving the integrity of sphincter complex and anal continence.

Treatment differs according to the different types of fistulous tracks. The following are the surgical techniques for fistula in ano [121].

- *Fistulotomy*: Incision of the entire fistulous tract which is laid open to allow healing by secondary intention. It is realized with a probe passing through the internal and external openings and it is indicated in most fistulae in ano (intersphincteric and low trans-sphincteric), especially at low levels of the anal canal, in which the internal and outer external sphincter fibers can be divided at right angles without affecting continence. These are not indicated in anterior fistulae in female patients.
  - Core fistulectomy: Excision of the fistulous tract without division of the muscle. It is realized with a probe passing through the internal and external openings, coring out the fistula. The internal opening can then be closed (possibly combined with an advancement flap).
- *Seton placement*: A foreign material (usually non-absorbable sutures or elastic bands) placed in a fistula tract to drain septic fluid or to slowly cut the sphincter slowly stimulating muscle fibrosis maintaining in the meanwhile the integrity of the sphincter mechanism. It is indicated in high, complex, recurrent, or multiple fistulae and in patients with demonstrated poor sphincter pressures or in female patients with anterior fistulae. It can also be performed in association with fistulotomy if the fistula has more complex branches going upward.

- *Mucosal advancement flap*: Curettage or abscission of the tract as a core fistulectomy closing the internal opening by a mucosal flap. It is used when traditional fistulotomy and setons have failed or are inappropriate due to complex disease.
- *Fibrin glue or collagen plugs*: The track is drained with a seton and prepared. Then is filled with fibrin glue or a collagen plug to allow scarring.

In patients with Crohn's disease-related fistulae, antibiotics are the first-line therapy, while purine analogs (azathioprine, 6-mercaptopurine) are used to maintain remission [181].

Anti-necrosis factor (TNF) antibodies (Infliximab) have been recently introduced with good clinical results inducing the rapid closure of the fistulous orifices. However, despite the closure of the openings, the fistula track persists causing chronic inflammation and recurrent fistula and abscesses [182].

MRI plays an important role also to confirm and evaluate recurrent disease after surgery and to assess response to medical therapy [183]. After surgical treatment MRI remains the best diagnostic tool, superior to endoanal ultrasound due the more wide evaluation of perianal region and to the capacity to differentiate active fistulae from healed tracts. For these reasons, for the follow-up of patients with perianal fistulous disease, ultrasound has a very limited role, reserved in case of contraindication to MRI.

A fistula successfully treated appears as a linear structure with low intensity in both T_2- and T_1-weighted images, corresponding to a scar and without enhancement after administration of contrast material. After seton placement, it is important to recognize the seton as a central low signal point traversing the high-signal fistula channel [169].

Early complications, after surgical treatment of perianal fistula, include urinary retention and bleeding. Hemorrhages can produce high signal intensity on unenhanced T_1-weighted images in the site of the surgery. Among late complications, recurrence occurs in 20% for low and simple fistulae and in 40% for high and complex fistulae. Recurrent fistulae appear as hyperintense tracks in T_2-weighted images surrounded by hypointense spicules due to fibrotic changes of the past surgical or medical therapy. Active recurrent fistulae or abscesses show intense enhancing of fistulous walls in post-enhanced fat-suppressed T_1-weighted images with central hypointensity due to fluid collection on pus and inflammatory fluids [121].

Incontinence usually only requires a clinical diagnosis and is due to damage to the anal sphincters can be of low entity (in up to 50%) mainly characterized by involuntary passage of flatus or of more severe entity (3%–5% of cases) with fecal incontinence. Another uncommon late complication is the anal stenosis secondary to fibrotic changes of the healing process [121].

References

1. Standring S (2008) *Gray's Anatomy: The Anatomical Basis of Clinical Practice*, 40th edn. Churchill, Livingston, Scotland.
2. Netter F (1973) In: Oppenheimer E, ed. *The CIBA Collection of Medical Illustrations*, volume 3: Digestive system.
3. Achiam MP, Chabanova E, Løgager V et al. (2007) Implementation of MR colonography. *Abdom Imaging* 32(4):457–462.
4. Korman LY, Overholt BF, Box T et al. (2003) Perforation during colonoscopy in endoscopic ambulatory surgical centers. *Gastrointest Endosc* 58(4):554–557.
5. Pickhardt PJ (2006) Incidence of colonic perforation at CT colonography: Review of existing data and implications for screening of asymptomatic adults. *Radiology* 239(2):313–316.
6. Debatin JF, Lauenstein TC (2003) Virtual magnetic resonance colonography. *Gut* 52(Suppl 4):iv17–22.
7. Florie J, Jensch S, Nievelstein RA et al. (2007b) MR colonography with limited bowel preparation compared with optical colonoscopy in patients at increased risk for colorectal cancer. *Radiology* 243(1):122–131.
8. Ajaj W, Lauenstein TC, Pelster G et al. (2004) MR colonography: How does air compare to water for colonic distention? *J Magn Reson Imaging* 19(2):216–221.
9. Bakir B, Acunas B, Bugra D et al. (2009) MR colonography after oral administration of polyethylene glycol-electrolyte solution. *Radiology* 251(3):901–909.
10. Rodriguez-Gomez S, Pages Llinas M, Castells Garangou A et al. (2008) Dark-lumen MR colonography with fecal tagging: A comparison of water enema and air methods of colonic distension for detecting colonic neoplasms. *Eur Radiol* 18(7):1396–1405.
11. Lauenstein T, Holtmann G, Schoenfelder D et al. (2001) MR colonography without colonic cleansing: A new strategy to improve patient acceptance. *AJR Am J Roentgenol* 177(4):823–827.
12. Achiam MP, Chabanova E, Løgager VB et al. (2008) MR colonography with fecal tagging: Barium vs. barium ferumoxsil. *Acad Radiol* 15(5):576–583.
13. Ajaj W, Lauenstein TC, Schneemann H et al. (2005a) Magnetic resonance colonography without bowel cleansing using oral and rectal stool softeners (fecal cracking)—A feasibility study. *Eur Radiol* 15(10):2079–2087.
14. Zalis ME, Barish MA, Choi JR et al. (2005) CT colonography reporting and data system: A consensus proposal. *Radiology* 236(1): 3–9.
15. Zijta FM, Bipat S, Stoker J et al. (2010) Magnetic resonance (MR) colonography in the detection of colorectal lesions: A systematic review of prospective studies. *Eur Radiol* 20(5):1031–1046.
16. Shah HA, Paszat LF, Saskin R et al. (2007) Factors associated with incomplete colonoscopy: A population-based study. *Gastroenterology* 132(7):2297–2303.
17. Achiam MP, Holst Andersen LP, Klein M et al. (2009a) Preoperative evaluation of synchronous colorectal cancer using MR colonography. *Acad Radiol* 16(7):790–797.
18. Ajaj W, Goyen M, Langhorst J et al. (2006) MR colonography for the assessment of colonic anastomoses. *J Magn Reson Imaging* 24(1):101–107.
19. Rottgen R, Herzog H, Lopez-Häninnen E et al. (2006) Bowel wall enhancement in magnetic resonance colonography for assessing activity in Crohn's disease. *Clin Imaging* 30(1):27–31.
20. Ajaj WM, Lauenstein TC, Pelster G et al. (2005c) Magnetic resonance colonography for the detection of inflammatory diseases of the large bowel: Quantifying the inflammatory activity. *Gut* 54(2):257–263.
21. Schreyer AG, Rath HC, Kikinis R et al. (2005) Comparison of magnetic resonance imaging colonography with conventional colonoscopy for the assessment of intestinal inflammation in patients with inflammatory bowel disease: A feasibility study. *Gut* 54(2):250–256.
22. Schreyer AG, Furst A, Agha A et al. (2004) Magnetic resonance imaging based colonography for diagnosis and assessment of diverticulosis and diverticulitis. *Int J Colorectal Dis* 19(5):474–480.
23. Ajaj W, Ruehm SG, Lauenstein T et al. (2005b) Dark-lumen magnetic resonance colonography in patients with suspected sigmoid diverticulitis: A feasibility study. *Eur Radiol* 15(11):2316–2322.
24. Scardapane A, Bettocchi S, Lorusso F et al. (2011) Diagnosis of colorectal endo-metriosis: Contribution of contrast enhanced MR-colonography. *Eur Radiol* 21:1553–1563.

25. Salerno G, Sinnatambi C, Branagan G et al. (2006) Defining the rectum: Surgically, radiologically and anatomically. *Colorectal Dis* 8:5–9.

26. Jorge JM, Wexner SD (1997) Anatomy and physiology of the rectum and anus. *Eur J Surg* 163:723–731.

27. Schäfer A-O, Langer M (2010) *MRI of Rectal Cancer Clinical Atlas*. Springer, New York.

28. Aigner F, Trieb T, Öfner D et al. (2007) Anatomical considerations in TNM staging and therapeutical procedures for low rectal cancer. *Int J Colorectal Dis* 22:1339–1342.

29. Sakorafas GH, Zouros E, Peros G (2006) Applied vascular anatomy of the colon and rectum: Clinical implications for the surgical oncologist. *Surg Oncol* 15:243–255.

30. Maier A, Fuchsjager M (2003) Preoperative staging of rectal cancer. *Eur J Radiol* 47:89–97.

31. Iafrate F, Laghi A, Paolantonio P et al. (2006) Preoperative staging of rectal cancer with MR Imaging: Correlation with surgical and histopathologic findings. *RadioGraphics* 26(3):701–714.

32. Harrison JC, Dean PJ, el Zeky F et al. (1994). From Dukes through Jass: Pathological prognostic indicators in rectal cancer. *Hum Pathol* 25: 498–505.

33. Jass JR, Love SB (1989) Prognostic value of direct spread in Dukes' C cases of rectal cancer. *Dis Colon Rectum* 32:477–480.

34. Tang R, Wang JY, Chen JS et al. (1995) Survival impact of lymph node metastasis in TNM stage III carcinoma of the colon and rectum. *J Am Coll Surg* 180:705–712.

35. Willett CG, Badizadegan K, Ancukiewicz M et al. (1999) Prognostic factors in stage T3N0 rectal cancer: Do all patients require postoperative pelvic irradiation and chemotherapy? *Dis Colon Rectum* 42:167–173.

36. Wolmark N, Fisher B, Wieand HS (1986) The prognostic value of the modifications of the Dukes' C class of colorectal cancer. *Ann Surg* 203:115–122.

37. Heald RJ, Moran BJ, Ryall RD et al. (1998) Rectal cancer: The Basingstoke experience of total mesorectal excision, 1978–1997. *Arch Surg* 133(8):894–899.

38. Kaur H, Choi H, You YN et al. (2012) MR imaging for preoperative evaluation of primary rectal cancer: Practical considerations. *RadioGraphics* 32(2):389–409.

39. Sauer R, Becker H, Hohenberger W et al. (2004) Preoperative versus postoperative chemoradiotherapy for rectal cancer. *N Engl J Med* 351(17):1731–1740.

40. Kapiteijn E, Marijnen CA, Nagtegaal ID et al. (2001) Preoperative radiotherapy combined with total mesorectal excision for resectable rectal cancer. *N Engl J Med* 345(9):638–646.

41. Akasu T, Kondo H, Moriya Y et al. (2000) Endorectal ultrasonography and treatment of early stage rectal cancer. *World J Surg* 24:1061–1068.

42. Blair S, Ellenhorn JD (2000) Transanal excision for low rectal cancers is curative in early-stage disease with favorable histology. *Am Surg* 66:817–820.

43. Gao JD, Shao YF, Bi JJ et al. (2003) Local excision carcinoma in early stage. *World J Gastroenterol* 9:871–873.

44. Langer C, Liersch T, Markus P et al. (2002) Transanal endoscopic microsurgery (TEM) for minimally invasive resection of rectal adenomas and *low-risk* carcinomas (uT1, G1–2). *Z Gastroenterol* 40:67–72.

45. MERCURY Study Group (2007) Extramural depth of tumor invasion at thin-section MR in patients with rectal cancer: Results of the MERCURY study. *Radiology* 243(1):132–139.

46. MERCURY Study Group (2006) Diagnostic accuracy of preoperative magnetic resonance imaging in predicting curative resection of rectal cancer: Prospective observational study. *BMJ* 333(7572):779.

47. Beets-Tan RG, Beets GL, Vliegen RF et al. (2001) Accuracy of magnetic resonance imaging in prediction of tumour-free resection margin in rectal cancer surgery. *Lancet* 357(9255):497–504.

48. Ho M-L, Liu J, Narra V (2008) Magnetic resonance imaging of rectal cancer. *Clin Colon Rectal Surg* 21:178–187.

49. Brown G, Daniels IR, Richardson C et al. (2005) Techniques and trouble-shooting in high spatial resolution thin slice MRI for rectal cancer. *Br J Radiol* 78(927):245–251.

50. Brown G, Richards CJ, Bourne MW et al. (2003) Morphologic predictors of lymph node status in rectal cancer with use of high-spatial-resolution MR imaging with histopathologic comparison. *Radiology* 227(2):371–377.

51. Figueiras RG, Goh V, Padhani AR et al. (2010) The role of functional imaging in colorectal cancer. *AJR Am J Roentgenol* 195(1):54–66.

52. Curvo-Semedo L, Lambregts DM, Maas M et al. (2012) Diffusion-weighted MRI in rectal cancer: Apparent diffusion coefficient as a potential noninvasive marker of tumor aggressiveness. *J Magn Reson Imaging* 35(6): 1365–1371.

53. Dzik-Jurasz A, Domenig C, George M et al. (2002) Diffusion MRI for prediction of response of rectal cancer to chemoradiation. *Lancet* 360(9329): 307–308.

54. Hein PA, Kremser C, Judmaier W et al. (2003) Diffusion-weighted magnetic resonance imaging for monitoring diffusion changes in rectal carcinoma during combined, preoperative chemoradiation: Preliminary results of a prospective study. *Eur J Radiol* 45(3):214–222.

55. Jung SH, Heo SH, Kim JW et al. (2012) Predicting response to neoadjuvant chemoradiation therapy in locally advanced rectal cancer: Diffusion-weighted 3 Tesla MR imaging. *J Magn Reson Imaging* 35(1):110–116.

56. Gu J, Khong PL, Wang S et al. (2011) Quantitative assessment of diffusion-weighted MR imaging in patients with primary rectal cancer: Correlation with FDG-PET/CT. *Mol Imaging Biol* 13(5):1020–1028.

57. Kierkels RG, Backes WH, Janssen MH et al. (2010) Comparison between perfusion computed tomography and dynamic contrast-enhanced magnetic resonance imaging in rectal cancer. *Int J Radiat Oncol Biol Phys* 77(2):400–408.

58. Gollub MJ, Gultekin DH, Akin O et al. (2012) Dynamic contrast enhanced-MRI for the detection of pathological complete response to neoadjuvant chemotherapy for locally advanced rectal cancer. *Eur Radiol* 22(4):821–831.

59. Nougaret S, Reinhold C, Mikhael HW et al. (2013) The use of MR imaging in treatment planning for patients with rectal carcinoma: Have you checked the "DISTANCE"? *Radiology* 268(2):330–344.

60. Tudyka V, Blomqvist L, Beets-Tan RG et al. (2014 Apr) EURECCA consensus conference highlights about colon & rectal cancer multidisciplinary management: The radiology experts review. *Eur J Surg Oncol* 40(4):469–475.

61. Taylor F, Mangat N, Brown G et al. (2010) Proforma-based reporting in rectal cancer. *Cancer Imaging* 10 Spec no A:S142–S150.

62. Moon SH1, Kim DY, Park JW et al. (2012) Can the new American Joint Committee on Cancer staging system predict survival in rectal cancer patients treated with curative surgery following preoperative chemoradiotherapy? *Cancer* 118(20):4961–4968.

63. Taylor FG, Quirke P, Heald RJ et al. (2011) Preoperative high-resolution magnetic resonance imaging can identify good prognosis stage I, II, and III rectal cancer best managed by surgery alone: A prospective, multicenter, European study. *Ann Surg* 253(4):711–719.

64. Merkel S, Mansmann U, Siassi M et al. (2001) The prognostic inhomogeneity in pT3 rectal carcinomas. *Int J Colorectal Dis* 16(5):298–304.

65. Shirouzu K, Akagi Y, Fujita S et al. (2011) Clinical significance of the mesorectal extension of rectal cancer: A Japanese multi-institutional study. *Ann Surg* 253(4):704–710.

66. Nagtegaal ID, van de Velde CJ, Marijnen CA et al. Low rectal cancer: A call for a change of approach in abdominoperineal resection. *J Clin Oncol* 2005;23(36):9257–9264.

67. Kao PS, Chang SC, Wang LW et al. (2010) The impact of preoperative chemoradiotherapy on advanced low rectal cancer. *J Surg Oncol* 102(7):771–777.

68. Weiser MR, Quah HM, Shia J et al. (2009) Sphincter preservation in low rectal cancer is facilitated by preoperative chemoradiation and intersphincteric dissection. *Ann Surg* 249(2):236–242.

69. Rouanet P, Saint-Aubert B, Lemanski C et al. (2002) Restorative and nonrestorative surgery for low rectal cancer after high-dose radiation: Long-term oncologic and functional results. *Dis Colon Rectum* 45(3):305–313.

70. Shihab OC, How P, West N et al. (2011) MRI staging system for low rectal cancer aid surgical planning? *Dis Colon Rectum* 54(10):1260–1264.

71. Shihab OC, Moran BJ, Heald RJ et al. (2009) MRI staging of low rectal cancer. *Eur Radiol* 19(3):643–650.

72. Zhang XM, Zhang HL, Yu D et al. (2008) 3-T MRI of rectal carcinoma: Preoperative diagnosis, staging, and planning of sphincter-sparing surgery. *AJR Am J Roentgenol* 190(5):1271–1278.

73. West NP, Anderin C, Smith KJ et al. (2010) European extralevator abdomino-perineal excision study group. Multicentre experience with extralevator abdominoperineal excision for low rectal cancer. *Br J Surg* 97(4):588–599.

74. MERCURY Study Group, Shihab OC, Taylor F et al. (2011) Relevance of magnetic resonance imaging-detected pelvic sidewall lymph node involvement in rectal cancer. *Br J Surg* 98(12):1798–1804.

75. Kim JH, Beets GL, Kim MJ et al. (2004) High-resolution MR imaging for nodal staging in rectal cancer: Are there any criteria in addition to the size? *Eur J Radiol* 52(1):78–83.

76. Koh DM, George C, Temple L et al. (2010) Diagnostic accuracy of nodal enhancement pattern of rectal cancer at MRI enhanced with ultra-small superparamagnetic iron oxide: Findings in pathologically matched mesorectal lymph nodes. *AJR Am J Roentgenol* 194(6): W505–W513.

77. Brown G, Richards CJ, Bourne MW et al. (2003) Morphologic predictors of lymph node status in rectal cancer with use of high-spatial-resolution MR imaging with histopathologic comparison. *Radiology* 227(2):371–377.

78. Moran B, Brown G, Cunningham D et al. (2008) Clarifying the TNM staging of rectal cancer in the context of modern imaging and neo-adjuvant treatment: 'y' 'u' and 'p' need 'mr' and 'ct'. *Colorectal Dis* 10(3):242–243.

79. Heijnen LA, Lambregts DM, Mondal D et al. (2013) Diffusion-weighted MR imaging in primary rectal cancer staging demonstrates but does not characterise lymph nodes. *Eur Radiol* 23(12):3354–3360.

80. Lambregts DM, Maas M, Riedl RG et al. (2011) Value of ADC measurements for nodal staging after chemoradiation in locally advanced rectal cancer-a per lesion validation study. *Eur Radiol* 21(2):265–273.

81. Taylor FG, Quirke P, Heald RJ et al. (2014) Preoperative magnetic resonance imaging assessment of circumferential resection margin predicts disease-free survival and local recurrence: 5-year follow-up results of the MERCURY study. *J Clin Oncol* 32(1):34–43.

82. Smith NJ, Barbachano Y, Norman AR et al. (2008) Prognostic significance of magnetic resonance imaging-detected extramural vascular invasion in rectal cancer. *Br J Surg* 95(2):229–236.

83. Lahaye MJ, Engelen SM, Beets-Tan RG et al. (2005) Imaging for predicting the risk factors—The circumferential resection margin and nodal disease—of local recurrence in rectal cancer: A meta-analysis. *Semin Ultrasound CT MR* 26(4):259–268.

84. Brown G, Radcliffe AG, Newcombe RG et al. (2003) Preoperative assessment of prognostic factors in rectal cancer using high-resolution magnetic resonance imaging. *Br J Surg* 90:355–364.

85. Madoff RD (2004) Chemoradiotherapy for rectal cancer—When, why, and how? *N Engl J Med* 351(17):1790–1792.

86. Patel UB, Blomqvist LK, Taylor F et al. (2012) MRI after treatment of locally advanced rectal cancer: How to report tumor response-the MERCURY experience. *AJR Am J Roentgenol* 199(4):W486–W495.

87. Nagtegaal I, Gaspar C, Marijnen C et al. (2004) Morphological changes in tumour type after radiotherapy are accompanied by changes in gene expression profile but not in clinical behaviour. *J Pathol* 204:183–192.

88. Yeo SG, Kim DY, Kim TH et al. (2010) Tumor volume reduction rate measured by magnetic resonance volumetry correlated with pathologic tumor response of preoperative chemo-radiotherapy for rectal cancer. *Int J Radiat Oncol Biol Phys* 78(1):164–171.

89. Kim YC, Lim JS, Keum KC et al. (2011) Comparison of diffusion-weighted MRI and MR volumetry in the evaluation of early treatment outcomes after preoperative chemoradiotherapy for locally advanced rectal cancer. *J Magn Reson Imaging* 34(3):570–576.

90. Kang JH, Kim YC, Kim H et al. (2010) Tumor volume changes assessed by three-dimensional magnetic resonance volumetry in rectal cancer patients after preoperative chemoradiation: The impact of the volume reduction ratio on the prediction of pathologic complete response. *Int J Radiat Oncol Biol Phys* 76(4):1018–1025.

91. Barbaro B, Fiorucci C, Tebala C et al. (2009) Locally advanced rectal cancer: MR imaging in prediction of response after preoperative chemotherapy and radiation therapy. *Radiology* 250(3):730–739.

92. Torkzad MR, Lindholm J, Martling A et al. (2007) MRI after preoperative radiotherapy for rectal cancer; correlation with histopathology and the role of volumetry. *Eur Radiol* 17(6):1566–1573.

93. Curvo-Semedo L, Lambregts DM, Maas M et al. (2011) Rectal cancer: Assessment of complete response to preoperative combined radiation therapy with chemotherapy—Conventional MR volumetry versus diffusion-weighted MR imaging. *Radiology* 260(3):734–743.

94. Kim DJ, Kim JH, Lim JS et al. (2010) Restaging of rectal cancer with MR imaging after concurrent chemotherapy and radiation therapy. *RadioGraphics* 30(2):503–516.

95. Kim SH, Lee JM, Hong SH et al. (2009) Locally advanced rectal cancer: Added value of diffusion-weighted MR imaging in the evaluation of tumor response to neoadjuvant chemo- and radiation therapy. *Radiology* 253(1):116–125.

96. Sun YS, Zhang XP, Tang L et al. (2010) Locally advanced rectal carcinoma treated with preoperative chemotherapy and radiation therapy: Preliminary analysis of diffusion-weighted MR imaging for early detection of tumor histopathologic downstaging. *Radiology* 254(1):170–178.

97. Lambregts DM, Vandecaveye V, Barbaro B et al. (2011) Diffusion-weighted MRI for selection of complete responders after chemoradiation for locally advanced rectal cancer: A multicenter study. *Ann Surg Oncol* 18(8):2224–2231.

98. Kim SH, Lee JY, Lee JM et al. (2011) Apparent diffusion coefficient for evaluating tumour response to neoadjuvant chemoradiation therapy for locally advanced rectal cancer. *Eur Radiol* 21(5):987–995.

99. Rao SX, Zeng MS, Chen CZ et al. (2008) The value of diffusion-weighted imaging in combination with T2-weighted imaging for rectal cancer detection. *Eur J Radiol* 65(2):299–303.

100. Feliu J, Calvilio J, Escribano A et al. (2002) Neo-adjuvant therapy of rectal carcinoma with UFT-leucovorin plus radiotherapy. *Ann Oncol* 13(5):730–736.

101. Fernandez-Martos C, Aparicio J, Bosch C et al. (2004) Preoperative uracil, tegafur, and con-comitant radiotherapy in operable rectal cancer: A phase II multicenter study with 3 years' follow-up. *J Clin Oncol* 22(15):3016–3022.

102. Barbaro B, Vitale R, Leccisotti L et al. (2010) Restaging locally advanced rectal cancer with MR imaging after chemoradiation therapy. *RadioGraphics* 30(3):699–716.

103. Prasad DS, Scott N, Hyland R et al. (2010) Diffusion-weighted MR imaging for early detection of tumor histopathologic downstaging in rectal carcinoma after chemotherapy and radiation therapy. *Radiology* 256(2):671–672; author reply 672.

104. Lezoche G, Baldarelli M, Guerrieri M et al. (2008) A prospective randomized study with a 5-year minimum follow-up evaluation of transanal endoscopic microsurgery versus laparoscopic total mesorectal excision after neoadjuvant therapy. *Surg Endosc* 22:352–358.

105. Habr-Gama A, Perez RO, Proscurshim I et al. (2006) Patterns of failure and survival for nonoperative treatment of stage c0 distal rectal cancer following neoadjuvant chemoradiation therapy. *J Gastrointest Surg* 10:1319–1328.

106. Dworak O, Keilholz L, Hoffmann A et al. (1997) Pathological features of rectal cancer after preoperative radiochemotherapy. *Int J Colorectal Dis* 12(1):19–23.

107. Koh DM, Chau I, Tait D et al. (2008) Evaluating mesorectal lymph nodes in rectal cancer before and after neo-adjuvant chemoradiation using thin-section T2-weighted magnetic resonance imaging. *Int J Radiat Oncol Biol Phys* 71:456–461.

108. Patel UB, Taylor F, Blomqvist L et al. (2011) Magnetic resonance imaging-detected tumor response for locally advanced rectal cancer predicts survival outcomes: MERCURY experience. *J Clin Oncol* 29:3753–3760.

109. King AD, Ahuja AT, Yeung DK et al. (2007) Malignant cervical lymphadenopathy: Diagnostic accuracy of diffusion-weighted MR imaging. *Radiology* 245:806–813.

110. Nakai G, Matsuki M, Inada Y et al. (2008) Detection and evaluation of pelvic lymph nodes in patients with gynecologic malignancies using body diffusion-weighted magnetic resonance imaging. *J Comput Assist Tomogr* 32:764–768.

111. Lahaye MJ, Beets GL, Engelen SM et al. (2009) Locally advanced rectal cancer: MR imaging for restaging after neoadjuvant radiation therapy with concomitant chemotherapy. Part II. What are the criteria to predict involved lymph nodes? *Radiology* 252(1):81–91.

112. Vliegen RF, Beets GL, Lammering G et al. (2008) Mesorectal fascia invasion after neoadjuvant chemotherapy and radiation therapy for locally advanced rectal cancer: Accuracy of MR imaging for prediction. *Radiology* 246(2): 454–462.

113. George ML, Dzik-Jurasz ASK, Padhani AR et al. (2001) Non-invasive methods of assessing angiogenesis and their value in predicting response to treatment in colorectal cancer. *Br J Surg* 88:1628–1636.

114. Kremser C, Trieb T, Rudisch A et al. (2007) Dynamic T1 mapping predicts outcome of chemoradiation therapy in primary rectal carcinoma: Sequence implementation and data analysis. *J Magn Reson Imaging* 26:662–671.

115. Sahani DV, Kalva SP, Hamberg LM et al. (2005) Assessing tumor perfusion and treatment response in rectal cancer with multi-section CT: Initial observations. *Radiology* 234:785.

116. Zissin R, Gayer G (2004) Postoperative anatomic and pathologic findings at CT following colonic resection. *Semin Ultrasound CT MR* 25(3): 222–238.

117. Tytherleigh MG, McC Mortensen NJ (2003) Options for sphincter preservation in surgery for low rectal cancer. *Br J Surg* 90:922–933.

118. Corman M (2005) Carcinoma of the rectum. In: Corman M, ed. *Colon and Rectal Surgery*, 5th ed. Lippincott Williams & Wilkins, Philadelphia, PA, pp. 905–1061.

119. Goldberg S, Klas JV (1998) Total mesorectal excision in the treatment of rectal cancer: A view from the USA. *Semin Surg Oncol* 15(2):87–90.

120. Dehni N, Tiret E, Singland JD et al. (1998) Long-term functional outcome after low anterior resection: Comparison of low colorectal anastomosis and colonic J-pouch-anal anastomosis. *Dis Colon Rectum* 41:817–822.

121. Brittenden J, Tolan DJM (2012) *Radiology of the Post Surgical Abdomen*. Springer, London.

122. Weinstein S, Osei-Bonsu S, Aslam R et al. (2013) Multidetector CT of the postoperative colon: Review of normal appearances and common complications. *RadioGraphics* 33(2):515–532.

123. Bell SW, Dehni N, Chaouat M et al. (2005) Primary rectus abdominis myocutaneous flap for repair of perineal and vaginal defects after extensive abdominoperineal resection. *Br J Surg* 92:482–486.

124. Taylor GI, Corlett R, Boyd JB (1983) The extended deep inferior epigastric flap: A clinical technique. *Plast Reconstr Surg* 72:751–765.

125. Lee JK, Stanley RJ, Sagel SS et al. (1981) CT appearance of the pelvis after abdomino-perineal resection for rectal carcinoma. *Radiology* 141:737–741.

126. Kelvin FM, Korobkin M, Heaston DK et al. (1983) The pelvis after surgery for rectal carcinoma: Serial CT observations with emphasis on nonneoplastic features. *AJR Am J Roentgenol* 141(5):959–964.

127. Even-Sapir E, Parag Y, Lerman H et al. (2004) Detection of recurrence in patients with rectal cancer: PET/CT after abdomino-perineal or anterior resection. *Radiology* 232(3):815–822.

128. Scardapane A, Brindicini D, Fracella MR et al. (2005) Post colon surgery complications: Imaging findings. *Eur J Radiol* 53(3):397–409.

129. DuBrow RA, David CL, Curley SA (1995) Anastomotic leaks after low anterior resection for rectal carcinoma: Evaluation with CT and barium enema. *AJR Am J Roentgenol* 165:567–571.

130. Nicksa GA, Dring RV, Johnson KH et al. (2007) Anastomotic leaks: What is the best diagnostic imaging study? *Dis Colon Rectum* 50(2):197–203.

131. Hoeffel C, Arrivé L, Mourra N (2006) Anatomic and pathologic findings at external phased-array pelvic MR imaging after surgery for anorectal disease. *RadioGraphics* 26(5):1391–1407.

132. So JB, Palmer MT, Shellito PC (1997) Postoperative perineal hernia. *Dis Colon Rectum* 40:954–957.

133. Jain KA (2000) Imaging of peritoneal inclusion cysts. *AJR Am J Roentgenol* 174:1559–1563.

134. Messiou C, Chalmers AG, Boyle K (2006) Surgery for recurrent rectal carcinoma. The role of magnetic resonant imaging. *Clin Rad* 61:250–258.

135. Heidary B, Phang TP, Raval MJ, Brown CJ (2014) Transanal endoscopic microsurgery: A review. *Can J Surg* 57(2):127–138.

136. Kunitake H, Abbas MA (2012) Transanal endoscopic microsurgery for rectal tumors: A review. *Perm J* 16(2):45–50.

137. Guerrieri M, Baldarelli M, Morino M et al. (2006) Transanal endoscopic microsurgery in rectal adenomas: Experience of six Italian centres. *Dig Liver Dis* 38(3):202–207.

138. Guerrieri M, Baldarelli M, Organetti L et al. (2008) Transanal endoscopic microsurgery for the treatment of selected patients with distal rectal cancer: 15 years experience. *Surg Endosc* 22(9):2030–2035.

139. Floyd ND, Saclarides TJ (2006 Feb) Transanal endoscopic microsurgical resection of pT1 rectal tumors. *Dis Colon Rectum* 49(2):164–168.

140. Bach SP, Hill J, Monson JR et al. (2009) Association of coloproctology of Great Britain and Ireland Transanal Endoscopic Microsurgery (TEM) Collaboration. A predictive model for local recurrence after transanal endoscopic microsurgery for rectal cancer. *Br J Surg* 96(3):280–290.

141. Endreseth BH, Wibe A, Svinsås M et al. (2005) Postoperative morbidity and recurrence after local excision of rectal adenomas and rectal cancer by transanal endoscopic microsurgery. *Colorectal Dis* 7(2):133–137.

142. Baatrup G, Elbrønd H, Hesselfeldt P et al. (2007) Rectal adenocarcinoma and transanal endoscopic microsurgery. Diagnostic challenges, indications and short term results in 142 consecutive patients. *Int J Colorectal Dis* 22(11):1347–1352.

143. Kreissler-Haag D, Schuld J, Lindemann W et al. (2008) Complications after transanal endoscopic microsurgical resection correlate with location of rectal neoplasms. *Surg Endosc* 22(3):612–616.

144. Stipa F, Burza A, Lucandri G et al. (2006) Outcomes for early rectal cancer managed with transanal endoscopic microsurgery: A 5-year follow-up study. *Surg Endosc* 20(4):541–545.

145. Maslekar S, Pillinger SH, Monson JR (2007) Transanal endoscopic microsurgery for carcinoma of the rectum. *Surg Encosc* 21(1):97–102.

146. Lezoche E, Guerrieri M, Paganini AM et al. (1998) Transanal endoscopic microsurgical excision of irradiated and non-irradiated rectal cancer. *Surg Laparosc Endocsc* 8(4):249–256.

147. Serra-Aracil X, Vallverdù H, Bombardó-Junca J et al. (2008) Long-term follow-up of local rectal cancer surgery by transanal endoscopic microsurgery. *World J Surg* 32(6):1162–1167.

148. Lezoche E, Guerrrieri M, Paganini AM et al. (2005) Long-term results in patient with T2 - 3 N0 distal rectal cancer undergoing radiotherapy before transanal endoscopic microsurgery. *Br J Surg* 92(12):1546–1552.

149. Gavagan JA, Whiteford MH, Swanstrom LL (2004) Full-thickness intraperitoneal excision by transanal endoscopic microsurgery does not increase short-term complications. *Am J Surg* 187(5):630–634.

150. Said S, Stippel D (1995) Transanal endoscopic microsurgery in large, sessile adenomas of the rectum. A 10-year experience. *Surg Endosc* 9(10):1106–1112.

151. Lee W, Lee D, Choi S et al. (2003) Transanal endoscopic microsurgery and radical surgery for T1 and T2 rectal cancer. *Surg End* 17(8):1283–1287.

152. Winde G, Nottberg H, Keller R et al. (1996) Surgical cure for early rectal carcinomas (T1). Transanal endoscopic microsurgery vs anterior resection. *Dis Colon Rectum* 39(9):969–976.

153. Lezoche E, Guerrieri M, Paganini AM et al. (2005) Transanal endoscopic versus total mesorectal laparoscopic resections of T2-N0 low rectal cancers after neoadjuvant treatment: A prospective randomized trial with 3-years 40. minimum follow-up period. *Surg Endosc* 19(6):751–756.

154. Duek SD, Issa N, Hershko DD et al. (2008) Outcome of transanal endo-scopic microsurgery and adjuvant radiotherapy in patients with T2 rectal cancer. *Dis Colon Rectum* 51(4):379–384.

155. Habr-Gama A, de Souza PM, Ribeiro U Jr et al. (1998) Low rectal cancer: Impact of radiation and chemotherapy on surgical treatment. *Dis Colon Rectum* 41:1087–1096.

156. Luna-Perez P, Rodriguez-Ramirez S, Rodriguez-Coria DF et al. (2001) Preoperative chemo-radiation therapy and anal sphincter preservation with locally advanced rectal adenocarcinoma. *World J Surg* 25:1006–1011.

157. Medich D, McGinty J, Parda D et al. (2001) Preoperative chemoradiotherapy and radical surgery for locally advanced distal rectal adenocarcinoma: Pathologic findings and clinical implications. *Dis Colon Rectum* 44:1123–1128.

158. Grann A, Minsky BD, Cohen AM et al. (1997) Preliminary results of preoperative 5-fluorouracil, low-dose leucovorin, and concurrent radiation therapy for clinically resectable T3 rectal cancer. *Dis Colon Rectum* 40:515–522.

159. Janjan NA, Khoo VS, Abbruzzese J et al. (1999) Tumor downstaging and sphincter preservation with preoperative chemoradiation in locally advanced rectal cancer: The M. D. Anderson Cancer Center experience. *Int J Radiat Oncol Biol Phys* 44:1027–1038.

160. Hiotis SP, Weber SM, Cohen AM et al. (2002) Assessing the predictive value of clinical complete response to neoadjuvant therapy for rectal cancer: An analysis of 488 patients. *J Am Coll Surg* 194:131–135.

161. Habr-Gama A, Perez RO, Nadalin W et al. (2004) Operative versus nonoperative treatment for stage 0 distal rectal cancer following chemoradiation therapy: Long-term results. *Ann Surg* 240(4):711–717.

162. Suppiah A, Hartley JE, Monson JR (2009) Advances in radiotherapy in operable rectal cancer. *Dig Surg* 26(3):187–199.

163. Sainio P (1984) Fistula-in-ano in a defined population: Incidence and epidemiological aspects. *Ann Chir Gynaecol* 73(4):219–224.

164. de Miguel Criado J, del Salto LG, Rivas PF et al. (2012) MR imaging evaluation of perianal fistulas: Spectrum of imaging features. *RadioGraphics* 32(1):175–194.

165. Lilius HG (1968) Fistula-in-ano, an investigation of human foetal anal ducts and intramuscular glands and a clinical study of 150 patients. *Acta Chir Scand Suppl* 383:7–88.

166. Parks AG (1961) Pathogenesis and treatment of fistula-in-ano. *BMJ* 1(5224):463–469.

167. Halligan S, Buchanan G (2003) MR imaging of fistula-in-ano. *Eur J Radiol* 47(2):98–107.

168. Ziech M, Felt-Bersma R, Stoker J (2009) Imaging of perianal fistulas. *Clin Gastroenterol Hepatol* 7(10):1037–1045.

169. Halligan S, Stoker J (2006) State of the art: Imaging of fistula in ano. *Radiology* 239(1):18–33.

170. Bartram C, Buchanan G (2003) Imaging anal fistula. *Radiol Clin North Am* 41(2):443–457.

171. Delfaut EM, Beltran J, Johnson G et al. (1999) Fat suppression in MR imaging: Techniques and pitfalls. *RadioGraphics* 19(2):373–382.

172. Haggett PJ, Moore NR, Shearman JD et al. (1995) Pelvic and perineal complications of Crohn's disease: Assessment using magnetic resonance imaging. *Gut* 36(3): 407–410.

173. Haramati N, Penrod B, Staron RB et al. (1994) Surgical sutures: MR artifacts and sequence dependence. *J Magn Reson Imaging* 4(2):209–211.

174. Yang RK, Roth CG, Ward RJ et al. (2010) Optimizing abdominal MR imaging: Approaches to common problems *RadioGraphics* 30(1):185–199.

175. Kim H, Lim JS, Choi JY et al. (2010) Rectal cancer: Comparison of accuracy of local-regional staging with two- and three-dimensional preoperative 3-T MR imaging. *Radiology* 254(2):485–492.

176. Lichy MP, Wietek BM, Mugler JP et al. (2005) Magnetic resonance imaging of the body trunk using a single-slab, 3-dimensional, T2-weighted turbo-spin-echo sequence with high sampling efficiency (SPACE) for high spatial resolution imaging: Initial clinical experiences. *Invest Radiol* 40(12): 754–760.

177. Proscia N, Jaffe TA, Neville AM et al. (2010) MRI of the pelvis in women: 3D versus 2D T2-weighted technique. *AJR Am J Roentgenol* 195(1):254–259.

178. Schaefer O, Lohrmann C, Langer M et al. (2004) Assessment of anal fistulas with high-resolution subtraction MR-fistulography: Comparison with surgical findings. *J Magn Reson Imaging* 19(1):91–98.

179. Morris J, Spencer JA, Ambrose NS (2000) MR imaging classification of perianal fistulas and its implications for patient management. *RadioGraphics* 20(3):623–635; discussion 635–637.

180. Parks AG, Gordon PH, Hardcastle JD (1976) A classification of fistula-in-ano. *Br J Surg* 63(1):1–12.

181. Bell SJ, Halligan S, Windsor AC et al. (2003) Response of fistulating Crohn's disease to infliximab treatment assessed by magnetic resonance imaging. *Aliment Pharmacol Ther* 17(3):387–393.

182. Karmiris K, Bielen D, Vanbeckevoort D et al. (2011) Long-term monitoring of infliximab therapy for perianal fistulizing Crohn's disease by using magnetic resonance imaging. *Clin Gastroenterol Hepatol* 9(2):130–136.

183. Keighley M, Williams N (2007) *Surgery of the Colon, Rectum and Anus*, 1st ed. Elsevier, London.

26

Posttraumatic and Postsurgical Abdomen

Jose Luis Moyano-Cuevas, Juan Maestre-Antequera, Diego Masjoan, José
Blas Pagador, and Francisco Miguel Sánchez-Margallo

CONTENTS

26.1 Introduction

Multiple systems (musculoskeletal, brain, and medulla, among others) have widely demonstrated the usefulness of MRI as noninvasive diagnostic and guided tool for surgical interventions. Mainly, these great outcomes of MRI are due to its high intrinsic contrast resolution and the potential application of multiple sequences to achieve excellent tissue characterization. However, abdominal application of MRI techniques has not been increased until recent years when MRI sequences have been evolved improving its performance, reducing its exposure time, and increasing its spatial resolution. Nowadays, these improved features of MRI allow to detect and to analyze the most of these abdominal organs and structures that are highly deformable and suffer usually complex morphological changes and movements all over the abdominal cavity. For this reason, its use on postsurgical applications, with rising interest in abdominal interventions, is fully implemented in some surgical protocols and is increasing in others. In this sense, depending on the surgical specialty some specific techniques such as MR urography, MR cholangiopancreatography, and MR angiography have been developed using MRI.

This chapter has been organized in two main topics. First, postsurgery MRI applications have been analyzed dividing liver and kidney pathologies. Furthermore, special interest on intra-operative assistance of MRI-based techniques has been described first in each section.

Later, postoperative monitoring of possible complications have been shown. Finally, although posttraumatic MRI applications are quite limited at this moment, mainly because of the pressing needs of patients who go to the hospital with serious traumatic injuries, a brief section about this topic is presented at the end of the chapter.

Traditionally, MRI was used in the liver to diagnose and characterize both focal lesions and diffuse diseases. Among others, maybe cirrhosis is one of the most common diffuse diseases of the liver that can be detected and diagnosed by using MRI. However, the liver section of this chapter will focus mainly on hepatocellular carcinoma (HCC) and metastasis because of the important guidance that MRI supposes for these diseases. Additionally, some considerations about cholelithiasis are also analyzed, comparing endoscopic retrograde cholangiopancreatography (ERCP) with MR cholangiopancreatography (MRCP).

In clinical practice, the most common imaging techniques in the kidney are ultrasonography (US) and computed tomography (CT) because of its wide availability, good diagnosis results, and further its affordable low cost. However, MRI has specially indicated to distinguish tumors when traditional techniques (US and CT) obtain non-concluding diagnoses. Furthermore, in those cases where iodinated contrast agents cannot be used, MRI lead on an alternative of diagnosis and a perfect monitoring technique to control postoperative problems. Hence, Section 26.2.2 will focus on renal tumors and renal failure.

According to consequences of minimally invasive surgery on different abdominal organs, some of the latest finding of alterations and changes caused by pneumoperitoneum during surgery to several organs is presented in each section. So, in these sections several morphological, hemodynamic, and functional changes are analyzed both for liver and kidney minimally invasive procedures.

Finally, other important advances that improve the quality and safety of the treatments are image-guided therapies (IGT). These kinds of therapies promote all developments that can increase surgical outcomes of patients. For this reason, MRI is one of the main medical imaging techniques that is used today for different IGT. Although some intra-operative problems (image registration and soft-tissues deformation, among others) are technically difficult and are still unsolved, IGT are raising its use on several specific fields. None of these technical aspects will be about in this chapter, but MRI implications in different IGT will be shown throughout this chapter.

26.2 Postsurgery MRI Applications

Minimally invasive surgery—more specifically laparoscopy—reduces the size of incisions when compared to open surgery, so shorter hospital stay for patients and optimized healthcare costs are achieved [1–4]. Although these surgical techniques have increased patient safety, there are multiple possible complications that can occur after surgery, which are highly dependent of the procedure performed [5,6]. For this reason, an early detection of these complications is essential to ensure patient safety, as they frequently require the patient to go under surgery again. In this sense, MRI plays a key role in soft tissues analysis (thanks to its high contrast and better definition) both for the early diagnosis of different pathologies and for the characterization and analysis of the surgical outcomes and postoperative complications. Hence, MRI has exceeded to other diagnosis techniques as CT or US, thanks to its high sensitivity and specificity in detecting abnormalities. Among other applications, some could be mainly utilized for the evaluation of tumor margin's resection in several abdominal organs, the internal bleeding detection after surgery, or the blockages identification in the bile ducts.

Some of the most important MRI functionalities and advantages to follow up monitoring several abdominal surgeries, focusing mainly on the liver and kidney, are described in this section.

26.2.1 Liver

The liver is the body's second largest organ; only the skin is larger and heavier. The liver performs many essential functions related to digestion, metabolism, immunity, and storage of nutrients within the body. These functions make it a vital organ, without which the tissues of the body would quickly die from lack of energy and nutrients.

Because of these features of the liver, MRI is considered as a crucial technique for the diagnostics of its pathologies. Although MRI was not completely accepted at the beginning, recent experiments have demonstrated its utility and effectiveness [7]. MRI is more sensitive when detecting and characterizing liver lesions compared to other medical image diagnosis such as CT. Furthermore, MRI allows a noninvasive exploration of the bile tree with similar sensitivity than other diagnosis techniques such as the endoscopic ERCP [8,9].

T_1- and T_2-weighted sequences with breath hold are used for the exploration of the liver (Figure 26.1). The

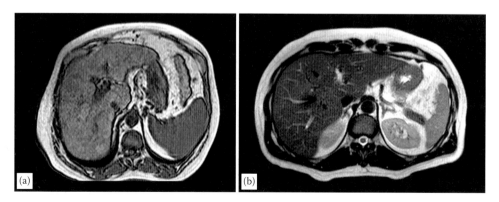

FIGURE 26.1
Example of an abdominal MRI: (a) T_1-weighted with breath-hold MR image obtained of the liver in a 45-year-old woman and (b) T_2-weighted with breath-hold MR image obtained of the liver in a 63-year-old man.

breath hold is used to reduce the image artifact because of the lungs movement. Variations of these sequences are used depending on the purpose of the study.

26.2.1.1 During Surgery

The rapid evolution of interventional and imaging techniques and its associated technologies carries out the inclusion of new therapies that improve the surgical outcomes. The interventional MRI is an example of this idea. *Interventional MRI* is the term used to refer all procedures guided by MRI [10]. These procedures can be performed by using either minimally invasive or open surgery. Because of space restrictions on conventional resonances, these image-guided techniques are usually performed in open resonances. In open resonances, because of their special features, it is difficult to catch high field intensity, what has a direct influence on the image quality. Because of these limitations and the special conditions needed in the operating room (non-ferromagnetic instruments and equipment), these image-guided techniques are not standardized yet.

In liver surgery, interventional MRI reaches a height for thermal ablation of different tumors such as HCC, because they are minimally invasive techniques with high patient safety and fast patient recovery. These techniques cause tissue necrosis by percutaneous needle access up to the location of the tumor in order to apply there the appropriate therapy, such as radio frequency (RF), cryogenics, or microwaves. By using either hot or cold thermal energies to coagulate or to devitalize the tumor tissue without removing it from the patient, resulting effects are comparable to those achieved by surgical resection [11]. Interventional MRI during this surgery provides information of the tumor localization to the surgeons, offering several advantages such as monitoring the therapies effect, visualization of soft tissues with high contrast, free selection of imaging planes, no requirement of iodinated contrast media, and absence of ionizing radiation. Besides, the effectiveness of HCC ablation guided by MRI has been demonstrated in scientific studies [7,13], performing better than other guided techniques such as US or CT in some cases. Clasen et al. [12] analyzed the effectiveness of RF therapy on 56 cases of HCC. They compared the effectiveness of the ablation technique using MRI and CT as guidance systems, obtaining better results with MRI guidance. In the particular case of cryoablation therapy, CT does not provide a clear image of the ice ball that covers the tumor, which limits the safety and effectiveness. US guidance is widely used in ablation therapies, but it has some disadvantages such as limited capability to visualize tumor tissue and monitor thermal effect due to air bubbles produced by vaporization [13]. These drawbacks increase the need for repeated sessions of RF in liver tumors, as it is not possible to precisely know the extent of therapy during the course of the same [14].

26.2.1.1.1 Effect of the Pneumoperitoneum

MRI is useful as image-guided technique in minimally invasive surgery, at the same time that it also increases the knowledge of abdominal organs during the laparoscopic procedure. Despite the advantages of laparoscopic surgery [15], it also presents a certain number of difficulties and complications because of the nature of this surgical technique. The negative effects of the required pneumoperitoneum are one of those complications.

In laparoscopic surgery, increasing the intra-abdominal pressure is necessary to achieve an adequate space in the abdomen [15]. In the case of the liver, the pneumoperitoneum causes alterations in the normal functionality of this organ, having being described alterations in the production of different hepatics enzymes as alanine aminotransferase or aspartate aminotransferase after a laparoscopic surgery [16,17]. Hemodynamic changes on the hepatic artery and the portal venous have also been observed due to the creation of the pneumoperitoneum. However, there is a discussion about these changes, because some authors show a reduced flow of the portal vein during the pneumoperitoneum [18,19], whereas others describe an increase of the flow [20]. Therefore, the scientific community continues investigating the changes and the chain of events that cause them.

There are some techniques available to analyze these morphological and functional changes such as esophagical probes and US probes, to analyze arterial flow or monitoring equipment based on patient pathways. They all consist on invasive techniques. On the other hand, MRI allows analyzing the morphological and hemodynamic changes of different abdominal structures noninvasively.

A T_2-weighted sequence with breath hold can be used to analyze the morphological changes of liver, but a T_1 sequence could also be used for this purpose. The suitable slice thickness can be between 1 and 4 mm without gap between the slices in order to extract changes that could be observed. A methodology based on performing MRI sequences before and after the insufflation of the pneumoperitoneum is used to analyze the morphological changes of the liver [21]. In this sense, Figure 26.2 shows several slices of a MRI study where both the liver and the abdominal anatomy show changes after increasing the intra-abdominal pressure with the pneumoperitoneum. These selected slices show a compression of the liver and other abdominal structures and distension of the abdominal wall to create the necessary space to perform the laparoscopic surgery.

Under experimental conditions using a porcine model, the mean liver volume is significantly increased in 57.67 cm³ during a laparoscopic procedure due to

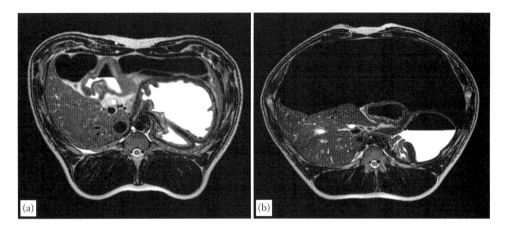

FIGURE 26.2

T_2-weigthed MRI obtained of swine model with a weight of the 35 Kg acquired in the supine position. (a) Images obtained before the pneumoperitoneum creation and (b) after of the pneumoperitoneum creation with a pressure of 14 mmHg. A deformation of the abdominal structures is observed.

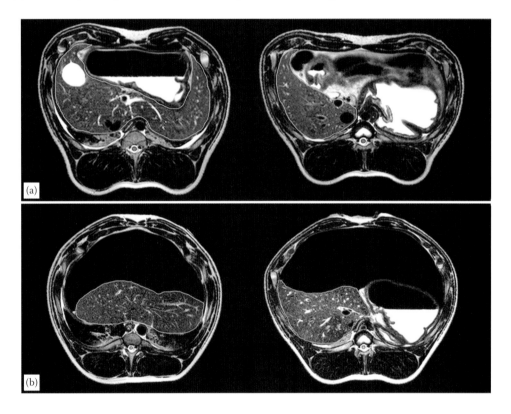

FIGURE 26.3

T_2-weighted image of swine model with a weight of 35 Kg acquired in the supine position. Liver manual segmentation is performed to evaluate the changes experimented by the liver volume with OXiris Visualization DICOM software (a) before and (b) after the pneumoperitoneum creation.

the increment of the intra-abdominal pressure up to 14 mmHg [21]. The reasons for this volume increment are not clear, as the portal artery flow was not analyzed in this study. Other studies analyze the hemodynamic effects in a porcine model at a pressure of 15 mmHg, showing an increase in the systemic vascular resistance, which can complicate the venous return [22]. Furthermore, several authors describe a decrease of portal blood flow in pigs when increasing pressure [18,19].

Therefore, it might be possible that a venous stasis lead to an increase in the liver volume.

Several image processing software of DICOM files are available to retrieve information associated with MRI studies and analyze parameters of a specific organ, such as calculating the volume of a selected structure. For this purpose, a previous identification of the liver using regions of interest (ROIs) in each of the slices making up the sequence is needed (Figure 26.3).

A liver deformation has also been observed when the intra-abdominal pressure increases [21]. In this case, the porcine liver undergoes an increase in the medio-lateral length and modifies its basal placement inside the abdomen. Figure 26.4 shows a coronal view of the same subject (porcine model) before and after the pneumoperitoneum insufflation, where a left medio-lateral extension of the liver can be observed. This reaction is due to the increased pressure. Hence, these parameters can be obtained by counting the number of slices where the liver is present, considering the slice thickness and the gap between each of them using the sagittal view.

The intra-abdominal pressure increase needed to perform laparoscopic procedures causes hemodynamic changes that have been widely described in the scientific literature. The MRI sequence phase contrast allows analyzing flows and other hemodynamic parameters in a noninvasive way, what has been validated against other gold standard techniques [23]. Although abdominal blood flows have not been analyzed during laparoscopic liver surgery, MRI can be used as a tool to analyze the morphological response of large hepatic vessels such as the portal vein. For this purpose, a high-resolution T_2-weighted sequence could be used to obtain images of the portal vein lumen where its morphological changes during surgery can be determined. Several images of porcine model before and after pressure increase are shown in Figure 26.5. These images show a narrowing of the portal vein lumen on the dorsal–ventral axis. For this study, three measures were taken along the vessel, from celiac trunk (first measure) up to 4 cm above of this (third measure).

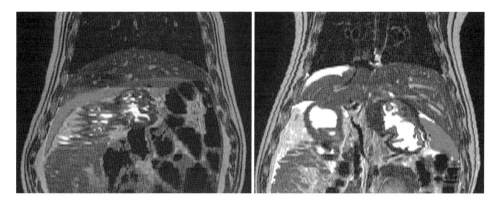

FIGURE 26.4
Coronal view of the T_2-weighted images of swine model with a weight of 35 Kg acquired in the supine position. Analysis of the medio-lateral expansion experienced by the liver by increasing the pressure up to 14 mmHg.

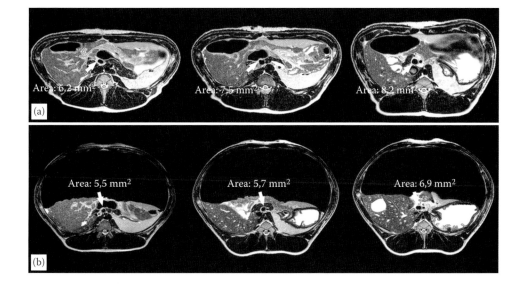

FIGURE 26.5
T_2-weighted images of a porcine model with a weight of 33 Kg acquired in the supine position. Analysis of the portal vein lumen in different position (a) before and (b) after the pneumoperitoneum creation. A reduction of the lumen in this vein was observed increasing the pressure to 14 mmHg.

26.2.1.2 Postoperative Finding

There are two main aspects related to the patient safety in the postoperative time: a monitoring of the surgical treatment efficiency and an early identification of complications derived from the surgical procedure itself. In this sense, MRI plays a crucial role in the tumor masses monitoring after surgical treatments or the identification of postsurgery problems derived from hepatic transplant surgeries. Subsequent sections will describe findings and complications detected by MRI, which have been classified according to the liver pathologies that cause the needs of these surgical interventions.

26.2.1.2.1 Hepatic Tumor

26.2.1.2.1.1 HCC   According to U.S. data, the primary liver cancer has increased its frequency in recent years: for the case of disease localized from 2.3 per 100.000 in 1999 to 4.2 per 100.000 in 2008 [24]. HCC is the most common primary liver cancer with 80% of the cases. HCC is the fifth and seventh most frequently diagnosed cancer in adult men and women, respectively, and the second leading cause of cancer death worldwide [25].

The initial diagnosis of HCC can be determined using a T_2-weighted sequence (Figure 26.6). This diagnostic study is complemented with a dynamic T_1 sequence with paramagnetic gadolinium contrast to discard benign lesions (such as angiomas), determine the lesion's vascularization and the presence of non-enhanced areas, what could mean intratumoral necrosis. These dynamic sequences are composed of four different images for each slice: images obtained before the administration of intravenous contrast together with images of the arterial, portal, and venous phases (Figure 26.7). The angioma shows a contrast absorption from external to the core of the tumor that remains along the time and constitutes a characteristic enhancement pattern of angiomas, because these lesions are composed by vascular cells with slow flow inside the tumor [26]. On the other hand, images of HCC lesions are hypointense in all phases in relation to the liver parenchyma, with small hypercaption focal points in the arterial phase, which decrease in later phases. Besides, the central areas of the tumor have decreased enhancement in all phases. Dysplastic nodules (precursor lesions to HCC) gradually replace the venous circulation by arterial vasculature due to the presence of newly formed arteries during the carcinogenesis (sinusoidal capillarization phenomenon) [27]. As it was introduced before, those non-enhanced areas in neither of the studied phases can be associated with tumorous necrosis areas.

Furthermore, a T_2 sequence with fat suppression is recommended to complete the diagnosis, discard the lipid nature of the detected lesions, and confirm the initial hypothesis (Figure 26.8).

There are different kinds of surgical treatment for the HCC according to the features and location of the tumor. Although hepatic resection is the first treatment, it is not an applicable option in high percentages of cases because of several reasons: the tumor placement, an insufficient remaining liver volume after resection, or possible tumor dissemination. On the other hand, liver transplant obtains high effectiveness, but it is highly restricted too, mainly due to very demanding criteria to select receptor patients, high associated costs, and limited availability of donors [28]. Other raising techniques, such as minimally invasive ablation therapies,

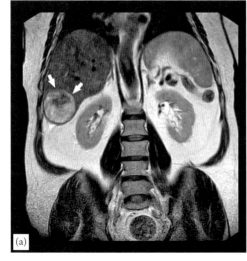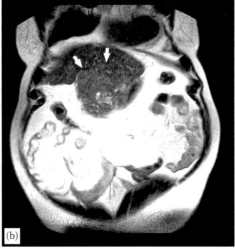

FIGURE 26.6

A 70-year-old man with two hepatic lesions corresponding with hepatocellular carcinoma. T_2-weighted coronal was acquired to diagnosis of the tumoral lesions. (a) The biggest lesion is located in the joint of the III and IV hepatic segments with a dimension of the 8.9 cm and the smallest lesion (b) with a dimension of the of the 5.9 cm is located in the V and VI segments. Both are solid lesions.

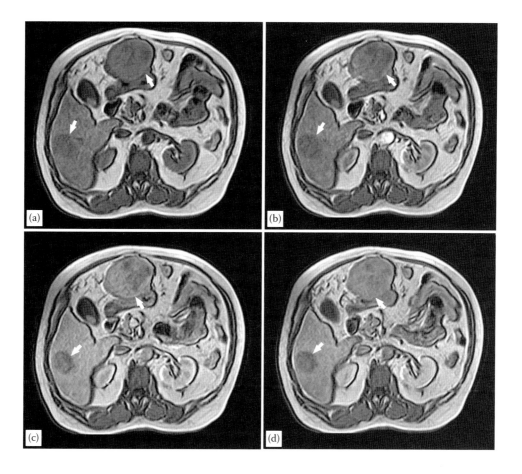

FIGURE 26.7

A 70-year-old man who is presented with two hepatic lesions corresponding with hepatocellular carcinoma. T_1 diffusion-weighted images of both lesions performed without using contrast agent (a), with gadolinium contrast in arterial phase (b), in portal phase (c) and in venous phase (d). Both lesions are hypointense in four phases relative to the hepatic parenchyma with small areas enhanced area in arterial phase.

offer many advantages to patients but increase technological demands and surgical skills for clinicians. These minimally invasive therapies are based on different physical principles to achieve tumor necrosis. Some of them use high temperatures (RF and microwaves ablations), while others use low temperatures (cryoablation) or high-intensity focused ultrasound (HIFU) or focused ultrasound (FUS). Although all these therapies have several advantages and drawbacks associated with them, RF and microwaves ablations are quite effective for HCC smaller than 5 cm and have a low rate of postsurgery complications. On the other hand, cryoablation and HIFU present higher rates of complications and long times of exposure, respectively, which limit their current use [29]. Some of the most common complications of cryoablation therapy are associated with hemorrhages, cold injuries to adjacent organs, or hepatic parenchyma fracture. However, recent studies show that these differences in the number of complications between cryoablation and RF therapies are not statistically significant (Figure 26.9) [30].

A proper monitoring of the liver tumors after applying these therapies is crucial to evaluate the tumor volume (decrease in tumor size), the need for additional sessions (repeating the therapy), or even to locate new tumorous areas in the same or adjacent organs. Therefore, contrast-enhanced multiphase CT or MRI is frequently performed to determine technical success of the tumor ablation.

Another critical use of MRI for evaluation is the intra-arterial therapies for tumor embolization. These therapies consist of obstructing the blood flow or reducing the vessels that feed the lesions, using different procedures such as transarterial embolization, drug-eluting bead, or radioembolization [31]. The liver tumors embolization has become a common procedure of treatment when the resection of the damaged segments and other ablation techniques cannot be performed due to the location or size of the tumor. Postoperative imaging studies are fundamental to determine the tumor response after the treatment, checking the successful obstruction performed to the vessels.

After performing the lesion embolization, it is critical to carry out several monitoring studies that help assessing the therapy effectiveness. So, it is possible through MRI to determine the obstruction level of the vessels that

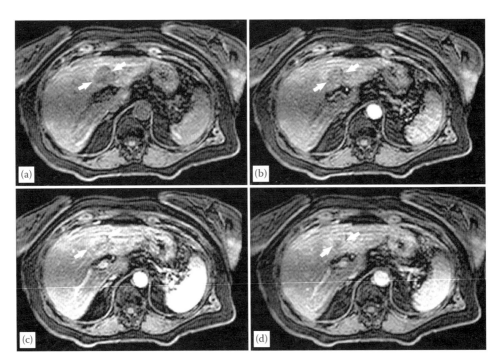

FIGURE 26.8
A 70-year-old woman who is presented with hepatic angioma. Breath-hold dynamic T_1-weighted dynamic imaging of the angioma before the administration of gadolinium contrast (a) and with gadolinium in arterial phase (b), in portal phase (c) and in venous phase (d). The lesion is localized in IV hepatic segment with a approximate dimension of 20–25 mm.

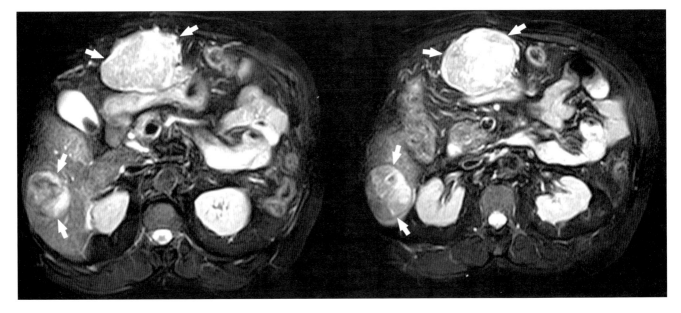

FIGURE 26.9
A 70-year-old man who is presented with two hepatic lesions corresponding with hepatocellular carcinoma. T_2-weighted fat-suppression images acquired to confirm the diagnostic of the tumor. The non-lipid nature of the lesion can be seen in the images.

feed lesions and analyze the size changes of the tumoral lesions. In these cases, a T_1 diffusion weighted (DW) sequence is the best choice. Specifically, DW imaging measures microscopic mobility of free water molecules and can be quantified by apparent diffusion coefficient [32]. Hence, this medical imaging technique allows a noninvasive measurement of the obstruction of the arteries that feed the tumoral lesions as well as the lesion size (high occlusion of vessels and decrease of tumor should be expected), because decrease uptake lesions

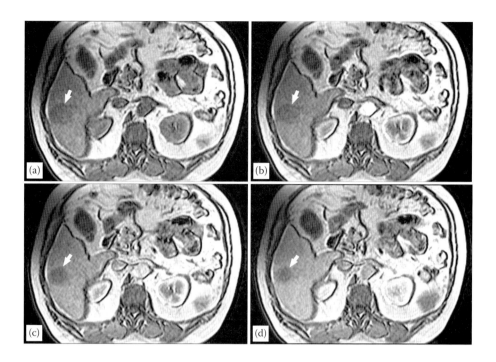

FIGURE 26.10

A 70-year-old man with two hepatic lesions corresponding with hepatocellular carcinoma has undergone a intra-arterial embolization therapy. Diffusion-weighted imaging of the lesion located in segments V and VI was acquired 2 months after embolization therapy. No difference between the lesion image without gadolinium contrast (a) and with gadolinium contrast in arterial, portal, and venous phase (b–d) are observed. These involve the artery obstruction is adequate due to the gadolinium administered don't irrigate of the tumor.

and changes in the lesion behavior after administration of contrast agent at different stages are easily detected in DW sequence. Therefore, no difference between the obtained images with and without contrast agent indicates a properly performed arteries obstruction. In this case, the contrast agent supplied through the endovascular system cannot achieve the tumor, producing the necrosis of this tumoral tissue (Figures 26.10 and 26.11). Although the tumor shows itself as a low-enhanced tissue with no differences in its behavior between presence and absence of contrast agent, a decrease of the tumoral lesion is not guaranteed (Figure 26.12b).

Another common sequence used to assess the embolization therapy outcomes, instead of DW imaging, is T_2 with fat suppression (Figure 26.12). This T_2 sequence allows observing changes in the tumor size and assessing the evolution of the tumor after the treatment too.

26.2.1.2.1.2 Metastasis Many metastasis produced by gastrointestinal primary tumors are frequently located in the liver [33]. Most of these metastasis come from colorectal cancer, being present in up to 55% of patients with this kind of primary cancer that eventually develop these tumoral lesions. However, other primary cancers can also provoke liver metastasis due to its proximity or due to other dissemination factors such as blood stream. Among others, some of the most common ones are breast, pancreas, stomach, or neuroendocrine

cancers. An example was presented by Weinrich et al., who correlated breast cancer with liver metastasis in a rate from 2% to 12% of the reviewed cases [34].

Liver metastasis resection increases the survival rates of patients with different kinds of primary cancers. According to the literature, resection of metastasis from breast cancer presents a survival rate ranging from 77% to 100% for the first year; between 50% and 86% for the second one; and between 9% and 61% for the fifth year [34]. On the other hand, liver metastasis resection from the colorectal cancer increases survival rates of patients from 0%–1% to 31%–58% [35].

Imaging acquisition systems are needed for both planning and monitoring liver metastasis resections. For this reason, effectiveness of these kinds of techniques has already been evaluated by several authors. Floriani et al. [34] presented a vast review of 6030 clinical cases where capacity of different imaging modalities to identify and characterize liver tumors were analyzed, providing results for US (63%–97%), CT scan (74.8%–95.6%), and MRI (80.1%–97.2%). According to these results, MRI effectiveness seems to be the highest, being exceeded only by FDG-PET that obtained a sensitivity of 93.8%–98.7% in all analyzed studies. However, chemotherapy before surgery causes changes in the delineation of tumors due to alterations in the liver parenchyma, the sensitivity of MRI being more effective than FDG-PET in these cases [36]. For these reasons, also according

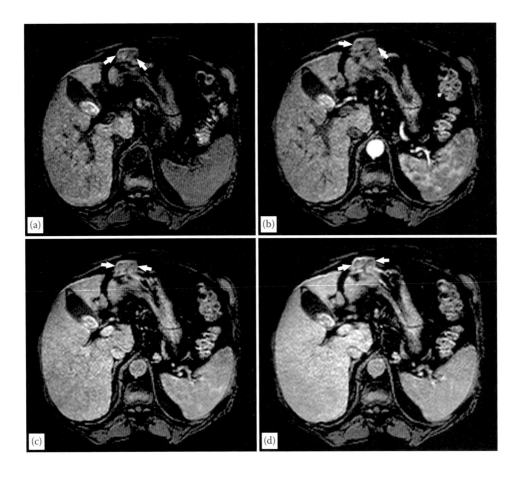

FIGURE 26.11

A 70-year-old man with two hepatic lesions corresponding with hepatocellular carcinoma underwent an intra-arterial embolization therapy. Diffusion-weighted imaging of the lesion located in segments III and IV were acquired 2 months after embolization therapy. No difference between the lesion image without gadolinium contrast (a) and with gadolinium contrast in arterial, portal, and venous phase (b–d) are observed. However, the size of the tumor has not been reduced.

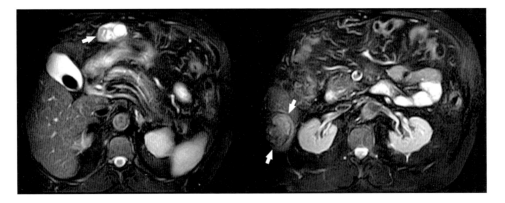

FIGURE 26.12

A 70-year-old man with two hepatic lesions corresponding with hepatocellular carcinoma has undergone an intra-arterial embolization therapy. T_2-weighted fat-suppression images after embolization was obtained. A reduction of 4 cm lesion located in segments III and IV was observed relative to the size of the Figure 3.10b. The lesion located in segment VI not experience significant reduction compared to the diagnostic imaging (Figure 3.10a).

to Legou et al. [37], MRI is the most optimal technique for the identification and characterization of the liver metastasis.

As for other interventions, liver metastasis resection requires a follow-up monitoring to determine that the tumoral tissue was correctly removed (Figure 26.13). In order to achieve this in a safe and noninvasive way, MRI sequences can be used to distinguish healthy and malignant tissues as well as hepatic collection resulting

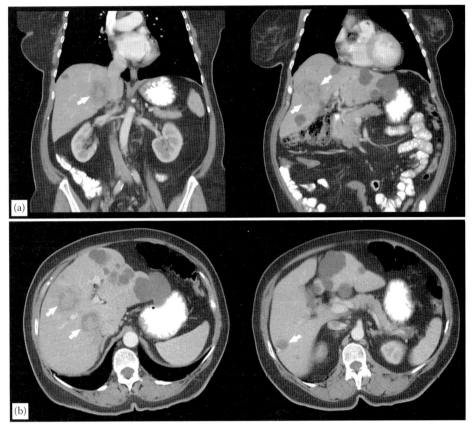

FIGURE 26.13
A 69-year-old woman who is presented with simple cyst and hepatic metastasis. The diagnostic CT in the coronal (a) and axial (b) views of the liver evidences the identification of the cysts in the left and right hepatic lobules, being the biggest cyst located in segment II with a dimension of the 5.3 cm. Besides, in the images, it is possible to identify two hepatic metastases with periferical enhance located in segment VIII of 5 cm, and the second located in a posterior region of the 4 cm. These metastases derived from colorectal carcinoma are evidenced in the colon.

from hepatic resection. After surgery, T_2 sequence can be used to identify these hepatic collections in those areas where metastasis were located, in order to assess surgical therapy effectiveness (Figure 26.14). Therefore, these sequences can be used to distinguish these hepatic collections from biliary cysts. While the biliary cysts have a homogenous signal, the posttherapeutic collections have an heterogeneous and made of a variable composition of several fluid substances that have different features.

26.2.1.2.1.3 Cholelithiasis Gallstone disease affects a large percentage of the population [38], and some of these patients also suffer from choledocholithiasis [39,40]. This pathology consists of the presence of lithiasis in the common bile duct and has serious health risks for patients, due to other more dangerous problems such as cholangitis or acute pancreatitis, which can be derived from it. Traditionally, endoscopic ERCP has been considered as the gold standard technique for the diagnosis and treatment of choledocholithiasis. Their

use is widespread and several authors are planning to include the MRCP in its daily routine for the choledocholithiasis diagnosis [41–43,44]. Despite presenting the highest degree of sensitivity and efficiency, both around 95% [39,44,45], this can cause severe complications. Furthermore, other authors consider their use as limited mainly due to its cost and availability [46].

MRCP allows obtaining static liquids signals after performing saturation of the background and other fluids as blood, but without requiring the administration of intravenous contrast agent. Both sensitivity and specificity of this technique is between 85%–92% and 93%–97%, respectively [47,48]. However, CT cholangiography has a sensitivity and specificity similar to MRCP, but CT requires the use of contrast agent and patients are exposed to ionizing radiation that can compromise its final use as choledocholithiasis identification method.

Other common non-preoperative use of MRCP is the assessment of complications derived from a cholecystectomy surgical procedure. Among others, the most

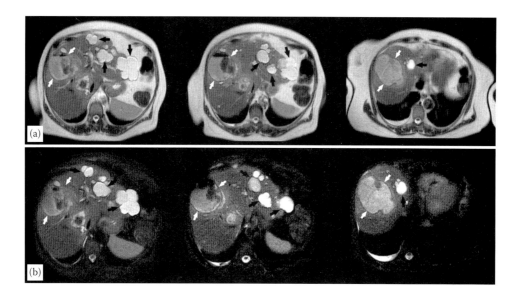

FIGURE 26.14
A 69-year-old woman with resection of the hepatic metastasis located in segment VIII. T_2-weighted (a) and T_2-weighted fat suppression (b) images was acquired after of the resection. Remaining lesions were not observed, only a collection postoperative liver 8 cm (white arrow). Cysts found (black arrow) not exhibit variations from the diagnostic study (Figure 3.14).

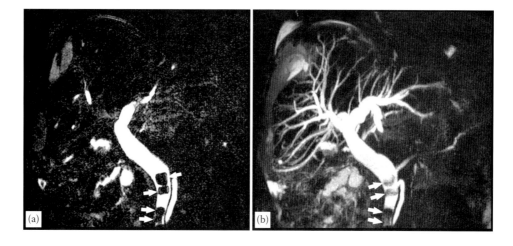

FIGURE 26.15
A 60-year-old man who is presented with a choledocholithiasis after a cholecystectomy. MRCP (a) and MIP reconstruction of MRCP (b) images was acquired after surgical intervention. A distension of the common bile duct of the 17 mm is identified due to a four stones with dimensions between 6 and 10 mm.

common complications of this surgical intervention are the bile leak or bile duct obstruction caused by cholelithiasis. These problems can usually appear several days after the surgical procedure was performed. Specifically, cholelithiasis can be derived from cholecystectomy when some stones are not detected in previous exams or are formed on the common bile duct after the surgical intervention. In these cases, an MRCP instead of ERCP is recommended as the preferred diagnostic method to perform the postcholecystectomy monitoring mainly due to its high precision based on the guideline of American Society of Gastrointestinal Endoscopy [49]. Dilatation of the common bile duct is a complication

usually identified using MRCP (Figure 26.15) that is described in the literature as a complication observed from the cholecystectomy surgical procedure. Furthermore, MRCP allows evaluating mild ectasia of pancreatic duct (Figure 26.16) and analyzing the nature of pancreatitis when MRCP is used in combination with T_2 axial sequences [50]. This work describes different dilatations of the common bile duct in several studies. Pavone et al. [51] confirm the accuracy of this technique, that also reveal the Winsurg duct dilatation and the stenotic tract for all studied cases of chronic pancreatitis. Other complication that can be analyzed with MRCP is ascites (accumulation of fluid into the abdomen).

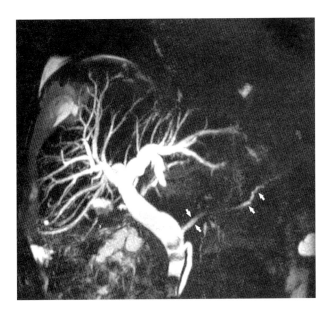

FIGURE 26.16
A 60-year-old man who is presented with a choledocholithiasis after a cholecystectomy. MIP reconstruction of MRCP image was acquired after surgical intervention. A distension of the common bile duct of the 17 mm is identified due to a four stone with dimensions between 6 and 10 mm.

This complication can be distinguished from other substances such as fat using T_2-weighted and T_2 with fat suppression (Figure 26.17).

26.2.2 Kidney

The kidney is a retroperitoneal bean-shaped organ located on either sides of the spine between the 12th thoracic vertebra and 4th lumbar vertebra. Some of the main functions of this organ are blood filtration and elimination of the waste and excessions.

MRI kidney exploration has increased its use as diagnosis method to evaluate several pathologies. Their main advantages against other diagnosis imaging techniques are the evaluation and characterization of tumoral lesions stages and its ability to characterize and distinguish cyst and solid lesions, even to allow monitoring lesions evolution after surgical treatment. Furthermore, it allows performing urographic techniques used to know the renal function.

Section 26.2.2.1 describes some of the possibilities of the MRI during surgery as surgeon assistance tool, and after surgery for evaluating and monitoring different complications derived from renal surgery procedures.

26.2.2.1 During the Surgery

Different renal surgery procedures are guided using MRI, what provides images with both high contrast and definition of soft tissues. However, research of interventional MRI is more focused on other surgical field such as liver or prostate tumor ablations.

26.2.2.1.1 Guidance Tool

Although its application on renal surgery is not extended, there are some examples of renal interventions that have used MRI as a guidance tool. Image-guided percutaneous nephrostomy is a well-established minimally invasive procedure that consists of placing a catheter into the renal collecting system to allow urine drain of patients with urinary tract obstruction. This surgical procedure is usually performed using a combined dual-imaging technique of US and fluoroscopy. However, both techniques present several limitations such as lower image quality due to the presence of air inside of the bowel, in the pleural space for US, or the use of intravenous contrast and iodinated radiation for fluoroscopy. On the other hand, MRI-guided nephrostomy is a feasible and safe procedure that can be performed both when there is dilatation in the renal pelvis [52] and when it is missing [53]. In this sense, Fischbach et al. [53] show the usefulness of this technique over US, while

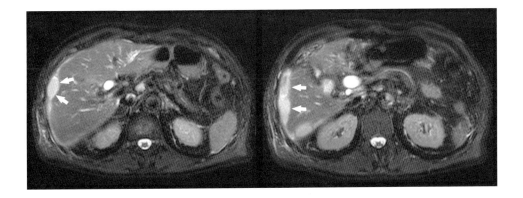

FIGURE 26.17
A 60-year-old man who is presented with a choledocholithiasis after a cholecystectomy. T_2-weighted fat-suppression images acquired after surgical intervention evidence the presence of ascites in the abdomen.

all patients of this study were rejected for US guidance due to several reasons: very poor general health condition that allowed only supine or semi-lateral positioning unusual for US guidance, no dilatation of the pelvic system or obesity reasons, it was possible to perform MRI guidance. Furthermore, MRI guidance has additional advantages such as images are not interfered with the presence of air in the bowel or contrast agents are not required.

Ablation therapies of kidney tumors can be also performed by means of MRI-guided procedures too. Likewise with liver tumors, kidney ablation therapies that are guided by MRI have advantages over other guidance techniques such as US or CT. Among these advantages are high-contrast images, multiplanar capabilities, and inherent sensitivity to temperature and blood flow. Furthermore, for thermal-based technologies such as RF, microwaves, or cryoablation, MRI allows monitoring the working area of thermal tissue destruction during therapy application and helps to adjust in situ the treatment according to intraoperative changes against of previous planned preoperative treatment [54]. In the case of cryoablation therapy, MRI allows identifying the generated ice ball so a more accurate tumor resection margins can be estimated, assuring that all resection margins are within the working area of the treatment. These MRI-guided procedures have been analyzed by several authors that show a high efficiency of the techniques (success rate from 91% to 100% the surgical interventions) [55,56]. In general, MRI sequences used to plan or assist surgical procedures can vary depending on the ablation technique and the nature of the tumor. In this sense, surgical planning is usually performed using T_2-weighted, T_1-weighted, or true fast sequences. On the other hand, intra-operative MRI-guided surgery requires a short echo time gradient-echo sequence (FSIP sequence in Siemens or FFE sequence in Philips) [55].

26.2.2.1.2 *Effect of Pneumoperitoneum*

The kidney is a retroperitoneal organ so changes in the intra-abdominal pressure requested during the laparoscopic procedures affects it indirectly. Despite this, the pressure increment provokes different alterations in the kidney functionality, with a significant decrease in renal blood flow and alterations in the urine production being the more important alterations described in the literature [57].

The analysis of these changes has been done through different techniques, such as intravenous catheters and monitoring the production of urine or biopsies [57]. All these techniques are carried out during the surgical procedures, so the maneuvers of the surgical procedure itself could also have an influence on the observed effects. Although these changes have been widely analyzed, studies are not focused on assessing the possible

morphological changes. As in the liver, MRI is a noninvasive, innocuous tool to analyze the morphological and functional changes of the kidney during laparoscopic procedures, thanks to its high contrast and sensitivity to soft tissues. This methodology has been used to analyze changes in animals [58] but not in humans yet.

T_2-weighted sequence is recommended for the analysis of these morphological changes, as it allows for a clear delimitation of both kidneys. The minimum slice thickness is desired, but due to the requirement of high acquisition times and due to the fact that the surface of the kidney is homogeneous, the slice thickness can be up to 4–5 mm. The methodology to assess the changes consists of the acquisition of two T_2 sequences. The first one should be done with the patient at ease, and the second one during the pneumoperitoneum, after a 20 min period for stabilization.

After the pressure increment up to 12 mmHg is used in laparoscopy, the placement of the abdominal structures suffers from important changes (Figure 26.18). Since the kidney is a more compact organ than the liver, deformations are apparently not appreciated in the images. But after detailed analysis of the images, a significant reduction in the volume of the kidney was observed. First, a manual segmentation of the kidney was accomplished in each of the slices where the kidney was visible. Then, using a DICOM visualization software such as Osirix, the volume of regions segmented is obtained (Figure 26.19). The mean value of the volume for five analyzed animals was reduced from 75.71 ± 11.73 cm^3 to 73.14 ± 12.39 cm^3 after the creation of the pneumoperitoneum [58]. This reduction might be caused for the compression that the viscera suffer and also due to the venous ecstasies caused by the vascular resistance described by other authors [22]. Regarding the location of the kidneys, there are no perceptible differences as they are retroperitoneal organs. Maximum area is also affected by the pressure increment.

Besides, the analysis of morphological changes in the abdominal organs, MRI also allows for a noninvasive means to analyze the blood flow in arteries and veins, such as the renal artery. During the pneumoperitoneum, the analysis of flows using MRI makes possible to isolate the effects of the pneumoperitoneum itself from other changes that might be caused by the surgical procedure, when flows are measured intraoperatively. Phase contrast is the sequence used for this purpose. This sequence does not require the use of contrast agents and after processing, it allows for the analysis of blood flow, speed, or stroke volume, among other hemodynamic parameters. Effectiveness of this sequence has been described by scientific studies that analyses the flow in different vessels [59]. For instance, usefulness of this sequence has been proved by analyzing the blood flow in the portal vein in systems where

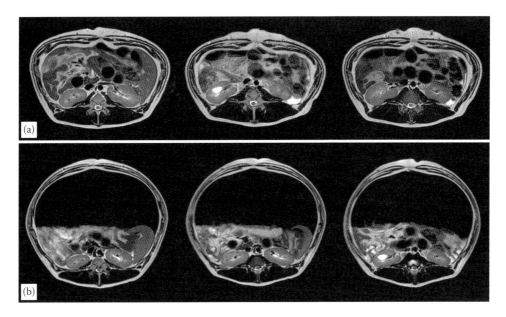

FIGURE 26.18
MRI performed in porcine model with weight of 30 kg. T_2-weighted images show the changes experimented by kidney before the neumoperitoneum creation (a) and after the pneumoperitoneum creation (b).

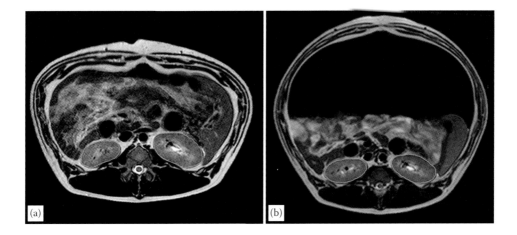

FIGURE 26.19
T_2-weighted image of a porcine model with a weight of 35 Kg acquired in the supine position. Kidney manual segmentation is performed to evaluate the changes experimented by the liver volume with OXiris Visualization DICOM software before (a) and after (b) the pneumoperitoneum creation.

the actual value of the flow is known beforehand, such as in phantom. A high correlation was shown between the obtained value using MRI and actual value of the flow [60]. In the case of the renal artery, the flow analysis through the phase contrast sequences has also been validated [61–63], comparing the obtained results to other already valid methods. Although the efficiency has been proved in a regular vascular anatomy, the presence of anatomical variations or when the artery is highly angular or tortuous causes variations on the results. Therefore, because of the reduced size of the renal artery, it is important to perform an angiography to accurately plan the sequence (Figure 26.20).

Figure 26.21 shows several images resulting from a phase contrast sequence on animal model before and after the creation of the pneumoperitoneum. The increment of the intra-abdominal pressure causes a reduction in the renal flow and a decrement in the lumen of the renal artery (Figure 26.22). These changes coincide with other finding in the literature, where a reduction in the renal flow is measured with other invasive methods such as ultrasound probes place around the artery or placement of intravenous catheters [57,64]. Besides, the reduction in the flow of the renal artery can be cause of functional alterations during laparoscopic surgery as described in the literature, such as a reduction in the

FIGURE 26.20
MR angiography images of the aorta and renal artery without contrast agent obtained in a porcine models with a weight of 33 and 32 kg. These images allow an adequate planning the phase contrast sequence of the renal artery.

FIGURE 26.21
Phase contrast image of the renal artery obtained of the porcine model with a weight of 31 Kg.

production of urine [57]. The observed increment of the volume of the kidney could also be a consequence of the reduction of the flow in the renal artery.

26.2.2.2 Postoperative Findings

Postoperative monitoring of kidney interventions is crucial to diagnose and evaluate different derived complications such as the identification of bleeding. As a noninvasive and harmless technique, MRI is a useful diagnosis and monitoring method for several surgical procedures such as tumor resection and postoperative complications, for instance, after kidney transplantation. Next section describes several kidney diseases, the advantages of MRI as a diagnosis tool and possible complications derived from the surgical procedure.

26.2.2.2.1 Renal Tumors

Renal cell carcinoma (RCC) is the eighth most common malignancy in adults [65] and the most common malignancy in the kidney [66]. The total number of detected RCC has recently increased due to the technical improvements in the medical imaging modalities such as US, CT, and MRI.

The gold standard technique to treat these tumors is a total nephrectomy, where the whole affected kidney is removed. However, depending on the tumor size and the kidney conditions of the particular patient, a partial nephrectomy could be indicated too. These surgical procedures require an exhaustive follow-up monitoring to assess and control the reproduction of lesions. For these reasons, using MRI as a monitoring method of this pathology is rising, thanks to its high contrast imaging and the features that allow distinguish tumoral and cystic tissues. In this sense, T_2- and T_1-weighted sequences allow showing the absence of tumoral tissues after surgery and distinguishing simple cysts (Figure 26.23). Furthermore, although kidney is a retroperitoneal organ, some anatomical changes in the abdominal area can be caused after its removal, as the renal basin can be occupied by abdominal structures, such as the pancreatic tail or small bowel loops (Figure 26.23).

26.2.2.2.1.1 Renal Failure
Renal failure consists of a reduction of the kidney ability to adequately filter waste products from the blood and balance body fluids and electrolytes. When the renal functions reach a critical malfunction and patients undergo haemodialysis, the only possible treatment for this disease is the kidney transplantation. This kind of transplantation is one of the most common methods nowadays, but it is not free of complications. Specifically, postsurgical complications are really hazardous, and they might require a reintervention to solve the postoperative problems in an early stage of the postoperative period. Unfortunately, a high percentage of these re-interventions lead to the

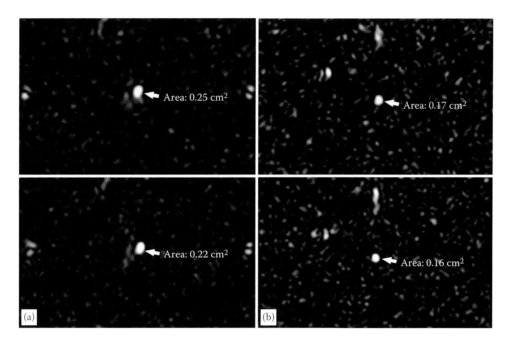

FIGURE 26.22
Phase contrast images of a porcine model of swine model with a weight of 30 Kg acquired in the supine position. Manual segmentation of the renal artery lumen (a) before and (b) after pneumoperitoneum creation. A reduction in the lumen artery renal is observed after increase of the intra-abdominal pressure to 14 mmHg.

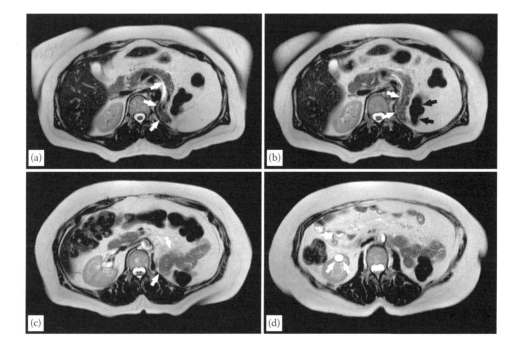

FIGURE 26.23
A 55-year-old woman with a renal cellular carcinoma in the left kidney subjected to a total nephrectomy. T_2-weighted images were acquired to evaluate the result of the surgery. Recurrence signs of the lesions are not seen in the images. Besides, the occupation of renal socket by the tail of the pancreas (a, b) and bowels (b black arrow and c) after the nephrectomy is observed. Simple cyst of 16 mm of dimension is also observed in the right kidney (d).

loss of the graft kidney [67]. According to the literature, vascular and urological problems are the most common complications [68,69]. For all these reasons, an exhaustive monitoring diagnosis of patients is needed in early postoperative stages to avoid possible complications and increase the graft survival rates.

Renal artery stenosis, renal vein thrombosis, renal artery thrombosis, and postoperative bleeding are the most common complications following a kidney transplantation. Among them, renal artery stenosis is the most frequent complication [70]. The gold standard technique to evaluate these vascular complications is the digital scripting arteriography (DSA), although other imaging techniques such as MR angiography are rising against DSA; since iodinated contrast agents are not needed in MR angiography, nephrotoxicity levels are decreased. This is a great advantage for patients with kidney malfunctions, because the use of iodinated contrast is too aggressive for them. On the other hand, MRI is a noninvasive method that avoids the use of catheters, decreasing the need of re-intervention in an early postoperative stage. Contrast-enhanced MR angiography is the MRI-based technique that obtains similar efficiency to DSA technique and its usefulness has been confirmed by effectively analyzing the state of the vascular anatomy in several studies [71,72]. Particularly, Huber et al. [73] designed a comparative study between effectiveness of MR angiography and DSA to determine the usefulness

and efficacy of MR angiography as monitoring technique for vascular postoperative complications. A sensitivity and a specificity of 100% and 95%, respectively, were found out to detect significant stenosis on the arterial vessels. Other recent studies show several findings using MR angiography in the early postoperative stages of kidney transplantation [74]. In this way, MR angiography can identify a very usual complication as the mild stenosis on the anastomoses of the renal vessels.

Complications in the vessels involved in a renal transplant have been also described using MR angiography. Similarly, MRI has demonstrated the presence of compressions of the renal vein, stenosis in the anastomosis of renal allograft and external iliac vein [73] or thrombosis of the renal vein [75].

Despite the advantages of MRI to monitorize vascular complications and to decrease radiation doses, as gadolinium contrast is less aggressive than ionized ones, some recent cases of nephrogenic systemic fibrosis (NSF) have been described [76,77]. These reactions suggest evaluating the risks and benefits of its use before employing such techniques.

Although urological complications are less frequent than vascular complications, they also have a high impact on the survival rate of patients with renal graft (Figure 26.24). According to the literature, complications rate varies between 2.5% and 27% of the cases [67,78–80]. Most frequent urological complications are

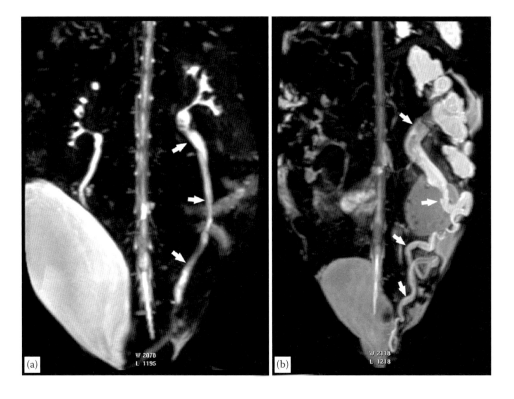

FIGURE 26.24
(a,b) MR urography images without contrast agent obtained in a porcine models with a weight of 33 Kg. Hydroneprhosis is observed in both images because of a vesicoutereteral obstruction.

vesicoureteral junction stenosis, urinary fistula, hydronephrosis, or lithiasis, among others. An early detection of these complications is critical to assure patient safety. In this sense, different medical imaging studies play a fundamental role to increase patient safety. Venous US is considered as a gold standard to monitor the urinary tract complications. However, this technique requires ionized contrast agents to be performed, and these agents can be toxic for certain patients. Thus, MRI studies are an alternative technique to analyze urological complications after surgery in an early postoperative stage of kidney transplantation. The efficiency of contrast-enhanced MR urography has been demonstrated to explore the urinary tract in standard conditions and to identify possible complications after kidney transplantation. Some of these complications are ureteral ducts stenosis or hydronephrosis due to ureteric compression (Figure 26.24). Furthermore, contrast-enhanced MR urography provides a higher image quality than other imaging techniques in maximum intensity projection reconstruction of renal graft and peritransplant region. Therefore, sensitivity and specificity of MR urography achieve values equal to 95.8% and 93.3%, respectively [81], with urinomas being one of the complications that present more difficulty to be detected with this technique. Although its nephrotoxicity is lower than iodinated contrast agents, the requested use of gadolinium contrast can cause other problems such as NSF. For these reasons, an alternative sequence to monitorize kidney complications is the static fluid T_2-weighted MR urography. This sequence is noninvasive as contrast agents are not required and allows an anatomical delineation of the ureteral graft without any kind of nephrotoxicity risk. However, it only allows the identification of complications when a hydronephrosis is presented due to this sequence is based on detecting static liquid. In this case, there are no significant differences between sensitivity and specificity of static fluid T_2-weighted sequence and contrast-enhanced MR urography to identify stenosis and ureteral obstructions [81].

26.3 Posttraumatic MRI Applications

The implementation of high-performance MR gradient systems, combined with fast-gradient ultrasound sequences, has allowed for a shortening of the images acquisition time, mainly using sequences with breath hold. Despite this, the role of MRI in the diagnosis of patients with abdominal trauma is limited. This is due to lower availability of this type of imaging systems in the emergency center, a higher imaging acquisition time, and the high cost of the diagnostic test [82].

Nevertheless, MR contrasted with intravenous gadolinium might be of utility during monitoring in case of allergy to iodized contrast agents or conservative treatment in hepatic, splenic, pancreatic, or renal injuries. Besides, the absence of ionizing radiation is an advantage to CT, especially in paediatric patients with lesions in solid intra-abdominal organs with conservative treatment, who require repeated controls [83].

In hepatic trauma, the utility of MRI in the detection of posttraumatic vascular shunts has been proved. Using intravenous contrast, image acquisition in the arterial and venous phases, and three-dimensional sequences in apnoea, the three signs that define posttraumatic vascular shunts are dilation of the afferent and efferent vessels of the shunt, parenchymal transient enhancement, and earlier enhancement of ipsilateral efferent vessels [84].

In the case of posttraumatic cysts in the spleen, MRI can determine the presence of recent bleeding within the cyst better than CT and ultrasound, by using hypersignal in a T_1 sequence and in order to distinguish it from secondary cysts, such as parasitic cysts [85].

In bruised injuries in the pancreas, CT is usually used as diagnostic method. MRI has proven its utility over other diagnostic methodologies, as it is able to show both the parenchymal lesion and the lesion of the main pancreatic duct (duct of Wirsung) [86]. To achieve this, specific sequences of MRCP, with 3D (MIP) reconstructions are used. Using these sequences, the disruption of the duct of Wirsung can be visualized. Furthermore, using T_1 and T_2 sequences with fat suppression, it is possible to observe the injuries in the parenchyma and associate collections.

MRI has been used over 20 years in the evaluation of obstetric pathologies without evidence of thermal injuries with magnetic fields up to 1.5 T. The literature is very limited with high field equipments (fields over 3 T) [87]. It is recommended that pregnant patients with abdominal trauma are not diagnosed using ionizing radiation, so US or RMI is recommended. US is usually the first imaging methodology and CT is the second one, bearing in mind always the ALARA (As Low As Razonable Achievable) criteria, referring to the ionizing radiation. Regarding the use of abdominal-pelvic MRI as imaging methodology in emergency cases, there are no enough studies that support its application, although it could be theoretically possible [87].

References

1. Wang MY, Cummock MD, Yu Y, Trivedi RA (2010) An analysis of the differences in the acute hospitalization charges following minimally invasive versus open posterior lumbar interbody fusion. *J Neurosurg Spine* 12(6):694–699.

2. Da Luz Moreira A, Kiran RP, Kirat HT et al. (2010) Laparoscopic versus open colectomy for patients with American Society of Anesthesiology (ASA) classifications 3 and 4: The minimally invasive approach is associated with significantly quicker recovery and reduced costs. *Surg Endosc* 24(6):1280–1286.

3. Lazzarino AI, Nagpal K, Bottle A et al. (2010) Open versus minimally invasive esophagectomy: Trends of utilization and associated outcomes in England. *Ann Surg* 252(2):292–298.

4. Wei B, Qi CL, Chen TF et al. (2011) Laparoscopic versus open appendectomy for acute appendicitis: A metaanalysis. *Surg Endosc* 25(4):1199–1208.

5. Yin J, Hou X (2014) Complications of laparoscopic versus open bariatric surgical interventions in obesity management. *Cell Biochem Biophys* 70(2):721–728. doi:10.1007/s12013-014-0041-2.

6. Sajid MS, Ahamd A, Miles WF et al. (2014) Systematic review of oncological outcomes following laparoscopic vs open total mesorectal excision. *World J Gastrointest Endosc* 166(5):209–219.

7. Wu B, Xiao YY, Zhang X et al. (2010) Magnetic resonance imaging-guided percutaneous cryoablation of hepatocellular carcinoma in special regions. *Hepatobiliary Pancreat Dis Int* 9(4):384–392.

8. Jendresen MB, Thorbøll JE, Adamsen S et al. (2002) Preoperative routine magnetic resonance cholangiopancreatography before laparoscopic cholecystectomy: A prospective study. *Eur J Surg* 168:690–694.

9. Nebiker CA, Baierlein SA, Beck S et al. (2009) Is routine MR cholangiopancreatography (MRCP) justified prior to cholecystectomy. *Langenbecks Arch Surg* 394:1005–1010.

10. Blanco RT, Ojala R, Kariniemi J et al. (2005) Interventional and intraoperative MRI at low field scanner—A review. *Eur J Radiol* 56(2):130–142.

11. Ni Y, Chen F, Mulier S et al. (2006) Magnetic resonance imaging after radiofrequency ablation in a rodent model of liver tumor: Tissue characterization using a novel necrosis-avid contrast agent. *Eur Radiol* 16(5):1031–1040.

12. Clasen S, Rempp H, Boss A et al. (2011) MR-guided radiofrequency ablation of hepatocellular carcinoma: Long-term effectiveness. *J Vasc Interv Radiol* 22(6):762–770.

13. Kim JE, Kim YS, Rhim H et al. (2011) Outcomes of patients with hepatocellular carcinoma referred for percutaneous radiofrequency ablation at a tertiary center: Analysis focused on the feasibility with use of ultrasonography guidance. *Eur J Radiol* 79(2):e80–84.

14. Tateishi R1, Shiina S, Teratani T (2005) Percutaneous radiofrequency ablation for hepatocellular carcinoma. An analysis of 1000 cases. *Cancer* 103(6):1201–1209.

15. Usón J, Sánchez FM, Sánchez MA, Pérez FJ, Hashizume M (2007) Principios básicos. In: Usón J, Sanchez FM, Pascual S, Climent S, (eds) *Formación en Cirugía Laparoscópica Paso a Paso*, 3rd edn. Centro de Cirugía de Mínima Invasión, Cáceres, Spain.

16. Guven HE, Oral S (2007) Liver enzyme alterations after laparoscopic cholecystectomy. *J Gastrointestin Liver Dis* 16(4):391–394.

17. Atila K, Terzi C, Ozkardesler S et al. (2009) What is the role of the abdominal perfusion pressure for subclinical hepatic dysfunction in laparoscopic cholecystectomy? *J Laparoendosc Adv Surg Tech A* 19(1):39–44.

18. Sáenz Medina J, Asuero de Lis MS, Galindo Alvarez J et al. (2007) Modification of the hemodynamic parameters and peripheral vascular flow in a porcine experimental of model of laparoscopic nephrectomy. *Arch Esp Urol* 60(5):501–518.

19. Smith MK, Mutter D, Forbes LE et al. (2004) The physiologic effect of the pneumoperitoneum on radiofrequency ablation. *Surg Endosc* 18(1):35–38.

20. Alexakis N, Gakiopoulou H, Dimitriou C et al. (2008) Liver histology alterations during carbon dioxide pneumoperitoneum in a porcine model. *Surg Endosc* 22(2):415–420.

21. Sánchez-Margallo FM, Moyano-Cuevas JL, Latorre R et al. (2011) Anatomical changes due to pneumoperitoneum analyzed by MRI: An experimental study in pigs. *Surg Radiol Anat* 33(5):389–396.

22. Bickel A, Loberant N, Bersudsky M, Goldfeld M, Ivry S, Herskovits M, Eitan A (2007) Overcoming reduced hepatic and renal perfusion caused by positive-pressure pneumoperitoneum. *Arch Surg* 142(2):119–124.

23. Gouya H, Vignaux O, Sogni P et al. (2011) Chronic liver disease: Systemic and splanchnic venous flow mapping with optimized cine phase-contrast MR imaging validated in a phantom model and prospectively evaluated in patients. *Radiology* 261(1):144–155.

24. Simard EP, Ward EM, Siegel R, Jemal A (2012) Cancers with increasing incidence trends in the United States: 1999 through 2008. *CA Cancer J Clin* 62(2):118–128. doi:10.3322/caac.20141.

25. Jemal A, Bray F, Center MM (2011) Global cancer statistics. *CA Cancer J Clin* 61(2):69–90.

26. SERAM (2009) *Actualizaciones Seram. Imagen en Oncología.* Médica Panamericana, Buenos Aires, Argentina.

27. Siegelman ES (2008) *Resonancia Magnetica Tórax Abdomen y Pelvis. Aplicaciones clínicas.* Médica Panamericana, Buenos Aires, Argentina.

28. Figueras J, Jaurrieta E, Valls C et al. (2000) Resection or transplantation for hepatocellular carcinoma in cirrhotic patients: Outcomes base on indicated treatment strategy. *J Am Coll Surg* 190:580–587.

29. McWilliams JP, Yamamoto S, Raman SS et al. (2010) Percutaneous ablation of hepatocellular carcinoma: Current status. *J Vasc Interv Radiol* 21(8 Suppl):204–213.

30. Dunne RM, Shyn PB, Sung JC et al. (2014) Percutaneous treatment of hepatocellular carcinoma in patients with cirrhosis: A comparison of the safety of cryoablation and radiofrequency ablation. *Eur J Radiol* 83(4):632–638.

31. Guo Y1, Yaghmai V, Salem R (2013) Imaging tumor response following liver-directed intra-arterial therapy. *Abdom Imaging* 38(6):1286–1299.

32. Malayeri AA, El Khouli RH, Zaheer A et al. (2011) Principles and applications of diffusion-weighed imaging in cancer detection, staging, and treatment follow-up. *RadioGraphics* 31(9):1773–1791.

33. Chatzifotiadis D, Buchanan JW, Wahl RL (2006) Positron emission tomography and cancer. In: Chang A, Ganz PA, Hayes DF et al. (eds) *Oncology: Ab Evidenced-Based Approach*, 1st edn. Springer, New York.

34. Weinrich M, Weiß C, Schuld J et al. (2014) Liver resections of isolated liver metastasis in breast cancer: Results and possible prognostic factors. *HPB Surg* 2014(4):1–6. doi:10.1155/2014/893829.

35. Floriani I, Torri V, Rulli E et al. (2010) A performance of imaging modalities in diagnosis of liver metastases from colorectal cancer: A systematic review and meta-analysis. *J Magn Reson Imaging* 31(1):19–31.

36. Van Kessel CS, Buckens CF, Van den Bosch MA et al. (2012) Preoperative imaging of colorectal liver metastases after neoadjuvant chemotherapy: A meta-analysis. *Ann Surg Oncol* 19(9):2805–2813.

37. Legou F, Chiaradia M, Baranes L et al. (2014) Imaging strategies before beginning treatment of colorectal liver metastases. *Diagn Interv Imaging* 95:505–512.

38. Shaffer EA (2006) Gallstone disease: Epidemiology of gallbladder stone disease. *Best Pract Res Clin Gastroenterol* 20(6):981–996.

39. Oneill CJ, Gillies DM Gani JS (2008) Choledocholithiasis: Overdiagnosed endoscopically and undertreated laparoscopically. *ANZ J Surg* 78:487–491.

40. Petelin JB (2003) Laparoscopic common bile duct exploration. *Surg Endosc* 17:1705–1715.

41. Jendresen MB, Thorbøll JE, Adamsen S et al. (2002) Preoperative routine magnetic resonance cholangiopancreatography before laparoscopic cholecystectomy: A prospective study. *Eur J Surg* 168:690–694.

42. Nebiker CA, Baierlein SA, Beck S et al. (2009) Is routine MR cholangiopancreatography (MRCP) justified prior to cholecystectomy. *Langenbecks Arch Surg.* 394:1005–1010.

43. Dalton SJ, Balupuri S, Guest J (2005) Routine magnetic resonance cholangiopancreatography and intra-operative cholangiogram in the evaluation of common bile duct stones. *Ann R Coll Surg Engl* 87:469–470.

44. Dumot JA (2006) ERCP: Current uses and less-invasive options. *Cleve Clin J Med* 73:418–425.

45. Mori T, Sugiyama M, Atomi Y (2006) Gallstone disease: Management of intrahepatic stones. *Best Pract Res Clin Gastroenterol* 20:1117–1137.

46. Al-Jiffry BO, Elfateh A, Chundrigar T et al. (2013) Non-invasive assessment of choledocholithiasis in patients with gallstones and abnormal liver function. *World J Gastroenterol* 19(35):5877–5882.

47. Romagnuolo J, Bardou M, Rahme E et al. (2003) Magnetic resonance cholangiopancreatography: A meta-analysis of test performance in suspected biliary disease. *Ann Intern Med.* 139(7):547–557.

48. Verma D, Kapadia A, Eisen GM et al. (2006) EUS vs MRCP for detection of choledocholithiasis. *Gastrointest Endosc.* 64(2):248–254.

49. ASGE Standards of Practice Committee, Maple JT, Ben-Menachem T et al. (2010) The role of endoscopy in the evaluation of suspected choledocholithiasis. *Gastrointest Endosc* 71(1):1–9.

50. Wang DB, Yu J, Fulcher AS et al. (2013) Pancreatitis in patients with pancreas divisum: Imaging features at MRI and MRCP. *World J Gastroenterol* 19(30):4907–4916

51. Pavone P, Laghi A, Catalano C et al. (1996) Non-invasive evaluation of the biliary tree with magnetic resonance cholangiopancreatography: Initial clinical experience. *Ital J Gastroenterol.* 28(2):63–69.

52. Kariniemi J, Sequeiros RB, Ojala R, Tervonen O (2009) MRI-guided percutaneous nephrostomy: A feasibility study. *Eur Radiol* 19(5):1296–1301.

53. Fischbach F1, Porsch M, Krenzien F, Pech M, Dudeck O, Bunke J, Liehr UB, Ricke J (2011) MR imaging guided percutaneous nephrostomy using a 1.0 Tesla open MR scanner. *Cardiovasc Intervent Radiol.* 34(4):857–863.

54. Nour SG, Lewin JS (2012) MRI-guided RF ablation in the kidney. In: Kahn T, Busse H (eds), *Interventional Magnetic Resonance Imaging*, 1st edn. Springer, Berlin, Germany.

55. Boss A, Clasen S, Kuczyk M et al. (2005) Magnetic resonance-guided percutaneous radiofrequency ablation of renal cell carcinomas: A pilot clinical study. *Invest Radiol* 40(9):583–590.

56. Grasso RF, Luppi G, Faiella E (2012) Radiofrequency ablation of renal cell carcinoma in patients with a solitary kidney: A retrospective analysis of our experience. *Radiol Med.* 117(4):606–615.

57. Demyttenaere S, Feldman LS, Fried GM (2007) Effect of pneumoperitoneum on renal perfusion and function: A systematic review. *Surg Endosc.* 21(2):152–160.

58. Sánchez Margallo FM, Moyano Cuevas JL, Maestre Antequera J et al. (2013) Efectos del neumoperitoneo en la morfología y hemodinámica renal analizado mediante resonancia magnética. In: *Proceeding Congreso Nacional de Urología*, Mexico.

59. Pelc LR, Pelc NJ, Rayhill SC et al. (1992) Arterial and venous blood flow: Noninvasive quantitation with MR imaging. *Radiology* 185(3):809–812.

60. Gouya H, Vignaux O, Sogni P et al. (2011) Chronic liver disease: Systemic and splanchnic venous flow mapping with optimized cine phase-contrast MR imaging validated in a phantom model and prospectively evaluated in patients. *Radiology* 261(1):144–155.

61. De Haan MW, van Engelshoven JM, Houben AJ et al. (2003) Phase-contrast magnetic resonance flow quantification in renal arteries: Comparison with 133Xenon washout measurements. *Hypertension* 41:114–118.

62. Park JB, Santos JM, Hargreaves BA et al. (2005) Rapid measurement of renal artery blood flow with ungated spiral phase-contrast MRI. *J Magn Reson Imaging* 21:590–595.

63. Bax L, Bakker CJ, Klein WM (2005) Renal blood flow measurements with use of phase-contrast magnetic resonance imaging: Normal values and reproducibility. *J Vasc Interv Radiol* 16:807–814.

64. Wiesenthal JD, Fazio LM, Perks AE et al. (2011) Effect of pneumoperitoneum on renal tissue oxygenation and blood flow in a rat model. *Urology* 77(6):1508.e9–1508.e15.

65. Jemal A, Siegel R, Ward E et al. (2007) Cancer statistics, 2007. *CA Cancer J Clin* 57:43–66.

66. Ng CS, Wood CG, Silverman PM (2008) Renal cell carcinoma: Diagnosis, staging, and surveillance. *AJR Am J Roentgenol* 191:1220–1232.

67. Barba Abad J, Rincón Mayans A, Tolosa Eizaguirre E et al. (2010) Surgical complications in kidney transplantation and their influence on graft survival. *Actas Urol Esp* 34(3):266–273.

68. Greco F, Fornara P, Mirone V (2014) Renal transplantation: Technical aspects, diagnosis and management of early and late urological complications. *Panminerva Med* 56(1):17–29.

69. Karam G, Maillet F, Braud G, Battis S (2007) Surgical complications of renal transplantation. *Annales d'urologie* 41:261–275.

70. Bruno S, Remuzzi G, Ruggenenti P (2004) Transplant renal artery stenosis. *J Am Soc Nephrol* 15:134–141.

71. Von Knobelsdorff-Brenkenhoff F, Gruettner H, Trauzeddel RF et al. (2014) Comparison of native high-resolution 3D and contrast-enhanced MR angiography for assessing the thoracic aorta. *Eur Heart J Cardiovasc Imaging* 15(6):651–658.

72. Makowski MR, Botnar RM (2013) MR imaging of the arterial vessel wall: Molecular imaging from bench to bedside. *Radiology* 269(1):34–51.

73. Huber A, Heuck A, Scheidler J, Holzknecht N, Baur A, Stangl M, Theodorakis J, Illner WD, Land W, Reiser M (2001) Contrast-enhanced MR angiography in patients after kidney transplantation. *Eur Radiol* 11(12):2488–2495.

74. Gufler H, Weimer W, Neu K et al. (2009) Contrast enhanced MR angiography with parallel imaging in the early period after renal transplantation. *J Magn Reson Imaging.* 29(4):909–916.

75. Di Felice A, Inguaggiato P, Rubbiani E (2002) Magnetic resonance in renal transplantation: Evaluation of post-surgery complications. *Transplant Proc* 34(8):3193–3195.

76. Cohan RH, Ellis JH, Hussain HK et al. (2008) Nephrogenic systemic fibrosis: Report of 29 patients. *AJR Am J Roentgenol* 190:736–741.

77. Marckmann P, Skov L, Rossen K et al. (2006) Nephrogenic systemic fibrosis: Suspected etiological role of gadodiamide used for contrast-enhanced magnetic resonance imaging. *J Am Soc Nephrol* 17:2359–2362.

78. Behzadi AH, Kamali K, Zargar M (2014) Obesity and urologic complications after renal transplantation. *Saudi J Kidney Dis Transpl* 25:303–308.

79. Kamali K, Zargar MA, Zargar H (2003) Early common surgical complications in 1500 kidney transplantations. *Transplant Proc* 35(7):2655–2656.

80. Koçak T, Nane I, Ander H (2004) Urological and surgical complications in 362 consecutive living related donor kidney transplantations. *Urol Int* 72(3):252–256.

81. Blondin D, Koester A, Andersen K et al. (2009) Renal transplant failure due to urologic complications: Comparison of static fluid with contrast-enhanced magnetic resonance urography. *Eur J Radiol* 69(2):324–330.

82. Weishaupt D, Grozaj AM, Willmann JK et al. (2002) Traumatic injuries: Imaging of abdominal and pelvic injuries. *Eur Radiol* 12(6):1295–1311.

83. Sivit CJ (2008) Contemporary imaging in abdominal emergencies. *Pediatr Radiol* 38(Suppl 4):S675–S678.

84. Semelka RC, Lessa T, Shaikh F et al. (2009) MRI findings of posttraumatic intrahepatic vascular shunts. *J Magn Reson Imaging* 29(3):617–620.

85. Acar M, Dilek ON, Ilgaz K, Çalışkan G, Kirpiko O, Tokyol C (2008) Posttraumatic splenic cyst Diagnostic value of MRI. *Eur J Gen Med* 5(4):242–244.

86. Yang L, Zhang XM, Xu XX (2010) MR imaging for blunt pancreatic injury. *Eur J Radiol* 75(2):97–101.

87. Katz DS, Klein MA, Ganson G et al. (2012) Imaging of abdominal pain in pregnancy. *Radiol Clin North Am* 50(1):149–171.

Index

Note: Locators followed by *'f'* and *'t'* refer to figures and tables, respectively